Concepts in Vaccine Development

Concepts in Vaccine Development

Editor
Stefan H. E. Kaufmann

Walter de Gruyter · Berlin · New York 1996

Editor

Professor Dr. Stefan H. E. Kaufmann

Universität Ulm
Abteilung Immunologie
Albert-Einstein-Allee 11
D-89081 Ulm

Max-Planck-Institut für Infektionsbiologie
Abteilung Immunologie
Monbijoustraße 2
D-10117 Berlin

Cover illustration
Scanning electron micrograph of biodegradable microspheres serving as antigen vehicles (courtesy of Gideon F.A. Kersten and Bruno Sander, see contribution pp. 265)

With 40 figures and 40 tables.

Library of Congress Cataloging-in-Publication Data

Concepts in vaccine development / editor, Stefan H. E. Kaufmann.
p. cm.
Includes bibliographical references and index.
ISBN 3-11-014815-3 (alk. paper)
1. Vaccines. I. Kaufmann, S. H. E. (Stefan H. E.)
QR189.C66 1996 96-11977
815'.372--dc20 CIP

Die Deutsche Bibliothek – CIP-Einheitsaufnahme

Concepts in vaccine development / ed. Stefan H. E. Kaufmann.
– Berlin; New York; de Gruyter, 1996
ISBN 3-11-014815-3
NE: Kaufmann, Stefan H. E. [Hrsg.]

♾ Printed on acid-free paper which falls within the guidelines of the ANSI to ensure permanence and durability.

Converting and typesetting by: Knipp Medien und Kommunikation, Dortmund. – Printing: Karl Gerike GmbH, Berlin. – Binding: Dieter Mikolei, Berlin. – Cover Design: Hansbernd Lindemann, Berlin.
Printed in Germany.

Preface

Although numerous vaccines provide highly cost efficient measures for disease control, thus far only one vaccine has achieved its ultimate goal, namely complete eradication of the infectious agent. Most likely the success of vaccination against small-pox is due to the unique characteristics of the virus, in particular its exclusive dependency on the diseased human host without animal reservoire or healthy human carrier. Hence, this success story will not be repeated soon. In 1796, exactly two-hundred years ago, Edward Jenner used vaccinia virus to protect against small-pox. This endeavour was purely empirical and it took more than hundred years until the mechanisms underlying vaccination became somewhat clearer. The art of vaccination, however, dates back before Jenner. Variolation was in use in China and India already one millenum ago and Lady Montagu is generally given credit for introducing active variolation into Europe in 1721. One of the earliest reports on deliberate attempts to protect against intoxication is that of Mithridates who protected himself against poisonous mushrooms by consuming subtoxic quantities of the fungi. Yet, this was probably some kind of biochemical adaption rather than acquisition of specific immunity. In the medieval ages it was widely accepted in Europe that those who had survived an epidemic were resistant against second infection and the survivers of bubonic plague were the people of choice for the care of sufferers of this disease. This notion, however, was apparently not taken as far as to employ deliberate vaccination.

Nowadays vaccination is generally regarded the applied arm of immunology representing its medically and economically most successful aspect. Unfortunately, this is not fully true and immunologists have to confess that almost all vaccines in use nowadays had been developed more or less independently from basic immunologic knowledge. Only rarely did vaccinology benefit from basic sciences directly. The golden ages of medical microbiology, dominated by Louis Pasteur and Robert Koch, were the cradle of immunology and laboratory designed vaccinology and strong ties existed between the newly emerging disciplines. The intimate interrelation of vaccinology and medical microbiology at this time is illustrated by the successful attenuation of anthrax bacilli by prolonged in vitro culture and of rabies virus by drying of infected nervous tissues by Louis Pasteur and Emile Roux. Both in vitro manipulations yielded effective vaccines. The close communication between vaccinology and immunology is best exemplified by the work of Emil v. Behring and Paul Ehrlich. The former scientist successfully introduced serum therapy (i.e., passive vaccination) against bacterial toxins, and the latter one deviced the standardized measurements for the quantification of antitoxic antisera.

Soon thereafter, vaccinology and immunology went their own ways. Vaccinology prospered through empiric approaches and immunology florished to become a successful basic research discipline. Thriving as the empiric approach has been for the available vaccines, it is no longer sufficient for vaccine design against infectious diseases which have thus far avoided control by conventional vaccination measures. Control by vaccination of major global scourges requires novel concepts and strategies which must be based on advances in the basic sciences, including molecular genetics, cell biology, protein chemistry, and immunology. Modern vaccinology, however, is even broader and, in addition, also rests on insights from epidemiology and public health. It is this interdisciplinary approach ranging from the laboratory bench to the field which renders vaccinology equally complex and fascinating.

The book "Concepts in Vaccine Development" brings together experts from various fields of vaccinology, and I deeply appreciate the enormous efforts of each contributor. It is hoped that the readers feel satisfied by such a broad coverage and do not restrict themselves to the more familiar chapters but also become interested in the more distantly related ones. Until a decade ago, infectious diseases were regarded as health threads that can be dealt with, at least in the industrialized world. Now we witness the appearance of new diseases, in particular AIDS; the reemergence of old diseases, such as tuberculosis; and the microbial etiology of diseases considered non-infectious until recently, such as *Helicobacter pylori* as cause of stomach ulcers. This has dramatically changed our comprehension and renewed interest in vaccine development as cost-efficient prevention measure. In the future, purely empiric efforts will be insufficient and new concepts and strategies, as they are described in this volume, will be needed to overcome current obstacles. In this way, vaccinology will evolve as a discipline of its own right.

Ulm and Berlin, January 1996 Stefan H.E. Kaufmann

Acknowledgements

I wish to thank the contributors for their time and effort, Rita Mahmoudi for her excellent secretarial help and my wife Elke and sons, Moritz and Felix, for their understanding that I spent so much time with the book.

Bicentennial of the first vaccination by Edward Jenner

This book on "Concepts in Vaccine Development" appears two-hundred years after Edward Jenner introduced vaccination. On May 14th, 1796, Jenner administered material to James Phipps which he had obtained from a pustular lesion of the dairymaid Sarah Nelmes who suffered from cow-pox infection. It was known to him and others that country people who had been infected with cow-pox, a mild disease transmitted from cattle, were generally immune against small-pox, a highly severe disease for man. Yet, Jenner was the first to use "humanized" cow-pox. Having observed that Phipps had recovered from slight inconvenience, he injected small-pox into Phipps' arms on July 1st with no adverse results. Some months later, small-pox were reinoc-

Figure 1: Edward Jenner (1749-1823)

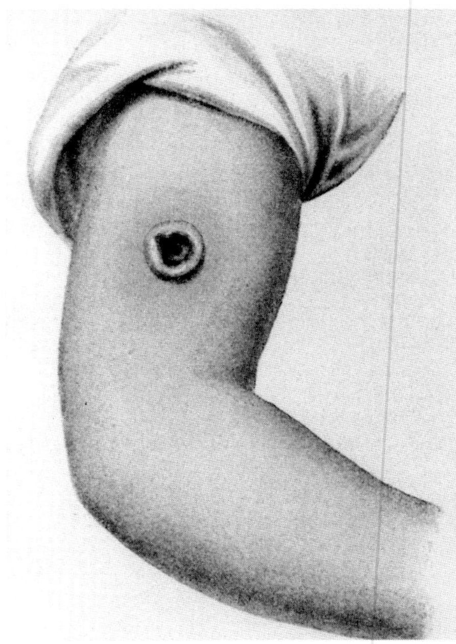

Figure 2: First immunization with "humanized" cow-pox by E. Jenner.

ulated without detrimental effects. This case is the most famous one amongst the 23 descriptions which were published by Jenner in June 1798 under the title "An inquiry into the causes and effects of the Variolae vaccinae, a Disease discovered in some of the western counties of England, particularly Gloucestershire, and known by the name of the Cow-Pox". Jenner hoped that his experiment would ultimately lead to the eradication of small-pox. Although he was proven correct by the World Health Assembly in May 8, 1980, he had hoped that this goal could be achieved much earlier.

Stefan H.E. Kaufmann

Contents

1. The Need for New Vaccines

1.1 Vaccines as Evolutionary Tools: The Virulence-Antigen Strategy

Paul W. Ewald

1.1.1 Introduction

Although vaccination has been used to combat diseases for two centuries, smallpox is the only disease that has been eradicated by vaccination. A few other diseases, such as diphtheria, have been virtually eliminated; several others, such as measles and whooping cough have been strongly suppressed but are still important threats. The preliminary successes at controlling diseases like smallpox and polio led researchers to overly optimistic predictions about the degree to which conventional approaches to vaccination would be able to control infectious diseases. Using the frequency of past successes of vaccination programs as a gauge for assessing future success will tend to lead to such over-optimism because the easiest pathogens will tend to be controlled first. Smallpox, for example, was easy to control because it was not antigenically variable and had an intrinsically slow rate of spread. As vaccination efforts continue, we can expect that control of additional pathogens will be progressively more difficult because of this winnowing process.

This difficulty results largely from the ability of pathogen populations to both multiply and evolve rapidly. Rapid multiplication allows pathogens to spread from small numbers of infected individuals through groups in which control efforts are weak. Rapid growth along with rapid generation of variation by mutation and genetic recombination may allow pathogens to evolve around vaccine-acquired immunity and spread through vaccinated or previously infected populations. New control strategies that pay particular attention to the resilience of pathogens may therefore be needed to maintain and heighten the levels of success that have been achieved in the past.

Traditionally, vaccination policies treat the resilience of pathogens as obstacles that make control and eradication of pathogens more difficult. An evolutionarily oriented perspective, in contrast, emphasizes the difference between control and eradication. A focus on eradication of pathogens will tend to pit vaccine against pathogen in an evolutionary arms race. A focus on control of disease without eradication of pathogens offers the possibility of guiding the evolutionary course of pathogens toward mild coexistence with people. Vaccines that move disease organisms closer to eradication may be different from vaccines that lead to maximal long-term control of disease.

The goal of traditional vaccination strategies has been to get as close as possible to eradication given the need to work within certain guiding criteria (summarized in the left circle of fig. 1.1.1).

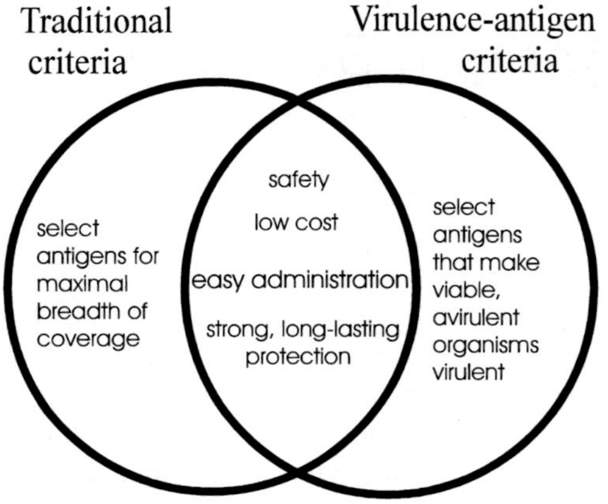

Fig. 1.1.1: Criteria for development of vaccines according to traditional strategies and the virulence-antigen strategy. Non-overlapping criteria occur because virulence-antigen criteria are based on evolutionary effects of the vaccine on the target organisms.

When eradication is realistic this traditional approach is justifiable. But the single success at eradication is not particularly encouraging. Evolutionary considerations alter these guidelines by emphasizing that without eradication we will be left with those variants of the disease organism that are not controlled by the vaccine. It is in our interest to have these variants be as mild as possible. I have suggested that this goal can be achieved by using a virulence-antigen strategy (Ewald, 1994).

The virulence-antigen strategy requires that policy makers resist the temptation to use vaccines with the broadest possible coverage of existing pathogens. It emphasizes that vaccines should preferentially incorporate antigens that contribute to virulence, that is, those antigens that make mild but transmissible organisms harmful (right circle of fig. 1.1.1). Such virulence-based vaccines should give a greater benefit per unit of investment than traditional vaccines because virulence-based vaccines disproportionately suppress the virulent variants, leaving mild variants to provide several benefits. The most obvious benefit is that mild variants cause less damage to those who are not vaccinated or who develop an insufficient immunity when vaccination programs are in full force. A less obvious benefit concerns the temporary lapses in coverage that occur even in countries with high quality health care. Standard vaccination of children in the U.S, for example, varied from 70 % to 98 % during the late 1970s and early 1980s (Warren, 1986). Evolutionary reductions in the virulence of circulating organisms will increase the chances that infections occurring during such lapses will be due to mild variants. A related benefit is that circulating mild variants may provide some protection against virulent variants. Thus, if virulent organisms arise by mutation or enter from other areas, they will be less able to spread because of the increased resistance attributable to mild natural infections. The mild variants circulating in the community thus act as free live vaccines for the unvaccinated.

The corollary of the virulence-antigen criteria (fig. 1.1.1) is that antigens should be excluded from vaccines if they do not contribute to virulence and if they occur on avirulent strains in addition to virulent strains. The presence of such antigens on mild variants may stimulate immunity that protects against virulent variants. This corollary holds so long as the other shared goals of the two approaches, such as the degree of vaccine-induced resistance (fig. 1.1.1), can be obtained from virulence antigens.

In theory, the virulence-antigen strategy should applicable to any virulence antigens that stimulate protective immunity and are not essential for reproduction and transmission from the host. Because such virulence antigens are often excellent candidates for vaccines according to traditional criteria, some have been included in vaccines strictly for nonevolutionary reasons. These vaccines offer insights into the feasibility of the virulence-antigen strategy, and the ways in which applications of the strategy may depend on details of pathogenicity and immunity. These issues are illustrated below for several pathogens.

1.1.2 Application to Specific Pathogens

1.1.2.1 Corynebacterium diphtheriae

Although the virulence-antigen strategy might seem unrealistic at first glance, the evolutionary effects of virulence-based vaccines are not simply hypothetical. The vaccine that has most fully met (albeit inadvertently) the virulence-antigen criteria is the diphtheria toxoid vaccine, which is based on a modified diphtheria toxin. To synthesize diphtheria toxin, *C. diphtheriae* must carry a viral *tox* gene, which is repressed by a DNA binding protein (DtxR) when iron is available. When iron is scarce, the dissociation of iron and DtxR allows transcription. The toxin blocks protein synthesis elongation factor 2, causing host cell death (Pappenheimer, 1977; 1982), which apparently liberates nutrients for *C. diphtheriae*'s use. By this mechanism *C. diphtheriae* may generate resources for itself when resource availability in its immediate vicinity is low (Schmitt and Holmes, 1991; Schmitt et al., 1992). In accordance with this argument, overt cases of diphtheria are more contagious than asymptomatic infections (Miller et al., 1972). In unimmunized people, the toxin therefore seems to provide its bearer with a fitness benefit.

If, however, a person is immunized with diphtheria toxoid, the toxin is impotent and therefore represents a drain of the bacterium's nutrients at a time when nutrients are particularly limited. Although both tox^+ and tox^- *C. diphtheriae* can infect vaccinated hosts (Miller et al., 1972), the toxin-producers should be at a competitive disadvantage in a vaccinated population because of this drain. Accordingly, diphtheria has vanished from areas with long-standing, thorough vaccination programs, while tox^- *C. diphtheriae* have been perpetuated, a change attributable to the selective pressure exerted by the vaccine (Uchida et al., 1971; Pappenheimer and Gill, 1973; Pappenheimer, 1982; Chen et al., 1985). The most detailed evaluation of the idea that the toxoid vaccine causes *C. diphtheriae* to evolve toward mildness came from Romania, where the precision of the positive association between doses of vaccine administered and the relative frequency of tox^- strains leave little room for alternative explanations (Pappenheimer, 1982).

When comprehensive vaccine efforts are maintained for prolonged periods, even the susceptibility to the phage carrying the *tox* gene may be a liability because production of the toxin would place the susceptible variants at a disadvantage. With prolonged high levels of vaccination one would therefore expect to see increased frequencies of corynebacteria that are refractory to the ß-corynephage that carry the *tox* gene. This possibility was investigated in an area of Siena, Italy, which has had compulsory vaccination of newborns since 1939: classic *C. diphtheriae* were absent but taxonomically unclassifiable corynebacteria that were closely related to *C. diphtheriae* and refractory to ß-corynephage infection were present (Mencarelli et al., 1992).

Next to the eradication of smallpox, the control of diphtheria has been the most successful vaccine program in terms of reductions in morbidity and mortality per unit of investment. A comparison with pertussis gives a sense of this success. In 1920, before the any vaccines were available, the annual incidences of pertussis and diphtheria in the US were similar, about two per 1000 people (Brooks, 1969; Brooks and Buchanan, 1970). Vaccines for both pathogens had become generally available during the second quarter of the century. Over the last quarter of the century, the incidence of pertussis has diminished in well-vaccinated populations to a level that is 100-fold less than the pre-vaccine incidence, but is still about 100-fold greater than the current incidence of diphtheria (Brooks, 1969; Brooks and Buchanan, 1970; Chen et al., 1985; Cherry et al., 1988). The measles trend has been like that of pertussis rather than diphtheria (Barkin, 1975; Poland and Jacobson, 1994), but the comparison between pertussis and diphtheria is particularly noteworthy because a single vaccine (DPT) has been used against both. The vaccinated population has therefore been virtually the same for most of century. The few outbreaks of diphtheria that still occur in the well-vaccinated regions, typically are attributable to foreign travel or limited circulation among small groups people who have compromised states of health, or live in poor, densely populated, urban areas (Kallick et al., 1970; Pappenheimer and Murphy, 1983; Harnisch et al., 1989). This disproportionate success is as expected from virulence-antigen strategy because control of disease does not result just from the administered vaccine but also from the remaining tox$^-$ *C. diphtheriae* which should generate some protection against tox$^+$ organisms.

The records on diphtheria also illustrates the stability of control by a virulence-antigen vaccine. In recent decades the proportion of people with protective immunity against diphtheria toxin has not been particularly high. In the United States about three-quarters of children and one-quarter of adults have protective immunity (Chen et al., 1985). If the diphtheria vaccine protected equally against all strains of *C. diphtheriae*, the large numbers of unprotected people might have generated a resurgence in diphtheria, much like the resurgence of other respiratory tract diseases. For example, even though diphtheria and pertussis vaccination are occurring at virtually the same rate, whooping cough has commonly reoccurred in areas where vaccination efforts have diminished (Hewlett, 1990; Pinchichero et al., 1992); diphtheria has reoccurred substantially only in areas with dramatically lowered rates of vaccination. In the Republic of Georgia, for example, coverage for the first dose of toxoid vaccine was only 37 % in 1992, down from 68 % in 1989; epidemic diphtheria began there in 1993 (Sasse et al., 1994). By inhibiting just the toxin-producing strains, the diphtheria vaccine allows the mild strains to be present to protect against the severe strains should vaccination efforts lapse slightly or miss a small percentage of a population.

1.1.2.2 Bordetella pertussis

For most of this century, the standard vaccine against pertussis was a suspension of killed *Bordetella pertussis*. The occasional encephalopathy associated with the vaccine (Howson and Fineberg, 1992) generated interest in developing safer alternative vaccines. Because the pertussis toxin (PT, also called lymphocytosis promoting factor, histamine sensitizing factor, pertussigen, or islet activating protein) is largely responsible for protective immune responses, it was used in the first acellular vaccines.

Pertussis toxin apparently benefits *B. pertussis* by inhibiting attack from various cells of the immune system (killer cells, phagocytic cells and mast cells) through interference with signal transduction (Weiss and Hewlett, 1986). Because the severe effects of pertussis are largely attributable to this toxin, a safe vaccine generated from deactivated toxin should help control pertussis by the same evolutionary process described above for diphtheria. Virulent *B. pertussis* should be replaced with benign *B. pertussis*, which will naturally immunize people against any remaining virulent strains. Different strains of *B. pertussis* are differentially attacked by the immune response to the various vaccines (Blumberg et al., 1992), suggesting the feasibility of a vaccine that selectively disfavors more virulent variants of *B. pertussis*.

The evolutionary bonus may be lost, however, if the vaccine is made to protect against mild as well as virulent strains of *B. pertussis*. Over the last decade researchers have been evaluating a wide variety of additional *B. pertussis* antigens to increase vaccine efficacy. Besides PT, the major antigen candidates are filamentous hemagglutinin (FHA), adenylate cyclase toxin (ACT), tracheal cytotoxin, dermonecrotic toxin, lippooligosaccharide endotoxins, agglutinogens, and pertactin (Cherry et al., 1988). Numerous studies have investigated different combinations and concentrations of these antigens in acellular vaccines, which tend to generate responses as good as or better than whole cell vaccines with fewer dangerous side effects (Petersen et al., 1991; Podda et al., 1991; Feldman, 1992; Pichichero et al., 1992; Stoersaeter and Olin, 1992; Marcinak et al., 1993; Anderson et al., 1994; Vanura et al., 1994; Seachrist, 1995). The virulence-antigen strategy emphasizes, however, that the candidate antigens should be carefully evaluated not just according to their contributions to short-term protection, but also on the basis of their contributions to virulence. Although knowledge is still too scanty to allow definite decisions about the relative merits of each antigen, the available evidence does allow some preliminary evaluations of their suitability for the virulence-antigen strategy.

Rather than having toxic effects, FHA contributes to the infectious process by facilitating adherence to cells and by allowing entry and internal survival in macrophages (Weiss and Hewlett, 1986; Locht et al., 1993). Although these characteristics foster infections by virulent *B. pertussis*, the characteristics themselves may not be harmful if products that directly cause pathology such as PT are not present.

Studies of immunological responses to FHA have yielded variable results. In one study, for example, FHA added an increment in protection against mild cases that ranged from a negligible amount within the first few months after vaccination to about 50% within the first few years (Storsaeter and Olin, 1992). In another study, addition of FHA reduced the serological response to PT (Anderson et al., 1994), a detrimental effect for a virulence-antigen vaccine. Respiratory IgG and IgA responses to FHA are long-lived following *B. pertussis* infection, perhaps because of prolonged survival of *B. pertussis* in macrophages or prolonged presence of FHA (Amsbaugh et al., 1993). Cellular immunity may also be important (Mills and Redhead, 1993), allowing the presence of mild intracellular *B. pertussis* to trigger protection against virulent strains. Studies of medical personnel indicate that continued exposure to natural infections keeps anti-FHA IgG (as well as anti-PT IgG) from waning (Tomoda et al., 1992), and antibodies to FHA are particularly apparent in subclinical reinfections (Long et al., 1990).

These findings indicate that long-lived resistance to pertussis is substantially maintained by infection-induced immunity, and that the FHA immunity generated by vaccines may be of only moderate importance. In light of the virulence-antigen hypothesis, this evidence suggests that pertussis would be better controlled if the FHA in acellular vaccines is replaced by a virulence antigen that provided similar short-term protection. Such a replacement would allow a greater prevalence of mild FHA^+ *B. pertussis*, which would in turn protect against virulent pertussis through the immune response triggered against their FHA. Assuming that infections with mild variants are effective in stimulating immunity, these results bolster the possibility of controlling pertussis by a virulence-antigen strategy in which FHA is not included in vaccines.

This advantage of replacing FHA in acellular vaccines is uncertain, however, because of gaps in knowledge. Specifically, we need a better understanding about the roles of FHA in pathogenesis to determine whether FHA is a virulence antigen and whether FHA on an avirulent *B. pertussis* would make a host more vulnerable to other pathogenic infections. We also need to know more about the degree of resistance to FHA induced by mild strains of *B. pertussis* to assess the benefits gained from vaccines that preferentially allow mild FHA^+ variants to circulate.

Even if further study shows that FHA does not directly contribute to virulence, acellular vaccines based on PT and FHA would probably still tip the competitive balance in favor of milder strains because PT is clearly a virulence determinant and FHA is intimately related to infection by virulent *B. pertussis*. In other words, although a vaccine using only virulence antigens may be better at long-term control of severe disease than a PT/FHA vaccine, the PT/FHA vaccine should be better than the whole cell vaccine because the former includes the most important virulence determinant (PT) and only one other antigen (FHA) that severe and mild strains of *B. pertussis* share.

The use of acellular vaccines in Japan offers a test. According to the preceding argument, the *B. pertussis* variants that are now present in Japan should be less severe than those that were present in 1981, when widespread use of acellular vaccines began in Japan. The lower incidence of pertussis in children less than two years old is consistent with this prediction because policy in Japan has been not to vaccinate children who are less than two years old (Christodoulides, 1990). However, this lower frequency could result from a lower overall exposure to *B. pertussis* rather than a lower virulence. More direct assessments of inherent virulence, such as measurements PT production, are needed to determine whether the *B. pertussis* circulating in Japan have become less virulent.

The remaining *B. pertussis* antigens include several candidates for virulence-antigen vaccines. Adenylate cyclase toxin contributes to virulence by inhibiting phagocytes (Cherry et al., 1988). The major biochemical mechanism apparently results from CAMP accumulation in the host cell, resulting from catalysis of the breakdown of ATP (Ehrmann et al., 1992). Like diphtheria toxin, ACT is not essential for bacterial growth, but ACT^- mutants grew to lower densities in a mouse model (Ehrmann et al., 1992), suggesting that ACT may confer a competitive benefit when not neutralized by an immune response. ACT therefore is a good candidate for further study to determine whether it can contribute to a virulence-antigen strategy. The presence of ACT in some avirulent strains (Weiss and Hewlett, 1986), however, suggests that the value of ACT may be minor relative to the more important determinants of virulence, such as PT.

The promise of the other antigens is unclear, but some contribute directly to virulence, at least in experimental settings. Pertactin may contribute to invasiveness and can trigger strong IgG responses in adults (Englund et al., 1992; Locht et al., 1993). Tracheal cytotoxin may inhibit cilia movement, which probably helps the bacteria avoid being cleared from the respiratory tract; it also damages epithelial cells apparently by inhibiting DNA synthesis (Weiss and Hewlett, 1986; Cherry et al., 1988). Dermonecrotic toxin causes skin necrosis in laboratory animals (Cherry et al., 1988).

Agglutinogens and lipooligosaccharides do not seem to contribute to virulence, and thus appear to be inappropriate for a virulence-antigen vaccine, even though they may stimulate protective responses and have therefore been added to acellular vaccines (e. g., Englund et al., 1992).

An evolutionary approach is particularly applicable to pertussis because pertussis vaccines tend to prohibit disease but not infection (Long et al., 1990). Even with high levels of vaccine coverage, *B. pertussis* continues to circulate (Herwaldt, 1993). Moreover, most damage by *B. pertussis* occurs in third world countries (Cherry, 1992), where intensive vaccination campaigns would be difficult. These factors indicate that chances for global eradication of *B. pertussis* by vaccination are remote. Virtual eradication of pertussis by virulence-antigen vaccines, however, should require less complete coverage (e. g., as with diphtheria), and therefore is a more realistic goal.

The changes that occur during individual *B. pertussis* infections lend credence to the possibility of driving *B. pertussis* to a milder state by using vaccine-induced protection against virulence antigens. As infections proceed, avirulent mutants tend to increase in frequency. This tendency indicates that mutations leading to mild infections occur readily and can be selected for on the basis of immune responses against the virulence antigens (Weiss and Hewlett, 1986).

Avirulent strains generated by this process revert back to virulent strains at a low frequency, apparently because sets of genes related to infections are controlled by a positive inducer locus (*bvg*), which experiences both deactivating and then restorative mutations (Weiss and Hewlett, 1986). This reversion to virulence along with the tendency for both subclinical infection in the immunized and prolonged infection after clinical recovery emphasize the need for a virulence-antigen strategy. Even a low frequency of reversion, can represent a substantial input of virulent variants. A virulence-antigen strategy should present a more formidable barrier to this reversion-based encroachment, by increasing the probability that unimmunized people will have developed a partial immunity due to colonization with mild strains. As with the vulnerability of *C. diphtheriae* to ß phage, if the virulence antigens are selected against strongly over long periods one would expect the vulnerability to reversion to eventually be reduced.

1.1.2.3 Haemophilus influenzae

Like the diphtheria toxoid vaccine, the *Hemophilus influenzae* vaccines have inadvertently conformed to the virulence-antigen criteria, and they too appear to have a disproportionately high effectiveness in suppressing severe disease. Although *H. influenzae* has been a major cause of bacterial meningitis, it has been almost eliminated as a factor where vaccination efforts have been intense (Musser et al., 1990; Bijlmer, 1991; Duclos, 1992; Eskola et al., 1992; Adams et al., 1993; Booy et al., 1993; 1994; Shapiro, 1993).

During vaccine development, the polysaccharide antigen that comprises the type b capsule (polyribosyl ribotol phosphate, abbreviated PRP) was identified as a promising vaccine component because type b *H. influenzae* (Hib) has been responsible for the vast majority of invasive disease due to *H. influenzae*; it typically causes more than 95 % of invasive disease and 99 % of the disease caused by the six typeable strains (Anonymous, 1992; Dajani et al., 1979; Granoff and Basden, 1980; Murphy et al., 1987; Bijlmer et al., 1992; Wenger et al., 1992). Simple PRP vaccines were introduced in the mid-1980s and replaced with polysaccharide conjugate vaccines during the late 1980s and early 1990s. During this period, rates of invasive disease dropped precipitously, often even more so than expected as a direct result of protection by the vaccine (Booy et al., 1992; Eskola et al., 1992; Peltola et al., 1992; Broadhurst et al., 1993; Murphy et al., 1993ab; Shapiro, 1993).

The conjugate vaccine is particularly responsible for the decline because it provides more powerful immunity in the infants and young children most prone to invasive disease and generates long-term T-cell immunity (Cartwright, 1992; Peltola et al., 1992). But one of the unexpected findings was that the onset of the decline occurred too early to be ascribable to the conjugate vaccine (Broadhurst et al., 1993; Murphy et al., 1993a). The early decline also seemed unascribable to the PRP vaccine because some of the decline occurred in children who were too young to receive the polysaccharide vaccine. Indirect effects were considered unlikely because empirical studies found no reduction in the rate of Hib colonization in response to the PRP vaccine (Murphy et al., 1993ab).

These findings, however, do not rule out the possibility of other kinds of indirect effects. Even if PRP vaccination has no discernable effects on the frequency of Hib colonization, it could have an effect by tipping the competitive balance against Hibs. Specifically, if dosages transmitted from colonized individuals are lower than those from more overtly infected individuals, and if the probability of generating Hib disease depends on dosage [both of which are implied by the association between high densities of bacteria in nasal mucosa and the occurrence of invasive disease (Liu and Smith, 1992), then PRP vaccination may suppress disease frequency over the long term even if it has no detectable indirect effect on the frequency of colonization. Any indirect effects would be enhanced by suppression due to competition between type b and other *H. influenzae*. Because serological responses to type b cross-react with nontypeable *H. influenzae* (Liu and Smith, 1992), this competition may not be limited to simultaneous coinfection with competing strains.

Such indirect effects could occur over a relatively large spatial scale as well as a long temporal scale. The large spatial scale requires that control populations be more removed from experimental populations than is necessary for more direct effects; declines in Hib disease among control populations chosen for detection of other indirect and direct effects may therefore be consistent with the evolutionary effects, even though the observed differences represent evidence against short-term effects of vaccination (e. g., the 1991 data described by Booy et al., 1992). Because the conjugate vaccines generate long-term immunity, their indirect suppression should be greater than that of the nonconjugate PRP vaccine.

These arguments assume that *H. influenzae* vaccines favor evolution toward less virulent *H. influenzae* variants. The available evidence supports this assumption. A multi-state study by the U.S. Centers for Disease Control (Wenger et al., 1992) showed that the absolute frequency of invasive disease caused by type b strains declined between 1986 and 1989 as PRP vaccines were being introduced, but the frequency of disease caused by other capsular serotypes and nontypeable strains did not. (The absolute frequency of non-b *H. influenzae* was about 15 % higher in 1989 than in 1986, but the rise was not statistically significant.) A study of school children showed that use of conjugate vaccine reduced the frequency of Hib carriage, confirming that fewer sources of Hib were present after vaccination (Murphy et al., 1993b). These findings indicate that evolutionary change (i. e., a change in the relative gene frequency) has occurred. *H. influenzae* has evolved toward a lower frequency of type b genotypes. This change probably involves an increase in the relative frequency of mild strains, because Hib tend to be more virulent than non-b *H. influenzae*.

The stability of the reduced frequency of Hib disease is a critical issue. The stability of protection afforded by virulence-antigen vaccines depends on whether the virulence mechanism provides a benefit to the disease organisms. If such benefits were not present, mild pathogens would not be under selective pressure to reevolve a virulence mechanism after virulent organisms were suppressed by a virulence-antigen vaccine. On the other hand, such benefits are generally expected because any pathogen paying the price of virulence without obtaining a compensating benefit would tend to be outcompeted by variants not paying the price. When virulence results from adaptive exploitation, virulence-antigen vaccines should eventually cause the pathogen to reevolve another mechanism of exploitation that does not rely on the antigen that was targeted by the vaccine. If a variety of exploitation mechanisms exist in the population or are readily generated by mutations or recombinations, then the success of a virulence-antigen vaccine will be relatively short-lived.

Various virulence-enhancing components occur among Hib and non-b strains. Some Hib strains are particularly mild, either not causing disease or causing it in a disproportionately low proportion of people harboring the organism (Musser et al., 1990; van Alphen et al., 1991; Bijlmer et al., 1992; Kroll et al., 1993). Studies of mucosal invasiveness show that different invasive mechanisms exist and are associated with differences in virulence; the dangerous *aegyptius* variant of Hib invades epithelial cells, whereas other Hib strains move between epithelial cells (Farley and Stephens, 1992; Farley et al., 1992). Although the type b capsule can enhance virulence, particularly after the bacteria are in the bloodstream, other virulence-enhancing genes are responsible for some of the virulence of type b *H. influenzae* (Liu and Smith, 1992).

The components responsible for this variation in virulence are candidates for virulence-antigen vaccines, especially when the components are produced by strains of both non-b and type b *H. influenzae*. One component is haemocin, which increases the ability of *H. influenzae* to infect hosts (LiPuma et al., 1992). It was present in 86 % of type b strains sampled, being absent in the most genetically divergent type b strains, which tend to be less virulent, and present in 95 % of the less divergent Hib (LiPuma et al., 1992). When equal mixtures of haemocin- and non-haemocin-producing strains were inoculated intranasally into infant rats, haemocin-producing strains became dominant in both nasopharyngeal and blood cultures, indicating an association between haemocin and increased infectious and invasive potential (LiPuma et al., 1992). Additional study will be needed to determine whether the enhancement of infection and invasion by haemocin results simply from interference with haemocin$^-$ strains or whether haemocin more directly contributes to virulence. Preliminary experiments suggest that some of the negative effects of haemocin may result directly from its interaction with host cells (J. J. LiPuma, personal communication). If so, haemocin would be an excellent candidate for inclusion in a virulence-antigen vaccine.

Lipopolysaccharide (LPS) variants offer other candidates. LPS damages the vascular endothelium and appears to contribute to damage at the blood-brain barrier (Liu and Smith, 1992). LPS variants that are associated with reduced virulence are still viable (Kimura and Hansen, 1986; Zwahlen et al., 1986), suggesting that targeting of particular LPS variants that contribute to virulence could favor mild variants. Variation in LPS may explain some of the particularly high virulence of type b strains because genes required for expression of LPS and capsule appear to be linked (Zwahlen et al., 1986). Variation in carbohydrate residues are responsible for differences in virulence, and LPS from Hib cross-react with those from nontypeable strains (Liu and Smith, 1992). In a study of two mild Hib strains, a larger LPS appeared to provide resistance to serum neutralization in one of two strains, and may thus be important for invasiveness (Kimura and Hansen, 1986). A vaccine directed against this LPS may protect against variants that are particularly capable of invasiveness, and thereby tip the competitive balance in favor of the less dangerous variants.

Outer membrane proteins (OMP) provide another set of candidates. The P2 outer membrane protein (OMP) of Hib contributes to shorter generation times and the ability to cause bacteremia in infant rats (Liu and Smith, 1992). OMP P2 also has the potentially valuable property of not being cross reactive (Liu and Smith, 1992). It might therefore selectively suppress the harmful Hib variants while allowing milder variants to be maintained. A 28 kDa OMP similarly contributes to bacteremia after intranasal inoculation in rats and occurs on many strains of Hib (Liu and Smith, 1992).

Conversely, although the P1 and 98 Kda OMPs induce protective antibodies, available evidence indicates that they do not contribute to virulence (Liu and Smith, 1992). According to the virulence-antigen strategy, these antigens should not be included in vaccines in spite of their immunogenicity.

Similarly, fimbria facilitate colonization of the buccal cavity and are serologically cross-reactive. Fimbrial antigens have therefore been considered as potential vaccine components, but their contribution to invasive disease is doubtful (Weber et al., 1991; Liu and Smith, 1992). If they contribute to colonization but not disease, fimbrial antigens are just the kind of antigen that should not be incorporated into vaccines, because they may suppress the circulation of mild variants in the community.

Although non-b strains are a less common cause of invasive disease, they do cause otitis media and sinusitis in infants and children, lower respiratory tract disease in the elderly, bronchitis patients, and children recovering from other infections, and sometimes invasive disease among neonates and postpartum women (Cox et al., 1991; Farley et al., 1992; Korppi et al., 1992; Murphy et al., 1992). One-quarter to one-half of the invasive *H. influenzae* disease among people ten years old or older is caused by non-b *H. influenzae* (Kostman et al., 1993; van Alphen et al., 1993). Although most of these cases involve compromised patients, the presence of virulence-enhancing products in non-b *H. influenzae* emphasizes both the need to protect against them and the potential for the use of virulence-antigen vaccines. Natural immunity to nontypeable strains is largely strain-specific and can select for new antigenic variants in chronic infections; yet cross-reactive antibodies can be elicited (van Alphen, 1992). Different variants of nontypeable *H. influenzae* therefore elicit the kind of partially overlapping immunity that is necessary for the virulence-antigen strategy.

Interestingly, when grown in cell culture, type b strains that lost capsules were associated with higher levels of growth and invasiveness (St. Geme and Falkow, 1992). Superficially this difference seems to be at odds with the epidemiological evidence suggesting that type b strains are particularly prone to causing invasive disease, but when analyzed more broadly the evidence is not contradictory. Capsules in and of themselves may reduce an organism's overall growth in cell culture by reducing the ability of the bacterium to adhere and by diverting resources that the bacterium might otherwise use for reproduction. The type b antigens, however, may favor increased virulence evolutionarily even though capsule production may divert resources away from production of other compounds related to virulence. When pathogens are durable in the external environment, natural selection should favor increased virulence because increased durability reduces the pathogen's reliance on host mobility for transmission (Ewald, 1987). This factor has recently been shown to explain statistically a great deal of the variation in mortality rates among pathogens of the human respiratory tract (Walther and Ewald unpublished manuscript). If the type b capsular antigen causes increased virulence evolutionarily by increasing durability in the external environment, the shift from Hib to other *H. influenzae* in response to the conjugate vaccine may be unstable over the long-term because durability in the external environment should not be particularly difficult to evolve.

In this context, invasive life-threatening disease is seen as an price of infection that a durable bacterium must occasionally pay as a result of its tendency to be more exploitative than its less durable competitors, but the price is outweighed by the fitness benefits associated with more extensive use of host resources. The increased virulence associated with capsule loss documented by St. Geme and Falkow (1992) therefore should not be interpreted as evidence that nonencapsulated strains are inherently more virulent than encapsulated strains, an interpretation that would be contrary to the body of evidence indicating that most invasive disease is due to *H.influenzae* type b. Rather, it suggests that type b *H. influenzae* has capabilities for virulence that may be reduced in the proximate mechanistic sense by the presence of a capsule, but increased in the ultimate sense because capsulated forms are selected to exploit the host to a greater degree.

The evolutionary hypothesis proposes that the increased capacity for causing disease that is associated with type b forms and is enhanced by capsule loss results from the selection for increased host exploitation among Hib. The evolutionary hypothesis that I propose above directly implicates a connection between durability and virulence. This connection could be tested by comparing the damage due to adherence and cell entry caused by type b capsule-minus mutants with the analogous damage caused by naturally unencapsulated *H. influenzae*. The evolutionary argument predicts that the capsule-minus mutants will have greater negative effects on cells than the naturally occurring unencapsulated *H. influenzae* (i. e., those not derived by recombination from type b organisms). Reciprocally, if naturally occurring unencapsulated *H. influenzae* suddenly obtain a type b capsule they should not be as severe.

The available evidence accords with this expectation. Phylogenetically divergent Hib apparently acquired their type b genes horizontally and are mild like the *H. influenzae* that have close affinities to the rest of their genome, rather than being highly virulent like the rest of the Hib (Musser, 1990; Kroll et al., 1993). Similarly, introduction of type b capsule genes results in enhanced virulence that falls short of the virulence of typical Hib, suggesting a linkage between type b capsular genes and other virulence genes (Zwahlen et al 1989, Kroll et al. 1993). This association is worrisome because it is a further indication that the current success of the PRP-conjugate vaccine will erode, as virulence components become unlinked from b capsule components. More generally, these evolutionary considerations indicate that even with the recent successes of Hib vaccines, we would be wise to determine other antigens that could form the basis of virulence-antigen vaccines to be used if the efficacy of the type-b conjugate vaccines begins to wane.

The strong geographic affinities of different types of *H. influenzae* suggests that spread by one type of *H. influenzae* into an areas dominated by another occurs slowly (Musser et al., 1990). This kind of evidence is encouraging because it suggests that a population of *H. influenzae* that has been shifted toward mildness by a virulence-antigen vaccine program may be resistant to reinvasion by more virulent pathogens from outside areas where such vaccine programs are not in place. Similarly if surveillance indicates that new virulent variants are beginning to emerge in an area, new virulence-antigen vaccines may be able to guard against them before the variants spreads regionally.

As the experience with *Haemophilus* conjugate vaccines unfolds, it will offer evidence illustrating the potential for short-to-intermediate-term success generated from vaccines that guard against the vast majority of virulent variants, but also the potential for reduced long-term efficacy that may result from use of antigens that failed to protect against what was originally a small proportion virulent variants. In the mean time, it would be wise to identify the action and importance of the other virulence antigens and to evaluate them in experimental vaccines.

1.1.2.4 Streptococcus pneumoniae

The virulence-antigen model is feasible for many vaccines that are in less advanced stages of development. *Streptococcus pneumoniae*, for example, is a major cause of pneumonia, meningitis, and severe ear infections, killing about 40,000 people each year in the United States (Lee et al., 1991). It causes its negative effect largely through production of pneumolysins, which inhibit the attack of white blood cells and rupture cell membranes, thereby liberating nutrients for the bacterium and expanding the scope of infection (Rubins et al., 1992; Berry et al., 1992). The similarity of the pneumolysins from different isolates of *S. pneumoniae* and the protection that a toxoid version elicited across *S. pneumoniae* capsular serotypes implicate pneumolysins as excellent vaccine candidates (Lock et al., 1992; Alexander et al., 1994). The currently used vaccine, which is a collection of complex sugar chains found on the capsules enclosing the different strains of the bacterium, often yields only weak protection (Lee et al., 1991; Lock et al., 1992). Besides providing better protection over the short term, pneumolysin-based vaccines should shift the competitive balance in favor of those *S. pneumoniae* without pneumolysins, thereby favoring evolution toward benignness.

1.1.2.5 Vibrio cholerae

The severe effects of *V. cholerae* on people are generated primarily by the cholera toxin, which causes a massive efflux of fluids into the intestinal lumen. The efflux probably benefits the organism first by flushing out competing organisms, and then by generating high densities of *V. cholerae* in easily dispersed fecal material (Ewald, 1994). The efflux may inflict great harm on the person, however, by causing dehydration and hypovolemic shock. Vaccines that block the effect of the cholera toxin could make cholera toxin a liability. The mild variants that infect without producing toxin (Minami et al., 1991) might then be favored by natural selection because they could channel the resources that would otherwise be wasted on toxin production into compounds that increase the bacterium's survival and reproduction.

Current development of vaccines to protect against cholera involve field tests of killed *V. cholerae* supplemented by portions of the cholera toxin; when evaluated by traditional criteria these supplemented vaccines appear to be about as effective as unsupplemented vaccines overall, but more effective during the six months after vaccination for young children (Sack et al., 1991; Holmgren et al., 1992). But the traditional criteria probably underestimate their effectiveness because the criteria do not include the evolutionary benefits. So long as the toxin-supplemented cholera vaccines inhibit cholera toxin, the net benefit of toxin production to *V.cholerae* declines. Although overall effects of the toxin-supplemented vaccine were comparable, there is some indication that the toxin-supplemented vaccine reduced the frequency of severe disease: individuals receiving the toxin-supplemented vaccine experienced significantly lower incidence of severe cholera than controls (73 % efficacy), whereas recipients of the unsupplemented vaccine did not (54 % efficacy) (Clemens et al., 1992); the incidence of mild and moderate cases, however, was higher in the toxin-supplemented group (38 % vs 62 % efficacy) (Clemens et al., 1992), perhaps indicating a tendency for the toxin-supplemented vaccine to make severe cases more mild. Although the short term measures of efficacy therefore are slight, even a slight short term effect is promising if it is associated with a long-term evolutionary disfavoring of the more toxigenic variants. Even if the short-term efficacy of such a vaccine is low, the long-term benefits might warrant its use, especially if other environmental factors that foster virulent *V. cholerae* variants, such as waterborne transmission (Ewald, 1994), could be simultaneously reduced.

On the basis of traditional criteria, researchers have suggested that cholera vaccines should include compounds that trigger immunity against the milder el tor *V. cholerae* but not against the more severe classical type (Osek et al., 1992). As with incorporation of FHA into acellular pertussis vaccines, evolutionary considerations suggest that over the long term these additional compounds may do more damage than good by disfavoring the milder variants.

1.1.3 Guidelines for the Virulence-Antigen Strategy

1.1.3.1 Virulence, Cross-Protection, and Antigen Selection

These vaccine programs against specific pathogens illustrate the importance of understanding better the relationships between pathogen products and virulence. If the spectrum of *H. influenzae* virulence, for example, encompasses many different mechanisms by which mild *H. influenzae* can be converted into virulent *H. influenzae* (biochemically through the use of products such as haemocin and/or evolutionarily through structures that increased durability in the external environment) then each mechanism needs to be understood and targeted by the vaccine to provide the highest stability of long-term control. If just one or a few mechanisms have evolved (as is the case with *C. diphtheriae* and *V. cholerae*) the job will be easier.

The virulence-antigen strategy emphasizes the need for increased study of the cross-protection that can be conferred against virulent variants by the mild variants that would continue to circulate during a virulence-antigen vaccine program. Assessment of this cross-protection will provide a sense of the long term benefits that can be expected from a virulence-antigen strategy. The cross-protection will likely be on the strong side for pathogens like toxigenic and nontoxigenic *C. diphtheriae* because they are exposing similar antigens whether or not the toxin is produced. The degree of cross-protection generated by the closely related *Corynebacterium* variants studied in Italy (Mencarelli et al., 1992) is more uncertain. The degree of cross-protection will be on the weak side for type b vs non-b *H. influenzae*, but should be very strong for strains that are from the same type but differ in whether they produce a virulence factor such as haemocin.

Virulence-antigen strategies should be particularly successful when the virulence enhancing antigens are costly for the bacterium to produce. The more costly they are, the more they will be a liability to the bacterium when their function is blocked by vaccination. In this regard, the diphtheria toxin is an excellent vaccine antigen because it accounts for 5 % of the protein production by toxigenic *C. diphtheriae* (Pappenheimer and Gill, 1973). Similarly, the type b capsular antigen appears to be expensive for *H. influenzae*. Hib isolated at sites of invasion often have extra copies of the type b encapsulation genes, which are thought to increase the degree of encapsulation, but these isolates revert to a reduced copy number on culturing in vitro (Kroll et al., 1993), indicating that the funnelling of resources into the capsular antigen becomes a liability in the fitness benefits of the antigen are reduced. *V. cholerae* infections in mice also suggest that virulent *V. cholerae* may be at a disadvantage relative to milder *V. cholerae* as infections develop; avirulent mutants tend to increase in frequency, presumably because toxin production facilitates establishment of the infection but reduces competitive ability within the host after establishment (Ewald, 1994). When choosing among virulence antigens, preference should be given to those antigens that are expensive for the organism to produce.

One approach of modern vaccinology has been almost the opposite of the virulence-antigen strategy: instead of using the virulence antigens, virulence genes are deleted in efforts to make a safe live vaccine. Such efforts may work well if the deleted virulence genes do not code for antigens that are recognized and attacked by the immune systems and if the pathogens have little evolutionary flexibility. If these situations do not hold however, this virulence deletion strategy could have negative evolutionary effects; the vaccines could preferentially select against the milder strains, leaving the virulent strains to repopulate the unimmunized. Lapses in vaccination efforts would then tend to generate outbreaks of severe rather than mild infections.

Most major vaccination programs, such as those for measles, rubella, and mumps, parallel the whooping cough experience; they have been dealt setbacks from vaccine-caused disease, outbreaks that spread among the unvaccinated, or reinfections as vaccine-induced immunity declines (Hewlett, 1990; Hilleman, 1992; Pinchichero et al., 1992; Weiss, 1992; Forsey et al., 1992). The incidence of measles in the US, for example, rose dramatically during the 1980s, with infections often occurring in previously vaccinated adults (Poland and Jacobson, 1994). This increase occurred in spite of a program initiated by the US Public Health service to eliminate measles from the US by 1982 (Centers for Disease Control, 1978). Changes in antigens suggest that this reappearnce might be at least partly due evolutionary changes in the virus; currently circulating measles viruses now contain new antigenic determinants that make them less strongly neutralized by the standard vaccine than by a vaccine generated from currently circulating virus (Tamin et al., 1994). Vaccines that are not based on virulence antigens select for pathogen variants that are merely different from the vaccine component, not more mild. As antigenically dissimilar viruses arise we can therefore expect these kinds of resurgences.

In this regard, researchers have recognized public health costs associated with having vaccines that do not incorporate a spectrum of antigens that protects against the spectrum of pathogenic variants (Kroll et al., 1993). The virulence-antigen strategy develops this line of reasoning by emphasizing the long-term benefits both of doing so and of restricting the vaccine components to virulence antigens. Traditional approaches propose that candidate antigens "must ... be conserved among strains so that immunization can protect against all or most strains" (Murphy et al., 1992). The virulence-antigen strategy does not advocate protection against all or most strains. Rather, it distinguishes the dangerous from the safe strains and protect against the former but not the latter. The virulence-antigen strategy therefore requires a much more complete understanding of the biochemical determinants of virulence than the traditional approach.

1.1.3.2 Compliance, Age, and Vaccination

The successes of vaccination programs against the most dangerous pathogens leave us with pathogens that cause disease in ever smaller percentages of infections. For such pathogens slight negative effects of vaccines may be sufficient to offset the positive effects because of the small probability of a serious infection in an unvaccinated individual. In such situations both the public policy impetus and the degree of compliance among individuals may be low, resulting in fewer vaccinations and more infections. The virulence-antigen strategy should offer extra protection by making these infections benign.

Similar issues of compliance arise from the age-dependence of vaccine responses. The substantial incidence of respiratory diseases in spite of stable vaccination programs has led researchers to recommend revaccination in older age groups to boost waning immunity (Cherry, 1992; Mencarelli et al., 1992; Poland and Jacobson, 1994). The virulence-antigen strategy reinforces but revises this recommendation. Without revaccination, people in older age groups can act as reservoirs. If adults experience infection but not disease, however, they may be unmotivated to become vaccinated, leading to low coverage. Even if these people are not adversely affected (e. g, because of partial immunity) and coverage is low, virulence-antigen vaccines should help protect all age groups, not only by reducing the spread of the virulent variants, but by favoring milder protective variants among the organisms circulating in the community.

The diphtheria vaccination experience in China illustrates the value of adult vaccination with virulence-antigen vaccines, even when adults are not severely affected. Vaccination coverage of infants in the Hubei province during the 1970s and 1980s rose above 80 %, an adequate level for controlling diphtheria if maintained across generations (Youwang et al., 1992). Adults, however, were not targeted because diphtheria does not affect adults as severely as children. The low vaccination levels among adults were insufficient to prevent an outbreak of toxigenic *C. diphtheriae*, which was controlled within a few weeks by targeting adults (Youwang et al., 1992). If data on nontoxigenic *C. diphtheriae* had been obtained, this study could have revealed whether adult vaccination generated increased spread of toxinless variants.

Age differences in immune responses to particular antigens can also be used to control pathogens with multiple virulence antigens. An acellular pertussis vaccine containing pertactin, for example, stimulated a strong immune response in adults but not in children when administered as part of an acellular DPT vaccine (Podda et al., 1991; Englund et al., 1992). If the adult response is protective, an adult booster using pertactin could be an effective part of a virulence-antigen strategy.

Concern has also focussed on reductions in infections in younger age groups that are caused by traditional vaccines. Such reductions could lead to increased disease in older age groups because immunity among older children may depend on repeated subclinical infections (Adams et al., 1993; Shapiro, 1993). The virulence-antigen strategy should reduce this problem by leaving mild variants to circulate in the community.

1.1.3.3 A Global Long-Term View

One problem with vaccines is that they require continued economic investment. Maintaining this investment in poor countries often requires good will from the wealthier countries. Virulence-based vaccines may foster this investment by more strongly benefitting both the wealthy countries that may contribute the bulk of the funds and the poorer countries that may receive help from the wealthy countries. Within wealthy countries with a high level of vaccine coverage, the predominant mild strains acts as buffers that inhibit the spread of virulent strains that enter from countries with poorer vaccine coverage. But wealthy countries also benefit by fostering such vaccination programs in countries with which they are connected by personal traffic. Such programs should make the disease organisms in these poorer countries more mild so that they will be less harmful when they inadvertently enter the wealthy countries.

The virulence antigen strategy emphasizes the need for continued development of virulence-antigen vaccines even after a successful non-virulence-based vaccine has been implemented. Just as development efforts continue when the best vaccines provide incomplete protection (e. g., the simple polysaccharide *H. influenzae* vaccines) or infrequently cause damaging side effects (e. g., the whole cell pertussis vaccine), vaccines that do not force the the locally and globally circulating pathogens toward mildness will be generating a higher frequency of illness than those that do generate such a shift. The virulence antigen strategy is therefore one that promises to drive the low numbers of residual cases left by traditional vaccines to even lower numbers, just as the incidences of residual cases of diphtheria are now about 100-fold lower than those of pertussis.

Once a safe and effective virulence-antigen vaccine has been developed, surveillance and vaccine development must continue because variants may reevolve virulence by a mechanism that is not blocked by the vaccine. Some *C. diphtheriae*, for example, can produce disease without producing a toxin (Kallick et al., 1970; Zuber et al., 1992). If the virulence of these variants is associated with a competitive superiority, these variants will spread. Vaccination efforts need to block this evolutionary process by incorporating the newly arising virulence antigens into vaccines. This kind of evolutionary control should be less difficult than the nonevolutionary control that has been the goal of conventional efforts. If the virulence-antigen vaccine protects against the existing spectrum of virulence mechanisms, breaching of evolutionary control requires that a pathogen evolve a new mechanism of virulence.

Modern health sciences has conquered or is well on its way to conquering the easy adversaries like the smallpox virus. We are now left with adversaries that will be more difficult to control because, unlike smallpox, they have a great potential for resisting and evolving around our interventions. The virulence-antigen strategy uses to our advantage just the characteristic that makes these organisms so difficult to control. Instead of allowing the evolutionary potential of these pathogenic adversaries to thwart our efforts, the virulence-antigen strategy uses the pathogen's evolutionary potential to generate mild strains that will be providing protection against the severe strains in the community.

In theory, virulence-antigen strategies could be applied to virtually all vaccines. The examples of virulence-antigen vaccines in use or in development involve bacterial diseases. Although the virulence-antigen strategy should be applicable to other kinds of pathogens, advancements will be slower because (i) determinants of virulence are more difficult to resolve and (ii) high rates of antigenic variation may make targeting of virulence antigens more complicated. The first test cases will therefore be bacterial. *C. diphtheriae* and *H. influenzae* are the first two pathogens for which effects of virulence-antigen vaccines can be evaluated, but even with these pathogens the degree to which mild variants interfere with virulent variants needs clarification. By the time research resolves the determinants of viral and protozoal virulence sufficiently to apply a virulence-antigen strategy, we should have a sufficient number of test cases among bacterial pathogens to understand the general effectiveness of the strategy.

Yet even now we can envision ways in which the strategy could be applied against some of the toughest adversaries. For example, as knowledge about the virulence determinants of HIV improve, the virulence-antigen strategy might be applicable in therapeutic vaccines. By using antigens that are associated with particularly virulent variants (e. g., antigens unique to syncytium inducing phenotypes) the progression of HIV infections to AIDS might be slowed and the transmitted variants (those not blocked by the vaccine) might be more mild.

Considering this potential breadth of applicability, it may seem surprising that we did not generalize from the diphtheria experience, even though researchers realized that the decline in diphtheria was associated with an evolution of *C. diphtheriae* toward more mild forms. The general reason is probably that researchers have focussed on eradication of pathogens rather than benign coexistence with them. They therefore did not see the shift from a severe toxigenic *C. diphtheriae* to nontoxigenic corynebacteria as being more beneficial than a comparable reduction of toxigenic *C. diphtheriae* in the absence of any nontoxigenic corynebacteria. Now that pathogens have better educated us about their potential for evolutionary change and for long term persistence in spite of our efforts to eradicate them, we are in a better position to recognize the general value of fostering benign coexistence through virulence-antigen strategies.

Acknowledgements

This study was supported by a grant from Leonard X. Bosack and Bette M. Kruger Charitable Foundation, a George E. Burch Fellowship in Theoretic Medicine and Affiliated Sciences awarded by the Smithsonian Institution, and an Amherst College Faculty Research Award. The ideas presented in this manuscript were improved by discussions with J. M. Musser, E. R. Moxon, and J. J. LiPuma, and were facilitated by participation at a June 1995 conference on emerging viruses sponsored by the New York Academy of Medicine. For help in obtaining articles I thank L. Bailey and S. Edelberg of the Robert Frost Library at Amherst College.

References

Adams, W. G., Deaver, K. A., Cochi, S. L., Plikaytis, B. D., Zell, E. R., Broome, C. V., Wenger, J. D. (1993) Decline of childhood *Haemophilus influenzae* type b (Hib) disease in the Hib vaccine era, J. Amer. Med. Assoc. *269*, 221-226.

Alphen, L. van (1992) Epidemiology and prevention of respiratory tract infections due to nonencapsulated *Haemophilus influenzae*, J. Infect. Dis. *165*, S177-S180.

Alphen, L. van, Bol, P., Kok, M. L. J., Geelen-van den Broek, L. (1991) Differences in subtype distribution of *Haemophilus influenzae* type b from carriers in the general population and patients with meningitis, J. Med. Entomol. *34*, 313-316.

Alphen, L. van, Spanjaard, L., Dankert, J. (1993) Non-typable *Haemophilus influenzae* invasive disease, Lancet *341*, 1536.

Amsbaugh, D. F., Li, Z. M., Shahin, R. D. (1993) Long-lived respiratory immune response to filamentous hemagglutinin following *Bordetella pertussis* infection, Infect. Immun. *61*, 1447-1452.

Anderson, E. L., Mink, C. M., Berlin, B. S., Shih, C. N., Tung, F. F., Belshe, R. B. (1994) Acellular pertussis vaccines in infants: evaluation of single component and two-component products, Vaccine *12*, 28-31.

Anonymous (1992) Hib, Hib, hooray, Lancet *340*, 845.

Barkin, R. M. (1975) Measles mortality. Analysis of the primary cause of death, American Journal of the Diseases of Children *129*, 307-309.

Berry, A. M., Paton, J. C., Hansman, D. (1992) Effect of insertional inactivation of the genes encoding pneumolysin and autolysin on the virulence of *Streptococcus pneumoniae* type 3, Microb. Pathog. *12*, 87-93.

Bijlmer, H. A. (1991) World wide epidemiology of *Haemophilus influenzae* meningitis: industrialized versus non- industrialized countries, Vaccine *9*, S5-S9.

Bijlmer, H. A., van Alphen, L., Geelen-van den Broek, L., Greenwood, B. M., Valkenburg, H. A., Dankert, J. (1992) Molecular epidemiology of *Haemophilus influenzae* type b in The Gambia, J. Clin. Microbiol. *30*, 386-390.

Booy, R., Hodgson, S., Carpenter, L., Mayonwhite, R. T., Slack, M. P. E., Macfarlane, J. A., Haworth, E. A., Kiddle, M., Shribman, S., Roberts, J. S. C., Moxon, E. R. (1994) Efficacy of *Haemophilus influenzae* type b conjugate vaccine PRP-T, Lancet *344*, 362-366.

Booy, R., Moxon, E. R., Macfarlane, J. A., Mayonwhite, R. T., Slack, M. P. E. (1992) Efficacy of *Haemophilus influenzae* type b conjugate vaccine in Oxford region, Lancet *340*, 847.

Broadhurst, L. E., Erickson, R. L., Kelley, P. W. (1993) Decreases in invasive *Haemophilus influenzae* diseases in US army children, 1984 through 1991, J. Amer. Med. Assoc. *269*, 227-231.

Brooks, G. F. (1969) Recent trends in diphtheria in the United States, J. Infect. Dis. *120*, 500-502.

Cartwright, K. A. V. (1992) Vaccination against *Haemophilus influenzae* b disease, Br. Med. J. *305*, 485-486.

Centers for Disease Control (1978) Goal to eliminate mesles fro the United States., Morbid. Mortal. Weekly Rep. *27*, 391.

Chen, R. T., Broome, C. V., Weinstein, R. A., Weaver, R., Tsai, T. F. (1985) Diphtheria in the United States, 1971-81, Am. J. Public Hlth. *75*, 1393-1397.

Cherry, J. D. (1992) Pertussis: the trials and tribulations of old and new pertussis vaccines, Vaccine *10*, 1033-1038.

Cherry, J. D., Brunell, P. A., Golden, G. S., Karzon, D. T. (1988) Report of the task force on pertussis and pertussis immunization-1988, Pediatrics *82*, S939-S984.

Christodoulides, M. (1990) Pertussis vaccines: present status, In A. Mizrahi (Editor). Bacterial Vaccines. Advances in Biotechnological Processes, Wiley-Liss, New York, pp. 169-199.

Clemens, J. D., Sack, D. A., Rao, M. R., Chakraborty, J., Khan, M. R., Kay, B., Ahmed, F., Banik, A. K., van Loon, F. P. L., Yunus, M., Harris, J. R. (1992) Evidence that inactivated oral cholera vaccines both prevent and mitigate *Vibrio cholerae* O1 infections in a cholera-endemic area, J. Infect. Dis. *166*, 1029- 1034.

Cox, R. A. (1991) Neonatal septicaemia due to non-capsulate *Haemophilus influenzae* in three siblings, Journal of Infection *23*, 317-320.

Dajani, A. S., Asmar, B. I., Thirumoorthi, M. C. (1979) Systemic *Haemophilus influenzae* disease: An overview, Journal of Pediatrics *94*, 355-364.

Duclos, P. (1992) Statement on *Haemophilus influenzae* type b conjugate vaccines for use in infants and children, Canadian Medical Journal Association *146*, 1363-1366.

Ehrmann, I. E., Weiss, A. A., Goodwin, M. S., Gray, M. C., Barry, E., Hewlett, E. L. (1992) Enzymatic activity of adenylate cyclase toxin from *Bordetella pertussis* is not required for hemolysis, FEBS Letters *304*, 51-56.

Englund, J. A., Decker, M. D., Edwards, K. M., Pichichero, M. E., Steinhoff, M. C., Anderson, E. L. (1994) Acellular and whole-cell pertussis vaccines as booster doses – a multicenter study, Pediatrics *93*, 37-43.

Englund, J. A., Glezen, W. P., Barreto, L. (1992) Controlled study of a new five-component acellular pertussis vaccine in adults and young children, J. Infect. Dis. *166*, 1436-1441.

Eskola, J., Peltola, H., Kayhty, H., Takala, A. K., Makela, P. H. (June 1992) Finnish efficacy trials with *Haemophilus influenzae* type b vaccines, J. Infect. Dis. *165*, S137-S138.

Ewald, P. W. (1994) Evolution of Infectious Disease, Oxford University Press, New York.

Farley, M. M., Stephens, D. S., Brachman, P. S., Harvey, R. C., Smith, J. D., Wenger, J. D. (1992) Invasive *Haemophilus influenzae* disease in adults. A prospective, population-based surveillance, Ann. Intern. Med. *116*, 806-812.

Farley, M. M., Stephens, D. S. (1992) Pathogenic events during *Haemophilus influenzae* type b infection of human nasopharyngeal mucosa, J. Infect. Dis. *165*, S109-S110.

Farley, M. M., Whitney, A. M., Spellman, P., Quinn, F. D., Weyant, R. S., Mayer, L., Stephens, D. S. (1992) Analysis of the attachment and invasion of human epithelial cells by Haemophilus influenzae biogroup aegyptius, J. Infect. Dis. *165*, S111-S114.

Feldman, S., Perry, C. S., Andrew, M., Jones, L., Moffitt, J. E., Abney, R., Carlyle, W., Freeman, E. E., Hendrick, J., Hopper, S., Ray, M., Sistrunk, W., Smith, W. H., Stone, L., Welch, P., Womack, N., Miller, J., Thompson, R. H., Simmons, L., Sherwood, J. A., Denney, S. J., Shaak, C., Cooke, D. T., Mccaslin, L. (1992) Comparison of acellular (B type) and whole-cell pertussis- component diphtheria-tetanus-pertussis vaccines as the first booster immunization in 15- to 24-month- old children, Journal of Pediatrics *121*, 857-61.

Forsey, T., Bentley, M. L., Minor, P. D., Begg, N. (1992) Mumps vaccines and meningitis, Lancet *340*, 980.

Granoff, D. M., Basden, M. (1980) Haemophilus influenzae infections in Fresno County, California: a prospective study of the effects of age, race, and contact with a case on incidence of disease, J. Infect. Dis. *141*, 40-46.

Harnisch, J. P., Tronca, E., Nolan, C. M., Turck, M., Holmes, K. K. (1989) Diphtheria among alcoholic urban adults., Ann. Intern. Med. *111*, 71-2.

Herwaldt, L. A. (1993) Pertussis and pertussis vaccines in adults, J. Amer. Med. Assoc. *269*, 93-94.

Hewlett, E. L. (1990) *Bordetella* species, In G. L. Mandell, R. G. Douglas, and J. E. Bennett (ed). Principles and practice of infectious disease, Wiley, New York, pp. 1757-1762.

Hilleman, M. R. (1992) Past, present, and future of measles, mumps, and rubella virus vaccines, Pediatrics *90*, 149-53.

Holmgren, J., Svennerholm, A. M., Jertborn, M., Clemens, J., Sack, D. A., Salenstedt, R., Wigzell, H. (1992) An oral b subunit whole cell vaccine against cholera, Vaccine *10*, 911-914.

Howson, C. P., Fineberg, H. V. (1992) Adverse events following pertussis and rubella vaccines. Summary of a report of the Institute of Medicine, J. Amer. Med. Assoc. *267*, 392-396.

Kallick, C. A., Brooks, G. F., Dover, A. S., Brown, M. C., Brolnitsky, O. (1970) A diphtheria outbreak in Chicago, Ill. Med. J. *137*, 505-512.

Kamiya, H., Nii, R., Matsuda, T., Yasuda, N., Christenson, P. D., Cherry, J. D. (1992) Immunogenicity and reactogenicity of takeda acellular pertussis-component diphtheria-tetanus- pertussis vaccine in 2- and 3-month-old children in Japan, Am. J. Dis. Child. *146*, 1141-1147.

Kimura, A., Hansen, E. J. (1986) Antigenic and phenotypic variations of *Haemophilus influenzae* type b lipopolysaccharide and their relationship to virulence, Infect. Immun. *51*, 69-79.

Korppi, M., Katila, M. L., Jääskeläinen, J., Leinonen, M. (1992) Role of non-capsulated *Haemophilus influenzae* as a respiratory pathogen in children, Acta Paediatr. Scand. *81*, 989-992.

Kostman, J. R., Sherry, B. L., Fligner, C. L., Egaas, S., Sheeran, P., Baken, L., Bauwens, J. E., Clausen, C., Sherer, D. M., Plorde, J. J., Stull, T. L., Mendelman, P. M. (1993) Invasive *Haemophilus influenzae* infections in older children and adults in Seattle, Clin. Infect. Dis. *17*, 389-396.

Kroll, J. S., Moxon, E. R., Loynds, B. M. (1993) An ancestral mutation enhancing the fitness and increasing the virulence of *Haemophilus influenzae* type b , J. Infect. Dis. *168*, 172- 176.

Lee, C. J., Banks, S. D., Li, J. P. (1991) Virulence, immunity, and vaccine related to *Streptococcus pneumoniae*, Critical Reviews in Microbiology *18*, 89-114.

LiPuma, J. J., Sharetzsky, C., Edlind, T. D., Stull, T. L. (1992) Haemocin production by encapsulated and nonencapsulated *Haemophilus influenzae*, J. Infect. Dis. *165*, S118-S119.

Liu, V. C., and Smith, A. L. Molecular Mechanism of Haemophilus- Influenzae Pathogenicity, H. Schonfeld, and H. Helwig Bacterial Meningitis, Vol 45, Karger, Postfach/CH-4009 Basel/ Switzerland, 30-51.

Locht, C., Bertin, P., Menozzi, F. D., Renauld, G. (1993) The filamentous haemagglutinin, a multifaceted adhesin produced by virulent *Bordetella* spp, Mol. Microbiol *9*, 653-660.

Long, S. S., Welkon, C. J., Clark, J. L. (1990) Widespread silent transmission of pertussis in families: antibody correlates of infection and symptomatology, J. Infect. Dis. *161*, 480-486.

Marcinak, J. F., Ward, M., Frank, A. L., Boyer, K. M., Froeschle, J. E., Hosbach, P. H. (1993) Comparison of the safety and immunogenicity of acellular (BIKEN) and whole-cell pertussis vaccines in 15- to 20-month-old children, Am. J. Dis. Child. *147*, 290-294.

Mencarelli, M., Zanchi, A., Cellesi, C., Rossolini, A., Rappuoli, R., Rossolini, G. M. (1992) Molecular epidemiology of nasopharyngeal corynebacteria in healthy adults from an area where diphtheria vaccination has been extensively practiced, Eur. J. Epidemiol. *8*, 560-567.

Miller, L. W., Older, J. J., Drake, J., Zimmerman, S. (1972) Diphtheria immunization. Effect upon carriers and the control of outbreaks, Am. J. Dis Child *123*, 197-199.

Mills, K. H. G., Redhead, K. (1993) Cellular immunity in pertussis, J. Med. Entomol. *39*, 163-164.

Minami, A., Hashimoto, S., Abe, H., Arita, M., Taniguchi, T., Honda, T., Miwatani, T., Nishibuchi, M. (1991) Cholera enterotoxin production in *Vibrio cholerae*-O1 strains isolated from the environment and from humans in Japan, Appl. Environ. Microbiol. *57*, 2152-2157.

Murphy, T. F., Nelson, M. B., Apicella, M. A. (1992) The P6 outer membrane protein of non-typeable *Haemophilus influenzae* as a vaccine antigen, J. Infect. Dis. *165*, S203- S205.

Murphy, T. V., Osterholm, M. T. (1987) Prospective surveillance of *Haemophilus influenza* type b disease in Dallas County, Texas, and in Minnesota, Pediatrics *79*, 173-180.

Murphy, T. V., Pastor, P., Medley, F., Osterholm, M. T., Granoff, D. M. (1993) Decreased *Haemophilus* colonization in children vaccinated with *Haemophilus influenzae* type b conjugate vaccine, Journal of Pediatrics *122*, 517-523.

Murphy, T. V., White, K. E., Pastor, P., Gabriel, L., Medley, F., Granoff, D. M., Osterholm, M. T. (1993) Declining incidence of *Haemophilus influenzae* type b disease since introduction of vaccination, J. Amer. Med. Assoc. *269*, 246- 248.

Musser, J. M., Kroll, J. S., Granoff, D. N., Moxon, E. R., Brodeur, B. R., Campos, J., Dabernat, H., Frederiksen, W., Hamel, J., Hammond, G., Hoiby, E. A., Jonsdottir, K. E., Kabeer, N., Kallings, I., Khan, W. N., Kilian, N., Knowles, K., Koornhof, H. J., Law, B., Li, K. I., Montgomery, J., Pattison, P. E., Piffaretti, J. C., Takala, A. K., Thong, M. L., Wall, R. A., Ward, J. I., Selander, R. K. (1990) Global genetic structure and molecular epidemiology of encapsulated *Haemophilus influenzae*, Rev. Infect. Dis. *12*, 75-111.

Osek, J., Svennerholm, A. M., Holmgren, J. (1992) Protection against *Vibrio cholerae* el tor infection by specific antibodies against mannose-binding hemagglutinin pili, Infect. Immun. *60*, 4961-4964.

Pappenheimer, A. M. (1977) Diphtheria toxin, Annu. Rev. Biochem. *46*, 69-94.

Pappenheimer, A. M. (1982) Diptheria: studies on the biology of an infectious disease, Harvey Lectures *76*, 45-73.

Pappenheimer, A. M., Gill, D. M. (1973) Diphtheria, Science *182*, 353-358.

Pappenheimer, A. M., Murphy, J. R. (1983) Studies on the molecular epidemiology of diphtheria, Lancet *322*, 923-926.

Peltola, H., Kilpi, T., Anttila, M. (1992) Rapid disappearance of *Haemophilus influenzae* type b meningitis after routine childhood immunisation with conjugate vaccines, Lancet *340*, 592-594.

Petersen, J. W., Ibsen, P. H., Bentzon, M. W., Capiau, C., Heron, I. (1991) The cell mediated and humoral immune response to vaccination with acellular and whole cell pertussis vaccine in adult humans, FEMS Microbiol. Immunol. *76*, 279-287.

Pichichero, M. E., Francis, A. B., Blatter, M. M., Reisinger, K. S., Green, J. L., Marsocci, S. M., Disney, F. A. (1992) Acellular pertussis vaccination of 2-month-old infants in the United States, Pediatrics *89*, 882-887.

Podda, A., Nencioni, L., Marsili, I., Peppoloni, S., Volpini, G., Donati, D., Ditommaso, A., Demagistris, M. T., Rappuoli, R. (1991) Phase-I clinical trial of an acellular pertussis vaccine composed of genetically detoxified pertussis toxin combined with FHA and 69-kDa, Vaccine *9*, 741-745.

Poland, G. A., Jacobson, R. M. (1994) Failure to reach the goal of measles elimination. Apparent paradox of measles infections in immunized persons, Arch. Intern. Med. *154*, 1815-1820.

Rubins, J. B., Duane, P. G., Charboneau, D., Janoff, E. N. (1992) Toxicity of pneumolysin to pulmonary endothelial cells in vitro, Infect. Immun. *60*, 1740-1746.

Sack, D. A., Clemens, J. D., Huda, S., Harris, J. R., Khan, M. R., Chakraborty, J., Yunus, M., Gomes, J., Siddique, O., Ahmed, F., Kay, B. A., Vanloon, F. P. L., Rao, M. R., Svennerholm, A. M., Holmgren, J. (1991) Antibody responses after immunization with killed oral cholera vaccines during the 1985 vaccine field trial in Bangladesh, J. Infect. Dis. *164*, 407-411.

Sasse, A., Malfait, P., Padrón, T., Erikashvili, M., Freixa, E., Moren, A. (1994) Outbreak of diphtheria in Republic of Georgia, Lancet *343*, 1358-1359.

Schmitt, M. P., Holmes, R. K. (1991) Characterization of a defective diphtheria toxin repressor (dtxR) allele and analysis of dtxR transcription in wild-type and mutant strains of *Corynebacterium diphtheriae*, Infect. Immun. *59*, 3903-3908.

Schmitt, M. P., Twiddy, E. M., Holmes, R. K. (1992) Purification and characterization of the diphtheria toxin repressor, Proc. Natl. Acad. Sci. USA *89*, 7576-7580.

Seachrist, L. (1995) New pertussis vaccines safer, more effetive, Science News *148*, 54.

Shapiro, E. D. (1993) Infections caused by *Haemophilus influenzae* type b. The beginning of the end?, J. Amer. Med. Assoc. *269*, 264-266.

St.Geme, J. W., Falkow, S. (1992) Capsule loss by *Haemophilus influenzae* type b results in enhanced adherence to and entry into human cells, J. Infect. Dis. *165*, S117-S118.

Storsaeter, J., Olin, P. (1992) Relative efficacy of two acellular pertussis vaccines during three years of passive surveillance., Vaccine *10*, 142-144.

Tamin, A., Rota, P. A., Wang, Z. D., Heath, J. L., Anderson, L. J., Bellini, W. J. (1994) Antigenic analysis of current wild type and vaccine strains of measles virus, Journal of Infectious Diseases *170*, 795-801.

Tomoda, T., Ogura, H., Kurashige, T. (1992) The longevity of the immune response to filamentous hemagglutinin and pertussis toxin in patients with pertussis in a semiclosed community, Journal of Infectious Diseases *166*, 908-910.

Uchida, T., Gill, D. M., Pappenheimer, A. M. (1971) Mutation in the structural gene for diphtheria toxin carried by temperate phage ß, Nature New Biol. *233*, 8-11.

Vanura, H., Just, M., Ambrosch, F., Berger, R. M., Bogaerts, H., Wynen, J., Vandevoorde, D., Wiedermann, G. (1994) Study of pertussis vaccines in infants: comparison of response to acellular pertussis DTP vaccines containing 25 mg of FHA and either 25 or 8 mg of PT with response to whole-cell pertussis DTP vaccine, Vaccine *12*, 210-214.

Warren, K. S. (1986) New scientific opportunities and old obstacles in vaccine development, Proc. Natl. Acad. Sci. USA *83*, 9275-9277.

Weber, A., Harris, K., Lohrke, S., Forney, L., Smith, A. L. (1991) Inability to express fimbriae results in impaired ability of *Haemophilus influenzae* b to colonize the nasopharynx, Infect. Immun. *59*, 4724-4728.

Weiss, R. (1992) Measles battle loses potent weapon, Science *258*, 546-7.

Wenger, J. D., Pierce, R., Deaver, K., Franklin, R., Bosley, G., Pigott, N., Broome, C. V. (1992) Invasive *Haemophilus influenzae* disease: a population-based evaluation of the role of capsular polysaccharide serotype, J. Infect. Dis. *165*, S34-S35.

Youwang, Y., Jianming, D., Yong, X., Pong, Z. (1992) Epidemiological features of an outbreak of diphtheria and its control with diphtheria toxoid immunization, Int. J. Epidemiol. *21*, 807-811.

Zuber, P. L. F., Gruner, E., Altwegg, M., von Graevenitz, A. (1992) Invasive infection with ynon-toxigenic *Corynebacterium diphtheriae* among drug users, Lancet *339*, 1359.

Zwahlen, A., Rubin, L. G., Moxon, E. R. (1986) Contribution of lipopolysaccharide to pathogenicity of *Haemophilus influenzae*: comparative virulence of genetically-related strains in rats, Microb. Pathog. *1*, 465-473.

Zwahlen, A., Kroll, J. S., Rubin, L. G. and Moxon, E. R. (1989) The molecular basis of pathogenicity in *Haemophilus influenzae*: comparative virulence of genetically related capsular transformants and correlation with changes at the capsulation locus cap. Microb. Pathog. *7*, 225-235.

1.2 Economic Perspectives on Vaccine Needs

Robert V. Ashley and Christopher J. L. Murray

1.2.1 Introduction

In a world of scarcity, not even the richest countries can afford to purchase every possible health intervention. In poor countries, where health needs are greatest, health resources are limited to less than $10 per capita (Murray et al., 1993). Given the difficult choices between competing health intervention priorities, some principles must be followed to guide the allocation of resources. One strategy is to seek to allocate health resources amongst the different interventions available to those that will make the greatest countributions to population health status. Economic analysis can then be used to quantify the costs and health benefits of different interventions so that resources can be used to improve maximally the health of the community.

This chapter will focus on how vaccine-related activities can be evaluated in terms of the health resources they require and the health benefits they yield. We must distinguish two different analytical questions: the appropriate role of existing vaccines and the priorities for vaccine research. To address both of these issues from an economic perspective will require four bodies of information: estimates of costs and benefits of health interventions; a comparative evaluation of disease burdens; speculative judgments on scientific opportunities in health research and development; and the likely cost-effectiveness of proposed research products. Considerable progress has been made in the past decade in developing compendiums of cost-effectiveness information (Jamison et al., 1993a; World Bank, 1993) and in the past three years the Global Burden of Disease study has provided comparable information on disease burden by cause (Murray and Lopez, 1994a; 1994b; 1996a; 1996b). Several studies have attempted to apply formalized methods to assess comprehensively the third required data-set – i.e., the likelihoods of success of future avenues of vaccine research – but such strict formulations may require infeasibly precise speculation (Institute of Medicine, 1985; 1986; Shepard, 1995). Finally the World Health Organization's Ad Hoc Committee on Health Research Relating to Future Intervention Options has attempted to evaluate the cost-effectiveness of different proposed research products for a number of diseases (World Health Organization, 1996).

In this chapter each of these areas will be discussed briefly. As essential background to this type of analysis, however, we will devote some attention to laying the foundations for the economic analysis of intervention and research priorities.

1.2.2 The Role for Existing Vaccines

Resources for health are limited and demands for health care exceed available resources. This is especially so in low-income countries, where resources are fewer and the burden of disease is greater (Murray et al., 1993a). For instance, total per capita health expenditure for the United States in 1990 has been estimated to be 1990 US$ 2763. For the same year, in sub-Saharan Africa, a region with five times the developed world's per-capita disease-burden, per-capita health expenditure is estimated to be only around 1990 US$ 25 (World Bank, 1993). Table 1.2.1 illustrates the contrast between health expenditures and the burden of disease for eight regions of the world; this contrast serves to emphasize the extra importance to developing countries of allocating health resources efficiently.

Table 1.2.1: Regional health expenditures and the burden of disease, 1990

Demographic region*	GDP 1990 (1990 US$ x 1 000 000)	Health expenditures, 1990			Burden of disease, 1990 (DALYs** per capita)
		Total (1990 US$ x 1 000 000)	As % GDP	Per capita (1990 US$)	
Established market economies	15974547	1483285	9.29	1869	0.13
Formerly socialist economies of Europe	1380409	49114	3.56	142	0.18
India	291561	17488	6.00	20	0.33
China	365557	12819	3.51	11	0.18
Other Asia and islands	817304	36817	4.51	53	0.25
Sub-Saharan Africa	275580	11607	4.21	22	0.55
Latin America and the Caribbean	1106035	43825	3.96	98	0.22
The Middle Eastern Crescent	1248990	44631	3.57	88	0.30

Notes: * These regions are standard regions defined by the World Bank's World Development Report 1993 (World Bank, 1993).
** Disability-adjusted life-years, discussed in detail in this chapter.

Source: Authors

Within such limits and with health-maximization as a goal, health planners should attempt to purchase the greatest health gain possible. Either of two modes might be used in order to achieve this gain: the market, or a health system where the state plays a role in controlling the set of interventions delivered through regulation of providers or direct provision of services.

A popular view with some health economists is that the allocation of resources between different health interventions, such as measles immunization and liver transplantation, should be left to consumers and providers interacting in a free competitive market. If such markets functioned according to textbook theory, they would be desirable mechanisms to maximize consumer satisfaction in the health sector. However, markets for health care are far from perfect and their limitations are various. Among them are: (1) Some products of the health sector are public goods, which would be undersupplied in a free market; public goods are goods the consumption of which by one individual does not lessen the value of which to others, such as public sanitation systems, health information campaigns, or alcohol control programs. (2) Externalities are present, which likewise leads to underprovision in a free market; an example of an externality is a situation in which an intervention has an effect on individuals other than the consumer – e.g., treatment for tuberculosis which decreases the risk of transmission of the disease for the entire community. (3) Information problems exist; e.g., consumers' inability to acquire complete information, asymmetries of information between providers and consumers, and the high costs of poor decisions to consumers. (4) Left alone the market is unable to address issues of equity.

Noting these problems with healthcare markets, it is the departure point of this analysis that health maximization cannot be achieved by a free market in healthcare, and there is some role to play for the quantitative analysis of the costs and benefits of different health interventions. Some forms of economic analyses, such as cost-effectiveness analysis, can serve as a guide to decision-makers to optimize this resource allocation.

While rational resource allocation oriented to achieving the greatest health gain given scarce resources may be a laudable goal, we must not lose sight of the fact that most government decision-makers are subject to a variety of influences which may impede the optimal allocation of health resources. Among these influences are: the onus of past capital investments demanding a repeated depletion of recurrent budget commitments year after year; international trends in health policy, driven by donor agencies; and political power of certain groups to garner support and services for interventions they value highly. While decisions in the real world will necessarily reflect all these influences, there is a role for rational resource planning to influence, even if only incrementally, the allocation of resources between interventions. Several recent examples, such as the Oregon Health Services Commission, illustrate that cost-effectiveness analysis of interventions can profoundly alter resource allocations (Oregon Health Services Commission, 1994; Kitzhaber, 1993).

The following sections will first describe the choice of cost-effectiveness analysis as a useful tool for rational resource allocation in the health sector and then provide an overview of the methods used in cost-effectiveness analyses. Specifically, the following sections will address the methods used in studies undertaken under the aegis of the Health Sector Priorities Review (Jamison et al., 1993a; World Bank, 1993) to arrive at an extensive data-base of cost-effectiveness calculations for vaccines and other health interventions. The use of such results in resource allocation will then be discussed.

1.2.2.1 Cost-Effectiveness Analysis and Cost-Benefit Analysis

Three quantitative methods are in wide use to analyze health interventions: decision analysis, cost-effectiveness analysis and cost-benefit analysis – see table 1.2.2. In decision analysis, the benefits of an intervention strategy, such as INH chemoprophylaxis following a positive PPD skin test, are quantified using the available literature. The intervention strategy that leads to greatest expected health improvement is then recommended. In decision analysis, which is primarily intended to inform clinical decisions, no consideration is made of the costs of different intervention strategies.

In cost-effectiveness analysis, both the costs and the benefits of an intervention are considered. Various types of cost-effectiveness analysis exist. In some cost-effectiveness studies, effectiveness is measured using process indicators such as the number of children fully immunized. There is an extensive literature on the cost per fully immunized child through the Expanded Programme on Immunization (e.g., Brenzel, 1990; Shepard et al., 1986). Other cost-effectiveness studies use measures of health outcome to denominate effectiveness. Most simply, cost-effectiveness analyses can express health outcomes in terms such as cases averted or deaths averted. More sophisticated health outcome measures include the quality-adjusted life-year (QALY) or its variant, the disability-adjusted life-year (DALY), both of which account for both years of life lost due to mortality and years of life lived with reduced quality of life due to morbidity or disability.

A third method, cost-benefit analysis, differs from cost-effectiveness only in that the benefits of health interventions are calculated in dollar terms. Cost-benefit analysis requires that we put a dollar value on human life and/or the improvement in the quality of human life. Denominating health gains in dollars has the advantage that it facilitates comparisons between health investments and investments in other sectors of the economy. Nevertheless, cost-benefit analysis remains quite unpopular because many people in public health feel uncomfortable attaching a specific dollar value to a year of healthy life or other health improvement.

Table 1.2.2: Summary comparison of decision analysis, cost-effectivenss analysis, and cost-benefit analysis

Method	Units of cost	Units of benefit
Decision	—	Health
Cost-effectiveness	Money	Process or health
Cost-benefit	Money	Money

The following discussion will focus on cost-effectiveness analysis as a less controversial (though not controversy-free) and more accessible tool. Recently, studies by the World Bank and Oregon Health Services Commission (Oregon state, USA) have made available compendia of cost-effectiveness data covering a wide range of health interventions (World Bank, 1993; Jamison et al., 1993a; Oregon Health Services Commission, 1994; Kitzhaber, 1993; Klevit et al., 1991). These data-bases indicate an increasing focus on cost-effectiveness in the health sector. While both the Oregon and World Bank studies provide cost-effectiveness figures, data from the World Bank are intended for broad international use. The next two sections provide an introduction to the methodology of cost-effectiveness analyses used by the World Bank and in similar studies.

1.2.2.2 Assessing Costs

1.2.2.2.1 Costing Perspectives

When interpreting a cost-effectiveness analysis, it is important to identify from what perspective the study has been carried out. The study's perspective affects what costs are evaluated and included in the results, therefore having a great impact on the interpretation and usefulness of the data. In the evaluation of the cost-effectiveness of vaccines or vaccine research, either of two perspectives will likely be taken: the societal perspective or the providers' perspective.

The societal perspective takes into consideration all costs incurred by the application of a health intervention. Costs incorporated into such an evaluation include costs of vaccines and their delivery systems and salaries of health workers, as well as costs to vaccine recipients – costs such as user fees, travel costs (including lost time) for persons traveling to hospitals or health centers to receive the vaccinations, and lost time for persons waiting on line. The societal perspective is the point of view preferred by economists because it leads one to make decisions that will maximize health for the community.

The providers' perspective, on the other hand, accounts only for costs incurred by the provider; this in itself can include just hospital costs or vaccination-programme costs or total costs to the Ministry of Health. From this perspective, costs to vaccine recipients are ignored. The cost if evaluated from the provider's perspective will appear lower than if evaluated from the societal perspective, obscuring the costs incurred by patients. If patient co-financing is extensive, such a study could be misleading. While the perspective taken by a cost-effectiveness study determines what costs are included or ignored in its analysis, studies from both perspectives use the same methods of evaluation of costs.

1.2.2.2.2 Opportunity Cost

Given the limited resources of any health sector, a decision to purchase more of one item is consequently a decision to purchase less of some other item. The economically relevant cost of any purchase is then its *opportunity cost*, i.e., the value of the best alternative that is not purchased. Economic analyses must focus specifically on opportunity costs, though the market price of an item may closely resemble its opportunity cost. In fact, the opportunity costs of goods and services that happen to be exchanged on the market are simply their respective market values – some adjustments might need to be made when there are externalities, government intervention in the market or other reasons for price distortions.

Some goods and services purchased by health sectors, however, are not exchanged on the market. These may include unpaid volunteer services or, as discussed in the preceding section, travel time and waiting time of patients. Such costs may or may not be evaluated, depending on the perspective of the study. In the following discussions of cost analysis, the cost referred to is invariably the opportunity cost.

1.2.2.2.3 Calculations of Costs

Analyses of costs generally calculate results in terms of annual costs of entire health programs or specific components of those programs. Data on resource consumption that are collected in the field are processed in order to express the value of these resources in terms of annual monetary costs.

All costs can be classified as recurrent or capital and the methods for transforming these two categories of costs into annual terms are different. Recurrent costs are costs incurred for items consumed over less than one year. Capital costs are incurred for items consumed over more than one year. While methods for such calculations are analytically straightforward, they can be computationally complex. However, with the help of computer spreadsheet programs or published annuity factor tables, these calculations are universally performed in analyses of intervention programmes. Summing the resulting annual costs for the recurrent and capital expenditure items gives the total annual cost of a health intervention.

When assessing the costs of individual resources consumed, care must be paid to the monetary units in which the costs are expressed. In any cost study, costs must be presented in useful units of currency. When importing to one setting cost data collected in another setting, it is important to adjust these costs to reflect temporal and geographic price differences. Two adjustments must be made: adjustments for inflation over time and conversions between different currencies.

Inflation is an issue in cost analyses because, as general price levels tend to increase over time, cost data from one year may not reflect the reality of costs for another year. A ten-dollar note in 1996 does not have the same value as a ten-dollar note in 1993. To interpret cost data meaningfully, costs are presented in standard units, denoting both the currency and the year of that currency. Such a standard unit denotes *real* or *constant* currency units. Comparisons of costs which have not been adjusted to standard units – i.e., *nominal* or *current* costs – can be virtually meaningless.

Adjustments for inflation are made accurately by applying consumer-price indices or gross domestic product (GDP) deflators, which are comparative measures of the inflation of specified currency units. These indices are calculated and published annually by individual countries and in cross-national compendia such as the International Financial Statistics Yearbook (International Monetary Fund, 1991). They facilitate free conversion of currency units from one year to another.

In comparing costs from one economy to those of another, adjustments to standardized currency units must also be made by converting data in one currency unit to that of another. This is facilitated by applying the exchange rate for the two currencies. Official exchange rates, reflective of the market for foreign exchange, can be found in annual publications such as the International Financial Statistics Yearbook (International Monetary Fund, 1991).

The market for foreign exchange, however, only accounts for currencies' comparative purchasing power for traded goods and services. The prices for non-traded goods and services, such as labor, can vary widely between countries. For example, in real terms, a nurse in Zimbabwe would be paid five or ten times less than would be an equivalently trained nurse in Brazil or Korea. Labor and other non-traded goods and services are undervalued by official exchange rates, which therefore exaggerate the purchasing power gap between richer and poorer countries. The International Comparison Project (ICP) has developed tools to assess more accurately the comparative purchasing power of currencies (Summers and Heston, 1991; Summers et al., 1980). The study produces purchasing-power parity (PPP) ratios to be used in place of official exchange rates in price conversions to US dollars. Costs exchanged using these ratios are frequently referred to as International dollars.

1.2.2.2.4 Interpretations of Cost Data

Processed total cost data are usefully disaggregated into three program-cost components. This categorization helps to illuminate how resources are being consumed by an existing program and how resources may be consumed if the program were expanded. These three categories are: variable costs, program fixed costs, and infrastructure fixed costs. A program's variable costs are a direct function of the number of patients treated; variable costs can include costs of vaccines, drugs, reagents for diagnosis, and food during hospitalization. Program fixed costs are those associated with an intervention program itself; for a vaccination program, for example, these costs may include costs for salaries of regional and district program coordinators, vehicles used solely by the vaccination program, and administrative costs of the program. Infrastructure fixed costs are the costs for resources used by the intervention program that are not variable costs, but that are incurred for resources used that are not associated with the intervention program itself. These costs include costs for the use of the fixed infrastructure of the health system, such as health clinics and hospitals.

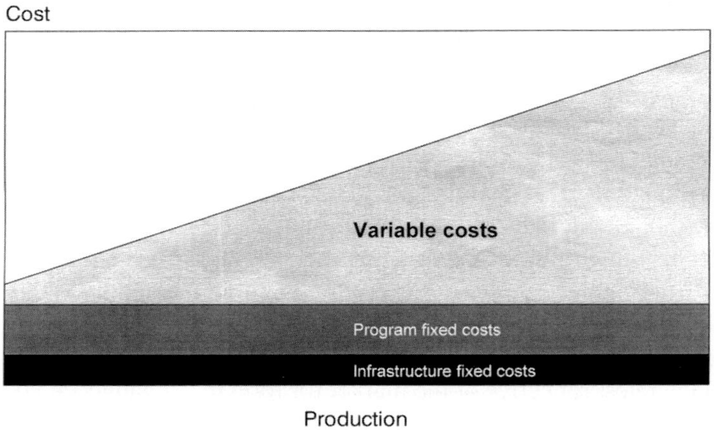

(Source: Authors)

Fig. 1.2.1: Cost categories: a schematic illustration of costs versus production.

Unlike variable costs, the two types of fixed costs are similar in that they represent purchases that do not vary as functions of program output. While output may fluctuate within a given year, the fixed costs cannot be altered to meet the changing activities of the program. Program fixed costs are incurred for items used specifically by the program in question. However, a program may also require the use of items purchased for the general health infrastructure. For example, a vaccination program may require the use of health clinic buildings and staff that are also used for other health activities. Both types of fixed costs might be thought of as start-up or overhead costs: no matter the level of productivity these costs incurred. The distinction between these two categories of fixed costs is often not made in cost-effectiveness studies. (For a fuller explanation of the necessity of this distinction, see Murray et al., 1994b)

Cost data are also usefully presented as unit costs. The unit costs of an intervention are presented as simple ratios of cost to output or, more tellingly, as functions of a program's output. Output here can be measured in terms of process-indicators – e.g., children vaccinated – or in terms of health outcomes – e.g., lives saved, or QALYs. The several types of unit-cost categories for a program are: average cost, marginal cost, and incremental cost.

Average cost is the total cost divided by the total output. Marginal cost is the cost to the program of gaining one additional unit of output. This is difficult to measure over a wide range and is therefore often approximated by the average variable cost, i.e., total variable costs divided by total output. This approximation is reasonable where the marginal cost rises nearly linearly. Incremental cost is the additional cost incurred to achieve a specified increase in output, divided by that increase in output. In analyzing the costs and benefits of expanding a program or of switching programs altogether, incremental cost is a useful indicator.

Figure 1.2.2 illustrates the relationships of average cost and marginal cost to output. The average cost curve takes a U-shape. Near the origin of the graph, average cost decreases as output increases; this occurs because, as output increases the fixed costs of the program are divided across a greater number of output units. Marginal cost tends to increase with output, as output units become more and more difficult to obtain. Intuitively, this is because easy-to-obtain, hence marginally less-expensive, outputs are obtained earlier in the production function. The marginal-cost curve happens always to cross the average-cost curve at the average-cost curve's lowest point. After crossing this point, the marginal cost rises above average cost. The average-cost curve thereby begins to increase.

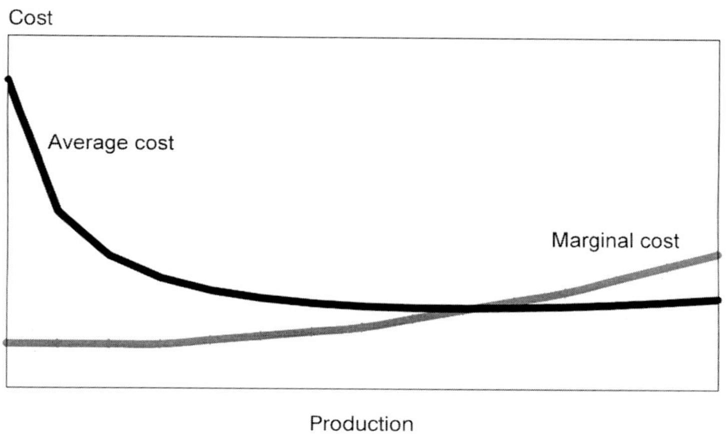

(Source: Authors)

Fig. 1.2.2: Unit costs: the relationship between average and marginal costs.

Having touched on unit costs, which are expressed as costs per unit of outcome, outcome measures should now be addressed.

1.2.2.3 Assessing Effectiveness

As discussed previously cost-effectiveness analysis assesses health benefits in terms of health outcomes. In cost-effectiveness analysis, the effectiveness of a health intervention is expressed simply as the difference between specified health outcomes with the intervention in place and without the intervention in place. Outcome measures span a wide range. A study's decision to use one of these effectiveness measures involves practical analytical issues as well as implicit value judgments. The interpretation of such studies should be made keeping in mind the processes and judgments involved. The outcome measures described in the next section are used in disease-burden analysis – which is a useful tool for evaluating research needs – as well as in cost-effectiveness analysis.

1.2.2.3.1 Mortality

Outcome measures used in studies range from specific process indicators – such as children fully immunized, or tuberculosis patients diagnosed and treated – to more general measures, such as deaths or years of life lost. Specific process indicators have the great advantage of being easy to measure, as they are generally recorded in routine program monitoring. However, their chief disadvantage is that they are not useful in comparisons across health interventions that do not involve the same process. For example, it is difficult to interpret the comparison of ten fully immunized children to six diagnosed and treated tuberculosis patients. More general measures of health outcome better facilitate such comparisons, which are necessary for health resource allocation.

The simplest general measure of effectiveness in the health sector is deaths-averted. Deaths-averted are frequently used as the measure of effectiveness in studies of immunization and oral rehydration therapy. However, society may value more greatly deaths averted at earlier ages than deaths averted at later ones. It is therefore often desirable to reflect the greater value of averting deaths at younger ages by weighting each death-averted by the number years of life lost (YLL) that are averted.

YLLs account for each death by assigning it a value in years, the number of years that a person could have been expected to live had he or she not died prematurely. The number of YLLs assigned to a particular death is a function of the age at which the individual dies.

Since Dempsey (1947) introduced the concept of measuring lost time due to mortality, a wide variety of methods for measuring years of life lost have been proposed. (For a full treatment of these methods, please see Murray, 1994) The World Bank study of cost-effectiveness analysis employed a method by which expectations of life are standardized for each age and are based on a published model life-table (Jamison et al., 1993a; World Bank, 1993; Murray, 1994). These expectations of life are illustrated in table 1.2.3.

1.2.2.3.2 Non-Fatal Health Outcomes

While YLL measures quantify mortality, other health outcomes are also addressed by health interventions. Measures for these non-fatal health outcomes include measures of morbidity, disability, or quality of life, all of which attempt to quantify non-fatal outcomes in terms which facilitate comparison to mortality measures.

Measuring non-fatal health outcomes in terms commensurate with time lost due to mortality has been the subject of extensive research for three decades. Beginning in the 1960s, a series of authors formulated models for composite indicators of mortality and morbidity (Sullivan, 1966; 1971; Chiang, 1965; Fanshel and Bush, 1970; Patrick et al., 1973; Berg, 1973). While each indicator had notable differences, they each defined a series of health states ranging from health to death and a set of weights

Table 1.2.3: Standard life expectancy at age of death

Age (years)	Life expectancy Females	Males	Age (years)	Life expectancy Females	Males
0	82.5	80	40	43.53	40.64
1	81.84	79.36	45	38.72	35.77
5	77.95	75.38	50	33.99	30.99
10	72.99	70.4	55	29.37	26.32
15	68.02	65.41	60	24.83	21.81
20	63.08	60.44	65	20.44	17.5
25	58.17	55.47	70	16.2	13.58
30	53.27	50.51	75	12.28	10.17
35	48.38	45.56	80	8.9	7.45

reflecting the comparative severities of those states. Since these pioneering studies, intellectual studies have evolved in three largely independent lines.

One line of development has been pursued by health economists interested in using the measures to evaluate the benefits of specific health interventions. The now familiar term "quality-adjusted life-year" (QALY) has become a standard tool in health program evaluation in developing countries. In the work on QALYs, the focus has been on developing sophisticated methods for measuring individuals' preferences for time spent in different health states. Nord (1992), for example, reviews five approaches to elicit utility weights for health states. For most cost-effectiveness studies, health states have been defined *ad hoc* for use in evaluating a specific health intervention. One recent variant of the QALY, the disability-adjusted life-year (DALY) employs standard health states and corresponding disability weights that are applicable to any disabling condition (Murray, 1994; 1996a).

The second school of development has been the burgeoning field of health status indicators pursued largely in North America (Lohr and Ware, 1987; Lohr, 1989; 1992). Rather than emphasizing utility weights as in the development of QALYs, the major thrust here has been defining the precise dimensions of health status and designing practical survey instruments for measurement. Beginning initially with a narrow vision of disease, the measures have progressively incorporated variables related to physical function, mental function, and, more recently, social function. The weights used in collapsing measurements of multiple variables into a single indicator have not been as great a topic for concern as in the QALY literature; frequently these weights are chosen on arbitrary grounds such as equal weighting.

The third cluster of work on measuring non-fatal health outcomes dates from the early 1970s. A World Health Organization (WHO) initiative in collaboration with the Paris WHO Centre for the Classification of Diseases and various non-governmental organizations led to the publication in 1980 of the *International Classification of Impairments, Disabilities, and Handicaps*, or ICIDH (World Health Organization,

1980). The conceptual framework that emerged from this collaboration differs substantially from the frameworks for the QALY and health status index approaches. The ICIDH work schematizes a linear progression from impairment to disability to handicap. Impairment is defined on the level of the organ system, disability is the consequent impact on the performance of the individual, and handicap represents the overall consequences for the individual. The ICIDH has been adopted by both WHO and the United Nations Statistics Division.

In cost-effectiveness analysis, the QALY approach is most often used. The QALY literature has measured the impact of disability or quality-of-life both in terms of severity of the health state in comparison with death and perfect health, and in terms of duration of the health state. These features of QALYs have been developed to facilitate comparisons between fatal and non-fatal health outcomes in shared units. A fatal health outcome is generally assigned the highest severity weight and a duration equal to the calculated years of life lost. A non-fatal health outcome is assigned a severity weight between death and perfect health and a duration equal to the duration of the health outcome – e.g., the disability or morbidity. The duration may range from a few days (e.g., for an episode of diarrheal disease) to the remaining length of life (e.g., for a permanent state such as uncorrectable blindness). QALYs from deaths are simply added to QALYs from non-fatal health outcomes to calculate the total health outcome.

The effectiveness of a health intervention then is usefully measured in terms of QALYs averted. The number of QALYs averted is simply the difference between the number of QALYs that would have resulted due to a specific disease had no intervention program been applied and the number of QALYs that actually resulted due to the disease after the intervention was applied. The cost-effectiveness of a health intervention is then calculated as the costs of the intervention divided by the health outcome averted.

1.2.2.3.3 Other Value Judgments in Health Outcome Measures

All measures of health outcome or disease burden incorporate specific value judgments, judgments made to address theoretical and practical concerns over what exactly should be measured. The widely used disability-adjusted life-year (DALY) is no exception. As a specific variant of the QALY, the DALY is calculated using explicit and largely standardized value judgments. As claculation methods for DALYs have been explicitly laid out, it is possible and perhaps desirable to alter these judgments in order to better reflect the values within a country setting (Murray, 1996a; Murray, 1994; World Bank, 1993; Jamison et al., 1993a).

Two specific judgments incorporated in the DALY indicator are particularly controversial and are discussed briefly below: (1) time preference, i.e. the value of future health outcomes is less than the value of equivalent present health outcomes; and (2) age weights, i.e. values of an individual's years-of-life vary with the individual's age.

At the simplest level, time preference is the economic concept that inividuals prefer benefits now rather than in the future. The value of goods or services today is greater than in one or ten years (even after these goods or services have been adjusted for inflation into real values). If offered a choice from a completely reliable source between $100 today or $100 one year from now (in real dollars), most indiviuals will prefer their money today. If offered $110 in one year or $100 today, some may choose the $110. The interest rate on a savings account is the rate at which individuals are willing to forego consumption today for consumption in the future. The market rate of interest is the aggregate rate at which individuals in society as a whole discount future consumption. It is standard practice in economic appraisal of projects to use the *discount rate* to discount benefits in the future (Dasgupta et al., 1972). The process of discounting future benefits converts them into present-value terms which can then be compared with program costs, some of which may also be discounted if their respective products or services are consumed in the future.

Despite the widespred use of discounting in cost-benefit and cost-effectiveness analyses, no consensus exists for either the conceptual justification for discounting health outcomes or the actual value of discount rate itself (for a fuller discussion of time preference and the discount rate, see Murray, 1994).

To choose the discount rate for a study, the social opportunity cost of capital can be used, as captured by the market rate of return on investment. Distortions of the market can complicate the determination of the social opportunity cost of capital. In practice, discount rates based on the social opportunity cost of capital are high (between 8% and 15%). The World Bank and the United States Congressional Budget Office have for many years used a 10% discount rate in project appraisal (Hartmann, 1990). Studies of long-term returns on investments, however, suggest a lower discount rate of between 1% and 3%. Rather than determining the social opportunity cost of capital, one can determine society's *social time preference*. This assumes the notion that societies are like individuals in having an inherent preference for gains today rather than equivalent gains in the future. Society's discount rate is thought to be lower than the market rate of interest (closer to the 1% to 3% range) (Lind, 1982). Discounting health benefits remains a controversial topic. DALYs incorporate a low, positive rate of 3%.

The second controversial value judgment incorporated into the DALY indicator is that values of an individual's years-of-life vary with the individual's age. In all societies social roles vary with age. The young and often the elderly depend on the rest of society for physical, emotional, and financial support. Given different roles and changing levels of dependency with age, it may be appropriate to value differentially the time lived at different ages. Higher valuation for a year of life at a particular age does not imply that the time lived at that age is in itself more important to the individual, but rather that because of social roles the social value of that time may be greater. Some studies have conceptualized age-weighting as a reflection of economic production and therefore determine age-weights accordingly (Prost and Prescott, 1984; Barnum, 1987). The DALY indicator attempts to capture the chaniging social roles filled by individuals as they age. Figure 1.2.3 illustrates the continuous mathematical function used in the DALY indicator to assign age weights.

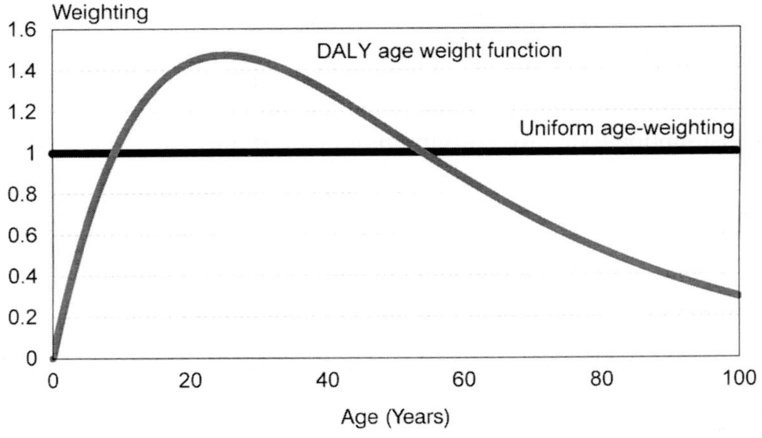

(Source: Authors)

Fig. 1.2.3: Age weights: The DALY age-weight function and uniform age-weighting.

1.2.2.4 The Cost-Effectiveness of Vaccines

1.2.2.4.1 Vaccine Cost-Effectiveness and Incidence of Disease

Generalizable estimates of the cost-effectiveness of any health intervention are difficult due to the varying costs in different environments and the various methods of assessing those costs. Studies should be interpreted with careful notice paid to the details of the program being evaluated and the assumptions and value judgments made.

Cost-effectivness studies of vaccines and screening programs are even more difficult to generalize due to the relationship between a vaccine's cost-effectivness and the incidence of the disease it is intended to prevent. This is because, as incidence declines, more individuals must be vaccinated for each case prevented.

As the incidence of a vaccine-preventable disease declines, all other things remaining equal, the ratio of cost to health outcome for the vaccine will decrease. If vaccination levels remain the same, the cost of a vaccination program will remain the same as the incidence rate declines. However, as the incidence rate declines, the benefits of vaccination will also decline. This is due to the fact that as incidence declines, so may the risk of infection and the incidence of disease. In an environment with a lower risk of infection and hence lower risk of death or disability, the benefits of the intervention program decrease. The benefits are the ill-health outcomes (e.g. DALYs) averted, which is calculated as the difference in health outcomes with the intervention and without the intervention. That is,

$$\text{Cost} - \text{effectiveness} = \frac{\text{Cost}}{\text{DALYs without program} - \text{DALYs with program}}$$

In the case of tuberculosis, it has been estimated that if the risk of infection declines from 2% to 1%, the cost per death averted will more than double (Murray et al., 1993). Figure 1.2.4 illustrates the relationship between the cost-effectiveness of BCG vaccine and the risk of infection.

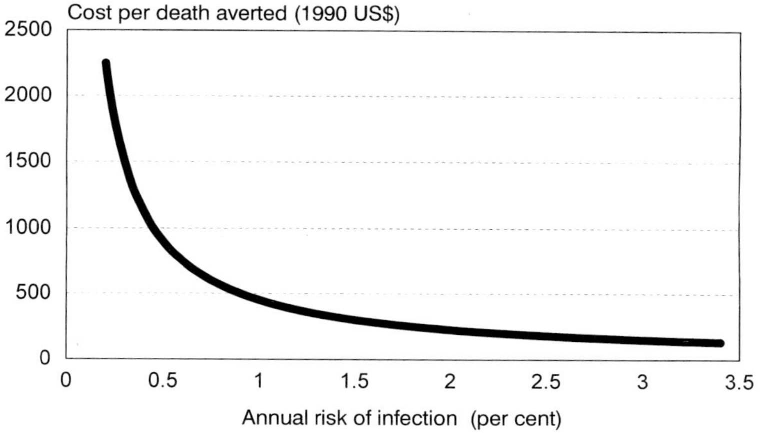

(Source: Murray et al., 1993)

Fig. 1.2.4: Cost-effectiveness of BCG vaccine against tuberculosis as a function of the annual risk of infection.

1.2.2.4.2 Results from the Health Sector Priorities Review

Using analytical methods as described in the previous sections, a wide range of cost-effectiveness analyses have been completed, each usually focusing on one or a few interventions for the same disease. The methods employed, however, vary across studies, from almost imperceptible nuances in the methods of costing capital to issues such as the severity weights used to assess non-fatal health outcomes to the choice of health indicator altogether. No universal standard methodology exists, complicating comparisons of interventions that are assessed in different studies.

A recent study, the World Bank's Health Sector Priorities Review, provides cost-effectiveness estimates for interventions spanning a range of diseases prevalent in developing countries. While the methods used vary from disease to disease, the authors' resulting evaluations are all expressed in a common measure, US dollars per disability-adjusted life-year (DALY), as described in earlier sections. This study provides reasonable estimates of the cost-effectiveness of the interventions it assesses and the methods used in the study are similar enough to facilitate meaningful, though not precise, comparisons across diseases (Jamison et al., 1993a).

Of the fifty-two interventions for which estimates are provided in Jamison et al. (1993a), the cost-effectiveness for twenty-two is less than US$ 25 per DALY. Five of these are vaccines: diphtheria-pertussis-tetanus plus polio vaccine, measles vaccine, BCG vaccine for tuberculosis, hepatitis B vaccine, and rotavirus vaccine. The two other vaccines assessed in the study, pneumococcal vaccine and cholera vaccine, are estimated to have cost-effectiveness values of between US$ 25 to US$ 75 per DALY and between US$ 75 and US$ 250 per DALY, respectively. Table 1.2.4, taken from Jamison et al. (1993a) illustrates the generally low cost-effectiveness of vaccines in the context of the Health Sector Prioirties Review.

Table 1.2.4: Cost-effectiveness of 55 interventions reviewed by the World Bank's Health Sector Priorities Review, by intervention type

Cost per DALY		Other public health interventions				Clinical interventions		
	Immunization	Environmental	Mass chemoprophylaxis	Screening and referral	Behavior change	Primary care	District hospital	Referral hospital
< $25	5	1	6	0	3	4	3	1
$25–$75	1	0	1	0	2	1	1	1
$75–$250	1	2	0	0	2	8	3	0
$250–$1000	0	0	0	1	1	0	0	0
> $1000	0	0	0	0	0	2	1	4
Total	7	3	7	1	8	15	8	6

Source: Adopted from Jamison et al. 1993

For the interventions evaluated by the Health Sector Priorities Review, both the costs and benefits of health interventions vary widely. The cost per intervention ranges from less than US$ 1 to nearly US$ 10 000, while benefits range from a few hours or days of healthy life to ten or more years of healthy life. Cost-effectiveness ratios have been estimated to range from around US $1 per DALY to upwards of US $10 000 per DALY (World Bank, 1993). Bobadilla et al. (1994), examining this data, note that there is little correlation between the cost of an intervention and the health gains it yields.

Figure 1.2.5 presents these findings and illustrates the range of cost-effectiveness ratios for the interventions evaluated. This graph is reproduced here using data from the Health Sector Prioirties Review and following the form of a graph published in World *Bank's World Development Report 1993: Investing in Health* (World Bank, 1993; Bobadilla et al., 1994).

Forty-seven interventions are shown on this scatter-plot as points, each plotted according to its cost and its effectiveness. The axes are scaled logarithmically. The x-axis represents the cost of one complete application of the intervention (or, for interventions that are applied continuously and indefinitely, the cost of one year of applying the intervention, as in the cases of environmental control, nutritional supplementation, or long-term case-management). The y-axis represents the effectiveness

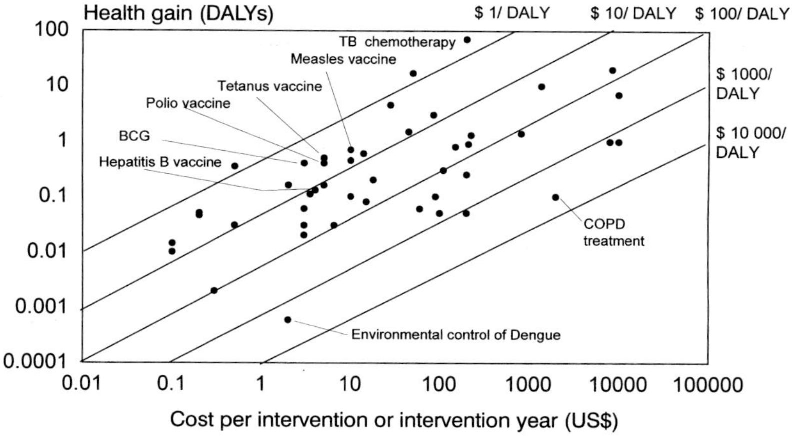

(Source: Bobadilla et al., 1994; Jamison et al., 1993; World Bank data)

Fig. 1.2.5: Costs and health gains from 47 interventions: results from the World Bank's World Development Report 1993.

of the intervention (either one complete application or one year of its contiuous application), denominated in terms of DALYs averted. Points farther to the left are less expensive. Points higher up yield greater benefits. The diagonal lines are contours connecting points of equal cost-effectiveness. The most cost-effective interventions lie near the highest of these diagonals, which represents a cost-effectiveness ratio of $1 per DALY. Less cost-effective interventions lie further toward the bottom right-hand corner, near the lowest diagonal, which represents cost-effectiveness ratios of $10,000 per DALY.

One striking observation from figure 1.2.5, is that preventive interventions are *not* universally more cost-effective than curative ones. One prime example of this is the case of tuberculosis. Chemotherapy for tuberculosis cases, represented as the highest point on the graph, is *more* cost-effective than BCG vaccine to prevent tuberculosis cases. Other highly cost-effective non-preventive interventions – costing between $5 and $30 per DALY – include cataract removal and treatments for leprosy, measles, and intestinal helminths. Some non-cost-effective preventive interventions include Dengue control programs, prevention of rheumatic fever, and management of hypertension and hypercholestermia. Also, primary care interventions are not universally more cost-effective than hospital care. Among the highly cost-effective non-primary-care interventions are cataract removal and rehabilitation for complications of leprosy.

As can be seen, vaccinations tend to be highly cost-effective. Four vaccines lie between the $1 per DALY diagonal and the $10 per DALY diagonal, ranking them among the most cost-effective of the interventions evaluated. Another, hepatitis B vaccine, lies near the $10 per DALY diagonal. Vaccination for tuberculosis costs around $7.50 per DALY, for tetanus around $10 per DALY, polio around $12.50, measles around $14, and hepatitis B around $25. Vaccines, along with nutritional supplements, are two groups of highly cost-effective interventions.

Conclusions from these results can be examined in more detail and in broader context by consulting the evalauvtions that arrived at them. The disease-specific evaluators for the Health Sector Priorities Review each present their methods and findings in a detailed chapter in Jamison et al. (1993a):

Evaluating BCG vaccination for tuberculosis, Murray et al. (1993) note the difficulties in generalizing the cost-effectiveness of BCG. They note as particular difficulties the relationship of cost-effectiveness with the annual risk of infection and the variation among BCG vaccination strategies in different countries. They note that only one study has attempted to evaluate the cost-effectiveness of a BCG program in a developing country (Barnum et al., 1980). Due to lack of evidence, it is difficult to draw any conclusions concerning adult BCG vaccination. The authors conclude that, at low levels of tuberculosis incidence, chemotherapy is more cost-effective than vaccination. At higher risks of infection, the two interventions have similar cost-effectiveness ratios and BCG may be a reasonable option. Some mix of treatment and prevention is likely most favorable.

Steinglass et al. (1993) review fifteen studies of the cost-effectiveness of tetanus immunization programs and present the data from each in a detailed table. They strongly conclude that prevention of neo-natal tetanus should be a priority in many developing countries, due to the large burden of the disease (see table 1.2.8), the severity of the disease even when treated, and the high cost of tetanus treatment when compared to immunization. The authors focus on immunization because of the lack of studies of the cost-effectiveness of prevention of neo-natal tetanus through asceptic delivery methods and the prevention of non-neonatal tetanus through wound management. The authors list the factors that could influence the choice of immunization strategy. Among these are: incidence rate, organization and utilization of health services and other channels of contact, and incremental costs of tetanus immunization strategies. Depending on these and other factors in a particular country, either mass immunization campaigns or continuous immunization through routine services may be appropriate.

Jamison et al. (1993b) conclude their evaluation of interventions against polio by stating that with sustained international support for the Expanded Program of Immunization and the strengthening of primary care in developing countries, "it is reasonable to hope that polio can be eradicated by the year 2000 or soon thereafter." While the costs and benefits of treatments and rehabilitation for polio-related disability are not determined, the cost-effectiveness of polio vaccines is clearly favorable. The authors note that the World Heath Organization recommends that schedules designed around Oral Polio Vaccine (OPV) are globally desirable, though they advise that with careful evaluation, other schedules may be more desirable in local settings.

Underscoring the global benefits of effective measles vaccination, Foster et al. (1993) state: "Almost all children unprotected by measles vaccine will eventually be infected with measles and 1 to 5 percent will die." The authors review a range of studies and employ a mathematical model to evaluate the immunization coverage achieveable with four different strategies. They conclude that the most cost-effective strategy appears to be adminstration of Edmonston-Zagreb or equivalent vaccine at age six months. The authors go on to note that, despite the highly favorable cost-effectiveness of measles vaccine and vaccination strategies, their use will not achieve the 1990 World Summit for Children goals for coverage and disease-burden reduction. The authors suggest eleven strategies to move toward these goals. These can be summarized as: vaccination in the first year of life, use of accelerated schedules when appropriate and two-dose regimens, better monitoring of intended vaccine recipients including newborns, vaccination of high risk groups, vitamin A dietary supplementation and case-treatment, treatment of measles complications, and development of new vaccine delivery points. The authors note two research priorities for measles: development of heat-stable vaccine with high effectiveness when administered before six months, and operations research to maximize the effectiveness of existing interventions.

Kane et al. (1993) state that, lacking any effective treatments for hepatitis, the most important step in the global control of hepatitis is the integration of hepatitis B vaccine into the Expended Program of Immunization. The authors report that the vaccine is 95 to 99 percent effective in studies of vaccinated infants. The vaccine is also highly flexible and can be administered effectively in accordance with any EPI schedule (Hadler et al., 1989), keeping minimal the incremental costs of integrating hepatitis B vaccine into EPI. The authors note that a study of EPI in the Gambia, the only African country with routine hepatitis B immunization for infants, calculated that the incremental cost of adding hepatitis B vaccine is $4.23 per fully immunized child (Robertson et al., 1992). In developed countries, where immunization costs are higher, immunization of high-risk individuals and screening or pregnant women and immunization of the newborns of carriers are reported to be cost-effective. The authors note that universal immunization in developed countries may also be cost-effective.

1.2.2.4.3 Results from the Center for Risk Analysis

A study from the Center for Risk Analysis at the Harvard School of Public Health provides results with a similarly broad spectrum of costs, effectiveness, and cost-effectiveness (Tengs et al., 1995). This study is not very useful for questions of health resource allocation: much of its energy is devoted to non-health-sector interventions (such as product safety and environmental pollution reduction) and to interventions that do not address causes of great disease burden, especially in developing countries. It make no attempt to be comprehensive in assessing interventions employable in the health sector. Its results, however, are methodologically precise and broadly interpretable in relation to the results from the Health Sector Prioirties Review. Some comparable, broad trends can be discerned.

The Center for Risk Analysis study discriminatingly reviewed the literature for economic analyses of life-saving interventions, ultimately evaluating 587 interventions, the data for which fit their strict criteria. The study measures costs in 1993 US dollars and benefits in years of life saved, both discounted at a rate of 5%. Cost-savings due to future health costs prevented were also taken into account as *negative* costs. While this methodology excludes disability from their assessments of effectiveness and differs from the DALY methodology in many other ways, some broad conclusions can be compared to those of the Health Sector Priorities Review. It should be noted that costs reported in the Tengs et al. study will fall into a higher range than those in the World Bank study because years lived with disablity are not included as benefits and the interventions tend to be for developed country settings.

The median cost for medical interventions reported by Tengs et al. was $19,000 per year of life saved. The median cost for all interventions evaluated was $42,000 per year of life saved. Figure 1.2.6 illustrates the eleven order-of-magnitude range of cost-effectiveness ratios compiled by the study, from less than zero up to nearly $100 billion per life-year saved.

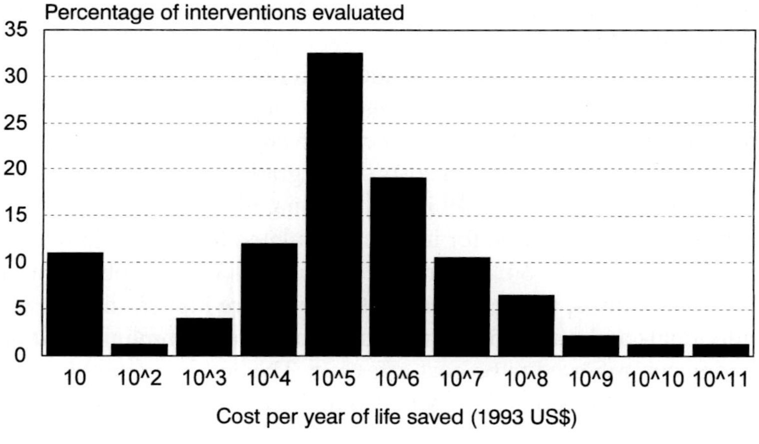

(Source: Tengs et al., 1994)

Fig. 1.2.6: The range of cost-effectiveness ratios: 587 interventions evaluated by the Harvard Life-Saving Project

Simlarly to the World Bank study, Tengs et al. found that immunizations were among the least expensive interventions, with a median cost per life-year of less than zero, signifying that the vaccine interventions not only save life-years but also avert the need to use other, more expensive health interventions.

The scope of the Tengs et al. study, while intentionally not comprehensive, is bemusingly inclusive. Other cost-saving interventions – i.e. those with cost-effectiveness of less than zero dollars per life-year – include: bans on residential growth in tsunami-prone areas, flammability standards for children's sleepwear sizes 0 to 6X, termination of sales of three-wheeled all-terrain vehicles, and smoking cessation advice for pregnant women. Among the least cost-effective interventions are:

benzene emisssion control at chemical manufacturing process vents ($526 million per life-year), and sickle-cell screening for non-black, low-risk newborns ($34.2 billion per life-year).

Five specific childhood immunizations evaluated had costs per year of life less than zero: DPT, measles-mumps-rubella, polio for ages 0 to 4 years, rubella for children under age 2, and measles vaccination within a national eradication program. In addition, influenza vaccination for all persons was found to have a favorable cost-effectiveness of $141 per life-year.

However, as was observed for the Health Sector Priorities Review results, preventive interventions are not universally more cost-effective than curative interventions, and primary-health interventions are not universally more cost-effective than hospital-based interventions. Among the cost-effective hospital-based curative interventions are: defribulators in emergency vehicles for resuscitation after cardiac arrest ($39 per life-year), coronary-care-units for patients under age 65 with cardiac arrest ($390 per life-year), heart transplantation for patients 55 years or younger with favorable prognosis ($3,583 per life-year), and neonatal intensive care units for infants weighing 1 to 1.5 kilograms ($5686 per life-year).

1.2.2.5 Resource Allocation

With knowledge of the comparative cost-effectiveness of a range of interventions, a society can analytically determine what interventions should be financed. In Oregon state, USA, the cost-effectiveness of each of 714 interventions was evaluated (Oregon Health Services Commission, 1991; Kitzhaber, 1993; Klevit et al., 1991; Østbye and Speechley, 1992; Dixon and Welch, 1991; Brown, 1991; Eddy, 1991; Garland and Hasnain, 1990). Having compiled this data, the Oregon Health Services Commission chose to finance these interventions in rank order from the most to least cost-effective until the entire health budget had been consumed.

In the World Bank's landmark *World Development Report 1993: Investing in Health* (WDR), data from the Health Sector Priorities Review are used and allocation recommendations are made to low-income and middle-income countries (which are assumed to have average annual per-capita incomes of US$ 350 and US$ 2500, respectively). Rather than a rank list approach as used in Oregon, the WDR recommends that countries finance a minimum *package* of essential services. Interventions included in the package are both cost-effective and directed toward major health threats (Bobadilla et al., 1994). These packages thereby exclude interventions which may be highly cost-effective, but that address health problems that are rare or that cause a negligible health loss to each afflicted individual. The respective packages recommended to lower and middle income countries both include the expanded programme of immunization (including immunization for hepatitis B and vitamin A nutritional supplements). Table 1.2.5 provides details of the recommendations of the WDR.

Table 1.2.5: Minimum packages of health services recommended in the World Development Report 1993

Intervention	Cost per beneficiary (US$)	Cost per capita (US$)	Cost per DALY (US$)	DALYs potentially gained*
Low Income Countries				
Public health interventions	14.6	0.5	12–17	45
Expanded program of immunization plus**	3.6	0.3	20–25	4
School health program	0.3	0.3	35–55	12
Tobacco and alcohol control program	112.2	1.7	3–5	35
Other public health interventions****	2.4	1.4	—	—
Sub-total	—	4.2	14	—
Clinical services				
Chemotherapy against tuberculosis	500.0	0.6	3–5	34
Integrated management of the sick child	9.0	1.6	30–50	184
Family planning	12.0	0.9	20–30	7
STD treatment	11.0	0.2	1–3	26
Prenatal and delivery care	90.0	3.8	30–50	57
Limited care*****	6.0	0.7	200–300	—
Sub-total	—	7.8	—	—
Total	12.0	—	—	
Middle-income countries				
Public health interventions				
Expanded program of immunization plus**	28.6	0.8	25–30	4
School health program	6.5	0.6	38–43	5
Tobacco and alcohol control program	0.3	0.3	45–55	9
AIDS prevention program***	132.3	2.0	13–18	15
Other public health interventions****	5.2	3.1	—	—
Sub-total	—	6.9	—	—
Clinical services				
Chemotherapy against tuberculosis	275.0	0.2	5–7	6
Integrated management of the sick child	8.0	1.1	50–100	21
Family planning	20.0	2.2	100–150	6
STD treatment	18.0	0.3	10–15	3.7
Prenatal and delivery care	255.0	8.8	60–110	25
Limited care*****	13.0	2.1	400–600	—
Sub-total	—	14.7	—	—
Total	—	21.5	—	—

Notes:
* Includes DALYs gained due to reduced transmission of disease
** EPI Plus includes vaccination against hepatitis B and nutritional supplementation for vitamin A
*** DALYs lost from AIDS includes reduced transmission only in the first year, which understates the value of preventing cases
**** Other public health interventions includes information, communication and education on selected risk factors and health behaviors, plus vector control and disease surveillance
***** Limited care includes treatment of infection and minor trauma and, for more complicated conditions, diagnosis, advice and pain relief and treatment as resources permit

Source: Adopted from Bobadilla et al 1994

Another study of resource allocation uses computer-aided modeling to process more detailed cost-effectiveness data in order to optimize health expenditures (Murray et al., 1994c). This Health Resource Allocation Model (HRAM) takes into account rising marginal costs and distinguishes among variable costs, program fixed costs, and infrastructure fixed costs. Fixed costs are accounted for in terms of the non-tradeable resources that they represent, and limits on the numbers of available hospital beds and other fixed resources are respected. Using the same Health Sector Priorities Review database used in the WDR, HRAM attempts to allocate optimally the health resources for the entire budget for a hypothetical sub-Saharan African country with an average annual income of 1990 US$ 340. This model allocates around 17% of the non-infrastructure costs in the health system to immunizations, including: BCG and DPT, measles, polio, and tetanus. Table 1.2.6 illustrates the prominence of immunizations in the model's output.

While HRAM provides allocation recommendations for only a hypothetical country, its results reflect that vaccine interventions are not only highly cost-effective, but, on further analysis, indeed demand prominence in a society's health expenditures. This study concurs to a great extent with the recommendations of the WDR.

The WDR package targets 95% of all born for the EPI Plus intervention (which includes EPI vaccinations and vitamin A supplementation). At a cost of around US$ 15 per fully immunized child, the intervention need for EPI Plus for a low-income country is around 0.14% of its gross domestic product (GDP). In middle-income countries, the cost per fully immunized child is around US$ 30, which gives a total intervention need of around 0.03% of GDP (Jamison and Saxenian, 1994). The HRAM computer model estimates an intervention need of less than 0.4% of GDP for non-infrastructure costs for BCG and DPT, hepatitis B, measles, polio, and tetanus.

While the proportion of resources in the WDR packages recommended for vaccines is relatively small, the health gains from vaccines are sizable, as is apparent from the highly favorable cost-effectiveness of vaccines. With low intervention needs, in low-income and middle-income countries, implementation of EPI Plus could avert 8% and 4%, respectively, of countries' total disease burdens (Jamison and Saxenian, 1994).

1.2.3 Economic Appraisal of Vaccine Research Priorities

Several groups have attempted to identify priorities for vaccine research or health research more generally based on a cost-effectiveness framework (Institute of Medicine, 1985; 1986; Shepard et al., 1995; World Health Organization, 1996). These efforts, all starting from the premise that the most cost-effective avenues of research should be pursued, have come to different conclusions on the limits of formal cost-

Table 1.2.6: Outputs from the Health Resource Allocation Model: allocations to specific interventions at varying health budget levels for a hypothetical sub-Saharan African country[(a)]

Intervention	At 3.0% of GDP		At 4.0% of GDP		At 5.0% of GDP	
	Spending ('000 $)	DALYS ('000s)	Spending ('000 $)	DALYS ('000s)	Spending ('000 $)	DALYS ('000s)
FIXED INFRASTRUCTURE[(b)]	32697		32699		32698	
ARI screening and referral	6879	233	11107	277	11107	277
ORT for Diarrheal Disease	789	12	4831	53	15661	123
BCG added to DPT for TB	920	71	1783	80	2291	84
Hepatitis B Immunization	391	8	505	9	505	9
Iodization of Salt or Water	249	32	249	32	249	32
Measles Immunization	4986	272	7686	298	12545	324
Poliomyelitis Immunization	907	30	1232	33	1503	34
Semiannual Vitamin A dose for children 0-5	577	38	881	41	881	41
Tetanus Immunization	1651	213	2042	222	2042	222
Breast-Feeding Promotion w/ education or hospital routine for Diarrheal Diseases	2564	74	2755	77	2755	77
Improved Weaning practices from education	1526	46	1526	46	1526	46
Oral Iron supplementation for duation of pregnancy	70	1	70	1	70	1
Chlamydia treatment w/ antibiotics	107	6	107	6	107	6
Gonorrhea Treatment w/ antibiotics	111	9	111	9	111	9
Syphilis Treatment w/ antibiotics	147	156	147	156	147	156
HIV Blood Screening	911	39	962	40	962	40
Annual breast exams	0	0	0	0	312	1
Antibiotics for Rheumatic Heart Disease	0	0	388	3	722	5
Cataract Surgery	860	9	935	10	935	10
CVD Preventive Program	0	0	0	0	287	2
Improved Domestic and Personal Hygiene	0	0	5306	47	9658	72
Injected Insulin and Health Education for IDDM	0	0	0	0	231	0
Leprosy multi-drug clinic	537	6	541	7	541	7
Low cost management of acute MI	0	0	0	0	780	3
Papsmear at 5-yr. intervals	246	1	227	2	422	2
Pneumococcal Vaccine	884	16	2543	32	3206	36
Schizophrenia	248	3	331	3	331	3
School based anthelminthic chemoprophylaxis	597	12	597	12	597	12
Short Course Chemotherapy for Sputum Negative Patients	6688	372	10208	415	14120	443
Short Course Chemotherapy for Sputum Positive Patients	3498	453	5018	484	5218	487
Sugar or Salt Fortified With Iron	124	19	124	19	124	19
Tetanus Referral Case Management	0	0	0	0	1176	3
Vector Control for Malaria	12123	304	16587	348	18942	364
ADDED INFRASTRUCTURE	20713		24502		27238	
TOTAL COSTS ('000 $)	102000	2435	136000	2762	170000	2950
TOTAL COST PER CAPITA ($)	10.20		13.60		17.00	

Notes: (a) Population is assumed to be 10,000,000 and GDP per capita $340.
(b) Fixed Infrastructure reports the costs of construction, maintenance and staffing of the clinics, district hospitals and referral hospitals which are assumed to have been constructed previously. Assumptions are based on infrastructure data for sub-Saharan African coutries.

Source: Murray et al, 1994

effectiveness analyses and the appraisal of health research. Each of these exercises will be briefly reviewed; the major emphasis, however, will be on the report of the more recent WHO Ad Hoc Committee on Health Research Relating to Future Intervention Options (World Health Organization, 1996).

1.2.3.1 The Instiute of Medicine Study

The Committee on Issues and Priorities for New Vaccine Development, of the USA's Institute of Medicine (IOM), attempted to use economic analysis to recommend best research opportunities for a wide range of diseases (Institute of Medicine, 1985; 1986). The IOM study focuses exclusively on vaccines, making it an encyclopedic compilation of assessments of vaccine research opportunities. However, it does not attempt formally to evaluate where research needs might better be met by pursuing research on non-vaccine interventions. The study nonetheless provided a pathbreaking first attempt at evaluating priorities for vaccine research.

The study released its findings in two volumes, the first of which focuses on vaccine research priorities for the United States and the second of which focuses on vaccine research priorities for developing countries (Institute of Medicine, 1985; 1986). The methodologies were largely the same, but different enough that results from one volume cannot be compared freely to those from the other. Assessments of vaccine research opportunities were made for a set of diseases determined by expert consultation.

The calculations of the IOM studies are explicit and completed in a clear, stepwise fashion. Costs are evaluated simply in terms of present-value (or real) US dollars. Benefits of pursuing a research avenue were evaluated in terms of infant-mortality equivalents (IME) saved. The IME, like the QALY and DALY, is a combined indicator of mortality and morbidity. As discussed earlier, while specific methods and values vary, QALYs and DALYs both denominate health outcomes in units of time (i.e., years of life) weighted for severity of morbidity or disability. IMEs denominate deaths and disability in units of infant-deaths. Deaths at different ages are weighted with respect to their values compared to infant deaths. Time spent in states of less than perfect health is likewise weighted with respect to its value in comparison with infant deaths. As in QALYs and DALYs, these weights vary with the severity of the morbidity or disability.

The methodology used is primarily that of cost-effectiveness analysis and the outputs are in the form of cost-effectiveness ratios. The formulae for costs and benefits can be summarized by two simple expressions. (These expressions are reproduced from the IOM evaluation of vaccine research opportiunities for developing countries (IOM, 1986). The evaluation for the United States uses more complicated, but largely similar formulae.)

For costs:

$$\mathrm{Cost} = \mathrm{C_{Dev}} + \frac{\mathrm{P_{Dev}C_v}}{(1+\mathrm{r})^{\mathrm{T_{use}}}}$$

where

Costs = annualized total cost of research for and implementation of vaccine;
C_{dev} = annualized cost of research and devlopment for vaccine;
P_{dev} = probability of successful vaccine development;
C_V = annual cost of vaccination program;
r = discount rate;
T_{use} = time until the resulting vaccine is implemented in a steady state fashion.

The denominator term $(1+r)^{T_{\text{Use}}}$ is included in order to discount the future costs of the vaccine program in order to account for the fact that they will not be incurred until time T_{use} has elapsed.

For benefits, both evaluations use the following formula:

$$\text{Benefits} = \frac{P_{\text{Dev}}B'}{(1+r)^{T_{\text{use}}+T_{\text{lag}}}}$$

where

Benefits = annualized potential benefits from implemention of vaccine;
P_{dev} = probability of successful vaccine development;
B' = potential annual benefits from vaccine
r = discount rate;
T_{use} = time until the resulting vaccine is implemented in a steady state fashion;
T_{lag} = lag between administration of vaccine and realization of health benefits.

The denominator term $(1+r)^{(T_{\text{Use}}+T_{\text{Lag}})}$ is included in order to discount the future benefits of the vaccine program in order to account for the fact that they will not be gained until time $T_{\text{use}} + T_{\text{Lag}}$ has elapsed.

These expressions can be combined to arrive at cost-effectiveness ratios. Such analysis can be completed for any intervention, not just for vaccines. The resultant cost-effectiveness ratios can be compared across a universe of research opportunities. If comparable methods were used, ostensibly, these ratios for as-yet non-existent interventions could be compared directly to the cost-effectiveness ratios for existing interventions.

One major difficulty of an endeavour such as the IOM study is the assembling of data. While cost and benefit data are in themselves difficult to acquire, they are in fact assessed in various ad hoc program evaluations and results have been usefully compiled for a range of diseases by the Health Sector Priorities Review. Where such existing data have not been made available, it may be possible from the disparate literature to estimate roughly costs and benefits. The additional data demanded by the IOM formulae are threatened by large practical limits on the data available and the reliability of estimates of such data.

Three inter-related terms in the above formulae pose great difficulties: P_{Dev} (probability of success), C_{Dev} (costs for research), and T_{Use} (time until implementation). The probability of success of a research opportunity is directly related to the costs of research and the time until implementation. Hence, the probability of success of a research opportunity can be increased by increasing the costs of that research and the time until use. To evaluate the possible values for these three variables imposes a greater analytical task.

For all of these variables, however, the greatest problem is that the speculation involved in predicting results of future research may be too difficult to perform to make it meaningful at all. Such speculation may actually be misleading and could result in a misallocation of resources. Shepard et al., (1995) have evaluated the predictions of the IOM studies and find them to have predicted greater success in vaccine development than was actually achieved. This points to the notion that, while the methods may be conceptually sound, the inherent uncertainty in the research process may make predictions of the probability of success largely meaningless.

1.2.3.2 The Children's Vaccine Initiative Study

Another attempt at formal cost-effectiveness analysis of vaccine research has been developed by Shepard et al. for the Children's Vaccine Initiative (CVI) (Shepard et al., 1995). The CVI was created following the 1990 World Summit for Children in New York. As part of the CVI's Task Force on Priorities and Strategic Plans, Shepard et al reviewed thirteen vaccine candidates using methodology similar to that of the IOM study discussed above. As the authors explain, besides updating data estimates, they modified the IOM's useful framework in four specific ways: (1) Rather than IMEs, they employ QALYs as a health measure, this breed of which are intended to be nearly equivalent to DALYs. (2) They consider not just novel vaccines but also improvements to presently existing vaccines. (3) They perform a sensitivity-analysis of their results to reveal how intervention costs and research costs could rise in order for the vaccine to remain cost-effective. (4) They incorporate data for life-spans of vaccines and consequences for disease-eradication.

Shepard et al. process their data using calculations similar to those described above for the IOM study. Table 1.2.7 illustrates the results of the model. The first column is the expected cost-effectiveness of research for and application of each vaccine candidate. These range from under \$5 per QALY for early single-dose administration of measles vaccine to over \$100,000 per QALY for pertussis vaccine.

These estimates, the authors acknowledge, are only as reliable as the data used to arrive at them. Shepard et al. note that the IOM study may have been overly optimistic in its evaluations; only four of the ten vaccines that it predicted would be liscensed by 1992 actually were. In order to arrive at more realistic estimates for their input data, Shepard et al. perform two sensitivity analyses of key parameters. In the first, they calculate how high the application-cost per dose can rise while the entire research and intervention endeavour remains cost-effective, which they define as under \$50 per QALY. The results of this analysis are presented in the second column of table 1.2.7. Application costs per dose for the bottom four vaccines shown would have to be close to zero in order for the research to fit the study's cost-effectiveness criteria. The sec-

Table 1.2.7: Results from the Children's Vaccine Inititaive Task Force on Prioirties and Strategic Plans: thirteen promising vaccine candidiates

Vaccine candidate	Cost per QALY (US$)	Maximum feasible cost per dose (US$)	Maximum feasible developmental cost (US$ x 10^9)
Measles: early single dose administration	5	5.26	4
Slow release tetanus toxoid	9	3.6	0.6
Measles: early 2-dose	13	1.9	4.7
Tyhoid	20	5.8	5.4
HBV-DPT combination	21	1.01	0.8
Rotavirus	39	2.6	1.3
Pneumococcal pneumonia	57	1.72	0
Hib-DPT combination	78	0.47	0
Enterotoxogenic E. coli	159	0.47	0
Dengue	399	0	0
Thermostable OPV	1005	0.07	0
Meningococcal meningitis: conjugate	1355	0.11	0
DT with acellular pertussis	113208	0	0

Source: Adopted from Shepard et al 1994

ond sensitivity analysis was performed similarly by holding all variables (including cost per dose) at their expected values and allowing research and development costs to rise until total costs for the endeavour reached the cut-off of $50 per QALY. These maximum feasible development costs are shown in the third column of table 1.2.7. For seven of the thirteen vaccines, research costs would have to be zero in order for the research and application of the vaccine to be cost-effective.

The authors note that their estimates of the cost-effectiveness of vaccine research are in general most sensitive to the following data: incidence and case-fatality rate of the disease, and the differences between proposed and existing vaccines in effectiveness, procurement costs and delivery costs.

1.2.3.3 The World Health Organization's Ad Hoc Committee on Health Research

A third study of research opportunities was undertaken from 1994 to 1996 by World Health Organization Ad Hoc Committee on Health Research Relating to Future Intervention Options, a major review of international health research priorities (World Health Organization, 1996). While initially starting with a method similar to the IOM and Shepard studies, this group of experts early on concluded that the analysis of health research priorities required three steps: identification of research needs based on a full analysis of present and future burden of disease, identification of research opportunities based on the collective judgment of scientists in each field, and specification of the characteristics of the product of the research endeavor that would be required to make the product cost-effective. The empirical basis of each of these steps will be discussed below.

1.2.3.3.1 Identifying Research Needs Based on Present and Future Burdens of Disease

Two factors may explain the existence of the disease-burden from any specific disease, injury, or risk factor. Either there is no cost-effective technology in existence that can be used to prevent, cure or ameliorate the problem; or cost-effective technologies exist, but are not fully implemented due to under-funding or program inefficiency. As a first estimate, a large disease-burden from a problem indicates the need for some form of research, either basic, strategic, developmental or operational. Given the rapid demographic and epidemiological shifts underway, research needs should be defined not simply with respect to the burden of diseases, injuries, or risk factors today, but also taking into consideration the likely trends in the burden of disease in the next decades. As part of the landmark Global Burden of Disease study, efforts have recently been made to provide such information on current and projected burden from specific disease, injuries, and risk factors.

1.2.3.3.2 The Global Burden of Disease Study

Despite the usefulness of knowledge of disease-burden, no comprehensive information system exists to assess it. Studies have, however, been completed for specific countries and, using rougher data, for the entire world. The Global Burden of Disease Study (GBD) provides global and regional estimates of the burden of disease and injury for nearly 500 conditions caused by 98 specific diseases and injuries (Murray and Lopez, 1996a; 1996b).

Only one attempt to measure the burden of disease in a comprehensive manner was made before the first version of the GBD in 1993. The Ghana Health Assessment Project Team estimated the burden of disease due to mortality and morbidity in Ghana for 48 causes, but little methodological detail was presented and the study was not followed up in Ghana or any other country (Ghana Health Assessment Project Team, 1981). In 1993 the World Bank's *World Development Report 1993: Investing in Health* presented results from the third version of the GBD. A fourth version is presented in papers published in the *Bulletin of the World Health Organization* (Murray, 1994; Murray and Lopez, 1994a; Murray and Lopez, 1994b; Murray et al., 1994a). Final results from the GBD project's fifth version are to be published in 1996 (World Health Organization, 1996; Murray and Lopez, 1996a; 1996b). Data from this most recent version are presented in summary here.

The GBD project's estimates were generated with the collaboration of over 100 experts, each providing consultation concerning his or her disease or diseases of expertise. Additional inputs included a wide range of published and unpublished data from around the world. Fatal and non-fatal health outcomes were included and incorporated into the DALY indicator, discussed more extensively in the preceding section concerning cost-effectiveness analysis. The GBD outputs are expressed in standard DALYs, which allows them to be easily interpreted with respect to the results of the Health Sector Priorities Review cost-effectiveness estimates. Results are generally presented by cause, region, and age-sex group for the years 1990 and 2020. Table 1.2.8 presents the rank order of diseases and injuries in terms of contribution to global burden as estimated for 1990 and Tables 1.2.9 and 1.2.10 provide summaries of the GBD results for 1990 and the baseline projections for 2020.

Measles and tuberculosis both rank within the top ten diseases in terms of global DALY loss in 1990. Tetanus, pertussis, polio, hepatitis and diphtheria rank 18th, 23rd, 61st, 69th, and 96th, respectively.

The two leading causes of DALY-loss are lower respiratory infections and diarrheal diseases, which, along with perinatal conditions, inflict far greater burdens than any other cause. Proportions of the DALY-loss from lower respiratory infections and diarrheal diseases may be precipitated by vaccine-preventable diseases, specifically measles and pertussis. While diphtheria ranks comparatively low, measles, tuberculosis, tetanus and pertussis are extremely important causes of disease-burden. These four each inflict greater burdens than some causes which presently receive great research support, among which are: HIV, diabetes mellitus, leukaemia, and lung, prostate and breast cancers (see tab. 1.2.8).

Other significant insights from the GBD study include the expected convergence of the epidemiological profiles in developed and developing countries. The predominance of infectious and parasitic diseases among the causes of disease-burden in the world is projected to drop from 23% in 1990 to 14% in 2020. By 2020, cardiovascular diseases will account for between fifteen and twenty percent of disease-burden in each demographic region of the world besides sub-Saharan Africa (SSA) and the formerly socialist economies of Europe (FSE). In the SSA region, the share of disease-burden attributable to cardiovascular disease is nevertheless projected to rise from around 4% in 1990 to nearly 6% in 2020. In FSE, that proportion is projected to have reached 27% by the year 2020 (see tabs. 1.2.9 and 1.2.10).

Some unexpected findings include the enormous share of global disease-burden accounted for by injuries, intentional and unintentional. Self-inflicted injuries, other-inflicted injuries (violence) and injuries from war each account for over 1.25% of the global disease-burden. In each demographic region, injuries accounted for between eleven and nineteen percent of disease-burden in 1990. These proportions are expected to rise to nearly 20% for all regions except the established market economies. The burden from neuro-psychiatric conditions are also unexpectedly high, with unipolar major depression ranking fourth globally among all causes of DALY-loss (see tabs. 1.2.9 and 1.2.10). Other disease accounting for two percent or more of the global burden of disease in 1990 are: ischaemic heart disease, cerebrovascular disease, motor-vehicle accidents, congenital anomalies, malaria, and chronic obstructive pulmonary disease.

Table 1.2.8: The Global Burden of Disease in 1990: rank list of specific causes evaluated by the GBD study, in descending order of magnitude

Rank order	Cause	Global Burden of Disease (DALYs)	
		DALYs ('000s)	As % Total
	All causes	1372344	100.00
1	Lower respiratory infections	111857	8.15
2	Diarrhoeal diseases	98714	7.19
3	Perinatal conditions	91457	6.66
4	Unipolar major depression	50342	3.67
5	Ischaemic heart disease	46269	3.37
6	Cerebrovascular disease	38168	2.78
7	Tuberculosis	38067	2.77
8	Measles	36184	2.64
9	Motor-vehicle accidents	34019	2.48
10	Congential anomalies	32329	2.36
11	Malaria	31414	2.29
12	Chronic obstructive pulmonary disease	28867	2.10
13	Falls	26434	1.93
14	Iron-deficiency anaemia	24386	1.78
15	Protein-energy malnutrition	20764	1.51
16	Self-inflicted injuries	18773	1.37
17	War	18559	1.35
18	Tetanus	17355	1.26
19	Violence	17309	1.26
20	Alcohol use	16507	1.20
21	Drownings	15552	1.13
22	Bipolar disorder	14125	1.03
23	Pertussis	13279	0.97
24	Osteoarthritis	13153	0.96
25	Cirrhosis of the liver	13061	0.95
26	Schizophrenia	12680	0.92
27	Fires	11899	0.87
28	HIV	11069	0.81
29	Diabetes mellitus	10887	0.79
30	Asthma	10676	0.78
31	Inflammatory heart disease	10243	0.75
32	Trachea, bronchus, and lung cancers	8790	0.64
33	Dementia and other degenerative and hereditary CNS disorders	8556	0.62
34	Nephritis and nephrosis	8538	0.62
35	Obsessive-compulsive disorders	8408	0.61
36	Stomach cancer	7623	0.56
37	Cataract	7441	0.54
38	Syphilis	6535	0.48
39	Liver cancer	6489	0.47
40	Obstructed labour	6402	0.47
41	Poisonings	6395	0.47
42	Bacterial meningitis and meningococcemia	6184	0.45
43	Abortion	6162	0.45
44	Rheumatic heart disease	6134	0.45
45	Chlamydia	5975	0.44
46	Endocrine disorders	5934	0.43
47	Drug use	5622	0.41
48	Epilepsy	5300	0.39

cont. ——>

Table 1.2.8 cont.

Rank order	Cause	Global Burden of Disease (DALYs) DALYs ('000s)	As % Total
49	Leukaemia	4940	0.36
50	Maternal sepsis	4868	0.35
51	Panic disorder	4722	0.34
52	Colon and rectum cancers	4575	0.33
53	Dental caries	4273	0.31
54	Breast cancer	4171	0.30
55	Lymphatic filariasis	3960	0.29
56	Gonorrhoea	3861	0.28
57	Vitamin A deficiency	3802	0.28
58	Mouth and oropharynx cancers	3709	0.27
59	Oesophagus cancer	3704	0.27
60	Maternal haemorrhage	3526	0.26
61	Polio	3340	0.24
62	Rheumatoid arthritis	3256	0.24
63	Lymphoma	3075	0.22
64	Cervix uteri cancer	2828	0.21
65	Peptic ulcer disease	2740	0.20
66	Edentulism	2729	0.20
67	Glaucoma	2554	0.19
68	Otitis media	2143	0.16
69	Hepatitis B and hepatitis C	2116	0.15
70	Leishmaniasis	2072	0.15
71	Post-traumatic stress disorder	1927	0.14
72	Benign prostatic hypertrophy	1801	0.13
73	Trichuriasiss	1771	0.13
74	Appendicitis	1747	0.13
75	Ascariasis	1734	0.13
76	Hypertensive disorders of pregnancy	1715	0.12
77	Pancreas cancer	1563	0.11
78	Iodine deficiency	1543	0.11
79	Schistosomiasis	1505	0.11
80	Ancylostomiasis and necatoriasis	1470	0.11
81	Trypanosomiasis	1454	0.11
82	Multiple sclerosis	1404	0.10
83	Ovary cancer	1390	0.10
84	Prostate cancer	1333	0.10
85	Upper respiratory infections	1287	0.09
86	Bladder cancer	1204	0.09
87	Parkinson disease	1040	0.08
88	Trachoma	1015	0.07
89	Onchocerciasis	876	0.06
90	Dengue	743	0.05
91	Japanese encephalitis	737	0.05
92	Corpus uteri cancer	638	0.05
93	Melanoma and other skin cancers	560	0.04
94	Chagas disease	548	0.04
95	Leprosy	380	0.03
96	Diphtheria	358	0.03
97	Periodontal disease	252	0.02

Sources: Murray and Lopez 1996a, World Health Organization 1996

Table 1.2.9: Summary results: the Global Burden of Disease in 1990

Condition	DALYs by cause, as percentage of regional total								
	WORLD	Sub-Saharan Africa	India	China	Other Asia and Islands	Latin America and the Caribbean	Middle Eastern Crescent	Formerly socialist economies of Europe	Established market economies
I. Communicable, maternal, perinatal, and nutritional conditions	44.4	66.3	56.8	25.1	45.0	35.6	48.9	10.6	7.5
A. Infectious and parasitic diseases	23.0	42.5	28.9	7.7	22.2	17.8	20.2	3.5	3.2
Tuberculosis	2.8	3.5	4.8	2.0	3.1	1.8	1.7	0.6	0.1
STDs excluding HIV	1.2	1.9	1.7	0.0	2.0	1.1	0.5	0.5	0.3
HIV infection	0.8	2.8	0.1	0.0	0.1	1.1	0.0	0.1	1.3
Diarrhoeal diseases	7.2	10.9	10.2	1.8	7.7	5.4	9.7	0.4	0.2
Childhood-cluster diseases	5.1	10.4	6.3	1.1	4.5	3.4	5.6	0.1	0.0
Bacterial meningitis and meningococcemia	0.5	0.3	0.5	0.6	0.5	0.5	0.5	0.4	0.2
Malaria	2.3	9.2	0.4	0.0	1.4	0.5	0.2	0.0	0.0
Tropical cluster diseases and leprosy	0.8	1.9	1.2	0.1	0.4	0.8	0.2	0.0	0.0
Intestinal nematode infections	0.4	0.2	0.3	0.7	0.9	0.7	0.1	0.0	0.0
Other infectious and parasitic diseases	2.0	1.4	3.5	1.4	1.6	2.5	1.7	1.6	1.0
B. Respiratory infections	8.4	10.5	11.8	6.0	8.7	4.9	10.6	2.0	1.3
Lower respiratory infections	8.2	10.3	11.4	5.7	8.5	4.7	10.3	1.9	1.2
Other respiratory infections	0.3	0.2	0.5	0.2	0.3	0.2	0.3	0.1	0.1
C. Maternal conditions	2.7	3.5	3.1	1.8	2.7	2.0	3.9	1.7	0.3
Obstructed labour	0.5	0.6	0.5	0.3	0.5	0.4	0.7	0.2	0.2
Abortion	0.4	0.7	0.7	0.0	0.5	0.6	0.3	0.3	0.0
Other maternal conditions	1.8	2.3	1.9	1.5	1.7	1.0	2.9	1.2	0.1
D. Perinatal conditions	6.7	6.6	8.8	4.9	6.9	7.3	9.6	2.1	1.8
E. Nutritional deficiencies	3.7	3.2	4.2	4.6	4.4	3.6	4.6	1.2	0.9
Protein-energy malnutrition	1.5	1.8	1.8	1.0	1.7	1.7	2.4	0.2	0.1
Vitamin A deficiency and iodine deficiency	0.4	0.4	0.4	0.4	0.5	0.3	0.6	0.0	0.0
Anaemia	1.8	0.9	2.1	3.2	2.3	1.7	1.6	0.6	0.7
Other nutritional deficiencies	0.0	0.0	0.0	0.0	0.0	0.0	0.0	0.4	0.0
II. Noncommunicable diseases	40.6	18.5	28.7	57.3	40.6	48.1	38.6	71.4	80.7
A. Malignant neoplasms	5.1	2.1	2.5	8.7	5.1	4.5	2.4	11.4	15.3
Stomach cancer	0.6	0.1	0.2	1.6	0.4	0.4	0.2	1.6	1.1
Colon and rectum cancers	0.3	0.1	0.1	0.5	0.3	0.2	0.1	1.1	1.6
Liver cancer	0.5	0.3	0.1	1.9	0.5	0.1	0.1	0.3	0.3
Trachea, bronchus and lung cancers	0.6	0.1	0.1	1.0	0.5	0.3	0.2	2.5	3.0
Breast cancer	0.3	0.1	0.2	0.2	0.2	0.4	0.1	0.8	1.4
Cervix uteri cancer	0.2	0.1	0.3	0.1	0.3	0.4	0.1	0.3	0.2
Lymphomas and multiple myeloma	0.2	0.2	0.1	0.2	0.2	0.3	0.1	0.4	0.7
Leukaemia	0.4	0.1	0.1	0.8	0.5	0.3	0.2	0.5	1.0
Other malignant neoplasms	2.0	1.1	1.3	2.4	2.2	2.0	1.2	4.0	6.0
B. Diabetes mellitus	0.8	0.2	0.8	0.5	0.7	1.4	0.9	1.0	2.3
C. Neuro-psychiatric conditions	10.4	4.0	6.9	13.6	10.8	16.0	8.6	16.7	24.8
Unipolar major depression	3.7	1.5	2.8	6.2	3.8	4.2	3.0	4.8	6.7
Bipolar disorder	1.0	0.4	0.8	1.8	1.1	1.2	0.9	1.3	1.7
Schizophrenia	0.9	0.2	0.6	1.3	1.3	1.3	0.9	1.3	2.2
Alcohol use	1.2	0.6	0.3	0.7	1.1	3.8	0.2	2.7	4.7
Dementia and other CNS disorders	0.6	0.1	0.3	0.7	0.5	0.8	0.2	1.5	2.9
Drug use	0.4	0.1	0.0	0.1	0.5	1.1	0.6	0.8	1.5
Epilepsy	0.4	0.2	0.3	0.4	0.5	0.7	0.3	0.6	0.5
Other neuro-psychiatric conditions	2.1	0.9	1.7	2.3	2.1	2.9	2.5	3.6	4.7

cont. ——>

Table 1.2.9 cont.

Condition	DALYs by cause, as percentage of regional total WORLD	Sub-Saharan Africa	India	China	Other Asia and Islands	Latin America and the Caribbean	Middle Eastern Crescent	Formerly socialist economies of Europe	Established market economies
II. Noncommunicable diseases, continued									
D. Glaucoma and cataracts	0.8	0.7	1.0	1.0	0.9	0.6	0.6	0.1	0.1
E. Cardiovascular diseases	9.6	3.9	8.1	11.0	10.1	7.9	11.0	22.6	18.3
Rheumatic heart disease	0.4	0.2	0.5	1.1	0.1	0.2	0.5	0.6	0.2
Ischaemic heart disease	3.4	0.8	3.5	2.9	2.2	3.0	3.5	11.1	8.9
Cerebrovascular disease	2.8	1.6	1.5	5.2	2.5	2.5	1.6	7.0	5.0
Inflammatory heart diseases	0.7	0.5	0.6	0.6	1.3	0.5	1.2	0.7	0.7
Other cardiovascular diseases	2.3	0.9	2.0	1.2	4.0	1.7	4.1	3.3	3.6
F. Respiratory diseases	4.3	2.5	2.6	10.6	2.6	3.9	4.0	4.7	4.8
COPD	2.1	0.6	0.9	8.6	0.7	1.0	0.9	1.7	2.3
Asthma	0.8	0.5	0.5	1.3	0.8	1.0	0.6	0.8	1.2
Other respiratory diseases	1.4	1.4	1.2	0.7	1.1	1.9	2.5	2.3	1.3
G. Digestive diseases	3.3	1.8	2.2	4.6	4.6	3.8	3.9	4.6	4.4
Cirrhosis of the liver	1.0	0.2	1.0	1.5	1.3	1.1	0.5	1.2	1.6
Other digestive diseases	2.4	1.6	1.2	3.1	3.3	2.6	3.4	3.4	2.8
H. Musculo-skeletal diseases	1.4	0.4	0.5	1.7	1.2	3.1	0.6	4.4	4.2
Rheumatoid arthritis	0.2	0.0	0.1	0.3	0.1	0.6	0.1	0.8	0.9
Osteoarthritis	1.0	0.3	0.4	1.0	0.9	2.1	0.4	3.1	2.7
Other musculo-skeletal diseases	0.2	0.0	0.0	0.4	0.2	0.4	0.1	0.5	0.5
I. Congenital anomalies	2.4	1.3	2.9	3.0	2.3	2.6	2.7	2.1	2.0
J. Oral conditions	0.5	0.1	0.4	0.5	0.7	1.0	0.9	0.8	0.9
K. Other noncommunicable diseases	2.0	1.5	0.9	2.1	1.6	3.3	3.0	3.0	3.5
III. Injuries	15.0	15.2	14.5	17.7	14.4	16.3	12.5	18.1	11.7
A. Unintentional injuries	11.0	9.4	13.0	13.0	12.1	11.8	6.7	12.5	8.6
Motor-vehicle accidents	2.5	1.9	2.1	2.1	2.7	4.0	1.7	4.3	4.3
Poisonings	0.5	0.4	0.3	0.7	0.6	0.2	0.2	1.4	0.3
Falls	1.9	0.7	3.5	2.2	2.3	1.7	1.1	1.7	1.3
Fires	0.9	1.2	2.0	0.3	0.3	0.3	0.5	0.3	0.3
Drownings	1.1	1.1	0.9	2.1	1.6	0.9	0.6	0.9	0.3
Other unintentional injuries	4.1	4.0	4.1	5.6	4.7	4.7	2.7	3.8	2.1
B. Intentional injuries	4.0	5.8	1.5	4.7	2.3	4.5	5.8	5.5	3.1
Self-inflicted injuries	1.4	0.2	1.0	3.9	1.1	0.6	0.9	2.5	2.1
Violence	1.3	2.2	0.5	0.8	0.9	3.2	0.8	1.3	1.0
War	1.4	3.4	0.0	0.0	0.3	0.6	4.1	1.7	0.0
Other intentional injuries	0.0	0.0	0.0	0.0	0.0	0.0	0.0	0.0	0.0

Source: Murray and Lopez, 1996a; World Health Organization, 1996

From these results, conclusions can be drawn concerning research needs. For causes for which there is a large present and projected disease-burden, there is a role for research. The next step in the determination of research needs is to identify whether the priority for research concerning that cause is (1) the development of new technologies, (2) operational research to enhance the effects of resources already invested in interventions for the cause, or (3) research and advocacy to enhance the priority attached to the existing cost-effective interventions for that cause. Toward the goal of discerning which type of research is of highest prioirty for a specific cause, the WHO Ad Hoc Committee on Health Research has made use of research-needs diagrams.

Table 1.2.10: Summary results: projections of the Global Burden of Disease in 2020

Condition	DALYs by cause, as percentage of regional total								
	WORLD	Sub-Saharan Africa	India	China	Other Asia and Islands	Latin America and the Caribbean	Middle Eastern Crescent	Formerly socialist economies of Europe	Established market economies
I. Communicable, maternal, perinatal, and nutritional conditions	21.9	41.2	26.3	5.7	18.3	13.9	21.1	3.7	6.0
A. Infectious and parasitic diseases	14.0	29.5	18.3	1.7	10.7	8.2	8.8	1.3	3.6
Tuberculosis	3.2	6.8	6.8	0.4	1.2	0.5	0.5	0.1	0.1
STDs excluding HIV	0.9	1.6	1.2	0.0	1.4	0.7	0.3	0.3	0.2
HIV infection	2.7	4.4	4.8	0.1	3.1	3.1	0.2	0.2	2.6
Diarrhoeal diseases	2.8	5.6	2.6	0.3	2.3	1.6	4.3	0.1	0.1
Childhood-cluster diseases	2.1	5.3	1.5	0.2	1.3	1.0	2.5	0.0	0.0
Bacterial meningitis and meningococcemia	0.1	0.2	0.1	0.1	0.1	0.1	0.2	0.1	0.1
Malaria	1.2	4.6	0.1	0.0	0.3	0.1	0.1	0.0	0.0
Tropical cluster diseases and leprosy	0.1	0.3	0.1	0.0	0.0	0.2	0.0	0.0	0.0
Intestinal nematode infections	0.2	0.1	0.2	0.3	0.5	0.3	0.1	0.0	0.0
Other infectious and parasitic diseases	0.6	0.6	0.8	0.3	0.5	0.6	0.7	0.4	0.6
B. Respiratory infections	3.3	5.6	3.3	1.2	2.9	1.6	4.8	0.9	1.2
Lower respiratory infections	3.2	5.4	3.2	1.1	2.9	1.6	4.6	0.9	1.1
Other respiratory infections	0.1	0.1	0.1	0.0	0.1	0.1	0.1	0.0	0.1
C. Maternal conditions	0.4	0.7	0.4	0.1	0.4	0.3	0.7	0.2	0.1
Obstructed labour	0.1	0.1	0.1	0.0	0.1	0.1	0.1	0.0	0.0
Abortion	0.1	0.1	0.1	0.0	0.1	0.1	0.1	0.0	0.0
Other maternal conditions	0.3	0.4	0.2	0.1	0.2	0.1	0.5	0.2	0.0
D. Perinatal conditions	2.6	3.8	2.5	0.9	2.3	2.3	4.7	0.7	0.7
E. Nutritional deficiencies	1.7	1.6	1.9	1.9	2.0	1.4	2.1	0.5	0.5
Protein-energy malnutrition	0.6	1.0	0.5	0.2	0.5	0.5	1.1	0.1	0.0
Vitamin A deficiency and iodine deficiency	0.1	0.2	0.1	0.1	0.1	0.1	0.3	0.0	0.0
Anaemia	1.0	0.4	1.3	1.7	1.4	0.8	0.8	0.4	0.5
Other nutritional deficiencies	0.0	0.0	0.0	0.0	0.0	0.0	0.0	0.1	0.0
II. Noncommunicable diseases	57.0	29.8	53.8	76.9	63.5	65.7	58.1	77.8	82.9
A. Malignant neoplasms	9.9	4.4	7.1	19.1	11.8	8.8	5.3	16.7	18.8
Stomach cancer	1.1	0.3	0.5	3.6	1.0	0.8	0.4	2.4	1.3
Colon and rectum cancers	0.6	0.1	0.2	1.0	0.6	0.5	0.2	1.5	1.9
Liver cancer	1.1	0.6	0.1	4.2	1.3	0.1	0.2	0.4	0.4
Trachea, bronchus and lung cancers	1.8	0.2	1.4	3.4	1.9	1.3	1.2	4.5	4.8
Breast cancer	0.5	0.2	0.4	0.3	0.5	0.8	0.3	0.8	1.4
Cervix uteri cancer	0.4	0.3	0.6	0.3	0.6	0.7	0.2	0.3	0.2
Lymphomas and multiple myeloma	0.4	0.4	0.2	0.3	0.4	0.4	0.2	0.5	0.8
Leukaemia	0.5	0.1	0.2	1.0	0.6	0.4	0.4	0.6	1.1
Other malignant neoplasms	3.7	2.2	3.2	4.9	4.8	3.7	2.3	5.7	6.9
B. Diabetes mellitus	0.8	0.2	0.8	0.4	0.9	1.6	1.0	0.8	2.2
C. Neuro-psychiatric conditions	13.3	7.6	11.7	14.0	16.0	19.4	14.0	15.0	23.5
Unipolar major depression	5.9	3.6	5.8	7.8	7.0	6.8	6.2	5.5	7.5
Bipolar disorder	1.6	1.0	1.6	2.1	1.9	1.8	1.7	1.4	1.8
Schizophrenia	0.6	0.1	0.5	0.6	1.1	1.0	0.8	0.8	1.2
Alcohol use	1.0	0.6	0.4	0.4	1.2	3.6	0.2	1.7	2.8
Dementia and other CNS disorders	0.8	0.1	0.5	0.8	0.7	1.0	0.2	1.6	3.8
Drug use	0.6	0.3	0.0	0.1	0.9	1.6	1.2	0.9	1.5
Epilepsy	0.3	0.2	0.3	0.2	0.4	0.5	0.3	0.4	0.3
Other neuro-psychiatric conditions	2.6	1.6	2.6	2.0	2.9	3.0	3.4	2.8	4.5

cont. ——>

Table 1.2.10 cont.

Condition	DALYs by cause, as percentage of regional total WORLD	Sub-Saharan Africa	India	China	Other Asia and Islands	Latin America and the Caribbean	Middle Eastern Crescent	Formerly socialist economies of Europe	Established market economies
II. Noncommunicable diseases, continued									
D. Glaucoma and cataracts	1.7	1.4	2.6	1.9	2.6	1.4	1.4	0.1	0.2
E. Cardiovascular diseases	14.5	5.9	17.7	16.3	15.7	13.5	17.5	27.1	20.1
Rheumatic heart disease	0.5	0.2	0.8	1.2	0.1	0.2	0.5	0.6	0.1
Ischaemic heart disease	5.8	1.4	8.8	4.7	5.0	5.7	7.3	13.7	10.1
Cerebrovascular disease	4.4	2.5	3.2	8.2	4.5	4.4	2.8	8.0	5.3
Inflammatory heart diseases	0.9	0.6	0.9	0.6	1.2	0.6	1.4	0.8	0.6
Other cardiovascular diseases	3.0	1.1	3.8	1.6	4.7	2.6	5.5	4.0	3.8
F. Respiratory diseases	7.3	4.4	6.3	16.5	4.3	6.6	6.5	8.6	5.7
COPD	4.1	1.4	2.7	14.6	1.7	2.6	2.2	3.4	3.0
Asthma	1.0	0.7	0.9	1.3	1.1	1.4	0.9	1.0	1.2
Other respiratory diseases	2.2	2.3	2.7	0.6	1.5	2.6	3.5	4.2	1.5
G. Digestive diseases	3.6	1.8	2.5	3.6	6.6	4.9	3.6	4.3	5.6
Cirrhosis of the liver	1.2	0.3	1.2	1.5	2.2	1.7	0.7	1.2	2.0
Other digestive diseases	2.4	1.5	1.2	2.1	4.4	3.1	2.9	3.1	3.6
H. Musculo-skeletal diseases	0.7	0.1	0.1	0.8	0.4	3.5	0.5	1.2	1.7
Rheumatoid arthritis	0.4	0.1	0.1	0.6	0.3	1.0	0.2	0.9	1.3
Osteoarthritis	0.2	0.0	0.0	0.0	0.0	2.2	0.1	0.0	0.0
Other musculo-skeletal diseases	0.1	0.0	0.0	0.3	0.2	0.3	0.1	0.3	0.4
I. Congenital anomalies	2.3	2.3	3.3	2.0	1.9	1.8	3.4	1.2	1.0
J. Oral conditions	0.9	0.2	0.8	0.7	1.6	1.4	1.8	0.9	1.1
K. Other noncommunicable diseases	1.9	1.7	0.9	1.6	1.8	2.8	3.0	2.0	3.2
III. Injuries	21.1	29.0	19.9	17.4	18.1	20.5	20.9	18.5	11.1
A. Unintentional injuries	13.7	16.0	17.1	11.7	14.3	14.0	10.2	12.5	7.5
Motor-vehicle accidents	5.4	5.3	6.7	5.1	5.4	6.8	3.8	5.7	4.0
Poisonings	0.5	0.6	0.4	0.5	0.6	0.1	0.3	1.2	0.2
Falls	1.6	1.0	2.7	1.5	2.1	1.5	1.3	1.5	1.2
Fires	0.9	1.6	2.1	0.2	0.2	0.2	0.6	0.2	0.2
Drownings	0.9	1.3	0.8	1.0	1.1	0.7	0.7	0.7	0.2
Other unintentional injuries	4.4	6.2	4.4	3.3	4.8	4.6	3.6	3.1	1.7
B. Intentional injuries	7.3	13.0	2.9	5.7	3.8	6.5	10.7	6.0	3.5
Self-inflicted injuries	1.9	0.4	1.9	4.8	1.8	0.9	1.7	2.8	2.5
Violence	2.4	5.2	0.9	0.9	1.5	4.7	1.4	1.4	1.1
War	3.0	7.4	0.1	0.0	0.5	0.9	7.6	1.8	0.0
Other intentional injuries	0.0	0.0	0.0	0.0	0.0	0.0	0.0	0.0	0.0

Source: Murray and Lopez, 1996a; World Health Organization, 1996

Figure 1.2.5 is a research-needs diagram devised by the WHO study from a model it developed and from data collected from disease-specific experts. Figure 1.2.5 illustrates the case of tuberculosis. The x-axis represents the aggregate coverage and efficiency from all interventions against tuberculosis. The y-axis represents the aggregate efficacy of all interventions against tuberculosis. The entire area of the box figure represents the total number of DALYs that would be lost due to tuberculosis if no interventions were applied. The black rectangle in the lower left corner of this box represents the total DALYs saved by the existing interventions. Hence, the total area of this diagram less the black rectangle representing DALYs averted is equal to the global burden of tuberculosis in 1990. The white band across the top represents the DALYs that can only be addressed by the development of new technologies. The dark grey area represents the gains in reducing the burden of tuberculosis that may be made through improvements in technical and allocative efficiency. These improvements may come about through operational research projects; e.g. improving the cold-chain for the Expanded Programme of Immunization. Technological breakthroughs also could be used to avert this share of the burden. For example, were a new ultra-short course chemotherapy regimen for tuberculosis developed and implemented, an accordingly updated version of this diagram would show the dark gray box as reduced in size and the black box increased. The lightest grey rectangle represents that share of burden that could be addressed with existing technology, the application of which at present would not be cost-effective.

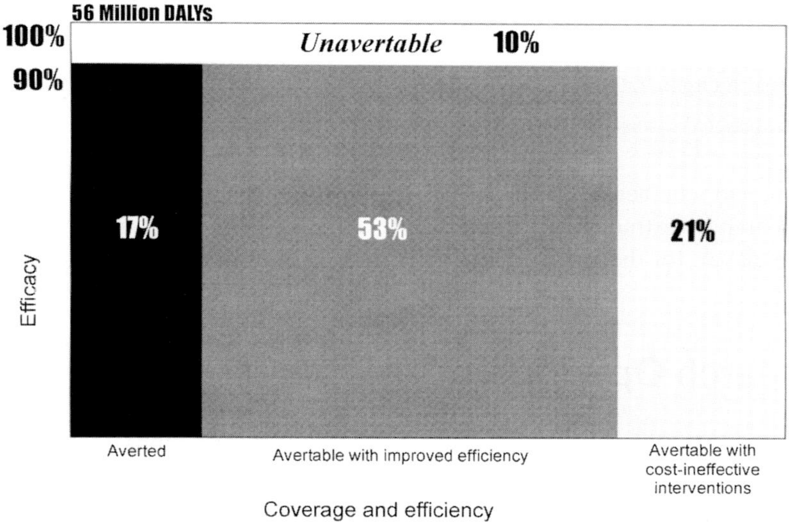

(Source: World Health Organization, 1996)

Fig. 1.2.7: Tuberculosis research needs: averted, avertable, and unavertable global burden of tuberculosis with existing interventions, 1990.

The data necessary to complete such an analysis for a disease include disease burden estimates as well as estimates of the coverage and costs for all existing interventions against the disease. Such data are not readily available and broad, aggregate estimates can be made. Due to the lack of precise data to complete such an analysis, results should be taken to give a rough picture of research needs.

Figure 1.2.8 is a research-needs diagram for pneumococcal disease.

This method helps to discrimante among research needs of various forms. Research needs, however, are not necessarily priorities for research investment. To define a priority for research investment, we must first examine both the opportunities

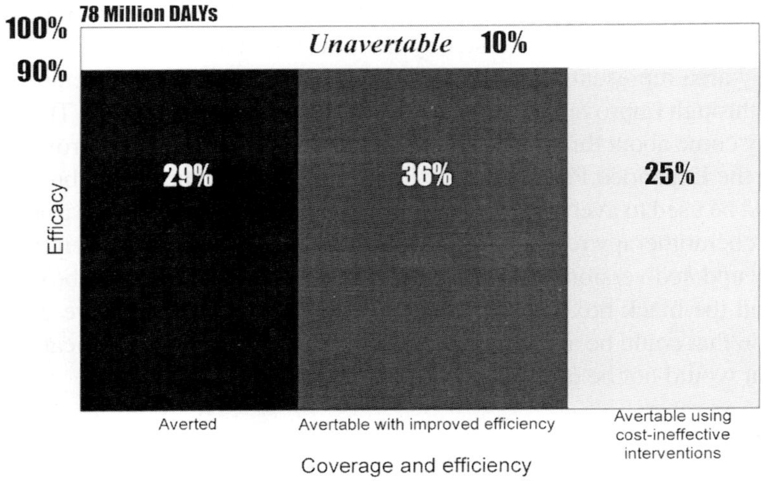

(Source: World Health Organization, 1996)

Fig. 1.2.8: Pneumococcal disease research needs, 1990.

for advancement, and whether the product of research would be cost-effective. If the product of a research investment were not cost-effective, the investment might yield no benefits whatsoever, for the intervention would likely never be applied.

1.2.4 Research Opportunities

The IOM and Shepard et al studies used expert judgement to attempt to estimate the probability of success of different research avenues (Institute of Medicine, 1985; 1986; Shepard et al., 1995). After undertaking an extensive year-long survey of several hundred researchers, the WHO Ad Hoc Committee on Health Research Relating to Future Intervention Options concluded that asking scientists to evaluate the probability of success of their own research areas was not likely to produce comparable or meaningful results. When forced to quantify such an uncertain activity as research, scientists are likely to be optimisitic. As the Shepard study itself points out, the IOM studies were extremely optimistic; only four of the ten vaccines candidates it predicted were liscenced by 1992. There is too much of a vested interest in many cases for scientists to be entirely objective when rendering this type of judgement. In any group of scientits, moreover, there will be a minority that are pessimists and will tend to underestimate the liklihood of success, therefore potentially biasing the evaluation of opporunities for research towards those activities that are proposed and evaluated by optimists.

Even though formal estimates of the probability of success cannot be made in a comparable and meaningful manner, there is no group in a better situation to judge the merits of different proposed avenues of research on a particular research topic than scientists themselves. The WHO study concluded that the process of identifying priority areas for research should be based on an assessment of research needs. The process of identifying specific approaches to researching a topic should be left to the standard mechansisms of peer review.

While formal assessments of the probability of success of a research project are probably futile, there is a clear informative role for economic analysis. Many research avenues are pursued without much consideration of the question of whether the product of the research will be use even if the research is successful. Before investing in a research project such as a vaccine for schistosomiasis, the characteristics of the proposed product that would be required in order that it would be cost-effective should be undertaken. Thus we can use cost-effectiveness for the intervention to estimate the efficacy, effectiveness and costs required of a specific intervention in order that when implemented it would cost less than $500 per DALY – or some other cutoff used to define a cost-effective intervention. The exact cutoff used to define a cost-effective intervention, and thus used for product specification, will depend on the income per capita of the country or region under consideration.

Through the interplay of the scientific communities' judgements of the technical merits of research projects and economic appraisals of the product specifications required for the outcome of reserach to be worthwhile, a more rational approach to research prioritization of vaccine research and health research in general can be envisaged.

Acknowledgments

The authors thank Kaniaru Wacieni for his research assistance. They gratefully acknowledge the assistance and support of Steven Goodreau, Joshua Salomon and Bonifasiyo Ssennyamantono.

References

Anonymous (1992) Discounting health care: only a matter of timing? Lancet *340*, 148-149.

Barnum, H. (1987) Evaluating healthy days of life gained from health projects. Social Science and Medicine *24*, 833-841.

Barnum, H., Tarantola, D. and Setaidy, I. F. (1980) Cost-effectiveness of an immunization programme in Indonesia. Bulletin of the World Health Organization *58*, 499-503.

Berg, R. L. (1973) Weighted life expectancy as a health status index. Health Services Research *8*, 153-156.

Bobadilla, J.-L., Cowley, P., Musgrove, P. and Saxenian, H. (1994) Design, content and financing od an essential package of health services. Bulletin of the World Health Organization *72*, 653-662.

Brenzel, L. (1990) The cost of EPI: a review of cost and cost-effectivness studies (1979-1987). Arlington, Virginia: Resources for Child Health.

Brenzel, L. (1991) Cost and financing of EPI. Arlington, Virginia: Resources for Child Health.

Brown, L. D. (1991) The national politics of Oregon's rationing plan. Health Affairs *10*, 28-51.

Chiang, C. L. (1965) An index of health: mathematical models. (Public Health Services Publications 1000 Series 2. No. 5) Washington DC: National Center for Health Statistics.

Cowen, T. and Parfit, D. (1992) Against the social discount rate. In: Justice between age groups and generations (Laslett, P. and Fishkin, J. S., eds.), New Haven, Yale University Press.

Cutts, F. T. and Smith, P. G. (1994) Vaccination and World Health. Chichester (UK), John Wiley and Sons Ltd.

Dasgupta, P., Marglin, S. and Sen, A. (1972) Guidelines for project evaluation. New York, United Nations.

Dempsey, M. (1947) Decline in tuberculosis. The death rate fails to tell the entire story. American Review of Tuberculosis *56*, 157-164.

Dixon, J. and Welch, H. G. (1991) Priority setting: lessons from Oregon. Lancet *337*, 891-894.

Drummond, M. F., Stoddart, G. L. and Torrance, G. W. (1987) Methods for the economic evaluation of health care programmes. Oxford, Oxford Medical Publications.

Drummond, M., Torrance, G. and Mason, J. (1993) Cost-effectivenss league tables: more harm than good? Social Science and Medicine *37*, 33-40.

Eddy, D. M. (1991) Oregon's methods – did cost-effectiveness analysis fail? Journal of the American Medical Association *266*, 2135-2141.

Fanshel, S. and Bush, J. W. (1970) A health-status index and its application to health services outcomes. Operations Research *18*, 1021-1066.

Foster, S. O., McFarland, D. A. and John, A. M. (1993) Measles. In: Disease Control Prioirties in Developing Countries (Jamison, D. T., Mosley, W. H., Measham, A. R. and Bobadilla, J.-L., eds.), New York: Oxford University Press for the World Bank.

Fox, D. M. and Leichter, H. M. (1991) Rationing care in Oregon: the new accountability. Health Affairs *10*, 7-27.

Garber, A. M. and Phelps, C. E. (1992) Economic foundations of cost-effectiveness analysis. Cambridge: National Bureau of Economic Research Working Paper 4164.

Garland, M. J. and Hasnain, R. (1990) Health care in common: setting priorities in Oregon. Hastings Center Report September-October 1990, 16-18.

Ghana Health Assessment Project Team (1981) A quantitative method of assessing the health impact of different diseases in less developed countries. International Journal of Epidemiology *10*, 73-80.

Greville, T. N. E. (1948) Comments on Mary Dempsey's articles on "decline in tuberculosis: the death rate fails to tell the entire story". American Review of Tuberculosis *57*, 417-419.

Hadler, S. C., de Monzan, M. A., Lugo, D. R. and Perez, M. (1989) Effect of timing of hepatitis B vaccine doses on response to vaccine in Yucpa Indians. Vaccine *7*, 106-110.

Hadorn, D. C. (1991) Setting health care priorities in Oregon: cost-effectiveness meets the Rule of Rescue. Journal of the American Medical Association *265*, 2218-2225.

Haenszel, W. (1950) A standardized rate for mortality defined in units of lost years of life. American Journal of Public Health *40*, 17-26.

Haaga, J. G. (1986) Cost-effectiveness and cost-benefit analyses of immunization programmes in developing countries. In: Advances in International Maternal and Child Health, volume 6 (Jelliffe, D. and Jelliffe, E. F. P., eds.), Oxford, Clarendon.

Hammit, J. (1993) Discounting health increments. Journal of Health Economics *12*, 117.

Hartman, R. W. (1990) One thousand points of light seeking a number: A case study of CBO's search for a discount rate policy. Journal of Environmental Economics and Management *18*, S3-S7.

Institute of Medicine (1985) New Vaccine Development: Establishing Priorities – Volume I Diseases of Importance in Developed Countries. Washington: National Academy Press.

Institute of Medicine (1986) New Vaccine Development: Establishing Priorities – Volume II Diseases of Importance in Developing Countries. Washington: National Academy Press.

International Monetary Fund (1991) International Financial Statistics Yearbook. Washington, International Monetary Fund.

Jamison, D. T. (1993) Disease control priorities in developing countries: an overview. In: Disease Control Prioirties in Developing Countries (Jamison, D. T., Mosley, W. H., Measham, A. R. and Bobadilla, J.-L., eds.), New York, Oxford University Press for the World Bank.

Jamison, D. T. and Saxenian, H. (1994) Investing in imminization: conclusions from the 1993 World Development Report. In: Vaccination and World Health (Cutts, F. T. and Smith, P. G., eds.), Chichester (UK), John Wiley and Sons Ltd.

Jamison, D. T., Mosley, W. H., Measham, A. R. and Bobadilla, J.-L. (1993a) Disease Control Prioirties in Developing Countries. New York, Oxford University Press for the World Bank.

Jamison, D. T., Torres, A. M., Chen, L. C. and Melnick, J. L. (1993b) Poliomyelitis. In: (1993) Disease Control Prioirties in Developing Countries (Jamison, D. T., Mosley, W. H., Measham, A. R. and Bobadilla, J.-L., eds.), New York, Oxford University Press for the World Bank.

Johannesson, M. (1992) On the discounting of gained life-years in cost-effectiveness analysis. International Journal of Technology Assessment in Health Care *8*, 359-364.

Kane, M., Clements, J. and Hu, D. (1993) Hepatitis B. In: Disease Control Prioirties in Developing Countries (Jamison, D. T., Mosley, W. H., Measham, A. R. and Bobadilla, J.-L., eds.), New York, Oxford University Press for the World Bank.

Keeler, E. and Cretin, S. (1983) Discounting of life-saving and other nonmonetary effects. Management Science *29*, 300.

Klevit, H. D., Bates, A. C., Castanares, T., Kirk, P., Sipes-Metzler, P. R. and Wopat, R. (1991) Prioritization of health care services – a progress report by the Oregon Health Services Commission. Archives of Internal Medicine *151*, 912-916.

Kitzhaber, J. A. (1993) Prioritising health services in an era of limits: the Oregon experience. British Medical Journal *307*, 373-377.

Krahn, M. and Gafni, A. (1993) Discounting in the economic evaluation of health care interventions. Medical Care *31*, 403.

Lind, R. (1982) Discounting for time and risk in energy policy. Baltimore, Johns Hopkins University Press.

Lohr, K. N. and Ware, J. E. Jr. (eds.) (1987) Proceedings of the advances in health assessment conference. Journal of Chronic Disease *40*, 1S.

Lohr, K. N. (ed.) (1989) Advances in health status assessment: conference proceedings. Medical Care *27*, S1.

Lohr, K. N. (1992) Advances in health status assessment: fostering the application of health status measures in clinical settings. Proceedings of a conference. Medical Care *30*, MS1-MS293.

Murray, C. J. L. (1994) Quantifying the burden of disease: the technical basis for disability-adjusted life years. Bulletin of the World Health Organization *72*, 429-445.

Murray, C. J. L. and Lopez, A. D. (1994a) Global and regional cause of death patterns in 1990. Bulletin of the World Health Organization *72*, 447-480.

Murray, C. J. L. and Lopez, A. D. (1994b) Quantifying disability: data, methods and results. Bulletin of the World Health Organization *72*, 481-494.

Murray, C. J. L. and Lopez, A. D. (1994c) Global Comparative Assessments in the Health Sector: Disease Burden, Expenditures and Intervention Packages. Geneva, World Health Organization.

Murray, C. J. L. and Lopez, A. D. (1996a) The Global Burden of Disease: a comprehensive assessment of mortality and disability from diseases, injuries, and risk factors in 1990 and projected to 2020. Cambridge, Massachusetts, Harvard School of Public Health for the World Health Organization and the World Bank.

Murray, C. J. L. and Lopez, A. D. (1996b) Global Health Statistics: a compendium of incidence, prevalence and mortality estimates for over 200 conditions. Cambridge, Massachusetts, Harvard School of Public Health for the World Health Organization and the World Bank.

Murray, C. J. L. and Lopez, A. D. (1996c) Health Dimensions of Sex and Reproduction: the global burden of sexually transmitted diseases, HIV, maternal conditions, perinatal disorders, and congenital anomalies. Cambridge, Massachusetts: Harvard School of Public Health for the World Health Organization and the World Bank.

Murray, C. J. L., Govindiraj, R. and Musgrove, P. (1994a) National health expenditures: a global analysis. Bulletin of the World Health Organization *72*, 623-637.

Murray, C. J. L., Kreuser, J. and Whang, W. (1994b) Cost-effectiveness analysis and policy choices: investing in health systems. Bulletin of the World Health Organization *72*, 663-674.

Murray, C. J. L., Styblo, K. and Rouillon, A. (1993) Tuberculosis. In: Disease Control Prioirties in Developing Countries (Jamison, D. T., Mosley, W. H., Measham, A. R. and Bobadilla, J.-L., eds.), New York, Oxford University Press for the World Bank.

Nord, E. (1992) Methods for quality adjustment of life years. Social Science and Medicine *34*, 559-569.

Olsen, J. (1993) On what basis should health be discounted. Journal of Health Economics *12*, 39.

Oregon Health Services Commission (1991) Prioritization of health services: A report to the Governor and Legislature. Portland, State of Oregon.

Østbye, T. and Speechley, M. (1992) The Oregon formula: a better method of allocating health care resources. Nordisk Medicin *107*, 92-95.

Patrick, D.L., Bush, J. W. and Chen, M. M. (1973) Methods for measuring levels of well-being for a health-status index. Health Services Research *8*, 228-245.

Patrick, D. L. and Erickson, P. (1993) Health status and health policy: allocating resources to health care. New York, Oxford University Press.

Phillips, M. A., Feachem, R. G. and Mills, A. (1987) Options for diarrhoea control. EPC publication 13. London, London School of Ygiene and Tropical Medicine.

Prost, A. and Prescott, N. (1984) Cost-effectiveness of blindness prevention by the Onchocerciasis Control Programme in Upper Volta. Bulletin of World Health Organization *62*, 795-802.

Robertson, R. L., Foster, S. O., Hull, H. F. and Williams, P. J. (1985) Cost-effectiveness of immunization in the Gambia. Journal of Tropical Medicine and Hygiene *88*, 343-351.

Robertson, R. L., Foster, S. O., Hull, H. F. and Williams, P. J. (1992) Cost-effectivenss analyses of the Expanded Porgramme on Immunization (EPI) and of the addition of hepatitis B virus vaccination to the EPI of the Gambia. Health Policy and Planning.

Robine, J. M., Mathers, C. D. and Bucquet, D. (1993) Distinguishing health expectancies and health-adjusted life expectancies from quality-adjusted life years. American Journal of Public Health *83*, 797-798.

Romeder, J. and McWhinnie, J. (1977) Potential years of life lost between ages 1 and 70: An indicator of premature mortality for health planning. International Journal of Epidemiology *6*, 143.

Shepard, D. S., Robertson, R. L., Cameron, C. S. M., Saturno, P., Pollack, M. and Manceau, J. (1987) The cost-effectiveness of immunization strategies in Ecuador. Arlington, Virginia, John Snow, REACH.

Shepard, D. S., Sanoh, L. and Coffi, E. (1986) Cost-effectiveness of the Expanded Programme on Immuization in the Ivory Coast: a preliminary assessment. Social Science and Medicine *22*, 369-377.

Shepard, D. S., Walsh, J. A., Kleinau, E., Stansfield, S. and Bhalotra, S. (1995) Setting prioirties for the Children's Vaccine Initiative: a cost-effectiveness approach. Vaccine *13*, 707-714.

Steinglass, R., Brenzel, L. and Percy, A. (1993) Tetanus. In: Disease Control Prioirties in Developing Countries (Jamison, D. T., Mosley, W. H., Measham, A. R. and Bobadilla, J.-L., eds.), New York, Oxford University Press for the World Bank.

Sullivan, D. F. (1966) Conceptual problems in developing an index of health. US Public Health Service Publication Series No. 1000. Vital and Health Statistics Series 2. No. 17. National Center for Health Statistics.

Sullivan, D. F. (1971) A single index of mortality and morbidity. Health Reports *86*, 347-354.

Summers, R. and Heston, A. (1991) The Penn World Table (mark 5): an expanded set of international comparisons, 1950-1988. Quarterly Journal of Economics *26*, 19-66.

Summers, R., Kravis, I. B. and Heston, A. (1980) International comparisons of real product and its composition, 1950-1977. Review of Income and Wealth *26*, 19-66.

Tengs, T. O., Adams, M., Pliskin, J. S., Safran, D. G., Siegel, J. E., Weinstein, M. C. and Graham, J. D. (1995) Five-hundred life-saving interventions and their cost-effectiveness. Risk Analysis *15*, 369-390.

Torrance, G. (1986) Measurement of health state utilities for economic appraisal. A review. Journal of Health Economics *5*, 1-30.

Torrance, G., Thomas, W. H. and Sackett, D. L. (1972) A utility maximization model for evaluation of health care programmes. Health Services Research *7*, 118-133.

Viscusi, K. and Moore, M. (1989) Rates of time preference and valuations of the durations of life. Journal of Public Economics *38*, 297.

Weinstein, M. C. and Stason, W. B. (1977) Foundations of cost-effectiveness analysis for health and medical practices. New England Journal of Medicine *296*, 716-721.

World Bank (1993) World Development Report 1993: Investing in Health. New York: Oxford University Press for the World Bank.

World Health Organization (1977) International Classification of Diseases, 1975 Revision. Geneva, World Health Organization.

World Health Organization (1980) International Classification of Impairment, Disability and Handicap. Geneva, World Health Organization.

World Health Organization (1992) International Classification of Diseases and Related Health Problems, Trenth Revision. Geneva, World Health Organization.

World Health Organization (1996) Investing in Health Research and Development. Geneva, World Health Organization.

1.3 Future Immunization Strategies – Considerations from the Public Health View

Sieghart Dittmann

1.3.1 Introduction

From the public health view, the health status of the people living in our world should be taken into consideration as the first priority as well as experiences gained in already implemented immunization programmes when future immunization strategies are being discussed. The World Health Organization (WHO), the United Nations Children's Fund (UNICEF) and other public health agencies focus on the potential impact of new vaccines on public health in all countries of the world and on the cost-effectiveness of immunization compared to other health interventions. However, private industrial efforts in the area of vaccine development have often been geared to the prevention of diseases which are considered as priorities for the people in the industrialized world.

1.3.2 Global Health Situation

Globally, about 51 million people of all ages died in 1993; some 39 million deaths took place in the developing world and about 12 million in the industrialized. Sixteen million deaths (31 % of the total) were due to infectious and parasitic diseases. Table 1.3.1 gives an overview of the most important infectious and parasitic diseases. There are five major killers: pneumonia, diarrhoea, malaria, measles, and malnutrition. Around 2 million malaria deaths occurring annually are caused directly or in association with acute respiratory infections and anemia, the vast majority among young children; some 500 million cases are estimated annually. Diarrhoeal diseases resulting from unsafe water and inadequate sanitation coupled with poor food-handling practice are responsible for 3 million deaths in children below 5 years in developing countries. Most of the deaths would be preventable by using oral rehydration salts. Acute respiratory infections, mainly pneumonia, killing 4 million children annually, could also be prevented in the majority of cases through treatment of the underlying bacterial infection with low-cost antibiotics for a few days. Following the worldwide implementation of immunization, 1993 saw the number of children dying from vaccine-preventable diseases reduced by 1.3 million compared to 1985. However, two million and four hundred thousand children aged under 5 are still dying every year from diseases preventable through current available vaccines. Major outbreaks of meningococcal meningitis are known to occur in Sub-Saharan Africa and China. Recurrent epidemics affecting 40 000 people, mainly children, are serious public health problems in Ethiopia causing case fatality rates of about 10 % (World Health Report, 1995; Programme Report, 1994).

In total, annually 12 million children under five years of age die, the greatest causes are extreme poverty, lack of clean water and sanitation, lack of vaccines, curative drugs and other treatment (World Health Report, 1995).

WHO estimates that in 1994 the prevalence of HIV infections among adults worldwide was over 13 million. Currently, some 2 million people are becoming infected each year. In parts of Africa and Asia the virus is advancing rapidly. The lethal relationship between tuberculosis and HIV infection is making the death toll many times worse. During the next 10 years in Asia alone it is estimated that tuberculosis and AIDS together will kill many millions of people. The highest rates for STD's are generally seen in the 20-24 age group, followed by those aged 15-19 and 25-29. HIV and AIDS have a devastating effect on young people. Half the global HIV infection have been in people aged under 25 (World Health Report, 1995; Murray and Lopez, 1995).

In terms of lost years of healthy life the major microbiological threats are: pneumococcal disease 3.5 %, tuberculosis 2.6 %, malaria 2.2 %, sexually transmitted diseases excluding HIV 1.1 %, HIV 0.8 % (estimates made 1990) (Murray and Lopez, 1995).

1.3.2.1 Health Situation in Industrialized Countries

In the period after World War II, the industrialized countries in Western Europe, North America, Australia, in Japan and some other parts of East Asia have achieved remarkable success in the prevention and control of infectious diseases. Improved living conditions, immunization programmes and the development of antimicrobial drugs made it appear, for a short while, as if human victory over microbes had been achieved. During the 60s and 70s of this century, the majority of the medical community in most industrialized countries had been ready to close the book on infectious diseases. The emergence of a completely new and very dangerous disease, HIV/AIDS, as well as the spread of drug-resistant strains of microbes such as *Mycobacterium tuberculosis* and *Staphylococcus aureus*, and the resurgence of once-controlled diseases, such as diphtheria in the Newly Independent States of the former USSR, have combined to deliver the message that the war against the microbes has not yet been won by far (WHO, 1995).

Table 1.3.1: Health impact of infectious and parasitic diseases

Disease	annual deaths	% of total annual deaths	currently affected	annual new cases
Tuberculosis	3 million	5.9 %		8–10 million
Malaria	2 million	3.9 %		500 million
Hepatitis B	1 million	2 %		
Diseases preventable by currently available vaccines such as	2.4 million	4.7 %		
– measles	1 million			45 million
– neonatal tetanus	500000			
Diarrhoeal diseases	3 million	5.9 %		
– cholera	7000			370000
Acute respiratory infections, mainly pneumonia	4 million	7.8 %		
HIV infection and other sexually transmitted diseases				
– HIV infection			13 million	2 million
– chlamydial infection				97 million
– trichomoniasis				94 million
– gonorrhea				78 million
– genital warts				32 million
– genital herpes				21 million
– syphilis				19 million
Schistosomiasis	200000		200 million	
Onchocerciasis			18 million	
Chagas disease	45000		17 million	
Leishmaniasis	80000		13 million	
Dracunculiasis			3 million	
Trypanosomiasis	55000			
Dengue/				millions
Dengue hemorrhagic fever	1000s			500000

The described global health situation applies only partially for industrialized countries. Infectious diseases, particularly acute respiratory diseases, contribute between 5-10 % to the total mortality. However, one third of all work absenteeism, two thirds of school absenteeism, and 80 % of absences from day-care centers are still due to acute infectious diseases (WHO, 1984). Deaths due to diseases preventable by vaccines are now a rarity although low or insufficient immunization coverage has left a considerable number of measles, mumps or pertussis cases in some countries.

1.3.3 Current Immunization Programmes – Programme Development and Design, Goals and Successes, Constraints and Lessons Learned

1.3.3.1 Programme Development and Design

The history of immunization started exactly 200 years ago. In 1796, Edward Jenner protected a boy against smallpox by inoculating material taken from a cow's pustule lesions caused by cowpox, an orthopoxvirus closely related to variola virus. A hundred and eighty one years later, after a very long, troublesome and finally extremely successful way, the last smallpox case occurred in Somalia. In 1980, the World Health Assembly announced that worldwide eradication of smallpox had been achieved and recommended to cease routine smallpox immunization in all countries of the world. Based on a large number of calculations made and experiences gained in many countries and projects by national health authorities and international organizations (e.g. World Bank, United Nations Children's Fund, World Health Organization), we know today that immunization is one of the least expensive and cost-effective of all health services. In addition to the savings of the lives of 100 000s of people, the smallpox eradication programme can be used as one of the first examples for an excellent cost-benefit ratio. The programme costs were estimated to be US$ 300 million, the savings (cessation of routine smallpox immunization, quarantine measures, costs of treatment) were estimated to be in excess of US$ 1 billion annually (Henderson, 1988).

During the second half of the 20th century, immunization is controlling in most parts of the world at least five major diseases through immunization: diphtheria, tetanus, pertussis, measles, and poliomyelitis. These five antigens, together with BCG protecting at least young children against severe forms of tuberculosis such as meningitis and miliary tuberculosis, are also the core of the World Health Organization's (WHO) Expanded Programme on Immunization (EPI). Recognizing that immunization programmes had been successful in controlling dangerous childhood diseases in industrialized countries, WHO established the EPI in 1974. The term 'expanded' recognized that immunization services of some sort already existed in many countries and indicated that expansion in terms of coverage of the childhood population as well as in terms of the antigens used would be addressed. During the last 20 years, EPI has grown to be an operational programme of Member States working in a broad coalition of UN agencies, multi- and bilateral development agencies, and private and voluntary groups (Galazka, 1994; Henderson, 1994).

To demonstrate different immunization programmes used in developing countries (mainly according to the EPI strategy) and in industrialized countries, tables 1.3.2 and 1.3.3 showing prototype immunization schedules may be useful.

Table 1.3.2: WHO recommended schedule, mainly used in developing countries

Age	Vaccine			
(1) Newborns	BCG		OPV_0	
(2) 6 weeks		DTP_1	OPV_1	
(3) 10 weeks		DTP_2	OPV_2	
(4) 14 weeks		DTP_3	OPV_3	
(5) 9 months				measles

Through five visits, six antigens (OPV considered as 1 antigen) are administered. Booster doses are not proposed. Additionally, some developing countries have already implemented hepatitis B vaccine, and yellow fever vaccine (yellow fever belt in central Africa and South America) and Japanese encephalitis vaccine (parts of South East and East Asia) are recommended according to the regional epidemiological situation.

In Western and Central European countries as well as in industrialized countries of North America, Asia and Australia, a broad variation of immunization schedules exists. However, the core of the schedules is similar and corresponds widely with the following prototype schedule (see tab. 1.3.3). Timing and spacing of primary and boster immunizations varies. Through 8-9 visits, 8-10 antigens are administered for both primary and booster immunization.

Table 1.3.3: Prototype schedule used in different variations in most industrialized countries

Age	Vaccine				
(1) Newborns	BCG				
(2) 2–3 months		DTP_1	OPV_1		Hib_1
(3) 4–5 months		DTP_2	OPV_2		
(4) 6–7 months		DTP_3	OPV_3		Hib_2
(5) 15 months				MMR_1	
(6) 15–24 months		DTP_4	OPV_4		Hib_3
(7) 4–8 years		DT			
(8) 6–12 years				MMR_2	
(9) 14–16 years		Td			

Either OPV (Oral Polio Vaccine) or IPV (Inactivated Polio Vaccine) is used, sometimes as sequential immunization (OPV followed by IPV). BCG routine immunization of newborns is scheduled in the majority of countries whereas some countries immunize only those children at risk. The number of countries introducing immunization against Haemophilus influenzae type b infections (Hib immunization) is increasing. Due to higher costs some countries have not yet been able to add rubella and mumps vaccine to their schedules. In general, the combined measles-mumps-rubella vaccine (MMR vaccine) is the vaccine of choice. Seven industrialized HBsAg-low prevalence countries are already known to have implemented universal hepatitis B immunization of children (and adolescents). In other HBsAg-low prevalence countries the implementation is under serious consideration. At least four combined vaccines (DTP, MMR, DTP/Hib, DTP/IPV) are already available commercially, further combined vaccines are under development or undergoing clinical trials.

For children, boosters are recommended against poliomyelitis, diphtheria and tetanus, in some countries against pertussis at school entry. In many countries, a second dose of measles (or MMR) is recommended to ensure that each child has a second chance to get immunized or to seroconvert if due to different reasons seroconversion has not occurred after the first dose. For adults, boosters are recommended in many countries, mainly against diphtheria, tetanus, and poliomyelitis.

Additionally, there are recommendations to immunize persons at risk, e.g.

– travellers	diphtheria, tetanus, poliomyelitis, hepatitis A, Japanese encephalitis, typhoid fever, yellow fever;
– older and chronically ill people	influenza;
– occupational risk	diphtheria, hepatitis B, influenza, varicella, measles, rubella;
– patients at risk	hepatitis B, Hib, pneumococcal disease;
– children with leucosis	varicella;
– lifestyle risk	hepatitis B.

The recommendations include both primary and booster immunization. The use of combined vaccines or the simultaneous administration of different vaccines is widely used (Dittmann, 1994).

1.3.3.2 Successes of Current Immunization Programmes

In 1974, when the EPI had been formed, the immunization coverage of children below one year of age with childhood vaccines was estimated to be lower than 5 % in developing countries. The 1990 goal to reach a global immunization coverage of 80 % triggered off a huge international pressure on Member States to reach the goal. Member States accelerated efforts to improve their immunization programmes and international donor agencies provided outstanding technical and material support. As a result, the overall global 80 % goal was reached by the end of 1990. In general, the currently achieved impact on morbidity and mortality of diseases preventable through immunization is impressive (Galazka, 1994).

Poliomyelitis

Since 1981, the number of reported cases of poliomyelitis has fallen sharply to 8549 in 1994, a decrease of 85 % compared with the period from 1974 to 1980 when on average 40 000 – 60 000 were reported annually. Provisional figures for 1995 show that the reported cases will be lower than 5 000 cases. The last laboratory-confirmed case of paralytic poliomyelitis due to wild poliovirus in the Americas occurred on 23 August 1991 in Peru. Three years later, in September 1994, the International Commission for the Certification of Poliomyelitis Eradication declared that the interruption of transmission of wild poliovirus had been achieved in the Americas.

The success was based on the strong political and social will of the Member States, strong leadership and coordination, large support from the donor community, and a strategy including the establishment of an extensive laboratory-based surveillance system, immunization campaigns carried out on National Immunization Days and supplemented by mopping-up operations in regions at risk. The experiences gained in the Americas will be used in other regions of the world to reach the goal established by the World Health Assembly in 1988: the global eradication of poliomyelitis by the year 2000 (Programme Report, 1994; Galazka, 1994; Henderson, 1994; de Quadros, 1994).

Measles

Most industrialized and many developing countries have reduced measles morbidity and particularly mortality due to measles to very low levels, and an estimated 90 million cases and 1.5 million deaths are now prevented annually. However, WHO estimates that 45 million cases and 1 million deaths due to measles still occur each year (Programme Report, 1994).

Tetanus and neonatal tetanus

Tetanus immunization protects only the vaccinee, even extreme high coverage does not lead to population immunity. Therefore, the reported number of non-neonatal tetanus is only slowly decreasing. WHO estimates that in 1993 over 700 000 deaths from neonatal tetanus were prevented as a result of maternal immunization and other interventions. However, half a million deaths due to neonatal tetanus are still occurring annually worldwide, with 80 % of these deaths in 12 developing Asian and African countries (Programme Report, 1994; Galazka, 1994).

Diphtheria

The high coverage achieved with DTP or other diphtheria-component-containing vaccines resulted in a considerable reduction of diphtheria incidence in industrialized and developing countries to an all-time-low of about 20 000 reported cases in

1992. The dramatic re-emergence of epidemic diphtheria in the Newly Independent States of the former USSR has reversed the trend. Universal childhood immunization with diphtheria toxoid began in the late 1950s throughout the Soviet Union, resulting in control of diphtheria for almost 30 years. In 1976, reported cases reached a nadir of only 198 (0.08 per 100 000 population). Epidemic diphtheria re-emerged in 1990 in the Russian Federation. Epidemics in Ukraine (1991), Tajikistan (1993) and in all other Newly Independent States of the former USSR followed (since 1994). The number of reported diphtheria cases increased to over 47 000 in 1994, a further increase occurred in 1995 when about 55 000 cases have been registered (provisional figures).

In the Russian Federation and some other regions, decreasing immunization coverage rates among infants and children in the second half of the 80s, and the gap of immunity among adults were important basic reasons for the development of epidemic diphtheria. Spread of epidemic diphtheria may have been facilitated by large population movements following the dissolution of the Soviet Union. Socioeconomic changes and a partially deterioriating health service infrastructure may have been other contributing factors. Adequate control measures were not implemented in the early phases of the epidemics, and lack of vaccines interrupted routine immunization activities following independence in some Newly Independent States.

A broad coalition of international donors is now supporting diphtheria control in the epidemic countries (Hardy et al., 1996).

Pertussis

Following the widespread use of DTP vaccine, the incidence of reported pertussis decreased dramatically initiating a trend that has continued up to now. In some countries having continously achieved very high coverages the pertussis incidence came down close to elimination (e.g. Czech Republic, Hungary). The epidemiological efficacy of the conventional whole-cell pertussis vaccine has also been demonstrated in the 80s when some countries (UK, Sweden) interrupted their pertussis immunization programme exaggerating side effects of the vaccine. Large pertussis epidemics reoccurred and have been controlled again by reimplementation of pertussis immunization (UK).

Rubella, mumps, Haemophilus influenzae type b

Immunization against rubella and mumps, preferably using a combined MMR vaccine, and against diseases caused by Haemophilus influenzae type b have been implemented in many industrialized countries. In countries where high coverages have been achieved over the years, these diseases have already come close to elimination (e.g. Finland, countries in Scandinavia).

1.3.3.3 Constraints and Lessons Learned

1.3.3.3.1 Maintenance of High Coverage

Since 1990, global immunization coverage has levelled and the steady progression to higher coverage rates has stopped. There have been dramatic declines in reported coverage rates, including Ethiopia (from 80 % to 34 %), Nepal (from 80 % to 60 %, Nigeria (from 56 % in 1990 to 29 % in 1993), Papua New Guinea (from 67 % to 37 %), and Yemen (from 89 % to 54 %). The following table 1.3.4 shows disparities among socioeconomically different groups of countries.

Table 1.3.4: Immunization coverage in %, based on data available in August 1994

Countries	BCG	DTP_3	OPV_3	HBV_3	Yellow Fever	Measles	TT_2*
Least developed	71	53	49			49	43
25 most populous developing	87	81	83			79	47
all developing	85	78	80	3	7	77	45
economics in transition	88	74	80			90	
industrialized	80	85	83	5		80	
global total	85	79	80	3	7	78	45

* 2 doses of tetanus toxoid for pregnant women

Countries which have achieved and sustained high coverage usually have health infrastructures capable of delivering immunization and educating the population regarding the benefits of vaccines as well as national health leaders strongly committed to disease prevention through immunization. By contrast, in countries with decreasing coverage, the infrastructure of the health service is not capable or has broken down, staff are not always motivated, there is often a shortage of vaccine and other supplies. Civil conflict or war often interrupted the programme.

Therefore, maintenance of a high level of immunization coverage is one of the major objectives set for the coming decades. As a basis for further action and necessary support, the specific reasons for not achieving or not maintaining high coverage should be analyzed carefully in the respective countries (Programme Report, 1994).

1.3.3.3.2 Disease Surveillance is Fundamental for Monitoring Immunization Programmes

Monitoring of immunization programmes includes the collection and evaluation of data regarding demography, vaccine supply and logistics, immunization coverage and, last but not least, incidence data of the diseases targeted through immunization. The majority of developing countries and even some highly developed countries have previously been unable to produce the accurate and complete data needed for disease control. In many other countries, especially in the Americas, parts of Western Europe and the Eastern Mediterranean and Western Pacific region, striking progress has been made in surveillance. The success of the polio eradication programme, particularly in the Americas, could only be achieved by developing a laboratory-based surveillance programme for cases of acute flaccid paralysis. Before implementing a new immunization programme, surveillance data of the target disease should be available and the surveillance programme should be designed in advance. In general, the decision to implement or reimplement a new immunization programme can only be made on the basis of surveillance data describing the size and the impact of the target disease and the necessity and possibility to reduce size and impact through immunization. During an implemented immunization programme, continuous monitoring must measure the impact of the programme on the target disease, possible changes of efficacy under field conditions, and try to detect adverse events following immunization (Programme Report, 1994; Begg and Cutts, 1994).

1.3.3.3.3 Laboratory Network

Laboratory-based surveillance plays an important role in evaluating the impact of vaccination programmes. Poliomyelitis, pertussis, Haemophilus influenzae, and measles are only few examples where the role of laboratory support is considered to be essential. The closer a programme comes to the final phase the more important laboratory confirmation or exclusion of cases becomes. The target of global eradication of poliomyelitis requires a global network including national, regional, and specialized (at global level) laboratories with different functions and responsibilities and considering the limited resources of many developing countries. Specialized laboratories supply improved reagents and receive strains of wild polioviruses for molecular characterization to show geographical relationships. Close cooperation between the management of immunization programmes and research groups is necessary to develop and to select standardized and appropriate diagnostic methods (Programme Report, 1994; Begg and Cutts, 1994).

1.3.3.3.4 Vaccine Demand, Supply, Financing and Quality Control

Today the medical community worldwide is faced with the question of how to sustain the high levels of immunization coverage achieved in the 80s and to make new vaccines accessible as well. While a shift away from donor spending on immunization is appropriate for most middle income countries, it can be damaging to existing programmes in low income countries. Low income countries are not able to include vaccines which have been additionally recommended for inclusion into EPI, such as hepatitis and yellow fever vaccines. An assessment of the global vaccine supply confirmed that poverty and population size were key factors influencing the possibility that a country become self-sufficient in vaccine supply. Small, low income countries should continue to receive donations of vaccines from the donor community; other countries should receive support to help them to become self-sufficient either through procurement or production. This strategy should also ensure that new vaccines will be made accessible.

Almost half of the world's vaccine supply for childhood immunization programmes is bought and provided through UNICEF; the vaccines purchased meet quality criteria set up by WHO. Today, vaccines are produced in 59 countries; 60 % of the vaccines used in childhood immunization programmes are produced in 14 developing countries. Evaluations in 10 of the 14 countries draw attention to the problem of poor quality of some of the locally produced vaccines. To improve the vaccine quality, strengthening of collaboration and technology transfer is urgently needed including establishing and strengthening national capabilities in quality control (Programme Report, 1994; Henderson, 1994).

1.3.3.3.5 Logistics and Cold Chain

Through an intensive development programme a new generation of freezers, refrigerators and cold boxes, adapted to the needs of immunization programmes in developing countries became available as a fundamental precondition for the implementation of the Expanded Programme on Immunization. Additionally, indicators which permit easy monitoring of temperatures during the transport and storage of vaccines have been developed and implemented into practice. UNICEF and WHO have worked closely and successfully together (Henderson, 1994). The developments even influenced the industrialized countries to consider the weaknesses of their own cold chain and logistic systems and to initiate improvements. This paper is not the place to present and to discuss the whole range of impressive developments and knowledge gained during the universal immunization campaign of the world's children in the field of cold chain and logistics. However, it can be considered that regarding cold chain and logistics the ground has been prepared for future immunization programmes.

1.3.4 Vaccine Research

Vaccine research represents the future of immunization programmes. Vaccine research is funded by industry, at national level in some countries, and by WHO or similar international institutions. International funding is limited compared to the considerable amounts invested by industry to develop just a few vaccines. However, the small amount of international funding promotes the development of some vaccines considered of primary importance for public health in developing countries. The report of the 11th session of the Scientific Group of Experts (SAGE) of WHO's Global Programme for Vaccines and Immunization summarizes the developments in this field. The following table 1.3.5 provides an overview. Several research groups involved in the work coordinated by WHO have already initiated active collaboration with vaccine producers in both the public and the private sector (Global Programme, 1994).

Other research projects coordinated by the WHO's Global Programme for Vaccines and Immunization and the Scientific Group of Experts (SAGE) include

- microencapsulation and other non-replicating delivery systems aimed at the reduction of doses necessary for primary immunization (single dose vaccines)
- development of live vaccine vectors and of nucleic acid vaccine
- mucosal immunization
- improvement of immunogenicity using adjuvants, cytokines and other approaches (Global Programme, 1994).

Vaccine development is also supported by other programmes of WHO such as the Global Programme on AIDS (HIV vaccine) and the Special Programme for Research and Training in Tropical Diseases (vaccines against malaria, leishmaniasis and schistomiasis – included in tab. 1.3.5). The HIV vaccine development strategy comprises three components: development, evaluation and future availability. Activities have been prioritized on the characterization of genetic, antigenic and biologic diversity of HIV strains in relation to vaccine development (Global Programme, 1994).

1.3.5 Future Immunization Programmes

Considering the global health situation as well as the successes and constraints of current immunization programmes, from the public health view the main objectives for future global immunization programmes are

- *maintenance of high coverage of immunization against diphtheria, tetanus, pertussis, tuberculosis, measles, and poliomyelitis achieved through the Expanded Programme on Immunization (EPI)*
- *implementation or strengthening of immunization programmes against hepatitis B in all countries of the world; implementation or strengthening of immunization programmes against yellow fever and Japanese encephalitis in countries at risk of these diseases*
- *development of new and improved vaccines against diseases which represent a significant morbidity and particularly mortality burden for the majority of countries in the world (see tabs. 1.3.1, 1.3.5, 1.3.6.)*
- *introduction of those newly developed or improved vaccines into the Expanded Programme on Immunization*
- *development of new and simplified approaches for vaccine administration, e.g., combined vaccines, fewer doses, alternative routes of administration*

Table 1.3.5: Vaccine research

Target disease	Vaccine candidates	Field trials, developments
Diseases due to encapsulated bacteria		
Meningococcal disease (serogroups A,C)	1) group A/C oligosaccharides-CRM_{197} conjugate vaccine 2) group A/C polysaccharide-protein conjugate vaccine	field trials in Gambia and Niger 1994/1995
Meningococcal disease (serogroup B)	outer membrane protein (OMP) vaccine developed in Cuba and Norway	comparative field trial has been carried out in Iceland; further development of this type of vaccine will be supported
Pneumococcal disease	common antigen vaccines and conjugate vacines, possibly also combined	production and evaluation supported
Diarrhoeal diseases		
Cholera (serotype 01)	1) whole-cell recombinant B subunit inactivated vaccine	field trials in USA, Chile, Peru have been carried out, vaccine is licensed in some countries
	2) CVD103HgR live vaccine	challenge studies in volunteers have been carried out
Cholera (serotypes 01 & 0139)	3) vaccine mentioned under 1) with addition of formalin-killed cells of V. cholerae 0139	field trials in Sweden and Thailand finished
	4) Live vaccine against V. cholerae 0139	challenge studies in volunteers finished
Shigellosis	Efforts to generate genetically-defined attenuated mutants both in Sh. flexneri and Sh. dysenteria type 1 or E. coli/Sh. flexneri hybrids	phase I trials
Escherichia coli (enterotoxigenic) disease	prototype vaccine consisting of recombinant cholera toxin B subunit and formalinized E. coli serotypes	phase III trials planned
Typhoid fever	1) Vi antigen of S. typhi conjugated to Pseudomonas toxin 2) Vi antigen conjugated to tetanus toxoid	studies in volunteers carried out
	3) attenuated S. typhi strains to serve as live oral vaccines	phase II clinical trials
Rotavirus infection	1) tetravalent rhesus-human reassortant vaccine 2) infectious DNA 3) nucleic acid vaccine	studies in animal models have been carried out, clinical trials prepared; licensing of rotavirus vaccine to be expected within 2,3 years
Tuberculosis and leprosy		
	research priorities are: – mycobacterial genome sequencing – identification of virulence-related genes as basis for rational attenuation of M. tuberculosis – identification of vaccine candidates and delivery systems – establishment of animal models	

cont. ——>

Table 1.3.5 cont.

Target disease	Vaccine candidates	Field trials, developments
Acute respiratory diseases including measles		
Measles	1) recombinant efficacy of vaccinia and canarypox viruses 2) ISCOM vaccine 3) nucleic acid vaccines	evaluation in animal models, construction of new vectors, subunit preparations
RSV infections	cold passage mutants of RSV	field trials in progress
Parainfluenza type 3 infections	1) bovine PIV3 vaccine 2) cold passage mutants of human PIV3	field trials in progress
Dengue and Japanese encephalitis		
	Research priorities are: – recombinant vaccines with rubella and vaccinia viruses as vectors – demonstrate nucleic acid feasibility – develop 17D chimeric viruses – develop dengue 4 mutants with mutations in the 5' + 3' regions of the genome – develop animal models for evaluation of dengue candidate vaccines	
Hepatitis and poliomyelitis		
Hepatitis A	several attenuated HA vaccines	further studies on the impact of attenuation on the ability of HAV to replicate and to cause disease in a primate model
Hepatitis E	constructing recombinant HEV antigens as an approach to candidate vaccine development	
Hepatitis C	no adequate culture system for HCV as a basis for vaccine development has yet been established; current research is aimed at developing new diagnostic assays to discriminate between carriers and disease state	
Poliomyelitis	research priorities are: – could the use of poliovirus-sensitive transgenic mice replace the monkey neurovirulence test – efforts to stabilize the poliovaccine	
*Tropical diseases**		
Leishmaniasis	– killed parasite – live vectors expressing Leishmania antigens – subunit vaccines	field trials in Brazil, Iran, Venezuela preclinical evaluation
Malaria	– pre-erythrocytic vaccine – asexual blood stage vaccine transmission-blocking vaccines	preclinical research field trials using Spf66 vaccinein Tansania, Colombia, Thailand, Gambia field trials started
Schistosomiasis	recombinant antigens (S. mansoni)	preclinical evaluation

* Vaccine research coordinated by WHO's Special Programme for Research and Training in Tropical Diseases

– *improvement of vaccine efficiency using new and alternative technologies, e.g., new adjuvants, controlled-release vaccines, live vectors for vaccines, nucleic acid vaccines*

– *improvement and implementation of programmes evaluating the efficacy of existing and newly introduced immunization programmes*

and last but not least

– *optimal use of 20 years EPI experience.*

Three of the goals mentioned should be discussed in more detail: the need for the development of new or improved vaccines, new or simplified approaches for vaccine administration, and the experience gained in carrying out WHO's Expanded Programme on Immunization.

1.3.5.1 Available and Needed Vaccines

Table 1.3.6 provides an overview on vaccines already available commercially, vaccines which need improvement, and new vaccines needed. Some new or improved vaccines are needed globally whereas others will be used according to the regional epidemiological situation or preventing disease in groups at risk. Protection through immunization should not only be provided in childhood but also in adolescence, such as vaccines against sexually transmitted diseases, and in adult life.

In subchapter 1.3.4 an overview on current research activities aiming at new or improved vaccines has been provided.

1.3.5.2 New or Simplified Approaches for Vaccine Administration

1.3.5.2.1 Combined Vaccines

In industrialized countries, the number of vaccine injections recommended for children, particularly to be administered during the first year of life, is already very high. New vaccines are likely to be introduced into immunization practice within the next ten years. Most mothers both in developing and industrialized countries will not accept that their babies receive an unlimited number of injections.

Combined vaccines are tools to enable reduced contact immunization schedules. During the coming years, the role of combined vacines will certainly increase, particularly for primary immunization. Development of combination of conventional DTP vaccine with hepatitis B, *Haemophilus influenzae type b* and enhanced inactivated poliovaccine had proceeded rapidly, and within a very short time combined vaccines based on DTaP (acellular pertussis) will be used in industralized countries. This development will radically change the global market. Because 60 % of DTP used in developing countries is produced locally, assistance to developing countries manufacturers is necessary to prevent a gap in technology.

Problems are connected with development, control and field testing regarding safety and efficacy of such combined vaccines but also with their application in the

Table 1.3.6: Vaccines available, vaccines needed, and vaccines which need improvement

Vaccines already available commercially	Improvement needed	New vaccines needed
Vaccines used routinely in EPI		*for infants*
BCG	more effective TB vaccine	E..coli-Infections
Diphtheria		Hepatitis C
Measles	for immunization of infants	Hepatitis E
Pertussis, whole-cell		RSVirus-Infections
Poliomyelitis, live attenuated	more stable poliovaccine	Rotavirus
Tetanus	single-dose tetanus toxoid	Parainfluenza
Vaccines recommended for inclusion into EPI but not yet widely implemented		Shigellosis
Hepatitis B, plasma-derived		Streptococcal disease
Hepatitis B, genetically engineered		
Japanese encephalitis (recommended for countries at risk)	expensive inactivated mouse brain vaccine to be replaced (?)	*for adolescents*
Yellow fever (recommended for countries at risk)		Chlamydia
Other vaccines licensed		Gonorrhoe
Cholera, whole-cell		Herpes (HSV2)
Haemophilus influenzae		Syphilis
Hepatitis A		
Influenza	vaccine capable of immunizing against antigenically drifted strains	*for adults*
Meningococcal meningitis (serogroups A and C)	conjugated vaccine for infants	(Helicobacter pylori ?)
Mumps		
Pneumococcal disease	conjugated vaccine for infants	*all age groups*
Poliomyelitis, inactivated		Dengue fever
Rabies		HFRS (Hantanvirus)
Rubella		HIV
Tick-borne encephalitis		Leishmaniasis
Typhoid fever	more effective vaccine	Leprosy
Varicella		Malaria
Vaccines developed and undergoing clinical trials , already licensed in some countries		Schistosomiasis
Cholera, whole-cell/recombinant B subunit	addition of V. cholerae 0139 antigen: clinical trials initiated	
Cholera live vaccine	addition of attenuated strains of V. cholerae 0139: clinical trials initiated	
Meningococcal meningitis (serogroup B)	more effective vaccine	
Pertussis, acellular		

doctors practice when due to different reasons schedules are interrupted or have been started later in childhood or if severe reactions have occurred after a previous dose. The development of further combined vaccines should be inaugerated after careful consideration of the practical consequences. In any case, the current available mono-, bi-, and trivalent vaccines should be kept as an alternative to four-, five-, and possibly six-component combined vaccines (Dittmann, 1994; Global Programme, 1994).

1.3.5.2.2 Controlled Release Vaccines

The number of injections necessary for primary immunization to achieve optimal protection against poliomyelitis, diphtheria, hepatitis B or tetanus is another reason for concern. High drop-out rates hamper the coverage of the population. Immunization programmes would very much benefit from vaccines capable to achieve the goal of primary immunization using just a single dose of vaccine. Approaches to simplify vaccine delivery include the development of single-dose vaccines based on the use of live vectors, of new adjuvants, or of controlled-release systems. Promising results have already been obtained with several tetanus candidate vaccines (Programme Report, 1994).

1.3.5.3 Optimal Use of EPI Experience

Immunization in combination with surveillance and containment had eradicated smallpox. The achievements of the Expanded Programme on Immunization are impressive. The developments made, the wealth of experience gained and the lessons learned during the last 30 years are of high value for the development of future immunization efforts. EPI experience should always be taken into consideration when new strategies will be elaborated.

References

Ada, G. L. (1995) The development of new vaccines. In: Vaccination and World Health (Cutts, F. T. and Smith, P. G., eds.), Wiley Publishers, Chichester, pp 67-80.

Begg, N. and Cutts, F. T. (1994) The role of epidemiology in the development of a vaccination programme. In: Vaccination and World Health (Cutts, F. T. and Smith, P. G., eds.), Wiley Publishers, Chichester, pp 123-138.

de Quadros, C. A. (1994) Strategies for disease control/eradication in the Americas. In: Vaccination and World Health (Cutts, T. C. and Smith, P. G., eds.), Wiley Publishers, Chichester, pp 3-16.

Dittmann, S. (1994) Booster doses and combined vaccines. Abstracts of the 2nd Conference on Vaccinology, organized by the European Vaccine Manufacturers, Brussels, 18-20 May 1994.

Galazka, A. M. (1994) Achievements, problems and perspectives of the Expanded Programme on Immunization. Zbl. Bakt. *281*, 353-364.

Global Programme for Vaccines and Immunization (1994) Research Priorities. Report of the 11th session of the Scientific Advisory Group of Experts (SAGE), WHO, Geneva, 22-24 June 1994. GPV/VRD/94.14.

Hardy, I. R. B., Dittmann, S., Sutter, R. W. (1996) Resurgence of diphtheria in the New Independent States of the former Soviet Union: Current situation and control strategies. Lancet (in press).

Henderson, D. A. (1988) Smallpox and vaccinia. In: Vaccines (Plotkin, S. A. and Mortimer, E. A., eds.), Saunders Company, Philadelphia, pp. 8-30.

Henderson, R. H. (1994) Vaccination: successes and challenges. In: Vaccination and World Health (Cutts, T. C. and Smith, P. G., eds.), Wiley Publishers, Chichester, pp. 3-16.

Murray, C. J. L. and Lopez, A. D. (1995) Global patterns of cause of death and burden of disease in 1990, with alternative projections to the year 2020. In: WHO Ad hoc committee on health research relating to future intervention options. Investing in health research and development (draft). WHO, Geneva.

Plotkin, S. A. (1995) Discussion of the development of new vaccines. In: Vaccination and World Health (Cutts, F. T. and Smith, P.G., eds.), Wiley Publishers, Chichester, pp 81-88.

Programme Report (1994) Global Programme for Vaccines and Immunization. WHO/GPV/95.1, Geneva.

Velimirovic, B. (1984) Infectious diseases in Europe. World Health Organization, Regional Office for Europe, Copenhagen, Denmark, pp. 25-29.

WHO Ad hoc committee on health research relating to future intervention options (1995) Investing in health research and development (draft). WHO, Geneva.

World Health Report (1995) Summary (A 48/3). Forty-eighth World Health Assembly. WHO Geneva.

1.4 The New Pertussis Vaccines

David L. Klein and Carole Heilman

1.4.1 Introduction

The development of new vaccines for the prevention of pertussis (whooping cough) remains a central focus of interest. The interest lies less in the immediacy of the disease burden and more in the use of this vaccine as a foundation for the development of new combination vaccines and the interest of the public in improving the safety/ reactogenicity profile of the currently licensed vaccines. As a result, a series of new pertussis vaccine candidates, collectively known as the "acellular pertussis vaccines", have been under intense clinical development throughout the world. The purpose of this review is provide an update on the current status of these new vaccines and to speculate on their future.

1.4.2 Background: The Disease

Pertussis, or whooping cough, is a very contagious and potentially devastating respiratory disease produced by the organism *Bordetella pertussis*. The early stages of the disease are marked by runny nose, slight fever, and a dry, irritating cough. Within two weeks, the disease progresses and the patient develops episodes of short rapid coughs followed by a quick, deep breath, the characteristic "whoop". These episodes can last up to two months and can cause vomiting, choking, and inability to breath. Approximately 20 % of infants with reported disease may develop pneumonia, a leading cause of pertussis related deaths. Other severe pertussis related complications include encephalopathy and seizures.

In countries where routine whole cell pertussis (DTwP) vaccines are mandatory, and compliance is high, pertussis is no longer a major public health concern. Although clinical pertussis has become a rare event in these countries, it continues to occur. For example, in the U.S. between 4,000-6,000 cases of pertussis are reported annually with fairly high complication rates ranging from pneumonia to encephalopathy (CDC, 1990; CDC, 1992; CDC, 1993; Farizo et al., 1992). A demographic evaluation of this group between 1980-1989 demonstrated that 67 % of disease occured in children less than five years of age and 45 % occured in infants. Of these children, 63 % were not appropriately immunized for their age with 34 % having no pertussis immunizations at all (Farizo et al., 1992). Worldwide, it is estimated that the disease continues to afflict more than 50 million people each year resulting in approximately 350,000 deaths annually. Most of this disease occurs in countries with poor immunization programs, or where the pertussis component (P) of the DTwP vaccines are no longer recommended or mandatory. In Sweden, for example, where mandatory vaccination for pertussis is not recommended, it is estimated that 16 % of their birth cohort experiences clinical pertussis by age 5 (Romanus et al., 1987). In India, where vaccine delivery remains problematic, The World Health Organization estimates that 185,000 lethal cases of whooping cough are recorded each year .

1.4.3 The Whole Cell Vaccine

1.4.3.1 Safety

Although use of DTwP has controlled pertussis in countries where it has been widely used, it causes side effects in over one-third of infants who receive it (Mortimer and Jones, 1979). The administration of DTwP has been temporally associated with a series of adverse events ranging from unpleasant (i. e. pain, redness and swelling at the site of injection) to severe such as serious allergic reactions, decreased consciousness, encephalopathy, and death (Cody et al., 1981; Long and DeForest, 1990). The reactogenicity associated with the DTwP vaccine has been blamed largely on the endotoxin subcomponent, a logical extension of its known biological activity (Robinson et al., 1985) as well as the undefined nature of the vaccine.

DTwP vaccines have been repeatedly demonstrated to be more reactogenic than candidate acellular pertussis vaccines (DTaPs). For example, in a prospective comparative evaluation of thirteen different DTaPs and two licensed DTwPs where vaccine was administered to infants using the 2, 4, 6 month schedule, the reactogenicity of the two DTwP arms were consistently and significantly shown to be greater than any of the DTaP arms for the following criteria: fever, redness, swelling, pain and fussiness. In addition, drowsiness, anorexia, and use of antipyretics were also significantly more prevalent in recipients of whole cell vaccine. These reactions followed a similar, and predicatable, pattern with adverse reactions being maximal within 12-24 hours post-vaccination and decreasing with time (Decker et al., 1995). These results were consistent with a phase 2 trial conducted in Great Britain between 1989-1990 to examine the safety and immunogenicity of three DTaPs manufactured by Porton (PT, FHA, AGG), Merieux (PT, FHA) and Lederle (PT, FHA, PRN, AGG) compared to the licensed Wellcome adsorbed DTwP in infants using a 3, 5, and 8-10 month schedule. The data also demonstrated significantly less local and systemic reactions in

DTaP vaccine recipients compared to DTwP vaccinees. Additionally, reactions such as unusual or prolonged crying, hypotonic/hyporesponsiveness and adverse events requiring hospitalization were reported less frequently following administration of the DTaP vaccines than the DTwP vaccines (Miller et al., 1990). Despite the transient reactions associated with DTwP vaccines, the risk/benefit ratio has been considered low enough to warrant mandatory or recommended vaccination in most countries.

During the last thirty years, public concern about vaccine safety has challenged the risk/benefit ratio argument due primarily to the decrease in pertussis disease seen in developed countries. Japan and the United Kingdom were forced to alter their recommendations in light of the public's concern over possible associations between DTwP and serious adverse events. In 1986, the United States passed legislation requiring the evaluation of serious vaccine associated reactions following pertussis vaccination. Seventeen potential adverse events were identified for evaluation including: infantile spasms, hypsarrhythmia, aseptic meningitis, encephalopathy, SIDS (Sudden Infant Death Syndrome), anaphylaxis, autism, erythema multiforme or other rashes, Guillain-Barre syndrome (polyneuropathy), peripherial mononeuropathy, hemolytic anemia, juvenile diabetes, learning disabilities and hyperactivity, protracted inconsolable crying or screaming, Reye syndrome, shock and unusual "shock-like" states, convulsions, and thrombocytopenia. Based on all available evidence, a causal relationship between DTwP vaccination and two of the seventeen evaluated adverse events were indicated: analyphylaxis and inconsolable crying. Evidence consistent with a causal relationship between DTwP vaccination and adverse reactions were identified for two additional events: acute encephalopathy and shock/unusual "shock-like" states. For all other catagories, evidence was either insufficient or did not show a relationship between DTwP vaccination and the adverse event (Howson et al. 1991). After reviewing the long-term outcome of children with acute encephalopathy, the Institute of Medicine recently concluded in a follow-up report, that the evidence was consistent, very rarely, with a causal relationship between DTwP and some forms of chronic nervous system disorders. However, the data remain insufficient to determine whether DTwP causes lasting brain damage in the general population. If it does, it is very rare (Howson and Fineberg, 1992). Based on these findings, the United States endorsed the continued use of DTwP.

1.4.3.2 Efficacy

The use of a killed preparation of *B. pertussis* as a vaccine to prevent and control epidemic outbreaks of pertussis dates back to the 1930's (Madsen, 1933; Kendrick and Eldering, 1939; Sauer, 1939) (see table 1.4.1 for historic perspective). Despite the publication of conflicting results on the efficacy and methodology used to demonstrate the efficacy of this approach (Lapin, 1943; Doull et al., 1936), interest in the widespread use of this preventative approach continued such that during the 1950's, vaccines made from inactivated whole cell *B. pertussis* were recommended for routine use in all children.

The association between widespread use of DTwP and the decline in pertussis disease is clear, although other socioeconomic factors, such as improvements in nutrition, living conditions, and overall preventative and theraputic health measures have contributed to this outcome. For example, a temporal association between availability and widespread use of DTwP in the United States and decrease in pertussis associated mortality is shown in figure 1.4.1. The most striking examples of a causal association between widespread pertussis vaccination and decreased morbidity/mortality due to pertussis are found in Japan, the United Kingdom and Sweden, which have (or had) discontinued the mandatory use of pertussis vaccines. In each of these countries, increases in pertussis related mobidity and mortality had been observed. For example, during the four years preceeding the withdrawal of DTwP from recommended usage in all infants, the number of recorded deaths due to pertussis in Japan was ten. This number increased ten fold during the four years in which DTwP was withdrawn from recommended usage, a measure that was taken following the large number of claims for damages caused by DTwP (Kanai, 1980; Nobel et al., 1987). Acellular vaccines completely replaced the old whole cell vaccines in 1981. Although little data were available from controlled clinical trials of the DTaPs, pertussis largely disappeared from Japanese children after they were introduced to the vaccine, beginning at the age of two, as part of the routine childhood immunization schedule (Kimura et al., 1991).

In England and Wales, public concern over the safety of whole cell vaccine resulted in a 75 % decrease in public acceptance of the vaccine which was shortly followed by a major epidemic of pertussis with associated increases in pertussis related mortality (Fine and Clarkson, 1982). More recent epidemiological assessments of pertussis disease in England and Wales have demonstrated a major reduction in the incidence of pertussis associated mortality and morbidity, and the prevalence of the disease in young children as a result of the increased utilization of DTwP from a low of 30 % in 1978 to the 91 % level in 1992 (Miller et al., 1992).

Similarly, Sweden stopped using DTwP in 1979 due to manufacturing difficulties in producing the vaccine and a loss of confidence in the product. By 1985 whooping cough was again a common childhood infection, with one in five Swedish children being treated for whooping cough (Romanus et al., 1987).

Despite data that demostrate a decrease in serious pertussis related events associated with the widespread use of DTwP, the central concern that continues to drive a country's decison to vaccinate or not is the ratio they calculate when they weigh the perceived risks and benefits. Much effort has been expended to define the efficacy of DTwPs. This has been a remarkably difficult task to accomplish prospectively, since the widespread use of the vaccine has reduced the prevalence of this disease. A variety of methods have been used to calculate DTwP efficacy. Using a "screening" method to determine age specific pertussis vaccine efficacy, Ramsay and collegues observed overall DTwP vaccine efficacies between 87 % and 93 %, with efficacy estimates inversely related to the prevalence of disease (Ramsay et al., 1993). Active surveillance and investigation of secondary attack rates have also been used to estimate the efficacy of DTwP. Using a series of clinical definitions, efficacy ranged from 64 % (less stringent case definitions) to 95 % (stringent case definition) in preschool children in the United States (Onorato et al., 1992). Most recently, several large Phase 3 trials of DTaPs have been concluded in countries which do not mandate DTwP vaccine usage and in which DTwP arms have been included for direct comparison. In the two largest randomized trials sponsored by the U.S. National Institute of Allergy and Infectious Diseases (NIAID), National Institutes of Health (NIH) and conducted in Sweden and Italy, pertussis efficacy for DTwP vaccine recipients was calculated to be 36 % in Italy and 48 % in Sweden (P. Olin and D. Greco, personal communication). The trial used the WHO consensus definition of pertussis cases (table 1.4.1), specifically a paroxysmal cough equal to or greater than 21 days plus a positive culture of *B. pertussis*, or a significant rise in IgG or IgA antibody to pertussis toxin or filamentous hemagglutinin, or a culture-confirmed case of pertussis in the child's household within 28 days (Sweden only).

Table 1.4.1: WHO Case Definition

- 21 days or more of spasmodic cough, and
- EITHER culture-confirmed *B. pertussis*
- OR serologic evidence of *B. pertussis* infection(100% IgG or IgA antibody rise in ELISA for FHA or PT)
- OR household contact of culture+ case with onset of cough within 28 days before or after study child.

1.4.4 New Candidate Vaccines

1.4.4.1 Acellular Vaccine Composition

Public perception and the concern of parents and pediatricians about the frequent and worrisome side effects stimulated the search for safer, but equally effective, alternatives to DTwP (table 1.4.2). Beginning in the early 1970's considerable effort was devoted to the development of "acellular" vaccines based on molecularly defined and purified antigens with little or no predicatable toxicity (Edwards and Karzon, 1990).

There are four major components of *B. pertussis* that have been shown in animal models and/or in vitro opsonic and functional assays to contribute to pertussis pathogenesis. All new DTaP vaccines contain Pertussis Toxin (PT), while most contain additional components including: Filamentous Hemalgglutinin (FHA), Agglutinogens (AGG), and Pertactin. (PRN).

Table 1.4.2: DTP Vaccines: Chronology

1906	Organism is isolated and grown in artificial media (Bordet-Gengou)
1912-14	Vaccine made from killed whole cell *B. pertussis* first introduced into children
1930's	Kendrick refines and uses whole cell vaccine in children
1942	Kendrick combines improved killed vaccine with Diphtheria and Tetanus Toxoids (DTP)
1947	DTP vaccine first recommended for routine administration in U.S.
1965	Many states in U.S. pass school-entry laws requiring DTP immunization
1974-77	Questions about the safety of whole cell vaccines in Great Britain and Japan. Vaccine uptake falls; cases increase dramatically
1979	Sweden discontinues use of whole cell vaccines due to safety issues and lack of efficacy
1981	The British National Childhood Encephalopathy Study is published suggesting rare association with acute neurologic reactions. Japan initiates routine immunization of two year-olds with several acellular vaccines
1986	National Childhood Vaccine Injury Act is passed by the U.S. Congress
1991-92	Several major efficacy studies begin in Europe and Africa
1993	Institute of Medicine publishes findings on the nature, frequency and circumstances of adverse events following pertussis
1994-95	Seven efficacy trials for evaluating eight acellular vaccines completed

Pertussis Toxin

To date, all candidate acellular vaccines contain a version of the PT protein. PT has been shown to be an immunogen which elicits a protective immune response against mouse intracerebral (Sato et al., 1981; Oda et al., 1984) or respiratory (Sato et al., 1981) challenge by *B. pertussis.* These studies, along with recent clinical data, indicate that PT is a dominant target of the immune response to pertussis infection. PT, thus, represents the major component in new generation vaccines against pertussis, either as an acellular vaccine candidate alone or combined with other *B. pertussis* immunogens. Unfortunately, in its active form, this toxin is also a major virulence factor in pertussis disease and trace amounts of this product in some DTwP vaccines are believed to be a contributing factor in severe vaccine reactions (Arai and Sato, 1976; Pittman, 1984; Pittman, 1986; Ui, 1988; Monack et al., 1989; Edwards and Karzon, 1990). PT displays a variety of biological effects which result from binding to cells, internalization, and subsequent ADP-ribosylation of a family of GTP-binding regulatory proteins (G proteins) (Katada and Ui, 1982). The effects of the toxin on cells of the immune system are multiple and include induction of lymphocytosis, inhibition of macrophage migration, adjuvant activity, and T-cell mitogenicity. PT is one of the most highly studied bacterial toxins. It is a member of a group of ADP-ribosylating exotoxins and consistes of six polypeptide chains assembled from five different subunits (S1-S5), and consists of two functional units, the A and B subunits (Tamura et al., 1982).

A-Subunit: The A subunit (active portion) contains the S1 subunit and catalyzes the ADP-ribosylation of the target proteins which damages the cells (Katada and Ui, 1982). This protein has been extensively characterized at the amino acid level, and several laboratories have identified critical amino acids for catalytic activity (Barbieri and Cortina, 1988; Pizza et al., 1989; Losemore et al., 1990). The PT S1 subunit catalyzes the transfer of ADP-ribose from NAD to several G-proteins. ADP-ribosylated G protein fails to interact with the G protein coupled receptor, effectively uncoupling signal transduction (91). In addition, other cellular functions, such as the regulation of ion channelling, are modified following ADP-ribosylation of G proteins (Katada and Ui, 1982).

B-Subunit. The catalytic portion of the toxin cannot function without the B subunit (binding moiety), containing S2-S5. This subunit allows S1 from the A subunit to bind and cross the membrane of the target cell. The B subunit of PT is a pentamer consisting of two heterodimers joined by S5. One dimer contains one copy of S4 associated with S2 and the other dimer contains a second copy of S4 associated with S3 (Tamura et al., 1982). Several lines of evidence suggest that unlike many toxins, the B-oligomer of PT has some toxin activity in the absence of the A-oligomer and ADP-ribosylation. Purified B oligomer, in the absence of ADP-ribosylation activity, has been shown to activate T lymphocytes, probably through its ability to interact with the T-cell receptor complex by binding the CD3 molecule (Gray et al., 1989). Tamura and coworkers (Tamura et al., 1983) were the first to demonstrate directly that the PT B-oligomer could elicit responses in cells independent of any ADP ribosyltransferase activity of the holotoxin. In that work, the PT B-oligomer was shown to stimulate the mitosis of lymphocytes and cause an insulin-like enhancement of glucose oxidation. Because of the high (ug/ml) concentrations of toxin required for these effects, it was considered unlikely that it had a substantial role in clinical pertussis. Nevertheless, the development of acellular pertussis vaccines containing recombinant forms of PT, which possess no ADP ribosyltransferase activity, but are still active in the assays for B-oligomer responses (Nenciani et al., 1991a), raised interest and concern about the possibility of untoward effects in vaccine recipients. Although there are, at present, no clinical data to suggest that biological effects of B-oligomer are occurring in vaccinees, studies are continuing into the mechanism of action and potential consequences of the B-oligomer activities which have been observed in vitro. Additional studies have demonstrated that PT-induced mitogenicity for lymphocytes is now clearly a B-oligomer effect (Kong and Morse, 1977; Tamura et al., 1983). The phenomenum of PT-induced platelet activation is also becoming more important since acellular vaccines, now being tested, contain PT which retains this activity (Banga et al., 1987).

Although DTaPs are generally treated with inactivating chemical agents such as formalin and glutaraldehyde, these procedures reduce immunogenicity and often permit recovery of measurable PT activity upon storage (Nencioni et al., 1991a). Further, even exhaustive purification cannot totally eliminate the possibility of contamination by other *B. pertussis* toxins. One of the newer approaches towards the development of an inactivated toxin (i. e. toxoid) for pertussis has relied upon advances in recombinant DNA technology (Tamura et al., 1982). The operon encoding the complex pertussis toxin was molecularly cloned and sequenced by J. M. Keith and colleagues, and a mutant-inactive pertussis toxin was engineered (Barnette et al., 1988; Locht et al., 1986). These early data set the stage for the development of PT mutants with properties desirable for a vaccine. The best characterized mutant (PT-9K/129G), a genetically detoxified derivative, was developed by Rappuoli and coworkers (Nencioni et al., 1991b; Pizza et al., 1989) and represents a potentially powerful vaccine component. This double mutant, with replacements of lysine for arginine at residue 9 and glycine for glutamic acid at residue 129 of S1, was found to be completely non-toxic but retained immunogenic properties similar to the wild type toxin and could protect immunized mice from challenge with virulent *B. pertussis* (Pizza et al., 1989; Losemore et al., 1990). PT9K/129G has been shown to be safe and immunogenic in human trials (Podda et al., 1992).

Filamentous Hemagglutinin

Four bioactive domains of FHA have been delineated. Two, designated Lectin Domains I and II, recognize carbohydrates and the other two, designated Integrin Recognition Domain (RGD) and Factor X, recognize integrins (i. e. adhesion molecules).

Lectin Domains: Ciliated cells and macrophages of the respiratory tract are the primary targets of *B. pertussis* in whooping cough. The attachment of pertussis to glycoconjugates is mediated by the lectin domains of PT and FHA (Tuomanen and Weiss, 1985). In animal models, adherence of *B. pertussis* to ciliated cells can be interrupted by intratracheal application of lactose-containing receptor analogues (Tuomanen et al., 1988). Antibodies against PT or FHA block bacterial adherence to ciliated cells suggesting that a subcomponent vaccine based on the adhesive domains of PT and FHA might be protective against colonization and disease. Lectin Domain 1, located at the amino terminus of FHA, is responsible for hemagglutination, a process which is dependent on recognition of sulfated species such as heparin (Menozzi et al., 1994). This activity may promote the association of the bacteria with respiratory mucous which is rich in sulfated glycoconjugates. Lectin Domain II, which spans amino acids 1141-1279 of the FHA, is responsible for all of the binding of FHA to cilia and approximately half of the binding to macrophages (Prasad et al., 1993). A truncated FHA molecule missing amino acids 1141-1279 is unable to bind carbohydrates of both cilated cells and macrophages. A homolog of this lectin domain is found in the S2 subunit of PT (Saukkonen et al., 1992). Residues 1141-1279, thus, represent a potential non-toxic candidate for a subcomponent vaccine for whooping cough that should prevent respiratory colonization.

Intergrin Binding Domains: FHA binding to integrins mediates the association and subsequent uptake of *B. pertussis* into macrophages (Relman et al., 1990). From the point of view of pathogenesis, this uptake may prolong the course of the infection (Saukkonen et al., 1991). In addition, however, the mechanism of binding to integrins is of interest since it involves mimicry by the bacteria of the natural ligands for the integrins. This suggests that fragments of FHA and anti-FHA antibodies may possess biological activities against natural leukocyte integrin-mediated events. The Integrin Recognition Domain RGD of FHA contains the signature triplet RGD (i. e Arg-Gly-Asp sequence) for binding to integrins (Relman, 1989; Relman et al., 1990). Although initially thought to directly mediate bacterial attachment to the leukocyte integrin CR3, more recent evidence suggests this region of FHA acts indirectly to enhance bacterial-CR3 binding, thereby enhancing entry of *B. pertussis* into the leukocyte. Based on potential mimicry, antibodies to the RGD region of FHA cross react with determinants on cerebral microvascular endothelium which are involved in leukocyte transmigration into the brain (Tuomanen et al., 1993).

The Integrin Recognition Domain Factor X contains sequences which mimic the sites on Factor X of the coagulation cascade that bind to leukocyte CR3 (Rozdzinski et al., 1995). These regions may mediate the direct binding of purified FHA to purified CR3 (Van Strijp et al., 1993; Rozdzinski et al., 1995). Peptides from these regions possess anti-coagulant and anti-inflammatory properties arising from inhibition of the activities of the adhesion molecule CR3.

Agglutinogens

Agglutinogens, which serve as structural proteins on the bacterium's surface, vary among strains of *B. pertussis* and have been used to type the three major serotypes of pertussis: 1-2, 1-3, and 1-2-3. This is based on the general belief that there are only three major agglutinogens (i. e. 1, 2 and 3) and three minor ones (i. e. 4, 5, and 6) (Preston et al., 1982). Studies now indicate that these agglutinogens are, in fact, the bacterial fimbriae (Robinson et al., 1990) and as such, agglutinogens are involved in the adherence of the pertussis organism to host respiratory tissue. However, the exact mechanism of protection conferred by antibodies to these antigens remain in question. Protection against disease has been seen in children both with and without high levels of antibodies directed against surface agglutinogens. For example, one early prototype acellular vaccine "Pillimer's antigenic fraction", was shown in the 1950's to confer protection, but induced no agglutinogen antibody response.

There is evidence for strain-specificity of response to agglutinogens contained in whole cell vaccines. Use of vaccines containing little or none of a given agglutinogen, or an agglutinogen that is less immunogenic that others in the vaccine, has been followed by increases in the prevalence of that agglutinogen in the vaccinated population (Robinson et al., 1989). With this in mind, most manufacturers use several strains of *B. pertussis* to ensure the presence of all three types of agglutinogens.

Pertactin

Pertactin is an outer membrane protein which was originally identified by a monoclonal antibody produced against a 68-kilodalton protein from *Bordetella bronchiseptica*. It was shown to be a protective immunogen in experimental *B. bronchiseptica* infections (Shahin et al., 1990). The corresponding protein in *B. pertussis* has a molecular mass of 69-kilodaltons by SDS-PAGE.

The gene for pertactin from *B. pertussis* has been cloned and sequenced. The open reading frame could encode a protein of 910 amino acids, with a predicted molecular mass of 93,478 (Charles et al., 1989), suggesting the mature protein is processed from a larger precursor, as is the case of FHA. Pertactin also contains a copy of the "RGD" consensus sequence, which appears to be essential for binding to target cells (Leininger et al., 1991), and thus represents a good candidate for genetic alteration. Pertactin also appears to be a protective antigen in the mouse model.

Results from the recently completed efficacy studies in Sweden (Stockholm) and Italy suggest an important role for pertactin in promoting protection against disease. The two-component SmithKline Beecham Biologicals (SKB) vaccine lacking pertactin was the only acellular vaccine that demonstrated low efficacy while the three acellular vaccines containing the antigen were highly efficacious.

1.4.4.2 Clinical Studies

The above four catagories of *B. pertussis* components have been purified and combined in various formulations by a variety of manufacturers for use as candidate acellular pertussis vaccines. To date, a subset of these candidates have been evaluated in various clinical trials for both safety and efficacy. These data provide us with a framework for determining the relative merit of various products.

Early Clinical Trials

Early efforts in the development of acellular pertussis vaccines focused on the isolation and characterization of two protein components of the *B. pertussis* organism, PT and FHA. Extensive animal studies showed both antigens played important roles in disease pathogenesis (Weiss et al., 1983). Furthermore, mice immunized with either PT or FHA were protected against lethal respiratory challenge with *B. pertussis*. However, only anti PT antibodies protected mice against lethal intracerebral challenge with live pertussis organisms. (Sato and Sato, 1984; Oda et al., 1984).

During the past nine years several phase 1/2 clinical studies have examined numerous experimental acellular vaccines, most containing PT and FHA (Edwards et al., 1986; Lewis et al., 1986; Blumberg et al., 1991; Anderson et al., 1987; Ad hoc, 1988; Kimura et al., 1991; Anderson et al., 1988). The immunogenicity and safety trends for each vaccine were remarkably similar. In general, vaccination with these acellular pertussis vaccines revealed an increase in the ELISA antibody titers to all antigens included in the vaccine. The extent of the response was proportional to the number of doses administered. When vaccines were administered at 2, 4 and 6 months of age, antibody to PT and FHA were significantly higher in recipients of DTaP when compared to DTwP controls and remained higher in infants previously primed with acellular vaccines and boosted with the same. Acellular vaccines all appeared to be safe, regardless of the vaccine administered for primary vaccination. In addition, the frequency of adverse events for the more common, as well as the more serious, reactions were consistently less in infants receiving DTaP than for those vaccinated with DTwP (for recent overview see Report of the Nationwide Multicenter Acellular Pertussis Trial, 1995).

In 1986-87, NIAID, NIH supported a large randomized, double-blind, placebo-controlled clinical trial of two Japanese acellular pertussis vaccines in Sweden. Approximately 1,400 infants between 5-11 months of age received a two-component PT/FHA vaccine and an equal number received a vaccine containing PT toxoid alone. Both vaccines, produced by Biken, were administered to infants starting at approximately 5-6 months of age and included a two dose regimen with a 2-3 month interval between doses. The pertussis attack rates in each vaccine group were compared to the attack rate in approximately 950 placebo recipients over a 15 month period. The two vaccines, referred to in the trial as JNIH-6 (a two-component PT/FHA vaccine) and JNIH-7 (a mono-component PT vaccine) were both moderately effective in preventing culture-confirmed disease including any duration of cough (Ad hoc, 1988; Storsaeter et al., 1988). Point estimates of efficacy for the vaccines were 69 % for JNIH-6 (95 % confidence interval, CI, 47-82 %) and 54 % for JNIH-7 (95 % CI 26-72 %) (Ad hoc, 1988). These outcomes were based on a 15 month blinded follow-up, starting 30 days post second trial dose. The point estimates of efficacy for the two acellular vaccines, as defined by the protocol case definition, were not considered statistically different. Efficacy was estimated to be lower than observed for DTwP, although no direct comparisons were made. Although these results provided confirmatory evidence of the type of protection achievable with acellular pertussis vaccines, for a variety of reasons, neither vaccine was licensed for use in infants, outside Japan..

In addition to efficacy measures, several pieces of additional information were obtained. Immediate side reactions (local and systemic), seen with both acellular vaccines, were less frequent than those commonly reported in association with whole cell vaccines. Local reactions were, in general, more common in vaccine than in placebo recipients and were more frequent following the second dose. The study was not large enough to detect rare adverse events. The occurrence of four deaths associated with severe invasive bacterial infection in vaccine recipients was greater than the number expected. However, investigation of children hospitalized with systemic bacterial infections showed no significantly increased risk associated with vaccination. These findings, coupled with the results of intensive epidemiological and immunological investigations, and the absence of a plausible biological explanation, led to the conclusion that it is unlikely that the vaccines played a casual role in the deaths.

Suprisingly, no clear relationship between the level and type of antibody to PT and FHA and protection against culture-proven disease was demonstrated (Ad hoc, 1988).

A post trial follow-up, involving three years of unblinded surveillance did not demonstrate any evidence of waning efficacy. The data indicated that the two acellular vaccines continued to provide protection nearly four and a half years beyond the initial immunization. However, when examining culture confirmed cases of disease of >30 days duration, a statistically significant difference of protection was observed between JNIH-6 and JNIH-7 (i. e. relative risk of 1.5), with the PT/FHA vaccine being more efficacious than the PT vaccine (Storsaeter et al., 1990).

Additional analyses of the original dataset also yielded new insights about the sensitivity and specificity of pertussis case definitions. In the post-trial follow-up, 22 cases of culture-confirmed pertussis with cough of any duration resulted in an efficacy of 65 % for JNIH-7 and 77 % for JNIH-6. For culture-confirmed cases with more than 30 days of cough, the efficacy for JNIH-7 was 79 % and JNIH-6 was 92 % (Olin, 1990). The difference between these two vaccines was determined to be statistically significant. Applying the current WHO case definition to the results of the 1986-87 Swedish trial reveals efficacy data for both vaccines greater than 80 %. However, one needs to examine this outcome with caution since the confidence intervals are wide and many of the cases used to determine these figures were identified during an unblinded phase of the study. Based on further analysis, the following conclusions were drawn for the above trial: 1) clinical diagnosis alone increases sensitivity, but with an accompanying loss of specificity; 2) a positive culture or significant rise in PT antibodies increases the specificity, but is associated with a differential sensitivity; and 3) diagnosis based on antibody rise against FHA seems to be less influenced by differential sensitivity, but does not discriminate between *B. pertussis and Bordetella parapertussis* infections.

The results of this trial raised more questions than answers including: 1) could efficacy be improved by adding additional antigens to an acellular vaccine or by administering three rather than two doses; 2) what is the efficacy of acellular vaccines in infants vaccinated before six months of age; 3) what is the relative efficacy of whole cell versus acellular pertussis vaccine; 4) is there a serologic correlate of protection which can be measured, and would the collection of nasopharyngeal secretions for measuring IgA activity provide any additional useful information; and 5) do acellular pertussis vaccines increase the risk of invasive bacterial infection?

Recent Clinical Trials

To both address many of the unresolved questions and to develop the information needed for licensure and acceptance of the DTaPs for use in infants, the NIAID sponsored a second efficacy trial. Prior to the conduct of this trial, vaccine manufacturers with formulated DTaP products were invited to participate in a phase I/II comparative evaluation of their products. Thirteen candidate acellular vaccines were evaluated in six contracted Vaccine Treatment and Evaluation Units (VTEUs) in the U.S., for their immunogenicity/safety compared to standard lots of whole cell pertussis vaccines. Over 2,300 infants ages two months and older were enrolled to receive three doses of vaccine at 2, 4, and 6 months of age (these children are currently being followed-up for a 4-6 year booster study). Evaluation of the preliminary results in the first year of life was essential to the selection of candidate DTaPs for use in phase III efficacy/safety trials outside the U.S. The results indicated that all the vaccines were of sufficient immunogenicity and less reactogenic than either of the two DTwPs used in the trial. After extensive evaluation by an independent Task Force, seven vaccines were selected for evaluation of the scientific, as well as the efficacy objectives, of the phase III trials. Two sites, Sweden and Italy, were competitively selected to conduct these trials which eventually involved 25,000 infants. The trials were designed to assure measurements of both absolute efficacy (i. e. comparing vaccines to a placebo control) and relative efficacy (i. e. comparing vaccines to one another). In determining absolute efficacy, the primary emphasis was to evaluate the effectiveness of DTaPs compared to a placebo (DT) control in preventing cases of pertussis. Secondary aims included: comparing the relative safety and efficacy of DTaPs with a DTwP, exploring serological correlates of protection among immunized infants (i. e. Swedish study only), looking at the effects of different case definitions on efficacy estimates, and evaluating the results of a household contact study as another measure of efficacy (i. e. Swedish study only).

The vaccines used in both studies were chosen to evaluate multi-component vs pauci-component DTaPs and the impact of different antibody responses and different vaccine formulations on vaccine efficacy. The Swedish investigators used a five-component acellular candidate vaccine manufactured by Connaught Laboratories, Ltd. (PT, FHA, PRN, and Agg 2 & 3), and a two-component vaccine manufactured by SmithKline Beecham Biologicals (PT, FHA). The Italian investigators used two different DTaP candidate vaccines manufactured by SmithKline Beecham Biologicals (PT, FHA, PRN) and Chiron/Biocine (PT, FHA, PRN). The PT portion of the latter vaccine is a novel recombinant DNA product.

The results confirm that all DTaP vaccines are safe, with levels of reactogenicity comparable to DT and significantly lower than the DTwP vaccine. In Sweden, serious adverse events were evenly distributed among the three vaccine groups except for hyptonic/hyporesponsive events in which five of the six events occurred in DTwP recipients. However, efficacy for the PT/FHA vaccine (i. e. 58 %) did not meet the present standards for efficacy, while the five-component vaccine did meet these standards with an efficacy of 85 %. Interestingly, the DTwP, though fulfilling U.S. licensing requirements, only had 48 % efficacy in Sweden The efficacy of the two DTaPs used in the Italian trial was 84 % while the whole cell vaccine was only 36 %. Local, systemic, and serious adverse events ocurred most frequently in the DTwP group and were comparably low for the DTaP and DT vaccine groups. Estimates of efficacy increased for both DTaPs with increasing severity of the clinical endpoints (e. g. type and length of cough) (Gustafsson, et al., 1995; Greco et al., 1995). A summary of these results is shown in table 1.4.3. Follow-up studies for long term safety and relative efficacy are underway in both Sweden and Italy.

Table 1.4.3: Results of NIAID Trials of Acellular Pertussis Vaccines in Sweden and Italy

Vaccine Manufacturer	Vaccine Composition	Efficacy	Trial Location
SmithKline Beecham	PT and FHA	58%	Sweden
Connaught Laboratories, Ltd.	PT, FHA, 69kD and 2 AGG	86%	Sweden
Connaught Laboratories, Inc.	Conventional WC	48% 36%	Sweden Italy
SmithKline Beecham	PT,FHA and 69kD	84%	Italy
Chiron Biocine	PT,FHA and 69kD	84%	Italy

WC = whole cell PT = pertussis toxin FHA = filamentous hemagglutinin
69kD = pertactin or 69kD protein AGG = agglutinogens

Table 1.4.4: International Efficacy Trials*

Vaccine Manufacturer	Site	Vaccine Composition†	Schedule	Vaccine Efficacy (95% C.I.) DTaP**	DTwP***
(1) Amvax	Sweden	PT	3, 5, 12	71 (63-78)	—
(2) Connaught, Inc.	Germany	PT, FHA (Biken)	2, 4, 6	96 (78-99)	97 (79-100)
(1) Chiron Biocine	Italy	PT, FHA, PRN	2, 4, 6	84 (76-90)	36 (14-52)
(3) Connaught, Ltd	Sweden	PT, FHA, PRN, FIM (2,3)	2, 4, 6	85 (81-89)	48 (37-58)
(4) Lederle	Germany	PT, FHA, PRN, FIM 2 (Takeda)	2, 4, 6 15-18 mos	84 (79-)	92 (89-)
(5) Pasteur/Merieux	Senegal	PT, FHA	2, 4, 6	86 (71-93)	96 (87-99)
(6) SKB	Germany	PT, FHA, PRN	3, 4, 5 15-19 mos	89 (77-95)	97 (83-100)
(3) SKB	Sweden	PT, FHA	2, 4, 6	59 (51-66)	48 (37-58)
(1) SKB	Italy	PT, FHA, PRN	2, 4, 6	84 (76-89)	36 (14-52)

* Absolute efficacy data based on WHO case definition
(> 21 days paroxysmal cough and: culture, or serology, or epidemiological link following three doses of vaccine)
† PT = pertussis toxin; FHA = filamentous hemagglutinin; PRN = 69kD outer membrane protein (pertactin); FIM = fimbriae (agglutinogen)
** DTaP = acellular vaccine
*** DTwP = whole cell vaccine
(2) Prospective case control study
(4) Randomized, double-blind multicenter study with an open non-randomized DT cohort
(5) Prospective, randomized study designed to demonstrate relative efficacy with a parallel DT non-study group
(6) Household contact study (surveillance in catchment area of an earlier immunogenicity study)
(1) Prospective, randomized, double-blind multicenter study
(3) Prospective, randomized, double-blind multicenter study – Trial I

Other large scale field trials were recently completed in several sites to obtain additional efficacy information on DTaP vaccines (table 1.4.4). The scope and objectives of the studies differed, but all sought to examine the safety and efficacy of the new DTaP vaccines. SmithKline Beecham Biologicals, Connaught Laboratories, Inc., and Wyeth/Lederle conducted studies of their own vaccines in various parts of Germany where DTwP vaccine is not used routinely. All three DTaP vaccines were included in the earlier NIAID multicenter phase I/II clinical trial. The vaccines were tested in the following respective ways: one was a family contact study, the second a case control study, while the third compared rates of whooping cough in the experimental DTaP to rates in a non-randomized DT control group. Preliminary data appear to indicate that both the DTwP and DTaP vaccines were efficacious, with the DTaP vaccines providing evidence of less reactogenicity for each of the common and rare serious adverse events measurable.

Specific data from SmithKline Beecham's family contact study demonstrated approximately 90 % protective efficacy against typical disease with their three component (i. e. PT, FHA, PRN) DTaP vaccine following household exposure to a confirmed case of pertussis (Schmitt et al., 1994). These data are consistent with a previous household contact study, conducted in Japan, in which Biken and Takeda-like acellular vaccines provided efficacy estimates of 77 % and 88 % respectively (Aoyama et al., 1988).

Pasteur/Merieux has recently completed a randomized, double blind, prospective efficacy trial to evaluate a DTaP containing PT and FHA compared to their DTwP in a rural area of Senegal with a high incidence of pertussis. Few cases were expected in both groups combined because all of the children in the trial received a pertussis vaccine. Preliminary results indicated that the two-component DTaP vaccine was safe and caused significantly fewer adverse events compared to the DTwP vaccine. Although the trial was designed to look at the relative risk of pertussis in the DTaP group compared to the DTwP group, the availability of a small matched cohort of non-immunized infants, not part of the protocol, allowed for an analysis of absolute efficacy. Based on the standard WHO case definition, efficacy of the two component DTaP (85 %) was similar to the DTwP vaccine (94 %) (Personal Communication, Dr. Michel Cadoz).

The National Institute of Child Health and Human Development, NIH has developed an acellular monovalent DTaP vaccine that contained only pertussis toxoid formulated with D and T. The vaccine, which was not included in the NIAID multicenter clinical trial, was evaluated for safety and efficacy in Goteborg, Sweden, in which approximately 3,400 infants received either the acellular or a DT control vaccine at 3, 5 and 12 months of age. The trial did not compare the acellular vaccine to DTwP. The vaccine is manufactured by North American Vaccines. The results of the trial were announced in 1994 and demonstrated that the vaccine was quite safe, with a point estimate of 71 % efficacy (95 % CI 63-78 %), based on a modified WHO case definition (Taranger et al., 1995). An important finding was that clinical disease among DTaP vaccinated children was significantly attenuated compared to DT recipients.

Efficacy, safety, and immunogenicity data are now available from seven large, prospective, doubled-blind trials containing eight different DTaP vaccines (table 1.4.3). Overall, the results indicate that several of the new vaccines are very effective in preventing whooping cough, while others provide less substantial levels of protection. It is likely that several of these vaccines will be licensed and broadly used by 1996. Extensions of these studies will allow investigators to evaluate the booster responses in children primed with either a DTwP or a DTaP vaccine. Extended surveillance will continue in many of the efficacy studies to determine the duration of protection following three doses of vaccine. Such studies may help establish new vaccine schedules and support the need for an additional dose of vaccine in countries relying on a primary series as the sole source of protection against pertussis disease. The continued use of DTwP in developing countries will also have to be re-evaluated in light of recent successes with DTaPs and the questionable levels of efficacy observed in certain studies in which DTwP was also evaluated. It should be emphasized that DTwP efficacies varied among studies. Whether this variation is a result of different DTwP compositions, force of infection observed during study period, schedule implemented, and/or endpoints evaluated is currently unclear.

Introduction of new DTaP vaccine into a developing country to replace the established and successful DTwP is one that must be carefully analyzed in terms of cost/benefit and the impact it may have on disrupting viable, ongoing immunization programs. Although the safety evidence remains consistently encouraging, none of the trials were large enough to eliminate the concern about serious, but rare, adverse neurological reactions. The answers to questions about these rare complications will require several years of careful post-marketing surveillance studies. Interesting preliminary data is just now becoming available from the Swedish Trial 2 study which was designed to examine safety and relative efficacy issues among three DTaPs and a DTwP (i. e. Evans Medeva) in a cohort of approximately 82,000 children. From a safety point of view, there were no overall statistically significant differences in the rates of serious adverse events between the vaccine groups. However, events contraindicating further vaccine administration, such as temperature of 40.5 °C or greater within 48 hours of vaccine administration and convulsions occurring within three days of any trial dose, were more frequently observed in the DTwP group. Meanwhile, additional studies are underway which will continue to evaluate the DTaPs alone and in combination with other routinely used vaccines to determine what effect these interactions have on overall safety and immunogenicity.

1.4.5 Future Impact

Newer Generation Vaccines

The expected value of new DTaPs will be in the increased compliance and interest of physicians and families for maintaining vaccine schedules. In addition, DTaPs, as foundations for combinination vaccines, are expected to simplify the development and delivery of these products. Nonetheless, questions remain regarding the possibilities of developing newer generations of DTaPs that have additional qualities, including protection against infection, as well as disease, and life-time immunity. The possibility of lifelong immunity derives it's strength from data that suggest that in contrast to vaccinated individuals, patients who recover from whooping cough seem to enjoy lifelong immunity from both disease and reinfection by the microorganism. This suggests that the ideal vaccine, one that prevents infection/reinfections, as well as disease, can by achieved. However, the mechanism for inducing this immunity has yet to be elucidated. The identification of new virulence factors with potential for inclusion in newer generation vaccines remains an important area for vaccine research. To date, several new virulence factors have been identified, which have the potential to be components in newer generation vaccines. Examples include the pertussis toxin liberation (Ptl) proteins, the *Bordetella* resistance to killing (Brk) proteins, the trachael cytotoxins (TCT), and the adenylate cyclase toxin (AC).

Ptl proteins have been shown to promote the secretion of assembled pertussis toxin across the outer membrane of *B. pertussis* (Weiss et al., 1993). Specific mutations in the Ptl pathway result in less virulent organisms, suggesting that it's virulence mechanism may be valuable as a vaccine component.

The Brk proteins are another group of virulence proteins which controls resistance to killing by human sera. Weiss and coworkers developed a mutant which was deficient in it's ability to promote a lethal infection in mice, yet retained the expression of all the known *B. pertussis* virulence factors (Weiss et al., 1989). These mutants were also 10 to 1000-fold more susceptible to killing by the antibody-dependent classical pathway of complement than wild type (Weiss and Goodwin, 1989). The parallels in humans are interesting. Immune sera from different sources has different killing capacities. Serum from some individuals can effectively kill even the resistant wild type strain, suggesting that acquired immunity can overcome this bacterial defense mechanism (Fernandez and Weiss, 1994). These results suggest the bacteria have a mechanism to resist killing by complement and that this mechanism can be overcome if the individual is able to generate the appropriate antibody. Evidence of this differential killing is seen in the human host. These proteins may play a role in bacterial carriage. While current vaccines protect the individual from severe disease, they can often harbor the bacteria in their respiratory tract and serve as a source of infection. However, some individuals, in particular those who have recovered from disease, often have a better immunity and higher killing titers than vaccinated individuals and are also protected from carriage as well as severe disease. Bactericidal titer may account for the ability of these individuals to prevent colonization of the bacteria.

TCT is the newest addition in the family of pertussis toxins which contribute significantly to the disease syndrome. It is produced from an innocuous peptidoglycan macromolecule found in nearly all bacteria, pathogenic or not, and is responsible for the specific respiratory tract damage that is the hallmark of pertussis disease (Cookson et al., 1989). TCT has been identified as a member of the "muramyl peptide" family, peptidoglycan fragments that are linked to a diverse array of biological activities including adjuvanticity, pyrogenicity, arthritogenicity, and somnogenicity (Adam and Lederer, 1984). The later could account for one of the more common side affects of pertussis vaccination, specifically hypotonic/hyporesponsiveness, which may have other neurological implications for complications associated with *B. pertussis* infection. These studies have had a major impact on the thinking in the pertussis field, and may have direct relevance to other infectious disease states and biological phenomena such as cellular responses in inflammation and regulation of fever and slow-wave sleep. *B. pertussis* demonstrates a propensity for proliferating exclusively on the surface of cilated epithelial cells in the human respiratory tract. The discovery of TCT-induced nitric oxide provides a mechanistic explanation for the lung pathology first observed in pertussis autopsy samples over 80 years ago. It is also likely that this specific pattern of cilated cell destruction is largely responsible for pertussis symptomatology and transmission (Heiss et al., 1989). The observation that nitric oxide synthase inhibitors can interrupt the pathway of respiratory tract damage, and quite possibly restore epithelial integrity, also has important consequences for clinical treatment. These findings suggest new possibilities for therapy of pertussis patients based on interfering with molecular signals involved in generating lung damage.

The AC toxin is a single 216 kD protein containing both enzymatic (the ability to convert ATP to cyclic AMP) and toxin (the ability to penetrate the membrane and introduce the enzyme into target mammalian cells) activity (Hewlett et al., 1993). Elevated intracellular cyclic AMP levels can lead to adverse consequences. The AC toxin has been shown to inhibit the function of several types of immune effector cells, such as polymorphonuclear leukocytes, macrophages, monocytes, natural killer cells, and lymphoma cell lines. This is probably the role of the AC toxin in disease, allowing the bacteria to resist the onslaught of the immune defenses. The toxin has been shown to be essential for lethal infection in infant mice (Goodwin and Weiss, 1990). Furthermore, antibodies to AC toxin have demonstrated protection in a mouse model of disease (Brezin et al., 1987; Guiso et al., 1989). Vaccinated individuals or individuals recovering from disease produce antibodies against AC toxin (Farfel et al., 1990). Several AC toxin mutants have been identified that appear to be antigenically intact, but lack toxin activity.

Crucial to the development of a safer and less reactogenic vaccine for *B. pertussis* is a more thorough understanding of the biology of the infecting organism. As previoulsy described, *B. pertussis* synthesizes a variety of factors which have been implicated in pathogenesis. Expression of all of these factors, except TCT, is positively regulated by the products of a single genetic locus (designated *BvgAS* or v*ir*) (Weiss and Falkow, 1984). This locus encodes two proteins, BvgA and BvgS, which act in trans to mediate a biphasic transition between two distinct phenotypic phases, a phenomenon known as phenotypic modulation, or phase switching from *vir*- to *vir*+. When *BvgAS* is active (*vir* + phase), all of the known virulence factors are expressed and *B. pertussis* is lethal in an infant mouse model (Weiss et al., 1984). The regulation of virulence factor expression is complex and appears to be transcriptionally modulated.

One hypothesis for how this pattern of gene activation might reflect the role of BvgAS during infection is that it may allow for the sequential synthesis of adhesins and toxins when the bacterium experiences an upshift in temperature upon entering a new host (Scarlato et al., 1991). Alternatively, or in addition, BvgAS may differentially regulate gene expression in response to incremental changes in conditions such as may be encountered at different sites within the host allowing the organism to precisely regulate virulence gene expression to meet its needs at specific locations. For example, adhesins regulated by BvgA might be required to mediate attachment to respiratory epithelial cells and establish a defined niche as the bacterium enters the nasal cavity of a susceptible host. Toxins, which might be required for evasion and/or persistence, might not be required until after the organism has reached deeper tissue. Such a virulence determinant becomes very useful in thwarting the host's immune defense system, thereby promoting bacterial survival. While this hypothesis is consistent with data regarding transcriptional control of virulence gene expression, the selective advantage provided by the ability of *B. pertussis* to switch between distinct phenotypic phases, and the role of BvgAS mediated signal transduction in pathogenesis remains unclear due, in part, to the lack of appropriate animal models.

In the *vir*- phase, a second class of virulence repressed genes (*vrgs*) are expressed resulting in a significantly less virulent stage. The environmental signal(s) provided by the host to induce phase switching has not been identified. Recently, investigators have demonstrated the importance of the *vir*+ phase in *B. bronchiseptica*, an organism closely related to *B. pertussis*, but having a broader host range, in the establishment of respiratory tract infection. (Cotter and Miller, 1994; Akerley et al, 1995). These results indicate that the *vir*+ phase is both necessary and sufficient for establishment of respiratory tract infections and strongly suggests that the Bvg- phase is not expressed in vivo, nor are antibodies against *vir*- phase specific antigens detectable. A similar phenomenum has been observed in convalescent serum from humans recovering from pertussis; antibodies which can be detected are directed exclusively against *vir*+ phase specific factors.

To determine if BvgAS dependent regulation is required for growth in response to nutrient limitation, studies were designed to examine the abilities of *vir*- phase locked mutants to survive in phosphate buffered saline. In contrast to the in vivo studies, the *vir*-phase appears to be advantageous for growth under extreme conditions. These results indicate an advantage for the *vir*-phase under nutrient limiting conditions such as may be encountered in the environmental reservoir, and suggest a possible role for the *vir-* phase during transmission (Cotter and Miller, 1994). Transmission, thus, can be divided into three steps: 1) release of the bacterium from an infected host; 2) survival between hosts; and 3) the initial interaction with the new host. Down-regulation of virulence factors and up-regulation of *vir-* phase specific factors could potentially contribute to any or all of these steps. Because pertussis is a highly infectious disease, an understanding of how transmission occurs, including the identification of specific required factors, would be useful in designing strategies aimed at limiting the spread of the organism. Although the experiments described above were done with *B. bronchiseptica*, the functional conservation of BvgAS within and across species suggests these results may be extrapolated to *B. pertussis*. Understanding the role and function of the BvgAS virulence regulon may have practical applications for vaccine development, particularly for vaccines which have potential to interupt transmission from human adult carriers.

Expanding the Use of Acellular Vaccines In other Populations

Unlike pertussis disease, which appears to confer life-long immunity, pertussis vaccines confer only partial and relatively transient protection. A high degree of protection persists from three to seven years, then decreases until little protection is evident after about 12 years. Adults who had childhood pertussis and who are re-exposed demonstrate variable antibody levels to antigens that do not correlate with clinical protection (Van Savage et al., 1990). During a recent outbreak of pertussis, the number of years since the last administration of vaccine correlated with clinical disease and provided evidence for waning immunity (Bass and Stephensen, 1987). Individuals who received their last immunization within three years of the outbreak experienced a 21 percent attack rate compared to a 95 percent attack rate in those who had an interval greater than 12 years.

The number of individuals susceptible to pertussis is increasing worldwide. The number of young adults without prior immunizations or with diminished post-vaccination immunity is growing. Individuals older than six years of age are not vaccinated routinely since pertussis morbidity in older children is not considered significant, and because these individuals routinely demonstrate increased rates of local reactions upon vaccination (Linneman et al., 1975). The reported annual incidence of disease continues to increase as this population of young adults, suceptible to pertussis, grows (Nelson, 1978; Mertsola et al., 1983). Because adults may serve as an important reservoir for pertussis, several studies have examined the feasibility of providing acellular pertussis vaccines to this population (Edwards et al., 1986; Edwards et al., 1992)). From the standpoint of immunogenicity and safety, the potential for introducing acellular pertussis vaccine into the general population on a routine basis when combined with adult formulations of diphtheria and tetanus toxoids is being discussed and will await the outcome of ongoing clinical studies.

Multicomponent Vaccines

A primary goal of the WHO's Childhood Vaccine Initiative (CVI) program is the selective eradication of various childhood diseases through the development of immunization programs which incorporate new emerging technologies for the creation of improved, safe, and efficacious vaccines. The concept of designing safe and effective multicomponent vaccines has been advanced by the CVI program as a modern and cost effective approach for providing fewer doses of vaccine to children to help facilitate delivery and broaden the coverage of immunization against multiple diseases (La Montagne and Rabinovich, 1994). Manufacturers are currently reformulating as many as four or five existing licensed vaccines into a single product.

DTaPs offer the prospect of improved immunogenicity and substantially reduced reactogenicity relative to DTwPs. In all likelihood, the success of multicomponent vaccines will depend on the availability and use of acellular vaccines since they are expected to be essential components and serve as the foundation for the development of these combination products. It is envisaged that DTaPs, when combined with other vaccines such as *Haemophilus influenzae* type b, hepatitis B vaccine, and inactivated polio, will have wide application in primary immunization schedules for infants and for boosting immunity in older-aged children (Edwards and Decker, 1994). Irrespective of the implied advantages, combined vaccines must have three minimal attributes that are essentially nonviolable. First, the individual components which make up the combined product must be as efficacious in preventing disease as the individual components when given separately in different sites. This means that there cannot be any substantial interference between the vaccine components and that the antibody and T cell responses must be quantitatively and qualitatively equivalent. Second, the duration of immunity for each component of a combination vaccine must be as long as the duration of immunity for that component when administered alone. Third, clinical reactions to the combined vaccines should not be significantly greater than the most reactive of the individual components.

A significant aspect of this new vaccine strategy would be to develop vaccines composed of relevant portions of multiple pathogens associated with a particular disease (e. g. pneumonia), or a particular need (e. g. children). The use of a vector system which expresses multiple foreign genes, or newly attenuated viruses with defined genetic mutations and a sufficient capacity to encode multiple foreign genes, may provide opportunities to evaluate this strategic approach. Combining vaccines for use in a single syringe will pose a unique challenge to manufacturers in understanding the immunologic basis for the compatibility and incompatibility of various vaccines as well as the infant immune response. The development of these products will greatly facilitate, and greatly challenge, the implementation of universal vaccination programs required to bring under control serious infectious diseases, such as pertussis, on a worldwide scale.

References

Ad hoc group for the study of pertussis vaccines (1988) Placebo-controlled trial of two acellular pertussis vaccines in Sweden: Protective efficacy and adverse events, Lancet *1*, 955-960.

Adam, A. and Lederer, E. (1984) Muramyl peptides: Immunomodulators, sleep factors, and vitamins, Med. Res. Rev. *4*, 111-152.

Akerley, B. J., Cotter, P. A., and Miller, J. F. (1995) Ectopic expression of the flagellar regulon alters the development of the Bordetella-host interaction, Cell *80*,611-620.

Anderson, E. L., Belshe, R. B., Bartram, J., Gurwith, M., Hung, P., Levner, M. and Vernon, S. K. (1987) Clinical and serologic responses to acellular pertussis vaccine in infants and young children, Am. J. Dis. Child. *141*, 949-953.

Anderson, E. L., Belshe, R. B. and Bartram, J. (1988) Differences in reactogenicity and antigenicity of acellular pertussis and standard pertussis vaccines combined with diphtheria and tetanus in infants, J. Infect. Dis. *157*, 731-737.

Anderson, E. L., Belshe, R. B., and Bartram J. (1988) Differences in reactogenicity and antigenicity of acellular and standard pertussis vaccines combined with diphtheria and tetanus in infants, J. Infect. Dis. *157*,731-7.

Aoyama, T., Marose, Y., Gonda, T., and Iwata, T. (1988) Type specific efficacy of acellular pertussis vaccine, Am. J. Dis. Child. *142*,40-42.

Banga, H. S., Walker, R. K., Winberry, L. K. and Rittenhouse, S. E. (1987) Pertussis toxin can activate human platelets: Comparative effects of the holotoxin and its ADP-ribosylation S_1 subunit, J. Biol. Chem. *262*,14871-14874.

Barbieri, J. T., and Cortina, G. (1988) ADP-ribosyltransferase mutations in the catalytic S-1 subunit of pertussis toxin, Infect. Immunol. *56*,1934-1941.

Bass, J. W. and Stephenson, S. R. (1987) The return of pertussis, Pediatr. Infect. Dis. J. *6*,141-144.

Blumberg, D. A., Mink, C. M., Cherry, J. D., Johnson, C., Garber, R., Plotkin, S. A. et al. (1991) Comparison of acellular and whole cell pertussis-component diphtheria-tetanus-pertussis vaccines in infants, J. Pediatr. *199*,194-203.

Brezin, C., Guiso, N., Ladant, D., Djavadi-Ohaniance, L., Megret F., Onyeocha, I., and Alonso, J.M. (1987) Protective effects of anti- adenylate cyclase antibodies against lethal respiratory infection of the mouse, FEMS Microb. Lett. *42*,75-80.

Burnette, W. N., Cieplak, W., Mar, V. L., Kaljot, K. T., Sato, H., and Keith, J. M. (1988) Pertussis toxin S1 mutant with reduced enzyme activity and a conserved protective epitope, Science *242*,72-74.

CDC (1990) Summary of notifiable diseases, United States, 1989, MMWR *38*, 33, 53-59

CDC (1992) Pertussis surveillance – United States, 1989-1991, MMWR *41*,11-19

CDC (1993) Resurgence of pertussis – United States, 1993, MMWR *42*,952-960

Charles, I. G., Dougan, G., Pickard, D., Chatfield, S., Smith, M., Novotny, P., Morrissey, P. and Fairweather, N. F. (1989) Molecular cloning and characterization of protective outer membrane protein P69 from *Bordetella pertussis*, Proc. Natl. Acad. Sci. *86*,3554-3558.

Cody, C. L., Baraff, L. J., Cherry, J. D., Marcy, S. M., and Manclark, C. R. (1981) The nature and rate of adverse reactions associated with DTP and DT immunization in infants and children, Pediatrics *68*,650-60.

Cotter, P. A. and Miller, J. F. (1994) BvgAS mediated signal transduction: Analysis of phase-locked regulatory mutants of *Bordetella bronchiseptica* in a rabbit model, Infect. Immun. *62*,3381-3390.

Decker, M., Edwards, K., Steinhoff, M., Rennels, M., Pichichero, M., Englund, J., Anderson, E., Deloria, M., and Reed, G. (1995) Composition of thirteen acellular pertussis vaccines: Adverse reactions, Pediatrics Supplement 96, 557-66.

Doull, J. A., Shibley, G. S., and McClelland, J. S. (1936) Active immunization against whooping cough: Interim report of the Cleveland experience, Am. J. Public Health *26*,1097-1105.

Edwards, K. M. and Decker, M. D. (1994) Combination vaccines: hopes and challenges, Pediatr. Infect. Dis. J. *13*,347-347.

Edwards, K. M., Decker, M. D., Graham, B. S., Mezzatesta, J., Scott, J., and Hackell, J. (1993) Immunization of adults with acellular pertussis vaccine, JAMA *269*,53-56.

Edwards, K. M. and Karzon, D. T., (1990) Pertussis vaccines, Pediatr. Clin. North Am. *37*,549-566.

Edwards, K. M., Lawrence, E., and Wright, P. F. (1986) Diphtheria, tetanus and pertussis vaccine. A comparison of the immune response and adverse reactions to conventional and acellular pertussis components, Am. J. Dis. Child. *140*, 867-871.

Farfel, Z., Konen, S., Wietz, E., Klapmuts, R., Addy, P., and Hanski, E. (1990) Antibodies to adenylate cyclase are produced in man during pertussis infection and after vaccination, J. Med. Microbiol. *32*,173-177.

Farizo, K. M., Cochi, S. L., Zell, E. R., Brink, E. W., Wassilak, S. G., and Patriarca, P. A. (1992) Epidemiological features of pertussis in the United States (1980-1989), Clin. Infect. Dis. *14*,708-719.

Fernandez, R. C. and Weiss, A. A. (1994) Cloning and sequencing of a serum resistance locus, Infect. Immun. *62* (11),4727-4738.

Fine, P. E. M. and Clarkson, J. A. (1982) The recurrence of whooping cough:Possible implications for assessment of vaccine efficacy, Lancet *1*,666-669.

Goodwin, M. S. and Weiss, A. A. (1990) Adenylate cyclase toxin is critical for bacterial colonization and pertussis toxin is critical for lethal infection by *Bordetella pertussis* in infants mice, Infect. Immun. *58*,3445-3447.

Gray, L. S., Huber, K. S., Gray, M. C., Hewlett, E. L., and Englehard, V. H. (1989) Pertussis toxin effects on T lymphocytes are mediated through CD3 and not by pertussis toxin catalyzed modification of a G protein, J. Immunol. *142*,1631-1638.

Greco, D., Salmaso, A., Mastrantonio, P., Giuliano, M., Tozzi, A. E. et al. (1996) A controlled trial of two acellular vaccines and one whole-cell vaccine against pertussis, NEJM 334, 341-48.

Gustafsson, L., Hallander, H. O., Olin, P., Reizenstein, E., and Storsaeter, J. (1996) A controlled trial of a two-component acellular, a five-component acellular, and a whole-cell pertussis vaccine, NEJM 334, 349-55.

Guiso, N., Rocancourt, M., Szatanik, M., and Alonso, J-M. (1989) *Bordetella* adenylate cyclase is a virulence associated factor and an immunoprotective antigen, Microb. Path. *7*,373-380.

Heiss, L. N., Flak, T. A., Lancaster, Jr., J. R., McDaniel, M. L., and Goldman, W. E. (1993) Nitric oxide mediates *Bordetella pertussis* tracheal cytotoxin damage to the respiratory epithelium, Infectious Agents and Disease 2,173-77.

Hewlett E. L., Gordon, V. M., McCaffery, J. D., Sutherland, W. M., and Gray, M. C. (1989) Adenylate cyclase toxin from *Bordetella pertussis*: Identification and purification of the holotoxin molecule, J. of Biol. Chem. *264*,19379-19384.

Howson, C. P. and Fineberg, H. V. (1992) Adverse events following pertussis and rubella vaccines. Summary of a report of the Institute of Medicine, JAMA *267*,92-396.

Howson, C. P., Howe, C. J., and Fineberg, H. V. (1991) Adverse effects of pertussis and rubella vaccines: A report of the committee to review the adverse consequences of pertussis and rubella vaccines. Washington, D.C., National Academy Press.

Kanai, K. (1980) Japan's experience in pertussis epidemiology and vaccination during the past thirty years, Jpn. J. Sci. Biol. *33*,107-143;49-51.

Katada, T. and Ui, M. (1982) Direct modification of the membrane adenylate cyclase system by islet-activating protein due to ADP-ribosylation of a membrane protein, Proc. Natl. Acad. Sci. USA *79*,3129-3133.

Kendrick, P. and Eldering, G. (1939) A study in active immunization against pertussis, Am. J. Hyg. *29*,133-153.

Kimura, M., Kuno-Sakai, H., Sato, Y., Kamiya, H., Nii, R., Isomura, S. et al., (1991) A comparative trial of the reactogenicity and immunogenicity of Takeda acellular pertussis vaccine combined with tetanus and diphtheria toxoids. Outcome in 3- to 8-month-old infants, 9- to 23-month-old infants and children, and 24- to 30-month-old children, Am. J. Dis. Child. *145*,734-741.

Kong, A. S. and Morse, S. J. (1977) The in vitro effect *Bordetella pertussis* lymphocytosis-promoting factor on murine lymphocytes. I. Proliferative response, J. Exp. Med. *145*,151-162.

La Montagne, J. R. and Rabinovich, N. R. (1994) The promise of new technologies, Internat'l, J. Technol. Assess. Health Care, *10*,7-13.

Lapin, L. H., Whooping Cough. Charles C. Thomas (1943), Springfield Ill

Leninger , E., Roberts, M., and Kenimer, J. G. (1991) Pertactin, an Arg-Gly-Asp containing *Bordetella pertussis* surface protein that promotes adherence of mammalian cells, Proc. Natl. Acad. Sci.U.S.A. *88*,345-349.

Lewis, K., Cherry, J. D., Holroyd, H. J., Baker, L. R., Dudenhoeffer, F. E., and Robinson, R. G. (1986) A double-blind study comparing an acellular pertussis-component DTP vaccine with a whole-cell pertussis-component DTP vaccine in 18-month-old children, Am. J. Dis. Child. *140*, 872-876

Linneman, C. C., Perlstein, Jr., P. G. H., Ramundo, N., Minton, S. D., Englender, G. S., McCormick, J. B. et al. (1975) Use of pertussis vaccine in an epidemic involving hospital staff, Lancet *2*,540-543.

Locht, C. and Keith, J. M. (1986) Pertussis toxin gene: Nucleotide sequence and genetic organization, Science *232*,1258-1264.

Long, S. S. and DeForest, A., Pennridge Pediatric Associates, Smith, D. G, Lazaro, C., Wassialk, S.G.F. (1990) Longitudinal study of adverse reactions following diphtheria-tetanus-pertussis vaccine in infants, Pediatrics *85*,294-302.

Losemore, S. M. et al. (1990) Engineering of genetically detoxified PT analogues for development of a recombinant whooping cough vaccine, Infect. Immun. *58*,3653-3662.

Madsen, T. (1933) Vaccination against whooping cough, JAMA *101*,187-188.

Medical Research Council (MRC) (1956) Vaccination against whooping-cough. Relation between protection in children and results of laboratory tests, Br. Med. J. *2*,44-462

Menozzi, F., Mutombo, R., Renauld, G., Gantiez, C., Hannah, J., Leininger, E., Brennan, M., and Locht, C. (1994) Heparin-inhibitable lectin activity of the filamentous hemagglutinin adhesin of *Bordetella pertussis*, Infect. Immun. *62*,769-778.

Mertsola, J., Ruuskanen, O., Eerola, E., and Viljanen, M. K. (1983) Intrafamilial spread of pertussis, J. Pediatr. *103*,359-363.

Miller, E., Miller, D. L., Asworth, L. A. E., Waight, P. A. and Harbert, K. A. Preliminary comparison of antibody responses and symptoms following primary immunization with British whole cell and three acellular DTP vaccines. In: Manclark, C.R., (ed.) 1990 Proceedings of the Sixth International Symposium on Pertussis. DHHS US Public Health Service, Bethesda, Maryland. DHHS Publication No. (FDA) 90-1164; 303-309.

Miller, E., Vurdien, J. E., and White, J. M. (1992) The epidemiology of pertussis in England and Wales, Commun. Dis. Rep. CDR Rev. *2*,152-154.

Monack, D., Munoz, J. J., Peacock, M. G., Black, W. J., and Falkow, S. (1989) Expression of pertussis toxin correlates with pathogenesis in *Bordetella* species, J. Infect. Dis. *159*,205-210.

Mortimer, E. A. and Jones, P. K. (1979) An evaluation of pertussis vaccine, Rev. Infect. Dis. *1*,927-32

Nelson, J. D. (1978) The changing epidemiology of pertussis in young infants. The role of adults as reservoirs of infection, Am. J. Dis. Child. *132*,371-373.

Nencioni, L., Volpini, G., and Peppoloni, S. (1991b) Properties of pertussis toxin mutant PT-9K/129G after formaldehde treatment, Infect. Immun. *59*,625-30.

Nencioni, L., Pizza, M., Volpini, G., De Magistris, M. T., Giovannoni, F., and Rappuoli, R. (1991a) Properties of the B. oligomer of pertussis toxin, Infect. Immun. *59*,4732-4734.

Noble, G. R., Bernier, R. H., Esber, E. C., Hardegree, M. C., Hinman, A. R., Klein, D., and Saah, A. J. (1987) Acellular and whole-cell pertussis vaccines in Japan: Report of a visit by US scientists, JAMA *257*,1351-1356.

Oda, M., Cowell, J. L., Burstyn, D. G., and Manclark, C. R. (1984) Protective activities of the filamentous hemagglutinin and the lymphocytosis-promoting factor of *Bordetella pertussis* in mice, J. Infect. Dis. *150*,823-833.

Olin, P. New conclusions and lessons learned from the vaccine trial in Sweden. In:Manclark, C. R.(ed.) (1990) Proceedings of the Sixth International Symposium on Pertussis, DHHS, USPHS, Bethesda, Maryland, DHHS Publication No. (FDA) 90-1164; 299-302.

Onorato, I. M., Wassilak, S. G., and Meade, B. (1992) Efficacy of whole-cell pertussis vaccine in preschool children in the United States, JAMA *267*,2788-2790.

Pittman, M. (1984) The concept of pertussis as a toxin-mediated disease, Pediatr. Inf. Dis. *3*,467-486

Pittman, M. (1986) Neurotoxicity of *Bordetella pertussis*, Neurotoxicology *7*,53-68.

Pizza, M., Covacci, A., Bartoloni, A., Perugini, M., Nencioni, L., de Magistris, M., Villa, L., Nucci, D., Manetti, R., Bugnoli, M., Giovannoni, F., Olivieri, R., Barbieri, J., Sato, H., and Rappuoli, R. (1989) Mutants of pertussis toxin suitable for vaccine development., Science *246*,497-500.

Podda, A., DeLuca, E. C., Titone, L., Casadei, A. M., Cascio, A., Peppoloni, S., Volpini, G., Marsilli, I., Nencioni, L., and Rappuoli, R. (1992) Acellular pertussis vaccine composed of genetically inactivated pertussis toxin: safety and immunology in 12-24 and 2- 4 month-old children, J. Pediatrics *120*(5),680-685.

Prasad, S., Tuomanen, E., and Masure, H. R. (1993) Identification of a carbohydrate recognition domain in filamentous hemagglutinin of *Bordetella pertussis*, Infect. Immun. *61*,2780-2785.

Preston, N. W., Surapatana, N., and Carter, E. J. (1982) A reappraisal of serotype factors 4, 5 and 6 of *Bordetella pertussis*, J. Hyg. Camb. *88*,39-46.

Ramsay, M. E., Farringon, C. P., and Miller, E. (1993) Age-specific efficacy of pertussis vaccine during epidemic and non-epidemic periods, Epidemiol. Infect. *111*,41-48.

Relman, D. A., Domenighini, M., Tuomanen, E., Rappuoli, R., and Falkow, S. (1989) Filamentous hemagglutinin of *Bordetella pertussis*; nucleotide sequence and crucial role in adherence, Proc. Natl. Acad. Sci. USA *86*,2637-2641.

Relman, D., Tuomanen, E., Falkow, S., Golenbock, D. T., Saukkonen, K., and Wright, S. D. (1990) Recognition of a bacterial adhesin by an eukaryotic integrin: CR3 (alpha M beta 2, CD11b/CD18) on human macrophages binds filamentous hemagglutinin of *Bordetella pertussis*, Cell *61*,1375-1382.

Report of the Nationwide Multicenter Acellular Pertussis Trial, 1995 In: Decker, M and Edwards, KM (eds) Supplement to Pediatrics 96, 547-603

Robinson, A., Gorringe, A. R., Funnell, S. G. P., and Fernandez, M. (1989) Serospecific protection of mice against intranasal infection with *Bordetella pertussis*, Vaccine *7*,321-324.

Robinson, A., Irons, L. I., Seabrook, R. N., Pearce, A., Matheson, M., and Funnell, S. G. P. (1990) Structure-Function studies of *Bordetella pertussis* fimbriae. In: Manclark, C. R. (ed.) Proceedings of the Sixth International Symposium on Pertussis, DHHS, USPHS, Bethesda, Maryland. DHHS Publication No. (FDA) 90,1164;126-135.

Romanus, V., Jonsell, R., and Bergquist, S. O. (1987) Pertussis in Sweden after the cessation of general immunization in 1979, Pediatr. Infect. Dis. J. *6*,364-371.

Rozdzinski, E., Sandros, J., van der Flier, M., Young, A., Spellerberg, B., Bharracharyya, C., Straub, J., Musso, G., Putney, S., Starzyk, R., and Tuomanen, E. (1995) Inhibition of leukocyte-endothelial cell interactions and inflammation by peptides from a bacterial adhesin which mimic coagulation factor X, J. Clin. Invest. *95*,1078-1085.

Sato, H. and Sato, Y. (1984) *Bordetella pertussis* infection in mice: Correlation of specific antibodies against two antigens, pertussis toxin, and filamentous hemagglutinin with mouse protectivity in an intracerebral or aerosol challenge system, Infect. Immun. *46*,415-421.

Sato, Y., Izumiya, K., Cowell, J. L., and Manclark, C. R. (1981) Role of antibody to leukocytosis promoting factor hemagglutinin and to filamentous hemagglutinin in immunity, Infect.Immun. *31*(3),1223-31.

Sato, Y., Izumiya, K., Sato, H., Cowell, J. L., and Manclark, C. R. (1981) Aerosol infection of mice with *Bordetella pertussis*, Infect. Immun. *31*,1223-1231.

Sauer, L. W. (1939) Whooping cough: New phases of the work of immunization and prophylaxis, JAMA *112*,305-308.

Saukkonen, K., Burnette, W. N., Mar, V., Masure, H. R., and Tuomanen, E. (1992) Pertussis toxin has eukaryotic-like carbohydrate recognition domains, Proc. Natl. Acad. Sci. USA *89*,118-122.

Scarlato, V., Arico, B., Prugnola, A., and Rappuoli, R. (1991) Sequential activation and environmental regulation of virulence genes in *Bordetella pertussis*, EMBO *10*,3971-3975.

Schmitt, H. J., Wirsing Von Konig, Neiss, A., Bogaerts, H., Bock, H. L., Schulte-Wissermann, H. et al. (1994) Protective efficacy of an acellular pertussis DTaP vaccine following household exposure to pertussis, 34th Interscience Conference on Antimicrobial Agents and Chemotherapy; Abstract #B/3; p.8.

Shahin, R. D., Brennan, M. J., Li, M. L., Meade, B. D. and Manclark, C. R. (1990) Characterization of the protective capacity and immunogenicity of the 69-kD outer membrane protein of *Bordetella pertussis*, J. Exp. Med. *171*,63-73.

Storsaeter, J., Hallander, H., Farrington, C. P., Olin, P., Mollby, R., and Miller, E. (1990) Secondary analysis of the efficacy of two acellular pertussis vaccines evaluated in a Swedish phase III trial, Vaccine *8*,457-461.

Storsaeter, J., Olin, P., Renemar, B., Langergard, T., Norberg, R., Romanus, V., and Tiru, M. (1988) Mortality and morbidity from invasive bacterial infections during a clinical trial of acellular pertussis vaccines in Sweden, Pediatr. Infect. Dis. J. *7*,637-645.

Tamura, M., Mogimori, K., Murai, S., Yakima. M., Ito, K., Katada, T., Ui, M., and Ishii, S. (1982) Subunit structure of pertussis toxin, Biochemistry *22*,5516-5522.

Tamura, M., Nogimori, K., Yajima, M., Ase, K., and Ui, M. (1983) A role of the B-oligomer moiety of islet-activating protein, pertussis toxin, in development of the biologic effects on intact cells, J. Biol. Chem. *258*,6756-6761.

Taranger, J. B., Trollfors, T., Lagergard, J., Robbins, G., Zackrisson, L., Lind, V., Sundh, Lowe, C., and Blackwelder, W. (1995) Efficacy of a pertussis toxoid vaccine in a randomized double-blind and placebo-controlled study, American Society for Microbiology; Abstract #E-93, p.297.

Tuomanen, E. and Weiss, A. (1985) Characterization of two adhesions of *Bordetella pertussis* for human ciliated respiratory epithelial cells, J. Infect. Dis. *153*,118-125.

Tuomanen, E., Towbin, H., Rosenfelder, G., Braun, D., Hansson, G., Larson, G., and Hill, R. (1988) Receptor analogs and monoclonal antibodies which inhibit adherence of *Bordetella pertussis* to human ciliated respiratory epithelial cells, J. Exp. Med. *168*,267-277.

Tuomanen, E., Prasad, S., George, J., Hoepelman, A. I. M., Ibsen, P., Heron, I., and Starzyk, R. (1993) Reversible opening of the blood brain barrier by anti-bacterial antibodies, Proc. Natl. Acad. Sci. USA *90*,7824-7828.

Ui, M. The multiple biological activities of pertussis toxin. In: Wardlaw A. C., Parton, R. (eds.) (1988) Pathogenesis and immunity in pertussis. Chichester, England, John Wiley & Sons Ltd; 121-134.

Van Savage, J., Decker, M. D., Edwards, K. M., Sell, S. H., and Karzon, D. T. (1990) Natural history of pertussis antibody in the infant and effect on vaccine response, J. Infect. Dis. *161*,487-492.

Van Strijp, J. A. G., Russell, D. G., Tuomanen, E., Brown, E. J., and Wright, S. D. (1993) Ligand specificity of purified complement receptor type 3: Indirect effects of an Arg-Gly-Asp sequence, J. Immunol. *151*,3324-3336.

Weiss, A. A. and Falkow, S. (1984) Genetic analyis of phase variation in *Bordetella pertussis*, Infect. Immun. *42*,263-269.

Weiss, A. A. and Goodwin, M. S. (1989) Lethal infection by mutants in the infant mouse model, Infect. Immun. *57*,3757-3764.

Weiss, A. A., Hewlett, E. L., Myers, G. A., and Falkow, S. (1983) Tn5-induced mutations affecting virulence factors of *Bordetella pertussis*, Infect. Immun. *42*,33-41.

Weiss, A. A., Hewlett, E. L., Myers, G. A., and Falkow, S. (1984) Pertussis toxin and extracytoplasmic adenylate cyclase as virulence factors of *Bordetella pertussis*, J. Infect. Dis. *150*,219-222.

Weiss, A. A., Johnson, F. D., and Burns, D. L. (1993) Molecular characterization of an operon required for pertussis toxin secretion, Proc. Natl. Acad. Sci. USA *90*,2970-2974.

Weiss, A. A., Melton, A. R., Walker, K. E., Andraos-Selim, C., and Meidl, J. J. (1989) Use of the promoter fusion transposon, Tn5 lac to identify mutants in vir-regulated genes, Infect. Immun. *57*,2674-2682.

2. General Principles of Immunology

2.1 Basic Principles of Immunity Against Intracellular Bacteria and Protozoa

Gudrun Szalay and Stefan H. E. Kaufmann

2.1.1 Introduction

Intracellular pathogens comprise bacteria as well as protozoa. The major common feature of these pathogens is their intracellular life-style (Hahn and Kaufmann, 1981; Moulder, 1985; Kaufmann, 1993; Kaufmann, 1994). Intracellular living implies invasion of and permanent residence inside host cells. An additional requirement for intracellular living is the low intrinsic toxicity of intracellular pathogens because toxins would impair their habitat. Intracellular pathogens are found in mononuclear phagocytes (MP) and various other tissue cells, in particular hepatocytes, epithelial cells, and endothelial cells (Kaufmann, 1993; Kaufmann, 1994). The intracellular bacteria can be further distinguished according to their capacity to survive outside of host cells:

a) Facultative intracellular bacteria predominantly live inside host cells but are also capable of replicating in the extracellular space. Their preferred habitat are MP. *Mycobacterium tuberculosis*, the causative agent of tuberculosis, is the medically most important representative of this group (Bloom, 1994). Worldwide, this disease causes enormous health problems, which increased in recent years in parallel with the emergence of multidrug resistant strains. Other facultative intracellular pathogens are *Salmonella typhi*, a gram negative rod causing typhoid fever, *Legionella pneumophila* responsible for Legionnaire‘s disease, and the gram positive rod, *Listeria monocytogenes*, the etiologic agent of listeriosis (Finlay and Falkow, 1989; Barbaree et al., 1993; Kaufmann, 1988). Disease is transmitted orally by contaminated food as is the case with *S. typhi* and *L. monocytogenes* or is air-borne as for *M. tuberculosis* and *L. pneumophila*.

b) Obligate intracellular bacteria fail to grow outside host cells because of their degenerate metabolism as a consequence of extreme adaption to the mammalian biotope. This group primarily encompasses the families rickettsiae and chlamydiae. Rickettsia organisms are transmitted by the bites of ticks, fleas, mites or lice. Infections with *Rickettsia prowazekii* and *Rickettsia typhi* cause typhus in man or mouse, respectively (Mandell et al., 1990; Winkler, 1995). *Rickettsia rickettsii* and *Rickettsia tsutsugamushi* are responsible for Rocky Mountain Spotted Fever or scrub thypus, respectively. Globally, all four types of disease are rare. In contrast, diseases caused by chlamydiae gain increasing importance worldwide (Mc Clarty, 1994). Depending on the serotyp, *Chlamydia trachomatis* causes conjunctivitis, trachoma, lymphogranuloma venerum, and urogenital infections. *Chlamydia psittaci* and *Chlamydia pneumoniae* cause atypical pneumoniae. *C. trachomatis* is spread by direct contact, whilst *C. pneumoniae* and *C. psittaci* are inhaled. Certain chlamydial infections have been implicated in the development of reactive arthritis. Obligate intracellular bacteria favour various non-professional phagocytes as habitat, in particular epithelial and parenchymal cells. An intermediate position between facultative and obligate intracellular bacteria is held by *Mycobacterium leprae*, the agent of leprosy (Hastings, 1985). Although it belongs to the family of mycobacteriaceae and shares many features with *M. tuberculosis*, attempts to grow *M. leprae in vitro* have failed thus far.

Development of intracellular protozoa and disease transmission depend on vectors. In its vector, the protozoan pathogen develops into the infectious phase for humans. Vectors for protozoan pathogens are mainly insects, and the major diseases occur in tropical and subtropical countries. Intracellular protozoa include *Plasmodium sp.* which cause malaria, a major infectious disease, affecting hundreds of millions of people (Miller et al., 1994). *Plasmodium sp.* are transmitted by bites of infected anopheline mosquitos. Other intracellular protozoa belong to the genus *Trypanosoma*. *Trypanosoma cruzi* is responsible for Chagas disease which is prevalent in South America (Takle and Hudson, 1989). Vectors for this pathogen are reduviid bugs. The third major tropical protozoan disease, leishmaniasis is transmitted by the bite of sand-flies carrying metacyclic larvae of *Leishmania sp.* (Locksley and Louis, 1992; Milon et al., 1995). In contrast, *Toxoplasma gondii* is a protozoan pathogen which is distributed worldwide (Subauste and Remington, 1993). Vectors for *T. gondii* are mice, and thus transmission is climate independent. Toxoplasmosis is a zoonotic disease with cats as regular hosts.

2.1.2 Specific Immune Response

Although nonspecific resistance mechanisms as well as innate immunity may interfere with microbial invasion and initiation of infection, infections with intracellular microbes are ultimately controlled by acquired immune responses. Dependency on specific immunity also forms the basis for vaccine induced protection. Acquired immunity is mediated by the humoral and cellular immune systems which specifically recognize foreign material entering the body. B lymphocytes are mediators of the humoral immune response. Upon direct recognition of foreign material they produce specific antibodies. T lymphocytes are responsible for the cellular part of the specific immune response. A major difference between B and T cells is the failure of T cells to recognize antigens directly as B cells do. Instead, T cells need adequate presentation of antigens by major histocompatibility complex (MHC) molecules expressed on antigen presenting cells (APC) (Townsend and Bodmer, 1989). T cells are distinguished according to their T cell receptor (TCR) and accessory molecules: The TCR is composed of an α and β chain (TCR α/β) or a γ and δ chain (TCR γ/δ). Distinguishing accessory molecules are the CD4 and CD8 markers. Exact functions of T cells will be discussed later in this chapter.

2.1.2.1 MHC and Antigen Presentation

MHC genes are organized in a gene complex of about 3500 kb size in man (Campbell and Trowsdale, 1993). Different classes of molecules are encoded in this gene complex, the MHC class I and the MHC class II genes being central to antigen presentation. MHC class I molecules consist of an α polypeptide chain in close association with β_2 microglobulin (β_2m), and MHC class II molecules are formed by an α and β polypeptide chain expressed as a dimer on the cell surface (Townsend and Bodmer, 1989). Association of antigen with either MHC molecule occurs inside the cell and processing of foreign antigens to peptide fragments is a prerequisite for sucessful association and presentation (Monaco, 1992; Neefjes and Ploegh, 1992; Neefjes and Momburg, 1993). The association product of MHC molecule and antigenic peptide is transported to the cell surface and is presented to T cells. MHC class I and MHC class II molecules can be further distinguished into classical and non-classical MHC molecules (Stroynowski and Fischer-Lindahl, 1994).

Classical MHC class I molecules are encoded by the genes H-2K, H-2D, H-2L in mouse, and by HLA-A, HLA-B, and HLA-C in man. These classical molecules present processed peptides to the vast majority of CD8 T cells. Non-classical or MHC class I like molecules, termed H-2Q, H-2T, and H-2M in the mouse, are also encoded in the MHC gene complex, and show homology to the classical MHC class I molecules (Shawar et al., 1994; Beckman and Brenner, 1995). Both, classical and non-classical MHC class I molecules are surface expressed in association with β_2m. Whilst for H-2Q molecules no corresponding T cell subset has been described so far, H-2M and H-2T gene products present a restricted set of peptides to CD8$^+$ TCR α/β cells (Pamer et al., 1992; Kurlander et al., 1992; Bouwer et al., 1994). Homologs of the murine non-classical MHC class I molecules in man are HLA-E, HLA-F and HLA-G with yet unidentified functions (Beckman and Brenner, 1995). The important difference of classical and non-classical MHC class I molecules is the high polymorphism of classical versus the low polymorphism of non-classical MHC molecules.

Classical MHC class II molecules are HLA-DP, HLA-DQ, HLA-DR in man, and H-2I-A and H-2I-E in mouse. Classical MHC class II molecules present processed peptides to CD4 T cells. Contrary to MHC class I like molecules, MHC class II like molecules are not expressed on the cell surface (Beckman and Brenner, 1995). Available evidence suggests that they support classical MHC class II molecules in antigen presentation (Roche, 1995). CD1 molecules represent a third class of molecules involved in antigen presentation (Beckman and Brenner, 1995; Bendelac, 1995). They share with MHC class I molecules the association with β_2m, although the genes are encoded outside the MHC. CD1 molecules have several isoforms in humans, CD1a-e, whilst in the mouse only the CD1d homologs CD1.1 and CD1.2 exist.

The task of APC is processing of endogenous and exogenous antigenic entities and their presentation to T cells. Virtually all host cells are able to present antigen together with MHC class I molecules. In contrast, only a restricted set of cells bears MHC class II molecules constitutively on the cell surface. Macrophages, dendritic cells, and B cells are paragons of professional APC, bearing MHC class I as well as MHC class II molecules. Two different pathways of antigen presentation are distinguished (Harding et al., 1995). In the endosomal processing pathway, phagocytozed exogeneous material is degraded within the phagosome and the resulting peptides of about 12-15 amimoacids (aa) length associate with MHC class II molecules. MHC class II molecules are synthesized in the endoplasmic reticulum (ER), from where they are transported to the endosomal compartment via the Golgi apparatus and the Trans-Golgi-network. They associate with peptides in the endosomal compartment, and then the MHC class II/peptide complex is presented at the cell surface.

Antigen processing and presentation through the cytosolic pathway represents the second way (fig. 2.1.1). This alternative is normally restricted to proteins located in the cytosol which are either of host-cell or viral origin. In the cytosol, proteasomes degrade proteins to peptides which are then transported by specialized proteins – transport-associated proteins (TAP) – into the ER. There, the peptides associate with the heavy chain of the MHC class I gene product. Prior to peptide association, the heavy chain is bound to β_2m. This complex is transported to and presented on the cell surface to stimulate CD8 T cells.

The large polymorphism of genes and pseudogenes coding for classical MHC class I and class II molecules enables the organism to bind and present an enormous diversity of peptides generated by the plethora of microbial pathogens. Diversity could be further increased by using longer peptides whereas shorter peptides would require a smaller T-cell repertoire with a higher risk of autoreactivity. Probably the 8-15 mer peptides presented by classical MHC molecules evolved as compromise between these two alternatives. The question how antigens from intracellular microbial pathogens enter the cytosolic pathway is only incompletely understood. Several possibilities are conceivable: First, the pathogen itself enters the cytosol, where synthesized proteins of the pathogen are degraded and transported to the ER. For ex-

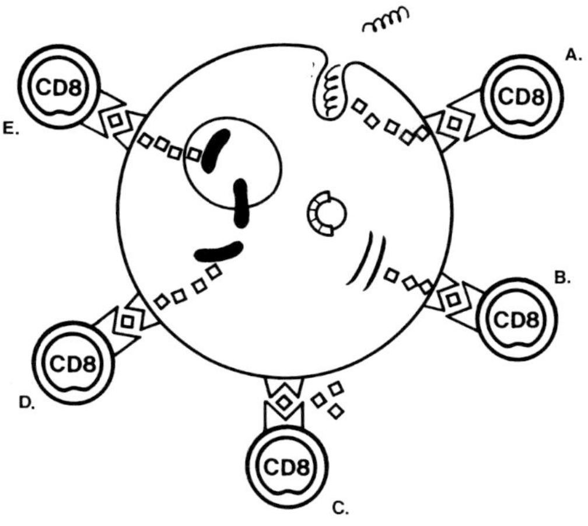

Fig. 2.1.1: Different antigen processing pathways leading to CD8 T-cell activation.

A. Phagocytosis of exogenous proteins and processing by an alternative pathway leads to MHC class I presentation.
B. Viral or host-cell proteins or those from naked DNA are present in the cytosol and processed and presented by MHC class I molecules.
C. Extracellular loading of MHC class I molecules.
D. Phagocytosis and egression of microbes into the cytosol.
E. Phagocytosis of microbes, degradation in the phagosome and introduction of microbial proteins or peptides into the cytosol.

ample, some intracellular organisms like hemolytic *L. monocytogenes* or *T. cruzi*, in response to the endosomal pH decrease, produce cytolysins which destroy the phagosomal membrane so that the whole pathogen egresses into the cytosol (Gaillard et al., 1986; Andrews et al., 1990). Second, proteins or peptides from microbes remaining in the endosome gain excess to the cytosol by still unknown mechanisms. For example, antigens of ahemolytic *L. monocytogenes*, *T. gondii*, or transfected *S. typhimurium* are presented by MHC class I molecules (Szalay et al., 1994; Denkers et al., 1993; Aggarwal et al., 1990; Hess et al., 1996). Further possibilities for MHC class I presentation are the extracellular loading of 'empty' MHC molecules with peptides by regurgitation of peptides from the endosome to the extracellular space.

The low polymorphism of non-classical MHC molecules drastically reduces the diversity of the corresponding T cells with concomitant risk of crossreactivity. This menace is by-passed by focussing on a restricted set of ligands such as N-formyl-methylated peptides which are abundant in bacteria and rare in mammals (Pamer et al., 1992; Kurlander et al., 1992; Wang et al., 1995). Specialization to unique peptides and low polymorphism promote development of T cells which are less dependent on a unique aa sequence. Instead, these T cells are directed against peptides which are characteristic for prokaryotes. CD1 molecules, as the last group of antigen presenting molecules, present not only proteinacious but also non-proteinacious antigens (Bendelac, 1995). The CD1d isotyp seems to preferentially present hydrophobic peptides, whereas the CD1b isoform appears specialized in binding glycolipids like mycolic acids and lipoarabinomannans from mycobacteria (Sieling et al., 1995; Castano et al., 1995; Beckman et al., 1994). Association of the CD1 molecule with its antigenic ligand is TAP-independent and chloroquine sensitive suggesting an endosomal processing pathway (Hanau et al., 1994). A specialized subset of αβ T cells lacking CD4 and CD8 is the prime target of the complex of CD1b plus glycolipid.

It can be envisaged, that focus on microbial glycolipids provides an additional advantage to the host, because glycolipid components are cleaved from the bacterial cell walls early during endosomal degradation. Therefore, presentation by CD1 molecules represents a more rapid event than antigen processing, association and presentation through either classical MHC processing pathway.

Increasing evidence suggests that nonproteinacious components like alkyl components, sugars, and nucleotides of bacterial origin are directly presented on the cell surface of APC and then stimulate γδ T cells (Tanaka et al., 1995). Common to these ligands is the presence of phosphate-residues (Constant et al., 1994; Schoel et al., 1994). For example, isopentenylpyrophosphat is presented and recognized directly without apparent involvement of any classical or non-classical MHC molecules or the CD1 molecules (Morita et al., 1995). Moreover, presentation of the prenyl pyrophosphate antigens seems to be processing independent. However, the risk of crossreactivity exists because isopentenylpyrophosphat represents a ubiquitous metabolite in pro- and eukaryotic cells.

The advantage of MHC class I presentation is obvious. MHC class I molecules are presented by all nucleated cells from all kinds of tissues. In contrast, MHC class II molecules are only expressed on professional phagocytes, either constitutively or after induction. So, triggering of T cells by MHC class II molecules is restricted to professional phagocytes. The spectrum of T-cell antigens is further extended by non-classical MHC class I and CD1 gene products as well as by direct antigen stimulation. These different ways of antigen presentation aim at the activation of specific T cells, so that they express specialized effector functions.

2.1.2.2 T Cells

Antigen recognition by T cells is mediated by the TCR with the help of the accessory molecules CD4 and CD8. The TCR is composed of two disulfide-linked chains, either a combination of an α and β chain or a γ and δ chain. The gene sequences of the chains are formed by rearrangements of multiple germline segments coding for variable (V), diversity (D), joining (J) and constant (C) regions (Leiden, 1993). Combinatorial association of the different segments during germline rearrangement as well as different chain combinations generate a high TCR diversity which is further increased by junctional diversification. This means the random addition of nucleotides to and imprecise joining of segments, as well as the existence of different reading frames for the D segments. Numbers of segments for the α and β chains are markedly higher than for γ and δ chains. A higher junctional diversification in the δ chain, however, compensates for the lower numbers of V,D,J segments. The high diversity of $>10^{12}$ T cells enables the host to respond to the plentitude of different antigenic ligands from microbial pathogens. In man and mouse, the T cell population in peripheral blood and lymphoid organs comprises $> 90\%$ α/β T cells and $< 10\%$ γ/δ T cells. Both, the TCR α/β heterodimer and the TCR γ/δ heterodimer are associated with the CD3 complex. The TCR binds to the hypervariable region of the MHC molecule which harbors the antigenic peptide, and the CD4 or CD8 molecules bind to conserved regions of the MHC molecule. Engagement of the TCR involves the CD3 complex leading to intracellular signal transduction and T cell activation. Additional costimulatory effects are induced by engagement of the accessory molecules CD4 or CD8. CD4 and CD8 T cells bind to monomorphic parts of the MHC class II or MHC class I molecules, respectively. T cells lacking both, CD4 and CD8 molecules, are called double negative (DN). DN T cells are found to a minor extent amongst TCR α/β cells and to a major extent amongst TCR γ/δ cells. DN T cells are independent from accessory molecules and classical MHC gene products. Potential ligands for DN TCR α/β cells are the CD1b molecules (Porcelli et al., 1989). Murine T cells recognizing CD1.1 and CD1.2 molecules express the phenotype $CD4^+CD8^-NK1.1^+$ or DN $NK1.1^+$ (MacDonald, 1995). Recognition by γ/δ T cells of antigen, direct or in the context of MHC class I like or CD1 molecules, remains incompletely understood (Beckman and Brenner, 1995; Kaufmann, 1995). Antigen specific T cell stimulation induces different effector functions like T Helper (T_H) or cytolytic activities.

2.1.2.2.1 T_H Functions

Differential processing and presentation of microbial pathogens has major consequences on the cellular immune response. Recognition of antigenic peptides in the context of MHC class II molecules activates CD4 T cells which generally express T_H functions characterized by specific cytokine production. Cytokines serve as transmitters between different leukocyte populations in which they activate appropriate effector functions. Cytokines play a decisive role in infections with intracellular pathogens, because they are involved in inflammation, T cell differentiation, granuloma formation, and activation of antimicrobial macrophage functions. According to their cytokine profiles, T cells are further devided into T_{H1} and T_{H2} cells (Mosmann and Coffman, 1989). T_{H1} cells primarily produce Interleukin 2 (IL-2) and Interferon γ (IFN-γ) which function as activators for T cells or macrophages, respectively. T_{H2} cells secrete IL-4, IL-5 which provide help to B cells. Inhibitory effects by secreted IL-10 is another function of T_{H2} cells. Separation into T_{H1} and T_{H2} cells is not stringent. Rather, T_{H1} and T_{H2} cells with their defined cytokine spectrum represent polar forms of a continuous spectrum. This is underlined by the recent observation, that T_{H1} cells can be driven to become T_{H2} cells. In contrast, conversion of T_{H2} cells into T_{H1} cells seems less likely. T_{H1} cells are central to control of infections with intracellular pathogens, whereas combat of helminth infections is primarily promoted by T_{H2} cells. Moreover, T_{H2} cells are involved in allergic reactions of the immediate type. The T_{H2}

cells are central to all humoral immune responses. IL-4 induces B cell maturation and immunoglobulin (Ig) E secretion while IL-5 induces IgA secretion. Yet, cytokine segregation into help for cell mediated versus humoral immunity is not strict and IFN-γ supports secretion of IgG isotypes with opsonizing capacities.

Differentation into T_{H1} and T_{H2} cells is mediated by cytokines, produced by the innate immune system (fig. 2.1.2). IL-12 is secreted by macrophages after infection and promotes differentiation into T_{H1} cells either directly or by activating natural killer (NK) cells. NK cells for their part produce IFN-γ. In contrast, IL-4 which drives T cells into the T_{H2} lineage is produced by basophils, mast cells, and unconventional T lymphocytes of $CD3^+CD4^+NK1.1^+$ phenotype. The development into either form is sustained by reciprocal negative regulation: IFN-γ downregulates T_{H2} cells, while IL-4 inhibits T_{H1} cells.

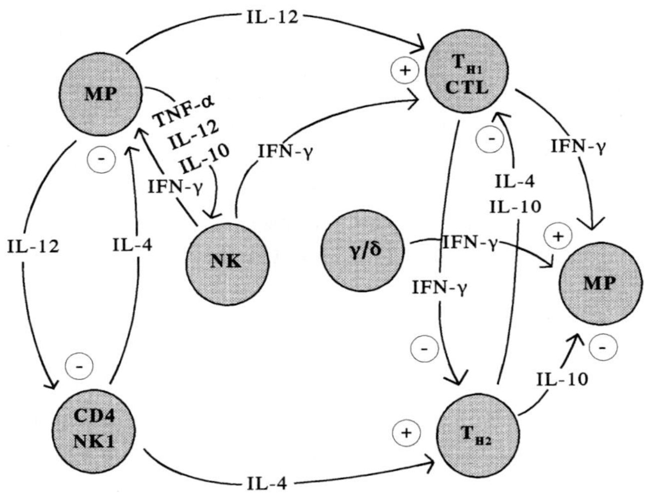

Fig. 2.1.2: Relationship between different immune cells and cytokines during microbial infections.

(+) indicates stimulatory and (-) inhibitory activities of cytokines. MP: mononuclear phagocyte; NK: natural killer cell, CTL: cytolytic T lymphocyte. T_{H1}: T helper cell type 1, T_{H2}: T helper cell type 2.

Although T_H cells predominantly belong to the CD4 TCR α/β set, CD8 TCR α/β cells as well as TCR γ/δ cells also express helper functions. The T_{H1} cytokines are essential for protection against intracellular pathogens with macrophage activation by IFN-γ being of upmost importance. In contrast, T_{H2} cytokines may induce detrimental effects. Accordingly, most infections with intracellular pathogens induce strong T_{H1} and weak T_{H2} activities. Benefical or detrimental effects of T_{H1} versus T_{H2} functions are most obvious in the murine model of leishmaniasis (Reiner and Locksley, 1995). In susceptible mouse strains, which cannot clear infection, T_{H2} cytokines are preponderant, whereas resistant mice primarily produce T_{H1} cytokines. IL-4 is also detected in patients suffering from *M. leprae* infection or *C. trachomatis* infection (Yamamura et al., 1991; Simon et al., 1993). Yet, an exclusive dependence on T_{H1} immune responses in infections with intracellular microbial pathogens is a black-and-white drawing (Kelso, 1995). Rather a sequential appearance of T_{H1} and T_{H2} cytokines during infection seems likely. From the onset to the height of infection, T_{H1} cytokines induce protective responses. Once the specific immune system has eliminated the pathogen, T_{H2} cytokines downmodulate the immune response to avoid uncontrolled tissue damage. Sequential appearance of T_H cytokines and their functioning as up- and downmodulators of the specific immune response is consistent with the finding that T_{H1} cell populations can be converted into T_{H2} cell populations (Perez et al., 1995). However, premature downmodulation of the T_{H1} response by T_{H2} cytokines prior to sterile pathogen elimination will have detrimental consequences for the host.

2.1.2.2.2 Cytolytic Functions

In addition to CD4 T cells, CD8 T cells participate in acquired immunity to intracellular pathogens (Kaufmann, 1988). Probably, the main function of CD8 T cells is the expression of cytolytic activities. Infected target cells are lysed by CD8 T cells via direct cell contact involving antigen presentation in the context of classical and non-classical MHC class I molecules. Target cell lysis encompasses two independent mechanisms. The first mechanism depends on perforin and granzymes (Kägi et al., 1995; Kägi et al., 1994). In the mouse seven different granzymes termed A-G with serine protease activity have been described and in man four granzymes have been identified (Ebner et al., 1995). Perforin and granzymes are contained within granules which degranulate after antigen specific recognition. Perforin induces disintegration of liposomes, and polymerized perforin spans in a tubular structure of the target cell membrane. This causes rapid Ca^+ dependent osmotic lysis of target cells. The role of granzymes is less clear. The only active enzymes are granzyme A and B. Granzyme B is required for cytotoxic activities, whereas granzyme A plays no apparent role in cell lysis of *L. monocytogenes* infected target cells (Ebner et al., 1995; Heusel et al., 1994). The mechanism of perforin action resembles active target cell killing. Apoptosis or programmed cell death, the second mechanism, resembles self-annihilation (Nagata and Goldstein, 1995). Cell death is induced by crosslinking of the Fas ligand on CTL, with the Fas antigen on target cells. This induces chromatin condensation, membrane blebbing, and DNA fragmentation, followed by cell shrinking, dilation of the ER, and cell fragmentation. Although the cytolytic effector function is prevailed in CD8 TCR α/β cells, to a minor extent also CD4 TCR α/β cells and TCR γ/δ cells are involved in target cell destruction. The major aim of all these cytolytic T cells is the destruction of infected cells. As a consequence, the habitat of the intracellular microbial pathogen is destroyed, and the organism can be taken up by more potent effector cells including activated mononuclear phagocytes.

2.1.3 T_{H1} Function and Cytolysis in Intracellular Microbial Infections

Both T cell functions – T_{H1} and cytolysis – are required for elimination of intracellular microbial pathogens (Farrar and Schreiber, 1993; Taniguchi and Minami, 1993). The benefical effects of a T_{H1} immune response with secretion of IFN-γ and IL-2 has been demonstrated for virtually all intracellular pathogens including *M. tuberculosis*, *M. leprae*, *L. monocytogenes*, *Rickettsia sp.*, *Chlamydia sp.*, *L. pneumophila* as well as for *Plasmodium sp.* and *Leishmania sp.* (Kaufmann, 1993; Kaufmann, 1994; Reiner and Locksley, 1995; Troye-Blomberg, 1994; Langhorne, 1994). In most cases CD4 T cells are responsible for T_{H1} activities. *T. gondii*, *T. cruzi*, *L. monocytogenes* and *C. trachomatis* live in non-professional phagocytes (Kaufmann, 1993). These host cells are devoid of constitutive MHC class II expression and therefore recognition of infected cells by CD8 T cells gains major importance. Accordingly, CD8 T cells are the principal IFN-γ secretors in these infections (Denkers et al., 1993; Shirahata et al., 1994; Tarleton et al., 1992). Induction of T_{H1} cytokines however appears insufficient for efficacious combat of intracellular infections. Rather, cytolytic activities are required in addition. Cytolytic capacities are mainly allocated to CD8 T cells. As shown in the listeriosis model, CD8 T cells dominate over CD4 T cells in protection against infection (Kaufmann, 1993; Ladel et al., 1994). Additionally, in *M. tuberculosis* infection and in *C. trachomatis* infection, protective, cytolytic CD8 T cells have been identified (DeLibero et al., 1988; Flynn et al., 1992; Magee et al., 1995; Igietseme et al., 1994; Starnbach et al., 1994). Infections with *Plasmodium sp.* and *T. cruzi* organisms also induce protective, cytolytic CD8 T cells (Troye-Blomberg, 1994; Langhorne, 1994; Hoffman and Franke, 1994; Nickell et al., 1993). Auxiliary help for the combat against intracellular pathogens is provided by DN α/β and γ/δ T cells (Beckman et al., 1994; Haas et al., 1993). These T cell populations have been shown to produce IFN-γ in response to microbial antigens.

2.1.4 Innate Immune Mechanisms

Prior to the specific immune response, the innate immune system controls microbial infection. Cells of the innate immune system act in an antigen-independent way and do not acquire memory. Lack of specificity and memory exclude them as major targets of vaccination. Yet, cytokines produced by the innate immune system profoundly influence the type of specific immune response, in particular the T_{H1}/T_{H2} balance, and hence rational vaccine design cannot neglect this early host response (see above). Although the innate immune system fails to cause sterile pathogen eradication, it is important for initial reduction of microbial numbers. The main mechanism of innate immunity is phagocytosis (Rabinovitch, 1995). Therefore, professional phagocytes namely macrophages, monocytes and polymorphonuclear granulocytes (PNG) represent the principle cell populations of the innate immune system. PNG are short-lived, whereas monocytes and tissue macrophages can survive several days up to months. Certain other cells like hepatocytes, fibroblasts or epithelial cells can also take up microbes. In contrast to professional phagocytes, the contribution of these so-called non-professional phagocytes to resistance is low because of their restricted capacity to engulf large numbers of particles and to express anti-microbial effector functions. Rather, they often serve as shelter for intracellular pathogens. However, after appropriate stimulation murine as well as human hepatocytes express antibacterial as well as antiprotozoan activities (Nüssler et al., 1991; Szalay et al., 1995; Nüssler et al., 1993).

After their adherence to phagocytes, pathogens are internalized into phagosomes. The intraphagosomal milieu of professional phagocytes is initially basic, and thus enables defensins to express their antimicrobial activities (Lehrer 1993; Selsted and Ouellette, 1995). Defensins are peptides of small molecular weight, which permeabilize bacterial membranes. After a short period of time, the phagosomal milieu becomes acidic. The low pH interferes with bacterial growth directly. Moreover, this acidic milieu is sustained after fusion of phagosomes with lysosomes to promote optimum activity of lysosomal enzymes. The various hydrolases degrade proteins, lipids, carbohydrates and nucleic acids. They are engaged in killing as well as in degradation of internalized pathogens. More importantly, phagocytozed bacteria and protozoa are attacked by reactive oxygen and nitrogen intermediates (ROI, RNI, respectively) (Babior, 1984; Nathan and Hibbs, 1991; Nüssler and Billiar, 1993).

The oxidative burst is stimulated as a consequence of receptor mediated phagocytosis via IFN-γ or Fcγ receptors. Molecular oxygen is converted to ROI: O_2^-, H_2O_2, OH^-, $^1O_2$, and °OH radical. These products are toxic for bacteria and protozoa. The RNI are synthesized by the inducible NO synthase (i-NOS), which transforms L-arginine to L-citrulline and NO. NO exists in three different redox forms, NO°, NO^+ (nitrosonium cation), and NO^- (nitroxyl anion). NO° is highly toxic for intracellular microbes as is OONO°, which is formed by NO° and O_2^- or by NO^+ and H_2O_2. Induction of i-NOS is cytokine-mediated, mainly by IFN-γ. Microbicidal effects of RNI are well documented for the murine system where RNI production is an important bactericidal effector mechanism for control of *M. tuberculosis*, *Plasmodium sp.* or *T. gondii* (Nüssler et al., 1991; Flesch and Kaufmann, 1991; Langermans et al., 1993). A different situation holds true for human macrophages which produce only minute amounts of RNI upon stimulation with cytokine combinations comprising granulocyte-macrophage colony stimulating factor (GM-CSF), TNF-α, and IFN-γ (Denis, 1994). Convincing evidence for profound effects of RNI produced by human macrophages on the growth of intracellular pathogens is lacking, although reduction of *L. major* growth by human macrophages due to NO action after ligation of CD23 (FcεRII) molecules has been described recently (Vouldoukis et al., 1995).

Another way of controlling intracellular microbial growth is limitation of nutrients. Supply with free iron is essential for microbial metabolism (Payne, 1993). Inside the cell, iron is complexed to ferritin and lactoferrin. Introduction of lactoferrin into the phagosome limits the amount of free iron in this compartment. Additionally, iron transport into the cell is receptor mediated. Transferrin is the major extracellular transport molecule for iron. Thus, downregulation of the transferrin receptor on the cell surface modulates microbial growth (Alford et al., 1991). Finally, lack of the aa tryptophan, caused by its degradation, restricts intracellular survival of *C. trachomatis* and *T. gondii* (McClarty, 1994; Pfefferkorn, 1984).

During infection, professional phagocytes are also a rich source of proinflammatory cytokines including IL-1, IL-6, TNF-α, and chemokines (Akira et al., 1993; Dinarello, 1992; Vassalli, 1992; Baggiolini et al., 1994). IL-1 and TNF-α cause leucocyte extravasation, and activate PNG and endothelial cells to secrete other cytokines; TNF-α also participates in macrophage activation. IL-6 together with IL-1 and TNF-α induces production of acute phase reactants in hepatocytes. Chemokines are produced by many different cell types. Chemokine activities include induction of inflammatory mediators at the site of microbial growth, attraction of leukocytes, and activation of microbicidal functions in phagocytes. Cell attraction is mediated by membrane-bound chemokines which induce adhesion to endothelial cells, rolling, and extravasation of leukocytes (Murphy, 1994). Chemokines activate a variety of cells including monocytes, neutrophils, eosinophils, basophils, and lymphocytes. According to the localization of conserved cysteine residues in their aa sequence, chemokines are separated into α and β chemokines. In α chemokines the first two cysteine residues are separated by one additional aa, while in β chemokines the two cysteines are adjacent. The α chemokines mainly attract neutrophils and induce the oxidative burst, while β chemokines mostly attract and activate monocytes, lymphocytes, eosinophils, and basophils. The production of chemokines depends on the presence of other proinflammatory cytokines, namely TNF-α, IL-1 and INF-γ.

The second major cell population of the innate immune system are NK cells (Bancroft, 1993; Bendelac, 1995). NK cells are a distinct lymphocyte population neither expressing the TCR/CD3 complex nor Ig on their surface. Specific receptors expressed by NK cells include members of the NKR-P1 and Ly 49 families in the mouse and the NKR-P1 and the p58 families in man (Raulet and Held, 1995; Yokoyama, 1995). Engagement of Ly 49 or p58 with particular MHC class I alleles on target cells causes a negative signal in NK cells while binding of NKR-P1 receptors to carbohydrate ligands on target cells induces NK effector functions. Additional activating receptors on NK cells are Fcγ receptors and a shorter version of the p58 receptor, the p50 receptor. NK cells lyse IgG-coated cells through binding to the Fcγ receptor, and interactions between p50 receptors and MHC class I molecules also induce cytotoxicity. In microbial infections, NK cells are potent IFN-γ producers, which activate antimicrobial capacities in macrophages (Bancroft, 1993). Macrophages on the other hand stimulate NK cells via IL-12. Moreover, NK cells have been shown to lyse host cells infected with various intracellular bacteria (Trinchieri, 1989). In contrast to T cells,

NK cells are insufficient for elimination of intracellular pathogens like *Trypanosoma sp.*, *T. gondii*, *Rickettsia sp.*, *L. pneumophila*, *Plasmodium sp.* (Bancroft et al., 1991; Mombaerts et al., 1993). In SCID mice or RAG mice, which are devoid of functional T and B cells but have apparently normal NK cell levels, microbial growth is controlled only initially, and these mice succumb to infection at later stages (Bancroft et al., 1991; Mombaerts et al., 1993).

2.1.5 Evasion Mechanisms of Pathogens

Induction of protective mechanisms against intracellular pathogens by specific T cells is ultimately aimed at the activation of potent antimicrobial effector mechanisms in professional phagocytes. An additional, often auxiliary, mean is starvation of the pathogen within non-professional phagocytes. Beside phagocyte activation through cytokine secretion by T_{H1} cells, target cell lysis releases pathogenic organisms from uncapacitated host cells so that more proficient phagocytes can attack the liberated organisms. Intracellular pathogens have developed various means to interfere with these antimicrobial mechanisms (Garcia-del Portillo and Finlay, 1995). A first possibility of escaping phagosomal attack is the egression into the cytosol where microbes are less sensitive to toxic products, acidic pH, and lysosomal enzymes. By producing cytolysins, the phagosomal membrane becomes porous and the microorganism evades into the cytosol. This mechanism is used by *L. monocytogenes*, *Rickettsia sp.*, and *T. cruzi* (Mandell et al., 1990; Gaillard et al., 1986; Mosmann and Coffman, 1989). *M. leprae* has also been found in the cytosol of host cells, but the underlying mechanism is not yet known (Mor, 1983). Egression into the cytosol has marked consequences on processing of microbial antigens because proteins from "cytosolic" pathogens are directly delivered to MHC class I molecules. Several microbes which remain in the endosomes such as *L. pneumophila*, *M. tuberculosis* and *Chlamydia sp.* inhibit phagosome-lysosome fusion, thus escaping attack by lysosomal enzymes (Yokoyama, 1995; Bancroft et al., 1991). *M. tuberculosis* and *T. gondii* developed resistance mechanisms against such lysosomal enzymes (Garcia-del Portillo and Finlay, 1995; Mor, 1983).

Attack by ROI can be avoided by inducing phagocytosis via complement receptors (CR). Phagocytosis mediated by CR1 and CR3 enhances uptake of *M. tuberculosis*, *M. leprae*, *L. pneumophila*, and *Leishmania sp.* without stimulating a respiratory burst as it is promoted by FcγR engagement (Kaufmann and Reddehase, 1989). Additionally, scavengers of ROI like superoxide dismutase, catalase or phenolic glycolipid which inactivate O_2^- and H_2O_2 are produced by *L. monocytogenes*, *L. pneumophila*, or *M. leprae* (Storz et al., 1990).

A further evasion strategy is the abuse of non-professional phagocytes as habitat which possess low antimicrobial potential. Epithelial cells are invaded by *Chlamydia sp.* and *Rickettsia sp.*; hepatocytes are targets for *Plasmodium sp.* and *L. monocyto-*

genes (Moulder, 1985; Kaufmann, 1988; Mandell et al., 1990; Miller et al., 1994). To overcome this obstacle the immune system has to destroy the infected host cells and release the pathogens so that uptake by more proficient phagocytes becomes possible.

2.1.6 Specific Problems

Specific T cell immunity against infectious agents bears the risk of autoimmune disease due to crossreactivity between microbial and self peptides (Kaufmann and Schoel, 1994). Heat shock proteins (hsp) which are induced by stress in all cells are highly conserved during evolution, so that cross-reactive T cells are generated frequently (Koga et al., 1989). Although T cells with specificity for epitopes shared by microbial and self hsp have been described, their relevance as a link between infection and autoimmune disease remains to be firmly established. In patients suffering from reactive arthritis, *Chlamydia sp.* and *Salmonella sp.* specific antigens have been detected in the joints with sequence homologies to HLA-B27, the haplotype prevalent in reactive arthritis patients (Daser et al., 1994; Hughes and Keat, 1994).

Oral vaccination is the route of choice because this way of administration minimizes the risk of potential side effects such as tissue reactions which are frequently caused by intravenous or intradermal vaccines (Staats et al., 1994; Service, 1994). Oral vaccination is simple, cheap, and easily delivered. However, the oral route causes specific problems for vaccines against intracellular pathogens because the antigens have to cross the intestinal epithelia and induce a systemic immune response. Low quantities of antigen, given orally, can induce tolerance rather than protective immunity (Weiner et al., 1994). Whilst vaccination against autoimmune disease requires tolerance induction, vaccine-induced tolerance to intracellular pathogens would have fatal consequences for the host. Many intracellular pathogens like *L. monocytogenes*, *Salmonella sp.*, *M. tuberculosis* and *M. leprae* invade the host via the mucosal epithelia. The first line of specific host defense to prevent pathogen invasion and dissemination is provided by IgA antibodies in the mucosa (Staats et al., 1994; Service, 1994). Although the mucosal immune response contributes to local protection at the site of entry, it frequently favours tolerance induction in the systemic immune system. Introduction of antigen into Peyer's patches preferentially results in T_{H2} responses which on the one hand are central to IgA secretion but on the other hand interfere with optimum T_{H1} cell induction. Oral vaccines should stimulate local IgA secretion. At the same time, vaccine antigens must be delivered to the systemic immune system in order to induce efficient T_{H1} immune responses. Currently, this is best achieved by viable vaccines which are not only introduced to Peyer's patches but are also spread from the mucosa to lymph nodes and spleen within monocytes.

Cytokines do not only support immune responses, but also cause negative side effects. Tissue necrosis is primarily triggered by proinflammatory cytokines. TNF-α produced by infected macrophages together with IFN-γ from specific T cells is central to granuloma formation (Vassalli, 1992). The formation of fibrotic walls around granulomas promotes microbial containment within a restricted area, but at the same time, damages the tissue. Moreover, high levels of TNF-α are responsible for cachexia and wasting, often seen in tuberculosis. Finally, cytokines may influence growth of protozoan pathogens directly (Barcinski and Costa-Moreira, 1994). Replication of *Leishmania sp.* promastigotes is enhanced in the presence of IL-2 and GM-CSF, whereas IFN-γ improves trypanosomal growth. In conclusion, production of cytokines stimulates not only the development and maintainance of specific cellular immunity, but may also favour survival of protozoan pathogens. These harmful sequelae underline the importance of downmodulatory immune mechanisms, such as induction of T_{H2} cytokines at later times during the immune response.

2.1.7 Vaccine Development

An effective vaccine has to be potent, long-lasting and safe (Rabinovich et al., 1994). In this chapter we have mainly focussed on strategies to improve vaccine potency. Our discussion of the immune mechanisms involved in protection emphasizes the need for vaccines capable of stimulating all T cells: CD4 T cells, CD8 T cells, DN TCR α/β and DN TCR γ/δ cells. Activation of specific CD4 T cells by exogenous antigens generally causes less severe problems, whereas effective induction of protective CD8 and DN T cells demands for more sophisticated approaches (see fig. 2.1.3). These include:

a) Immunization with live recombinant carriers expressing a unique protective antigen. Possible carriers are attenuated strains of *M. bovis*, *S. typhi/S. typhimurium* and *L. monocytogenes* (Sadoff et al., 1988; Ikonomidis et al., 1994). Although unlikely, these strains may convert to virulent forms. Second, *M. bovis BCG* and *Salmonella sp.* are "endosomal pathogens" which may fail to stimulate protective CD8 T cells in an efficient way. These two organisms mainly induce CD4 T cells which are insufficient for full protection. During the course of infection with an intracellular pathogen, the appearance of secreted antigens generally precedes that of somatic antigens because microbial killing and degradation is a prerequisite for the latter but not the former type of antigen display. Therefore, viable vaccines should secret antigens in order to achieve rapid T cell activation. Specific secretion systems are available which allow active export of foreign antigens from various carrier systems (Hess et al., 1996). In this way, the peptide is no longer stored in the cytosol of the carrier organism but is secreted in high amounts into the extracellular milieu. Carriers expressing cytolysins per se, e.g. *L. monocytogenes* or recombinant carriers transfected with cytolysin genes, e.g. *r-S. typhimurium* can egress into the host cytosol and deliver antigenes directly to MHC class I molecules (Hess et al., 1995).

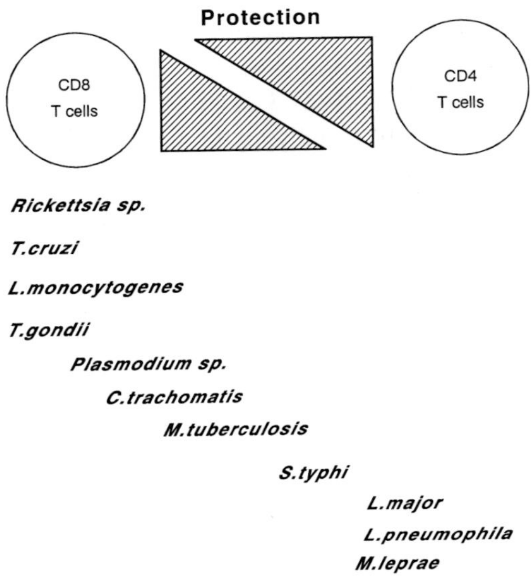

Fig. 2.1.3: Relative contribution of CD4 and CD8 T cells to protection against various intracellular microbial pathogens.

b) Immunization with defined peptides (Rabinovich et al., 1994). Experiments using peptides thus far gave poor protective immune responses. One reason may be, that internalized peptides are restricted to the endosomal processing pathway, thus activating CD4 T cells only. The administration of peptides or proteins with adjuvants bears the problem that currently available adjuvants often induce undesired inflammation.
c) Immunization with denatured proteins or killed pathogens (Schirmbeck et al., 1994; Szalay et al., 1995). This is another promising way to induce protective CD8 T cells. Denaturation is chemically achieved by SDS or heat treatment. Repeated immunization with heat killed *L. monocytogenes* organisms has been shown to stimulate CD8 T cells which protect against subsequent listeriosis (Szalay et al., 1995). As a major advantage of this strategy immunogenic antigens need not be identified because whole killed microorganisms can be used for immunization.
d) Vaccination with "naked" DNA coding for immunogenic antigens (Fynan et al., 1993). This is one of the latest approaches of vaccine development. The introduction of naked DNA into the muscle stimulates protective CD8 T-cell mediated immune responses. Effective protection has been obtained with DNA of influenza virus of *Plasmodium sp.* (Ulmer et al., 1993; Sedegah et al., 1994). The mechanisms leading to the stimulation of protective T cells by DNA injection into muscle cells remain to be clarified (Pardoll and Beckerleg, 1995). Because myocytes lack costimulatory molecules it appears unlikely that they induce protective T cell responses directly. Rather, antigen transfer from myocytes to macrophages seems to be the likely mechanism. Influx of macrophages into the muscular tissue is sup-

ported by local inflammation at the site of injection and is further improved by introduction of the GM-CSF gene into the vaccine (Xiang and Ertl, 1995).

e) Coadministration of cytokines. Induction of T_{H1} responses can be supported by coadministration of appropriate cytokines. For example, leishmania-infected mice, which have been treated with IL-12, develop a T_{H1} response and become resistant to disease (Muray and Hariprashad, 1995). Cointroduction of DNA encoding both, protective antigen and IL-12, yielded higher protection than vaccination with antigen DNA alone (Muray and Hariprashad, 1995). Modification of the immune response by exogenous cytokines holds promise not only for preventive but also for therapeutic vaccination strategies.

2.1.8 Concluding Remarks

The aim of vaccine development against intracellular pathogens has to focus on the induction of the "right" T cell response with CD8 T cells playing a central role. This population expresses similar functions as the CD4 T cell subset, namely cytokine production and cytolytic activities. However, CD8 T cells have a broader scope of target cells. Recognition of antigens in the context of MHC class I molecules, which are present on all nucleated cells, enables the CD8 T cells to detect infected host cells of any type. In contrast, CD4 T cells are restricted to MHC class II expressing host cells and thus they are more limited in their scope. In addition to classical peptide presentation molecules, antigen presentation by non-classical MHC class I molecules and CD1 molecules broadens the antigenic spectrum to lipids and carbohydrates. This type of antigen recognition – which is the domain of DN TCR α/β and TCR γ/δ cells – is just being uncovered and its contribution to antimicrobial immunity remains to be determined. Yet, stimulation of the appropriate blend of all T cell subsets is the desired goal of an optimum vaccine against intracellular pathogens.

Acknowledgements

S. H. E. Kaufmann acknowledges financial support from the Sonderforschungsbereich 322 "Lympho-Hämopoese", the German Science Foundation, project Ka 573/3-1/2 and the BMBF Verbundprojekt "Mykobakterielle Infektionen". We are grateful to Mrs. R. Mahmoudi for excellent secretarial assistance.

References

Aggarwal, A., Kumar, S., Jaffe, R., Hone, D., Gross, M., and Sadoff, J. (1990) Oral Salmonella: Malaria circumsporozoite recombinants induce specific $CD8^+$ cytotoxic T cells. J. Exp. Med. *172*, 1083-1090.

Akira, S., Taga, T., and Kishimoto, T. (1993) Interleukin-6 in biology and medicine. Adv. Immunol. *54*, 1-78.

Alford, C. E., King, T. E. J., and Campbell P. A. (1991) Role of transferrin, transferrin receptor, and iron in macrophage listericidal activity. J. Exp. Med. *174*, 459-466.

Andrews, N. W., Abrams, C. K., Slatin, S. L., and Griffiths, G. (1990) A T. cruzi secreted protein immunologically related to the complement component C9: Evidence for membrane pore-forming activity at low pH. Cell *61*, 1277-1287.

Babior, B. M. (1984) Oxidants from phagocytes: Agent of defense and destruction. Blood *64*, 959-966.

Baggiolini, M., Dewald, B., and Moser, B. (1994) Interleukin 8 and related chemotactic cytokines – CXC and CC chemokines. Adv. Immunol. *55*, 97-179.

Bancroft, G. J. (1993) The role of natural killer cells in innate resistance to infection. Curr. Opin. Immunol. *5*, 503-510.

Bancroft, G. J., Schreiber, R. D., and Unanue, E. R. (1991) Natural immunity: A T-cell independent pathway of macrophage activation defined in the SCID-mouse. Immun. Rev. *124*, 5-24.

Barcinski, M. A. and Costa-Moreira, M. E. (1994) Cellular response of protozoan parasites to host-derived cytokines. Parasitol. Today *10*, 352-355.

Beckman, E. M. and Brenner, M. B. (1995) MHC class I-like, class II-like and CD1 molecules: distinct roles in immunity. Immunol. Today, *16*, 349-352.

Beckman, E. M., Porcelli, S. A., Morita, C. T., Behar, S. M., Furlong, S. T., and Brenner, M. B. (1994) Recognition of a lipid antigen by CD1-restricted $\alpha\beta^+$ T cells. Nature *372*, 691-694.

Bendelac, A. (1995) CD1: Presenting unusual antigens to unusual T lymphocytes. Science *269*, 185-186.

Bendelac, A. (1995) Mouse $NK1^+$ T cells. Curr. Opin. Immunol. *7*, 367-374.

Bloom, B. R. (1994) Tuberculosis. Pathogenesis, protection, and control, edn, American Society for Microbiology, Washington, DC.

Bouwer, H. G. A., Fischer-Lindahl, K., Baldridge, J. R., Wagner, C. R., Barry, R. A., and Hinrichs, D. J. (1994) An H2-T MHC class Ib molecule presents Listeria monocytogenes-derived antigen to immune $CD8^+$ cytotoxic T cells. J. Immunol. *152*, 5352-5360.

Britton, W. J., Roche, P. W., and Winter, N. (1994) Mechanisms of persistence of mycobacteria. Trends in Microbiol. *2*, 284-288.

Campbell, R. D. and Trowsdale, J. (1993) Map of the human MHC. Immunol. Today *14*, 349-352.

Castano, A. R., Tangri, S., Miller, J. E. W., Holcombe, H. R., Jackson, M. R., Huse, W. D., Kronenberg, M., and Peterson, P. A. (1995) Peptide binding and presentation by mouse CD1. Science *269*, 223-226.

Constant, P., Davodeau, F., Peyrat, M.-A., Poquet, Y., Puzo, G., Bonneville, M., and Fourie, J.-J. (1994) Stimulation of human $\gamma\delta$ T cells by nonpeptidic mycobacterial ligands. Science *264*, 267-270.

Daser, A., Urlaub, H., and Henklein, P. (1994) HLA-B27 binding peptides derived from the 57kD heat shock protein of Chlamydia trachomatis: novel insights into peptide binding rules. Mol. Immunol. *31*, 331-336.

DeLibero, G., Flesch, I. E. A., and Kaufmann, S. H. E. (1988) Mycobacteria reactive Lyt2^{+} T cell lines. Eur. J. Immunol. *18*, 59-66.

Denis, M. (1994) Human monocytes/macrophages: NO or no NO? J. Leuko. Biol. *55*, 682-684.

Denkers, E. Y., Sher, A., and Gazzinelli R. T. (1993) CD8^{+} T cell interactions with Toxoplasma gondii: implications for processing of antigen for class I restricted recognition. Res. Immunol. *144*, 51-57.

Dinarello, C. A. (1992) Role of interleukin-1 in infectious diseases. Immunol. Rev. *127*, 119-146.

Ebner, K., Hausmann, M., Lehmann-Grube, F., Müllbacher, A., Kopf, M., Lamers, M., and Simon, M. M. (1995) Granzyme A-deficient mice retain potent cell-mediated cytotoxicity. EMBO, *14*, 4230-4239.

Farrar, M. A. and Schreiber, R. D. (1993) The molecular cell biology of interferon-γ and its receptor. Annu. Rev. Immunol *11*, 571-612.

Finlay, B. B. and Falkow, S. (1989) Salmonella as an intracellular parasite. Mol. Microbiol. *3*, 1833-1841.

Flesch, I. E. A. and Kaufmann, S. H. E. (1991) Mechanisms involved in mycobacterial growth inhibition by gamma-interferon activated bone marrow macrophages: Role of reactive nitrogen intermediates. Infect. Immun. *59*, 3213-3218.

Flynn, J. L., Goldstein, M. M., Triebold, K. J., Koller, B., and Bloom, B. R. (1992) Major histocompatibility complex class I-restricted T cells are required for resistance to Mycobacterium tuberculosis infection. Proc. Natl. Acad. Sci. USA *89*, 12013-12017.

Fynan, E. F., Webster, R. G., Fuller, D. H., Haynes, J. R., Santoro, J. C., and Robinson, H. L. (1993) DNA vaccines: Protective immunizations by parental, mucosal, and gene-gun inoculations. Proc. Natl. Acad. Sci. USA *90*, 11478-11482.

Gaillard, J. L., Berche, P., and Sansonetti, P. (1986) Transposon mutagenesis as a tool to study the role of hemolysin in virulence of Listeria monocytogenes. Infect. Immun. *522*, 50-55.

Garcia-del Portillo, F. and Finlay, B. B. (1995) The varied life-styles of intracellular pathogens within eukaryotic vacuolar compartments. Trends in Microbiol. *3*, 373-380.

Haas, W., Pereira, P., and Tonegawa, S. (1993) γ/δ T-cells. Annu. Rev. Immunol. *11*, 637-685.

Hahn, H. and Kaufmann, S. H. E. (1981) The role of cell mediated immunity in bacterial infections. Rev. Infect. Dis *3*, 1221-1250.

Hanau, D., Fricker, D., Bieber, T., Esposito-Farese, M.-E., Bausinger, H., Cazenavem, J.-P., Donato, L., Tongio, M.-M., and de la Salle, H. (1994) CD1 expression is not affected by human peptide transporter deficiency. Human Immunol. *41*, 61-68.

Harding, C. V., Song, R., Griffin, J., France, J., Wick, M. J., Pfeifer, J. D., and Geuze, H. J. (1995) Processing of bacterial antigens for presentation to class I and II MHC-restricted T lymphocytes. Infectious Agent and Disease, Raven Press, Ltd. NY *4*, 1-12.

Hastings, R. C. (1985) Leprosy. Edinburg, London, Melbourne, Churchill Livingstone.

Hess, J., Gentschev, I., Miko, D., Welzel, M., Ladel, C., Goebel, W., and Kaufmann, S. H. E. (1996) Superior efficacy of secreted over somatic antigen display in recombinant Salmonella: Vaccine induced protection against listeriosis. PNAS 93, 1458-1463.

Heusel, J. W., Wesselschmidt, R. L., Shresta, S., Russell, J. H., and Ley, T. J. (1994) Cytotoxic lymphocytes require granzyme B for the rapid induction of DNA fragmentation and apoptosis in allogeneic target cells. Cell *76*, 977-987.

Hoffman, S. L. and Franke, E. D. (1994) Inducing protective immune responses against the sporozoite and liver stages of Plasmodium. Immun. Lett. *41*, 89-94.

Hughes, R. A. and Keat, A. C. (1994) Reiter‘s syndrome and reactive arthritis: a current view. Semin. Arthritis-Rheum. *24*, 190-210.

Igietseme, J. U., Magee, M. D., Williams, D. M., and Rank, R. G. (1994) Role for CD8^{+} T cells in antichlamydial immunity defined by chlamydia-specific T-lymphocyte clone. Infect. Immun. *62*, 5195-5197.

Ikonomidis, G., Peterson, Y., Kos, F. J., and Portnoy, D. A. (1994) Delivery of a viral antigen to the class I processing and presentation pathway by Listeria monocytogenes. J. Exp. Med. *180*, 2209-2218.

Kägi, D., Ledermann, B., Bürki, K., Hengartner, H., and Zinkernagel, R. M. (1994) $CD8^+$ T cell-mediated protection against an intracellular bacterium by perforin-dependent cytotoxicity. Eur. J. Immunol. *24*, 3068-3072.

Kägi, D., Ledermann, B., Bürki, K., Zinkernagel, R. M., and Hengartner, H. (1995) Lymphocyte-mediated cytotoxicity in vitro and in vivo: Mechanisms and Significance. Immunol. Rev. *146*, 95-115.

Kaufmann, S. H. E. (1988) $CD8^+$ T lymphocytes in intracellular microbial infections. Immunol. Today *9*, 168-174.

Kaufmann, S. H. E. (1988) Listeriosis: new findings – current concern. Microbial. Pathogen. *5*, 225-231.

Kaufmann, S. H. E. (1993) Immunity to intracellular bacteria. Annu. Rev. Immunol. *11*, 129-163.

Kaufmann, S. H. E. (1994) Immunity to intracellular bacteria. In: Fundamental Immunology, 3rd ed. Paul W. E. (ed), New York, Raven Press, 1287-1308.

Kaufmann, S. H. E. (1996) $\gamma\delta$ and other unconventional T lymphocytes: What do they see and what do they do? Proc. Nat. Acad. Sci. USA 93: 2272-2279.

Kaufmann, S. H. E. and Reddehase, M. J. (1989) Infection of phagocytic cells. Curr. Opin. Immunol. *2*, 43-49.

Kaufmann, S. H. E. and Schoel, B. (1994) Heat shock proteins as antigens in immunity against infection and self. The biology of heat shock proteins and molecular chaperones. Morimoto, R. I, Tissieres, A., and Goergopoulos, C. (eds.), Cold Spring Harbor Press 495-531.

Kelso, A. (1995) Th1 and Th2 subsets: paradigms lost? Immunol.Today *16*, 374-378.

Koga, T., Wand-Württemberger, A., DeBruyn, J., Munk, M. E., Schoel, B., and Kaufmann, S. H. E. (1989) T cells against a bacterial heat shock protein recognize stressed macrophages. Science, *245*, 1112-1115.

Kurlander, R. J., Shawar, S. M., Brown, M. L., and Rich, R. R. (1992) Specialized role for a murine class I-b MHC molecule in prokaryotic host defenses. Science *257*, 678-679.

Ladel, C. H., Flesch, I. E. A., Arnoldi, J., and Kaufmann, S. H. E. (1994) Studies with MHC deficient Knock-out mice reveal impact of both MHC I and MHC II dependent T cell responses on Listeria monocytogenes infection. J. Immunol. *153*, 3116-3122.

Langermans, J. A. M., Van der Hulst, M. E. B., Nibberling, P. H., Hiemstra, P. S., Fransen, L., and Van Furth, R. (1992) IFN-gamma induced L-arginine-dependent toxoplasmastatic activity in murine peritoneal macrophages is mediated by endogenous tumor necrosis factor. J. Immunol. *148*, 568-574.

Langhorne, J. (1994) The immune response to the blood stages of Plasmodium in animal models. Immun. Lett. *41*, 99-102.

Barbaree, J. M., Breimann, R. F. and Dufour, A. P. (1993) *Legionella. Current Status and Emerging Perspectives.* AMS, Washington,

Lehrer, R. I. (1993) Defensins: Antimicrobial and cytotoxic peptides of mammalian cells. Annu. Rev. Immunol. *11*, 105-128.

Leiden, J. M. (1993) Transcriptional regulation of T cell receptor genes. Annu. Rev. Immunol. *11*, 539-570.

Locksley, R. M. and Louis, J. A. (1992) Immunology of leishmania. Curr. Opin. Immunol. *4*, 413-418.

MacDonald H. R. (1995) $NK1.1^+$ T cell receptor-α/β^+ cells: New clues to their origin, specificity and function. J. Exp. Med. *182*, 633-638.

Magee, D. M., Williams, D. M., Smith, J. G., Bleicker, C. A., Grubbs, B. G., Schachter, J., and Rank, R. G. (1995) Role of CD8 T cells in primary Chlamydia infection. Infect. Immun. *63*, 516-521.

Mandell, G. L., Douglas, R. G., and Bennett, J. E. (1990) Section D: Rickettsiosis. In: Principles and Practice of Infectious Diseases. 3rd ed. Mandell, G. L., Douglas, R. G., Bennett, J. E. (eds.), New York, Churchill Livingstone.

McClarty, G. (1994) Chlamydiae and the biochemistry of intracellular parasitism. Trends in Microbiol. *2*, 157-164.

Miller, L.H., Good, M.F., and Milon, G. (1994) Malaria Pathogenesis. Science, *264*, 1878-1883.

Milon, G., Del Giudice, G., and Louis, J. A. (1995) Immunobiology of experimental cutaneous Leishmaniasis. Parasitol. Today, *11*, 244-247.

Mombaerts, P., Arnoldi, J., Russ, F., Tonegawa, S., and Kaufmann, S. H. E. (1993) Differential roles of α/β and γ/δ T cells in immunity against an intracellular bacterial pathogen. Nature *365*, 53-56.

Monaco, J. J. (1992) A molecular model of MHC class I restricted antigen processing. Immunol. Today, *13*, 173-179.

Mor, N. (1983) Intracellular location of *Mycobacterium leprae* in macrophages of normal and immune-deficient mice and effect of rifampicin. Infect. Immun. *42*, 802-811.

Morita, C. T., Beckman, E. M., Bukowski, J. F., Tanaka, Y., Band, H., Bloom, B. R., Golan, D. E., and Brenner, M. B. (1995) Direct presentation of nonpeptide prenyl pyrophosphate antigens to human $\gamma\delta$ T cells. Immunity *3*, 495-507.

Mosmann, T. R. and Coffman, R. L. (1989) TH1 and TH2 cells: Different patterns of lymphokine secretion lead to different functional properties. Annu. Rev. Immunol. *7*, 145-173.

Moulder, J. W. (1985) Comparative biology of intracellular parasitism. Microbio. Rev. *49*, 298-337.

Muray, H. W. and Hariprashad, J. (1995) Interleukin 12 is effective treatment for an established systemic intracellular infection: Experimental visceral leishmaniasis. J.Exp.Med. 181, 387-391.

Murphy, P. M. (1994) The molecular biology of leukocyte chemoattractant receptors. Annu. Rev. Immunol. *12*, 593-633.

Nagata, S. and Goldstein, P. (1995) The Fas death factor. Science *267*, 1449-1456.

Nathan, C. F. and Hibbs J. B. Jr. (1991) Role of nitric oxide synthesis in macrophage antimicrobial activity. Curr. Opin. Immunol. *3*, 65-70.

Neefjes, J. J. and Momburg, F. (1993) Cell biology of antigen presentation. Curr. Opin. Immunol. *5*, 27-34.

Neefjes, J. J. and Ploegh, H. L. (1992) Intracellular transport of MHC class II molecules. Immunol. Today, *13*, 179-184.

Nickell, S. P., Stryker, G. A., and Arevalo, C. (1993) Isolation from Trypanosoma cruzi-infected mice of $CD8^+$, MHC-restricted cytotoxic T cells that lyse parasite-infected target cells. J. Immunol. *150*, 1446-1457.

Nüssler, A. K. and Billiar, T. R. (1993) Inflammation, immunoregulation, and inducible nitric oxide synthase. J. Leuc. Biol. *54*, 171-178.

Nüssler, A. K., Drapier, J.-C., Renia, L., Pied, S., Miltgen, F., Gentilini, M., and Mazier, D. (1991) L-arginine-dependent destruction of intrahepatic malaria parasites in response to tumor necrosis factor and/or interleukin 6 stimulation. Eur. J. Immunol. *21*, 227-230.

Nüssler, A. K., Renia, L., Pasquetto, V., Miltgren, F., Matile, H., and Mazier, D. (1993) In vivo induction of the nitric oxide pathway in hepatocytes after injection with irradiated malaria sporozoites, malaria blood parasites or adjuvants. Eur. J. Immunol. *23*, 882-887.

Pamer, E. G., Wang, C.-R., Flaherty, L., Fischer-Lindahl, K., and Bevan, M. J. (1992) H-2M3 presents a Listeria monocytogenes peptide to cytotoxic T lymphocytes. Cell *70*, 215-223.

Pardoll, D. M. and Beckerleg, A. M. (1995) Exposing the immunology of naked DNA vaccines. Immunity *3*, 165-169.

Payne, S. M. (1993) Iron acquisition in microbial pathogenesis. Trends in Microbiol. *1*, 66-69.

Perez, V. L., Lederer, J. A., Lichtman, A. H., and Abbas, A. K. (1995) Stability of Th1 and Th2 populations. Int. Immunol. *7*, 869-875.

Pfefferkorn, E. R. (1984) Interferon blocks the growth of *Toxoplasma gondii* in cultured fibroblasts by inducing the host cells to degrade tryptophan. Proc Natl Acad Sci USA *81*, 908-912.

Porcelli, S., Brenner, M. B., Greenstein, J. L., Balk, S. P., Terhorst, C., and Bleicher, P. A. (1989) Recognition of cluster of differentation 1 antigens by human $CD4^-CD8^-$ cytolytic T lymphocytes. Nature *341*, 447-450.

Rabinovich, N. R., McInnes, P., Klein, D. L., and Hall, B. F. (1994) Vaccine technologies: view to the future. Science *265*, 1401-1404.

Rabinovitch, M. (1995) Professional and non-professional phagocytes: an introduction. Trends in Cell Biol. *5*, 85-88.

Raulet, D. H. and Held, W. (1995) Natural Killer Cell Receptors: The Offs and Ons of NK Cell Recognition. Cell *82*, 697-700.

Reiner, S. L. and Locksley, R. M. (1995) The regulation of immunity to Leishmania major. Annu. Rev. Immunol. *13*, 151-177.

Roche, P. A. (1995) HLA-DM: In vivo facilitator of MHC class II peptide loading. Immunity *3*, 259-262.

Sadoff, J. C., Ballou, W. R., Baron, L. S., Majarian, W. R., Brey, R. N., Hockmeyer, W. T., Young, J. F., Cryz, S. J., Ou, J., Lowell, G. H., and Chulay, J. D. (1988) Oral Salmonella typhimurium vaccine expressing circumsporozoite protein protects against malaria. Science *240*, 336-338.

Schirmbeck, R., Böhm, W., and Reimann. J. (1994) Injection of detergent-denatured ovalbumin primes murine class I restricted cytolytic T cells in vivo. Eur. J. Immunol. *24*, 2068-2072.

Schoel, B., Sprenger, S., and Kaufmann, S. H. E. (1994) Phosphate is essential for stimulation of Vγ9Vδ2 T lymphocytes by mycobacterial low molecular weight ligands. Eur. J. Immunol. *24*, 1886-1892.

Sedegah, M., Hedstrom, R., Hobart, P., and Hofman, S. L. (1994) Protection against malaria by immunization with plasmid DNA encoding circumsporozoite protein. Proc. Natl. Acad. Sci. USA *91*, 9866-9870.

Selsted, M. E. and Ouellette, A. J. (1995) Defensins in granules of phagocytic and non-phagocytic cells. Trends in Cell Biol. *5*, 114-119.

Service, R. F. (1994) Triggering the first line of defense. Science *265*, 1522-1524.

Shawar, S. M., Vyas, J. M., Rodgers, J. R., and Rich, R. R. (1994) Antigen presentation by major histocompatibility complex Ib molecules. Annu. Rev. Immunol. *12*, 839-880.

Shirahata, T., Yamashita, T., Ohata, C., Goto, H., and Nakane, A. (1994) $CD8^+$ T lymphocytes are the major cell population involved in the early gamma interferon response and resistance to acute primary Toxoplasma gondii infection in mice. Microbiol. Immunol. *38*, 789-796.

Sieling, P. A., Chatterjee, D., Porcelli, S. A., Progozy, T. I., Mazzaccaro, R. J., Soriano, T., Bloom, F. R., Brenner, M. B., Kronenberg, M., Brennan, P. J., and Modlin R. L. (1995) CD1-restricted T cell recognition of microbial lipoglycan antigens. Science *269*, 227-230.

Simon, A. K., Seipelt, E., Wu, P., Wenzel, B., Braun, J., and Sieper, J. (1993) Analysis of cytokine profiles in synovial T cell clones from chlamydial reactive arthritis patients: predominance of the Th1 subset. Clin. Exp. Immunol. *94*, 122-126.

Staats, H. F., Jackson, R. J., Marinaro, M., Takahashi, I., Kiyono, H., and McGhee, J. R. (1994) Mucosal immunity to infection with implications for vaccine developement. Curr. Opinion. Immunol. *6*, 572-583.

Starnbach, M. N., Bevan, M. J., and Lampe, M. F. (1994) Protective cytotoxic T lymphocytes are induced during murine infection with Chlamydia trachomatis. J. Immunol. *153*, 5183-5189.

Storz, G., Tartaglia, L. A., Farr, S. B., and Ames, B. N. (1990) Bacterial defenses against oxidative stress. Trends in Gen. *6*, 363-368.

Stroynowski, I. and Fischer-Lindahl, K. (1994) Antigen presentation by non-classical I molecules. Curr. Opin. Immunol. *6*, 38-44.

Subauste, C. S. and Remington, J. S. (1993) Immunity to Toxoplasma gondii. Curr. Opin. Immunol. *5*, 532-537.

Szalay, G., Hess, J., and Kaufmann, S. H. E. (1994) Presentation of *Listeria monocytogenes* antigens by major histocompatibility complex class I molecules to CD8 cytotoxic T lymphocytes independent of listeriolysin secretion and virulence. Eur. J. Immunol. 24, 1471-1477.

Szalay, G., Hess, J., and Kaufmann, S. H. E. (1995) Restricted replication of Listeria monocytogenes in a gamma interferon-activated murine hepatocyte line. Infect. Immun. *63*, 3187-3195.

Szalay, G., Ladel, C. H., and Kaufmann, S. H. E. (1995) Stimulation of protective CD8 T lymphocytes by vaccination with non-living bacteria. Proc. Natl. Aca. Sci. USA 92, 12389-12392.

Takle, G. B. and Hudson, L. (1989) Autoimmunity and Chagas disease. Curr. Top. Microb. Immunol. *145*, 79-92.

Tanaka, Y., Morita, C. T., Tanaka, Y., Nieves, E., Brenner, M. B., and Bloom, B. R. (1995) Natural and synthetic non-peptide antigens recognized by human γδ T cells. Nature *375*, 155-158.

Taniguchi, T. and Minami, Y. (1993) The IL-2/IL-2 receptor system: A current overview. Cell, 73, 5-8.

Tarleton, R. L., Koller, B. H., Latour, A., and Postan, M. (1992) Susceptibility of beta 2 microglobulin deficient mice to Trypanosoma cruzi infection. Nature *356*, 338-340.

Townsend A. and Bodmer, W. (1989) Antigen recognition by class I restricted T lymphocytes. Annu. Rev. Immunol. *7*, 601-624.

Trinchieri, G. (1989) Biology of natural killer cells. Adv. Immunol. 47, 187-376.

Troye-Blomberg, M. (1994) Human-T cell responses to blood stage antigens in Plasmodium falciparum malaria. Immun.Lett. *41*, 103-107.

Ulmer, J. B., Donnelly, J. J., Parker, S. E., Thodes, G. H., Felgner, P. L., Dwarki, V. J., Gromkowski, S. H., Randall Deck, R., DeWitt, C. M., Friedman, A., Hawe, L. A., Leander, K. R., Martinez, D., Perry, H. C., Shiver, J. W., Montgomery, D. L., and Liu, M. A. (1993) Heterologous protection against influenza by injection of DNA encoding a viral protein. Science *259*, 1745-1749.

Vassalli, P. (1992) The pathophysiology of tumor necrosis factors. Annu. Rev. Immunol. *10*, 411-452.

Vouldoukis, I., Riveros-Moreno, V., Dugas, B., Ouaaz, F., Becherel, P., Debre, P., Moncada, S., and Mossalayi, M. D. (1995) The killing of Leishmania major by human macrophages is mediated by nitric oxide induced after ligation on the FcεRII/CD23 surface antigen. Proc. Natl. Acad. Sci. USA, *92*, 7804-7808.

Wang, C. R., Castano, A. R., Peterson, P. A., Slaughter, C., Fischer-Lindahl, K., and Deisenhofer, J. (1995) Nonclassical binding of formylated peptide in crystal structure of the MHC class Ib molecule H2-M3. Cell *82*, 655-664.

Weiner, H. L., Friedman, A., Miller, A., Khoury, S. J., Al-Sabbagh, A., Santos, L., Sayegh, R. B., Nussenblatt, R. B., Trentham, D. E., and Hafler, D. A. (1994) Oral tolerance: Immunologic mechanisms and treatment of animal and human organ-specific autoimmune diseases by oral administration of autoantigens. Ann. Rev. Immunol. *12*, 809-837.

Winkler, H. H. (1995) Rickettsia prowazekii, ribosomes and slow growth. Trends in Microbiol. *3*, 196-198.

Xiang, Z. and Ertl, H. E. J. (1995) Manipulation of the immune response to a plasmid-encoded viral antigen by coinoculation with plasmids expressing cytokines. Immunity *2*, 129-13.

Yamamura, M., Uyemura, K., Deans, R. J., Weinberg, K., Rea, T. H., Bloom, B. R., and Modlin, R. L. (1991) Defining protective responses to pathogens: Cytokine profiles in leprosy. Science, *254*, 277-279.

Yokoyama, W. M. (1995) Natural killer cell receptors. Curr. Opin. Immunol. *7*, 110-120.

2.2 Viral Immunity and Vaccine Strategies

David L. Woodland, Geoffrey A. Cole and Peter C. Doherty

The value of any immunization protocol to prevent, or ameliorate, the consequences of a viral infection is a function of the characteristics of the pathogen, the effectiveness of the recall response and the nature of memory in the T and B cell compartments (Doherty et al., 1994). The aim here is to consider each of these parameters in a very general way, then develop some more specific ideas addressing possible T cell-directed vaccine strategies. The discussion does not deal in any depth with the particular problem of the human immunodeficiency viruses, which are the subject of a separate chapter in this volume. The focus is on basic principles influenced, particularly, by our recent studies with respiratory viruses in mouse model systems.

2.2.1 Characteristics of the Pathogen

The efficacy of any vaccine will be influenced by how and where a pathogen replicates and causes pathology. Viruses can obviously evolve much more rapidly than their mammalian hosts. The nature of the host-parasite relationship is thus dictated by the evolutionary imperative of the virus. Fortunately for us, this generally means that single host pathogens will evolve a balanced parasitism that allows optimal transmission but does not normally lead to the demise of the sustaining, target population. However a virus that may not threaten the group as a whole (e.g., poliovirus) may be unacceptable to contemporary society as a cause of individual death and disability.

2.2.1.1 Multi-Host Viruses with a Systemic Pathogenesis

The "balanced parasitism" argument breaks down when infection of humans is an occasional feature for a virus that is normally maintained in some other species. Many examples are available for viruses that depend on a sylvatic cycle, transmission by biting or blood-sucking insects (fig. 2.2.1) and replication within the vector. A prominent pathogen of this type is the flavivirus, yellow fever virus. The attenuated 17D yellow fever virus has proved to be a superb vaccine.

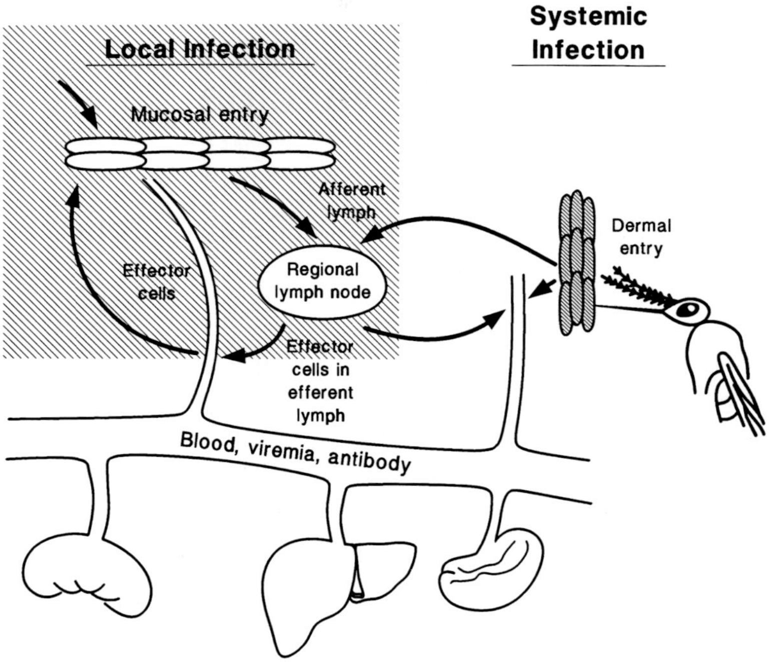

Fig. 2.2.1: The efficacy of any vaccine is, to some extent, a function of the nature of the particular infectious process. Viruses will generally enter via the mucosae or the dermis. Viral antigen will then be carried to the regional lymph nodes via afferent lymph, either as free virions or in the cytoplasm of antigen presenting cells, such as dendritic cells or Langerhans' cells. Pathogens that cause severe pathology at these sites of entry are the most difficult to counter by vaccination, as there is a need to maintain a high level of neutralising antibody in the local environment. Most successfull vaccines are directed against viruses that cause pathology as a consequence of systemic spread, and are thus exposed to serum antibody at a stage prior to entry into the target organ.

The experience with yellow fever illustrates four important points about vaccine strategies. These are:

(A) A virus that can replicate in alternative hosts (including the mosquito vector) does not need to develop a capacity for variation to avoid immune elimination, though it will have to cause a sufficiently severe infection to induce high titer viremia if it is to transmit.

(B) Viruses that do not change their surface glycoproteins, are readily neutralized by antibody, and depend on a systemic phase to cause pathology (fig. 2.2.1) are dealt with effectively by vaccines that promote long-term humoral immunity.

(C) Would 17D yellow fever virus, which was developed in the 1930s by attenuation following serial passage in mice, pass current guidelines for neurovirulence testing required for the licensing of new vaccines?

(D) The experience with yellow fever and Japanese encephalitis indicates that it would probably be relatively straightforward to develop effective attenuated, killed or recombinant vaccines for many of the mosquito and tick-borne diseases. Obvious examples are Murray Valley encephalitis, St Louis encephalitis or the polyarthritis and rash caused by Ross River virus. The problem is essentially one of economics. The judgement has been that diseases affecting relatively small numbers of people on a sporadic basis do not merit the development of a specific vaccine. Perhaps this perception will change as some of the current, novel strategies for multi-component vaccines are developed further. Hopefully, regulatory requirements can take account of these new opportunities.

When discussing vaccines directed at flaviviruses there is also the caveat raised by experience with the dengue viruses. Hemorrhagic dengue is thought to result from sequential infection with different variants of the virus. Any effective vaccine will probably need to involve simultaneous administration of the three major dengue serotypes in a single formulation, a strategy that is being pursued vigorously in Thailand where this disease has been a major pediatric problem. Though the physiological nature of dengue hemorrhagic shock is understood, the underlying immunopathogenesis has still not been clarified. This is a classical example of the difficulty of dissecting an immunologically-based human disease in the absence of a good experimental animal model.

2.2.1.2 Viruses with a Mucosal Entry but a Systemic Pathogenesis

Many of the viruses for which we have very effective vaccines infect via mucosae (fig. 2.2.1), but cause little problem if pre-existing antibody prevents viremic spread to distal sites of organ pathology. Classical examples are poliomyelitis and measles. Though waning immunity to polioviruses at a mucosal surface may not prevent recurrent, local infection when these viruses are circulating in a human community, the lack of involvement of large motor neurons means that there is no apparent disease. Such sporadic challenge may also serve to boost immune memory.

One virus in this category that is still a major human problem is the gammaherpesvirus, Epstein Barr virus (EBV). Initial entry via the respiratory mucosa or oropharynx leads to persistent, latent infection of B lymphocytes. Evaluation of antibody-directed vaccine to the major gp340 of EBV in primates indicates that an immunization strategy that relies on antibody has some promise though it may not provide "sterilizing" immunity (Ragot et al., 1993; Finerty et al., 1994). More recently, a multi-peptide polytope vaccine directed at the $CD8^+$ T cells is also being evaluated (Thomson et al., 1995). The use of this vaccine may, however, be more to boost immunity than to prevent infection, as discussed below in the section on immune memory. A promising development is the relatively recent availability of a murine gammaherpesvirus model (Nash and Sunil-Chandra, 1994) that mimics many of the features of, at least, the B lymphocyte-directed phase of EBV infection. This should allow experimental screening of potential vaccine strategies.

There is no major reason for thinking that effective vaccines to prevent herpesvirus infections cannot be developed. Even if a vaccine against *Varicella zoster* does not completely prevent infection, limiting viremia (and thus the extent of chicken-pox) should greatly diminish the extent of latent infection of neurons and the consequent, later development of shingles as the immune system declines in effectiveness with age. An excellent, attenuated living vaccine against the avian Marek's disease herpesvirus has been available for many years. However, humans differ from chickens in the sense that both biology and the nature of animal husbandry ensure that long-term survival is generally not a characteristic of the latter group.

2.2.1.3 Viruses That Cause Severe Pathology at Mucosal Surfaces

Setting aside the problem of antibody-selected variation, the most difficult pathogens to deal with by vaccination protocols are those that replicate mainly in superficial mucosal cells. In order to prevent such an infection completely, it is probably necessary to have neutralizing antibody present constantly in the fluid bathing the mucosal surface. This can be achieved either by locally-secreted antibody produced by plasma cells in mucosal sites, or by serum Ig that is at sufficiently high titer (and of the appropriate isotype) to gain access to the mucosa. Influenza infection in mice can be prevented by systemic administration of large amounts of neutralizing monoclonal antibody. Furthermore, even after influenza infection is established, passive antibody treatment can (at least for a time) terminate the disease process (Scherle et al., 1992; Palladino et al., 1995).

Evolving better protocols for promoting local immunity would seem to be highly desirable. The caveat is that strategies for inducing the development of extensive sites of lymphoid organization in, particularly, the respiratory tract could be dangerous in other senses. Any approach that increases the likelihood of local IgE production might not be useful overall because of the increased risk of allergic disease (Alwan et al., 1993).

2.2.2 The Nature of Immune Memory

The analysis of immune memory is complex, as it involves the study of populations of lymphocytes operating *in vivo* over the very long-term (Ahmed, 1992; Doherty et al., 1994). Most understanding of memory in the vaccine context is derived from measuring serum antibody. However, the definition of memory for the B cell/plasma cell compartment presents, if anything, even greater challenges than for the T cells because of the greater ease of quantifying T cell memory by *in vitro* limiting dilution analysis (LDA).

2.2.2.1 The Essential Difference Between T Cell and B Cell Memory

Operationally, the desirable scenarios for T cell and B cell memory are somewhat different. Optimal B cell memory is concerned with maintaining the persistent presence of the effector product, the Ig molecule, that binds to (predominantly) tertiary structures on proteins and has the capacity to neutralize virions (Colman et al., 1987). This requires the continued functioning of the Ig-secreting plasma cells, many of which seem to be sequestered (Tew et al., 1992; Bachmann et al., 1994; Sangster et al., 1995) in the bone marrow (BM). Perhaps the "hemopoietic" growth factors in the BM also operate to sustain the Ig-secreting plasma cells (Doherty, 1995).

The effector CD4+ and CD8+ T cells are not, however, maintained much past the initial phase of elimination of a viral pathogen (Sarawar et al., 1993; Sarawar and Doherty, 1994; Mo et al., 1995; Allan et al., 1990; Doherty et al., 1992). Lymphocytes that have the capacity to kill other cells, or secrete large amounts of cytokines that may have systemic effects, are no longer required once the job is done. The key characteristic of T cell memory is to maintain a population of "poised" precursors (p) that can be rapidly recalled to effector function (Tripp et al., 1995).

2.2.2.2 The Initiation of T Cell Memory

The basis of immune memory is established during the acute phase of the host response (fig. 2.2.2). The extent of clonal expansion of $CD8^+$ cytotoxic T lymphocyte precursors (CTLp) is such that the prevalence determined by LDA increases from something in the range of $1{:}10^5$ to $1{:}10^6$ in whole mouse lymph node or spleen to 1:3,000 or so within seven days of first encounter with antigen (Doherty et al., 1994). Recent estimates (Tripp et al., 1995) of the duration of this proliferative phase for the $CD8^+$ CTLps indicate that the 30-300x increase in frequency probably represents a balance between clonal expansion, the provision of terminally-differentiated CTLs in sites of pathology, and cell loss as a consequence of (perhaps) destruction in the spleen or excretion via mucosal sites in the respiratory or gastrointestinal tracts. The $CD8^+$ effectors that, for example, clear influenza virus from the lung within seven or eight days, are no longer detectable in bronchoalveolar lavage (BAL) populations by day 20 after the initial exposure to the pathogen.

Similarly, cytokine-producing cells in the BAL that can be detected by single cell ELISPOT analysis are at highest frequency during the time virus is being eliminated, and decrease rapidly in prevalence thereafter. Most of those producing IL-2, IL-4 and IFN-γ are probably $CD4^+$ effectors. Frequency estimates of $CD4^+$ T helper precursors

(Thp) are, from an as yet much less substantial body of LDA data (Ewing et al., 1995), somewhat similar to those for the CTLps. It is not, however, clear at this stage that the pattern of enormous proliferation and concomitant cell loss inferred (Doherty et al., 1994) for the virus-specific CTLps is also true for the $CD4^+$ subset. Such effects are, however, characteristic for $CD4^+$ T cells exposed to superantigens (Webb et al., 1994; Marrack et al., 1993).

The prevalence of both $CD4^+$ Thps and $CD8^+$ CTLps achieved by 21 days after initial infection with a respiratory virus seems to remain reasonably stable for the life of a laboratory mouse. Frequencies generally vary by a factor of < 2-3 fold and may even increase in very old animals, perhaps reflecting that T cell memory is being preserved in the face of declining numbers of naive Thps and CTLPs.

2.2.2.3 The Nature of B Cell Memory

Analysing the establishment of virus-specific B cell memory (fig. 2.2.2) has depended mainly on determining the prevalence of antibody-producing plasma cells using the single cell ELISPOT assay. These effectors can be shown to undergo the expected increase in prevalence and class-switching in both regional lymph nodes and spleen, then to disappear progressively from secondary lymphoid tissue. Furthermore, these cells can be shown to be maintained in the BM as antibody-secreting cells for the long term. Both the magnitude of the acute IgM response and Ig class switching depend on cognate help from antigen-specific $CD4^+$ T cells. The prevalence of memory B cell precursors can be assessed by the spleen focus assay, or by transfer into SCID mice (Linton et al., 1992). However, these approaches are much less convenient than the LDA protocols for T cells. Furthermore, much of the debate about the nature of B cell responses is framed more in the context of affinity maturation and somatic diversification, factors that do not play an obvious part for the T lymphocyte subsets. The combined use of ELISPOT analysis of various tissue sites and measurement of both serum and mucosal antibody is, at least in experimental animal models, the likely path of development for the further analysis of humoral immunity in the context of vaccine development.

2.2.2.4 The Antigen Persistence Debate

The idea (Gray and Matzinger, 1991) that persistence of the inducing antigen is mandatory for the maintenance of memory had considerable if (at least for $CD8^+$ T cells) somewhat brief currency (Gray and Matzinger, 1991; Lau et al., 1994; Hou et al., 1994). Adoptive transfer experiments with antigen-primed T cells transferred to antigen-negative, irradiated recipients indicate that memory $CD8^+$ T cells survive in the long term in a situation where the continued presence of the inducing virus (or components thereof) is highly improbable. These studies did not, however, exclude an alternative possibility that cross-reactive (Beverley, 1990), low affinity/avidity interactions with either self or other foreign antigens may help to maintain the more

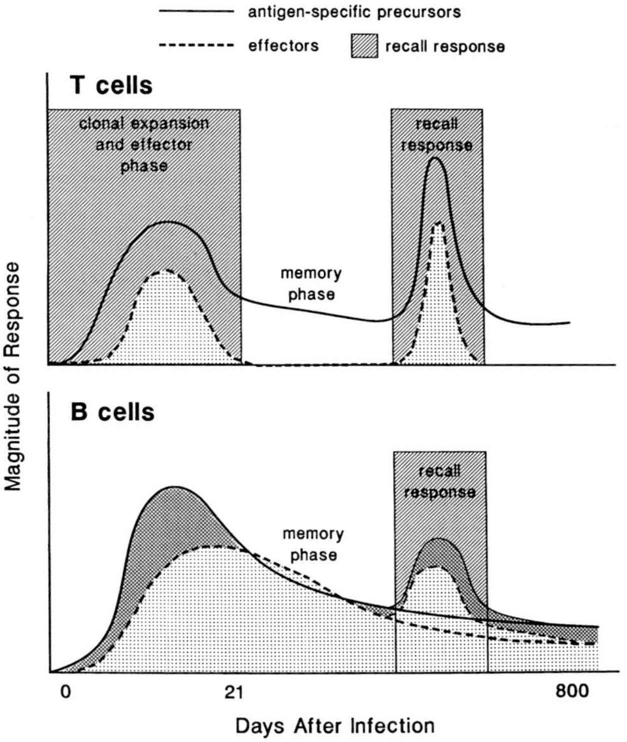

Fig. 2.2.2: Clonal expansion during the antigen-driven phase of the primary response leads to expanded pools (>30x) of $CD8^+$ CTLps and $CD4^+$ Thps. Some of these lymphocytes differentiate further to become CTL and Th effectors, which are probably removed as a consequence of apoptotic cell death within a week or so of elimination of the antigen. Present indications are that the CTLps may be produced in vast excess, with many being eliminated in (perhaps) the liver or exiting via the gastrointestinal or respiratory tracts. This is less clear for virus-specific Thps. Secondary challenge with an antigen that is not substantially removed by pre-existing antibody leads to further proliferation in the expanded pool of precursors, more rapid emergence of effector T cells and enhanced elimination of the pathogen. Clonal expansion is also characteristic of the B cell response, accompanied by maturation and Ig class switching. The essential difference between the T and B cell systems is that the effector cells, the plasma cells, remain activated and may continue to produce antibody for the life of a laboratory mouse. A primary site of plasma cell persistence is the bone marrow. Seconday challenge may cause a transient increase in plasma cell prevalence in the regional lymph nodes and spleen, but little perturbation of the size of the population maintained in the bone marrow.

readily-induced memory T cells. In the vaccine context, this is an academic argument of little (if any) relevance.

The situation for both $CD4^+$ T cells and B cells is much less clear. The argument here is that antigen/antibody complexes persist on specialized follicular dendritic cells in germinal centers (Kosco-Vilbois et al., 1993; Gray, 1994). The possibility is that, from time to time, the Ig receptor on the surface of memory B cells "picks off" one or more of these complexes. Processing through the exogenous pathway then leads to the expression of the appropriate peptide+class II MHC glycoprotein complex, which may in turn engage the T cell receptor (TCR) on the $CD4^+$ memory set. The proof, or otherwise, of this model should rest with current experiments using Ig^- μMT mice (Kitamura and Rajewsky, 1992). The analysis to date indicates that virus-specific $CD4^+$ T cell memory is maintained in these animals, at least for several months (D. Topham and P. C. Doherty, manuscript in preparation).

2.2.2.5 The Differentiation State of Memory T Cells

Though the numbers of virus-specific Thps and CTLps may remain relatively stable in the very long-term, it is clear that the activation state of these lymphocytes changes with time (Ewing et al., 1995; Webb et al., 1994; Tripp et al., 1995). During the acute response, FACS separation followed by LDA has been used to show (fig. 2.2.3) that the Thps and CTLps tend to assume the CD44-high, CD62L-low, α4-integrin-high phenotype characteristic of effector T cells. At least in the C57BL/6 mouse, the CD44-high characteristic is maintained for life (Tabi et al., 1988). The extent of the switch to the CD62L-low form during the acute phase varies for different viruses and, after a time, progressively reverts to the naive CD62L-high phenotype. The rapidity of this change is directly related to the profile seen during the acute phase, as illustrated in figure 2.2.3. The CD62L, or L-selectin, molecule is the lymph node homing receptor detected by the Mel-14 mAb. Lymphocytes that are CD62L-high will transit from blood to lymph node via the high endothelial venules, while the CD62L-low set is thought to recirculate through somatic tissues and return only to the node via afferent lymphatics. This means that the memory T cells tend, in the relative sense, to be excluded from the lymph nodes.

The CD62L gating mechanism does not operate in the spleen. Treatment with the Mel-14 mAb to CD62L modulates the molecule from the surface of naive T cells, but does not kill the lymphocytes. This leads to a very substantial diminution in the size of the lymph nodes, and the switching of the major site of primary response to the spleen. No analysis has been published on the effect, if any, on the recall of T cell memory.

2.2.3 Secondary Stimulation and the Recall Response

Immune memory is thus ideally characterized by the persistent presence of neutralizing antibody, both in serum and at mucosal sites, and by the availability of expanded

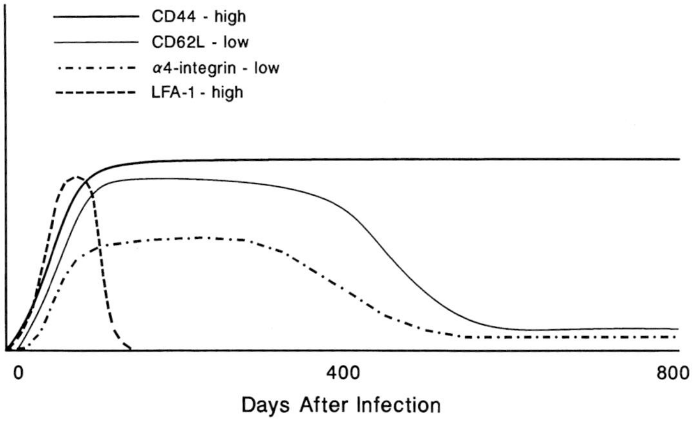

Fig. 2.2.3: The profile of cell-surface marker expression on influenza virus-specific memory $CD8^+$ CTLps is shown for B6 mice. The Sendai virus model has also been analysed in this way, the difference being that the switch to the activated CD62L-low form is never as absolute and the return to the naive CD62L-high phenotype occurs much more rapidly (5-8 months cf 12-15 months). CD62L (L-selectin) is the lymph node homing receptor. The central point is that both the activation and recirculation status of memory T cells change with time. Questions concerning the effectiveness of recall responses for $CD8^+$ T cells need to be addressed in this context, and in terms of T cell frequencies.

pools of "poised" Thps and CTLps that can rapidly be recalled to effector function (fig. 2.2.3).

2.2.3.1 The Effect of Antibody

The extent of antigen stimulation in any secondary challenge will depend on the availability of virus-specific Ig at the site of viral entry. If the levels are high enough to neutralize all input virus immediately, the likely result is that little perturbation of established immunity will occur unless a substantial amount of antigen is given. Virus complexed with Ig will be processed through the exogenous pathway of potential antigen-presenting cells and, if the dose is sufficient, this may lead to secondary stimulation of the $CD4^+$ set. Further promotion of humoral immunity presumably requires that at least some of the native protein be available to bind the surface Ig on the memory B cells: these complexes should in turn be interiorized and serve to focus the Thp population to deliver cognate help.

An alternative scenario is that the availability of neutralizing Ig at the site of viral entry may be insufficient to prevent the infection of epithelial cells and dendritic cells. The respiratory tract is, for example, lined by dendritic cells (Schon-Hegrad et al., 1991; Holt et al., 1994). Even if, as with the influenza A viruses and Sendai virus, the infection of dendritic cells is defective in the sense that infectious viral progeny are not produced (Horimoto and Kawaoka, 1995), this will be sufficient to cause secondary stimulation of both $CD4^+$ and $CD8^+$ T cells following transit via the lymph to the regional nodes or the blood to the spleen.

2.2.3.2 Characteristics and Limitations of the Recall Response for Memory T Cells

Secondary stimulation of T cell subsets independent of possible interference from antibody has been analyzed for situations where the T cell, but not the B cell, compartment has been primed. The first experiments of this type were done with influenza A viruses that share the internal nucleoprotein determinants that target $CD8^+$ T cells, but differ for the hemagglutinin and neuraminidase proteins recognized by neutralizing antibodies (Doherty et al., 1977). Later studies have utilized recombinant viruses (particularly vaccinia) incorporating particular proteins (Bennink and Yewdell, 1990), or peptides in various adjuvants. Such protocols can, for example, be shown to protect mice from a lethal challenge with Sendai virus (Kast et al., 1991). In general, however, the efficacy of T cell vaccines has generally been better for viruses that have a substantially systemic rather than mucosal pathogenesis (Doherty et al., 1989; Andrew et al., 1987; Oldstone et al., 1988; Castrucci et al., 1994; Connors et al., 1992).

The inherent problem is that the Thps and CTLps are in a "poised", but not an effector, state. The time to recall of CTL activity seems to be somewhere between three and five days. This may vary with the activation status of the memory T cells, though the progressive changes in levels of activation marker expression illustrated in figure 2.2.3 constitute a new story that has yet to be analyzed systematically in the context of possible consequences for virus challenge. Even so, it is clear that, if only the T cell compartment is primed, viruses will have at least 3-4 days to become established prior to the onset of effective cell-mediated immunity.

2.2.4 Maintaining Effective T Cell Memory

Though the evidence from laboratory mice indicates that T cell memory (characterized by expanded pools of virus-specific Thps and CTLps) can be maintained for at least two years , it is also the case that the inherent activation status of these memory T cells may decrease with time (fig. 2.2.3). Also, any generalization derived from the dissection of memory in mice has the inherent problem that the species is short-lived. The rules for memory over a human lifespan may be very different. Analysis of $CD8^+$ T cell memory to the influenza viruses indicates that people lose evidence of any protective effect after about three years (McMichael, 1994).

2.2.4.1 Secondary Stimulation

Though antigen persistence may not be mandatory for the maintenance of enhanced numbers of memory T cells in mice, sporadic re-stimulation may help to maintain the Thps and CTLps in an optimal state of activation (Tripp et al., 1995a; Tripp et al., 1995b). Furthermore, secondary challenge can be shown to cause substantial clonal expansion of the appropriate $CD8^+$ CTLp pool. Even so, it is not clear that this leads to enhanced T cell numbers in the longer term if the CTLps are already at the 1:3,000 or so level (in lymph nodes and spleen) that seems to be a recurring theme for T cell memory in the mouse.

The new DNA strategies (Webster et al., 1994; Zarozinski et al., 1995) may be particularly important for maintaining the prevalence and activation of memory T cells that have already been primed by infection, or administration of a more traditional vaccine. Though "immune exhaustion" (Moskophidis et al., 1993) may be a factor with viruses that cause persistent, massive stimulation of CTLps, there is good reason to think that this will not be a problem with more modest, though regular, antigenic exposure (Doherty, 1993). The prime example is EBV, which is never eliminated from the body but, as a consequence of continuing control by $CD8^+$ T cells, is generally maintained in a latent form in infected B lymphocytes (Moss et al., 1992; Doherty et al., 1994). Drastic immunosuppression, or simply *in vitro* culture, can readily lead to the emergence of lymphomas or lympoblastoid cell lines. Even in normal individuals infected with EBV, there is sporadic (or constant) excretion from the pharyngeal mucosa (Gan et al., 1993). This is essential for the epidemiology of the virus, and presumably reflects that some epithelial cells and (perhaps) B cells are regularly entering the lytic cycle that stimulates memory T cells. All the evidence suggest that potent EBV-specific T cell memory is maintained well into old age, and that this virus is not a common cause of lymphoma in the elderly. Old people are, however, very susceptible to periodic re-exposure to a variety of respiratory viruses, an effect that has generally been attributed to declining immune competence and, as all the viruses that are problematic have been encountered before, memory (Effros and Walford, 1983; Powers, 1993).

2.2.4.2 Bystander Activation and Memory T Cell Loss

Recent experiments indicate that infection with viruses that are not known to cross-react in the conventional sense can cause "bystander" activation of memory T cells (Yang et al., 1989). As many as 20 % of the $CD8^+$ T cells in the regional lymphoid tissue of mice responding to an influenza A virus can be shown to be in cycle at the peak of the response (Tripp et al., 1995). Many of these lymphocytes are probably memory T cells specific for other antigens, and there is some evidence that exposure in the cytokine-rich environment of the activated lymph node (Sarawar and Doherty, 1994) can cause both a measure of clonal expansion and the return to a higher activation state (Tripp et al., 1995b; Tripp et al., 1995a). This raises the possibility that periodic encounters with (often) minor pathogens may be important to maintain the total memory T cell compartment.

Other experiments with lymphocytic choriomeningitis virus (LCMV) indicate the contrary: intercurrent infection may lead to the generalized loss of memory T cells (Selin and Welsh, 1994). However, LCMV is known to replicate in a proportion of $CD4^+$ T cells (Ahmed et al., 1987) and some isolates cause severe damage in both the lymphoid and hemopoietic compartments. The possibility needs to be analyzed further with other infectious agents that cause substantial disease processes.

2.2.5 Vaccine Strategies Based on Priming $CD8^+$ T Cells

The preceding sections addressed the interface between vaccine strategies and viral immunity in a very general way. The remainder of this review concentrates on the much more specific topic of priming $CD8^+$ T cell-mediated immunity. Particular emphasis is placed on the idea that expanding T cells specific for sub-dominant, but conserved, epitopes may offer a strategy for immunizing against viruses that vary antigenically to escape immune elimination.

CTL epitopes are composed of processed viral peptides complexed with MHC class I glycoproteins on the surface of virus-infected cells (Townsend et al., 1986; Madden, 1995). Recognition of the class I/peptide complex is mediated by a cell surface $\alpha\beta$ T cell antigen receptor (TCR) which contacts amino acid side-chains of both the MHC molecule and the bound peptide (Madden, 1995; Ajitkumar et al., 1988; Chien and Davis, 1993; Jorgensen et al., 1992; Cole et al., 1995). This mode of recognition ensures that CTL are targeted to infected cells despite the presence of potentially large amounts of extracellular viral protein. Furthermore, all proteins of the virus, including conserved internal proteins, are potential targets for CTL-mediated immunity.

The biochemical details of antigen processing into peptides and loading of peptides onto class I molecules are the focus of intense investigation. Endogenously synthesized proteins are proteolytically degraded in the cytoplasm into small peptides which are translocated into the endoplasmic reticulum and associate with class I heavy chains, promoting their proper folding and association with β2-microglobulin (Townsend et al., 1989; Ortmann et al., 1994; Cox et al., 1990; Hosken and Bevan, 1990; Monaco et al., 1990; Yewdell and Bennink, 1992). Stable peptide-MHC complexes are then transported to the cell surface, where they are displayed for recognition by CTL. Crystallographic studies with both human and murine class I molecules have shown that the antigenic peptide occupies a groove composed of two α-helical regions and a β-pleated sheet at the membrane distal portion of the molecule (Bjorkman et al., 1987a; Bjorkman et al., 1987b; Matsumura et al., 1992; Fremont et al.,

1992; Zhang et al., 1992). Peptides eluted from purified class I molecules tend to be eight or nine amino acids long and to contain one or two conserved anchor residues with structural complementarity to pockets in the antigen binding groove (Zhang et al., 1992; Rotzscke et al., 1990a; Rotzscke et al., 1990b; Falk et al., 1991a; Rammensee et al., 1993; Falk et al., 1991b; Van Bleek and Nathenson, 1990). Individual class I molecules can therefore present a vast number of chemically distinct peptides by binding those which contain the appropriate binding motif.

2.2.6 Cytotoxic T Lymphocytes as Targets for Vaccines

The ability of CTL to recognize conserved internal proteins of viruses makes them potential targets for vaccines directed against latent viral infections, or against viruses that evade antibody responses. For example, CTL vaccines could be included as a component of a more complex vaccine engineered to induce both humoral and CTL priming.

The appropriate priming of CTL and the establishment of $CD8^+$ T cell memory is not a trivial issue. Inactivated virus vaccines or subunit vaccines generally fail to prime CTL because the antigen is routed primarily through the class II processing pathway, although there may be exceptions (Liu et al., 1995; Reis e Sousa and Germain, 1995). Live vaccines tend to circumvent this problem because the target proteins are synthesized endogenously and access the class I processing system, but have the disadvantage that they can themselves cause morbidity. An alternative approach is to prime CTL directly using free peptides (Kyburz et al., 1993; Melief and Kast, 1994; Mandelboim et al., 1994; Melief and Kast, 1995). This approach offers several potential advantages for CTL induction. First, peptides bind directly to class I molecules thereby bypassing any requirements for antigenic processing. Second, peptides generally do not induce antibody responses. Third, peptide-based vaccines should be safe and highly specific. The disadvantages of this type of approach are that peptides have short half-lives *in vivo* unless stabilized in appropriate adjuvants, and that protective epitopes need to be defined for many different MHC haplotypes.

Using murine models, it has been possible to directly test the ability of peptides to induce protective CTL responses against virus infection. Schulz et al. were able to show that a peptide derived from the LCMV nucleoprotein (NP) was able to protect mice from infection with a normally lethal dose of virus (as measured by virus replication in the spleen) (Schulz et al., 1991). Similarly, Kast et al. demonstrated that an immunodominant, K^b binding, NP peptide of Sendai virus ($NP_{324-332}$) could provide significant protection against a subsequent lethal challenge with Sendai virus (Kast et al., 1991). More recently, Sastry et al. have vaccinated mice with an immunodominant peptide derived from the influenza nucleoprotein and shown that they were partially protected against a lethal dose of influenza virus (Sastry et al., 1994). Interestingly, the kinetics of viral clearance was not affected despite the fact that there was a significant delay in the death of the mice. In all of these cases the peptide was delivered in the presence of incomplete or complete Freund's adjuvant to boost the immune response to the peptide and to extend the half-life of the peptide *in vivo*. Peptide immunizations have also been shown to be beneficial for the control of tumor cells. Feltkamp et al. have identified a peptide epitope from the E7 oncogene of human papilloma virus (HPV) that is present on an HPV-16-transformed murine tumor (Feltkamp et al., 1993). Immunization of mice with this peptide gave complete protection against a subsequent challenge with the syngeneic tumor. Other studies have reported similar success for peptide-induced tumor rejection (Joa and Kranz, 1994).

Peptide immunization does not always induce strong CTL responses for reasons that are, as yet, unclear (Rock et al., 1993; Bevan, 1989; Carbone et al., 1988; Carbone and Bevan, 1990). One possibility is that the concentration of β2-microglobulin (β2M) *in vivo* is too low to stabilize peptide association with empty class I molecules. To overcome this problem, Rock et al. have immunized mice with peptides in the presence of excess human β2M (human β2M binds to mouse class I heavy chains with a higher affinity than the mouse β2-microglobulin) (Rock et al., 1993). This protocol greatly enhanced the priming of CTL immunity to several different peptides. The effect was not due to the induction of T cell help since other human proteins could not substitute for β2M. An exciting alternative may be to bypass cell surface peptide loading by employing other molecules to ferry antigen into the appropriate presenting cells. Arnold et al. have suggested that heat shock proteins (HSP) may function in this manner since HSP from tumor cells prime strong CTL response to tumor antigens (Arnold et al., 1995; Blachere et al., 1993). A second possible difficulty in generating peptide specific CTL *in vivo* is the quality of T cell help in the form of maturational and proliferative lymphokines. Peptide immunization in the complete absence of helper activity induces T cell tolerance to the peptide (Aichele et al., 1995). Thus, in the murine models described above, T cell help is induced by the coinjection of adjuvant. However, an alternative, and very promising, approach is to take advantage of dendritic cells to prime peptide-specific immune responses. Dendritic cells are potent stimulators of primary T cells in both mice and humans and they are able to induce antigen-specific CTL from naive precursors *in vitro* and *in vivo* (Nair et al., 1993; Takahashi et al., 1993; Macatonia et al., 1989; Inaba et al., 1987; Nair et al., 1992; Mehta-Damani et al., 1994). For example, Porgador and Gilboa have used dendritic cells pulsed with an ovalbumin peptide to induce ovalbumin-specific CTL *in vivo* (Porgador and Gilboa, 1995). These CTL were primarily $CD8^+$ and dependent on $CD4^+$ T cells for induction. Clearly, dendritic cells represent a powerful alternative to classical adjuvants for the induction of MHC class I restricted T cells. A third possible difficulty in generating CTL *in vivo* is related to the strength of peptide recognition by T cells. Interestingly, T cells that recognize peptide epitopes with apparently higher avidity may be much more efficient at secreting IL-2 which regulates CTL proliferation (Heath et al., 1993). This may be further influenced by the relative efficiency of peptide loading onto class I molecules. In this regard, antigen processing-defective cells that express empty class I molecules can be used to enhance peptide loading. This approach has been shown to be extremely efficient at inducing both primary and secondary CTL activity (De Bruijn et al., 1991; Franksson et al., 1993).

2.2.7 Epitope Selection

The studies described above demonstrate the potential utility of peptides in priming protective CTL responses to either infectious agents or tumor cells. Thus, a central issue becomes the selection of appropriate peptide antigens. Indeed, a great deal of effort has been directed at defining natural CTL epitopes and understanding the mechanisms that control epitope selection. Epitopes can generally be defined experimentally using truncated proteins of various sources or peptide scans of whole proteins (Kast et al., 1991). However, the process has been greatly simplified by advances in our understanding of the molecular interactions between peptides and MHC molecules. In the case of MHC class I-associated peptides, the side-chains of two or three residues of the peptide fit into pockets in the peptide binding cleft of the MHC molecule (Zhang et al., 1992; Rammensee et al., 1993; Falk et al., 1991b; Van Bleek and Nathenson, 1990). Since only a limited number of amino-acid residues will allow functional peptide binding, it is possible to identify potential class I-binding peptides from the primary sequence of any given protein (Pamer et al., 1991; Rammensee et al., 1995). While this is a major benefit in the identification of CTL epitopes, it should be noted that not all such predicted peptides will actually bind the appropriate class I molecule (Deres et al., 1991). Secondary peptide/MHC interactions can influence peptide binding such that up to 50 % of sequences that express the appropriate binding motif, do not actually bind, or bind only weakly, to the class I molecule (Jameson and Bevan, 1992; Wipke et al., 1993; Schonbach et al., 1995; Ressing et al., 1995). Furthermore, not all peptides that can be shown to bind to class I molecules are immunogenic in the context of an infection, since the availability of different peptides in the cell may be strongly influenced by the processing machinery (Restifo et al., 1995; Eisenlohr et al., 1992).

While it is clear that multiple peptide/MHC epitopes can be derived from a complex antigen such as a virus, very few such epitopes appear to function as immune targets *in vivo*. Furthermore, there appears to be a hierarchy among epitopes in terms of their contribution to an immune response. For example, the $CD8^+$ T cell response of C57BL/6 mice to Sendai virus is directed exclusively against a single peptide derived from the nucleoprotein ($NP_{324-332}$) presented in the context of K^b (Kast et al., 1991; Dave et al., 1994). Studies by Cole et al. (manuscript in preparation) have shown that other potential CTL epitopes reside in the hemagglutinin and nucleoprotein molecules, but that these do not induce effector cells during the acute response to infection unless the $NP_{324-332}/K^b$ epitope is not available. Similar hierarchies have been described in LCMV infection (Gegin and Lehmann-Grube, 1992), minor histocompatibility antigen systems (Yin et al., 1993; Wolpert et al., 1995; Wettstein and Bailey, 1982; Johnson et al., 1981) and tumor systems (Sijts et al., 1994). The preferential involvement of some epitopes in an immune response (to the exclusion of other epitopes) has been termed immunodominance. This phenomenon was initially described in the late 1970's when it was noted that there was differential involvement of restriction elements in the CTL responses to virus infection. For example, mice of the K^bD^b haplotype responded to vaccinia and influenza A virus infection in the context of D^b. In contrast, mice of the K^kD^b haplotype preferentially responded to these viruses in the context of K^k (Zinkernagel et al., 1978; Doherty et al., 1978). Thus, there appeared to be a hierarchy of distinct epitopes presented by different class I molecules where $K^k > D^b > K^b$.

2.2.8 Factors that Control Immunodominance

Two key factors appear to determine the immunodominance of class I-restricted epitopes, namely, antigen density and the T cell repertoire. Antigen density is controlled by the relative ability of different peptides to be loaded onto MHC class I molecules. This depends on the relative efficiency of distinct antigen processing steps (including protease digestion and transport into the endoplasmic reticulum) (Eisenlohr et al., 1992; Yewdell and Bennink, 1992; Bergmann et al., 1994) and the affinity of the peptide for class I molecules (or its ability to form stable peptide/class I complexes). Sette et al. have analyzed over 100 potential peptide epitopes from the hepatitis B virus for class I binding and ability to induce CTL in HLA-A2 transgenic mice (Sette et al., 1994). It was found that a minimal affinity of 500nM was required to generate a CTL response, suggesting that peptide immunogenicity is correlated with high affinity for class I. Other studies have reached similar conclusions (Chen et al., 1994; Al-Ramadi et al., 1995). It is also possible that competition between peptides will influence the relative availability of peptides for presentation (Pala et al., 1988; Bodmer et al., 1989).

The T cell repertoire also influences immunodominance and epitope selection in the course of an infection. Studies by Daly et al. (1995) have shown that limiting the available T cell repertoire profoundly affects the immunodominance of class I epitopes following influenza virus infection. Infection of normal H-2^k mice with influenza virus induced strong immunodominant T cell responses to nucleoprotein and basic polymerase epitopes. In contrast, infection of syngeneic TCR transgenic mice expressing a single TCR β-chain paired with a full array of endogenous TCR α-chains induced T cell responses directed exclusively against non-structural and matrix derived epitopes. These data suggest that the relative frequencies of naive T cells specific for each of these epitopes differ substantially between wild type and transgenic mice. This idea was supported by the observation that the repertoire of T cells responding to the non-structural and matrix protein epitopes in transgenic mice was extremely limited (Daly et al., 1995). Despite the potential role of the T cell repertoire in immunodominance, some subdominant epitopes are able to induce strong T cell responses under appropriate conditions (see discussion below) (Oukka et al., 1992).

The repertoire of T cells able to recognize a given peptide antigen can be influenced by two factors. First, tolerance induced by self peptides analogous to the antigenic peptide, may drastically reduce the frequency of T cells able to recognize a particular epitope. Maryanski and colleagues (Casanova and Maryanski, 1993) have suggested that tolerance of this nature may affect not only the precursor frequency, but also the diversity of the T cell response to a given antigen. Second, individual peptides may differ in their immunogenicity for T cells. For example, a detailed study of the T cell response to the $NP_{324-332}/K^b$ epitope has shown that the TCR focuses on a central asparagine residue (residue 5) which is positioned at the top of a bulge in the peptide (Cole et al., 1995; Matsumura et al., 1992; Fremont et al., 1992). However, despite this constraint, multiple patterns of fine-specificity could be discerned depending on whether the TCR additionally recognized solvent exposed residues at positions 1 or 8. Thus, this epitope can be recognized in many distinct ways by different TCR and, consistent with this, the repertoire of T cells responding to this epitope is extraordinarily diverse (Cole et al., 1994). It follows that the frequency of naive precursor T cells able to recognize this epitope is likely to be high and be a predominant factor in the immunodominance of this epitope. Crystallographic analysis of several other immunodominant peptide/class I complexes have shown that the center of the peptide bulges out of the peptide binding cleft (Zhang et al., 1992; Madden et al., 1993; Guo et al., 1992). Thus the predominant recognition of the peptide center with degenerate recognition of other peptide residues may reflect a general paradigm in TCR interactions with immunodominant peptide/class I complexes (Young et al., 1994; Madden et al., 1992).

2.2.9 Subdominant Epitopes as Vaccine Targets?

One of the most intriguing aspects of the immunodominance phenomenon is that it appears to reflect a degeneracy in the T cell response to antigen. In many circumstances, the removal of an immunodominant epitope does not seem to affect the overall strength of the immune response, suggesting that subdominant epitopes are able to readily compensate for the loss of dominant epitopes. The clearest example of this comes from the previously mentioned studies of Daly et al. who demonstrated that the severely limited T cell repertoire of TCR β-chain transgenic mice forced a complete switch in CTL epitopes during influenza infection, as compared to wild-type mice (Daly et al., 1995). Despite the effect on epitope selection, transgenic mice mounted strong CTL responses and there was only slight delay in viral clearance. Furthermore, the transgenic mice were not more susceptible to Sendai or LCMV infection despite the severe constraints on TCR diversity (Doherty et al., 1994; Ewing et al., 1994). Other studies have shown that LCMV infection can be controlled by T cell responses to subdominant epitopes that normally cannot be detected *in vitro* (Gegin and Lehmann-Grube, 1992). In contrast, another report has suggested that the removal of an immunodominant epitope can sometimes immunocompromise the host. Thus, an early study by de Waal et al. suggested that the absence of the immunodominant $NP_{324-332}/K^b$ epitope in $H\text{-}2^{bm1}$ mice rendered them significantly more susceptible to Sendai virus infection (de Waal et al., 1983). Although this observation has not been supported by subsequent studies (Cole et al. unpublished data), the data suggest that subdominant epitopes may not always fully compensate for the loss of a dominant epitope.

It has now been well established that subdominant epitopes can elicit very strong immune responses if appropriately primed. For example, Oukka et al. have investigated the relationship between the immunodominant $NP_{366-374}/D^b$ and the subdominant NP_{55-63}/D^b epitopes of influenza virus (Oukka et al., 1992). Although influenza infection did not induce CTL against the subdominant epitope, direct immunization with the NP_{55-63} peptide induced a strong CTL response. Furthermore, these CTL were lytic for syngeneic target cells that were infected with a high dose of influenza virus. Subdominance in this case appears to be the result of relatively low expression of the NP_{55-63} epitope on influenza virus-infected cells. This level of expression is sufficient to allow CTL lysis, but is insufficient to induce CTL effectors. In the Sendai virus system, Cole et al. have examined the presentation and recognition of the $NP_{324-332}$ peptide in C57BL/6 mice (Cole et al., 1995; Cole et al., 1994). This peptide binds to both the K^b and D^b class I molecules with approximately equal affinity, but only $NP_{324-332}/K^b$-specific CTL were induced by Sendai virus infection (Dave et al., 1994; Deres et al., 1991). Immunization of $B6^{bm1}$ mice (K^{bm1}, D^b) with the $NP_{324-332}$ peptide, or infection of the same mice with Sendai virus, induced a strong $NP_{324-332}/D^b$-specific T cell response (Cole et al., manuscript in preparation). These D^b-restricted CTL were subsequently used to demonstrate that the $NP_{324-332}/D^b$ epitope was present on the surface of Sendai virus-infected C57BL/6 cells although there was evidence that the epitope density was lower than that of $NP_{324-332}/K^b$. This is an interesting system because the same peptide is being presented by the two different class I molecules suggesting that the generation of the peptide by the processing machinery is not a factor in this pattern of immunodominance. However, it has not been ruled out that the peptide is initially generated as a longer peptide that is trimmed to size after associating with the class I molecule. Thus, the longer peptide may associate more readily with K^b than D^b affecting the ultimate level of expression of these two epitopes. It should be noted that a longer version of this peptide ($NP_{321-336}$) may be recognized by K^b-restricted CTL suggesting that such a mechanism is possible (Kast et al., 1991; Schumacher et al., 1991). While the $NP_{324-332}/D^b$ epitope is clearly subdominant in terms of effector CTL in the lungs, there is effective priming of memory CTL to this epitope. Mediastinal lymph node $CD8^+$ T cells were isolated from C57BL/6 mice that had fully recovered from Sendai virus infection several weeks earlier. The cells were stimulated in vitro with $NP_{324-332}$ peptide and the resulting CTL were found to be specific for both $NP_{324-332}/K^b$ and $NP_{324-332}/D^b$ (Cole et al., manuscript in preparation). In terms of bulk CTL activity, the response to the D^b-restricted epitope at least equivalent to that of the K^b-restricted epitope. This phenomenon was only observed in long term memory cells since similar D^b-restricted activity was not readily generated from MLN immediately after the peak of infection. It is possible that D^b-restricted cells are swamped by K^b-restricted effector cells at early stages of the infection, an issue that can be resolved by LDA. Taken together, these studies demonstrate that subdominant epitopes can be readily primed by peptides and are fully capable of participating in an immune response (Albouche et al., 1982).

The relative potency of subdominant epitopes makes them potential targets for peptide vaccines. Increasing the number of epitopes involved in a vaccination protocol may result in a broader degree of protection and eliminates the requirement for positively identifying immunodominant epitopes, a significant problem in out-bred human populations. This idea has been successfully used in the "string of beads" vaccine approach adopted by Whitton et al. in which multiple class I-restricted epitopes were linked in a minigene expressed by vaccinia virus (Whitton et al., 1993). In addition, subdominant epitopes may confer protection in situations where persistent viruses escape immune surveillance by altering the immunodominant T cell epitopes, as has been described for LCMV (Pircher et al., 1990; Bertoletti et al., 1994a; de Campos-Lima et al., 1994; Bertoletti et al., 1994b; Klenerman et al., 1994). While the protective capacity of subdominant epitopes still needs to be evaluated, at least one study has shown that a subdominant epitope can confer significant protection against tumor cells. As mentioned earlier, immunization of mice with an E7 derived peptide protected mice against an HPV-16 transformed syngeneic tumor (C3)(Feltkamp et al., 1993). Despite the success of this system, it should be noted that this type of vaccination did not effectively eradicate existing tumors, although growth retardation was frequently observed (Feltkamp et al., 1993).

2.2.10 Concluding Remarks

Vaccines have proven to be extremely effective at controlling some virally-mediated diseases and have led to the virtual eradication of certain diseases associated with significant mortality and morbidity. Most of these successful antiviral vaccines depend on strong induction of neutralizing antibody. However, this may not be the most effective approach for viruses that have evolved strategies to evade antibody responses. An alternative, or complementary, approach for vaccine development is to specifically prime CTL responses. The advantage of this approach is that the CTL response is directed against internal components of the virus that are likely to be highly conserved between viral variants. Such vaccines would be unlikely to confer complete protection since CTL responses tend to be slow to develop and are only able to eradicate virus after infection has already been established. However, a memory CTL response may effectively control the viral load long enough for other mechanisms of viral clearance to develop. The studies discussed in this review demonstrate the feasibility of using peptides to induce CTL responses against viral infections and tumor cells. In some cases, these strategies induce significant protection. For example, peptide immunization gave significant protection against Sendai virus infection (Kast et al., 1991). However, further success in this area depends on a more complete understanding of the relationship between the virus and the immune system. It is essential that we understand the biochemical basis underlying the primary activation of T cells and the establishment of T cell memory. In particular, methods for inducing CTL memory in humans without the use of adjuvants and without having to define the immunodominant epitope need to be developed. Given the rate of progress in understanding immunity in general, the outlook for this area of science is extremely good.

Acknowledgements

This study was supported by National Institutes of Health Grants CA21765, AI29579, AI31596, and by the American Syrian Lebanese Associated Charities (ALSAC).

References

Ahmed, R. (1992) Immunological memory against viruses. Semin. Immunol., *4*, 105-109.

Ahmed, R., King, C. C., and Oldstone, M. B. (1987) Virus-lymphocyte interaction: T cells of the helper subset are infected with lymphocytic choriomeningitis virus during persistent infection in vivo. J. Virol., *61,* 1571-1576.

Aichele, P., Brduscha-Riem, K., Zinkernagel, R. M., Hengartner, H., and Pircher, H. (1995) T cell priming versus T cell tolerance induced by synthetic peptides. J. Exp. Med., *182*, 261-166.

Ajitkumar, P., Geier, S. S., Kesari, K. V., Boriello, F., Nakagawa, M., Bluestone, J. A., Saper, M. A., Wiley, D. C., and Nathenson. S. G. (1988) Evidence that multiple residues on both the a-helices of the class I molecule are simultaneously recognized by the T cell receptor. Cell, *54*, 47-56.

Al-Ramadi, B. K., Jelonek, M. T., Boyd, L. F., Margulies, D. H., and Bothwell, A. L. (1995) Lack of strict correlation of functional sensitization with the apparent affinity of MHC/peptide complexes for the TCR. J. Immunol., *155*, 662-673.

Allan, W., Tabi, Z., Cleary, A., and Doherty, P. C. (1990) Cellular events in the lymph node and lung of mice with influenza. Consequences of depleting $CD4^+$ T cells. J. Immunol., *144,* 3980-3986.

Allouche, M., Owen, J. A., and Doherty, P. C. (1982) Limit-dilution analysis of weak influenza-immune T cell responses associated with H-2Kb and H-$2D^b$. J. Immunol., *129*, 689-693.

Alwan, W. H., Record, F. M., and Openshaw, P. J. (1993) Phenotypic and functional characterization of T cell lines specific for individual respiratory syncytial virus proteins. J. Immunol., *150,* 5211-5218.

Andrew, M. E., Coupar, B. E., Boyle, D. B., and Ada, G. L. (1987) The roles of influenza virus haemagglutinin and nucleoprotein in protection: analysis using vaccinia virus recombinants. Scand. J. Immunol. *25,* 21-28.

Arnold, D., Faath, S., Rammensee, H., and Schild, H. (1995) Cross-priming of minor histocompatibility antigen-specific cytotoxic T cells upon immunization with the heat shock protein gp96. J. Exp. Med., *182*, 885-889.

Bachmann, M. F., Kündig, T. M., Odermatt, B., Hengartner, H., and Zinkernagel, R. M. (1994) Free recirculation of memory B cells versus antigen-dependent differentiation to antibody-forming cells. J. Immunol., *153,* 3386-3397.

Bennink, J. R. and Yewdell, J. W. (1990) Recombinant vaccinia viruses as vectors for studying T lymphocyte specificity and function. Curr. Top. Microbiol. Immunol., *163,* 153-184.

Bergmann, C. C., Tong, L., Cua, R., Sensintaffar, J., and Stohlman, S. (1994) Differential effects of flanking residues on presentation of epitopes from chimeric peptides. J. Virol., 68, 5306-5310.

Bertoletti, A., Costanzo, A., Chisari, F. V., Levrero, M., Artini, M., Sette, A., Penna, A., Giuberti, T., Fiaccadori, F., and Ferrari, C. (1994a) Cytotoxic T lymphocyte response to a wild type hepatitis B virus epitope in patients chronically infected by variant viruses carrying substitutions within the epitope. J. Exp. Med., *180*, 933-943.

Bertoletti, A., Sette, A., Chisari, F. V., Penna, A., Levrero, M., De Carli, M., Fiaccadori, F., and Ferrari, C. (1994b) Natural variants of cytotoxic epitopes are T-cell receptor antagonists for antiviral cytotoxic T cells. Nature, *369*, 407-410.

Bevan, M. J. (1989) Stimulating killer cells. Nature, *342*, 478-479.

Beverley, P. C. (1990) Is T-cell memory maintained by crossreactive stimulation? Immunol. Today, *11,* 203-205.

Bjorkman, P. J., Saper, M. A., Samraoui, B., Bennett, W. S., Strominger, J. A., and Wiley, D. C. (1987a) Structure of the human class I histocompatibility antigen, HLA-A2. Nature, *329*, 506-512.

Bjorkman, P. J., Saper, M. A., Samraoui, B., Bennett, W. S., Strominger, J. L., and Wiley, D. C. (1987b) The foreign antigen binding site and T cell recognition regions of class I histocompatibility antigens. Nature, *329*, 512-518.

Blachere, N. E., Udono, H., Janetzki, S., Li, Z., Heike, M., and Srivastava, P. K. (1993) Heat shock protein vaccines against cancer. J. Immunother., *14*, 352-356.

Bodmer, H. C., Bastin, J. M., Askonas, B. A., and Townsend, A. R. (1989) Influenza-specific cytotoxic T-cell recognition is inhibited by peptides unrelated in both sequence and MHC restriction. Immunology, *66*, 163-169.

Carbone, F. R. and Bevan, M. J. (1990) Class I-restricted processing and presentation of exogenous cell-associated antigen in vivo. J. Exp. Med., *171*, 377-387.

Carbone, F. R., Moore, M. W., Sheil, J. M., and Bevan, M. J. (1988) Induction of cytotoxic T lymphocytes by primary in vitro stimulation with peptides. J. Exp. Med., *167*, 1767-1779.

Casanova, J. L. and Maryanski, J. L. (1993) Antigen-selected T-cell receptor diversity and self-nonself homology. Immunol. Today, *14*, 391-394.

Castrucci, M. R., Hou, S., Doherty, P. C., and Kawaoka, Y. (1994) Protection against lethal lymphocytic choriomeningitis virus (LCMV). infection by immunization of mice with an influenza virus containing an LCMV epitope recognized by cytotoxic T lymphocytes. J. Virol. *68,* 3486-3490.

Chen, W., Khilko, S., Fecondo, J., Margulies, D. H., and McCluskey, J. (1994) Determinant selection of major histocompatibility complex class I- restricted antigenic peptides is explained by class I-peptide affinity and is strongly influenced by nondominant anchor residues. J. Exp. Med., *180*,1471-1483.

Chien, Y. and Davis, M. M. (1993) How ab T-cell receptors 'see' peptide/MHC complexes. Immunol. Today, *14*, 597-602.

Cole, G. A., Hogg, T. H., and Woodland, D. L. (1994) The MHC class I-restricted T cell response to Sendai virus infection in C57BL/6 mice: a single immunodominant epitope elicits an extremely diverse repertoire of T cells. Int. Immunol., *6*, 1767-1775.

Cole, G. A., Hogg, T. L., and Woodland, D. L. (1995) T cell recognition of the immunodominant Sendai virus $NP_{324-332}/K^b$ epitope is focused on the center of the peptide. J. Immunol., *155*, 2841-2848.

Colman, P. M., Laver, W. G., Varghese, J. N., Baker, A. T., Tulloch, P. A., Air, G. M., and Webster, R. G. (1987) Three-dimensional structure of a complex of antibody with influenza virus neuraminidase. Nature, *326,* 358-363.

Connors, M., Kulkarni, A. B., Collins, P. L., Firestone, C.-Y., Holmes, K. L., Morse, H. C., III, and Murphy, B. R. (1992) Resistance to respiratory syncytial virus (RSV) challenge induced by infection with a vaccinia virus recombinant expressing the RSV M2 protein (Vac-M2) is mediated by $CD8^+$ T cells, while that induced by Vac-F or Vac-G recombinants is mediated by antibodies. J. Virol., *66,* 1277-1281.

Cox, J. H., Yewdell, J. W., Eisenlohr, L. C., Johnson, P. R., and Bennink, J. R. (1990) Antigen presentation requires transport of MHC class I molecules from the endoplasmic reticulum. Science, *247*, 715-718.

Daly, K., Nguyen, P., Woodland D. L., and Blackman, M. A. (1995) Immunodominance of MHC class I restricted influenza virus epitopes can be influenced by the T cell receptor repertoire. J Virol., (in press).

Dave, V. P., Allen, J. E., Slobod, K. S., Smith, F. S., Ryan, K. W., Takimoto, T., Power, U. F., Portner, A., and Hurwitz, J. L. (1994) Viral cross-reactivity and antigenic determinants recognized by human parainfluenza virus type 1-specific cytotoxic T cells. Virology, *199*:376-383.

De Bruijn, M. L., Schumacher, T. N., Nieland, J. D., Ploegh, H. L., Kast, W. M., and Melief, C. J. (1991) Peptide loading of empty major histocompatibility complex molecules on RMA-S cells allows the induction of primary cytotoxic T lymphocyte responses. Eur. J. Immunol., *21,* 2963-2170.

de Campos-Lima, P. O., Levitsky, V., Brooks, J., Lee, S. P., Hu, L. F., Rickinson, A. B., and Masucci, M. G. (1994) T cell responses and virus evolution: loss of HLA A11-restricted CTL epitopes in Epstein-Barr virus isolates from highly A11-positive populations by selective mutation of anchor residues. J. Exp. Med., *179*, 1297-1305.

de Waal, L. P., Kast, W. M., Melvold, R. W., and Melief, C. J. M. (1983) Regulation of the cytotoxic T lymphocyte response against Sendai virus analyzed with H-2 mutants. J. Immunol., *130*, 1090-1096.

Deres, K., Schumacher, T. N., Wiesmuller, K. H., Stevanovic, S., Greiner, G., Jung, G., and Ploegh, H. L. (1991) Preferred size of peptides that bind to H-2 Kb is sequence dependent. Eur. J. Immunol., *22*, 1603-1608.

Doherty, P. C. (1993) Immune exhaustion: driving virus-specific $CD8^+$ T cells to death. Trends Microbiol., *1,* 207-209.

Doherty, P. C. (1995) Anatomical environment as a determinant in viral immunity. J. Immunol., *155,* 1023-1027.

Doherty, P. C., Allan, W., Boyle, D. B., Coupar, B. E., and Andrew, M. E. (1989) Recombinant vaccinia viruses and the development of immunization strategies using influenza virus. J. Infect. Dis. *159,* 1119-1122.

Doherty, P. C., Allan, W., Eichelberger, M., and Carding, S. R. (1992) Roles of alpha beta and gamma delta T cell subsets in viral immunity. Annu. Rev. Immunol., *10,* 123-151.

Doherty, P. C., Biddison, W. E., Bennink, J. R., and Knowles, B. B. (1978) Cytotoxic T-cell responses in mice infected with influenza and vaccinia viruses vary in magnitude with H-2 genotype. J. Exp. Med., *148*, 534-543.

Doherty, P. C., Effros, R. B., and Bennink, J. (1977) Heterogeneity of the cytotoxic response of thymus-derived lymphocytes after immunization with influenza viruses. Proc. Natl. Acad. Sci. U.S.A., *74,* 1209-1213.

Doherty, P. C., Hou, S., and Tripp, R. A. (1994) CD8$^+$ T-cell memory to viruses. Curr. Opin. Imm., *6,* 545-552.

Doherty, P. C., Hou, S., Evans, C. F., Whitton, J. L., Oldstone, M. B., and Blackman, M. A. (1994) Limiting the available T cell receptor repertoire modifies acute lymphocytic choriomeningitis virus-induced immunopathology. J. Neuroimmunol., *51,* 147-152.

Doherty, P. C., Sarawar, S. R., Tripp, R., and Hou, S. (1994) Characteristics of virus-specific CD8$^+$ T cell memory. In: Strategies in Vaccine Design, Ada, G. L. (ed.), R.G. Landes Company, Austin, 51-67.

Doherty, P. C., Tripp, R. A., and Sixbey, J. W. (1994) Evasion of host immune responses by tumours and viruses. Ciba Found. Symp., *187,* 245-256.

Effros, R. B. and Walford, R. L. (1983) Diminished T-cell response to influenza virus in aged mice. Immunology, *49,* 387-392.

Eisenlohr, L. C., Yewdell, J. W., and Bennink, J. R. (1992) Flanking sequences influence the presentation of an endogenously synthesized peptide to cytotoxic T lymphocytes. J. Exp. Med., *175,* 481-487.

Ewing, C., Allan, W., Daly, K., Hou, S., Cole, G. A., Doherty, P. C., and Blackman, M. A. (1994) Virus-specific CD8$^+$ T-cell responses in mice transgenic for a T-cell receptor beta chain selected at random. J. Virol., *68,* 3065-3070.

Ewing, C., Topham, D. J., and Doherty, P. C. (1995) Prevalence and activation phenotype of Sendai virus-specific CD4$^+$ T cells. Virology, *210,* 179-185.

Falk, K., Rotzschke, O., Stevanovic, S., Jung, G., and Rammensee, H.-G. (1991a) Allele-specific motifs revealed by sequencing of self peptides eluted from MHC molecules. Nature, *351,* 290-296.

Falk, K., Rotzscke, O., Deres, K., Metzger, J., Jung, G., and Rammensee. H.-G. (1991b) Identification of naturally processed viral nonapeptides allows their quantification in infected cells and suggests an allele-specific T-cell epitope forecast. J. Exp. Med., *174,* 1059-1071.

Feltkamp, M. C., Smits, H. L., Vierboom, M. P., Minnaar, R. P., de Jongh, B. M., Drijfhout, J. W., Schegget, J., Melief, C. J., and Kast, W. M. (1993) Vaccination with cytotoxic T lymphocyte epitope-containing peptide protects against a tumor induced by human papillomavirus type 16- transformed cells. Eur. J. Immunol., *23,* 2242-2249.

Finerty, S., Mackett, M., Arrand, J. R., Watkins, P. E., Tarlton, J., and Morgan, A. J. (1994) Immunization of cottontop tamarins and rabbits with a candidate vaccine against the Epstein-Barr virus based on the major viral envelope glycoprotein gp340 and alum. Vaccine, *12,* 1180-1184.

Franksson, L., Petersson, M., Kiessling, R., and Karre, K. (1993) Immunization against tumor and minor histocompatibility antigens by eluted cellular peptides loaded on antigen processing defective cells. Eur. J. Immunol., *23,* 2606-2613.

Fremont, D. H., Matsumura, M., Stura, E. A., Peterson, P. A., and Wilson, I. A. (1992) Crystal structures of two viral peptides in complex with murine MHC class I H-2k^b. Science, *257,* 919-927.

Gan, Y., Shirley, P., Zeng, Y., and Sixbey, J. W. (1993) Human oropharyngeal lesions with a defective Epstein-Barr virus that disrupts viral latency. J. Infect. Dis., *168,* 1349-1355.

Gegin, C. and Lehmann-Grube, F. (1992) Control of acute infection with lymphocytic choriomeningitis virus in mice that cannot present an immunodominant viral cytotoxic T lymphocyte epitope. J. Immunol., *149,* 3331-3338.

Gray, D. (1994) Regulation of immunological memory. Curr. Opin. Immunol., *6,* 425-430.

Gray, D. and Matzinger, P. (1991) T cell memory is short-lived in the absence of antigen. J. Exp. Med., *174,* 969-974.

Guo, H.-C., Jaradetzky, T. S., Garrett, T. P. J., Lane, W. S., Strominger, J. L., and Wiley, D. C. (1992) Different length peptides bind to HLA-Aw68 similarly at their ends but bulge out in the middle. Nature, *360*, 364-366.

Heath, W. R., Kjer-Nielsen, L., and Hoffmann, M. W. (1993) Avidity for antigen can influence the helper dependence of $CD8^+$ T lymphocytes. J. Immunol., *151*, 5993-6001.

Holt, P. G., Haining, S., Nelson, D. J., and Sedgwick, J. D. (1994) Origin and steady-state turnover of class II MHC-bearing dendritic cells in the epithelium of the conducting airways. J. Immunol., *153,* 256-261.

Horimoto, T. and Kawaoka, Y. (1995) The hemagglutinin cleavability of a virulent avian influenza virus by subtilisin-like endoproteases is influenced by the amino acid immediately downstream of the cleavage site. Virology, *210,* 466-470.

Hosken, N. A. and Bevan, M. J. (1990) Defective presentation of endogenous antigen by a cell line expressing class I molecules. Science, *248*, 367-370.

Hou, S., Hyland, L., Ryan, K. W., Portner, A., and Doherty, P. C. (1994) Virus-specific $CD8^+$ T-cell memory determined by clonal burst size. Nature, *369,* 652-654.

Inaba, K., Young, J. W., and Steinman, R. M. (1987) Direct activation of $CD8^+$ cytotoxic T lymphocytes by dendritic cells. J. Exp. Med., *166*, 182-194.

Jameson, S. C. and Bevan, M. J. (1992) Dissection of major histocompatibility complex (MHC) and T cell receptor contact residues in a Kb-restricted ovalbumin peptide and an assessment of the predictive power of MHC-binding motifs. Eur. J. Immunol., *22*, 2663-2667.

Joa, B. A. and Kranz, D. M. (1994) Generation of cytotoxic T-lymphocytes to a self-peptide/class I complex: a model for peptide-mediated tumor rejection. Cancer Res., *54*, 204-208.

Johnson, L. L., Bailey, D. W., and Mobraaten, L. E. (1981) Antigenic competition between minor (non H-2) histocompatibility antigens. Immunogenetics, *13*, 451-456.

Jorgensen, J. L., Esser, U., Fazekas de St. Groth, B., Reay, P. A., and Davis, M. M. (1992) Mapping T-cell receptor-peptide contacts by variant peptide immunization of single-chain transgenics. Nature, *355*, 224-230.

Kast, W. M., Roux, L., Curren, J., Blom, H. J., Voordouw, A. C., Meloen, R. H., Kolakofsky, D., and Melief, C. J. (1991) Protection against lethal Sendai virus infection by in vivo priming of virus-specific cytotoxic T lymphocytes with a free synthetic peptide. Proc. Natl. Acad. Sci. U.S.A., *88,* 2283-2287.

Kitamura, D. and Rajewsky, K. (1992) Targeted disruption of mu chain membrane exon causes loss of heavy- chain allelic exclusion. Nature, *356,* 154-156.

Klenerman, P., Rowland-Jones, S., McAdam, S., Edwards, J., Daenke, S., Lalloo, D., Koppe, B., Rosenberg, W., Boyd, D., and Edwards, A. (1994) Cytotoxic T-cell activity antagonized by naturally occurring HIV-1 Gag variants. Nature, *369*, 403-407.

Kosco-Vilbois, M. H., Gray, D., Scheidegger, D., and Julius, M. (1993) Follicular dendritic cells help resting B cells to become effective antigen-presenting cells: Induction of B7/BB1 and upregulation of major histocompatibility complex class II molecules. J. Exp. Med., *178,* 2055-2066.

Kyburz, D., Aichele, P., Speiser, D. E., Hengartner, H., Zinkernagel, R. M., and Pircher, H. (1993) T cell immunity after a viral infection versus T cell tolerance induced by soluble viral peptides. Eur. J. Immunol., *23*, 1956-1962.

Lau, L. L., Jamieson, B. D., Somasundaram, T., and Ahmed, R. (1994) Cytotoxic T-cell memory without antigen. Nature, *369,* 648-652.

Linton, P. J., Lo, D., Lai, L., Thorbecke, G. J., and Klinman, N. R. (1992) Among naive precursor cell subpopulations only progenitors of memory B cells originate germinal centers. Eur. J. Immunol., *22,* 1293-1297.

Liu, T., Zhou, X., Orvell, C., Lederer, E., Ljunggren, H. G., and Jondal, M. (1995) Heat-inactivated Sendai virus can enter multiple MHC class I processing pathways and generate cytotoxic T lymphocyte responses in vivo. J. Immunol., *154*, 3147-3155.

Macatonia, S. E., Taylor, P. M., Knight, S. C., and Askonas, B. A. (1989) Primary stimulation by dendritic cells induces antiviral proliferative and cytotoxic T cell responses in vitro. J. Exp. Med., *169*, 1255-1264.

Madden D. R (1995) The three-dimensional structure of peptide-MHC complexes. Annu. Rev. Immunol., *13*, 587-622.

Madden, D. R., Garbozy, D. N., and Wiley, D. C. (1993) The antigenic identity of peptide-MHC complexes: a comparison of the conformations of five viral peptides presented by HLA-A2. Cell, *75*, 693-708.

Madden, D. R., Gorga, J. C., Strominger, J. L., and Wiley, D. C. (1992) The three dimensional structure of HLA-B27 at 2.1A resolution suggests a general mechanism for tight peptide binding to MHC. Cell, *70*, 1035-1048.

Mandelboim, O., Berke, G., Fridkin, M., Feldman, M., Eisenstein, M., and Eisenbach, L. (1994) CTL induction by a tumour-associated antigen octapeptide derived from a murine lung carcinoma. Nature, *369*, 67-71.

Marrack, P., Hugo, P., McCormack, J., and Kappler, J. (1993) Death and T cells. Immunol. Rev., *133*, 119-129.

Matsumura, M., Fremont, D. H., Peterson, P. A., and Wilson, I. A. (1992) Emerging principles for the recognition of peptide antigens by MHC class 1 molecules. Science, *257*, 927-934.

McMichael, A. (1994) Cytotoxic T lymphocytes specific for influenza virus. In: Cytotoxic Tlymphocytes in Human Viral and Malaria Infections, Vol. 189, Oldstone, M. B. A. (ed), Springer-Verlag, New York, 75-91.

Mehta-Damani, A., Markowicz, S., and Engleman, E. G. (1994) Generation of antigen-specific $CD8^+$ CTLs from naive precursors. J. Immunol., *153*, 996-1003.

Melief, C. J. M. and Kast, W. M. (1994) Prospects for T cell immunotherapy of tumours by vaccination with immunodominant and subdominant peptides. In: Vaccines against virally induced cancers. Chadwick, D. J. and Marsh, J. (eds.), John Wiley & Sons., 97-104.

Melief, C. J. M. and Kast, W. M. (1995) T-cell immunotherapy of tumors by adoptive transfer of cytotoxic T lymphocytes and by vaccination with minimal essential epitopes. Immunol. Rev., *146*, 167-177.

Mo, X. Y., Sarawar, S. R., and Doherty, P. C. (1995) Induction of cytokines in mice with parainfluenza pneumonia. J. Virol., *69*, 1288-1291.

Monaco, J. J., Cho, S., and Attaya, M. (1990) Transport protein genes in the murine MHC: possible implications for antigen processing. Science, *250*, 1723-1726.

Moskophidis, D., Lechner, F., Pircher, H., and Zinkernagel, R. M. (1993) Virus persistence in acutely infected immunocompetent mice by exhaustion of antiviral cytotoxic effector T cells. Nature, 362, 758-761.

Moss, D. J., Burrows, S. R., Khanna, R., Misko, I. S., and Sculley, T. B. (1992) Immune surveillance against Epstein-Barr virus. Semin. Immunol., *4*, 97-104.

Nair, S., Babu, J. S., Dunham, R. G., Kanda, P., Burke, R. L., and Rouse, B. T. (1993) Induction of primary, antiviral cytotoxic, and proliferative responses with antigens administered via dendritic cells. J. Virol., *67*, 4062-4069.

Nair, S., Zhou, F., Reddy, R., Huang, L., and Rouse, B. T. (1992) Soluble proteins delivered to dendritic cells via pH-sensitive liposomes induce primary cytotoxic T lymphocyte responses in vitro. J. Exp. Med., *175*, 609-612.

Nash, A. A. and Sunil-Chandra, N. P. (1994) Interactions of the murine gammaherpesvirus with the immune system. Curr. Opin. Immunol., *6*, 560-563.

Oldstone, M. B., Whitton, J. L., Lewicki, H., and Tishon, A. (1988) Fine dissection of a nine amino acid glycoprotein epitope, a major determinant recognized by lymphocytic choriomeningitis virus-specific class I-restricted H-2Db cytotoxic T lymphocytes. J. Exp. Med., *168*, 559-570.

Ortmann, B., Androlewicz, M. J., and Cresswell, P. (1994) MHC class I/b2-microglobulin complexes associate with TAP transporters before peptide binding. Nature, *368*, 864-867.

Oukka, M., Riche, N., and Kosmatopoulos, K. (1992) A nonimmunodominant nucleoprotein-derived peptide is presented by influenza A virus-infected H-2^b cells. J. Immunol., *152*, 4843-4851.

Pala, P., Bodmer, H. C., Pemberton, R. M., Cerottini, J. C., Maryanski, J. L., and Askonas, B. A. (1988) Competition between unrelated peptides recognized by H-2-K^d restricted T cells. J. Immunol., *141*, 2289-2294.

Palladino, G., Mozdzanowska, K., Washko, G., and Gerhard, W. (1995) Virus-neutralizing antibodies of immunoglobulin G (IgG) but not of IgM or IgA isotypes can cure influenza virus pneumonia in SCID mice. J. Virol., *69*, 2075-2081.

Pamer, E. G., Harty, J. T., and Bevan, M. J. (1991) Precise prediction of a dominant class I MHC-restricted epitope of Listeria monocytogenes. Nature, *353*, 852-855.

Pircher, H., Moskophidis, D., Rohrer, U., Burki, K., Hengartner, H., and Zinkernagel, R. M. (1990) Viral escape by selection of cytotoxic T cell-resistant virus variants in vivo. Nature, *346*, 629-633.

Porgador, A. and Gilboa, E. (1995) Bone marrow-generated dendritic cells pulsed with a class I- restricted peptide are potent inducers of cytotoxic T lymphocytes. J. Exp. Med., *182,* 255-260.

Powers, D. C. (1993) Influenza A virus-specific cytotoxic T lymphocyte activity declines with advancing age. J. Am. Geriatr. Soc., *41,* 1-5.

Ragot, T., Finerty, S., Watkins, P. E., Perricaudet, M., and Morgan, A. J. (1993) Replication-defective recombinant adenovirus expressing the Epstein-Barr virus (EBV) envelope glycoprotein gp340/220 induces protective immunity against EBV-induced lymphomas in the cottontop tamarin. J. Gen. Virol., *74,* 501-507.

Rammensee, H. G., Friede, T., and Stevanoviic, S. (1995) MHC ligands and peptide motifs: first listing. Immunogenetics, *41*, 178-228.

Rammensee, H.-G., Falk, K., and Rotzchke, O. (1993) Peptides naturally presented by MHC class I molecules. Ann. Rev. Immunol., *11*, 213-244.

Reis e Sousa, C. and Germain, R. N. (1995) Major histocompatibility complex class I presentation of peptides derived from soluble exogenous antigen by a subset of cells engaged in phagocytosis. J. Exp. Med., *182*, 841-851

Ressing, M. E., Sette, A., Brandt, R. M., Ruppert, J., Wentworth, P. A., Hartman, M., Oseroff, C., Grey, H. M., Melief, C. J., and Kast, W. M. (1995) Human CTL epitopes encoded by human papillomavirus type 16 E6 and E7 identified through in vivo and in vitro immunogenicity studies of HLA- A*0201-binding peptides. J. Immunol., *154*, 5934-5943.

Restifo, N. P., Bacik, I., Irvine, K. R., Yewdell, J. W., McCabem, B. J., Anderson, R. W., Eisenlohr, L. C., Rosenberg, S. A., and Bennink, J. R. (1995) Antigen processing in vivo and the elicitation of primary CTL responses. J. Immunol., *154*, 4414-4422.

Rock, K. L., Fleischacker, C., and Gamble, S. (1993) Peptide-priming of cytolytic T cell immunity in vivo using beta 2- microglobulin as an adjuvant. J. Immunol., *150*, 1244-1252.

Rotzscke, O., Falk, K., Deres, K., Schild, H., Norda, M., Metzger, J., Jung, G., and Rammensee, H.-G. (1990b) Isolation and analysis of naturally processed viral peptides as recognized by cytotoxic T cells. Nature, 348, 252-254.

Rotzscke, O., Falk, K., Wallny, H.-J., Faath, S., and Rammensee, H.-G. (1990a) Characterization of naturally occurring minor histocompatibility peptides including H-4 and H-Y. Science, *249*, 283-287.

Sangster, M., Hyland, L., Sealy, R., and Coleclough, C. (1995) Distinctive kinetics of the antibody-forming cell response to Sendai virus infection of mice in different anatomical compartments. Virology, *207,* 287-291.

Sarawar, S. R. and Doherty, P. C. (1994) Concurrent production of interleukin-2, interleukin-10, and gamma interferon in the regional lymph nodes of mice with influenza pneumonia. J. Virol., *68,* 3112-3119.

Sarawar, S. R., Carding, S. R., Allan, W., McMickle, A., Fujihashi, K., Kiyono, H., McGhee, J. R., and Doherty, P. C. (1993) Cytokine profiles of bronchoalveolar lavage cells from mice with influenza pneumonia: consequences of $CD4^+$ and CD^+ T cell depletion. Reg. Immunol., *5,* 142-150.

Sastry, K. J., Bender, B. S., Bell, W., Small P. A. Jr., Arlinghaus, R. B. (1994) Effects of influenza virus-specific cytotoxic T-lymphocyte responses induced by a synthetic nucleoprotein peptide on the survival of mice challenged with a lethal dose of virus. Vaccine, *12*, 1281-1287.

Scherle, P. A., Palladino, G., and Gerhard, W. (1992) Mice can recover from pulmonary influenza virus infection in the absence of class I-restricted cytotoxic T cells. J. Immunol., *148,* 212-217.

Schon-Hegrad, M. A., Oliver, J., McMenamin, P. G., and Holt, P. G. (1991) Studies on the density, distribution, and surface phenotype of intraepithelial class II major histocompatibility complex antigen (Ia)-bearing dendritic cells (DC) in the conducting airways. J. Exp. Med., *173,* 1345-1356.

Schonbach, C., Ibe, M., Shiga, H., Takamiya, Y., Miwa, K., Nokihara, K., and Takiguchi, M. (1995) Fine tuning of peptide binding to HLA-B*3501 molecules by nonanchor residues. J. Immunol., *154*, 5951-5958.

Schulz, M., Zinkernagel, R. M., and Hengartner, H. (1991) Peptide-induced antiviral protection by cytotoxic T cells. Proc. Natl. Acad. Sci., USA, *88*, 991-993.

Schumacher, Y. N. M., de Bruijn, M. L. H., Vernie, L. N., Kast, W. M., Melief, C. J. M., Neefjes, J. J., and Ploegh, H. L. (1991) Peptide selection by MHC class I molecules. Nature, *350*, 703-706.

Selin, L. K. and Welsh, R. M. (1994) Specificity and editing by apoptosis of virus-induced cytotoxic T lymphocytes. Curr. Opin. Immunol., *6,* 553-559.

Sette, A., Vitiello, A., Reherman, B., Fowler, P., Nayersina, R., Kast, W. M., Melief, C. J., Oseroff, C., Yuan, L., Ruppert, J. et al. (1994) The relationship between class I binding affinity and immunogenicity of potential cytotoxic T cell epitopes. J. Immunol., *53*, 5586-5592.

Sijts, A. J., De Bruijn, M. L., Ressing, M. E., Nieland, J. D., Mengede, E. A., Boog, C. J., Ossendorp, F., Kast, W. M., and Melief, C. J. (1994) Identification of an H-2 Kb-presented Moloney murine leukemia virus cytotoxic T-lymphocyte epitope that displays enhanced recognition in H-2 Db mutant bm13 mice. J. Virol., *68*, 6038-6046.

Tabi, Z., Lynch, F., Ceredig, R., Allan, J. E., and Doherty, P. C. (1988) Virus-specific memory T cells are Pgp-1^+ and can be selectively activated with phorbol ester and calcium ionophore. Cell Immunol., *113,* 268-277.

Takahashi, H., Nakagawa, Y., Yokomuro, K., and Berzofsky, J. A. (1993) Induction of $CD8^+$ cytotoxic T lymphocytes by immunization with syngeneic irradiated HIV-1 envelope derived peptide-pulsed dendritic cells. Int. Immunol., *5*, 849-857.

Tew, J. G., DiLosa, R. M., Burton, G. F., Kosco, M. H., Kupp, L. I., Masuda, A., and Szakal, A. K. (1992) Germinal centers and antibody production in bone marrow. Immunol. Rev., *126,* 99-112.

Thomson, S. A., Khanna, R., Gardner, J., Burrows, S. R., Coupar, B., Moss, D. J., and Suhrbier, A. (1995) Minimal epitopes expressed in a recombinant polyepitope protein are processed and presented to $CD8^+$ cytotoxic T cells: Implications for vaccine design. Proc. Natl. Acad. Sc., USA, *92,* 5845-5849.

Townsend, A. R., Rothbard, J., Gotch, F. M., Bahadur, G., Wraith, D., and McMichael, A. J. (1986) The epitopes of influenza nucleoprotein recognized by cytotoxic T lymphocytes can be defined with short synthetic peptides. Cell, *44*, 959-968.

Townsend, A., Ohlen, C., Bastin, J., Ljunggren, H. G., Foster, L., and Karre, K. (1989) Association of class I major histocompatibility heavy and light chains induced by viral peptides. Nature, *340*, 443-448.

Tripp, R. A., Hou, S., and Doherty, P. C. (1995a) Temporal loss of the activated L-selectin-low phenotype for virus-specific $CD8^+$ memory T cells. J. Immunol., *154,* 5870-5875.

Tripp, R. A., Hou, S., McMickle, A., Houston, J., and Doherty, P. C. (1995b) Recruitment and proliferation of $CD8^+$ T cells in respiratory virus infections. J. Immunol., *154,* 6013-6021.

Tripp, R. A., Lahti, J. M., and Doherty, P. C. (1995a) Laser light suicide of proliferating virus-specific $CD8^+$ T cells in an in vivo response. J. Immunol., *155,* 3719-3721.

Van Bleek, G. M. and Nathenson, S. G. (1990) Isolation of an endogenously processed immunodominant viral peptide from the class I H-$2K^b$ molecule. Nature, *348*, 213-216.

Webb, S. R., Hutchinson, J., Hayden, K., and Sprent, J. (1994) Expansion/deletion of mature T cells exposed to endogenous superantigens *in vivo*. J. Immunol., *152,* 586-597.

Webster, R. G., Fynan, E. F., Santoro, J. C., and Robinson, H. (1994) Protection of ferrets against influenza challenge with a DNA vaccine to the haemagglutinin. Vaccine, *12,* 1495-1498.

Wettstein, P. J. and Bailey, D. W. (1982) Immunodominance in the immune response to "multiple" histocompatibility antigens. Immunogenetics, *16*, 47-58.

Whitton, J. L., Sheng, N., Oldstone, M. B., and McKee, T. A. (1993) A "string-of-beads" vaccine, comprising linked minigenes, confers protection from lethal-dose virus challenge. J. Virol., *67*, 348-352.

Wipke, B. T., Jameson, S. C., Bevan, M. J., and Pamer, E. G. (1993) Variable binding affinities of listeriolysin O peptides for the H-$2K^d$ class I molecule. Eur. J. Immunol., *23*, 2005-2010.

Wolpert, E., Franksson, L., and Karre, K. (1995) Dominant and cryptic antigens in the MHC class I restricted T cell response across a complex minor histocompatibility barrier: analysis and mapping by elution of cellular peptides. Intl. Immunol., *7*, 919-928.

Yang, H. Y., Dundon, P. L., Nahill, S. R., and Welsh, R. M. (1989) Virus-induced polyclonal cytotoxic T lymphocyte stimulation. J. Immunol., *142,* 1710-1718.

Yewdell, J. W. and Bennink, J. R. (1992) Cell biology of antigen processing and presentation to major histocompatibility complex class I molecule-restricted T lymphocytes. Adv. Immunol., *52*, 1-123.

Yin, L., Poirier, G., Neth, O., Hsuan, J. J., Totty, N. F., and Stauss, H. J. (1993) Few peptides dominate cytotoxic T lymphocyte responses to single and multiple minor histocompatibility antigens. Intl. Immunol., *5*, 1003-1009.

Young, A. C. M., Zhang, W., Sacchettini, J. C., and Nathenson, S. G. (1994) The three-dimensional structure of H-$2D^b$ at 2.4A resolution: implications for antigen-determinant selection. Cell, *76*, 39-50.

Zarozinski, C. C., Fynan, E. F., Selin, L. K., Robinson, H. L., and Welsh, R. M. (1995) Protective CTL-dependent immunity and enhanced immunopathology in mice immunized by particle bombardment with DNA encoding an internal virion protein. J. Immunol., *154,* 4010-4017.

Zhang, W., Young, A. C. M., Imarai, M., Nathenson, S. G., and Sacchettini, J. C. (1992) Crystal structure of the major histocompatibility complex class I H-$2K^b$ molecule containing a single viral peptide: Implications for peptide binding and T-cell receptor recognition. Proc. Natl. Acad. Sci. USA, *89*, 8403-8407.

Zinkernagel, R. M., Althage, A., Cooper, S., Kreeb, G., Klein, P. A., Sefton, B., Flaherty, L., Stimpfling, J., Shreffler, D., and Klein, J. (1978) Ir-genes in H-2 regulate generation of anti-viral cytotoxic T cells. Mapping to K or D and dominance of unresponsiveness. J. Exp. Med., *148*, 592-606.

2.3 Broadly Reactive HLA Restricted T Cell Epitopes and Their Implications for Vaccine Design

John Sidney, Ralph T. Kubo, Peggy A. Wentworth, Jeff Alexander, Robert W. Chesnut, Howard M. Grey, and Alessandro Sette

2.3.1 General Introduction

HLA-class I and class II molecules bind antigenic peptides and present them for recognition by T lymphocytes (see Townsend and Bodmer, 1989; Germain and Margulies, 1993; Engelhard, 1994 for reviews). The formation of a trimolecular complex between antigenic peptides, class I or class II MHC molecules, and the antigen-specific TCR of CD4+ or CD8+ T cells represents one of the most crucial events in the generation of immune responses to viral infections and tumors. Both the induction and functionality of helper and cytotoxic T cells are dependent on this event. In fact, the normal functions of the immune response, such as, induction of DTH responses against parasites, CTL responses against tumors and viruses, as well as induction of specific antibodies against bacteria, are all intimately tied to the formation of peptide MHC complexes and their recognition by T cells.

A remarkable expansion of our understanding of the molecular mechanisms involved in peptide-MHC class I and class II interactions has occured in the last few years. Crystallographic structures of several MHC molecules and MHC-peptide complexes have been solved (Madden, 1995). The sequencing of pools of naturally processed and endogenously bound peptides isolated from both class I and class II molecules have uncovered peptide sequence motifs recognized by various HLA molecules (see Rammensee et al., 1995, for review).

This increased understanding has generated considerable interest in the potential application of antigen-specific cytotoxic T cells to prophylatic and therapeutic treatments for chronic viral diseases and cancer (Boon et al., 1994; Chesnut et al., 1995; Hill et al., 1992; Vitiello et al., 1995). Strategies to rapidly identify antigenic peptides based on the combined use of allele specific motifs, quantitative peptide-HLA class

I binding assays, and biological assays have been developed. Peptide-based vaccines offer several advantages compared to live or recombinant vaccines. For example, it should be possible to specifically direct the immune system towards specific epitopes, focusing on regions that are highly conserved among different strains. One would also be able to direct the immune response towards subdominant epitopes in cases where tolerance might exist to immunodominant epitopes. Or, a response might be elicited against MHC/peptide complexes that are not present on the cell surface in amounts sufficient to induce an immune response, but may be adequate for the target recognition phase (De Bruijin et al., 1991).

The focus of this chapter will be to describe concepts in the binding of peptides to class I molecules. We will also discuss recent data on the generation of broadly reactive class II epitopes. Thoughout this review we will particularly emphasize the implications of broadly reactive class I and class II epitopes for vaccine design.

2.3.2 A General Strategy for Vaccine Design

a. Motif Definition

Perhaps the most formidable obstacle to the development of broadly efficacious peptide-based immunotherapeutics is the extreme polymorphism of HLA class I molecules. More than 100 alleles and isotypes are already known. Because MHC polymorphism is concentrated in the peptide binding cleft (Bjorkman and Parham, 1990; Parham et al., 1995), it has been generally believed that this high degree of polymorphism allowed class I molecules to bind unique repertoires of peptide ligands. Thus, effective coverage of the general human population by a given vaccine would be a task of considerable complexity.

One initial approach to the problem has been to characterize the binding specificity of the most common HLA-A alleles (Kubo et al., 1994). For example, HLA-A2 is present with a minimum phenotypic frequency of 30 % in every major ethnic group. Detailed understanding of the mechanism of peptide binding to A2 would thus be of considerable benefit for the development of peptide-based vaccines. To increase the potential population coverage achieved by a particular vaccine, a "cocktail" containing multiple epitopes could be designed. For example, a "cocktail" of five epitopes, each specific for one of the five most common HLA-A alleles (A1, A2.1, A3, A11, and A24) would allow for coverage of 80 to 90 % of the general population, at least of caucasian and oriental descent.

In this light, Kubo et al. determined the peptide binding motifs for A1, A2.1, A3, A11, and A24. The allele-specific motifs were characterized by three complementary approaches: 1) direct amino acid sequencing by automated Edman degradation of the mixture of naturally processed peptides eluted from affinity-purified class I mixtures, 2) sequencing of individual naturally processed peptides by tandem mass spectrometry, and 3) the analysis of binding of polyalanine analog peptides. This three-pronged approach allowed a broad characterization of the peptide motifs recognized by each molecule, and also enabled the identification of individual naturally processed peptides.

b. Motif Validation

The reliability of peptide motifs to predict CTL epitopes was addressed by Kast et al. (Kast et al., 1994). In that study the binding capacity of an overlapping set of nonamer peptides from the human papilloma virus type 16 E6/E7 protein was determined. The expanded motifs derived by Kubo et al. were present in 73 % of the high affinity binders, compared to just 27 % for the pool sequencing derived motifs.

On the other hand, the majority of motif containing peptides failed to bind to the relevant class I molecule. This result confirmed observations made earlier for A*0201 (Ruppert et al., 1993) that the presence of motif-specific anchor residues was a necessary, but not sufficient, condition to determine binding to class I molecules. It is also apparent that class I-peptide binding is not an all-or-none phenomenon. For example, when several hundred different peptides, all containing primary anchor residues conforming to the consensus A*0201 binding motifs were analyzed for their HLA-A*0201 binding affinity, it was found that A*0201 binding affinity varied, in an apparent continuum, over at least a 10,000-fold range (Ruppert et al., 1993). This effect was attributed to the important roles that secondary anchor residues play in peptide-MHC interactions. The contributions of secondary anchor interactions to peptide binding have been now confirmed for a number of HLA-A alleles besides A*0201 (see below and Sidney et al., 1996).

c. Immunogenicity and Binding

The data described above demonstrate how detailed allele-specific motifs aid in the identification of peptides which can bind to any given HLA molecule. The relationship between peptide binding affinity for HLA class I molecules and immunogenicity of discrete peptide epitopes has also been analyzed. Sette et al. (1994) utilized two different experimental approaches to investigate whether a discrete affinity threshold might be associated with immunogenicity in the class I system. In the first approach, the immunogenicity of potential epitopes ranging in MHC binding affinity over a 10,000-fold range was analyzed in HLA-A*0201 transgenic mice. In the second approach, the antigenicity of approximately 100 different hepatitis B virus (HBV) derived potential epitopes, all carrying A*0201 binding motifs, was assessed by measuring recall responses using peripheral blood lymphocytes (PBL) from patients with acute hepatitis B infection. The results obtained indicate that only peptides with a relatively high binding affinity for MHC are immunogenic. In both cases, it was found that the majority (60-90 %) of immunogenic peptides corresponded to those with an A*0201 binding affinity of 50 nM or less. All of the remaining immunogenic peptides were found to have binding A*0201 affinities in the 50 to 500 nM range. Thus, an affinity threshold of approximately 500 nM (preferably 50 nM or less) apparently determines the capacity of a peptide epitope to elicit an A*0201-restricted CTL response. These data correlate well with class I binding affinity measurements of either naturally processed peptides or previously described T cell epitopes, and emphasize how MHC binding affinity has a dramatic impact on immunogenicity.

Taken together, the data discussed above illustrate an efficient method to select peptides with high HLA class I binding affinity, and which, therefore, have a high likelihood of being immunogenic for CTL responses.

d. Cellular Assays

The study of Sette et al. also showed that for a peptide to be immunogenic, binding to HLA class I with high (ICSO $\leq$ 50 nm) or intermediate (ICSO 50-500 nm) affinity is a necessary, but not sufficient criteria. Thus, in addition to motif identification and MHC binding analysis, it is necessary to demonstrate that a candidate peptide is capable of eliciting a CTL response in the context of the appropriate class I molecule.

Rapid screening of potential epitopes for their in vitro immunogenicity for human T cells can be accomplished using the primary induction protocol described by Wentworth et al. (Wentworth et al., 1995). This protocol, which employs activated, peptide-loaded human peripheral blood mononuclear cells (PBMCs) as the antigen presenting cells (APC), results in effective and reproducible in vitro induction of primary antigen-specific CTL in humans. In brief, SAC-I activated PBMCs are acid stripped of endogenous peptides, resulting in the transient expression of empty class I molecules. These empty class I molecules can subsequently be stabilized with exogenous peptide and ß2-microglobulin. Exogenously loaded PBMCs are then used to stimulate cultures enriched for CD8+ T cells in the presence of recombinant Interleukin-7. After 12 days cells are restimulated with autologous, peptide-pulsed, adherent cells and tested for CTL activity seven days later. Specific positive responses are defined as lysis of peptide-sensitized target cells at least 15 % above lysis of unsensitized target cells at both the 90:1 and 30:1 E:T ratios. This protocol is generally applicable to different antigens and class I alleles.

e. Lipopeptide Constructs

Most CTL peptide epitopes are poor immunogens. However, if they are specifically modified by attaching a T helper peptide epitope and two lipid molecules, they have been shown to elicit powerful and specific CTL responses in mice (Chesnut et al., 1995) and in phase I clinical trials conducted in healthy volunteers (Vitiello et al., 1995).

Using the murine influenza virus CTL epitope NP 147-155 as a model system, Vitiello et al. found such a lipidated construct to be highly immunogenic. A single injection of the construct resulted in memory CTL induction for over one year. Based on these animal studies, an HBV vaccine for use in human subjects was designed. This vaccine included three components: 1) the HBV core antigen peptide 18-27 as the CTL epitope, 2) the class II promiscuous tetanus toxoid peptide 830-843 as the T helper peptide, and 3) two palmitic acid molecules as the lipids. A dose escalation trial (5, 50, 500 μg) carried out in 26 normal subjects showed that the vaccine was safe and able to induce a primary HBV-specific CTL response. A dose response curve was observed and five out of five subjects responded to the 500 μg dose.

2.3.3 The Concept of Class I Supertypes

a. Introduction

Developing an immunotherapeutic composed of a large number of epitopes specific for each of the common HLA alleles, is a potentially complex and expensive task. However, it may be possible to invert this logical framework, and instead attempt the identification of epitopes capable of binding multiple HLA types.

Data generated in the course of the last few years have suggested that HLA motifs, and resultant peptide repertoires, recognized by class I molecules are neither as narrowly specific, or unique as originally believed. For example, in the analysis of peptide binding to class I molecules encoded by the most frequent HLA-A (Kubo et al., 1994; Kast et al., 1994; del Guercio et al., 1995) and B alleles (Sidney et al., 1995; Huczko et al., 1993) it was shown that class I molecules could tolerate a wider range of residues at their anchor positions than identified by pool sequencing analyses. It has also been demonstrated that many alleles share somewhat overlapping peptide binding motifs. For example, Hill and co-workers have shown that HLA-B*3501 and B*5301 both prefer proline in position 2 of their peptide ligands (Hill et al., 1992). Other studies have indicated that several different HLA types, such as A*0301, A*1101 and A*6801, are associated with specificity for peptides carrying small or hydrophobic residues in position 2 and positively charged residues at the C-terminus (see Rammensee et al., 1995 for review). Further experiments demonstrated that A*0301 and A*1101 did indeed share overlapping peptide specificities. In fact, testing of a large panel of motif-carrying peptides revealed that while certain peptides bound (specifically) only A*0301 or A*1101, other peptides were capable of binding both alleles with high affinities (Kast et al., 1994).

On the basis of these results, it appears that HLA Class I molecules that share somewhat similar peptide binding motifs can be grouped into different HLA-supertypes, defined by a broad peptide binding motif recognized by all members of the supertype. We have recently defined four HLA-supertypes: the A2-like (del Guercio et al., 1995), B7-like (Sidney et al., 1995), A3-like (Sidney et al., submitted), and the B44-like (Sette and Sidney, unpublished observations).

b. The A2-Like Supertype

Del Guercio et al. have examined the degree of cross-reactivity of the A*0201-restricted hepatitis B virus core 18-27 (HBVc 18-27) peptide with other A2 subtypes (del Guercio et al., 1995). With direct MHC binding assays utilizing radiolabeled peptides and HLA class I-expressing mammalian cells, it was determined that this peptide epitope also binds A*0202, A*0205, and A*0206, but not A*0207. Subsequent experiments have shown that HBVc 18-27 also binds A*0203 (M.-F. del Guercio and A. Sette, unpublished observations). Similar observations have been reported by Tosi and co-workers (Tanigaki et al., 1994), who noted broad cross-binding of peptides between A*0201, A*0204, A*0205, and A*0206.

According to X-ray crystallographic data (Saper et al., 1991; Madden et al., 1993), the side chain of the residue at position 2 and the C-terminus of bound peptides normally occupy, respectively, the B and F pockets in the binding groove of HLA class I molecules. Analysis of the polymorphic residues that form the B and F pockets of various HLA alleles allowed prediction of binding of the hepatitis B virus core 18-27 epitope to two other HLA alleles (HLA-A*6802 and A*6901).

Inhibition experiments with panels of synthetic peptide analogues underlined the similar ligand specificities of the HLA-A*0201, A*0202, and A*0205 alleles (del Guercio et al., 1995). Subsequent analyses using synthetic peptides representing naturally occurring sequences extended these observations to A*0203, A*0206, and A*6802 (Sette et al., unpublished observations). The peptide binding motifs recognized by these molecules are characterized by a preference for peptide ligands with small or aliphatic amino acids (L, I, V, M, A, and T) at position 2 and their C-terminus.

Thus, it appears that at least seven different HLA-A molecules (A*0202, A*0203, A*0204, A*0205, A*0206, A*6802, and A*6901) share overlapping ligand specificities with A*0201. These alleles have been designated as the A2-like supertype. As a group, they are represented in very high frequency in the major ethnic populations.

c. The A3-Like Supertype

The HLA-A3-like supertype includes five common HLA-A alleles: A3, A11, A31, A*3301 and A*6801, and is defined by a shared preference for peptides carrying A, L, I, V, M, S, or T at position 2 and positively charged residues, such as R and K, at their C-termini. Based on the analysis of the binding capacity of large collections of motif-carrying peptides derived from important viruses and tumor antigens, an A3-like supermotif was also derived. The A3-like supermotif is defined by the primary and secondary anchor residues which are important for selecting peptides capable of cross-reacting amongst most or all molecules of the A3-like supertype.

The structural basis of these striking similarities in ligand specificities are apparent when the amino acid polymorphisms which make up the B and F pockets of the HLA class I binding grove are analyzed. For example, analysis of the B pocket revealed a consensus MHC motif which was shared between A*0301 and A*1101. This consensus motif was also shared among other HLA alleles which bind peptides carrying somewhat hydrophobic residues in position 2, but not by HLA alleles with different binding specificities (tab. 2.3.1). A similar consensus was also defined in the case of the F pocket of HLA and the corresponding C-terminal anchor residue of bound peptides (Sidney et al., 1996).

Interestingly, several other alleles such as A*3101, A*3301, A*3401, A*6601, A*6801, and A*7401 also share these consensus B and F pocket motifs. On this basis, this set of alleles (A3, A11, A31, A*3301, A*3401, A*6601, A*6801, and A*7401)

Table 2.3.1: B pocket analysis of various HLA-A and B alleles

Allele	B pocket binding motif[a]	B pocket residue[b]				
		45	66	67	70	99
A*0101	TSM	M	N	M	H	Y
A*0201	LM	M	K	V	H	Y
A*0301	LMV	M	N	V	Q	Y
A*1101	VT	M	N	V	Q	Y
A*6801	VT	M	N	V	Q	Y
A*6802	LM	M	N	V	Q	Y
Consensus	LMVTS(IA)	M	N/K	M/V	Q/H	Y
A*2401	YF	M	K	V	H	F
B*2705	RK	E	I	C	K	Y
B*3701	DE	T	I	S	N	S
B*0701	P	E	I	Y	Q	Y
B*3501	P	T	I	F	N	Y
B*5301	P	T	I	F	N	Y
B*5401	P	G	I	Y	Q	Y
B*0801	—	E	I	F	N	Y

a Residues listed represent those preferred in the peptide binding motif. See Rammensee et. al., 1995 for a review and listing of motifs. Other residues may be tolerated, though typically with lower affinity (Kubo et. al., 1994).

b Residues of class I molecule hypothesized to form and influence peptide binding to the B pocket; derived from Saper et. al., 1991, and Madden et. al., 1993.

has been designated as the A3-like supertype. Because of the importance of the B and F pockets in peptide binding (Madden, 1995), it was predicted that these HLA molecules would recognize overlapping peptide motifs. Experimental evidence from our own (Sidney et al., 1996) and others' (see Rammensee et al., 1995 for review) laboratories has so far demonstrated that this prediction is indeed correct in the case of HLA A3, A11, A31, A*3301, and A*6801.

Figure 2.3.1 indicates the frequencies, in five major ethnic groups, of the alleles confirmed to be part of the A3-like supertype. It is striking that while the frequency of any particular allele can vary widely amongst ethnic groups, the overall frequency of HLA-A3-like alleles is remarkably conserved.

d. The B7-Like Supertype

The B7-like supertype is characterized by molecules that recognize peptides bearing proline in position 2 and hydrophobic or aliphatic amino acids (L, I, V, M, A, F, W, and Y) at their C-terminus (Sidney et al., 1995). Compiling data from a num-

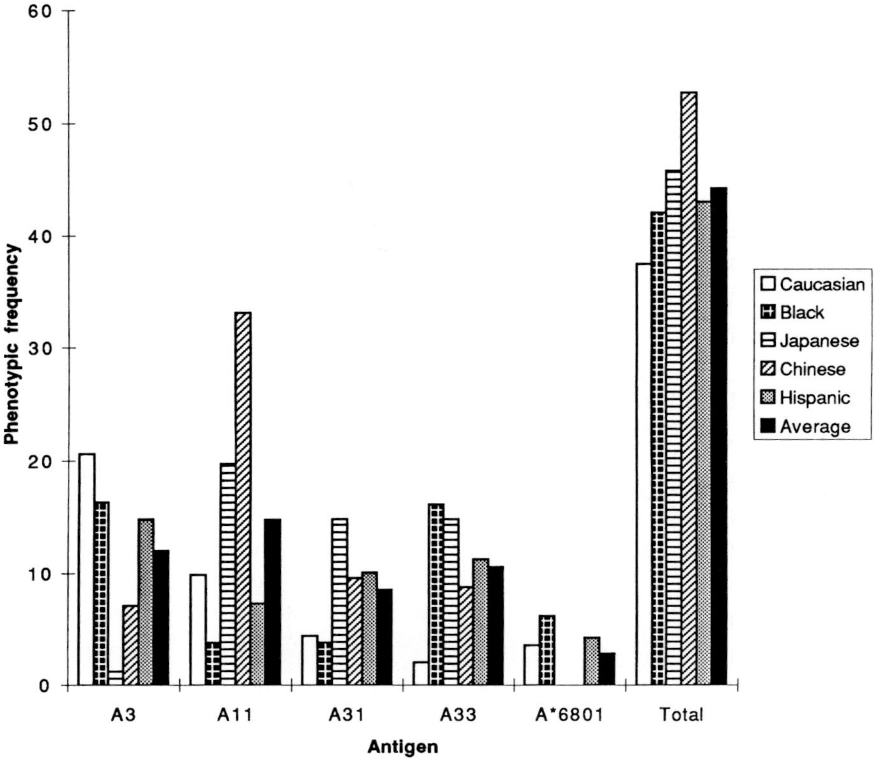

Fig. 2.3.1: Phenotypic frequencies of the A3-like supertype antigens and alleles. Phenotypic frequencies are compiled from Imanishi et al.1992a and Fernandez-Viña et al., 1992. Total coverage was calculated assuming Hardy-Weinberg equilibrium, and considers only those antigens or alleles confirmed to share the supertype binding preference. As more peptide binding motifs become available, it is conceivable that the total coverage confired by a particular supertype will increase. Phenotypic data to date does not have resolution at the level of alleles as defined by DNA sequences, and, therefore, do not distinguish between subtypes. A one to one correspondence between the serologically defined antigen and subtype allele was assumed, however, when comparison of the peptide binding specificities of subtypes, as determined by either binding data, published motifs, or sequence analysis, suggested that subtypes have overlapping peptide binding specificities.

ber of sources suggest that the B7-like supertype is comprised of products from at least a dozen HLA-B alleles: B7, B*3501, B*3502, B*3503, B51, B*5301, B*5401, B*5501, B*5502, B*5601, B*6701, and B*7801 (Sidney et al., 1995; Barber et al., 1995; Rammensee et al., 1995).

As in the cases of the A2- and A3-like supertypes, analysis of the sequences of various HLA class I alleles suggested that the B7-like supertype molecules also shared key consensus residues in their B and F pockets (Sidney et al., 1995). These observations were verified by prediction, on the basis of this consensus B and F pocket structure, of the peptide-binding specificity of HLA-B*5401. Quantitative binding assays demonstrated that, indeed, many (25 %) of the peptide ligands carrying proline in position 2 and hydrophobic/aromatic residues at the C-terminus (the B7-like supermotif) were capable of binding at least three of five HLA-B7-like supertype molecules tested (Sidney et al., 1995). Parham and co-workers have also reported significant overlap in the repertoires of peptides bound by a group of HLA-B molecules which recognize the B7-like motif (Barber et al., 1995). Similar to the A3-like supertype, the B7-like supertype is expressed in a large portion (>40 %) (fig. 2.3.2) of all major ethic groups.

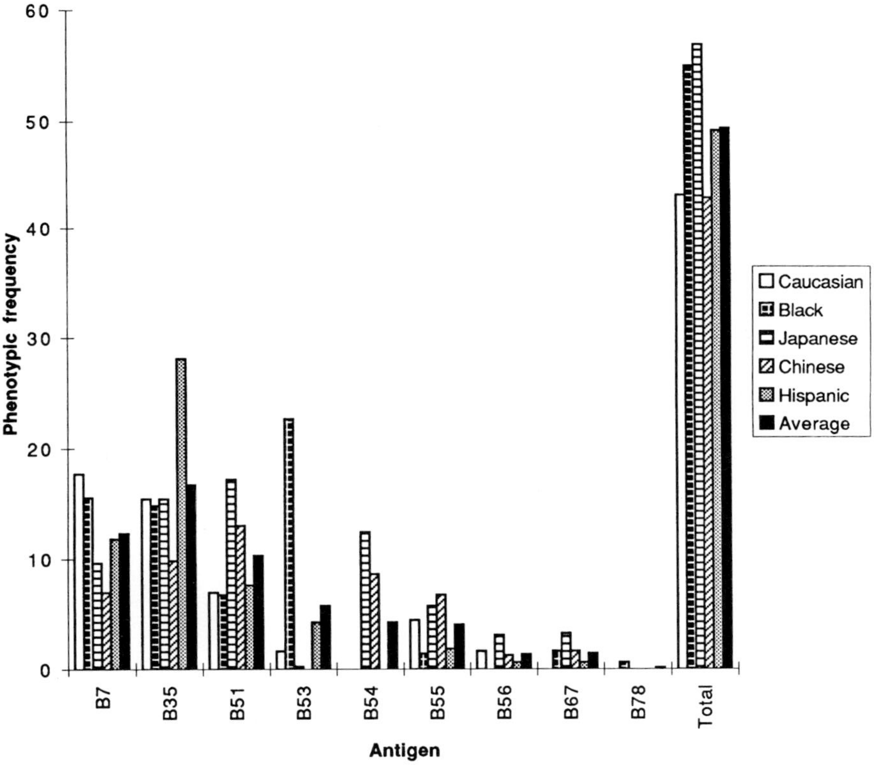

Fig. 2.3.2: Phenotypic frequencies of the B7-like supertype alleles. Phenotypic frequencies were determined as described in fig. 2.3.1.

e. The B44-Like Supertype

A review of the literature indicates that a significant number of HLA-B molecules share preferences with HLA-B44 for peptides bearing acidic residues (D, E) in position 2 and hydrophobic or aromatic residues at their C-terminus (see Rammensee et al., 1995 for a listing of motifs). The structural reasons for these shared preferences can again be inferred from sequence analysis of the B and F pockets of various HLA molecules. On this basis, we have proposed the existence of the B44-like supertype. Examining the population representation of the alleles which are believed to comprise this group suggests that the B44-like supertype, similar to the other supertypes, is very common (fig. 2.3.3). The degree of overlap of peptide binding repertoires within this supertype will be the subject of future experiments.

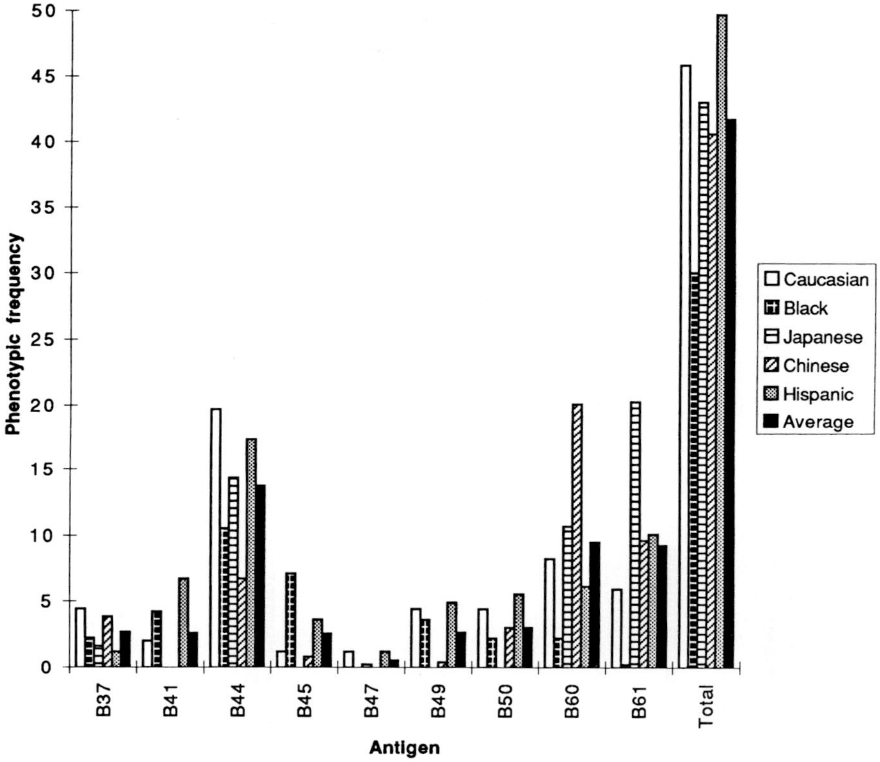

Fig. 2.3.3: Phenotypic frequencies of the B44-like supertype alleles. Phenotypic frequencies were determined as described in fig. 2.3.1.

f. Validation of Supertypes: Supermotifs

To demonstrate that supertypes represent a powerful tool for developing peptide based vaccines, it is necessary to show that supertype molecules share significantly overlapping peptide binding repertoires. The degree to which the peptide binding repertoires of A3-like supertype molecules overlapped was examined using a library of 200 different synthetic peptides corresponding to naturally occurring sequences from either tumor or viral antigens. When this library was screened for binding to the five most common A3-like alleles (Sidney et al., submitted), highly crossreactive peptides could readily be identified, with a relatively large fraction (10 %) of the peptides binding to 80 % to 100 % of the five A3-like alleles tested.

An analysis of the structural requirements for high affinity binding was performed to generate detailed secondary anchor maps for each molecule. These data were then used to derive an extended A3-like supermotif which describes secondary anchor requirements shared within the A3-like supertype. For example, it was found that the presence of aromatic residues at positions 3, 6, or 7 of peptide ligands was associated with highly crossreactive binding capacities. Other residues, such as D and E in position one, were associated with poor binding capacity, or poor crossreactivity (fig. 2.3.4). The a priori prediction of the crossreactivity of an independent set of peptides verified the predictive power of the A3-like supermotif. The biological relevance of these observations was underlined by the fact that peptides capable of binding to several A3-like alleles were also capable of eliciting CTL responses restricted by these different A3-like alleles (Sidney et al., 1996).

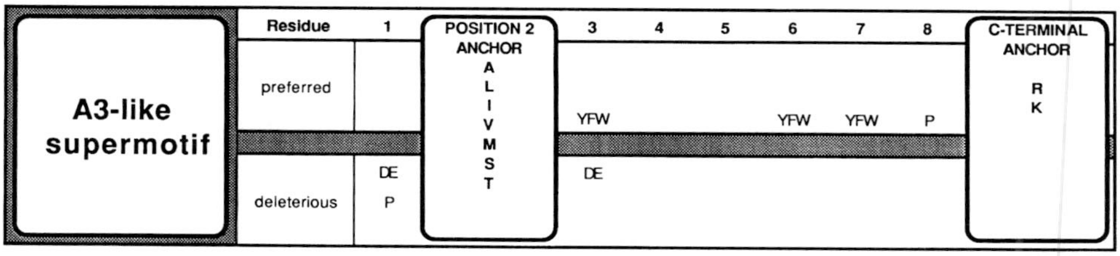

Fig. 2.3.4: The A3-like supermotif. The A3-like supermotif describes features of peptide ligands associated with a high degree of degeneracy amongst A3-like supertype alleles. The occurrence of certain residues at specific non-anchor positions has been demonstrated to be associated with either degeneracy (supermotif preferred residues) or poor binding in general (supermotif deleterious residues). The A3-like supermotif was derived from the analysis of the binding of a set of 200 peptide ligands representing naturally occurring sequences and each bearing the A3-like supertype primary anchor motif (A, I,V, L,M, S, or T at position 2, and R or K at the C-terminus) to A3, A11, A31, A*3301, and A*6801.

g. Evolution of Class I MHC Supertypes and Supermotifs

The potential relevance of class I supertypes at the level of the general human population becomes most evident when the incidence of the various supertype alleles in different ethnic backgrounds is examined. While it is apparent that the frequency of each individual allele varies drastically between ethnic groups (Imanishi et al., 1992a) (see figs. 2.3.1-2.3.3 and tab. 2.3.2), the cumulative frequency of the supertype is remarkably constant (in the 40 to 60 % range). For example, A3 is common in Caucasians, North American Blacks, and Hispanics, but almost absent in Japanese. Conversely, A31 is frequent in Japanese but rare in Caucasians and North American Blacks. By contrast, in each of the five populations examined, the A3-like supertype was present in at least 37 % of the individuals.

Calculation of genotype frequencies from overall phenotypic frequencies indicates that at least 50 % to 60 % of all HLA-A and -B genes in existence today are members

Table 2.3.2: Summary of four known HLA-A and B supertypes

Supertype	Supermotif B pocket	Supermotif F pocket	Average Allelic Frequency
A2-like	AILMVT	AILMVT	43.2
A3-like	AILMVST	RK	44.2
B7-like	P	AILMVFWY	49.5
B44-like	DE	AILMVFWY	41.9
Total			94.2

of one of the four HLA supertypes described above. The remarkable conservation of the overall frequency of each supertype across widely different ethnicities, together with their very high overall frequencies, may be reflective of important biological phenomena related to HLA supertypes.

In this regard, both common ancestry and convergent evolution need to be considered. For example, examination of the frequency of genetic variations among HLA-A and -B alleles (Imanishi et al., 1992b; Kato et al., 1989) suggests that all members of the A2-like and B44-like supertypes are phylogenetically closely related. Conversely, representatives of the A3-like and B7-like supertypes can be found in most major HLA-A or B lineages. Indeed, members of the A3-like supertype are found in four of the five major HLA-A evolutionary branches (Imanishi et al., 1992b; Ishikawa et al., 1994; Kato et al., 1989), a distribution compatible with a convergent evolution explanation. Yet, the alternative explanation of common ancestry is also still possible. This would raise the intriguing possibility that the A3-like and B7-like supertype specificities represent primeval A and B specificities present before polymorphism began to shape the structure and function of MHC molecules.

Another hypothesis is that the high phenotypic frequency of HLA supertypes may be related to the generation of an efficient peptide binding repertoire by focusing on residue types which are either frequent (such as hydrophobic or basic residues), highly resistant to proteolysis (such as proline), or available due to the specificities of the peptide processing and transport machineries. Along this line, it is interesting to consider the peptide specificity of human TAP molecules. TAP molecules preferentially transport peptides with certain sequence features such as hydrophobic, aromatic or positively charged C-termini (see Howard, 1995 for review). An extended TAP binding motif has been described by van Endert and associates, in collaboration with our own group, by evaluating the relative affinities for TAP of a large collection of peptides. Strikingly, this motif contains many of the features associated with the A3-like supermotif, such as preference for aromatic residues at positions 3 and 7 of 9-mer peptides, and absence of negatively charged residues at positions 1 and 3, and P at position 1 (van Endert et al., 1995).

These findings suggest that Class I and the peptide processing and transport machineries may have coevolved in order to optimize effective peptide presentation. According to this view, transported peptides are both the product of the "transporter enzymes" (TAP molecules), and the substrate of the "peptide binding enzymes" (HLA-A and -B molecules). It makes good biological sense that enzymes participating in the same chain of reactions, and expressed on contiguous locations of the same chromosome, also share coordinate substrate specificities.

h. Summary

In both practical and theoretical, terms, the discovery of HLA supertypes and supermotifs could have important consequences. Because it might allow broad, effective, and non-ethnically biased population coverage with only a handful of peptide epitopes, the capacity to identify broadly crossreactive peptides represents, in practical terms, a major advance towards the development of CTL peptide-based immunotherapeutics. Also, as hypothesized by Itescu et al. in a recent manuscript (Itescu et al., 1995), grouping HLA specificities on the basis of shared structure and peptide binding motifs may give insight into disease susceptibility and resistance associations.

In theoretical terms, HLA supertypes and supermotifs demonstrate that MHC polymorphism may be, in functional terms, more limited than generally thought. While it remains unknown how many supertypes will be identified, and how inclusive they will be, the available data demonstrate that the phenomenon of degeneracy of peptide binding specificities, previously thought to be restricted to HLA-class II (Panina-Bordignon et al., 1989; O'Sullivan et al., 1990; Busch et al., 1990), is also a feature of peptide binding to HLA-class I. These results also underline the point that the pressure to maximize MHC polymorphism may be balanced by the necessity for HLA molecules to be compatible with a highly complex and coordinate series of events. Analysis of HLA supertypes might, as described above, provide useful insights into the actual functional specificity of primeval MHC types. Finally, in terms of classification and nomenclature, the time may be near when it will be possible, and even desirable, to reclassify MHC polymorphisms not on the basis of serological reactivity, sequence, or evolutionary relationships, but on the basis of their biological function represented by their specific peptide binding motifs.

2.3.4 Pan DR Class II Epitopes (PADRE)

As is the case with class I binding peptides, there is a good correlation between class II binding affinity and immunogenicity of class II restricted epitopes (Schaeffer et al., 1989). The determination of class II allele specific peptide binding motifs has been the subject of a number of studies in recent years (see Sinigaglia and Hammer, 1994; Rammensee et al., 1995 for reviews). It is clear from these studies that there is a significant degree of overlap in the peptide motifs recognized by class II molecules. Indeed, a number of epitopes that are recognized in the context of multiple DR types are known from the literature. The possibility that peptide epitopes could be identified that are recognized in the context of multiple, or even all, DR alleles has profound implications for the use of peptide antigens in the design of subunit vaccines.

The development of high potency, universal, DR-restricted helper epitopes by the modification of high affinity DR-blocking peptides has recently been demonstrated (Alexander et al., 1994). When the degeneracy of previously reported class II epitopes was examined with respect to their DR binding capacity it was found that although they indeed have the capacity to bind to several DR alleles, they are by no means universal DR binders. By introducing anchor residues for different DR motifs within a poly-alanine backbone peptide, these investigators were able to engineer peptides with broader DR-binding capacity than known degenerate epitopes. These pan DR-binding peptides bound 10 of 10 DR molecules tested with affinities, in most cases, in the nanomolar range. They also found that these pan DR-binding peptides are effective blockers of DR-restricted antigen presentation.

Because of the limited intermolecular binding energy of the small methyl side chain of alanine that would be involved in T cell recognition, these degenerate binders were poor immunogens. However, by introducing bulky and/or charged residues at positions accessible for T cell recognition extremely powerful pan DR epitope peptides (PADRE) were obtained. These peptides elicited powerful responses in vitro from human PBMC. In one example of their capacity to elicit T help, PADRE were approximately 1000 times more powerful than natural T cell epitopes. In mouse studies, these peptides were also shown to be active immunogens in vivo.

Peptides such as PADRE have potential use as inducers of T help for a variety of vaccine applications. The need for added helper epitopes is suggested by the fact that many peptide-based vaccines are not universally efficacious (Vitiello et al., 1995). It is also possible that such highly immunogenic epitopes might be superior to the currently used carrier protein systems for polysaccharide antigens. A more speculative application of synthetic T cell epitopes might be in the regulation of cytokine production and the type of T cell response generated by such peptides. There is growing evidence that the fine antigenic structure of peptide epitopes may be important in differentially inducing the expression of Th1- versus Th2-like cytokines (Kumar et al., 1995; Pfeiffer et al., 1995). If this can be substantiated and reduced to practice, it might be possible in the future to specifically induce a T cell response that can optimally stimulate a humoral immune response or a cell-mediated immune response, depending upon the therapeutic application intended.

2.3.5 Concluding Remarks

The demonstration that cytotoxic T cells can confer protection against infectious diseases and cancer, opens up the possibility of manipulating the immune response through the use of antigenic peptides. Peptides are an especially attractive component of therapeutics and vaccines in that they are safe and relatively cheap to manufacture.

They also offer a large degree of flexibility in the engineering of immunotherapeutics. For example, in the treatment of tumors or chronic viral infection the use of subdominant peptides may allow the shifting of the immune response to epitopes that have not induced a state of tolerance. Similarly, they offer the ability to target immune responses to epitopes that are conserved in a targeted polymorphic pathogen (e.g., HIV or HCV).

The dependence of the immune response on the highly polymorphic class I and class II molecules encoded by the MHC presents a potential stumbling block in the the development of efficacious peptide-based vaccines and immunotherapeutics. However, the data presented herein suggests a solution to the problem. The polymorphism of MHC molecules may be much more limited than previously believed when considered at the level of functionality. It is clearly apparent that many MHC molecules share overlapping peptide binding specificities. Similarly, peptides which are capable of binding to multiple class I and class II molecules have been identified from natural sources, and high affinity peptides with broad specificity can also be easily engineered. Most importantly, many of these broadly reactive peptides are potent immunogens in the context of multiple MHC molecules.

Although in its infancy, the field of immunotherapeutics based on CTL-inducing epitopes is rapidly maturing to the point at which its real potential in therapy and prophylaxis against human diseases will be determined.

Acknowledgements

We would like to thank Glenna Marshall for assistance in manuscript preparation. This work was supported in part with Federal funds from the National Institute of Allergy and Infectious Diseases, National Institutes of Health, under contract NO1-AI-45241.

References

Alexander, J., Sidney, J., Southwood, S., Ruppert, J., Oseroff, C., Maewal, A., Snoke K., Serra, H. M., Kubo, R. T., Sette, A., and Grey, H. M. (1994) Development of high potency universal DR-restricted helper epitopes by modification of high affinity DR-blocking peptides. Immunity *1*, 751-761.

Barber, L. D., Gillece-Castro, B., Percival, L., Li, X., Clayberger, C., and Parham, P. (1995) Overlap in the repertoires of peptides bound in vivo by a group of related class I HLA-B allotypes. Curr. Biol. *5*, 179-190.

Bjorkman, P. J. and Parham, P. (1990) Structure, function, and diversity of class I major histocompatibility complex molecules. Annu. Rev. Biochem. *59*, 253-288.

Boon, T., Cerottini, J.-C., Van den Eynde, B., van der Bruggen, P., and Van Pel, A. (1994) Tumor antigens recognized by T lymphocytes. Annu. Rev. Immunol. *12*, 337-366.

Busch, R., Strang, G., Howland, K., and Rothbard. J. B. (1990) Degenerate binding of immunogenic peptides to HLA-DR proteins on B cell surfaces. Int. Immunol. 2, 443-451.

Chesnut, R. W., Sette, A., Celis, E., Wentworth, P., Kubo, R. T., Alexander, J., Ishioka, G., Vitiello, A., and Grey, H. M. (1995) Design and testing of peptide-based cytotoxic T-cell mediated immunotherapeutics to treat infectious diseases and cancer. Ch. 38 in Vaccine Design: The subunit and adjuvant approach. (Powell, M. F. and Newman, M. J. eds.) Plenum Press, New York , 847-874.

De-Bruijn, M. L. H., Schumacher, T. N. M., Nieland, J. D., Ploegh, H. L., Kast, W. M., and Melief, C. J. M. (1991) Peptide loading of empty major histocompatibility complex molecules on RMA-S cells allows the induction of primary cytotoxic T lymphocyte responses. Eur. J. Immunol. *21*, 2963-2970.

del Guercio, M.-F., Sidney, J., Hermanson, G., Perez, C., Grey, H. M., Kubo, R. T., and Sette, A. (1995) Binding of a peptide antigen to multiple HLA alleles allows definition of an A2-like supertype. J. Immunol. *154*, 685-693.

Engelhard, V. H. (1994) Structure of peptides associated with class I and class II MHC molecules. Annu. Rev. Immunol. *12*, 181-207.

Fernandez-Viña, M. A., Falco, M., Sun, Y., and Stastny, P. (1992) DNA typing for HLA class I alleles: I. Subsets of HLA-A2 and of -A28. Hum. Immunol. *33*, 163-173.

Germain, R. N. and Margulies, D. H. (1993) The biochemistry and cell biology of antigen processing and presentation. Annu. Rev. Immunol. *11*, 403-450.

Hill, A. V. S., Elvin, J., Willis, A. C., Aidoo, M., Allsopp, C. E. M., Gotch, F. M., Gao, X. M., Takiguchi, M., Greenwood, B. M., Townsend, A. R. M., McMichael, A. J., and Whittle, H. C. (1992) Molecular analysis of the association of HLA-B53 and resistance to severe malaria. Nature *360*, 434-439.

Howard, J. C. (1995) Supply and transport of peptides presented by class I MHC molecules. Curr. Opin. Immunol. *7*, 69-76.

Huczko, E. L., Bodnar, W. M., Benjamin, D., Sakaguchi, K., Zhu, N. Z., Shabanowitz, J., Henderson, R. A., Appella, E., Hunt, D. F., and Engelhard, V. H. (1993) Characteristics of endogenous peptides eluted from the class I MHC molecule HLA-B7 determined by mass spectrometry and computer modeling. J. Immunol. *151*, 2572-2588.

Imanishi, T., Akaza, T., Kimura, A., Tokunaga, K., and Gojobori, T. (1992a) Allele and haplotype frequencies for HLA and complement loci in various ethnic groups. In HLA 1991: Proceedings of the Eleventh International Histocompatibility Workshop and Conference, Vol. 1. K. Tsuji, M. Aizaqa, and T. Sasazuki, eds. Oxford University Press Tokyo, Japan, pp. 1065-1074.

Imanishi, T. and Gojobori, T. (1992b) Patterns of nucleotide substitutions inferred from the phylogenies of the class I major histocompatibility complex genes. J. Mol. Evol. *35*, 196-204.

Ishikawa, Y., Tokunaga, K., Lin, L., Imanishi, T., Saitou, S., Kimura, A., Kashiwase, K., Akaza, T., Tadokoro, K., and Juji. T. (1994) Sequences of four splits of HLA-A10 Group: Implications for serologic cross-reactivities and their evolution. Hu. Immunol. *39*, 220-224.

Itescu, S., Rose, S., Dwyer, E., and Winchester, R. (1995) Grouping HLA-B locus serologic specificities according to shared structural motifs suggests that different peptide-anchoring pockets may have contrasting influences on the course of HIV-1 infection. Hum. Immunol. *42*, 81-89.

Kast, W. M., Brandt, R. M. P., Sidney, J., Drijfhout, J.-W., Kubo, R. T., Grey, H. M., Melief, C. J. M., and Sette, A. (1994) The role of HLA-A motifs in identification of potential CTL epitopes in human papillomavirus type 16 E6 and E7 proteins. J. Immunol. *152*, 3904-3912.

Kato, K., Trapani, J. A., Allopenna, J., Dupont, B., and Yang, S. Y. (1989) Molecular analysis of the serologically defined HLA-Aw19 antigens: A genetically distinct family of HLA-A antigens comprising the A29, A31, A32, and Aw33, but probably not A30. J. Immunol. *143*, 3371-3378.

Kubo, R. T., Sette, A., Grey, H. M., Appella, E., Sakaguchi, K., Zhu, N.-Z., Arnott, D., Sherman, N., Shabanowitz, J., Michel, H., Bodnar, W. M., Davis, T. A., and Hunt, D. F. (1994) Definition of specific peptide motifs for four major HLA-A alleles. J. Immunol. *152*, 3913-3924.

Kumar, V., Bhardwaj, V., Soares, L., Alexander, J., Sette, A., and Sercarz, E. (1995) MHC-binding affinity of an antigenic determinant is crucial for the differential secretion of IL-4 or IFN-g by T cells. Proc. Natl. Acad. Sci. (USA) in press.

Madden, D. R., Garboczi, D. N., and Wiley, D. C. (1993) The antigenic identity of peptide/MHC complexes: a comparison of the conformations of five viral peptides presented by HLA-A2. Cell *75*, 693-708.

Madden, D. R. (1995) The three-dimensional structure of peptide-MHC complexes. Annu. Rev. Immunol. *13*, 587-622.

O'Sullivan, D., Sidney, J., Appella, E., Walker, L., Phillips, L., Colón, S. M., Miles, C., Chesnut, R. W., and Sette, A. (1990) Characterization of the specificity of peptide binding to four DR haplotypes. J. Immunol. *145*, 1799-1808.

Panina-Bordignon, P., Tan, A., Termijtelen, A., Demotz, S., Corradin, G., and Lanzavecchia, A. (1989) Universally immunogenic T cell epitopes: promiscuous binding to human MHC class II and promiscuous recognition by T cells. Eur. J. Immunol. *19*, 2237-2242.

Parham, P., Adams, E. J., and Arnett, K. L. (1995) The origins of HLA-A,B,C polymorphism. Immunol. Rev. *143*, 141-180.

Pfeiffer, C., Stein, J., Southwood, S., Ketelaar, H., Sette, A., and Bottomly, K. (1995) Altered peptide ligands can control CD4 T lymphocyte differentiation in vivo. J. Exp. Med. *181*, 1569-1574.

Rammensee, H.-G., Friede, T., and Stevanovic, S. (1995) MHC ligands and peptide motifs: first listing. Immunogenetics *41*, 178-228.

Ruppert, J., Sidney, J., Celis, E., Kubo, R. T., Grey, H. M., and Sette, A. (1993) Prominent role of secondary anchor residues in peptide binding to HLA-A2.1 molecules. Cell *74*, 929-937.

Saper, M. A., Bjorkman, P. J., and Wiley, D. C. (1991) Refined structure of the human histocompatibility antigen HLA-A2 at 26 Å resolution. J. Mol. Biol. *219*, 277-319.

Schaeffer, E. B., Sette, A., Johnson, D. L., Bekoff, M. C., Smith, J. A., Grey, H. M., and Buus, S. (1989) Relative contribution of "determinant selection" and "holes in the T-cell repertoire" to T-cell responses. Proc. Natl. Acad. Sci. (USA) *86*, 4649-4653.

Sette, A., Vitiello, A., Reherman, B., Fowler, P., Nayersina, R., Kast, W. M., Melief, C. J. M., Oseroff, C., Yuan, L., Ruppert, J., Sidney, J., del Guercio, M.-F., Southwood, S., Kubo, R. T., Chesnut, R. W., Grey, H. M., and Chisari, F. V. (1994) The relationship between class I binding affinity and immunogenicity of potential cytotoxic T cell epitopes. J. Immunol. *153*, 5586-5592.

Sidney, J., del Guercio, M.-F., Southwood, S., Engelhard, V. H., Appella, E., Rammensee, H.-G., Falk, K., Rötzschke, O., Takiguchi, M., Kubo, R.T., Grey, H. M., and Sette, A. (1995) Several HLA alleles share overlapping peptide specificities. J. Immunol. *154*, 247-259.

Sidney, J., Grey, H. M., Southwood, S., Celis, E., Wentworth, P. A., del Guercio, M.-F., Kubo, R. T., Chesnut, R. W., and Sette, A. (1996) Definition of an HLA-A3-like supermotif demonstrates the overlapping peptide binding repertoires of common HLA molecules. Hu. Immunol. *45*, 79-93.

Sinigaglia, F. and Hammer, J. (1994) Defining the rules for the peptide-MHC class II interaction. Curr. Op. Immunol. *6*, 52-56.

Tanigaki, N., Fruci, D., Chersi, A., Falasca, G., Tosi, R., and Butler, R. H. (1994) HLA-A2 binding peptides cross-react not only within the A2 subgroup but also with other HLA-A locus allelic products. Hum. Immunol. *39*, 155-162.

Townsend, A. and Bodmer, H. (1989) Antigen recognition by class I-restricted T lymphocytes. Annu. Rev. Immunol. *7*, 601-624.

van Endert, P. M., Riganelli, D., Greco, G., Fleischhauer, K., Sidney, J., Sette, A., and Bach, J.-F. (1995) The peptide-binding motif for the human transporter associated with antigen processing. J. Exp. Med. *182*, 1883-1895.

Vitiello, A., Ishioka, G., Grey, H. M., Rose, R., Farness, P., LaFond, R., Yuan, L., Chisari, F. V., Furze, J., Bartholomeuz, R., and Chesnut, R. W. (1995) Development of a lipopeptide-based therapeutic vaccine to treat chronic HBV infection: I. Induction of a primary CTL response in man. J. Clin. Invest. *95*, 341-349.

Wentworth, P. A., Celis, E., Crimi, C., Stitely, S., Hale, L., Tsai, V., Serra, H. M., del Guercio, M.-F., Livingston, B., Alazard, D., Fikes, J., Kubo, R. T., Grey, H. M., Chesnut, R. W., Chisari, F. V., and Sette, A. (1995) In vitro induction of primary, antigen-specific CTL from human peripheral blood mononuclear cells stimulated with synthetic peptides. Molec. Immunol. *32*, 603-610.

2.4 Quantitative Considerations in the Design of Vaccination Strategies Against Pathogens Uniquely Susceptible to Cell-Mediated Attack

Peter Bretscher

2.4.1 Introduction

Some infections appear to be contained only by a cell-mediated, Th1 attack. The induction of antibody at the expense of the cell-mediated response, and the generation of Th2 cells antagonistic to the action of Th1 cells, usually leads to chronic or progressive and fatal disease. The purpose of vaccination in these cases must be to ensure a stable and adequately dominant cell-mediated response upon natural infection. Effective vaccination must generate an imprint upon the immune system such that an otherwise "susceptible" animal or individual mounts a protective, cell-mediated, Th1 response upon infection. We have achieved this in one case, namely that of mice "susceptible" to *L. major*, a parasite able to cause cutaneous leishmaniasis. Quantitative considerations were essential to our approach. I elaborate here on these quantitative considerations and discuss their general validity. In addition, I describe some ideas relevant to broader and more speculative issues that have contributed to my thinking on the conditions that might be required to achieve such vaccination. These more speculative ideas try to address why it is that some infections can only be contained by a cell-mediated response and to relate this requirement to the similar requirements for cell-mediated immunity to contain cancers, to reject transplants containing minor histocompatibility antigens, and for some forms of autoreactivity to be evident as damaging autoimmunity.

2.4.2 Pathophysiological Significance of Distinct Classes of Immunity and Their Exclusive Regulation

A correct understanding of the pathophysiological significance of immune class regulation might be expected to be central to understanding the regulation governing the induction of different classes. Such regulation should ensure that an "optimal" immune response is *usually* induced. This "optimal" response must be defined in pathophysiological terms. In many cases we can identify what response is optimal, such as cell-mediated immunity in leishmaniasis, as such immunity is required to contain the infection.

I have recently discussed (Bretscher, 1992a) an old view (Bretscher, 1974; Bretscher, 1981) of the pathophysiological significance of immune class regulation in terms of modern findings. I therefore outline this view to provide a context for discussion, without attempts at a full justification (Bretscher, 1992). There are about 100 to 1000 peptides generated in a cell and able to bind to class I MHC molecules. There are roughly 10^5 class I MHC molecules on a cell surface. Thus there are expected to be approximately 100 to 1000 molecules of each peptide-MHC complex on a cell surface, representing a potential minor histocompatibility antigen. Cells bearing such minor antigens are lysable by specific cytoxic T cells. Thus target cells bearing only roughly 100 recognisable sites are effectively attacked by cell-mediated mechanisms. This contrasts dramatically with what is known about antibody-mediated mechanisms. Target cells must have upwards of several 100,000 sites recognisable by IgG antibody to be lysed by the IgG antibody-dependent complement-mediated, or by the IgG antibody-dependent cell, mechanism. I have argued that cells with few recognisable "foreign" sites are generally not susceptible to IgG-dependent effector mechanisms, and that IgG antibody is not usually induced against such cellular antigens. In contrast, cells with many recognisable "foreign" sites can be attacked/lysed by antibody-dependent effector mechanisms, and such cellular antigens can under appropiate conditions induce antibody. Autoreactivity can be induced by antigens that crossreact with self, and cells infected with intracellular parasites would appear to represent the most common form of crossreactive antigen. The advantage to an individual of the exclusive induction of antibody against cells with many foreign sites (according to the view I have developed) (Bretscher, 1992a; Bretscher, 1974; Bretscher, 1981) is that antibody can successfully attack the infected cells upon recognition of the foreign sites, and yet any anti-self autoantibody induced is *unlikey* to cause damage to uninfected self-cells recognised by the autoantibody due to the requirement that so many sites have to be recognised by IgG antibody to activate effector functions. The induction of cell-mediated immunity is not required to eliminate the cell bearing many foreign sites, so long as sufficient antibody is induced, and the absence of a cell-mediated response minimises the damage caused by any autoreactivity induced.

In contrast, cells with few recognised sites are only susceptible to cell-mediated attack, and any cell-mediated autoreactivity generated will be more damaging than the "corresponding" antibody response. I have therefore argued that the broad features of the regulation governing cell-mediated and humoral responses serve the purpose of providing effective immunity, in that the effector mechanisms generated can successfully attack the inciting foreign invader, and at the same time generates immunity that minimises the destructive consequences of any autoreactivity induced (Bretscher, 1974; Bretscher, 1981).

This view naturally explains why transplants bearing minor histocompatibility antigens are only suscetible to cell-mediated attack, as cells of such transplants bear so few foreign sites. The nature of the events leading to the establishment of some cancers has been clarified in recent years (Fearon and Vogelstein, 1990) and this, together with the characterisation of tumour-associated transplantation antigens (Boon et al., 1994), would again suggest that tumour cells differ relatively slightly from normal cells. The view elaborated here would therefore naturally account for the general requirement for cell-mediated responses to contain tumours. I have suggested that the characteristic of those intracellular parasites that can cause chronic disease and that are only contained by a cell-mediated response is their slow replication. This relatively slow growth rate means that a relatively small fraction of the proteins synthesised in an infected host cell are parasite-specific, and consequently relatively few parasite-dependent antigens are expressed on the surface of the infected cell. I have suggested that for these reasons such cells are only susceptible to cell-mediated attack (Bretscher, 1992a). These ideas on why cell-mediated immunity is required to contain certain infections have influenced our approach to vaccination, as I hope will become apparent later. Finally, we might expect on the general view outlined here that autoreactivity would often be damaging if it existed at the cell-mediated but not at the antibody level (Bretscher, 1974; Bretscher, 1981). This is a subject to which we shall return when we discuss observations that support this view and that suggest strategies for prevention of autoimmune disease.

2.4.3 Basic Studies on Immune Class Regulation

I give here a brief outline of my understanding of immune class regulation as an introduction to our discussion of strategies of vaccination. I emphasise those aspects where I find myself at odds with the most currently held views and which seem to me pertinent to vaccination.

Studies in the mid-60's showed that rodents immunised to produce antibody to an antigen could no longer be induced to express delayed type hypersensitivity (DTH), a major class of cell-mediated immunity, to this antigen (Asherson and Stone, 1962). Such a state can be referred to as a state of humoral immune deviation, meaning that the immune response has been locked into a humoral mode. Other studies in the late

60's and early 70's demonstrated the converse phenomenon, whereby the induction of cell-mediated immunity rendered rodents unresponsive for the induction of antibody (Parish, 1972). The immune response to the antigen employed appeared to have been locked into a cell-mediated mode. In the early 70's I proposed a conceptual scheme to explain the pathophysiological significance and mechanisms of immune class regulation (Bretscher, 1974; Bretscher, 1981). I summarise experiments carried out in the mid-70's to test some critical predictions of this scheme, as they are relevant to current ideas and controversies on the role of Th1 and Th2 cells in immune class regulation. Moreover, implications drawn from these observations are central to our vaccination strategy.

We wished to examine why mice immunised to produce antibody were unresponsive for the induction of DTH. The conceptual scheme proposed had postulated that this unresponsiveness is due to the generation of antigen-specific T cells that suppress the induction of DTH. We found that the transfer of Ly1+ Ly2- T cells, from mice immunised to produce antibody and unresponsive for the induction of DTH, to normal mice, rendered these recipients unresponsive for the induction of DTH. We refer to the transferred T cells able to suppress DTH as TsDTH cells. I should like to stress two aspects of our findings. (i) DTH-mediating T cells had the same Ly1+Ly2- phenotype as TsDTH cells, so we knew, in more modern notation, that there were at least two classes of CD4+ T cells, one mediating DTH and another, not doing so, but capable of suppressing the induction of DTH (Ramshaw et al., 1976; Ramshaw et al., 1977). These observations are still relevant to the proposal, not yet formally demonstrated, that Th2 cells inhibit the induction of Th1 cells. Secondly, we carried out experiments to examine some features of the mechanism of suppression of the induction of DTH by TsDTH cells. Animals immunised to produce antibody to a protein P could not be induced to express DTH to P. The transfer of (CD4+) TsDTH cells from this mouse to a normal mouse inhibited the induction of DTH to a foreign red blood cell F when the immunising antigen was P-F, but not when the antigen was F, or when the antigens administered were F and P-Q, where Q was an antigen that did not crossreact with F (Ramshaw et al., 1976; Ramshaw et al., 1977). These observations parallel those showing that the induction of anti-hapten B cells requires the hapten to be physically linked to the carrier for which the helper T cells are specific. Thus, just as B cell-Th cell cooperation in the induction of B cells requires the linked recognition of antigenic determinants, our observations demonstrated that the suppression of the induction of DTH by TsDTH cells is also mediated by the linked recognition of antigenic determinants. This conclusion is somewhat paradoxical in terms of current concepts, as CD4+ T cells are known to recognise peptides of the nominal antigen seen in the context of class II MHC molecules, and so recognition of linked epitopes on the nominal antigen by CD4+ T cells appears paradoxical. Nevertheless, the fact that TsDTH cells operationally act by the linked recognition of antigenic epitopes is central to our vaccination strategy, as further discussed later.

We also tested the prediction of the conceptual scheme that the unresponsive state for the induction of antibody, sometimes associated with the induction of cell-mediated immunity, is due to Ly1-Ly2+ (i.e. CD8+) T cells. The transfer of CD8+ T cells from mice expressing potent DTH and unresponsive for the induction of antibody rendered the recipent mice unresponsive for the induction of antibody. We refer to these CD8+ T cells able to suppress the induction of antibody as TsAb cells (Ramshaw et al., 1977; Ramshaw et al., 1977). Three points are worth stressing concerning these CD8+ T cells. (i) It seems likely, as originally suggested, that not only do these CD8+ T cells inhibit the induction of antibody responses but also the generation of the CD4+ TsDTH cells, and conversely that the CD4+ TsDTH cells inhibit not only the induction of DTH responses but the generation of the CD8+ TsAb cells. Such additional activities would explain the exclusive generation of CD4+ TsDTH cells and CD8+ TsAb cells (Bretscher, 1974; Bretscher, 1981). There is as yet, however, no direct evidence for this mutual inhibition. (ii) It is current dogma that Th1 cells inhibit the generation of Th2 cells, and vice versa, thus explaining the exclusivity of cell-mediated and antibody responses. This proposition is based upon observations made in potentially non-physiological systems. For example, the kind of evidence leading to this proposal is that Th1 clones make IFNγ and that this cytokine inihibts the multiplication of cells from Th2 clones, and that the IL10, made by cells of Th2 clones, can inhibit the production of IFNγ by cells of Th1 clones. This proposal predicts that the unresponsiveness of mice for the induction of antibody, sometimes associated with the induction of a substantial DTH response, is due to CD4+ Th1-like cells, a result inconsistent with the transfer studies just outlined, demonstrating the role of CD8+ T cells (Ramshaw et al., 1976; Ramshaw et al., 1977). One feature of the observations made with clones is that they do not address how the cytokines such as IL10 produced by Th2-like cells might be delivered with the requisite accuracy to the specific Th1-like cells, and how IFNγ produced by Th1-like cells might be delivered to Th2-like cells. I feel the phenotype of the "TsAb" cell is central for the following reason. In proposing a scheme of immune class regulation over twenty years ago, I was struck by the fact that the four conditions of administering antigen parenterally to immunocompetent rodents that resulted in the induction of DTH were just the same four conditions that resulted in the generation of antigen-specific TsAb cells, with the Ly1-Ly2+ (CD4-CD8+) phenotype, in those systems where the phenotype was established. This led me to suggest that the physiological role of these CD8+ T cells was to ensure that antibody was not induced during the course of a strong cell-mediated response, and led to the prediction, subsequently verified as described above, that animals induced to express cell-mediated immunity, and unresponsive for the induction of antibody, bore antigen-specific CD8+ TsAb cells. This interpretation of the correlation between the conditions under which DTH and TsAb were generated appears to be both valid and significant. (iii) Lastly, observations by others demonstrated that CD8+ TsAb cells act by the linked recognition of antigenic epitopes (Basten, 1974).

Finally, the scheme envisaged also proposed what factors determine whether the primary response an antigen induces is predominantly cell-mediated or humoral. These factors I refer collectively to as the *decision criterion*. It seemed that an appropiate proposal for nature of the the decision criterion should satisfy two conditions.

Firstly, according to the pathophysiological considerations discussed above, and on the basis of experimental evidence (Pearson, 1971), one variable that must affect the decision is the density of foreign sites on a cellular antigen, ensuring that cells with a low density can usually only induce cell-mediated immunity, whereas those with a high density can induce antibody. Secondly, a correct decision criterion *must* account for those variables of immunisation that critically affect the class of immunity induced. The decision criterion proposed has three components. (i) It is based upon the assumption that there are helper/inducer CD4 T cells specific for foreign but not for self epitopes. This assumption is now supported by much evidence, as detailed elsewhere (Bretscher, 1992b). (ii) It assumes that the primary induction of all lymphocytes requires CD4+ Th cells that act by the linked recognition of antigenic epitopes. Much evidence supports this assumption (Bretscher, 1992). (iii) It postulates that the induction of lymphocytes involved in cell-mediated responses require the generation of fewer helper T cell-dependent signals than the induction of lymphocytes involved in antibody responses. In summary, this decision criterion postulates that the threshold of helper T cell-dependent signals required to induce lymphocytes involved in cell-mediated immunity is lower than the threshold of those involved in antibody responses. I refer to it as *The Threshold Hypothesis*. I have recently reviewed the substantial evidence supporting this hypothesis (Bretscher, 1994), and so will not do so here. Figure 2.4.1 summarises the salient features of immune class regulation that I have discussed.

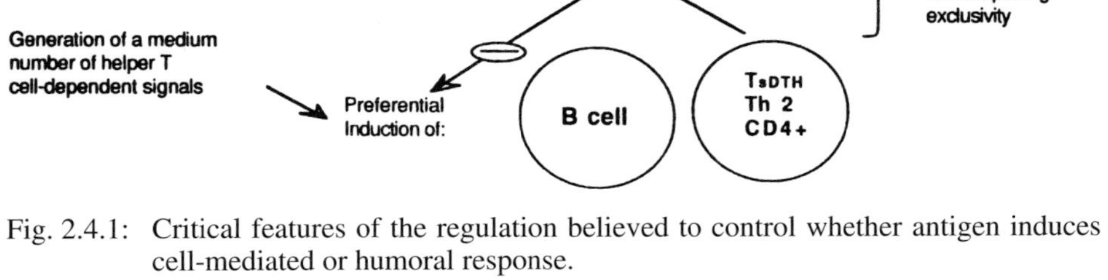

Fig. 2.4.1: Critical features of the regulation believed to control whether antigen induces a cell-mediated or humoral response.

2.4.4 Paradoxes of this View of Immune Class Regulation in Terms of Current Models for the Induction of Th Cells and a Potential Resolution

A number of observations and considerations concerning immune class regulation are difficult to reconcile with the simplest kind of model most frequently envisaged for the induction of helper T cells. According to this model, a precursor of an activated helper T cell must bind to peptides themselves bound to class II MHC molecules of an antigen presenting cell (APC) in order to be induced. In addition, this precursor cell probably requires costimulatory signals to be induced. The generation and delivery of the costimulatory signals by APC may or may not be regulated.

I shall describe two related difficulties of this view and a potential solution to them. A component of the above model is that all CD4+ T cells recognise peptides derived from nominal antigens that are bound to class II MHC molecules. There is evidence, however, that some interactions between two antigen-specific CD4+ T cells operationally require the linked recognition of antigenic epitopes. These include the induction of precursor cells that express DTH upon activation (Tucker and Bretscher, 1982), and the suppression of the induction of these precursors by TsDTH cells, as described above. How can there be such a requirement when the antigen, in order to be recognised by the CD4+ T cells, must be broken down into small peptide fragments? The only possibility I have been able to envisage is based upon the solution to the similar problem in B cell-Th cell collaboration, in which a requirement exists for linked recognition of hapten by the B cell and of the carrier for which the T cell is specific, even though the T cell recognises a carrier-derived peptide to which the hapten is not attached. Logically, the requirement for linkage must occur before the degradation of the nominal antigen. I would suggest that there is an obligatory step in these processes, that involve CD4+ T cell-CD4+ T cell interactions operationally mediated by the linked recognition of antigenic epitopes, in which the intact nominal antigen must be taken up by specific B cells that can then act as an APC to present two unrelated peptides to the two interacting CD4+ Th cells. These two peptides will only be present on the same B cell if they were initially linked by being derived from the same nominal antigen. This solution postulates an obligatory role for antigen-specific B cells as APC in different kinds of response. Some evidence supports such requirements (Bottomly and Janeway, 1989; Lanzavecchia, 1990).

A more general, very broad class of observations, is also difficult to reconcile with the minimal model for the induction of Th cells. This is the phemenon that the class of immune response to the different peptides of simple protein molecules (Parish, 1972), to the different components of foreign red blood cells (Lagrange et al., 1974; Wortis et al., 1968), or to the different components of the protozoan parasite *L. major*, (Bretscher et al., 1992), for example, tend to be coordinately regulated such that either cell-mediated, Th1- or antibody, Th2-responses are induced under a given set of circumstances. Although this is a tendency and not an absolute feature of the response, it is extraordinarily dramatic in various situations. This again suggests that the regulation of the immune response to the different parts of an antigen are interdependently regulated and that the kind of response an antigen induces depends upon the intact properties of the antigen. Indeed, we have discussed above the proposal that antigens with few foreign sites induce cell-mediated immunity and Th1-like cells, whereas cells with many foreign sites can induce antibody and Th2-like cells. This old proposal was originally made on the basis of observation (Pearson and Raffel, 1971). How can we reconcile this suggestion with the idea that the first significant interaction of antigenic material occurs after the antigen has been processed into small peptides? It is impossible to distinguish at this stage whether a peptide has been derived from an antigen with few or many foreign sites. These considerations again suggest that the intact antigen *must be* recognised by antigen-specific B cells at an early step in the inductive process. I refer to the tendency for there to be coordinate regulation of the class of imunity to the different components of simple and complex antigens as *coherence.* Coherence is a phenomenon central to our strategy of vaccination, as will become apparent shortly. Coherence is most easily explained if the regulation governing immune class regulation involves T cell-T cell interactions mediated by the linked recognition of antigenic epitopes.

2.4.5 Dependence of the Class of Immunity Induced on Dose of Antigen Administered and on Time After Immunisation: Relevance to Establishing Cell-Mediated Immune Deviation

It has been reported that low doses of antigens with many foreign sites induce cell-mediated immunity whereas higher doses induce antibody. In addition, many doses induce both classes, but in this case cell-mediated immunity, at least in the form of DTH, is induced first and decays as antibody is produced (Salvin, 1958). I have argued that a similar kinetic pattern probably governs the induction of cytotoxic T lymphocytes (Bretscher, 1994). These observations on the induction of DTH are explicable on The Threshold Hypothesis, which was proposed in part to account for such observations. Thus antigens with many foreign sites are expected to induce Th cells less efficiently at low and suboptimal doses, and the activated Th cells will also deliver the antigen-mediated helper T cell dependent signals to precursor cells less efficiently at such low doses, thus activating those precursor lymphocytes requiring the generation of the fewest helper T cell-dependent signals to be induced, i.e. those involved in cell-mediated responses. In addition, antigen is known to activate helper T cells, and so the number of helper T cell-dependent signals is expected to increase with time. This can explain why the immune response to an antigen often first goes through a cell-mediated phase that declines as antibody is produced (Salvin, 1958).

Parish showed in the early 70's that the administration of low doses of a protein antigen to rats not only induced DTH but made the rats unresponsive for the induction of antibody (Parish, 1972). The immune response was locked into a cell-mediated mode. This state is referred to as cell-mediated low-zone immune deviation. It is interesting that such a state has been observed only after administering low doses of antigen for several weeks. This seems very important to me for several reasons. Firstly, we know that the induction of DTH does not automatically result in the generation of sufficient TsAb cells to lock the immune response into a cell-mediated mode, as the induction of DTH occurs before antibody production in many responses, as just noted. Secondly, the chronic administration of low doses of a non-replicating antigen, that causes a lock of the immune response into a cell-mediated mode, is very close to the pattern of antigen stimulation expected from infection with a fairly low dose of a replicating antigen *that replicates relatively slowly.* We have argued above that it is against just such slowly growing intracellular pathogens that a cell-mediated attack is required to contain them, and so it is reasonable to suppose that natural selection has ensured that the immune response in these cases is locked into a cell-mediated mode. Thus the conditions required to establish cell-mediated low-zone immune deviation fit very nicely with the idea that this lock serves the purpose of usually ensuring a stable cell-mediated response to slowly-growing intracellular parasites.

2.4.6 Establishing Low-Zone Cell-Mediated Immune Deviation to *L. major* in "Susceptible Mice"

Different strains of mice produce different kinds of immune response upon infection with a standard and substantial number of *L. major* parasites, a protozoan that causes cutaneous leishmaniasis in humans. Resistance is associated with the generation of a cell-mediated, Th1-like response and susceptibility with an antibody, Th2-like response (Locksley, 1991). BALB/c are the prototypic susceptible strain. We have shown that this susceptibility is not an absolute but a conditional trait, as BALB/c mice infected with a number of parasites that is well below the standard number resist the infection and mount an exclusive and stable cell-mediated, Th1-like response. Furthermore, such mice become resistant with time to a standard, normally pathogenic challenge. This resistance to a standard challenge is associated with a prolonged and stable cell-mediated, Th1-like response (Bretscher et al., 1992; Menon and Bretscher, 1996).

The two cytokines IFNγ and IL4 are charactistically produced by Th1 and Th2 cells respectively, and have opposing effects upon macrophage activation and the consequent killing/containment of intracellular *L. major* parasites (Lehn et al., 1989; Essner et al., 1989). IFNγ is involved in activating macrophages, whereas the delivery of IL4 to an infected macrophage is believed to counteract such activation. The ability of Th1-derived cytokines to contain the parasite may therefore be related not only to their absolute level of production but also to the ratio of their production to that of IL4. We measured at various times post-infection the parasite antigen-dependent production by spleen cells of IFNγ and IL4 following challenge of normal "susceptible" mice with a standard, normally pathogenic number of parasites and of mice made resistant by pre-exposure to low dose infection. We calculated the ratio in the production of IFNγ to that of IL4 at different times post infection to give a measure of the relative size of the Th1- and Th2-components of the immune response. Two points are noteworthy. The IFNγ/IL4 ratio in normal mice given a standard challenge is moderate at around three weeks post infection, and then declines over successive weeks. This corresponds to the well known fact that a cell-mediated response often precedes a humoral one (Salvin, 1958; Howard, 1986). Secondly, the ratio in mice pre-exposed to a low dose differs little from normal mice shortly after infection, but increases dramatically over several weeks (Menon and Bretscher, 1996). This suggests that low dose imprinting ensures that the immune response following high dose challenge evolves with time such that the Th1 component becomes ever more dominant and presumably ever more effective against the parasite.

2.4.7 The Dependence of the Generation of Th1 and Th2 Cells on Parasite Dose Appears to be General: Potential Implications for Universally Efficacious Vaccination

Different strains of mice show different susceptibilities to a standard and substantial infection of *L. major*. CBA mice are the prototypic "resistant" strain, BALB/c mice the prototypic "susceptible" strain, and A/J mice are of "intermediate susceptibility". Our observations on low dose vaccination of "susceptible" BALB/c mice show that "susceptibility" is a trait conditional upon parasite dose, as BALB/c mice mount a protective, cell-mediated response upon low dose infection. Furthermore, the susceptibility to the standard infection is modifyable by pre-exposure to low doses. These observations led to studies in other strains of mice. We infected CBA and A/J mice with different numbers of parasites. Infection with very high doses of parasites caused progressive infection in "resistant" CBA mice. Infection of A/J mice with a number of parasites below the standard dose led consistently to a non-progressive infection and infection with a number above the standard dose consistently led to progressive disease. These observation suggest that infection with relatively high doses can cause progressive disease in all strains of mice, and that infection with relatively low doses induces a cell-mediated response that can contain the infection. In this sense, all strains of mice conform to the same rule. It is just that the absolute number of parasites at which the response changes from a mode associated with protection to one associated with progression is different. We examined how the generation of Th1 and Th2 cells differed in two groups of mice of each strain, one infected with a low dose that was non-progressive and the other infected with a high progressive dose, by assessing the parasite-antigen dependent production of IFNγ and IL4 by spleen cells from infected mice obtained at various times post-infection. We found the ratio of IFNγ/IL4 to differ at about 10 weeks post infection by about a 100 to a 1000 fold in the two groups of mice, with the ratio being higher in the mice infected with the lower dose in all the different strains (Menon and Bretscher, in preparation). These studies show the very dramatic effect on the generation of Th1 and Th2 cells of parasite dose.

Is there a similar dependency of the generation of Th1 and Th2 cells on dose of infecting microorganism in other systems? We are particularly interested in mycobacteria because of the relevance of tuberculosis. We have found a similar dependence of cell-mediated, Th1 and antibody, Th2 responses on the dose of mycobacteria given either intravenously or subcutaneously. Furthermore, pre-exposure to a low dose subcutaneous infection can dramatically bend the anti-mycobacterial response to a high dose intravenous challenge of BCG away from a Th2 towards a Th1 pole, and such modulation of the response is associated with more rapid clearance of the mycobacteria from the spleen (Bretscher and Wei, in preparation). These results support the general validity of the low dose vaccination strategy.

I believe these observations allow us to envisage a standard vaccination protocol that may be universally efficacious despite the genetic diversity of people. Consider what we would expect to happen if a hereogenous population of individuals were infected with a low dose of BCG, chosen so that it did not induce antibody in any individual. We would expect it to either induce a cell-mediated response, and thus lead to the generation of the imprint required to confer protection, or to be below the threshold required to induce a cell-mediated response. In the latter case, the original mycobacteria inoculated would be expected to grow unimpeded, until they reached the threshold at which a cell-mediated response is induced. We anticipate that they would then induce the imprint required to confer protection. Thus vaccination with a low dose that does not induce antibody in any individual has the potential to provide universally efficacious protection (Bretscher, 1992c).

2.4.8 Further Quantitative Considerations

Our observations clearly show the importance of the dose of a viable vaccination agent in achieving a protected state. However, protection takes time to be established.

Mice challenged shortly after low dose infection with a normally pathogenic dose are not protected but suffer progressive disease. This is consistent with the idea that the lock of the immune response into a cell-mediated mode takes time to be established, and with the arguments elaborated above that such a lock should only be induced upon infection with a slowly-growing intracellular parasite, as only in this case is a cell-mediated response required to contain the infection. Suppose a low dose infection occurs with a rapidly replicating microorganism. In this case, the microoganism is likely to have reached the high levels normally required to induce antibody formation before the imprint has been established that locks the immune response into a cell-mediated mode, and consequently antibody will be induced. These considerations make it clear that there is a relationship between the dose of the vaccinating agent, its rate of replication, and the time needed to generate the imprint, in determining whether a lock of the immune response into a cell-mediated mode is established. In particular, we would expect that relatively high doses would be ineffective unless such a dose is initially below the threshold required to induce antibody and unless the immunising agent replicated relatively *slowly.*

Certain observations are particularly interesting from this view point. I consider the observations suggesting that cell-mediated immunity is protective against HIV-1 plausible, and I therefore consider the equivalent of low dose infection as an interesting possibility for efficacious vaccination (Salk et al., 1993). Some evidence suggests that infection with low doses of SIV can protect in the monkey model of AIDS (cited in Salk et al., 1993). However, the most dramatic form of protection that has been reported is that achieved by immunising with a moderate dose of a mutant form (nef$^-$) of SIV that is partially deficient in replication and hence replicates slowly (Daniel et al., 1992). This finding seems to make eminent sense in terms of the considerations just elaborated.

2.4.9 The Other Side of the Coin: Organ-Specific Immunity

In some diseases, such as leprosy, the leishmaniases, and most probably tuberculosis (Surcel, 1994), the induction of antibody and associated cells at the expense of (and/or in antagonism to) a cell-mediated response results in pathology. The existence of this pathology naturally leads to the physiological question of what useful purpose does the antibody arm of the immune system serve, as such pathology would not occur in these infectious diseases if strong and exclusive cell-mediated responses were always generated. The humoral arm most likely serves many purposes. However, a proposal for a major advantage in having a humoral arm has already been discussed above, and observations lend some credence to this proposal, as I shall now discuss. The induction of autoreactivity is likely to occur when the immune system is stimulated by antigens that crossreact with self cells (Bretscher, 1992a). Autoreactivity at the cell-mediated level is more likely to be damaging than a corresponding antibody response, due to the fact that damaging cell-mediated effector mechanisms are activated against cells bearing only a very few recognised sites. According to this view, antibody is induced against cells that have many recognisable foreign sites, as antibody-dependent effector mechanisms can be effective against such cells. The advantage gained by the generation of an antibody response over a corresponding cell-mediated response is that any autoreactivity generated and directed against self cells is less likely to be damaging to these self cells than if cell-mediated autoreactivity were generated. A series of experiments support these ideas. In vivo depletion of lymphocytes or CD4+ T cells in mice can switch the class of response induced from a humoral, Th2 to a cell-mediated, Th1 mode (Mitchell et al., 1981; Bretscher, 1983; Sadick et al., 1987). This is understable in terms of The Threshold Hypothesis, as this hypothesis states that the induction of lymphocytes involved in antibody responses requires the generation of more helper T cell-dependent signals than the induction of lymphocytes involved in cell-mediated responses. Interestingly, depletion of lymphocytes in healthy rats can result in the appearance of various forms of organ-specific cell-mediated autoimmunity, and this appearance can be suppressed by giving the rats at the time of lymphocyte depletion a small number of CD4+ T cells obtained from normal rats. The T cells from normal rats that are effective in suppressing the appearance of autoimmunity belong to a subset that can both help secondary antibody responses and produce IL4 upon stimulation (Penhale et al., 1973; Fowell and Mason, 1993). I favour the following interpretation of these observations. Normal rats have innocuous antibody responses directed against organ-specific antigens, probably due to crossreactions between such antigens and gut flora (Penhale and Young, 1988). Depletion of lymphocytes, CD4+ T cells in particular, results in a modulation of this humoral response to a cell-mediated one, with the consequence that the autoreactivity becomes damaging and pathologically evident. The parallel with immune responses to slowly growing intracellular parasites appears simple. In both cases, cell-mediated immunity is effective. In the case of organ-specific autoimmunity, the cell-mediated response is seen as self-destructive damage, and in the case of intracellular parasites as successful containment of an infection.

2.4.10 Establishment of Resistance to Tumours by Excision Priming and its Potential Relationship to Cell-Mediated Low-Zone Immune Deviation

I would like to point out what appear to me to be some extraordinarily striking similarities between findings in the immune response to, and resistance against, tumours and slowly growing intracellular parasites. These parallels support the validity of our low dose vaccination strategy.

(i) Both tumours and such parasites are believed to be contained only by cell-mediated immune responses, a conclusion understandable in terms of the patho-physiological significance of distinct classes of immunity, as discussed at the beginning of this article.

(ii) Infection with a substantial and pathogenic number of parasites results first in the generation of an exclusive cell-mediated response, that is most likely protective, and an increase in net parasite numbers occurs more dramatically once this response declines and antibodies are produced (Howard, 1986). The induction of antibody is associated with the generation of CD4+ T cells, presumably of the Th2 variety, that suppress the generation of a protective response (Howard, 1986). Injection of a lethal dose of tumour cells to mice results first in the generation of a cell-mediated response that is protective.This protective response can be detected by the passive transfer of lymphocytes from the infected mouse to a naive mouse given a normally lethal challenge of tumour cells; the transfer of cells can result in protection. Studies show that this "concomitant immunity", generated in animals given a lethal dose of tumour cells, is ineffective in completely protecting the host because "too little immunity is generated too late". This deficiency is at least partly because the generation of concomitant immunity is inhibited by the generation of tumour specific CD4+ T cells (North, 1986), presumably of the Th2 variety.

(iii) A classical procedure has been used for decades to make mice specifically resistant to tumours. A normally lethal dose of tumour cells is injected into the skin, and once the tumour has reached a diammeter of roughly a millimeter, about ten days after injection, it is excised. The mouse survives and can be shown to have acquired resistance about two months later to a normally lethal challenge of the same tumour (Foley, 1953). Interestingly, the tumour is excised at roughly the time when concomitant immunity has reached maximal expression and before the CD4+ T cells that can suppress this immunity have been generated. Excision priming thus appears to be a way of inducing a protective cell-mediated response and preventing the normal progression of the response (into what I would suggest is a Th2 mode) by removal of most of the antigen by excision of the tumour at a judicious time. In this respect, excision priming seems very similar to our low dose vaccination strategy, as in the latter case our dose of agent is chosen to be sufficiently low to induce an exclusive cell-mediated response. A second parallel is the length of time, of the order of a month or two, it takes to firmly establish resistance. I would argue that our strategy of establishing low-zone cell-mediated immune deviation to intracellular parasites is a quite parallel process to that resulting in resistance to tumours by excision priming. Since excision priming has been achieved in very many tumour systems, this parallelism suggests that cell-mediated immune deviation can be generally established.

2.4.11 Problems in Vaccine Design Against Tuberculosis and Speculation on Potential Solutions

I would like to summarise the important points discussed above by looking at the recognised problems in achieving efficacious vaccination in humans against tuberculosis (Fine, 1988), and discussing how the perspective developed here might be relevant in addressing them.

i) *The Nature of Protective Antigens*
A full understanding of the protective immune response will involve a definition of those protective antigens against which immunity is required to contain/eliminate the pathogen. Much effort has therefore been made in various systems to define such antigens as a first step in raising protective immunity. Our strategy attempts to avoid the complexities associated with the need to define protective antigens and is based upon the coherence of the immune response. We surmise that a strategy, ensuring that the immune response against most of the pathogen's antigens is of the effective, cell-mediated class, will automatically ensure that this same class is induced against the protective antigens. This surmise is likely to be as good as coherence is a real characteristic of the immune response.

ii) *The Nature of the Protective Response*
We need to know what response is protective, or at least what are the correlates of a protected state, if we are to rationally attempt to achieve such a state. Expression of mycobacterial-specific DTH is found in healthy individuals infected with *M. tuberculosis,* and so such a state might appear to be an appropiate correlate. However, those ill with tuberculosis usually express such hypersensitivity as well, and it is often argued that these observations collectively show that we do not know what correlates with protection. Although the detailed mechanisms responsible for containing *M. tuberculosis* are an involved subject that I shall not enter into here, it seems to me that a plausible and consistent view of what correlates with a protective response can be made. The low dose infection of BALB/c mice with *L. major,* that results in the establishment of resistance, induces an exclusive cell-mediated response that disappears at three months post-infection (Bretscher et al., 1992). These resistant mice display a memory state for a Th1, cell-mediated response, as they mount a rapid and substantial response of this kind upon rechallenge. We interpret these observations to mean that the response induced upon low dose infection has been effective in reducing the parasite and hence antigen load to such a low level by three months post-infection that parasite-specific lymphocytes are no longer activated in sufficient numbers for a cell-mediated, Th1 response to be evident (Menon and Bretscher, 1996). Infection with slightly higher doses gives rise to borderline leishmaniasis with chronic parasitemia

and a chronic Th1/Th2 immune response (Bretscher and Wei, in preparation). I would therefore argue that the induction of an effective immune response would be a cell-mediated immune response exclusive of an antibody, Th2 component. Vaccination with BCG should induce mycobacterial-specific DTH that should disappear after immunisation, a disappearance that is associated with resistance (Bretscher, 1992). Prolonged expression of mycobacterial-specific DTH following BCG immunisation, in areas where such expression is not due to stimulation by enviromental mycobacteria, is a sign that BCG immunisation has induced a mixed Th1/Th2 response ineffective in eliminating the BCG. In other words, we draw the same conclusions as has been drawn from observations made on patients with leprosy. An exclusive cell-mediated response is optimally protective. The expression of cell-mediated immunity by individuals with borderline leprosy does not invalide the view that an exclusive cell-mediated response is protective.

Ensuring a Protective Response upon Natural Infection

We have argued extensively above that this can be achieved by low dose immunisation.

Avoiding Disadvantageous Priming by Atypical Mycobacteria

The reported efficacy of BCG trials against tuberculosis varies widely. It appears that the efficacy of trials is lower when they are carried out in countries closer to the equator, where atypical mycobacteria are present in the environment (Fine, 1988). Such mycobacteria may cause advantageous Th1- or disadvantageous Th2-imprints upon the immune system. It would appear that disadvantageous priming could be minimised by neonatal immunisation, so that an effective imprint would be usually established before enviromental priming occurs.

Problems Posed by the Genetic Diversity of the People for Achieving Standard Vaccination that Is Universally Efficacious

Genetic polymorphisms, including HLA polymorphisms (Singh et al., 1983; van Eden et al., 1985 and other articles in this volume), affect immune responses generally and those against mycobacteria in particular. Some HLA loci partially control susceptibility to leprosy and tuberculosis (van Eden et al., 1985; Singh et al., 1983). Such genetic polymorphisms mean that standard vaccination will usually induce different kinds of response in genetically different people. We are currently testing in mice our view that the traits of "resistance/susceptibility" to mycobacteria, that are due to MHC polymorphisms, are not absolute traits but rather traits that are conditional on the dose of mycobacteria employed for infection. Should this turn out to be a correct hypothesis, it will support the feasibility of the low dose vaccination strategy outlined above, according to which immunisation of a heterogenous population with a low dose of BCG, that does not induce antibody in any individual, results in universal efficacious protection.

References

Asherson, C. R., Stone, S. H. (1962) Immunol. *9*, 205.

Basten, A. (1974) In: Immunological Tolerance, D. H. Katz, B. Benacerraf, Eds. Academic Press New York, 107-117.

Boon, T., Cerottini, J. C., den Eynde, B. V., Van den Burggess, P., and Pel, A. V. (1994) Ann. Rev. Immunol. *337*, 337.

Bottomly, K. and Janeway, C. A. (1989) Nature *337*, 24.

Bretscher, P. A. (1974) Cell Immunol *13*, 171.

Bretscher, P. A. (1981) Fed Proc. *40*, 1473.

Bretscher, P. A. (1983) Cell Immunol. *81*, 345.

Bretscher, P. A. (1992a) Immunol Cell Biol. *70*, 343.

Bretscher, P. A. (1992b) Immunol Today *13*, 74.

Bretscher, P. A. (1992c) Immunol Today *13*, 342.

Bretscher, P. A., Wei, G., Menon, J. N., and Bielefeldt-Ohmann, H. (1992) Science *257*, 539.

Bretscher, P. A. (1994) In: Strategies in Vaccine Design, G. L. Ada, Ed. RG Landes Co. Anston, Texas, 99-100.

Daniel, M. D. et al. (1992) Science *258*, 1938.

van Eden, W., Gonzale, N. M., de Vries, R. K., Convit, J., and van Road, J. J. (1985) J. Inf. Dis. *151*, 9.

Essner, R., Rhoades, K., McBride, W. H., Morton, D. L., and Economon, J. S. (1989) J. Immunol. *142*, 3857.

Fearon, E. R. and Vogelstein, B. (1990) Cell *61*, 759.

Fine, P. E. M. (1988) Br. Med. Bull. *44*, 691.

Foley, E. J. (1953) Cancer Res. *3*, 326.

Fowell, D. and Mason, D. W. (1993) J. Exp. Med. *177*, 627.

Howard, J. G. (1986) Int. Rev. Exp. Pathol. *28*, 79.

Lagrange, P. H., Mackaness, G. B., and Miller, T. E. (1974) J. Exp. Med. *139*, 528.

Lanzavecchia, A. (1990) Ann. Rev. Immunol. *8*, 773.

Lehn, M., Weiser, W. Y., Engelhorn, S., Gillis, S, and Remold, H. G. (1989) J. Immunol. *143*, 3020.

Locksley, R. M. and Scott, P. (1991) In: C. Ash and R. G. Gallagher, Eds., Immunoparasitol Today. Cambridge:Elsevier Trends Jounals A58-A61.

Mitchell, G. F., Curtis, J. M., Scollay, R. G., and Handman, E. (1981) Aus. J. Exp. Biol. Med. Sci. *59*, 539.

Menon, J. N. and Bretscher, P. A. (1966) Eur. I. Immunol. 26, 243.

North, R. J. (1986) J. Exp. Med. *164*, 1652.

Parish, C. R. (1972) Transplant Rev. *13*, 35.

Pearson, M. N. and Raffel, S. (1971) J. Exp. Med. *133*, 494.

Penhale, W. J., Farmer, A., McKenna, R. P., and Irvine, W. J. (1973) Clin. Exp. Immunol. *25*, 6.

Penhale, W. J. and Young, R. P. (1988) Clin. Exp. Immunol. *72*, 288.

Ramshaw, I. A., Bretscher, P. A., and Parish, C. R. (1976) Eur J Immunol. *6*, 674.

Ramshaw, I. A., Bretscher, P. A., and Parish, C. R. (1977) Eur J Immunol. *7*, 180.

Ramshaw, I. A., Bretscher, P. A., McKenzie, I. F. C., and Parish, C. R. (1977) Cell Immunol *31*, 364.

Sadick, M. D., Heinzel, F. P., Shigekane, M., Fisher, W. L., and Locksley, R. H. (1987) J. Immunol. *139*, 1303.

Salk, J., Bretscher, P. A., Salk, P., Clerici, M., and Shearer, G. M. (1993) Science *260*, 1740.

Salvin, S. B. (1958) J. Exp. Med. *107*, 109.

Singh, S. P., Mehra, N. K., Dingley, H. B., Pande, J. N., and Vaida, M. C. (1983) Tissue Antigens *21*, 380.
Surcel, H. M. (1994) Immunol *81*, 171.
Tucker, M. J. and Bretscher, P. A. (1982) J. Exp. Med. *155*, 1037.
Wortis, H. H., Taylor, R. B., and Dresser, D. W. (1968) Immunology *14*, 69.

2.5 The Impact of the Type 1 and Type 2 T Helper Cell Concept on Novel Vaccine Design with Emphasis on Protection Against *Leishmania* Parasites

Christian Bogdan and Martin Röllinghoff

2.5.1 Introduction

Approaching the year 2000 leishmaniasis remains one of the ten most threatening infectious disease complexes in the world for which we still lack a molecularly defined, efficient and safe vaccine approved for human use. *Leishmania* are flagellated protozoan parasites which are transmitted to mammal hosts by sand-flies. They are readily taken up by phagocytic cells with subsequent transformation into amastigote forms. Depending on the parasite species as well as the immune status and the genetic background of the host organisms, *Leishmania* can lead to a broad spectrum of clinical manifestations. These include single cutaneous ulcers which are mostly self-healing (e.g. *Leishmania tropica, L. major, L. aethiopica, L. braziliensis, L. mexicana*), diffuse cutaneous leishmaniasis with characteristic chronicity (*L. amazonensis*, *L. mexicana*), muco-cutaneous lesions which are locally progressive and tissue-destructive (*L. braziliensis*) as well as visceral disease (kala azar) affecting liver, spleen and bone marrow (e.g. *L. donovani, L. infantum*) which is fatal if left untreated (Pearson and de Queiroz Sandoz, 1995). For centuries, injection of whole, virulent *Leishmania* strains at covered sites of the body ("leishmanization") has been the only means for preventing disfiguring cutaneous lesions after natural exposure to the parasite. This method, although proven efficient in many field trials in former Soviet Union, Iran and Israel, is hampered by severe side effects and is not suitable for protection against the most severe form of leishmaniasis, i.e. kala azar (Greenblatt, 1988). However, considerable advances in our understanding of the immune response to *Leishmania*, the biochemical characterization of leishmanial antigens and the advent of modern molecular techniques have now paved the way towards real anti-*Leishmania* vaccines, which provide a high degree of protection and do not entail injection of whole parasites.

In this chapter we will introduce the type 1/type 2 T helper-cell (Th1/Th2) concept which is valuable for understanding the key requirements and components of a protective Th1-type immune response against *Leishmania*. We will also delineate potential antagonistic, disease-mediating mechanisms including the expansion of Th2 cells, the induction of which needs to be avoided during vaccination based on animal studies and analysis of infected patients. Finally, an overview of the currently available molecularly defined *Leishmania* antigens and vaccine candidates will be provided. Despite the recent progresses it is important to bear in mind that the majority of encouraging results presented has been obtained in mouse models and not yet confirmed in humans.

2.5.2 T Helper Cell Subpopulations and Infections with Intracellular Parasites

2.5.2.1 Basic Aspects of the Th1/Th2 Concept

In 1986 Mosmann and Coffmann reported on a functional diversity and different cytokine secretion pattern amongst two groups of murine $CD4^+$ T helper cell clones, which ever since have been referred to as type 1 and type 2 T helper (Th1, Th2) cells (Mosmann et al., 1986).

Classification and Functions

Th1 cells are characterized by the synthesis of interleukin 2 (IL-2), interferon-γ (IFN-γ) and tumor necrosis factor-β (lymphotoxin, TNF-β), whereas Th2 cells are producers of IL-4, IL-5, IL-6, IL-10 and IL-13. Both types of cells were found to secrete IL-3, TNF-α and granulocyte-macrophage colony-stimulating factor (GM-CSF) and are believed to develop from a common precursor (designated Thp cells) via a Th0 population, which coexpresses variable sets of Th1 and Th2 cytokines (Mosmann et al., 1991; Kamogawa et al., 1993) (fig. 2.5.1). Some of the divergent functions of Th1 and Th2 cells can be deducted from the known effects of the produced cytokines. Thus, Th1 cells induce cytostatic or antimicrobial activity in macrophages due to their production of IFN-γ and mediate a strong delayed type hypersensitivity (DTH) reaction, whereas Th2 cytokines downregulate macrophage functions (IL-4, IL-10) or stimulate various aspects of B cell growth and differentiation including the isotype switching to IgE (IL-4, IL-5, IL-6). However, it would be a crude oversimplification to assign macrophage activation solely to Th1 cells and B cell help exclusively to Th2 cells. For example, Th2 cells have repeatedly been reported to activate macrophages via membrane-bound TNF-α and cell-cell contact (Stout, 1993), and Th1 cells also provide B cell help which results in the secretion of IgG2a (Coffman et al., 1988). The complexity of the system will become evident during the further discussion.

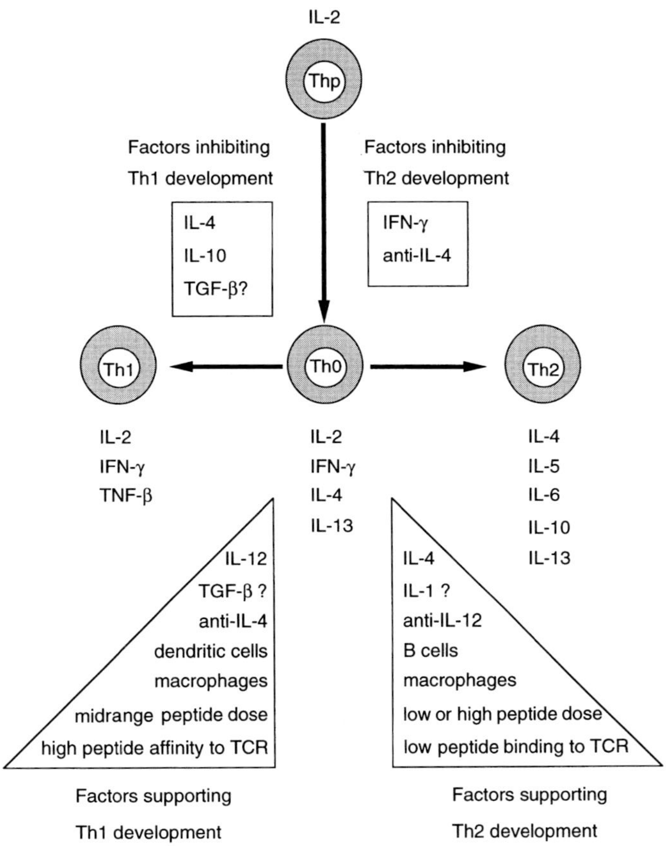

Fig. 2.5.1: Development of Th1 and Th2 cells from a common precursor

Induction, Expansion and Interregulation

The signals which will lead to differentiation of Thp and Th0 cells into Th1 or Th2 cells include cytokines, antigen (peptide) dose, peptide affinity for the T cell receptor (TCR) and interaction with certain subsets of antigen-presenting cells. Th1 cell development is preferentially driven by macrophages, dendritic cells and midrange peptide concentrations, whereas B cells, mast cells and (very) low or very high peptide doses promote the generation and proliferation of Th2 cells (Fitch et al., 1993; Germann et al., 1992; Macatonia et al., 1993; Huels et al., 1995; Macatonia et al.,

1995; Hosken et al., 1995; Constant et al., 1995). Both (splenic) macrophages and dendritic cells, but apparently not primary resting B cells, produce IL-12 (Macatonia et al., 1993; Macatonia et al., 1995). Th1 development of naive $CD4^+$ T cells *in vitro* and *in vivo* is strongly enhanced by IL-12 and supported by neutralization of endogenous IL-4 (Macatonia et al., 1995; Manetti et al., 1993; Hsieh et al., 1993; McKnight et al., 1993; Gately et al., 1994). *In vitro*, optimal differentiation into Th1 cells only occurs in the presence of IL-12 *and* IFN-γ and is not efficiently triggered by IL-12 or IFN-γ alone (Schmitt et al., 1994). *In vivo*, stimulation and expansion of Th1 cells by IL-12 does not require IFN-γ (McKnight et al., 1994; Gately et al., 1994; Schmitt et al., 1994; Wynn et al., 1995), but appears to be partly dependent on natural killer (NK) cells (McKnight et al., 1994). IL-12 stimulates the production of IFN-γ by natural killer (NK) cells, naive $CD4^+$ T cells and cloned Th1 cells (reviewed in Trinchieri, 1994; Germann and Rüde, 1995).

The induction and propagation of Th2 cells, in contrast, requires IL-4 in addition to IL-2 (Le Gros et al., 1990; Swain et al., 1990). There is also evidence that IL-1 stimulates established Th2 rather than Th1 clones (reviewed in Fitch et al., 1993). However, under some circumstances IL-1 was found to promote the generation of IFN-γ-producing T-helper cells (Schmitz et al., 1993). Similarly, transforming growth factor-β (TGF-β), a family of cytokines produced, for example, by platelets, macrophages, fibroblasts and T cells, was both reported to facilitate (Swain et al., 1991; Sad and Mosmann, 1994) or to inhibit the differentiation of Th1-like memory or effector cells (Schmitt et al., 1994).

As it is unlikely that *in vivo* Th1- or Th2-promoting stimuli are present in a mutually exclusive manner, the question arises how a preponderance of either T helper cell subtype will develop, for example during the course of an infection. Considering the known requirements for Th1 and Th2 differentiation *in vitro*, it is likely that the cytokine milieu in the microenvironment will govern the development of the Thp and Th0 cells. Potential early sources of IL-4 are mast cells/basophils (Gordon et al., 1990), γ/δ T cells (Ferrick et al., 1995) or a subpopulation of α/β, memory-like T cells ($CD4^+$, $NK1.1^+$, $CD44^{high}$, $CD62L^{low}$) (MacDonald, 1995). IFN-γ can be provided by NK cells (Naume and Espevik, 1994) as well as macrophages (Fultz et al., 1993; Di Marzio et al., 1994). After induction further propagation of Th2 cells can be achieved by (Th2- or macrophage-derived) IL-10 which (a) downregulates IL-12 production and IL-12-mediated IFN-γ production by NK cells (D'Andrea et al., 1993; Tripp et al., 1993) and (b) inhibits cytokine secretion of Th1, but not of Th2 cells (Mosmann et al., 1991; Fiorentino et al., 1989). Conversely, IFN-γ inhibits the proliferation of Th2, but not of Th1 cells (Fitch et al., 1993) (fig. 2.5.1).

Caveats

Although the Th1/Th2 concept has proven useful in dissecting the immune responses to various microbes, a number of findings clearly indicate its limitations which should also be considered in the context of vaccine design.

- Coexpression of Th1 and Th2 cytokines in various combinations by one T cell is not uncommon. In fact, several studies have demonstrated the existence of far more cytokine patterns than Th1 and Th2 (Schmitz et al., 1993; Kelso and Gough, 1988; Firestein et al., 1989). Th1 and Th2 clones appear to be only two of the possible extreme phenotypes within a wide spectrum (Kelso, 1995). Furthermore, *in vivo* Th1 or Th0 cells are present even in situations, which are dominated by Th2 cells, and vice versa (Lohoff et al., 1989; Powrie et al., 1994).
- A number of cytokines are not readily incorporated into the Th1/Th2 concept, but strongly influence Th1 and Th2 functions *in vitro* and *in vivo*. TGF-β, for example, downregulates IFN-γ mediated macrophage activation, exerts inconsistent (stimulatory or inhibitory) effects on the development of Th1 cells (see above), suppresses the proliferation of T cells and induces IgA switching of B cells (in cooperation with IL-10) (Wahl, 1994; Bogdan and Nathan, 1993). In *Candida (C.) albicans*-infected mice TGF-β delayed progression of disease and was required for optimal Th1 development (Spaccapelo et al., 1995), whereas in *Trypanosoma cruzi* or *L. braziliensis* infected mice TGF-β abrogated a protective Th1 response (Silva et al., 1991; Barral-Netto et al., 1992).
- Th1 and Th2 cytokines can upregulate each others production and are not necessarily antagonistic in their functions. Thus, IL-2 propagates Th1 *and* Th2 cells (Fitch et al., 1993; Le Gros et al., 1990), IL-4 enhances the IFN-γ-induced production of TNF-α by macrophages (Bogdan et al., 1994), and IL-13 synergizes with IL-2 in the induction of IFN-γ by NK cells (Minty et al., 1993). Furthermore, IL-12, which is generally thought to only act on Th1 cells and their production of IFN-γ, can also increase the production of IL-4 by established Th2 clones *in vitro* (Schmitt et al., 1994; Jeannin et al., 1995) and *in vivo* (Wang et al., 1994).
- For quite some time $CD4^+$ Th1 and Th2 cells were thought to be the major source of cytokines amongst T lymphocytes. More recently, however, it was found that $CD8^+$ precursor T cells can differentiate into cytotoxic $CD8^+$ T cells secreting Th1 or Th2 cytokines or into non-cytolytic $CD8^-CD4^-$ T cells, which produce IL-4, IL-5 and IL-10 and provide help to B cells towards antibody production (Croft et al., 1994; Sad et al., 1995; Erard and Le Gros, 1994). Similarly, γ/δ T cells can respond *in vivo* to different pathogens in a biased manner, i.e. release IFN-γ or IL-4, with subsequent development of Th1 or Th2 cells (Ferrick et al., 1995). These results underline that T cells other than $CD4^+$ helpers can influence or account for the cytokine balance in a host organism.

Relevance for Infectious Diseases

As the majority of the findings discussed above were obtained *in vitro* with murine cells, the question needs to be addressed whether these also apply for an *in vivo* situation, especially in humans. A strongly polarized T helper cell response compatible with the expansion of Th1 or Th2 cells has been observed in mice after infection with the protozoan *L. major* (Bogdan et al., 1993; Reiner and Locksley, 1995), the fungus

C. albicans (Puccetti et al., 1995) or the helminth *Trichuris muris* (Else et al., 1994). In these model systems the course of the disease is host strain-dependent and the type of Th response (Th1 or Th2) in the parasitized organs clearly correlates with susceptibility or resistance to disease (tab. 2.5.1). In several other cases (e.g. murine toxoplasmosis, trypanosomiasis, schistosomiasis, listeriosis and tuberculosis) IL-12 also plays a pivitol role for the induction of IFN-γ and protection (Biron and Gazzinelli, 1995), but resistance and disease-mediating responses do not strictly fall into Th1 or Th2 categories. In patients suffering from infectious diseases, mixed T lymphocytes or T cell clones with a Th1- or Th2-like cytokine pattern have been repeatedly isolated leaving no doubt about the appearance of skewed T-helper cell responses in humans (Romagnani, 1994). In patients with lepromatous leprosy, for example, skin lesions are characterized by the expression of Th2 cytokines (IL-4, IL-5, IL-10) and a lack of IFN-γ, which presumably accounts for the presence of exuberant amount of mycobacteria. In contrast, in tuberculous (resistant) lesions IL-12 and Th1 cytokines predominate (Yamamura et al., 1991; Sieling et al., 1994). Thus, in infectious diseases, where the antimicrobial defense mechanisms are known and a protective or disease-promoting response is characterized by the expansion of a defined T cell subset, the Th1/Th2 cytokine concept represents a valuable basis for the rational design of immunomodulatory treatments as well as vaccines. This will be exemplified in the following discussion of immunity against *Leishmania* .

Table 2.5.1: Protective and counter-protective T cell populations in various murine parasite models

parasite model	protective T cell population	non-protective T cell population
L. major	Th1	Th2
C. albicans	Th1	Th2
T. muris	Th2	Th1

2.5.2.2 Principles of the Immune Response to *Leishmania*

Infections with *L. major* and *L. donovani* not only differ significantly in the clinical course of the disease, but also in several parameters governing susceptibility and resistance. Therefore they will be discussed separately. Most of our insights into the immunopathogenesis of leishmanial infections are derived from studies with inbred mice. The results from these models will be presented along with data obtained in patients.

2.5.2.2.1 Mouse Model of *L. major* Infection

In the context of vaccination against *L. major* three questions need to be addressed. First, what are the cellular and cytokine requirements for the development of a protective immune response during primary infection. Second, how can protection be achieved in a host organism which is otherwise unable to control the infection? Third, what components of the immune system are necessary for maintaining resistance to a challenge infection after cured primary disease or successful vaccination? For a more detailed analysis of the immunobiology of leishmaniasis the reader is referred to a number of recent reviews (Bogdan et al., 1993; Reiner and Locksley, 1995; Liew and O'Donnell, 1993).

Resistance-Promoting T Cells and Cytokines

Leishmania reside and replicate within macrophages and dendritic cells (reviewed in Bogdan et al., 1993). Thus, control of the parasite depends first and foremost on the induction of efficient killing mechanisms in its host cells. The macrophage-activating cytokine IFN-γ is indispensable in this regard (fig. 2.5.2). In *L. major*-infected resistant mouse strains (e.g. CBA, C3H, C57BL/6, B10.D2) the spontaneous healing of the cutaneous lesions is paralleled by the expansion of IFN-γ producing, Th1-type $CD4^+$ T lymphocytes. In contrast, the exulceration of the skin lesion and the (lethal) spread of the parasite to visceral organs (spleen, liver, bone marrow) in susceptible mice (BALB/c, SWR/J, DBA/2) correlates with a predominance of IL-4-producing, Th2-type $CD4^+$ T cells and a lack of IFN-γ at least in the later stages of the disease (Sadick et al., 1986; Solbach et al., 1987; Heinzel et al., 1989; Heinzel et al., 1991; Morris et al., 1993; Moll and Röllinghoff, 1990). In the (complete) absence of IFN-γ *L. major* disseminates as uniformly documented by anti-IFN-γ treatment of *L. major*-infected resistant mice (Belosevic et al., 1989; Sadick et al., 1990; Scott, 1991; Leiby et al., 1993) or infection of mice with a genetic deletion of the IFN-γ- or IFN-γ receptor-gene (Wang et al., 1994; Swihart et al., 1995). However, a number of observations demonstrated that IFN-γ alone is not sufficient for the induction of protection. IFN-γ treatment of BALB/c mice led only to a transient improval of the course of infection (Sadick et al., 1990; Scott, 1991). $CD4^+$, *L. major*-specific T cell lines were described, which exhibited a Th1-type cytokine production profile, but still caused exacerbation of *L. major* infection in BALB/c mice (Titus et al., 1991). Transfectants of *L. major* promastigotes expressing IFN-γ were able to activate macrophages *in vitro*, but did not impede the progression of disease in BALB/c mice (Tobin et al., 1993). These results suggest that the resolution of cutaneous leishmaniasis requires the expression of additional cytokines other than IFN-γ (fig. 2.5.2). One of these is TNF-α as documented by the appearance of more severe or even non-healing lesions in *L. major*-infected resistant mice after treatment with anti-TNF-α (Liew et al., 1990a; Titus et al., 1989; Liew et al., 1990b) or soluble TNF-receptor (Garcia et al., 1995).

Although the long-established concept that both disease-promoting (Th0- or Th2-like) and resistance-mediating (Th1-like) T lymphocytes carry the $CD4^+$ surface marker still holds true, a number of recent studies add some complexity to the picture:

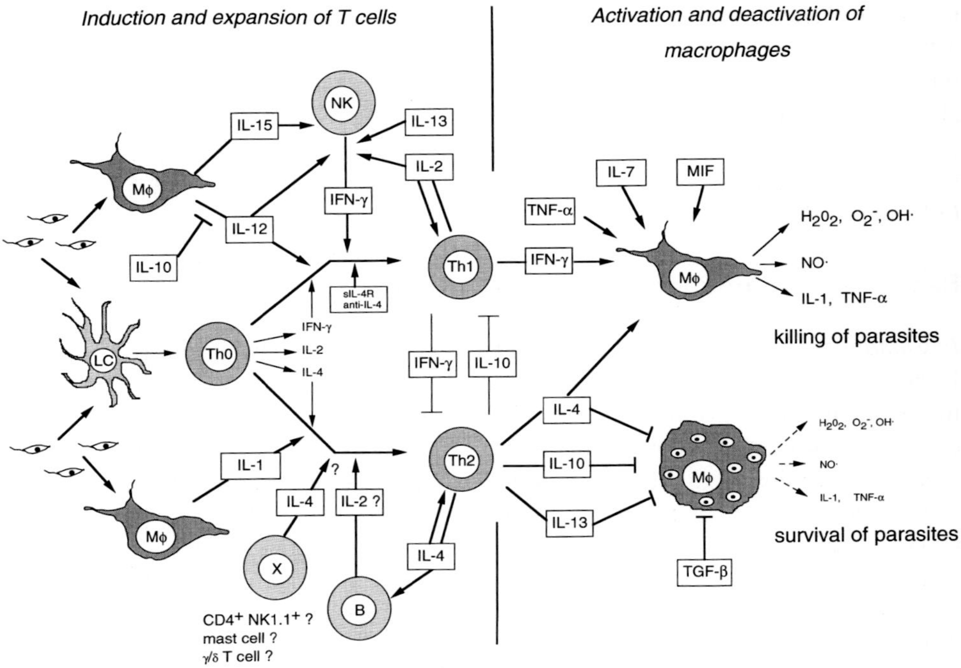

Fig. 2.5.2: Cellular and cytokine network in *L. major* infections
L. major promastigotes are phagocytosed by macrophages (mφ) and Langerhans cells (LC), which present leishmanial antigens to T cells. The development of T helper cell precursors or Th0 cells into IFN-γ- and IL-2-producing Th1 cells is governed (a) by macrophage-derived IL-12; (b) the secretion of IFN-γ by NK cells early after infection, which is triggered by IL-2, IL-12 and IL-13; (c) in the presence of anti-IL-4 or soluble IL-4 receptor (sIL-4R). In contrast, Th0 will preferentially differentiate into Th2 cells (a) if there is a lack of sIL-4R, IL-12 and/or NK-cell-derived IFN-γ; (b) in the presence of a putative early IL-4 source (X), e.g. $CD4^+$ $NK1.1^+$ T cells, mast cells or γ/δ- T cells. Macrophage-derived IL-1 as well as B cells and/or B-cell-derived IL-2 are also implicated in this process. The activation of NK cells by IL-15 has been demonstrated in the human, but not yet in the murine system. TNF-α, IFN-γ, IL-4, IL-7 and migration inhibitory factor (MIF) (synergistically) activate macrophages for the kill of intracellular *Leishmania* via induction of nitric oxide, reactive oxygen intermediates and/or TNF-α. TGF-β, IL-10, and – under certain conditions – IL-4 and IL-13 are able to antagonize these effector pathways. The macrophage-deactivating effect of IL-10 can also result from suppression of Th1 development and Th1 cytokine secretion. → denotes stimulation/induction, ⊥ denotes inhibition. Modified from Bogdan et al., 1993.

- Studies with gene-targeted resistant mice revealed that MHC class II-restricted antigen presentation, but not the expression of the costimulatory molecule CD4 are required for the development of a protective T helper cell population and the control of *L. major* in resistant mice (Chakkalath et al., 1995; Locksley et al., 1993).

- Analysis of the lymph nodes of BALB/c mice chronically infected (4-6 weeks) with *L. major* revealed the existence of two subpopulations of $CD4^+$ T cells, $CD45RB^{high}$ (40-50 %) and $CD45RB^{low}$ (15-20 %). The purified $CD45RB^{high}$ population produced IFN-γ and conferred resistance to congenic immunodeficient *scid* mice, whereas the $CD45RB^{low}$ population exhibited a Th2 cytokine pattern, caused non-healing disease upon adoptive transfer and inhibited the activity of the $CD45RB^{high}$ subpopulation *in vitro* and *in vivo* (Powrie et al., 1994). Thus, in infected BALB/c mice the Th1 $CD4^+$ T cell trait is functionally suppressed rather than *a priori* deleted which makes immunomodulatory vaccination approaches feasible.
- Anti-CD8 treatment during primary infection with *L. major* was reported to lead to a significant exacerbation of the disease in both resistant (CBA) and susceptible (BALB/c) mice (Titus et al., 1987). These earlier findings, however, could not be reproduced in a different strain of mice (129Sv) which lacked $CD8^+$ T cells due to a deletion of the β2-microglobulin gene (Overrath and Harbecke, 1993; Wang et al., 1993). Furthermore, immunization with peptides which led to the induction of cytotoxic T lymphocytes failed to alter the course of infection in BALB/c mice (Wang et al., 1993). Currently, it seems that $CD8^+$ T cells are not critical for the resolution of primary *L. major* infection as long as $CD4^+$ T cells are present. In thymectomized and CD4-depleted BALB/c mice, in contrast, $CD8^+$ T cells were described to control the dissemination of parasites to visceral organs (Hill et al., 1989).

Mechanisms of Susceptibility and Induction of Protection

The strikingly different outcome of *L. major* infection in susceptible BALB/c mice and various strains of resistant mice has sparked off tremendous research efforts over the years to unravel potential underlying mechanisms for this phenomenon. Susceptibility to *L. major* is largely independent of major histocompatibility complex (H-2) genes (Howard et al., 1980), but closely linked to a gene locus (Scl-1) on chromosome 11 (Blackwell et al., 1994). The exact genetic basis for the susceptible phenotype of BALB/c mice is not yet known, but the comparative analysis of various immunological parameters in resistant and susceptible mice following the infection with *L. major* as well as the discovery of a number of immunoprophylactic strategies to induce a protective immune response in BALB/c mice has provided us with plausible hypotheses. Although these concepts focus on different aspects of the induction of a non-protective Th2 response in BALB/c mice, they do not exclude each other. All of them are relevant for the design of an effective vaccine in a susceptible host organism.

- Hypothesis 1: Altered Innate Resistance
  During the natural cause of *L. major* infection the first detectable difference between BALB/c mice and resistant strains (C57BL/6, CBA/J, C3H/HeJ) relates to the kinetics of parasite dissemination. As early as 10 h after cutaneous infection *Leishmania* are already detectable in visceral organs in BALB/c mice, whereas in

resistant strains the parasite remains confined to the local site of infection and the first draining lymph node. Parasite containment in resistant mice occurred in the absence of $CD4^+$ and $CD8^+$ T cells, but was dependent on IL-12, NK cells (NK1.1.$^+$) and IFN-γ (Laskay et al., 1995). Cure of the disease in resistant mice was delayed by depletion of NK cells (Laskay et al., 1993; Scharton and Scott, 1993) and abolished by anti-IL-12 treatment (Heinzel et al., 1993; Sypek et al., 1993; Scharton-Kesten et al., 1995). Conversely, in BALB/c mice containment, NK cell activity and a self-curing course of infection was inducible by treatment with IL-12 (Laskay et al., 1995; Heinzel et al., 1993; Sypek et al., 1993). Activation of NK cells by poly-I:C amelioriated the disease in BALB/c mice (Laskay et al., 1993). In the light of the striking effect of exogenous IL-12 it was unexpected to find that during the early phase of infection (day 1-7) BALB/c mice do not lack IL-12 in the draining lymph node (Scharton-Kesten et al., 1995). However, in the skin lesion and lymph node of *L. major*-infected BALB/c mice there is evidence for enhanced expression of TGF-β (Scharton-Kersten et al., 1995; Stenger et al., 1994), which acts as a functional antagonist of IL-12 and seems to account for the early reduction of IFN-γ production and NK cell activity in BALB/c as compared to resistant mice (Scharton-Kesten et al., 1995). Furthermore, the upregulation of TGF-β in BALB/c mice at the site of infection was accompanied by a reduced amount of inducible nitric oxide synthase (iNOS), which generates leishmanicidal reactive nitrogen intermediates (Stenger et al., 1994). Finally, *in vitro* macrophages from resistant mice were consistently superior to macrophages from BALB/c mice in their ability to release NO in response to IFN-γ and to kill *L. major* (Stenger et al., 1994; Liew et al., 1991). Together, these results underline the importance of innate immunity (macrophages, IL-12, NK cells, IFN-γ, iNOS) in the early phase of infection and suggest that TGF-β and genetically determined differences in macrophage effector functions contribute to the rapid dissemination of *L. major* in BALB/c mice (fig. 2.5.2).

- Hypothesis 2: Upregulation of Early IL-4 Production
  Inhibition of IL-12 and macrophage functions by TGF-β as described above will favor the development of Th2 cells. Alternatively, excess amounts of IL-4 are also able to prime the expansion of Th2 cells. One potential early source of IL-4 is a newly described subclass of $CD4^+$ T cells with a very limited $V_β$ T cell receptor repertoire, which are characterized by the additional presence of NK1.1 and CD44 on the cell surface (MacDonald, 1995). These cells account for ca. 90 % of the IL-4 produced by $CD4^+$ T cells in reponse to anti-CD3 or superantigen (Yoshimoto et al., 1994). As BALB/c mice do not express the NK1.1 marker (Trinchieri, 1989), the function and possible hyperactivity of $CD4^+$ $NK1.1^+$ T cells cannot be directly evaluated in BALB/c mice in the absence of a known NK1.1-homologue. A recent study demonstrated that the early IL-4 response to *L. major* in BALB/c or anti-IFN-γ-treated C57BL/6 mice did not occur amongst $CD4^+$ T cells with a restricted TCR repertoire or positive for the NK1.1 marker, respectively (Launois et al., 1995). Therefore, other sources of IL-4 in the pre-T cell phase of infection have to be considered as discussed in section 2.5.2.1. BALB/c are efficiently pro-

tected against disseminating leishmaniasis by application of anti-IL-4 (Sadick et al., 1990), and all other known preventive treatments in BALB/c mice lead to the attentuation of the normally observed Th2 expansion (reviewed in Bogdan et al., 1993). Thus, it is attractive to postulate that an exuberant production of IL-4 in the early phase of infection accounts for the lethal disease in BALB/c mice.

- Hypothesis 3: Altered Antigen-Presentation and Costimulatory Pathways
  The representation of V_β TCR-elements in the T cell population expanding after infection with *L. major* is very similar in resistant and susceptible mice (Reiner et al., 1993; Lohoff et al., 1994). Thus, there is no convincing evidence that antigen-presenting cells (APC) from resistant and BALB/c mice process different *Leishmania* peptides and thereby preferentially activate Th1 or Th2 cells, respectively. The development of Th1 and Th2 cells, however, could arise from a differential expression of costimulatory cytokines and surface molecules. *L. major* parasitized macrophages from BALB/c mice produced substantially more IL-1 than macrophages from resistant mice (Wagner et al., 1991; Chakkalath and Titus, 1994). IL-1 is known to participate in the induction and proliferation of Th2 rather than Th1 cells (Fitch et al., 1993). *In vitro* experiments with APC and soluble *L. mexicana* antigen suggested that B cells from BALB/c were better in inducing Th2 responses than B cells from resistant C57BL/6 mice, whereas macrophages from C57BL/6 mice were superior to BALB/c macrophages in triggering a Th1 response (Rossi-Bergmann et al., 1993). *In vivo* blockade of the interaction between CD28 and B7 did not alter the course of infection in C57BL/6 mice, but completely abrogated the progressive disease in BALB/c mice (Corry et al., 1994). It appears that different costimulatory pathways are utilized between T cells and APC in resistant as compared to susceptible mice.
- Hypothesis 4: Inherent Defect in the $CD4^+$ Compartment of BALB/c Mice
  Recent experiments with T cell receptor-transgenic mice derived in different genetic backgrounds (but congenic for H-2^d) have demonstrated that under neutral stimulation conditions (i.e. in the absence of exogenous IL-4 or IL-12) T cells from resistant B10.D2 mice will acquire a Th1 phenotype, whereas T cells from susceptible BALB/c mice develop into Th0-like cells (producing 2- to 3-fold less IFN-γ and 3- to 5-fold more IL-4 than the B10.D2 cells). This effect was observed upon restimulation with APC from both strains indicating that the impact of the genetic background on Th phenotype development resides within the T cell and not the APC compartment (Hsieh et al., 1995). As upregulation of IL-12 production by macrophages does not occur after phagocytosis of the infective, promastigote stage of *L. major* (Reiner et al., 1994; Vieira et al., 1994), but only in response to the appearance of intracellular amastigotes (Vieira et al., 1994), T cells might encounter a "cytokine neutral" environment in the earliest phase of infection and therefore succumb to the genetically determined Th pathway. On the other hand, T cells from both genetic backgrounds developed a Th1 or Th2 phenotype after addition of exogenous IL-12 or IL-4, respectively. Thus, a host organism with a genetic predisposition of its T cells to turn into Th2 cells following an infection with

L. major should still mount a Th1 response after treatment with IL-12. As discussed already above, this form of immunoprophylaxis indeed appears to work, at least in the mouse model.

That these concepts of the pathogenesis of non-healing leishmaniasis are complementary rather than mutually exclusive, is nicely demonstrated by recent studies with adult thymectomized lethally irradiated, bone marrow-reconstituted (ATXBM), T cell-chimeric, H-2 congenic mice. Transfer of C57BL/6 $H2^d$ T cells into BALB/c ATXBM mice led to a Th1 response and protection against *L. major* infection. On the other hand, C57BL/6 ATXBM hosts given naive BALB/B6-H-2^d T cells progressed from a Th0 to a Th2 phenotype, but still resolved their infections (Shankar and Titus, 1995). These results and those summarized above clearly indicate that T cell and non-T cell compartments can independently mediate genetically determined resistance to *L. major*.

Maintainance of Resistance to a Challenge Infection After Cured Primary Disease or Successful Vaccination

During primary *L. major* infection the activity of $CD4^+$ T cells, but not of $CD8^+$ T cells is critical for the healing of the disease as well as for the further control of residual parasites persisting in the lymphoid tissue (Müller et al., 1989). Genetically resistant mice, which have resolved a primary infection, only develop small lesions upon reinfection. This resistance to challenge requires not only $CD4^+$, but also $CD8^+$ T cells (Müller, 1992). The protective function of $CD8^+$ T cells in this setting can be explained by their ability to secrete IFN-γ (Müller et al., 1993). In genetically susceptible mice, $CD8^+$ T cells are indispensable for the induction of protection by i.v. immunization (Farrell et al., 1989), anti-IL-4 treatment or partial $CD4^+$ depletion (Müller et al., 1991); in the absence of $CD8^+$ T cells these immunoprophylactic regimes do not confer resistance to BALB/c mice. Furthermore, in contrast to genetically resistant mice the protection of immune BALB/c mice against a challenge infection is primarily mediated by IFN-γ producing $CD8^+$ and not by $CD4^+$ T cells (Müller et al., 1993; Müller et al., 1991). Induced immunity in BALB/c mice was at least partially dependent on TNF-α (Liew et al., 1990a; Theodos et al., 1991). In fact, after vaccination with an avirulent clone of *L. major*, the induced resistance correlated with a reduced production of IL-4 and enhanced levels of TNF-α, but there was no increase of IFN-γ production (Boom et al., 1990). These results suggest that immunization of susceptible hosts should not only aim to induce *L. major*-specific $CD4^+$ Th1 cells, but also $CD8^+$ T cells producing IFN-γ and TNF-α.

2.5.2.2.2 Human Cutaneous Leishmaniasis

Infection of humans with *L. tropica*, *L. major* , *L. braziliensis* or *L. mexicana* usually leads to single, localized, and self-healing lesions with few parasites (LCL). In rare cases, diffuse cutaneous leishmaniasis (DCL) characterized by multiple parasite-rich lesions or a chronic destructive mucocutaneous form of leishmaniasis (MCL) arising years after the healing of the primary skin lesions will develop (Pearson and de Queiroz Sousa, 1995). A kala azar-like disease has also been described after infection with *L. major* or *L. tropica* (Pearson and de Queiroz Sousa, 1995; Mebrahtu et al., 1989; Magill et al., 1993). Several *in situ* studies have documented a normal number of Langerhans cells, $CD4^+$ and $CD8^+$ T cells (CD4/CD8 ratio ca. 1.5) as well as the expression of IFN-γ, IL-2, TNF-α, IL-1β, IL-6 and macrophage-chemoattractant protein-1 (MCP-1) in LCL lesions, whereas IL-4, IL-5, IL-10 and TGF-β were present only at low levels or in lesions with prolonged duration. There is evidence that intralesional $CD4^+$ T cells mostly exhibit a Th1 cytokine pattern, whereas $CD8^+$ T cells account for the IL-4 and IL-10 expression within the T cell compartment (Uyemura et al., 1993). In DCL lesions, in contrast, the number of Langerhans cells, the MHC class II expression of keratinocytes and the CD4/CD8 ratio was decreased, IL-6, TNF and MCP-1 was absent or weakly expressed and Th2 cytokines (IL-4, IL-5, IL-10) and parasite-loaded macrophages were more prominent. MCL granuloma exhibited a strong infiltration of $CD4^+$ T cells and a mixture of Th1 and Th2 cytokines, consistent with the hypersensitivity state of this form of leishmaniasis (Cáceres-Dittmar et al., 1993; Pirmez et al., 1993; Ritter et al., 1996). Peripheral blood T cells from LCL patients responded well to antigen with a high production of IFN-γ, especially in less severe cases and during the healing phase of infection (Murray et al., 1984; Gaadar et al., 1995; Kemp et al., 1994). The response of blood T cells from DCL patients to mitogens or leishmanial antigens was suppressed (Castés et al., 1984; Akuffo et al., 1984). The risk to develop MCL following infection with *L. braziliensis* is increased in patients with certain allelic polymorphisms in the promoter region of the TNF genes affecting the production of TNF-α and TNF-β (Cabrera et al., 1995). Taken together, upregulation of IFN-γ and control of Th2 cytokines and TNF expression appear to be necessary for the cure of at least some leishmanial skin lesions in humans.

2.5.2.2.3 Mouse Model of *L. donovani* Infection

L. donovani infection in mice differs from murine *L. major* infections in that the course of the disease is under the independent control of both non-H-2- and H-2 genes in the early and late phase of infection, respectively (Blackwell et al., 1985). The gene within the non-H2-locus (Ity/Lsh/Bcg locus) on chromosome 1 governing resistance to *L. donovani* (and other intracellular microbes) in the first 2-3 weeks of infection has been cloned (Nramp1, natural resistance associated macrophage protein-1; Lsh^r) and its mutated allele conferring susceptibility is known, but its function is still under investigation (Blackwell et al., 1994; Vidal et al., 1995). It is important to note that *L. donovani* infected mice, which carry the susceptibility allele (Lsh^s, $Nramp1^{Asp169}$) and therefore are unable to restrict the parasite growth during the early phase of infection, can still overcome the disease later on under the influence of their H-2 immune response genes (Vidal et al., 1995).

Unlike the *L. major* model, resolution of primary *L. donovani* infection requires not only $CD4^+$ T cells, but also $CD8^+$ T cells (Stern et al., 1988). Acquisition of resistance involves the secretion of IL-2, IFN-γ and TNF-α. The depletion of each of these cytokines will promote the visceralization of the parasite and delay or even abrogate its control in the liver. However, nude mice will not be rendered resistant by combined treatment with these cytokines in the absence of T cells (Squires et al., 1989; Murray et al., 1993; Murray et al., 1995). Similar to cutaneous leishmaniasis, resistance of *L. donovani*-immune mice to rechallenge was strongly dependent on $CD8^+$ T cells (Murray et al., 1992).

A characteristic feature of visceral leishmaniasis in susceptible hosts is the profound suppression of T cell responses to both mitogens and leishmanial antigen. Parasitized macrophages are likely to be responsible for this effect (Reiner and Finke, 1983; Nickel and Bonventre, 1985). Recent *in vitro* and *in vivo* studies have demonstrated that infection of macrophages with *L. donovani* leads to a reduced expression of MHC class II antigens and to a downregulation of costimulatory molecules (B7-1, heat stable antigen) involved in the activation of T cells (Reiner et al., 1988; Kaye et al., 1994; Saha et al., 1995).

The most remarkable difference between *L. major* and *L. donovani* infection relates to the significance of the Th1/Th2 concept for the pathogenesis of the disease. In visceral leishmaniasis, resistance and susceptibility are determined by the level and duration of IFN-γ production. There is no compelling evidence that over-expansion of Th2 cells contributes to the course of infection. IL-4 and IL-10 was expressed in both curing and non-curing mice and anti-IL-4 treatment did not accelerate control over visceral infection (Kaye et al., 1991; Miralles et al., 1994). The importance of the Th1 cytokines in overcoming any Th2 activity is further underlined by the striking effect of IL-12 in this system. In contrast to *L. major* infection, where IL-12 was unable to cure BALB/c mice with an established Th2 response unless a leishmanicidal drug (pentostam) was coadministered (Nabors et al., 1995), IL-12 was highly effective in accelerating the cure of *L. donovani*-infected BALB/c mice (Lsh^s) whether it was applied at the beginning or at two weeks of infection (Murray and Hariprashad, 1995). IL-12 is likely to be also active in mice strain (e.g. B10.D2/n) which develop a chronic, *non-curing* disease in the absence of an overwhelming Th2 response as the residual Th1 activity should be accessible for recovery.

2.5.2.2.4 Human Visceral Leishmaniasis (kala azar)

Infection of humans with *L. donovani* does not inevitably lead to visceral leishmaniasis. In fact, inapparent infections are probably ten times more common than clinical disease (Evans et al., 1992). Any deficiency of cell mediated immunity will predispose to fulminant visceral leishmaniasis, but nothing is known about potential susceptibility or resistance genes determining the outcome of *L. donovani* infection in humans. The role of the recently cloned human homologue of *Nramp1* in this context remains to be analysed (McLeod et al., 1995).

Visceral leishmaniasis is characterized by an antigen-specific and profound immunosuppression during the acute stage of infection, which is reversible upon successful treatment. Peripheral blood mononuclear cells (PBMNC) of kala-azar patients do not proliferate and do not secrete IFN-γ, IL-2 and IL-12 (p40) in response to leishmanial antigen *in vitro*. Instead, they upregulate IL-10 mRNA upon restimulation (Carvalho et al., 1985; Sacks et al., 1987; Holaday et al., 1993; Ghalib et al., 1993; Ghalib et al., 1995). In one study suppression was transferable by $CD8^+$ T cells, but it is unclear how these cells mediate suppression (production of IL-10 or induction of IL-10 release by other cells) (Holaday et al., 1993). In lymph nodes and bone marrow from acute-stage patients IL-10 mRNA is also prominently expressed, whereas IL-4 and IL-5 mRNA are much less elevated or not detectable. Importantly, whereas the expression of IL-10 mRNA dropped after treatment, IFN-γ and IL-2 mRNA were found at similar levels before and after chemotherapy (Ghalib et al., 1995; Karp et al., 1993). IFN-γ production and proliferation of PBMNC from untreated kala-azar patients are restored by the addition of IFN-γ plus IL-2, anti-IL-10 or IL-12 to the *in vitro* cultures. Successful treatment also leads to the reappearance of antigen-driven IFN-γ and IL-12 secretion (Ghalib et al., 1995; Carvalho et al., 1994). Administration of IFN-γ promotes cure in patients, which failed to respond to pentostam alone (Badaró and Johnson, 1993). As IL-10 is an important downmodulator of various macrophage functions including antimicrobial activity (Bogdan and Nathan, 1993), it is likely that the upregulation of IL-10 during acute kala-azar contributes to the uncontrolled parasite replication. In humans, IL-10 is a product of monocytes/macrophages as well as Th0, Th1 and Th2 cells (del Prete, G. et al., 1993). Fulminant visceral leishmaniasis therefore does not reflect a classical Th2 situation, especially as the Th1 cytokine response is functionally suppressed, but not absent in these patients. Further induction of IFN-γ (e.g. via IL-12) and downregulation of IL-10, however, seem to be crucial for the resolution of the disease. Whether vaccines when mediating this form of immunomodulation will ultimately protect against infection and disease, remains to be demonstrated.

2.5.3 Vaccination Against *Leishmania*

2.5.3.1 General Considerations

In the previous sections we have focussed on the immunological parameters which direct a protective or non-protective immune response against *Leishmania* species. Experimental and clinical studies on cytokines, costimulatory molecules and T cell subpopulations have helped to partially unravel the mechanisms which *amplify* and finally *deviate* the immune response leading to a resistant or susceptible phenotype.

In terms of the development of an efficient vaccination regime, however, we need to exactly know the immunological and non-immunological factors which will or will not allow for the *initiation* of protective responses. As with all other parasites, the complexity of *Leishmania* entails a multitude of potential problems in the design of a vaccine. Most of the following considerations arise from results in the mouse model of cutaneous leishmaniasis discussed above. Thus, it is uncertain, whether they are valid for vaccination against visceral leishmaniasis and the human system.

2.5.3.1.1 Type of Antigen

Whole Parasites

A long-established, historical method for protection of humans against cutaneous leishmaniasis in the Old World has been the injection of live, virulent promastigote parasites (e.g. *L. major*, *L. tropica*, *L. arabica*). Although this procedure is epidemiologically successful (Katzenellenbogen, 1944; Naggan et al., 1972; Modabber, 1989) and, in the absence of an alternative, still persued by the World Health Organization in a number of countries, it is not without problems. First, it is not a classical vaccination, but entails infliction of natural disease with possible risks (e.g. Leishmaniasis recidivans, spreading of the live parasites, reactivation of persisting parasites later during life). Second, it requires long-term culture and *in vivo* propagation of a *Leishmania* clone which has to maintain its virulence over time and needs to be routinely evaluated for its clinical safety. Third, whole *Leishmania* promastigotes do not trigger a protective Th1 pathway per se (Reiner et al., 1994). Thus, the outcome of the immunization very much depends on the immune response and the genetic background of the hosts leaving the possibility that susceptible individuals remain unprotected.

Some of these concerns could be dispelled by using avirulent or inactivated (killed) parasites. Virulent *L. major* promastigotes rendered non-infective by irradiation or live avirulent clones of *L. major* have been shown to protect BALB/c or CBA mice against a challenge with virulent *Leishmania* (Howard et al., 1984; Mitchell et al., 1985; Kimsey et al., 1993). In the case of avirulent *Leishmania* the protection involved control of IL-4 release and upregulation of the production of TNF-α rather than IFN-γ (Kimsey et al., 1993). There is also evidence that i.v. injection of radioattenuated *L. major* promastigotes results in tolerization rather than immunization (Aebischer et al., 1994). In humans, killed *Leishmania* were safe, highly immunogenic (even in the absence of adjuvant) and have shown some protection in two trials in Brazil, although the incidence of cutaneous leishmaniasis was already quite small in the control group (Nascimento et al., 1990; Mayrink et al., 1985; Antunes et al., 1986). There are observations in experimental systems indicating that killed *Leishmania* do not provide the same spectrum of antigens as live parasites (Kimsey et al., 1993; Müller and Louis, 1989). Whether this can negatively affect the efficiency of a killed vaccine in humans, is unknown.

Further parameters which have to be taken into account during the design of whole *Leishmania* vaccine include the parasite strain, its growth phase and developmental stage. In order to improve protection, it might be advisable to combine several *Leishmania* isolates from the living area of the vaccinees. Second, there is experimental evidence that in the logarithmic growth phase promastigote *Leishmania* are less virulent and immunogenic than in the stationary, metacyclic phase (Sacks, 1989; Jarecki-Black et al., 1986). Third, disease is initiated by promastigotes, but entertained by amastigotes. As several genes and antigens are differentially expressed in promastigotes and amastigotes (Zilberstein et al., 1991; Soong et al., 1995; Joshi et al., 1993), immunization with promastigotes alone might be insufficient to confer complete clinical protection.

Protein Fractions, Purified or Cloned Proteins, Peptides

In order to circumvent the problems with whole parasite vaccines, considerable effort has been put into the characterization of subunit vaccines which comprise immunodominant, protective T cell epitopes (tab. 2.5.2). This strategy aims to combine two opposing principles. On the one hand side it is necessary to minimize antigenic complexity in order to eliminate potential disease-promoting components. On the other hand, one requirement for a candidate vaccine is a sufficiently high degree of antigenic promiscuity in order to induce an immune response in individuals of multiple MHC haplotypes. A number of *Leishmania* antigen fractions, purified proteins and peptides have been described which upon immunization lead to partial or complete protection against a subsequent *Leishmania* infection (McConville et al., 1987; Russell and Alexander, 1988; Kahl et al., 1990; Scott et al., 1987; Frommel et al., 1988; Jardim et al., 1990; Steinberger et al., 1984; Yang et al., 1990; White and McMahon-Pratt, 1990; Rachamin and Jaffe, 1993; Mougneau et al., 1995; Symons et al., 1994; Wilson et al., 1995) or which act in an opposite manner, i.e. drastically worsen the course of disease (Jardim et al., 1990; Mitchell and Handman, 1986; Bogdan et al., 1990; Rodrigues et al., 1987; Liew et al., 1990). In one of the major leishmanial surface proteins (gp63) several protective and counter-protective peptide epitopes were mapped (Jardim et al., 1990; Soares et al., 1994). A single purified protein, and even more a fraction of proteins, is therefore unpredictable with regard to its immunizing effect *in vivo*. Nevertheless, immunization of humans with *Leishmania* fractions is a currently persued approach with promising results (Monjour et al., 1986; Monjour et al., 1992).

What makes a peptide mediate resistance or promote disease? The earlier hypothesis that Th1 and Th2 cells differ in their T cell receptor repertoire and therefore recognize distinct peptides was not confirmed by analysis of the V_β usage of *L. major*-infected resistant and susceptible mice (Reiner et al., 1993; Lohoff et al., 1994), but finds some support by studies on intralesional T cells from patients with American cutaneous leishmaniasis (Uyemura et al., 1993). Experimental evidence exists for an alternative model, which suggests that the binding affinity of a peptide ligand to the

Table 2.5.2: Present anti-*Leishmania* vaccines and future candidates

category	examples	references	human trials
whole leishmania	virulent live parasites (leishmanization)	Naggan et al., 1972; Modabber, 1989	GUS, Iran, Israel
	virulent killed parasites	Mayrink et al., 1985; Antunes et al., 1986	Brazil, Venezuela Iran
	avirulent live *Leishmania*	Kinsey et al., 1993	
antigen mixtures	purified antigen fractions	Monjour et al., 1986; Monjour et al., 1992	in progress
purified molecules	LPG	McConville et al., 1987; Russell and Alexander, 1988; Moll et al., 1989	
recombinant molecules	*L. donovani* dp72 *	Rachamin and Jaffe, 1993	
	L. major gp63 *	Russell and Alexander, 1988; Jardim et al., 1990	
	L. mexicana gp46/M-2	McMahon-Pratt et al., 1993	
	L. major PSA-2 *	Symons et al., 1994; Handman, 1995	
	L chagasi Lcr1	Wilson et al., 1995	
	L. major LACK*	Mougneau et al., 1995	
	L. braziliensis LeIF	Skeiky, 1994; Skeiky et al., 1995	
recombinant *Leishmania* DNA (transfected into live vector)	Salmonella/gp63 *vaccinia*/gp46 *BCG*/gp63	Yang et al., 1990; Xu et al., 1995 McMahon-Pratt et al., 1993 Connell et al., 1993	
recombinant *Leishmania* DNA (in DNA vector)	pCMV/gp63	Xu and Liew, 1995	
recombinant (transfected) *Leishmania*	*L. major*/IFNγ *L. major*/DHFR-TS -/- **	Tobin et al., 1993 Cruz et al., 1993; Titus et al., 1995	
combination of antigen plus cytokine	SLA plus IL-12 LACK plus IL-12	Afonso et al., 1994 Mougneau et al., 1995	

* demonstrated to be expressed in several other *Leishmani*a species

** attenuated (or virulent?) *L. major* parasites which a homozygous deletion of the dihydrofolate reductase-thymidylate synthase locus (required for infectivity *in vivo* but not for growth *in vitro*)

MHC class II and to the T cell receptor directs the differentiation of naive $CD4^+$ T cells into Th1-like or Th2-like cells independently of the T cell specificity (Evavold et al., 1993; Windhagen et al., 1995; Pfeiffer et al., 1995) (fig. 2.5.1). A final mechanism how a peptide/protein could modulate the development of Th1 and Th2 cells is via selective induction of cytokines. For at least one *Leishmania* product, the *L. braziliensis* homologue of the eukaryotic initiation factor 4A (Skeiky et al., 1995), this seems to be the case (see below).

Biochemical Nature of Antigen

An important discovery in immunology was the observation that T lymphocytes can also respond to non-protein antigens. A number of mycobacterial lipids have already been defined, which stimulate $CD4^+$ $NK1.1^+$ α/β-T cells or γ/δ-T cells in a CD1-restricted manner. The CD1-antigen presentation pathway is particularly attractive in the context of vaccination as the CD1 proteins unlike classical MHC molecules are not polymorphic, but rather comprise a limited number of different isotypes conserved between species (Bendelac, 1995). Thus, a protective antigen presented by one of the human CD1 isotypes should be highly effective in an MHC-independent manner. In the *Leishmania* mouse model lipophosphoglycan, a glycolipid on the surface of promastigotes and amastigotes, was shown to mediate protection and to stimulate T cells (Moll et al., 1989). Whether this occurs via CD1 presentation is unknown, but possible.

2.5.3.1.2 Route of Antigen Application

In the *L. major* mouse model the site of application appeared to be critical for the nature of the immune response induced by some antigen preparations. Local (e.g. subcutaneous) injection of avirulent, irradiated, heat-killed, sonicated or frozen-thawed *L. major* promastigotes was ineffective or led to exacerbation of a subsequent infection with live parasites, whereas systemic immunization (i.p. or i.v.) conferred complete protection (reviewed in Bogdan et al., 1993; Liew and O'Donnell, 1993). There is also evidence that the course of *L. major* infection and the type of immune response is strikingly dependent on the site of cutaneous inoculation. Skin temperature, lymphatic drainage, microvasculature and the density of tissue dermal macrophages and epidermal Langerhans cells might all influence the development of anti-leishmania immunity (Nabors and Farrell, 1994). These findings raised considerable concerns in terms of local immunization of humans. However, in the limited number of studies performed an exacerbated course of infection as a result of previous injection of killed *Leishmania* was not observed (Modabber, 1989). As will be discussed below the simultaneous application of cytokines as adjuvants might completely eliminate the risk of disease aggravation after local immunization.

2.5.3.1.3 Antigen Dose

Elegant studies and thoughts reviewed elsewhere in this book demonstrated that immunity to *L. major* /*L. tropica* can be achieved by low-dose infection with virulent promastigotes (100-1000) in both resistant (CBA) *and* susceptible (BALB/c) mice (Preston and Dumonde, 1976; Bretscher et al., 1992). This opens up a new way of "leishmanization" in humans. However, factors like the virulence of the used parasite isolate might influence the outcome of such a procedure. In at least one study, BALB/c mice still succumbed to disease after infection with as little as 20 *L. tropica* promastigotes (Howard et al., 1980).

2.5.3.1.4 Host Factors

The impact of the MHC- and non-MHC immune response genes of the host or vaccinee on the course of infection or on the development of protective immunity against *Leishmania* has been repeatedly touched in the previous sections. The MHC polymorphism within a population is certainly a major limiting factor for the efficiency of peptide vaccines. However, recent prototypic analysis of peptides spanning the complete *Leishmania* gp63 molecule has demonstrated that the design of protective and MHC-promiscous peptides is feasible (Soares and Barcinski, 1992). In a reverse approach, protective peptides can be isolated from the MHC class II molecules of infected macrophages and used for the cloning of the respective *Leishmania* peptide donor protein(s) (Campos-Neto et al., 1995). Although we know little about non-MHC genes influencing resistance to *Leishmania*, there is so far no evidence for a form of susceptibility to leishmaniasis which is irreversible and could not be tackled by immunomodulators.

2.5.3.2 Biochemical and Immunological Characterization of Protective *Leishmania* Antigens

Based on immunization experiments with mice there are now a number of well characterized *Leishmania* molecules, which are candidates for vaccines in humans (tab. 2.5.1).

Lipophosphoglycan (LPG)

LPG, a polymer of phosphodiester-linked oligosaccharide units linked via a hexasaccharide glycan core to a novel lyso-1-O-alkylphosphatidylinositol lipid anchor, is the dominant surface glycoconjugate on *Leishmania* promastigotes and essential for parasite infectivity and intracellular survival (Turco and Descouteaux, 1992). LPG is crucial for the binding of non-infective (procyclic) promastigotes to the midgut of the sand-fly vector and subsequently undergoes extensive modifications leading finally to the release of metacyclic, infective promastigotes (Sacks et al., 1995). Amastigotes also exhibit LPG on their surface, but it appears to be antigenically different (Turco and Descouteaux, 1992). Immunization with LPG prevented disease in resistant mice after a challenge with *L. major* and also partially protected susceptible BALB/c. In order to achieve protection it was necessary to combine LPG with an adjuvant (*C. parvum* or complete Freund's adjuvant) or to incorporate it into liposomes. In resistant mice LPG-containing liposomes were also effective after local (subcutaneous) injection. The protective effect of LPG appears to be T cell-dependent (McConville et al., 1987; Russell and Alexander, 1988; Moll et al., 1989). It is possible that proteins tightly associated with LPG contribute to its effect (Bates, 1995). As the internalization of both promastigote and amastigote *Leishmania* is significantly inhibited by anti-LPG-antibodies (Kelleher et al., 1995), part of the protection achieved by LPG might also be antibody-mediated. Whether antibodies against LPG in the blood of vaccinated individuals could interfere with the adhesion of procyclic promastigotes to the midgut of sand-flies after their blood-meal and thereby interrupt the development of infective parasites in the vector, is a speculation which remains to be tested.

Leishmania Major Surface Protease (Glycoprotein gp 63)

Similar to LPG, gp 63 is abundantly expressed on the surface of *Leishmania*, is developmentally regulated in the promastigote and amastigote stage (Schneider et al., 1992; Roberts et al., 1995), participates in the uptake of *Leishmania* by macrophages (Russell and Talamas-Rohana, 1989) and is immunogenic in mice and men (Russell and Alexander, 1988; Jardim et al., 1990; Yang et al., 1990; Xu et al., 1995; Xu and Liew, 1995; Russo et al., 1993; Yang et al., 1993). There is also evidence that the function of gp 63 as a metalloproteinase contributes to the survival of *Leishmania* in their host (Hey et al., 1994). Resistant mice immunized with gp63 encapsulated in liposomes did not develop any lesions after challenge with *L. major*, independent of the route of vaccination. In susceptible mice, the same gp63 preparation was largely ineffective, but significantly delayed the course of infection when combined with LPG and applied intraperitoneally (Russell and Alexander, 1988). Recently, promising results were obtained when gp63 was delivered orally via a transformed and attenuated *Salmonella typhimurium* strain, which constitutively expresses gp63, or when a DNA expression vector containing the gp63 cDNA was injected intramuscularly prior to *L. major* infection. In both cases, the development of cutaneous lesions in BALB/c mice was significantly slower and the lymph node and spleen cells of these mice released much higher levels of IFN-γ compared to the controls, compatible with a Th1 response (Xu et al., 1995; Xu and Liew, 1995). Comparable effects were achieved in a related approach, where recombinant bacille Calmette-Guerin expressing gp63 was inoculated intravenously or subcutaneously prior to challenge with *L. major* or *L. mexicana*. Consistent with the reported low level expression of gp63 on *L. major* as compared to *L. mexicana* amastigotes the protection of BALB/c against *L. major* was weaker than against *L. mexicana* (Connell et al., 1993). Whether BALB/c mice immunized with gp63 by any of these novel techniques will ultimately be able to cure a primary infection and develop strong immunity against rechallenge, remains to be demonstrated. In patients recovered from cutaneous leishmaniasis, Th1-like T cell responses to gp63 were observed, whereas T cells from patients cured of visceral leishmaniasis (*L. donovani*) were mostly non-responsive or released IL-4 upon restimulation with gp63 (Kemp et al., 1994; Kurtzhals et al., 1994). These findings suggest that gp63 is not an immunodominant antigen in kala azar and therefore question its use for immunoprophylaxis.

72 kDa *L. donovani* Protein (dp72)

If applied together with adjuvant (*C. parvum*), dp72, a non-glycosylated protein purified from *L. donovani* membranes by affinity chromatography, partially protected BALB/c mice against a challenge with *L. donovani* and completely prevented lesion development in *L. major*-infected BALB/c during a 6 month observation period. The antigen is expressed in both promastigote and amastigote stages (White and McMahon-Pratt, 1990). The protection is dependent on $CD4^+$ and $CD8^+$ T cells as well as on the presence of IFN-γ. dp72 appears to have a 60 kDa homologue in other *Leishmania* species (*L. tropica*, *L. major*, *L. aethiopica*) and therefore is an excellent candidate for a pan-*Leishmania* vaccine (Rachamin and Jaffe, 1993).

L. mexicana 46 kDa Surface Membrane Glykoprotein (GP46/M-2)

GP46/M-2 is a GPI-anchored membrane glycoprotein with a single carbohydrate chain, which is expressed on the surface of *Leishmania* promastigotes. Apparently it is expressed in all major lineages of the genus except *L. braziliensis*. Treatment of susceptible mice with GP46/M-2 plus *C. parvum* conferred an impressive degree of protection against a low dose challenge with *L. mexicana*. More recently, a recombinant GP46/M-2 vaccinia virus was constructed, which after two i.p. injections was similarly effective. The cytokine secretion pattern of T cells from protected mice was indicative of a mixed $CD4^+$ T helper cell response (Th1 plus Th2 or Th0) (McMahon-Pratt et al., 1993).

The Promastigote Surface Antigen 2 (PSA-2) Complex of *L. major*

The PSA-2 complex of *L. major* consists of 3 major polypeptides of approximate M(r) 96,000, 80,000 and 50,000. The proteins are GPI-anchored, expressed in both promastigotes and amastigotes and present in *L. major*, *L. tropica*, *L. donovani* and *L. mexicana*. The affinity-purified PSA-2 complex plus *C. parvum* completely protected C3H/HeN and conferred partial protection of BALB/c mice against a challenge with *L. major* (Symons et al., 1995; Handman, 1995). Similar to dp72 the PSA-2 complex is a candidate vaccine which might be active against several *Leishmania* species.

L. chagasi Lcr1

LcR1 is a >200 kDa protein which was isolated from a *L. chagasi* amastigote cDNA library after initial immunoscreening (with sera from Brazilian kala-azar patients) followed by secondary T cell screening (with T cells from *L. chagasi*-infected mice). The antigen is present in promastigotes and amastigotes and related to the flagellar repeat antigen of *T. cruzi*. Upon subcutaneous immunization together with complete Freund's adjuvant LcR1 significantly lowered the parasite burden in BALB/c during a subsequent *L. chagasi* infection (Wilson et al., 1995).

Leishmania Homolog of Receptors for Activated C Kinase (LACK)

LACK is a 36 kDa protein that exhibits homology with intracellular receptors for activated protein kinase C (RACK), members of an ancient family of regulatory proteins containing regularly spaced Trp-Asp amino acid sequence motifs. The protein was obtained from a *L. major* cDNA library by expression cloning with a protective Th1 clone specific for soluble leishmanial antigen (SLA). LACK is expressed in both promastigotes and amastigotes and highly conserved within various species of *Leishmania* including *L. major*, *L. donovani*, *L. amazonensis* and *L. chagasi*. Subcutaneous immunization of BALB/c mice with p24, a 24 kDa fragment of LACK, failed to protect BALB/c mice against a challenge with *L. major*, but prevented progressive disease in 72 % of all mice when it was applied in combination with IL-12. Mice, which continued to control parasite replication, exhibited a Th1-like cytokine pattern in their lymph nodes. Interestingly, in the same series of experiments vaccination with gp63 plus IL-12 did not exert any protective effect (Mougneau et al., 1995).

Leishmania Homologue of Eukaryotic Initiation Factor 4 A (LeIF)

LeIF is a 45kDa protein which was cloned from a *L. braziliensis* genomic expression library by screening with sera from a mucocutaneous leishmaniasis patient followed by analysis in T cell proliferation assays. LeIF contains sequence elements which are characteristic of the so-called "DEAD box" family of ATP-dependent RNA helicases. Both promastigotes and amastigotes of *L. braziliensis* and other *Leishmania* species express LeIF. LeIF was able to stimulate proliferation and cytokine secretion of peripheral blood mononuclear cells (PBMC) from the majority of patients with localized cutaneous, mucocutaneous or diffuse cutaneous leishmaniasis. A salient feature of LeIF is its ability to induce the release of biologically active IL-12 (p70) in cultured PBMC from both patients and uninfected individuals. Part of the IL-12 release was dependent on endogenous IFN-γ. LeIF also downregulated the production of IL-10 by patient PBMC and stimulated the generation of IFN-γ (Skeiky et al., 1995). These results are exciting considering earlier observations that whole *Leishmania* promastigotes are poor inducers of macrophage IL-12 *in vitro* (Reiner et al., 1994; Vieira et al., 1994). Overall, LeIF elicited a Th1-like cytokine pattern from leishmaniasis patient

PBMC, whereas total *L. braziliensis* lysate induced a mixed Th1/Th2 profile (Skeiky et al., 1995). *L. braziliensis* LeIF also partially protected BALB/c mice against a challenge with *L. major* (Skeiky, 1994).

2.5.3.3 Induction of Protective CD4$^+$ T Lymphocytes by Combined Immunization with Antigens and Immunomodulators

Based on the *in vitro* and *in vivo* results discussed above, immunization against leishmaniasis should be particularly effective, if the selected antigen(s) do not only contain immunodominant, MHC promiscuous T cell epitopes, but also lead to induction of the IL-12/NK-cell/Th1 pathway. The latter requirement can be easily achieved whenever the antigen itself triggers the production of cytokines (IL-12, IFN-γ) favoring the development of Th1 cells (Skeiky, 1994; Skeiky et al., 1995). In other cases, the inclusion of adjuvants (e.g. *C. parvum*, complete Freund's adjuvant) might provide the necessary stimulatory signals. An interesting, although less practicable approach is the immunization with autologous macrophages presenting *L. major* antigen. Resistance against *L. major* developed in 70 % of BALB/c mice after subcutaneous injection of GM-CSF-derived, antigen-pulsed bone marrow-macrophages 24 h prior to infection at the same site. The same procedure was ineffective, if the macrophages originated from M-CSF-stimulated bone marrow cells (Doherty and Coffman, 1993). This observation is consistent with earlier findings that GM-CSF- but not M-CSF-activated macrophages release IL-12 during their interaction with T cells (Germann et al., 1992). The presumed adjuvant effect of IL-12 was directly demonstrated in the *L. major*-model, where BALB/c mice immunized locally with IL-12 plus soluble leishmanial antigen (SLA) or LACK remained clinically resistant for up to 20 weeks after infection (Mougneau et al., 1995; Afonso et al., 1994). These results are promising as to the use of IL-12 for redirecting the immune response in susceptible individuals. However, it is important to bear in mind that IL-12 might upregulate already established Th2-like responses when there is a lack of IFN-γ (Wynn et al., 1995; Jeannin et al., 1995). This may limit its use for vaccination in patients with a Th2-dependent pathology.

2.5.4 Concluding Remarks

The advances in immunology as well as molecular biology have improved our chances to develop molecularly defined vaccines against cutaneous and visceral leishmaniasis, the resolution of which require an effective Th1-like T cell response. It has become feasible to detect immunodominant T cell antigens by modern expression cloning techniques and to systemically analyse complex molecules for protective T cell epitopes. While it is certainly possible to design a true subunit (peptide) vaccine which is effective in a heterogenous population with diverse MHC backgrounds, it is questionable whether this is also desirable. It might very well be that immunity to leishmaniasis requires persistence of antigen, which is probably hard to achieve after injection of a single peptide or protein. In this context the availability of live recombinant vectors (BCG, *S. typhimurium*) or DNA vaccines, which release antigens for prolonged periods of time, might offer new perspectives. An alternative, and for the moment more practical approach, would be the use of rather complex *Leishmania* antigen preparations (e.g. SLA, avirulent *Leishmania*, combination of several cloned protective antigens expressed in many or all *Leishmania* species), which contain a broad spectrum of T cell epitopes and thereby should increase the percentage of responders amongst vaccinees. In order to assure a T cell response dominated by IFN-γ-producing T cells, the simultaneous treatment with novel immunomodulators (e.g. IL-12) might be warranted. In the future, the use of avirulent *Leishmania* transfected with IL-12 could facilitate to achieve both aims.

Acknowledgements

The preparation of this chapter was supported by the Deutsche Forschungsgemeinschaft (SFB 263 A5). We apologize to those researchers whose original publications we overlooked or only referred to by citing reviews due to the limited space.

References

Aebischer, T., Morris, L., and Handman, E. (1994) Intravenous injection of irradiated *Leishmania major* into susceptible BALB/c mice: immunization or protective tolerance. Int. Immunol. *6,* 1535-1543.

Afonso, L. C. C., Scharton, T. M., Vieira, L. Q., Wysocka, M., Trinchieri, G., and Scott, P. (1994) The adjuvant effect of interleukin-12 in a vaccine against *Leishmania major*. Science *263,* 235-237.

Akuffo, H. O., Fehninger, T. E., and Britton, S. (1984) Differential recognition of *Leishmania aethiopica* antigens by lymphocytes from patients with local and diffuse cutaneous leishmaniasis. J. Immunol. *141,* 2461-2466.

Antunes, C. M., Mayrink, W., Magalhaes, P. A., Costa, C. A., Melo, M. N., Dias, M., Michalick, M. S., Williams, P., Lima, A. O., and Vieira, J. B. (1986) Controlled field trials of a vaccine against New World cutaneous leishmaniasis. Int. J. Epidemiol. *15,* 572-580.

Badaró, R. and Johnson, W. D. (1993) The role of interferon-γ in the treatment of visceral and diffuse cutaneous leishmaniasis. J. Infect. Dis. *167,* S13-17.

Barral-Netto, M., Barral, A., Brownell, C. E., Skeiky, Y. A. W., Ellingsworth, L. R., Twardzik, D. R. and Reed, S. G. (1992) Transforming growth factor-β in leishmanial infection: a parasite escape mechanism. Science *257,* 545-548.

Bates, P. A. (1995) The lipophosphoglycan-associated molecules of *Leishmania*. Parasitol. Today *11*, 317-318.

Belosevic, M., Finbloom, D. S., van der Meide, P. H., Slayter, M. V., and Nacy, C. A. (1989) Administration of monoclonal anti-IFN-γ antibodies in vivo abrogates natural resistance of C3H/HeN mice to infection with *Leishmania major*. J. Immunol. *143*, 266-274.

Bendelac, A. (1995) CD1: presenting unusual antigens to unusual T lymphocytes. Science *269*, 185-186.

Biron, C. A. and Gazzinelli, R. T. (1995) Effects of IL-12 in immune responses to microbial infections: a key mediator in regulating disease outcome. Curr. Opin. Immunol. *7*, 485-496.

Blackwell, J. M., Barton, C. H., White, J. K., Roach, T. I. A., Shaw, M.-A., Whitehead, S. H., Mock, B. A., Searle, S., Williams, H., and Baker, A.-M. (1994) Genetic regulation of leishmanial and mycobacterial infections: the Lsh/Ity/Bcg gene story continues. Immunol. Letters *43*, 99-107.

Blackwell, J. M., Roberts, B., and Alexander, J. (1985) Response of BALB/c mice to leishmanial infection. Curr. Top. Microbiol. Immunol. *122*, 97-106.

Bogdan, C. and Nathan, C. (1993) Modulation of macrophage function by transforming growth factor-β, interleukin 4 and interleukin 10. Ann. NY. Acad. Sci. *685*, 713-739.

Bogdan, C., Gessner, A., and Röllinghoff, M. (1993) Cytokines in Leishmaniasis: a complex network of stimulatory and inhibitory interactions. Immunobiol. *189*, 356-396.

Bogdan, C., Schröppel, K., Lohoff, M., Röllinghoff, M., and Solbach, W. (1990) Immunization of susceptible hosts with a soluble antigen fraction from Leishmania major leads to aggravation of murine leishmaniasis mediated by CD4+ T cells. Eur. J. Immunol. *20*, 2533-2540.

Bogdan, C., Vodovotz, Y., Paik, J., Xie, Q.-W., and Nathan, C. (1994) Mechanism of suppression of nitric oxide synthase expression by interleukin 4 in primary mouse macrophages. J. Leukoc. Biol. *55*, 227-233.

Boom, W. H., Liebster, L., Abbas, A. K., and Titus, R. G. (1990) Patterns of cytokine secretion in murine leishmaniasis: correlation with disease progression or resolution. Infect. Immun. *58*, 3863-3870.

Bretscher, P. A., Wei, G., Menon, J. N., and Bielefeldt-Ohmann, H. (1992) Establishment of stable, cell-mediated immunity that makes "susceptible" mice resistant to *Leishmania major*. Science *257*, 539-542.

Cabrera, M., Shaw, M.-A., Sharples, C., Williams, H., Castes, M., Convit, J., and Blackwell, J. M. (1995) Polymorphism in tumor necrosis factor genes associated with mucocutaneous leishmaniasis. J. Exp. Med. *182*, 1259-1264.

Cáceres-Dittmar, G., Tapia, F. J., Sánchez, M. A., Yamamura, M., Uyemura, K., Modlin, R. L., Bloom, B. R., and Convit, J. (1993) Determination of the cytokine profile in american cutaneous leishmaniasis using the polymerase chain reaction. Clin. Exp. Immunol. *91*, 500-505.

Campos-Neto, A., Soong, L., Cordova, J. L., Sant'Angelo, D., Skeiky, Y. A. W., Ruddle, N. H., Reed, S. G., Janeway, C., and McMahon-Pratt, D. (1995) Cloning and expression of a *Leishmania donovani* gene instructed by a peptide isolated from major histocompatibility complex class II molecules of infected macrophages. J. Exp. Med. *182*, 1423-1433.

Carvalho, E. M., Bacellar, O., Brownell, C., Regis, T., Coffman, R. L., and Reed, S. L. (1994) Restoration of IFN-γ production and lymphocyte proliferation in visceral leishmaniasis. J. Immunol. *152*, 5949-5956.

Carvalho, E. M., Badaro, R., Reed, S. G., Jones, T. C., and Johnson, W. D. (1985) Absence of gamma interferon and interleukin 2 production during active visceral leishmaniasis. J. Clin. Invest. *76*, 2066-2069.

Castés, M., Agnelli, A., and. Rondón, A. J (1984) Mechanisms associated with immunoregulation in human american cutaneous leishmaniasis. Clin. Exp. Immunol. *57*, 279-286.

Chakkalath, H. R. and Titus, R. G. (1994) *Leishmania major*-parasitized macrophages augment Th2-type T cell activation. J. Immunol. *153*, 4378-87.

Chakkalath, H. R., Theodos, C. M., Markowitz, J. S., Grusby, M. J., Glimcher, L. H., and Titus, R. G. (1995) Class II major histocompatibility complex-deficient mice initially control an infection with *Leishmania major* but succumb to the disease. J. Inf. Dis. *171,* 1302-1308.

Coffman, R. L., Seymour, B. W., Lebman, D. A., Hiraki, D. D., Christiansen, J. A., Shrader, B., Cherwinski, H. M., Savelkoul, H. F., Finkelman, F. D., Bond, M. W., and Mosmann, T. R. (1988) The role of helper T cell products in mouse B cell differentiation and isotype regulation. Immunol. Rev. *102,* 5.

Connell, N. D., Medina-Acosta, E., McMaster, R. W., Bloom, B. R., and Russell, D. G. (1993) Effective immunization against cutaneous leishmaniasis with recombinant bacille-Calmette-Guerin expressing the *Leishmania* surface proteinase gp63. Proc. Natl. Acad. Sci. USA *90,* 11473-11477.

Constant, S., Pfeiffer, C., Woodard, A., Pasqualini, T., and Bottomly, K. (1995) Extent of T cell receptor ligation can determine the functional differentiation of naive $CD4^+$ T cells. J. Exp. Med. *182*, 1591-1596.

Corry, D. B., Reiner, S. L., Linsley, P. S., and Locksley, R. M. (1994) Differential effects of blockade of CD28-B7 on the development of Th1 or Th2 effector cells in experimental leishmaniasis. J. Immunol. *153,* 4142-4148.

Croft, M., Carter, L., Swain, S. L., and Dutton, R. W. (1994) Generation of polarized antigen-specific CD8 effector populations: reciprocal action of interleukin (IL)-4 and IL-12 in promoting type 2 versus type 1 cytokine profiles. J. Exp. Med. *180,* 1715-1728.

Cruz, A. K., Titus, R., and Beverley, S. M. (1993) Plasticity in chromosome number and testing of essential genes in *Leishmania* by targeting. Proc. Natl. Acad. Sci. USA *90,* 1599-1603.

D'Andrea, A., Aste-Amezaga, M., Valiante, N. M., Ma, X., Kubin, M., and Trinchieri, G. (1993) Interleukin 10 (IL-10) inhibits human lymphocyte interferon γ-production by suppressing natural killer cell stimulatory factor/IL-12 synthesis in accessory cells. J. Exp. Med. *178,* 1041-1048.

del Prete, G., de Carli, M., Almerigogna, F., Giudizi, M. G., Biagiotti, R., and Romagnani, S. (1993) Human IL-10 is produced by both type 1 helper (Th1) and type 2 helper (Th2) T cell clones and inhibits their antigen-specific proliferation and cytokine production. J. Immunol. *150*: 353-360.

Di Marzio, P., Puddu, P., Conti, L., Belardelli, F., and Gessani, S. (1994) Interferon-γ upregulates its own gene expression in mouse peritoneal macrophages. J. Exp. Med. *179,* 1731-1736.

Doherty, T. M. and Coffman, R. L. (1993) Leishmania antigens presented by GM-CSF-derived macrophages protect susceptible mice against challenge with *Leishmania major*. J. Immunol. *150,* 5476-5483.

Else, K. J., Finkelman, F. D., Maliszewski, C. R., and Grencis, R. K. (1994) Cytokine-mediated regulation of chronic intestinal helminth infection. J. Exp. Med. *179,* 347-351.

Erard, F. and Le Gros, G. (1994) Th2-like CD8 T cells: their role in protection against infectious diseases. Parasitol. Today *10,* 313-315.

Evans, T. G., Teixeira, M. J., McAuliffe, I. T., de Alencar Barros Vasconcelos, I., Vasconcelos, A. W., de Queiroz Sousa, A., de Oliveira Lima, J. W., and Pearson, R. D. (1992) Epidemiology of visceral leishmaniasis in northeast Brazil. J. Infect. Dis. *166,* 1124-1132.

Evavold, B. D., Sloan-Lancaster, J., and Allen, P. M. (1993) Tickling the TCR: selective T-cell functions stimulated by altered peptide ligands. Immunol. Today *14,* 602-609.

Farrell, J. P., Müller, I., and Louis, J. A. (1989) A role for Lyt-2+ T cells in resistance to cutaneous leishmaniasis in immunized mice. J. Immunol. *142,* 2052-2056.

Ferrick, D. A., Schrenzel, M. D., Mulvania, T., Hsieh, B., Ferlin, W. G., and Lepper, H. (1995) Differential production of interferon-γ and interleukin-4 in response to Th1- and Th2-stimulating pathogens by γδ T cells in vivo. Nature *373,* 255-257.

Fiorentino, D. F., Bond, M. W., and Mosmann, T. R. (1989) Two types of mouse T helper cell. IV. Th2 clones secrete a factor that inhibits cytokine production by Th1 clones. J. Exp. Med. *170,* 2081-2095.

Firestein, G. S., Roeder, W. D., Laxer, J. A., Townsend, K. S., Weaver, C. T., Hom, J. T., Linton, J., Torbett, B. E., and Glasebrook, A. L. (1989) A new murine CD4+ T cell subset with an unrestricted cytokine profile. J. Immunol. *143,* 518-525.

Fitch, F. W., McKisic, M. D., Lancki, D. W., and Gajewski, T. F. (1993) Differential regulation of murine T lymphocyte subsets. Ann. Rev. Immunol. *11,* 29-48.

Frommel, D., Ogunkolade, B. W., Vouldoukis, I., and Monjour, L. (1988) Vaccine-induced immunity against cutaneous leishmaniasis in BALB/c mice. Infect. Immun. *56,* 843-848.

Fultz, M. J., Barber, S. A., Dieffenbach, C. W., and Vogel, S. N. (1993) Induction of IFN-γ in macrophages by lipopolysaccharide. Int. Immunol. *5,* 1383-1392.

Gaadar, A., Kharazmi, A., Ismail, A., Kemp, A., Hey, A., Christensen, C. B. V., Dafalla, M., El Kadaro, A. Y., El Hassan, A. M., and Theander, T. G. (1995) Dichotomy of the T cell response to Leishmania antigens in patients suffering from cutaneous leishmaniasis; absence or scarcity of Th1 activity is associated with severe infections. Clin. Exp. Immunol. *100,* 239-245.

Garcia, I., Miyazaki, Y., Araki, K., Lucas, R., Grau, G. E., Milon, G., Belkaid, Y., Montixi, C., Lesslauer, W., and Vassalli, P. (1995) Transgenic mice expressing high levels of soluble TNF-R1 fusion protein are protected from lethal septic shock and cerebral malaria, and are highly sensitive to *Listeria monocytogenes* and *Leishmania major* infections. Eur. J. Immunol. *25,* 2401-2407.

Gately, M. K., Warrier, R. R., Honasoge, S., Carvajal, D. M., Faherty, D. A., Connaughton, S. E., Anderson, T. D., Sarmiento, U., Hubbard, B. R. and Murphy, M. (1994) Administration of recombinant IL-12 to normal mice enhances cytolytic lymphocyte activity and induces production of IFN-γ in vivo. Int. Immunol. *6,* 157-167.

Germann, T. and Rüde, E. (1995) Interleukin 12. Int. Arch. Allergy Immunol. *108,* 103-112.

Germann, T., Mattner, F., Partenheimer, A., Schmitt, E., Reske-Kunz, A. B., Fischer, H.-G., and Rüde, E. (1992) Different accessory function for Th1 cells of bone marrow derived macrophages cultured in granulocyte macrophage colony stimulating factor or macrophage colony stimulating factor. Int. Immunol. *4,* 755-764.

Ghalib, H. W., Piuvezam, M. R., Skeiky, Y. A. W., Siddig, M., Hashim, F. A., El-Hassan, A. M., Russo, D. M., and Reed, S. G. (1993) Interleukin 10 production correlates with pathology in human *Leishmania donovani* infections. J. Clin. Invest. *92,* 324-329.

Ghalib, H. W., Whittle, J. A., Kubin, M., Hashim, F. A., El-Hassan, A. M., Grabstein, K. H., Trinchieri, G., and Reed, S. G. (1995) IL-12 enhances Th1-type responses in human *Leishmania donovani* infections. J. Immunol. *154,* 4623-4629.

Gordon, J. R., Burd, P. R., and Galli, S. J. (1990) Mast cells as a source of multifunctional cytokines. Immunol. Today *11,* 458-464.

Greenblatt, C. L. (1988) Cutaneous Leishmaniasis: the prospects for a killed vaccine. Parasitol. Today *4,* 53-54.

Handman, E., Symons, F. M., Baldwin, T. M., Curtis, J. M., and Scheerlinck, J.-P. Y. (1995) Protective vaccination with promastigote surface antigen 2 from *Leishmania major* is mediated by a Th1 type of immune response. Infect. Immun. *63*, 4261-4267.

Heinzel, F. P., Sadick, M. D., Holaday, B. J., Coffman, R. L., and Locksley, R. M. (1989) Reciprocal expression of interferon-γ or interleukin 4 during the resolution or progression of murine leishmaniasis. Evidence for expansion of distinct helper T cell subsets. J. Exp. Med. *169,* 59-72.

Heinzel, F. P., Sadick, M. D., Mutha, S. S., and Locksley, R. M. (1991) Production of interferon-γ, interleukin 2, interleukin 4 and interleukin 10 by CD4+ lymphocytes in vivo during healing and progressive murine leishmaniasis. Proc. Natl. Acad. Sci. USA *88,* 7011-7015.

Heinzel, F. P., Schoenhaut, D. S., Rerko, R. M., Rosser, L. E., and Gately, M. K. (1993) Recombinant interleukin 12 cures mice infected with *Leishmania major*. J. Exp. Med. *177,* 1505-1509.

Hey, A. S., Theander, T. G., Hviid, L., Hazrati, S. M., Kemp, M., and Kharazmi, A. (1994) The major surface glycoprotein (gp63) from *Leishmania major* and *Leishmania donovani* cleaves CD4 molecules on human T cells. J. Immunol. *152,* 4542-4548.

Hill, J. O., Awwad, M., and North, R. J. (1989) Elimination of CD4+ suppressor T cells from susceptible BALB/c mice releases CD8+ T lymphocytes to mediate protective immunity against *Leishmania*. J. Exp. Med. *169,* 1819-1827.

Holaday, B. J., de Lima Pompeu, M. M., Jeronimo, S., Texeira, M. J., de Queiroz Sousa, A., Vasconcelos, A. W., Pearson, R. D., Abrams, J. S., and Locksley, R. M. (1993) Potential role for interleukin-10 in the immunosuppression associated with kala-azar. J. Clin. Invest. *92,* 2626-2632.

Hosken, N. A., Shibuya, K., Heath, A. W., Murphy, K. M., and O'Garra, A. (1995) The effect of antigen dose on $CD4^+$ T helper cell phenotype development in a T cell-receptor-αβ-transgenic model. J. Exp. Med. *182,* 1579-1584.

Howard, J. G., Hale, C., and Chan-Liew, W. L. (1980) Immunological regulation of experimental cutaneous leishmaniasis. I. Immunogenetic aspects of susceptibility to *Leishmania tropica* in mice. Parasite Immunol. *2,* 303-314.

Howard, J. G., Hale, C., and Liew, F. Y. (1980) Genetically determined susceptibility to *Leishmania tropica* infection is expressed by haematopoietic donor cells in mouse radiation chimeras. Nature *288,* 161-162.

Howard, J. G., Liew, F. Y., Hale, C., and Nicklin, S. (1984) Prophylactic immunization against experimental leishmaniasis II. Further characterization of the protective immunity against fatal Leishmania tropica infection induced by irradiated promastigotes. J. Immunol. *132,* 450-455.

Hsieh, C.-S., Macatonia, S. E., O'Garra, A., and Murphy, K. M. (1993) Pathogen-induced Th1 phenotype development in CD4+ αβ-TCR transgenic T cells is macrophage dependent. Int. Immunol. *5,* 371-382.

Hsieh, C.-S., Macatonia, S. E., O'Garra, A., and Murphy, K. M. (1995) T cell genetic background determines default T helper phenotype development in vitro. J. Exp. Med. *181,* 713-721.

Huels, C., Germann, T., Giedert, S., Hoehn, P., Koelsch, S., Hültner, L., Palm, N., Rüde, E., and Schmitt, E. (1995) Co-activation of naive CD4+ T cells and bone marrow-derived mast cells results in the development of Th2 cells. Int. Immunol. *7,* 525-532.

Jardim, A., Alexander, J., Teh, H. S., Ou, D., and Olafson, R. W. (1990) Immunoprotective *Leishmania major* synthetic T cell epitopes. J. Exp. Med. *172,* 645-648.

Jarecki-Black, J. C., James, E. R., Kirshtein, J. W., Kirshtein, J. D., and Glassman, A. B. (1986) *Leishmania donovani*: immunization against infection as a function of parasite growth phase. Am. J. Trop. Med. Hyg. *35,* 1117-1120.

Jeannin, P., Delneste, Y., Life, P., Gauchat, J.-F., Kaiserlian, D., and Bonnefoy, J.-Y. (1995) Interleukin-12 increases interleukin-4 production by established human Th0 and Th2-like T cell clones. Eur. J. Immunol. *25,* 2247-2252.

Joshi, M., Dwyer, D. M., and Nakhasi, H. L. (1993) Cloning and characterization of differentially expressed genes from in vitro-grown 'amastigotes' of *Leishmania donovani*. Mol. and Biochem. Parasitol. *58,* 345-354.

Kahl, L. P., Lelchuk, R., Scott, C. A., and Beesley, J. (1990) Characterization of Leishmania major antigen-liposomes that protect BALB/c mice against cutaneous leishmaniasis. Infect. Immun. *58,* 3233-3241.

Kamogawa, Y., Minasi, L.-A. E., Carding, S. R., Bottomly, K., and Flavell, R. (1993) The relationship of IL-4- and IFN-γ-producing T cells studied by lineage ablation of IL-4 producing cells. Cell *75,* 985-995.

Karp, C. L., El-Safi, S. H., Wynn, T. A., Satti, M. M. H., Kordofani, A. M., Hashim, F. A., Hag-Ali, M., Neva, F. A., Nutman, T. B., and Sacks, D. L. (1993) In vivo cytokine profiles in patients with kala-azar. Marked elevation of both interleukin-10 and interferon-g. J. Clin. Invest. *91,* 1644-1648.

Katzenellenbogen, I. (1944) Vaccination against oriental sore. Report of results of five hundred and fifty-five inoculations. Arch. Dermatol. Syph. *50,* 239-242.

Kaye, P. M., Curry, A. J., and Blackwell, J. M. (1991) Differential production of Th1- and Th2-derived cytokines does not determine the genetically controlled or vaccine-induced rate of cure in murine visceral leishmaniasis. J. Immunol. *146,* 2763-2770.

Kaye, P. M., Rogers, N. J., Curry, A. J., and Scott, J. C. (1994) Deficient expression of co-stimulatory molecules on *Leishmania*-infected macrophages. Eur. J. Immunol. *24,* 2850-2854.

Kelleher, M., Moody, S. F., Mirabile, P., Osborn, A. H., Bacic, A., and Handman, E. (1995) Lipophosphoglycan blocks attachment of *Leishmania major* amastigotes to macrophages. Infect. Immun. *63,* 43-50.

Kelso, A. (1995) Th1 and Th2 subsets: paradigms lost? Immunol. Today *16,* 374-379.

Kelso, A. and Gough, N. M. (1988) Coexpression of granulocyte-macrophage colony-stimulating factor, γ interferon, and interleukin 3 and 4 is random in murine alloreactive T-lymphocyte clones. Proc. Natl. Acad. Sci. USA *85,* 9189-9193.

Kemp, M., Hey, A. S., Kurtzhals, J. A. L., Christensen, C. B. V., Gaafar, A., Mustafa, M. D., Kordofani, A. A. Y., Ismail, A., Kharazmi, A., and Theander, T. G. (1994) Dichotomy of the human T cell response to Leishmanial antigens. I. Th1-like response to Leishmania major promastigote antigens in individuals recovered from cutaneous leishmaniasis. Clin. Exp. Immunol. *96,* 410-415.

Kimsey, P. B., Theodos, C. M., Mitchen, T. K., Turco, S. J., and Titus, R. G. (1993) An avirulent lipophosphoglycan-deficient *Leishmania major* clone induces CD4+ T cells which protect susceptible BALB/c mice against infection with virulent *L. major*. Infect. Immun. *61,* 5205-5213.

Kurtzhals, J. A. L., Hey, A. S., Jardim, A., Kemp, M., Schaefer, K., Odera, E. O., Christensen, C. B. V., Githure, J. I., Olafson, R. W., Theander, T. G., and Kharazmi, A. (1994) Dichotomy of the human T cell response to *Leishmania* antigens. II. Absent or Th2-like response to gp63 and Th1-like response to lipophosphoglycan-associated protein protein in cells from cured visceral leishmaniasis patients. Clin. Exp. Immunol. *96,* 416-421.

Laskay, T., Diefenbach, A., Röllinghoff, M., and Solbach, W. (1995) Early parasite containment is decisive for resistance to *Leishmania major* infection. Eur. J. Immunol. *25,* 2220-2227.

Laskay, T., Röllinghoff, M., and Solbach, W. (1993) Natural killer cells participate in the early defense against *Leishmania major* infection in mice. Eur. J. Immunol. *23,* 2237-2241.

Launois, P., Ohteki, T., Swihart, K., MacDonald, H. R., and Louis, J. A. (1995) In susceptible mice, *Leishmania major* induce very rapid interleukin-4 production by $CD4^+$ T cells which are $NK1.1^-$. Eur. J. Immunol. *25,* 3298-3307.

Le Gros, G., Ben-Sasson, S. Z., Seder, R., Finkelman, F. D., and Paul, W. E. (1990) Generation of interleukin-4 (IL-4)-producing cells in vivo and in vitro: IL-2 and IL-4 are required for in vitro generation of IL-4 producing cells. J. Exp. Med. *172,* 921-929.

Leiby, D. A., Schreiber, R. D., and Nacy, C. A. (1993) IFN-γ produced in vivo during the first two days is critical for resolution of murine *Leishmania major* infections. Mircobial Pathog. *14,* 495-500.

Liew, F. Y. and O'Donnell, C. A. (1993) Immunology of leishmaniasis. Adv. Parasitol. *32,* 161-259.

Liew, F. Y., Li, Y., and Millott, S. (1990b) Tumor necrosis factor (TNF-α) in leishmaniasis. II. TNF-α induced macrophage leishmanicidal activity is mediated by nitric oxide from L-arginine. Immunology *71,* 556-559.

Liew, F. Y., Li, Y., Moss, D., Parkinson, C., Rogers, M. V., and Moncada, S. (1991) Resistance to *Leishmania major* infection correlates with the induction of nitric oxide synthase in murine macrophages. Eur. J. Immunol. *21,* 3009-3014.

Liew, F. Y., Millott, S. M., and Schmidt, J. A. (1990) A repetitive peptide of *Leishmania* can activate T helper type 2 cells and enhance disease progression. J. Exp. Med. *172,* 1359-1365.

Liew, F. Y., Parkinson, C., Millott, S., Severn, A., and Carrier, M. (1990a) Tumor necrosis factor (TNFα) in leishmaniasis. I. TNFα mediates host protection against cutaneous leishmaniasis. Immunol. *69,* 570-573.

Locksley, R. M., Reiner, S. L., Hatam, F., Littman, D. R., and Killeen, N. (1993) Helper T cells without CD4: control of leishmaniasis in CD4-deficient mice. Science *261,* 1448-1451.

Lohoff, M., Sommer, F., Solbach, W., and Röllinghoff, M. (1989) Coexistence of antigen-specific Th1 and Th2 cells in genetically susceptible BALB/c mice infected with *Leishmania major*. Immunobiology *179,* 412-421.

Lohoff, M., Steinert, M., Weiss, A., Röllinghoff, M., Balderas, R. S., and Theofilopoulos, A. N. (1994) Vβ gene repertoires in T cells expanded in local self-healing and lethal systemic murine cutaneous leishmaniasis. Eur. J. Immunol. *24,* 492-495.

Macatonia, S. E., Hosken, N. A., Litton, M., Vieira, P., Hsieh, C.-S., Culpepper, J. A., Wysocka, M., Trinchieri, G., Murphy, K. M., and O'Garra, A. (1995) Dendritic cells produce IL-12 and direct the development of Th1 cells from naive CD4+ T cells. J. Immunol. *154,* 5071-5079.

Macatonia, S. E., Hsieh, C.-S., Murphy, K. M., and O'Garra, A. (1993) Dendritic cells and macrophages are required for Th1 development of CD4+ T cells from αβ TCR transgenic mice: IL-12 substitution for macrophages to stimulate IFN-γ production is IFN-γ-dependent. Int. Immunol. *5,* 1119-1128.

MacDonald, R. (1995) NK1.1+ T cell receptor-α/β+ cells: new clues to their origin, specificity, and function. J. Exp. Med. *182,* 633-638.

Magill, A. J., Grögl, M., Gasser Jr., R. A., Sun, W., and Oster, C. N. (1993) Visceral infection caused by *Leishmania tropica* in veterans of operation desert storm. New Engl. J. Med. *328,* 1383-1387.

Manetti, R., Parronchi, P., Giudizi, M. G., Piccinni, M.-P., Maggi, E., Trinchieri, G., and Romagnani, S. (1993) Natural killer cell stimulatory factor (Interleukin 12 [IL-12]) induces T helper type 1 (Th1)-specific immune responses and inhibits the development of IL-4-producing Th cells. J. Exp. Med. *177,* 1199-1204.

Mayrink, W., Williams, P., da Costa, C. A., Magalhães, P. A., Melo, M. N., Dias, M., Lima, A. O., Michalick, M. S. M., Carvalho, E. F., Barros,G. C., Sessa, P. A., and de Alencar, J. T. A. (1985) An experimental vaccine against American dermal leishmaniasis: experience in the State of Espirito Santo, Brazil. Annals Trop. Med. Parasitol. *79,* 259-269.

McConville, M. J., Bacic, A., Mitchell, G. F., and Handman, E. (1987) Lipophosphoglycan of Leishmania major that vaccinates against cutaneous leishmaniasis contains an alkylglycerophosphoinositol lipid anchor. Proc. Natl. Acad. Sci. USA *84,* 8941-8945.

McKnight, A. J., Zimmer, G. J., Fogelman, I., Wolf, S. F., and Abbas, A. K. (1994) Effects of IL-12 on helper T cell-dependent immune responses in vivo. J. Immunol. *152,* 2172-2179.

McLeod, R., Buschman, E., Arbuckle, L. D., and Skamene, E. (1995) Immunogenetics in the analysis of resistance to intracellular pathogens. Curr. Opin. Immunol. *7,* 539-552.

McMahon-Pratt, D., Rodriguez, D., Rodriguez, J.-R., Zhang, Y., Manson, K., Bergman, C., Rivas, L., Rodriguez, J. F., Lohman, K. L., Ruddle, N. H., and Esteban, M. (1993) Recombinant vaccinia viruses expressing GP46/M-2 protect against *Leishmania* infection. Infect. Immun. *61,* 3351-3359.

Mebrahtu, Y., Lawyer, P., Githure, J., Were, J. B., Muigai, R., Hendricks, L., Leeuwenburg, J., Koech, D., and Clifford, R. (1989) Visceral leishmaniasis unresponsive to pentastam caused by *Leishmania tropica* in Kenya. Am. J. Trop. Med. Hyg. *41,* 289-294.

Minty, A., Chalon, P., Derocq, J.-M., Dumont, X., Guillemot, J.-C., Kaghad, M., Labit, C., Leplatois, P., Liauzun, P., Miloux, B., Minty, C., Casellas, P., Loison, G., Lupker, J., Shire, D., Ferrara, P., and Caput, D. (1993) Interleukin 13 is a new human lymphokine regulating inflammatory and immune responses. Nature *362,* 248-250.

Miralles, G. D., Stoeckle, M. Y., McDermott, D. F., Finkelman, F. D., and Murray, H. W. (1994) Th1 and Th2 cell-associated cytokines in experimental visceral leishmaniasis. Infect. Immun. *62,* 1058-1063.

Mitchell, G. F. and Handman, E. (1986) The glycoconjugate derived from a Leishmania major receptor for macrophages is a suppressogenic, disease-promoting antigen in murine cutaneous leishmaniasis. Parasite Immunol. *8,* 255-263.

Mitchell, G. F., Handman, E., and Spithill, T. W. (1985) Examination of variables in the vaccination of mice against cutaneous leishmaniasis using living avirulent cloned lines and killed promastigotes of *Leishmania major*. Int. J. Parasitol. *15,* 677-684.

Modabber, F. (1989) Experiences with vaccines against cutaneous leishmaniasis: of men and mice. Parasitology *98,* S49-S60.

Moll, H. and Röllinghoff, M. (1990) Resistance to murine cutaneous leishmaniasis is mediated by Th1 cells, but disease-promoting CD4+ cells are different from Th2 cells. Eur. J. Immunol. *20,* 2067-2074.

Moll, H., Mitchell, G. F., McConville, M. J., and Handman, E. (1989) Evidence that T-cell recognition in mice of a purified lipophosphoglycan from *Leishmania major*. Infect. Immun. *57,* 3349-3356.

Monjour, L., Monjour, E., Vouldoukis, I., Ogunkolade, B. W., and Frommel, D. (1986) Protective immunity against cutaneous leishmaniasis achieved by partly purified vaccine in a volunteer. Lancet *i,* 1490.

Monjour, L., Silva, O. A., Vouldoukis, I., Neogy, A. B., Brito, M. E., Filho, S. P., and Jardim, M. L. (1992) Immunoprophylaxis in cutaneous leishmaniasis. Lancet *340,* 1098-1099.

Morris, L., Troutt, A. B., McLeod, K. S., Kelso, A., Handman, E., and Aebischer, T. (1993) Interleukin-4, but not interferon-γ production correlates with the severity of murine cutaneous leishmaniasis. Infect. Immmun. *61,* 3459-3465.

Mosmann, T. R., Cherwinski, H., Bond, M. W., Giedlin, M. A., and Coffman, R. L. (1986) Two types of murine helper T cell clone. I. Definition according to profiles of lymphokine activities and secreted proteins. J. Immunol. *136,* 2348-2357.

Mosmann, T. R., Schumacher, J. H., Street, N. F., Budd, R., O'Garra, A., Fong, T. A. T., Mond, M. W., Moore, K. W. M., Sher, A., and Fiorentino, D. F. (1991) Diversity of cytokine synthesis and function of mouse CD4+ T cells. Immunol. Reviews *123,* 209-229.

Mougneau, E., Altare, F., Wakil, A. E., Zheng, S., Coppola, T., Wang, Z.-E., Waldmann, R., Locksley, R. M., and Glaichenhaus, N. (1995) Expression cloning of a protective *Leishmania* antigen. Science *268,* 563-566.

Müller, I. (1992) Role of T cell subsets during the recall of immunologic memory to *Leishmania major*. Eur. J. Immunol. *22,* 3063-3069.

Müller, I. and Louis, J. A. (1989) Immunity to experimental infection with *Leishmania major*: generation of protective L3T4+ T cell clones recognizing antigen(s) associated with live parasites. Eur. J. Immun. *19,* 865-871.

Müller, I., Garcia-Sanz, J. A., Titus, R., Behin, R., and Louis, J. (1989) Analysis of the cellular parameters of the immune responses contributing to resistance and susceptibility of mice to infection with the intracellular parasite, *Leishmania major*. Immunol. Reviews *112,* 95-113.

Müller, I., Kropf, P., Etges, R. J., and Louis, J. A. (1993) Gamma interferon response in secondary *Leishmania major* infection: role of CD8+ T cells. Infect. Immun. *61,* 3730-3738.

Müller, I., Pedrazzini, T., Kropf, P., Louis, J., and Milon, G. (1991) Establishment of resistance to *Leishmania major* infection in susceptible BALB/c mice requires parasite-specific CD8+ T cells. Int. Immunol. *3,* 587-597.

Murray, H. W. and Hariprashad, J. (1995) Interleukin 12 is effective treatment for an established systemic intracellular infection: experimental visceral leishmaniasis. J. Exp. Med. *181,* 387-391.

Murray, H. W., Hariprashad, J., and Aguero, B. (1995) Antimicrobial response of a T cell-deficient host to cytokine therapy: effect of interferon-γ in experimental visceral leishmaniasis in nude mice. J. Infect. Dis. *171,* 1309-1316.

Murray, H. W., Miralles, G. D., Stoeckle, M. Y., and McDermott, D. F. (1993) Role and effect of interleukin-2 in experimental visceral leishmaniasis. J. Immunol. *151,* 929-936.

Murray, H. W., Rubin, B. Y., Carriero, S., and Acosta, A. M. (1984) Reversible defect in antigen-induced lymphokine and g-interferon generation in cutaneous leishmaniasis. J. Immunol. *133,* 2250-2254.

Murray, H. W., Squires, K. E., Miralles, C. D., Stoeckle, M., Granger, A. M., Granelli-Piperno, A., and Bogdan, C. (1992) Acquired resistance and granuloma formation in experimental visceral leishmaniasis. Differential T cell and lymphokine roles in initial versus established immunity. J. Immunol. *148,* 1858-1863.

Nabors, G. S. and Farrell, J. P. (1994) Site-specific immunity to *Leishmania major* in SWR mice: the site of infection influences susceptibility and expression of the antileishmanial immune response. Infect. Immun. *62,* 3655-3662.

Nabors, G. S., Afonso, L. C. C., Farrell, J. P., and Scott, P. (1995) Switch from a type 2 to a type 1 T helper cell response and cure of established *Leishmania major*infection in mice is induced by combined therapy with interleukin 12 and pentostam. Proc. Natl. Acad. Sci. USA *92,* 3142-3146.

Naggan, L., Gunders, A. E., and Mochaeli, D. (1972) Follow-up study of a vaccination programme against cutaneous leishmaniasis. II. Vaccination with a recently isolated strain of *L. tropica* from Jericho. Trans. Roy. Soc. Trop. Med. Hyg. *66,* 239-243.

Nascimento, E., Mayrink, W., da Costa, C. A., Michalick, M. S. M., Melo, M. N., Barros, G. C., Dias, M., Antunes, C. M. F., Lima, M. S., Taboada, D. C., and Liu, T. Y. (1990) Vaccination of humans against cutaneous leishmaniasis: cellular and humoral immune responses. Infect. Immun. *58,* 2198-2203.

Naume, B. and Espevik, T. (1994) Immunomodulatory effects of cytokines on natural killer cells. Scand. J. Immunol. *40,* 128-134.

Nickol, A. D. and Bonventre, P. F. (1985) Visceral leishmaniasis in congenic mice of susceptible and resistant phenotypes: immunosuppression by adherent spleen cells. Infect. Immun. *50,* 160-168.

Overath, P. and Harbecke, D. (1993) Course of *Leishmania* infection in β2-microglobulin-deficient mice. Immunol. Letters *37,* 13-17.

Pearson, R. D. and de Queiroz Sousa, A. (1995) Leishmania species: visceral (kala-azar), cutaneous, and mucosal leishmaniasis. In: Mandell, Douglas and Bennett's principles and practice of infectious diseases. Mandell, G. L., Bennett, J. E., and Dolin, R. (eds.). Churchill Livingstone, New York, 2428-2442

Pfeiffer, C., Stein, J., Southwood, S., Ketelaar, H., Sette, A., and Bottomly, K. (1995) Altered peptide ligands can control CD4 T lymphocyte differentiation in vivo. J. Exp. Med. *181,* 1569-1574.

Pirmez, C., Yamamura, M., Uyemura, K., Paes-Oliveira, M., Conceição-Silva, F., and Modlin, R. L. (1993) Cytokine patterns in the pathogenesis of human leishmaniasis. J. Clin. Invest. *91,* 1390-1395.

Powrie, F., Correa-Oliveira, R., Mauze, S., and Coffman, R. L. (1994) Regulatory interactions between CD45RBhigh and CD45RBlow CD4+ T cells are important for the balance berween protective and pathogenic cell-mediated immunity. J. Exp. Med. *179,* 589-600.

Preston, P. M. and Dumonde, D. C. (1976) Experimental cutaneous leishmaniasis V. Protective immunity in subclinical and self-healing infection in the mouse. Clin. Exp. Immunol. *23,* 126-138.

Puccetti, P., Romani, L., and Bistoni, F. (1995) A Th1-Th2-like switch in candidiasis: new perspectives for therapy. Trends in Microbiol. *3,* 237-240.

Rachamin, N. and Jaffe, C. L. (1993) Pure protein from *Leishmania donovani* protects mice against both cutaneous and visceral leishmaniasis. J. Immunol. *150,* 2322-2331.

Reiner, N. E. and Finke, J. H. (1983) Interleukin-2 deficiency in murine leishmaniasis and its relationship to depressed spleen cell responses to phytohemagglutinin. J. Immunol. *131,* 1487-1491.

Reiner, N. E., Ng, W., Ma, T., and McMaster, W. R. (1988) Kinetics of γ interferon binding and induction of major histocompatibility complex class II mRNA in *Leishmania*-infected macrophages. Proc. Natl. Acad. Sci. *85,* 4330-4334.

Reiner, S. L. and Locksley, R. M. (1995) The regulation of immunity to *Leishmania major.* Annu. Rev. Immunol. *13,* 151-177.

Reiner, S. L., Wang, Z.-E., Hatam, F., Scott, P., and Locksley, R. M. (1993) Th1 and Th2 cell antigen receptors in experimental leishmaniasis. Science *259,* 1457-1460.

Reiner, S. L., Zheng, S., Wang, Z.-E., Stowring, L., and Locksley, R. M. (1994) *Leishmania* promastigotes evade interleukin 12 (IL-12) induction by macrophages and stimulate a broad range of cytokines from CD4+ T cells during initiation of infection. J. Exp. Med. *179,* 447-456.

Ritter, U., Moll, H., Laskay, T., Bröcker, E.-B., Castrejon, O. V., Becker, I., and Gillitzer, R. (1996) Differential expression of chemokines in skin lesions of patients with localized and diffuse American cutaneous leishmaniasis. J. Infect. Dis. (in press).

Roberts, S. C., Wilson, M. E., and Donelson, J. E. (1995) Developmentally regulated expression of a novel 59 kDa product of the major surface protease (Msp or gp63) gene family of *Leishmania chagasi.* J. Biol. Chem. *270,* 8884-8892.

Rodrigues, M. M., Mendonca-Previato, L., Charlab, R., and Barcinski, M. A. (1987) The cellular immune response to a purified antigen from *Leishmania mexicana* subsp. *amazonensis* enhances the size of the leishmanial lesion on susceptible mice. Infect. Immun. *55,* 3142-3148.

Romagnani, S. (1994) Lymphokine production by human T cells in disease states. Annu. Rev. Immunol. *12,* 227-257.

Rossi-Bergmann, B., Müller, I., and Godinho, E. B. (1993) Th1 and Th2 T-cell subsets are differentially activated by macrophages and B cells in murine leishmaniasis. Infect. Immun. *1993,* 2266-2269.

Russell, D. G. and Alexander, J. (1988) Effective immunization against cutaneous leishmaniasis with defined membrane antigens reconstituted into liposomes. J. Immunol. *140,* 1274-1279.

Russell, D. G. and Talamas-Rohana, P. (1989) Leishmania and the macrophage: a marriage of inconvenience. Immunol. Today *10,* 328-333.

Russo, D. M., Jardim, A., Carvalho, E. M., Sleath, P. R., Armitage, R. J., Olafson, R. W., and Reed, S. G. (1993) Mapping of human T cell epitopes in *Leishmania* gp63. J. Immunol. *150,* 932-939.

Sacks, D. L. (1989) Metacyclogenesis in *Leishmania* promastigotes. Exp. Parasitology *69,* 100-103.

Sacks, D. L., Lal, S. L., Shrivastava, S. N., Blackwell, J., and Neva, F. A. (1987) An analysis of T cell responsiveness in indian kala-azar. J. Immunol. *138,* 908-913.

Sacks, D. L., Pimenta, P. F. P., McConville, M. J., Schneider, P., and Turco, S. J. (1995) Stage-specific binding of *Leishmania donovani* to the sand fly vector midgut is regulated by conformational changes in the abundant surface lipophosphoglycan. J. Exp. Med. *181,* 685-697.

Sad, S. and Mosmann, T. R. (1994) Single IL-2 secreting precursor CD4 T cell can develop into either Th1 or Th2 cytokine secretion phenotype. J. Immunol. *153,* 3514-3522.

Sad, S., Marcotte, R., and Mosmann, T. R. (1995) Cytokine-induced differentiation of precursor mouse CD8+ T cells into cytotoxic CD8+ T cells secreting Th1 or Th2 cytokines. Immunity *2,* 271-279.

Sadick, M. D., Heinzel, F. P., Holaday, B. J., Pu, R. T., Dawkins, R. S., and Locksley, R. M. (1990) Cure of murine leishmaniasis with anti-interleukin 4 monoclonal antibody. Evidence for a T cell-dependent, interferon-γ-independent mechanism. J. Exp. Med. *171,* 115-127.

Sadick, M. D., Locksley, R. M., Tubbs, C., and Raff, H. V. (1986) Murine cutaneous leishmaniasis: resistance correlates with the capacity to generate interferon-γ in response to leishmania antigens in vitro. J. Immunol. *136,* 655-661.

Saha, B., Das, G., Vohra, H., Ganguly, N. K., and Mishra, G. C. (1995) Macrophage-T cell interaction in experimental visceral leishmaniasis: failure to express costimulatory molecules on *Leishmania*-infected macrophages and its implications in the suppression of cell-mediated immunity. Eur. J. Immunol. *25,* 2492-2498.

Scharton, T. M. and Scott, P. (1993) Natural killer cells are a source of IFN-γ that drives differentiation of CD4+ T cell subsets and induces early resistance to *Leishmania major* in mice. J. Exp. Med. *178,* 567-578.

Scharton-Kersten, T., Afonso, L. C. C., Wysocka, M., Trinchieri, G., and Scott, P. (1995) IL-12 is required for natural killer cell activation and subsequent T helper 1 cell development in experimental leishmaniasis. J. Immunol. *154,* 5320-5330.

Schmitt, E., Hoehn, P., Germann, T., and Rüde, E. (1994) Differential effects of interleukin-12 on the development of naive mouse CD4+ T cells. Eur. J. Immunol. *24,* 343-347.

Schmitt, E., Hoehn, P., Huels, C., Goedert, S., Palm, N., Rüde, E. and Germann, T. (1994) T helper type 1 development of naive CD4+ T cells requires the coordinate action of interleukin 12 and interferon-γ and is inhibited by transforming growth factor-β. Eur. J. Immunol. *24,* 793-798.

Schmitz, J., Assenmacher, M. and Radbruch, A. (1993) Regulation of T helper cell cytokine expression: functional dichotomy of antigen-presenting cells. Eur. J. Immunol. *23,* 191-199.

Schneider, P., Rosat, J. P., Bouvier, J., Louis, J., and Bordier, C. (1992) *Leishmania major*: differential regulation of the surface metalloprotease in amastigote and promastigote stages. Exp. Parasitol. *75,* 196-206.

Scott, P. (1991) IFN-γ modulates the early development of Th1 and Th2 responses in a murine model of cutaneous leishmaniasis. J. Immunol. *147,* 3149-3155.

Scott, P., Pearce, E., Natovitz, P., and Sher, A. (1987) Vaccination against cutaneous leishmaniasis in a murine model. II. Immunologic properties of protective and non-protective subfractions of a soluble promastigote extract. J. Immunol. *139,* 3118-3125.

Shankar, A. H. and Titus, R. G. (1995) T cell and Non-T cell compartments can independently determine resistance to *Leishmania major*. J. Exp. Med. *181,* 845-855.

Sieling, P., Wang, X.-H., Gately, M. K., Oliveros, J. L., McHugh, T., Barnes, P. F., Wolf, S. F., Golkar, L., Yamamira, M., Yogi, Y., Uyemura, K., Rea, T. H., and Modlin, R. L. (1994) IL-12 regulates T helper type 1 cytokine responses in human infectious disease. J. Immunol. *153,* 3639-3647.

Silva, J. S., Twardzik, D. R., and Reed, S. G. (1991) Regulation of Trypanosoma cruzi infections in vitro and in vivo by transforming growth factor β (TGF-β). J. Exp. Med. *174,* 539-545.

Skeiky, Y. A. W. (1994) 8th International Congress of Parasitology, Izmir, Turkey.

Skeiky, Y. A. W., Guderian, J. A., Benson, D. R., Bacelar, O., Carvalho, E. M., Kubin, M., Badaro, R., Trinchieri, G., and Reed, S. G. (1995) A recombinant *Leishmania* antigen that stimulates human peripheral blood mononuclear cells to express a Th1-type cytokine profile and to produce interleukin-12. J. Exp. Med. *181,* 1527-1537.

Soares, L. R. B. and Barcinski, M. A. (1992) Differential production of granulocyte-macrophage colony-stimulating factor by macrophages from mice susceptible and resistant to *Leishmania mexicana amazonensis*. J. Leukocyte Biol. *51,* 220-224.

Soares, L. R. B., Sercarz, E. E., and Miller, A. (1994) Vaccination of the *Leishmania major* susceptible BALB/c mouse. I. The precise selection of peptide determinant influences CD4+ T cell subset expression. Int. Immunol. *5,* 785-794.

Solbach, W., Lohoff, M., Streck, H., Rohwer, P., and Röllinghoff, M. (1987) Kinetics of cell-mediated immunity developing during the course of *Leishmania major* infection in 'healer' and 'non-healer' mice: progressive impairment of response to and generation of interleukin-2. Immunol. *62,* 485-492.

Soong, L., Duboise, M., Kima, P., and McMahon-Pratt, D. (1995) *Leishmania pifanoi* amastigote antigens protect mice against cutaneous leishmaniasis. Infect. Immun. *63,* 3559-3566.

Spaccapelo, R., Romani, L., Tonnetti, L., Cenci, E., Mencacci, A., Del Sero, G., Tognelli, R., Reed, S. G., Pucetti, P., and Bistoni, F. (1995) TGF-β is important in determining the in vivo patterns of susceptibility or resistance in mice infected with Candida albicans. J. Immunol. *155,* 1349-1360.

Squires, K. E., Schreiber, R. D., McElrath, M. J., Rubin, B. Y., Anderson, S. L., and Murray, H. W. (1989) Experimental visceral leishmaniasis: role of endogenous IFN-γ in host defense and tissue granulomatous response. J. Immunol. *143,* 4244-4249.

Steinberger, A., Slutzky, G. M., El-On, J., and Greenblatt, C. L. (1984) *Leishmania tropica*: protective response in C3H mice vaccinated with excreted factor crosslinked with the synthetic adjuvant, muramyl dipeptide. Exp. Parasitol. *58,* 223-229.

Stenger, S., Thüring, H., Röllinghoff, M., and Bogdan, C. (1994) Tissue expression of inducible nitric oxide synthase is closely associated with resistance to *Leishmania major*. J. Exp. Med. *180,* 783-793.

Stern, J. J., Oca, M. J., Rubin, B. Y., Anderson, S. L., and Murray, H. W. (1988) Role of L3T4+ and Lyt-2+ cells in experimental visceral leishmaniasis. J. Immunol. *140,* 3971-3977.

Stout, R. D. (1993) Macrophage activation by T cells: cognate and non-cognate signals. Current Opinion in Immunology *5,* 398-403.

Swain, S. L., Weinberg, A. D., English, M. and Huston, G. (1990) IL-4 directs the development of Th2 like helper effectors. J. Immunol. *145,* 3796-3806.

Swain, S., Huston, G., Tonkonogy, S. and Weinberg, A. (1991) Transforming growth factor-β and IL-4 cause helper T cell precursors to develop into distinct effector helper cells that differ in lymphokine secretion pattern and cell surface phenotype. J. Immunol. *147,* 2991-3000.

Swihart, K., Fruth, U., Messmer, N., Hug, K., Behin, R., Huang, S., Giudice, G. D., Aguet, M., and Louis, J. A. (1995) Mice from genetically resistant background lacking the interferon-γ receptor are susceptible to infection with *Leishmania major* but mount a polarized T helper cell 1-type CD4+ T cell response. J. Exp. Med. *181,* 961-971.

Symons, F. M., Murray, P. J., Ji, H., Simpson, R. J., Osnorn, A. H., Cappai, R., and Handman, E. (1994) Characterization of a polymorphic family of integral membrane proteins in promastigotes of different *Leishmania* species. Mol. Biochem. Parasitol. *67,* 103-113.

Sypek, J. P., Chung, C. L., Mayor, S. H. E., Subramanyam, J. M., Goldman, S. J., Sieburth, D. S., Wolf, S. F., and Schaub, R. G. (1993) Resolution of cutaneous leishmaniasis: interleukin 12 initiates a protective T helper type 1 immune response. J. Exp. Med. *177,* 1797-1802.

Theodos, C. M., Povinelli, L., Molina, R., Sherry, B., and Titus, R. G. (1991) Role of tumor necrosis factor in macrophage leishmanicidal activity in vitro and resistance to cutaneous leishmaniasis in vivo. Infect. Immun. *59,* 2839-2842.

Titus, R. G., Milon, G., Marchal, G., Vassalli, P., Cerottini, J.-C., and Louis, J. A. (1987) Involvement of specific Lyt-2+ T cells in the immunological control of experimentally induced murine cutaneous leishmaniasis. Eur. J. Immunol. *17,* 1429-1433.

Titus, R. G., Müller, I., Kimsey, P., Cerny, A., Behin, R., Zinkernagel, R. M., and Louis, J. A. (1991) Exacerbation of experimental murine cutaneous leishmaniasis with CD4+ *Leishmania major*-specific T cell lines or clones which secrete interferon-γ and mediate parasite-specific delayed-type hypersensitivity. Eur. Immunol. *21,* 559-567.

Titus, R. G., Sherry, B., and Cerami, A. (1989) Tumor necrosis factor plays a protective role in experimental murine cutaneous leishmaniasis. J. Exp. Med. *170,* 2097-2104.

Titus, R. D., Gueiros-Filho, F. J., de Freitas, L. A. R., and Beverley, S. M. (1995) Development of a safe live *Leishmania* vaccine line by gene replacement. Proc. Natl. Acad. Sci. USA *92,* 10267-10271.

Tobin, J. E., Reiner, S. L., Hatam, F., Zheng, S., Leptak, C. L., Wirth, D. F., and Locksley, R. M. (1993) Transfected *Leishmania* expressing biologically active IFN-γ. J. Immunol. *150,* 5059-5069.

Trinchieri, G. (1989) Biology of natural killer cells. Adv. Immunol. *47,* 187.

Trinchieri, G. (1994) Interleukin-12: a cytokine produced by antigen-presenting cells with immunoregulatory functions in the generation of T helper cells type 1 and cytotoxic lymphocytes. Blood *84,* 4008-4027.

Tripp, C. S., Wolf, S. F., and Unanue, E. R. (1993) Interleukin 12 and tumor necrosis factor α are costimulators of interferon-γ production by natural killer cells in severe combined immunodeficiency mice with listeriosis, and interleukin 10 is a physiologic antagonist. Proc. Natl. Acad. Sci. USA *90,* 3725-3729.

Turco, S. J. and Descouteaux, A. (1992) The lipophosphoglycans of *Leishmania* parasites. Annu. Rev. Microbiol. *46,* 65-94.

Uyemura, K., Pirmez, C., Sieling, P. A., Kiene, K., Paes-Oliveira, M., and Modlin, R. L. (1993) CD4+ type1 and CD8+ type 2 T cell subsets in human leishmaniasis have distinct T cell receptor repertoires. J. Immunol. *151,* 7095-7104.

Vidal, S., Tremblay, M. L., Govoni, G., Gauthier, S., Sebastiani, G., Malo, D., Skamene, E., Olivier, M., Jothy, S., and Gros, P. (1995) The Ity/Lsh/Bcg locus: natural resistance to infection with intracellular parasites is abrogated by disruption of the NRAMP1 gene. J. Exp. Med. *182,* 655-666.

Vieira, L. Q., Hondowicz, B. D., Afonso, L. C. C., Wysocka, M., Trinchieri, G., and Scott, P. (1994) Infection of *Leishmania major* induces interleukin 12 production in vivo. Immunol. Letters *40,* 157-161.

Wagner, H. M., Beuscher, H. U., Röllinghoff, M., and Solbach, W. (1991) Interferon-γ inhibits the efficacy of interleukin 1 to generate a Th2-cell biased immune response induced by *Leishmania major*. Immunobiol. *182,* 292-306.

Wahl, S. M. (1994) Transforming growth factor β: the good, the bad, and the ugly. J. Exp. Med. *180,* 1587-1590.

Wang, Z.-E., Reiner, S. L., Hatam, F., Heinzel, F. P., Bouvier, J., Turck, C. W., and Locksley, R. M. (1993) Targeted activation of CD8 cells and infection of β2-microglobulin-deficient mice fail to confirm a primary protective role for CD8 cells in experimental leishmaniasis. J. Immunol. *151,* 2077-2086.

Wang, Z.-E., Reiner, S. L., Zheng, S., Dalton, D. K., and Locksley, R. M. (1994) CD4+ effector cells default to the Th2 pathway in interferon-γ-deficient mice infected with *Leishmania major*. J. Exp. Med. *179,* 1367-1371.

Wang, Z.-E., Zheng, S., Corry, D. B., Dalton, D. K., Seder, R. A., Reiner, S. L., and Locksley, R. M. (1994) Interferon-γ-independent effects of interleukin-12 administered during acute or established infection due to *Leishmania major*. Proc. Natl. Acad. Sci. USA *91,* 12932-12936.

White, A. C. and McMahon-Pratt, D. (1990) Prophylactic immunization against experimental *Leishmania donovani* infection by use of a purified protein vaccine. J. Infect. Dis. *161,* 1313.

Wilson, M. E., Young, B. M., Andersen, K. P., Weinstock, J. V., Metwall, A., Ali, K. M., and Donelson, J. E. (1995) A recombinant *Leishmania chagasi* antigen that stimulates cellular immune responses in infected mice. Infect. Immun. *63,* 2062-2069.

Windhagen, A., Scholz, C., Hollsberg, P., Fukaura, H., Sette, A., and Hafler, D. A. (1995) Modulation of cytokine patterns of human autoreactive T cell clones by a single amino acid substitution of their peptide ligand. Immunity *2,* 373-380.

Wynn, T. A., Jankovic, D., Hieny, S., Zioncheck, K., Jardieu, P., Cheever, A. W., and Sher, A. (1995) IL-12 exacerbates rather than suppresses T helper 2-dependent pathology in the absence of endogenous IFN-γ. J. Immunol. *154,* 3999-4009.

Xu, D. and Liew, F. Y. (1995) Protection against leishmaniasis by injection of DNA encoding a major surface glycoprotein, gp63, of *L. major*. Immunology *84,* 173-176.

Xu, D., McSorley, S. J., Charfield, S. N., Dougan, G., and Liew, F. Y. (1995) Protection against *Leishmania major* infection in genetically susceptible BALB/c mice by gp63 delivered orally in attenuated *Salmonella typhimurium* (AroA-AroD-). Immunology *85,* 1-7.

Yamamura, M., Uyemura, K., Deans, R. J., Weinberg, K., Rea, T. H., Bloom, B. R., and Modlin, R. L. (1991) Defining protective responses to pathogens: cytokine profiles in leprosy lesions. Science *254,* 277-279.

Yang, D. M., Fairweather, N., Button, L. L., McMaster, W. R., Kahl, L. P., and Liew, F. Y. (1990) Oral *Salmonella typhimurium* (AroA-) vaccine expressing a major leishmanial surface protein (gp63) preferentially induces T helper 1 cells and protective immunity against leishmaniasis. J. Immunol. *145,* 2281-2285.

Yang, D., Rogers, M. V., Brett, S. J., and Liew, F. Y. (1993) Immunological analysis of the zinc-binding peptides of surface metalloproteinase (gp63) of *Leishmania major*. Immunology *78,* 582-585.

Yoshimoto, T. and Paul, W. E. (1994) CD4+, NK1.1+ T cells promptly produce interleukin 4 in response to in vivo challenge with anti-CD3. J. Exp. Med. *179,* 1285-1295.

Zilberstein, D., Blumenfeld, N., Liveanu, V., Gepstein, A., and Jaffe, C. L. (1991) Growth at acidic pH induces an amastigote stage-specific protein in *Leishmania* promastigotes. Mol. Biochem. Parasitol. *45,* 175-178.

3. General Principles of Vaccinology

3.1 Modern Adjuvants. Functional Aspects

Bror Morein, Karin Lövgren-Bengtsson and John Cox

3.1.1 Introduction

Despite the enormous number of publications dealing with adjuvants, only aluminium salts are currently registered for human use, although there are a number of adjuvants in phase 1, 2 or 3 human trials. In contrast, there are several adjuvants registered for animal vaccines notably aluminium salts, water in oil emulsions, oil in water emulsions and adjuvants based on saponin products.

During the last decade the capacity to produce antigens by recombinant DNA techniques, expressing single proteins in various cell systems, has dramatically increased the number of antigens available for vaccine use and reduced the price of production. Their limitation is their low immunogenicity and for that reason they require adjuvants both to improve their physical antigen presentation and, in most cases, as immune potentiators to convert these antigens into protective vaccines. This is particularly the case for antigens aimed at inducing protection against pathogens causing persistant infections, for which, in general, we lack protective vaccines. This review does not aim at an extensive coverage of the adjuvant field, as has been done by others (Cox and Coulter, 1992), but rather to consider various categories of adjuvants from a functional point of view. The use of adjuvants in relation to mucosal vaccine administration, vaccine adjuvants in relation to disease and immunization of infants during the period of maternal immunity are selected for special attention.

3.1.1.1 Experience with Conventional Adjuvant Formulations

The combined experience from decades of adjuvant and vaccine research has taught us that adjuvant activity is a result of multiple factors and that an adjuvant effect, in terms of the immune responses obtained with one antigen cannot as a rule be extrapolated to another antigen. Consequently, an immunomodulator must be chosen based on what type of immune response is desired and it must be formulated with the antigen in such a way that both are optimally distributed and presented to relevant lymphatic tissues.

Among the many immunomodulators and adjuvant formulations explored, two in particular have been extensively studied and used. One is based on mineral oil emulsions, with or without the addition of killed mycobacyteria, i.e., Freund's complete and incomplete adjuvants; the other is based on adsorbtion of antigens to a gel formed by aluminium salts, particularly aluminium hydroxide. Freund's complete adjuvant formulation is generally very effective with most types of antigens ranging from small soluble oligopeptides up to killed whole microorganisms. However, the side effects of Freund's adjuvant are unacceptable and prohibit its use in humans and domestic animals. In addition its use in laboratory animals is strongly discouraged, thus efficacious and safe alternatives are urgently needed.

The potential of aluminium hydroxide ($Al(OH)_3$) to increase antibody responses to some vaccine antigens, e.g., toxoids and hepatitis B surface antigen, is well documented. Although we cannot deny their benefits, the aluminium salt adjuvants have limitations. Not all antigens adsorb properly to the gel, and they induce poor cell mediated immune (CMI) responses and essentially no cytotoxic T-lymphocyte (CTL) response. The booster effect afforded by a secondary administration is generally weak and the antibodies induced are predominantly of the non complement-fixing type (IgG1 isotype in the mouse, IgG 2 in humans). In addition, aluminium salts promote the formation of IgE antibodies which may lead to hypersensitivity reactions.

The above two examples of adjuvant formulations are chemically different and have different immunological properties, yet they share one common feature – they interact with antigens to formulate them into larger entities. The aluminium salts adsorb antigen to a gel while in Freund's adjuvant the antigens are in the aqueous phase droplets of the water-in-oil emulsion. Both formulations exhibit a depot effect. According to Osebold (1982), antigen persists in aluminium gel granulomas for 2 to 3 weeks although earlier data showed excision of the depot 14 days post injection did not reduce the final antibody titre (Holt, 1949) suggesting that antigen persisting beyond this time was not immunologically useful. Similarly, antigen in Freunds adjuvant persists at the injection site for months (Sacco et al., 1989) although Freund observed (Freund, 1951) that excision eight days after injection decreased but did not eliminate the adjuvant effect and excision after 14 days failed to modify it at all. These observations suggest that the long-term depot effect of Freunds is not useful and may perhaps be detrimental (Lascelles et al., 1989).

3.1.2 Categories of Adjuvants

Individual antigens vary in their physical and biological properties and antigens may have different needs for help from an adjuvant. Hydrophobic or amphipathic antigens that form multimeric structures or particles are themselves comparatively strong antigens and adjuvants readily potentiate their immune responses in contrast to hydrophilic monomeric antigens.

Efforts have been made to categorize adjuvants through the years, but the terminology is still unclear. The term adjuvant or adjuvant formulation covers a whole range of different activities which lead to increased immunogenicity of an antigen. Allison and Byars (1990) introduced some structure into the terminology by defining an adjuvant as an agent that augments specific immune stimulation to antigens, a vehicle as the substance used for the delivery of antigen and an adjuvant formulation the combination of adjuvant in a suitable vehicle. An even broader definition was advanced by Cox and Coulter (1992) who defined adjuvant as "any substance or procedure which results in a specific increase in the immunogenicity of a vaccine component". Though it is not always feasible to make distinctions for all systems, there are three major areas in which adjuvants may exert their activities:

1. Physical presentation of the antigen, defined by the physical appearance of the antigen in the vaccine. This includes stabilization and exposure of native conformational epitopes in the antigen and also the ability of the adjuvant to formulate the antigen into small soluble particles or aggregates or by some other mechanism bring the antigen into an organized multimeric structure.
2. Antigen/adjuvant uptake and distribution (targeting), which covers a range of activities including slow release of antigen from a depot at the injection site, initiation of the immune response by attracting appropriate antigen presenting cells and other activities leading to increased antigen uptake and transport to relevant lymphatic organs.
3. Immune potentiation/modulation which includes activities that regulate both quantitative and qualitative aspects of the ensuing immune responses. These activities may include the intracellular trafic of antigens, their proteolytic processing, their association with MHC class I or II molecules and the expansion of T cells with different profiles of cytokine production.

3.1.2.1 Characteristics of Some Adjuvant Formulations

Despite the fact that there are few adjuvant formulations commercially available there are numerous adjuvants or adjuvant formulations described in the literature. The more important examples are listed in table 3.1.1; more extensive lists are available in Cox and Coulter (1992) and Powell and Newman (1995).

Table 3.1.1: Categories of Adjuvant Activity

Adjuvant	Presentation	Targeting	Immunomodulation
Oil-in-water emulsions	Good	Moderate	No
Liposomes	Good	Moderate	No
ISCOMs	Good	Moderate	Good
Non-ionic block copolymers	Good	No	?
Water-in-oil emulsions	No	Good	No
Aluminium salts	Moderate	Good	Moderate
Micro-and nano-particles	No	Good	No
Carbohydrates	No	No/yes[1)]	No
Saponins	No	No	Good
MDP derivates	No	No	Good
Lipid A	No	No	Good
Cytokines	No	No	Good

[1)] receptors on macrophages are targets for certain carbohydrates, e.g. manos

Aluminium hydroxide adsorbs soluble antigens to the gel structure and thereby improves the physical presentation of many antigens, particularly monomeric antigens. It is also reported to enhance antigen uptake and presentation by macrophages and upregulate the latter as measured by an increased IL-1 production (Mannhalter et al., 1985). The immune modulation afforded by aluminium salts is moderate and predominantly T_H2 and the CMI responses are poor (Bomford, 1980). Whether the T_H2 type of response is an immune modulatory effect of $Al(OH)_3$ or merely a potentiation of the response generated by the antigen itself is not clear since poorly immunogenic antigens will induce a T_H2 type of response (Hu and Kitagawa, 1990). However, the induction of IgE implies the former.

Saponins, in contrast to $Al(OH)_3$, induce good cellular immune responses, especially with particulate antigens. Monomeric antigens, however, frequently perform poorly when adjuvanted with saponins, most likely because the saponins do not provide a good physical antigen presentation. However, a combination of saponin and $Al(OH)_3$ can yield an efficient formulation, inducing both cellular and humoral responses (Dalsgaard, 1978).

Oil-in-water (o/w) emulsions can be very effective adjuvants for amphipathic molecules, provided the particle size of the discrete oil phase is small (preferably around 200 nm) (WO 90/14837) and the antigen is present during the emulsification process. Careful selection of emulsifiers is also important as was demonstrated by the choice of a mixture of Tween 80 and Span 85 to formulate the Chiron MF-59 adjuvant (Ott et al. 1995). An o/w emulsion without added immunomodulators confers no advantage to an adjuvant formulation if the antigen is not associated with the oil phase.

Another adjuvant group giving efficient physical antigen presentation is the surface active molecules such as non-ionic block polymers (Hunter et al., 1994). The precise mode of action of this class of molecule is unknown but the association of antigen with the polymer seems to be essential, since copolymers with low antigen binding capacity have a low adjuvant activity. Copolymers also augment the expression of MHC class II on macrophages (Howerton et al., 1990) hence it seems likely that a major activity exerted by the block co-polymers relates to a favorable physical presentation of the antigen in a condensed two-dimensional matrix in an environment of activated antigen presenting cells (Hunter et al., 1994).

Addition of substances with good immunomodulating properties such as lipid A or MDP derivatives, along with the proper presentation of the antigen given by the copolymers, can further increase adjuvant activity (Takayama et al., 1991). This concept was used by Allison and Byars (1986; 1991) to make SAF-1 (Syntex adjuvant formulation) which consists of threonyl-MDP (N-acetylmuramyl-L-threonyl-D-isoglutamine) in an emulsion of squalane, non-ionic block polymer (L121) and Tween 80 in phosphate buffered saline.

Ribi Chemical Co. also has several o/w emulsion-based adjuvant formulations designed according to a similar general principle; to formulate a mixture of selected substances with complementary adjuvant activities to yield desired immune stimulation to an antigen (Ribi et al., 1984).

Although physically fundamentally different to the emulsion-based formulations mentioned above, immunostimulating complexes (ISCOM™ adjuvant – Iscotec) (Morein et al., 1984) were created with similar aims in mind. Iscoms are complexes in which antigens are optimally exposed in multimeric presentation in a particulate structure composed of immunomodulating Quillaja saponins (Morein et al., 1995). The iscom is a small (about 40 nm) physically stable particle, with a hydrophilic surface created by the branched sugar chains of the Quillaja saponins. They are highly immunogenic, inducing strong antibody and CMI responses of a T_H1 type.

Lipid vesicles are vehicles for vaccine adjuvant formulations that have attracted the interest of many groups due to their versatility. Liposomes with physically diverse properties can be made by altering the composition of the lipids and by the method used for their production (Alving, 1992). The size of the liposomes may range from 20 nm up to more than 10 μm, they can be uni-or multilamellar, rigid or fusogenic, charged or non-charged. Hydrophilic antigen can be contained in the interior or between lipid bilayers while hydrophobic or amphipathic antigens can be integrated in the lipid membrane during preparation. Due to the great number of variations both of the composition of liposomes and their interactions with different antigens, it is difficult to generalize their effects. It is clear that liposomes are potent in presenting antigens, especially hydrophobic or amphipathic antigens presented on the liposomal surface. However, liposomes by themselves do not efficiently enhance and modulate the immune response and require supplementation with immunomodulators. Almost every possible combination of liposome with other adjuvants has been tested (Kersten and Crommelin, 1995).

A recently described antigen presentation form related to liposomes are the so-called protein cochleate formulations. These are protein-lipid (cholesterol/phosphatidyl serine)-calcium precipitates consisting of a large continuous solid lipid bilayer sheet rolled up in a "jelly roll" fashion (Gould-Fogerite et al., 1994). Amphipathic or hydrophobic antigens are integrated into the lipid bilayer of the cochleates. The calcium maintains the cochleates in their rolled-up form and removal of the calcium by diffusion or addition of chelating agents such as EDTA allows the cochleate to unroll and form large, mainly unilamellar, liposomes.

3.1.2.2 Encapsulation and Slow-Release Formulations

Microencapsulation describes any procedure for the manufacture of small, solid particles, generally in the range 10 nm to 100 μm. Microparticles are ≥ 1 μm; nanoparticles are < 1 μm.

Particles < 5 μm diameter offer excellent targeting to macrophages and dendritic cells, especially if they are composed of a hydrophobic polymer or alternatively have selected carbohydrates bound to their surface. This property has led to the use of such particles for oral vaccine delivery (O'Hagan et al., 1989; Eldridge et al., 1990).

Perhaps more importantly, microencapsulation offers the opportunity to produce vaccines which can give a pulsed (as opposed to trickle) release of antigens at consistent and defined times post-injection. As a result, a single vaccine inoculation can contain a priming dose and one or more booster doses of vaccine. This concept was first described by Eldridge (1991) and subsequent, technological improvements (WO 94/15636) have permitted large-scale preparation of microcapsules which contain both antigen and adjuvant by a process which avoids exposure of biological materials to organic solvents. The most common polymer used in microencapsulation is polylactide coglycolide (PLG), the basis of resorbable sutures. This polymer degrades by hydrolysis, the length of time for breakdown being determined both by chain length and ratio of the lactic to glycolic acid units.

3.1.2.3 Immunomodulators – One Part of the Adjuvant Formulation

Immunomodulatory molecules act upon the cytokine network to cause an overall upregulation which is frequently selective for T_H1 or T_H2 (types of responses). Their activity *in vivo* is of short duration and is not dependent upon association with antigen. Optimal efficiency, however, is generally obtained when both immunomodulator and antigen are formulated into multimeric particles. For this reason immunomodulators are frequently incorporated into liposomes and o/w formulations. Conversely, certain particulate adjuvants are also immunomodulatory e.g. iscoms, aluminium salts. The activity of a range of immunomdulators is summarised in table 3.1.2.

Table 3.1.2: Principal Actions of Immunomodulatory Adjuvants

T_H1 Induction:	lipophilic MDP eg MTP-PE lipid A and derivatives (eg MPL) avridine, DDA TDM DHEA, DHEAS
T_H2 Induction:	hydrophilic MDP eg GMDP vitamin D3 poly A:U aluminium salts
T_H1/T_H2 Induction:	saponins poly 1:C and poly ICLC

3.1.3 Antigen Presentation and Targeting

3.1.3.1 Uptake and Intracellular Distribution of Ag in Antigen Presenting Cells

The first stage by which the adjuvant can influence the processing of the antigen is its attachment to APC and its internalization. Amine containing compounds like DDA and avridine are reported to act by their positive charge causing electrostatic attachment of Ag or by hydrophobic interaction (Snippe et al., 1981). Similarly the SAF-1 formulation might attach Ag by the block polymer component. These compounds are poorly soluble in water but well suited for incorporation into o/w emulsions.

Macrophages are both APC and professional scavenger cells, two potentially antagonistic activities. Limitation of their proteolytic activity on the Ag may, therefore, enhance their capacity to present Ag. This has been postulated as the mode of adjuvant activity of dextran sulphate (DXS).

While there is a vast literature on Ag processing and presentation by APC to T-cells, there is a limited number of reports of how adjuvants influence these activities in APC. One reason for this is that most adjuvants are toxic *in vitro* to cells, as is the case with oil adjuvants or $Al(OH)_3$. In contrast, iscoms are well suited for cell culture work as well as for immuno-electron- microscopy where their stability makes it possible to follow by EM for about 30 minutes their internalization in cells (Watson et al., 1992). In contrast, micelles disintegrate and cannot be visualized in intracellular vesicles. In the studies of Villacres-Eriksson (1993) using immuno-EM on biotinylated influenza virus Ag in iscoms, these antigens were traced in about equal amounts to both cytosol and vesicles. These results were further supported by quantitative studies which determined the amount of biotinylated Ag in subcellular fractions obtained by differential centrifugation. A quantitative ELISA using a polyclonal antibody for capture and streptavidin peroxidase was used for the detection. While macrophages take up 50 %, DC 16 % and B-cells 13 % of iscom borne Ag the corresponding values for influenza virus micelles were 25 to 50-fold lower. Enhancement resulting in efficient Ag uptake by APC is probably an important function of an adjuvant. The capacity to deliver Ag to the cytosol is likely to pave the way for MHC class I restricted Ag presentation resulting in cytotoxic T lymphocyte (CTL) response (Takahashi et al., 1990). This is a feature largely confined to acid sensitive liposomes (Harding et al., 1991), iscoms and synthetic lipopeptide vaccines among the non-viable Ag delivery systems (Deres et al., 1989). With iscoms, long-lived cytotoxic memory T-cells are induced in mice after one s.c. immunization (Takahashi et al., 1990). One reason why iscoms induce CTL is their strong capacity to stimulate lymphocytes producing IFN-γ and IL-2 i.e. a T_H1 type of response (for ref. see Morein et al., 1995). There are a number of reports of other adjuvants which are claimed to enhance CTL, but generally high doses of Ag and several immunizations are required.

3.1.3.2 Adjuvant Influences the Transport of Ag and Localization of B and T Cell Responses Following Parenteral Immunization

The uptake of non-adjuvanted influenza virus Ag and influenza iscoms by the spleen was compared. Following intraperitoneal administration of influenza iscoms there was a significant though transient increase of neutrophils in the peritoneal fluid and a subsequent higher amount of radioactive virus Ag was found in the spleen up to eight days (Watson et al., 1989). Recently, an elegant fluorochrome labelling method was introduced (Claassen et al., 1995) where lipophilic fluorescent carbocyanine dye, which readily incorporates into lipids and liposomes was used to study the kinetics of liposome uptake by macrophages. Liposomes are taken up by marginal zone macrophages (MZM) and red pulp macrophages in the spleen following i.v. administration. MZM are known to be important for uptake of particulate Ag e.g. liposomes and bacteria. Rabies virus (RV) and RV glycoprotein iscoms were labelled by this technique and their localization followed after i.v., i.p. and s.c. administration. Two hours after i.v. and i.p. injections RV was found in MZM as expected. Conversely, iscoms were found mainly in the marginal metallophilic macrophages (MMM) located at the border of the marginal zone, and to a lesser extent, in MZM, follicular dendritic cells and B cells. MMM are characterized by slender processes protruding into periarteriolar lymphoid sheat (PALS) and are the main site of antibody production in the spleen and hence a strategic localization for antibody stimulation.

The role of MMM in Ag handling is unclear (Claassen et al., 1995) but it has been suggested that they are involved in Ag processing (Kraal et al., 1988), while MZM function as a means of rapid removal and elimination of Ag thus removing it from immune responses. This difference of distribution pattern between RV and RV iscoms may partly explain why RV iscoms were 20 to 30 times more immunogenic than RV. When injected s.c. 0.08 μg of RV iscoms would induce a RV-neutralizing antibody response while 10 μg of RV was required to obtain a similar neutralizing antibody response. Full protection to intracerebral challenge was obtained with 0.4 μg of RV iscoms.

Ag's are transported from the site of injection to the draining lymph nodes and subsequently specific T and B cell responses can be detected in various lymphatic tissues e.g. lymph nodes, spleen and bone marrow (BM). This process can be positively or negatively influenced by adjuvants. For example, CFA can cause a delay in transfer of the antibody producing cells from draining lymph nodes to BM due to the granulopoiesis induced in the spleen (Benner et al., 1981a). In the early 70's, pioneering studies were performed to locate specific sites for antibody producing cells in mice (for review see Benner et al., 1981a). In general, these studies showed that BM is an important site for antibody production. The antibody formation was concluded to be dependent on migration of Ag-activated lineage cells from elsewhere. The use of CFA or high doses of LPS interfered with or even abolished the ongoing Ig synthesis in BM. In contrast, $Al(OH)_3$ neither caused granulopoiesis, nor interfered with antibody formation in the BM and $Al(OH)_3$ adjuvanted antigen induced increased numbers of antibody producing cells in the BM after booster.

Recently, Sjölander et al. (1996) studied the distribution of B and T-cell responses after parenteral immunization with influenza virus envelope Ag as micelles incorporated into iscoms or adjuvanted with CFA. The T-cell response, measured by proliferation and production of IL-2, interferon gamma (IFN-γ) and IL-4, was first confined to the draining lymph node both for iscom borne Ag and Ag adjuvanted with CFA. It was transient for iscom but comparatively long lasting for CFA adjuvanted Ag. In the spleen, however, the T-cell response was prominent for iscom-borne Ag but low for CFA adjuvanted Ag.

The B-cell response after immunization with iscoms (Sjölander et al., 1996) measured as Ab producing cells, was first detected in draining lymph nodes, and was low in the spleen with a late but prominent response in BM. The implication of a strong BM response seems to be that Ab production is retained there for a long period of time. Moreover an increasing proportion of the Ab producing cells of all isotypes are located there with increasing age (Benner et al., 1981a; Benner et al., 1981b). Possibly the BM as an organ producing antibodies is particularly important for elderly people and animals. The mechanisms behind distribution of B-memory response in BM and effects of various adjuvants on that distribution need to be further explored.

3.1.3.2.1 Induction of CTL

The induction of CTL responses generally requires that antigens are processed within the cytosol (the endogenous pathway) where peptides, generally nonamers, become incorporated within the closed-end groove of the MHC class I molecule and are then expressed on the cell surface. This processing contrasts with the exogenous or class II pathway. Current evidence suggest that the turnover of cellular proteins occurs in a 26S multi-enzyme complex. The proteolytic component of which is the proteasome, a highly conserved 20S structure comprising 24-28 subunits (Rivett, 1993; Rechsteiner et al., 1993; Driscoll, 1994; Goldberg and Rock, 1992). The majority of proteins which pass through this complex exit as peptides which are further processed by exopeptidases to amino acids. However a small proportion are selectively transported to the ER (endoplasmic reticulum) by specialized transporter proteins (TAP1 and TAP2) where they are incorporated into the MHC I groove of the forming MHC class I and passed via the Golgi to the cell surface (Neefjes et al., 1995). There is increasing evidence that a low molecular mass protein (LMP) is a specialized proteasome where two (or more) of the subunits are MHC encoded (Rivett, 1993). The presence of these MHC-encoded subunits within a proteasome may modify the proteolytic cleavage towards peptides which are MHC compatible. Production of LMP is upregulated by IFN-γ (Goldberg and Rock, 1992) and it is tempting to speculate that this is a mechanism whereby the proportion of peptide capable of insertion into MHC-I can be increased in response to a cytokine warning signal.

For an adjuvant to be useful for CTL induction it must facilitate incorporation of appropriate peptides into MHC-I. The most effective way to achieve this is for the adjuvant to interact in some way with cell membranes so that antigen associated with the adjuvant is deposited within the cytosol in a form suitable for normal processing in the proteasome. This may occur by fusion with the external membrane or by endocytosis/pinocytosis followed by endosome membrane fusion or rupture (endosomal escape). Incorporation of an immunomodulator within this adjuvant formulation, especially one which induces IFN-γ production, could be expected to increase relevant MHC-I peptide expression. Although all cells express MHC-I the most effective target for CTL induction is an APC (Macatonia, 1989).

An alternative mechanism for CTL induction is by direct attachment of peptides to empty, externally exposed MHC-I. Suitable adjuvants can enhance this process by creating a depot containing peptides and foreign antigen to attach to APC. This is best achieved with a w/o formulation where the aqueous phase contains peptide and an ubiquitous protein e.g., tetanus toxoid. The w/o emulsion creates a depot which will attract APC and at the same time protect the peptide from proteolysis. The tetanus toxoid will cause the APC to migrate to lymphoid tissue and, whilst presenting the tetanus-derived peptides, will attract T cells some of which will be CD8+ cells which recognize the original peptide (Scalzo et al., 1995).

3.1.3.3 Adjuvants and Delivery Systems for Induction of Mucosal Immunity

In recent years there has been an increasing interest in adjuvants and vaccine delivery systems for induction of mucosal immune responses especially by the oral and to a lesser extent by respiratory tract routes. There are three problems to overcome for oral vaccines; the acid pH in the stomach, the mucosal barrier and the induction of tolerance which is clearly observed where oral exposure precedes parenteral immunization.

The requirements of adjuvants are different for mucosal and parenteral modes of administration. In contrast to parenterally administered antigens which contact APC at the site of injection, mucosally administered antigen must cope with and penetrate the mucus to reach APC. Antigens for mucosal administration also require adjuvants targeting to Peyer's patches (PP) or lymphoid tissues in the lamina propria (LP). Further they demand activation of Th cells expressing a different array of regulatory cytokines above all, IL-5, IL-4, TGF-β and IFNγ.

Perhaps the most successful mucosal adjuvant is CT which induces a strong secretory antibody response and a long-term immunological memory in mice when added to an unrelated Ag e.g. keyhole limplet hemocyanin (KLH). Memory B and T cells are detected locally in PP in the intestinal LP and mesenteric lymphnodes, and also systemically in the spleen. The B-cell memory initially encompasses IgM producing cells but on re-exposure to recall antigen they switch to IgG and IgA (Vajdy and Lycke, 1993). Memory T cell responses were dominated by IL-4, IL-5 but IL-2 was also present (Vajdy and Lycke, 1993). In contrast, CTB seems not to induce memory cells. The thermolabile enterotoxin of E coli (LT) is very similar to CT in structure and mode of action.

The B subunits of CT or LT are good for targeting but they have a weak adjuvant activity in contrast to the whole toxin. Therefore, considerable efforts were expended to modulate the A subunit to abolish toxicity but to retain its antigenicity and adjuvant activity. In very recent results, Fontana et al. (1995) constructed a potentially efficient CT vaccine where, by site directed mutagenesis, the toxic activity was eliminated but the antigenicity and strong immunogenicity remained and the construct was able to induce neutralizing antibodies against both A and B subunits. Other groups (Dickinson and Clements, 1995; Douce et al., 1995) have constructed LT's with strong immunogenicity but no toxicity.

Iscoms have also been shown to prevent induction of immunological tolerance and to exert adjuvant activity in the digestive tract. Low but repeated oral doses of iscoms induce secretory IgA, CTL and systemic immune responses (Mowat and Maloy, 1994). Using fluorochrome labelled iscoms containing the G protein of rabies virus, Claassen et al. (1995) showed that iscoms target PP more effectively than rabies virus particles. Another possible route for targeting to the lymphatic system in the gut is through the enterocytes which may act as APC (Santos et al., 1990). A strong indication that iscoms may use enterocytes as APC was shown by Lazarova et al. (1995). Incorporated influenza virus Ag was transported in an apical to basolateral direction across a monolayer of Caco-2 intestinal epithelial cells and collected for further analysis. In contrast to non-adjuvanted influenza virus Ag in micelle form, the iscom borne Ag was processed and induced in vitro a dose dependent proliferative response in T cells from primed mice. Ag fragments collected were 10-20 amino acids smaller than those collected from non-adjuvanted micelles. Iscom matrix added to the micelle Ag preparation enhanced significantly the transport through the Caco-2 epithelial cells. These results indicate that both iscoms and iscom matrix influence Ag processing in the gut, and may target two routes for induction of immune responses.

Microspheres of suitable size (< 10 μm in diameter) target PP and can remain there for extended periods. Microspheres were also detected in mesenteric lymph nodes from 1 to 35 days but only those with a diameter < 5 μm (Eldrich et al., 1990). Other particles like liposomes (Adams et al., 1987; Labbe et al., 1991) and VLPs enhance immune responses to various Ag by targeting PP.

The respiratory tract is the second desired target for a mucosal vaccine. Locally applied Ag may induce tolerance possibly by γ/δ T cells (McMenamin et al., 1991), an observation which requires consideration for prospective vaccines. Various liposome preparations have been tested and in general low secretory IgA titers along with serum antibody responses are obtained after two immunizations (Wilschut et al., 1994; De Haan et al., 1995). Liposomes co-presented as a separate entity to the Ag in an experimental vaccine were as effective as liposomes used as a carrier for influenza virus Ag (De Haan et al., 1995) possibly by abolishing a state of tolerance or anergy (Weiner et al., 1994). It is also possible that the mechanism for the observed mucosal adjuvant activity of liposomes in the respiratory tract involves an inhibition of the immune suppression exerted by alveolar macrophages by their saturation with liposomes (Thepen et al., 1989). If this is correct, the use of liposomes in this way has little practical value. It should be noted that CT and LT are equally efficient for delivery to respiratory tract as for the digestive tract (Staats et al., 1994). Iscoms have been used for respiratory tract delivery to mice of a number of Ags mainly from enveloped proteins e.g. corona virus and various herpesvirus and respiratory syncytial virus (RSV) (for references see Morein et al., 1995). Recently iscoms were shown to induce good serum antibody responses after one intranasal (IN) immunization, and two IN immunizations with 5 μg RSV-iscoms induced high secretory IgA titers in the lungs as well as in uterus-vagina of mice (not published data). Iscoms containing the envelope protein of influenza virus induced protection upon challenge (Lövgren et al., 1990, and Jones et al., 1988) and IgA producing cells and CTL memory cells.

3.1.4 Vaccine-Disease Relationships

3.1.4.1 Requirement of Adjuvant for Parenterally Administered Vaccines for Prevention of Disease Caused by Mucosal Infections

Generally, infectious agents have a preferred site of invasion and colonisation which most commonly is a mucosal surface consisting of mucous layered over epithelial cells. In the mucous, a first line of defence is exerted mainly by secretory IgA. However, most vaccines are designed for parenteral immunization. This discrepancy has practical reasons but is obviously not ideal. Local mucosal immunity is in general best stimulated by replicating antigens, whereas systemic immune responses are readily induced parenterally by non-replicating antigens. However, both routes of administration require adjuvants for non-replicating antigens to avoid multiple injections and to reach the levels of immune responses required for protection.

Killed poliovaccine used in Scandinavian countries and in the Netherlands is a good example of the use of a subcutaneous immunisation procedure to protect against a disease which develops following natural oral infection. The immune response to natural infection and oral vaccination is characterised by IgA and TH1 cells, a response which is low or non-existant after parenteral immunization (Mahon et al., 1995). The parenteral poliovaccine is either unadjuvanted or uses $Al(OH)_3$ and requires, an initial three dose regimen and subsequent, regular boosting. Although the killed vaccine does not protect against oral infection, it protects against paralytic polio (Böttiger, 1984). It induces a TH2 response in contrast to what would be considered an optimal immune response as indicated by the work of Mahon et al. (1995) who used a poliovirus-receptor transgenic BALB/c-(H2-d) model which measured protection to i.v. challenge. The protection was conferred with primed B-cells in combination with polyclonal T cells or by B-cells in the presence of a VP-4 specific Th1 clone. B or T cells on their own would not confer protection.

The currently used inactivated poliovaccine will not prevent an infection in the gut and virus is excreted following natural infection and can be traced to sewage (Böttiger, 1973; Böttiger, 1992). Most countries use the live orally administered polio vaccine which induces a high degree of protection to infection.

Foot and mouth disease virus (FMD) is another picornavirus classified as an aftaevirus, which is widely spread in the world mainly among cattle but also among pigs. It infects via the respiratory route and tonsils but following viremia, the virus is spread to the skin causing blisters on the nose and between the cloves. The goal of vaccination is to protect the animal but more importantly to prevent the spread of virus. For many years, vaccination programmes based on adjuvanted killed virus have been used. While these killed vaccines to various degrees were protective, live virus vaccines administered by various routes were not. For FMDV, circulating immunity based on antibody is important in prevention of viremia and subsequent establishment of virus infection in various epithelial cells which in turn result in the disease and become the source for spread to other animals. Interestingly, protective immunity was induced by most adjuvants tested including $Al(OH)_3$, $AlPO_4$, water-in-oil adjuvants and dextran. Experimental FMD-vaccines have been used as models for development of new adjuvants, the most efficient being the saponin adjuvant (Dalsgaard, 1978). The use started with crude saponins from various plants, and progressed to more defined products such as saponin from quality controlled batches (Bomford, 1980; Strobbe et al., 1976) and finally defined purified fractions of Quillaja saponaria Molina, e.g. Quil A (Dalsgaard, 1974). Perhaps the most efficient formulation is Quil A used with $Al(OH)_3$ (Dalsgaard, 1978; Dalsgaard et al., 1977) which gives a relatively high and long lasting IgG serum antibody response in cattle inducing both IgG1 and IgG2.

Influenza vaccine has its own sphere of interest and application. Although it causes infection in the respiratory tract, there is no proof that local administration of vaccine via the respiratory tract induces a better protection than parenteral administration.

Such studies have been carried out in animal systems based on challenge with pathogenic strains. Protection was as good after s.c. or i.m. immunizations with subunit vaccines or whole virus vaccines as after respiratory tract immunization (Lövgren et al., 1990; Ben-Ahmeida et al., 1994). The explanation may depend on the fact that the respiratory tract consists of at least two compartments. The upper part, encompassing the sinuses and nasal cavities, has a mucosa where secretory IgA is the dominant antibody induced by local administration. In contrast, the mucus of the lower respiratory tract has a mixed antibody composition with both IgG and secretory IgA, the former derived from the serum pool of IgG. The most obvious reason for induction of secretory IgA response to influenza virus is that a broader reacting immune response encompassing a larger cross reactivity between isolates and subserotypes would be achieved (Liew et al., 1984; Asanuma et al., 1995).

An important arm of the protective immune response to influenza virus is the more broadly reacting CTL response which recognizes epitopes from heterologous subserotypes both in the membrane proteins (HA and NA) as well as in the nucleoprotein. The target for CTL is the virus infected cell hence the CTL exerts its effect later than preformed antibodies which are active as a first line of immune protection (Yap et al., 1978). Influenza virus antigen in iscoms was shown to induce CTL in a mouse model by both i.n. (Jones et al., 1988) and by s.c. immmunization (Takahashi et al., 1990). Recent studies in monkeys (Osterhaus, manuscript in preparation) show that influenza virus iscoms after one immunization with 15 μg, induce high ELISA serum antibody titres, virus neutralizing antibodies and hemagglutination inhibition titres. A strong booster effect was obtained after revaccination one month later with more than tenfold titre increases in all tests. In contrast, monkeys showed no response with the Ag alone as a split vaccine. It is a general feature in man and horses as well as in monkeys that envelope proteins from influenza virus have low immunogenicity in contrast to what is the case in mice and guinea pigs, emphasizing a need for a suitable adjuvant. Aluminium salts are unsatisfactory because they show very low immunoenhancing effect on influenza virus antigen, and promote a Th2 instead of the desired Th1 type of response. From published data it appears that iscoms containing the envelope proteins of influenza virus will induce a comparatively long lasting protective immune response up to 18 months when tested in the natural host horse (Mumford et al., 1994).

3.1.4.2 Vaccines for Infants Requires That the Interference of Maternal Antibodies is Overcome

Live measles virus vaccine has been a success in the Western World for the last decade, although an increasing number of cases of measles have occurred in vaccinees in their late teens or soon thereafter. In contrast, in developing countries, measles is a major threat to young children, and vaccination of the young population of children at threat is ineffective because of the presence of maternal antibodies. In a recent project sponsored by the WHO, a number of measles vaccine formulations encompassing the commercial live measles vaccines, poxvirus constructs and adjuvant formulations were evaluated in vaccination experiments in cynomolgous macaques. Some experimental groups had received anti-measles antibodies to mimic the presence of maternal antibodies. The iscom vaccine induced high VN-titres, 10- to 100-fold higher than the other vaccines. In particular, live vaccines were unable to induce good responses in the presence of passively transferred antibodies (Osterhaus, manuscript in preparation).

Recently, studies in horses show that vaccination of two week old foals with iscoms containing envelope antigen from Equine Herpes 2 virus, a member of the gammavirinae subfamily, induce virus neutralizing antibodies in the presence of maternal antibodies and protection to disease when exposed to natural infection. Previous attempts with conventional vaccines were unsuccessful (Nordengrahn et al., 1995).

Passive immunization to enteric diseases in the newborn is transferred from mother to child by maternal antibodies. Immunization of mothers using a cholera vaccine induced increased levels of IgA in the milk (Svennerholm et al., 1980). Likewise an iscom matrix adjuvanted rotavirus vaccine delivered s.c. induced increased secretory IgA levels in the milk of baboon mothers which was reflected in the serum of the children (Snodgrass et al., 1995). In both cases vaccines were used to boost an existing immune response evoked by natural infection of the mothers.

In conclusion it is likely that the inhibitory effects of maternal antibodies will be overcome by certain vaccine formulations which will make it possible to immunize younger individuals. Possibly, local mucosal administration with suitable vaccine formulations may further overcome the maternal antibody problem.

3.1.5 Future Adjuvant Formulations

Vaccines of the future will be subject to three potentially conflicting requirements; they must give maximum efficacy, require the minimum number of doses and be delivered without pain or inconvenience. The achievement of these goals is dependent upon continuing developments in adjuvant studies.

The efficacy of a vaccine formulation will be determined by the nature of the immunogen and the nature of the immune response to that immunogen that is required for protection or in the case of a therapeutic vaccine, for cure. Knowledge of the immunogen and the response required will determine the choice of adjuvant. Certain antigens have in-built immunomodulatory activity e.g. CT, LT, the G protein of RSV (Bright et al., 1995) and certain proteins from *Trypanosoma cruzi* (Hansen et al., 1996). This activity may be compatible with the desired response or alternatively may be contrary in which case an adjuvant with powerful immunomodulatory activity of the desired type will be required to redirect the response. Adjuvants with a range of immunomodulatory activities are identified in table 3.1.2, and are described in more detail in Cox and Coulter (1996). Multimeric presentation of antigen is frequently beneficial to the immune response. This can readily be achieved for amphipathic molecules by incorporation into liposomes, iscoms or o/w emulsions, and can be similarly achieved for hydrophobic molecules but with more difficulty. Hydrophilic molecules may be best formulated with the appropriate aluminium salt or in a w/o emulsion if multimeric presentation is considered important. In most cases, antigens which naturally aggregate into particles such as HepBsAg and other VLPs will be best when formulated without disruption. One important exception to this is when CTL induction is required for optimal efficacy. In these situations, antigen needs to be incorporated into an adjuvant formulation which is able to deliver antigen to the cell cytosol for class I processing. This is best achieved with liposomes or iscoms although amphipathic adjuvants such as DDA and P_3CSS can also act in this way.

The number of vaccine doses can be minimised by use of either multicomponent vaccines, single dose delayed release vaccines or both. Convenience to the recipient results from either approach and single dose vaccines have the added advantage of guaranteeing compliance because a full vaccine course is administered. Multicomponent vaccines require powerful adjvuants because of the increased antigenic load. In addition, there are risks of antigenic competition, conflicting responses required for different antigens (e.g. T_H1 vs T_H2) and problems such as incompatibilities of buffers, pH etc. The use of dried vaccines may help to overcome these component incompatibilities. Single dose vaccines are best achieved by microencapsulation. The lack of total control over the time of release may be an advantage because the antigens of a multicomponent vaccine will be released to the immune system over the period of several weeks, thus minimizing competition for APC.

There is little argument that orally-delivered vaccines would be highly desirable if problems of efficacy and efficiency of delivery could be overcome. Despite promising data on microspheres (O'Hagan et al., 1989; Eldridge et al., 1991; Powell and Newman, 1995) most hopes still rest with the various mutants of LT and CT. Whether these can be used repeatedly without diminishing their efficacy is not known. If the problems of oral delivery cannot be solved, intranasal or pulmonary delivery may be an acceptable and achievable alternative. Lipsomes, iscoms, LTB and CTB have all been used successfully for this application.

Mechanisms to overcome the inhibitory effects of maternal antibody upon vaccination of the newborn are important in human and veterinary medicine. Iscoms offer promise in this area.

Further studies in adjuvant research need to concentrate on the mechanisms of action of immunomodulatory adjuvants, non-parenteral delivery and the development of effective single dose vaccines. This knowledge should permit the rational selection of and adjuvant formulation to meet the specific needs of any vaccine or vaccine combination.

References

Adams, S. E., Dawson, K. M., Gull, K., Kingsman, S. M., and Kingsman, A. J. (1987) The expression of hybrid HIV: Ty virus-like particles in yeast. Nature *329*, 68-70.

Allison, A. C. and Byars, N. E. (1986) An Adjuvant formulation that selectively elicits the formation of antibodies of protective isotypes and cell-mediated immunity. J. Immunol. Methods *95*, 157-168.

Allison, A. C. and Byars, N. E. (1990) Adjuvant formulations and their mode of action. Semin Immunol. 2 (5), 369-374.

Allison, A. C. and Byars, N. E. (1991) Immunological adjuvants: desirable properties and side-effects. Mol. Immunol. 28(3), 279-284.

Alving, C. R. (1992) Immunological aspects of liposomes: Presentation and processing of liposomal protein and phospholipid antigens. Biochim. Biophys. Acta *1113*, 307-322.

Asanuma, H., Koide, F., Suzuki, Y., Nagamine, T., Aizawa, C., Kurata, T., and Tamura, S. (1995) Cross-protection against influenza virus infection in mice vaccinated by combined nasal/subcutaneous administration. Vaccine *13*, 3-5.

Ben-Ahmeida, E. T. S., Potter, C. W., Gregoriadis, G., Adithan, C., and Jennings, R. (1994) IgG subclass response and protection against challenge following immunisation of mice with various influenza A vaccines.J Med Microbiol. *40*, 261-269.

Benner, R., Hijmans, W., and Haaijman, J. J. (1981b) The bone marrow: the major source of serum immunoglobulins, but still a neglected site of antibody formation. Clin.Exp. Immunol. *46*, 1-8.

Benner, R., van Oudenaren, A., and Koch, G. (1981a) Chapter 14. Induction of antibody formation in the mouse bone marrow. Immunol. methods II, 247-261.

Bomford, R. (1980) The comparative selectivity of adjuvants for humoral and cell-mediated immunity, Effect on the antibody response to bovine serum albumin and sheep red blood cells of Freund's incomplete and complete adjuvants, alhydrogel, Corynebacterium parvum, Bordetella pertussis, muramyl dipeptide and saponin. Clin. Exp. Immunol. *39,* 426-434.

Böttiger, M. (1973) Experiences from investigations of virus isolations from sewage over a 2-year period with special regard to polioviruses. Arch. Gesamte Virusforschung *41,* 80-85.

Böttiger, M. (1984) Long-term immunity following vaccination with killed poliovirus vaccine in Sweden, a country with no circulating poliovirus. Rev. infect. Dis. *6*, 548-551.

Böttiger, M. and Herrström, E. (1992) Isolation of poliovirues from sewage and their characteristics: Experience over two decades in Sweden. Scand. J. Infect. Dis. *24,* 151-155.

Bright, H., Turnbull, T., Toms, G. L., and Scott, R. (1995) Comparison of the T helper cell response induced by respiratory syncytial virus and its fusion protein in BALB/c mice. Vaccine *13,* 915-922.

Claassen, I. J. T. M., Osterhaus, A. D. M. E., and Claassen, E. (1995) Antigen detection in vivo after immunization with different presentation forms of rabies virus antigen: involvement of marginal metallophilic macrophages in the uptake of immune-stimulating complexes. Eur. J. Immunology *25,* 1446-1452.

Claassen, I., Osterhaus, A., Boersma, W., Schellekens, M., and Claassen, E. (1995) Fluorescent labelling of virus, bacteria and iscoms: In vivo systemic and mucosal localisation patterns. In: Advances in Mucosal Immunology. Mestecky, J. et al. (eds). Plenum Press, New York, 1485-1489.

Cox, J. and Coulter, A. (1992) Advances in adjuvant technology and application. In: Animal Parasite Control Utilizing Biotechnology. Yong, W. K. (ed.), CRC Press Boca Raton, Ann Arbor, London, Tokyo. 49-112.

Cox, J. and Coulter, A. (1996) Adjuvants – structure and function. Submitted.

Dalsgaard, K. (1974) Saponin adjuvants III Isolation of a substance from Quillaja saponaria molina with adjuvant activity in foot-and-mouth disease vaccines. Arch. ges. virus forsch. *44*, 243-254.

Dalsgaard, K. (1978) A study on the isolation and characterization of the saponin Quil A. Evaluation of its adjuvant activity with special reference to the application in foot-and-mouth disease. Acta Vet. Scand. (suppl) **69**, 1-40.

Dalsgaard, K., Jensen, M. H., and Sörensen, K. J. (1977) Saponin adjuvants IV. Evaluation of the adjuvant Quil A in the vaccination of cattle against foot-and-mouth disease. Acta vet. Scand. *18,* 349-360.

De Haan, A., Geerligs, H. J., Huchshorn, J. P., Van Scharrenburg, G. J. M., Palache, A. M., and Wilsschut, J. (1995) Mucosal immunoadjuvant activity of liposomes: induction of systemic IgG and secretory IgA responses in mice by intranasal immunization with an influenza subunit vaccine and coadministrated liposomes. Vaccine *13*, 155-162.

Deres, K., Schild, H., Wiesmuller, K. H., Jung, G., and Rammensee, H. G. (1989) In vivo priming of virus-specific cytotoxic T lymphocytes with synthetic lipopeptide vaccine. Nature *342*, 561-564.

Dickinson, B. L. and Clements, J. D. (1995) Dissociation of Escherichia coli heat-labile enterotoxin adjuvanticity from ADP-ribosyltransferase activity. Infect. Immun. *63*, 1617-1623.

Douce, G., Turcotte, C., Cropley, I., Roberts, M., Pizza, M., Domenghini, M., Rappuoli, R., and Dougan, G. (1995) Mutants of Escherichia coli heat-labile toxin lacking ADP-ribosyltransferase activity act as nontoxic, mucosal adjuvants. Proc. Natl. Acad. Sci. USA, *92*, 1644-1648

Driscoll, J. (1994) The role of the proteasome in cellular protein degradation. Histol. Histophathol. *9*, 197-202.

Eldridge, J. H., Hammond, C. J., Meulbroek, J. A., Staas, J. K., Gilley, R. M., and Tice, T. R. (1990) Controlled vaccine release in the gut-associated lymphoid tissues. I. Orally administered biodegradable microspheres target the Peyer's patches. J. controlled Rel. 11, 205-214.

Eldridge, J. H., Staas, J. K., Meulbroek, J. A., McGhee, J. A., Tice, T. R., and Gilley, R. M. (1991) Biodegradable microspheres as a vaccine delivery system. Mol. Immunol. 28, 287-294.

Fontana, M. R., Manetti, R., Giannelli, V., Magagnoli, C., Marchini, A., Olivieri, R., Domenighini, M., Rappuoli, R., and Pizza, M. (1995) Construction of nontoxic derivatives of cholera toxin and characterization of the immunological response against the A subunit. Infect. Immun. *63*, 2356-2360.

Freund, J. (1951) The effect of paraffin oil and mycobacteria on antibody formation and sensitization. a review. Am. J. Clin. Pathol. *21*, 645-656.

Goldberg, A. L. and Rock, K. L. (1992) Proteolysis, proteasomes and antigen presentation. Nature, *357*, 375-379.

Gould-Fogerite, S., Edghill-Smith, Y., Kheiri, M., Wang, Z., Das, K., Feketeova, E., Canki, M., and Mannio, R. J. (1994) Lipid matrix-based subunit vaccines: A structure-function approach to oral and parenteral immunization. Aids Research and human retroviruses 10 suppl.2, S99-S103.

Hall, C. B. (1994) Prospects for a respiratory syncytial virus vaccine. Science *265*, 1393-1394.

Hansen, D. S., Villacres-Eriksson, M., Åkerblom, L., Hellman, U., Segura, E., Carlomagno, M., and Morein, B. (1996) An immunoaffinity purified *Trypanosoma cruzi* antigen suppresses cellular proliferation through a TGF-β mediated mechanism. Submitted.

Harding, C. V., Collins, D. S., Slot, J. W., Geuze, H. J., and Unanue, E. R. (1991) Liposome-encapsulated antigens are processed in lysosomes, recycled, and presented to T cells. Cell *64,* 393-401.

Holt, L. B. (1949) Quantitative studies in diphtheria prophylaxis: the primary response. Br. J. Exp. Pathol. *30*, 289-297.

Howerton, D. A., Hunter, R. L., Ziegler, H. K., and Check, I. J. (1990) Induction of macrophage Ia expression in vivo by a synthetic block copolymer, L81. J. Immunol. *144*, 1578-1584.

Hu, J. G. and Kitagawa, T. (1990) Studies on the optimal immunization schedule of experimental animals. VI. Antigen dose-response of aluminium hydroxide-aided immunization and booster effect under low antigen dose. Chem. Pharm. Bull. *38,* 2775-2779.

Hunter, R. L., McNicholl, J., and Lal, A. (1994) Mechanism of action of nonionic block copolymer adjuvants. Aids Research and human retroviruses *10* suppl.2, S95-S98.

Jones, P. D., Tha Hla, R., Morein, B., Lövgren, K., and Ada, G. L. (1988) Cellular immune responses in the murine lung to local immunization with influenza A virus glycoproteins in micelles and immunostimulatory complexes (iscoms). Scand. J. Immunol. *27*, 645-652.

Kersten, G. F. A. and Crommelin, D. J. A. (1995) Liposomes and iscoms as vaccine formulations. Biochim. Biophys Acta *1241*, 117-138.

Kraal, G., Janse, M., and Claassen, E. (1988) Marginal metallophilic macrophages in the mouse spleen: effects of neonatal injections of MOMA-1 antibody on the humoral immune response. Immunol. Lett. *17*, 139-144.

Labbe, M., Charpilienne, A., Crawford, S. E., Estes, M. K., and Cohen, J. (1991) Expression of rotavirus VP2 produces empty corelike particles. J. Virol. *65,* 2946-2952.

Lascelles, A. K., Eagleson, G., Beh, K. J., and Watson, D. L. (1989) Significance of Freund's adjuvant/antigen injection granuloma in the maintenance of serum antibody response. Vet. Immunol. Immunopathol. *22*, 15-27.

Lazorova, L., Artursson, P., Lövgren Bengtsson, K., and Sjölander, A. (1995) Influence of a particulate Quillaja saponin-containing adjuvant, iscom-matrix, on the handling and transport of influenza virus antigens in human intestinal epithelial cells (Caco-2). Submitted.

Liew, F. Y., Russell, M., Appleyard, G., Brand, C. M., and Beale, J. (1984) Cross-protection in mice infected with influenza A virus by the respiratory route is correlated with local IgA antibody rather than serum antibody or cytotoxic T cell reactivity. Eur. J. Immunol. *14*, 350-356.

Lövgren, K., Kåberg, H., and Morein, B. (1990) An experimental influenza subunit vaccine (iscom): induction of protective immunity to challenge infection in mice after intranasal or subcutaneous administration. Clin. Exp. Immunol. *82*, 435-439.

Macatonia, S. E., Taylor, P. M., Knight, S. C., and Askonas, B. A. (1989) Primary stimulation by dendritic cells induces antiviral proliferative and cytotoxic T-cell responses in vitro. J. Exp. Med. 169, 1255-1264.

Mahon, B. P., Katrak, K., Nomoto, A., Macadam, A. J., Minor, P. D., and Mills, K. H. (1995) Poliovirus-specific CD4+ Th1 clones with both cytotoxic and helper activity mediate protective humoral immunity against a lethal poliovirus infection in transgenic mice expressing the human poliovirus receptor. J. Exp. Med. *181*, 1285-1292.

Mannhalter, J. W., Neychev, H. O., Zlabinger, G. J., Ahmad, R., and Eibl, M. M. (1985) Modulation of the human immune response by the non-toxic and non-pyrogenic adjuvant aluminium hydroxide: effect on antigen uptake and antigen presentation. Clin. Exp. Immunol. *61*, 141-151.

McMenamin, C., Oliver, J., Girn, B., Holt, B. J., Kees, U. R., Thomas, W. R., and Holt, P. G. (1991) Regulation of T-cell sensitization at epithelial surfaces in the respiratory tract:

suppression of IgE responses to inhaled antigens by CD3+ Tcr alpha-/beta- lymphocytes (putative gamma/delta T cells). Immunology *74*, 234-249.

Morein, B., Lövgren, K., Rönnberg, B., Sjölander, A., Villacrés-Eriksson, M. (1995) Immunostimulating Complexes. Clinical Potential in Vaccine Development. Clinical Immunotherapeutics *3*, 461-475. ADIS International.

Morein, B., Sundquist, B., Höglund, S., Dalsgaard, K., and Osterhaus, A. (1984) Iscom, a novel structure for antigenic presentation of membrane proteins from enveloped viruses. Nature *308*, 457-460.

Mowat, A. McI. and Maloy, K. (1994) Chapter 3. Immune Stimulating Complexes as Vectors for Oral Immunization. In: Novel Delivery Systems for Oral Vaccines. O'Hagan, D. T. (ed.), CRC Press, Inc. Boca Raton, USA.

Mumford, J. A., Jesset, D., Rollinson, E. A., Hannant, D., and Draper, M. E. (1994) Duration of protective efficacy of equine influenza immunostimulating complex/tetanus vaccines. The Veterinary Record, *12*, 158-162.

Neefjes, J., Gottfried, E., Roelse, J., Gromme, M., Obst, R., Hammerling, G. J., Momburg, F. (1995) Analyis of the fine specificity of rat, mouse and human TAP peptide transporters. Eur. J. Immunol. *25*, 1133-1136.

Nordengrahn, A., Merza, M., Rusvai, M., Ekström, J., Morein, B., and Belák, S. (1995) Equine Herpesvirus type 2 (EHV-2) as a predisposing factor for Rhodococcus equi pneumonia in foals: Prevention of the bifactorial disease with EHV-2 immunostimulating complexes. Submitted.

O'Hagan, D. T., Palin, K. J., and Davis, S. S. (1989) Poly(butyl-2-cyanoacrylate) particles as adjuvants for oral immunization. Vaccine *7*, 213-216.

Osebold, J. W. (1982) Mechanisms of action by immunologic adjuvants. J. Vet. Med. Assoc. *181*, 983-987.

Ott, G., Barchfeld, G. L., Chernoff, D., Radhakrishnan, R., van Hoogevest, P., and Van Nest, G. (1995) Design and evaluation of a safe and potent adjuvant for human vaccines. In: Vaccine design: The subunit and adjuvant approach. Powell, M. F. and Newman, M. J. (eds) Plenum Press, New York.

Powell, M. F. and Newman, M. J. (eds) (1995) Vaccine Design: The subunit and adjuvant approach. Plenum Press, New York.

Rechsteiner, M., Hoffmann, L., and Dubiel, W. (1993) The multicatylytic and 26S proteases. J. Biol. Chem. *268*, 6065-6068.

Ribi, E., Cantrell, J. L., Takayama, K., Qureshi, N., Peterson, J., and Ribi, H. O. (1984) Lipid A and immunotherapy. Rev. Infect. Dis. *6*, 567-572.

Rivett, J. (1993) Proteasomes: multicatylytic proteinase complexes. Biochem. J. *291*, 1-10.

Sacco, A. G., Yurewicz, E. C., and Subramanian, M. G. (1989) Effect of varying dosages and adjuvants on antibody response in squirrel monkeys (Saimiri sciureus) immunized with the porcine zona pellucida Mr = 55,000 glycoprotein (ZP3). Am. J. Reprod. Immunol. *21*, 1-8.

Santos, L. M., Lider, O., Audette, J., Khoury, S. J., and Weiner, H. L. (1990) Characterization of immunomodulatory properties and accessory cell function of small intestinal epithelial cells. Cell. Immunol. *127*, 26-34.

Scalzo, A. A., Elliot, S. L., Cox, J. C., Gardner, J., Moss, D. J., and Suhrbier, A. (1995) Induction of cytotoxic T cells to murine cytomegalovirus by using a nanopeptide and a human-compatible adjuvant (Montanide LSA-720). J. Virol. *69*, 1306-1309.

Sjölander, A., Lövgren Bengtsson, K., and Morein, B. (submitted) Kinetics, localization and cytokine profile of T cell responses to immune stimulating complexes (iscoms) containing human influenza virus envelope glycoproteins.

Sjölander, A., Lövgren Bengtsson, K., Johansson, M., and Morein, B. (1996) Kinetics, localization and isotype profile of antibody responses to immune stimulating complexes (iscoms) containing human influenza virus envelope glycoproteins. Scand. J. Immunol., 43, 164-172.

Snippe, H., De Reuver, M. J., Strickland, F., Willers, J. M. N., and Hunter, R.L. (1981) Adjuvant effect of nonionic block polymer surfactants in humoral and cellular immunity Int. Arch. Allergy Appl. Immunol. *65*, 390-398.

Snodgrass, D. R., Campbell, I., Mwenda, J. M., Chege, G., Seleman, M. A., Morein, B., and Hart, C. A. (1995) Stimulation of rotavirus IgA, IgG and neutralising antibodies in baboon milk by parenteral vaccination. Vaccine *13*, 408-413.

Staats, H. F., Jackson, R. J., Marinaro, M., Takahashi, I., Kiyono H., and McGhee, J. R. (1994) Mucosal immunity to infection with implications for vaccine development. Curr. Opin. Immunol. *6*, 572-583.

Strobbe, R., Charlier, G, Debecq, J., and van Aert, A. (1976) Studies about the adjuvant activity of saponin fractions in foot-and-mouth disease vaccine. III. Comparison of the irritant, adjuvant and hemolytic activities of six commercial saponins and their hemolytic fractions obtained by chromatography on Sephadex G100. Arch. Exp. Vet. Med. *30*, 173-181.

Svennerholm, A.-M., Hanson, L. A., Holmgren, I., Lindblad, B. S., Nilsson, B., and Quereshi, F. (1980) Different secretory immunoglobulin A antibody responses to cholera vaccination in Swedish and Pakistani women. Infect. Immun. *30*, 427-430.

Takahashi, H., Takeshita, T., Morein, B., Putney, S., Germain, R. N., and Berzofsky, J. (1990) Induction of CD8+ cytotoxic T cells by immunization with purified HIV-1 envelope proteins in iscoms. Nature *344*, 873-875.

Takayama, K., Olsen, M., Datta, P., and Hunter, R. L. (1991) Adjuvant activity of non-ionic block copolymers. V. Modulation of antibody isotype by lipopolysaccharides, lipid A and precursors. Vaccine *9*, 257-265.

Thepen, T., van Rooijen, N., and Kraal, G. (1989) Alveolar macrophage elimination in vivo is associated with an increase in pulmonary immune response in mice. J. exp. Med. *170*, 499-509.

Vajdy, M. and Lycke, N. (1993) Stimulation of antigen-specific T- and B-cell memory in local as well as systemic lymphoid tissues following oral immunization with cholera toxin adjuvant. Immunology *80*, 197-203.

Villacres-Eriksson, M. (1993) Induction of immune responses by iscoms. Ph. D. Thesis, Swedish University of Agricultural Sciences.

Watson, D. L., Lövgren, K., Watson N. A., Fossum, C., Morein, B., and Höglund, S. (1989) The inflammatory response and antigen localization following immunization with influenza virus iscoms. Inflammation *13*, 641-649.

Watson, D., Watson, N., Fossum, C., Lövgren, K., and Morein, B. (1992) Interactions between immune-stimulating complexes (ISCOMS) and peritoneal mononuclear leucocytes. Microbiol. and Immunol. *36*, 199-203.

Weiner, H. L., Friedman, A., Miller, A., Khoury, S. J., al-Sabbagh, A., Santos, L., Sayegh, M., Nussenblatt, R. B.,Trentham, D. E., and Hafler, D. A. (1994) Oral tolerance: immunologic mechanisms and treatment of animal and human organ-specific autoimmune diseases by oral administration of autoantigens. Annu. Rev .Immunol. *12*, 809-837.

Wilschut, J., De Haan, A., Geerligs, H. J., Huchschorn, J. P., van Scharrenburg, G. J. M., Palache, A. M., Renegar, K. B., and Small Jr., P. A. (1994) Liposomes as a mucosal adjuvant system: An intranasal liposomal influenza subunit vaccine and the role of IgA in nasal anti-influenza immunity. J. Liposome Res. *4*, 301.

WO 90/14837. Adjuvant formulation comprising a submicron oil droplet emulsion. van Nest, G., Ott, G., Barchfield.

WO 94/15636. Vaccine preparations. Cox, J., Sparks, R., Jacobs, I., Mason, N.

Yap, K. L., Ada, G. L., and McKenzie, I. S. C. (1978) Transfer of specific cytotoxic T cells protects mice inoculated with influenza viruses. Nature *273*, 238-239.

3.2 Biodegradable Microspheres as Vehicles for Antigens

Gideon F.A. Kersten and Bruno Gander

3.2.1 Introduction

Biodegradable microspheres have been studied widely as a drug delivery system (Langer, 1990). Until recently, their use was thought to be limited to low molecular weight drugs like steroids and peptides. Sustained release of macromolecules was considered impossible because of restricted diffusion and the sensitivity of proteins to the often harsh microencapsulation processes requiring organic solvents, elevated temperature or freeze-drying. Thanks to the development of advanced preparation techniques and new polymers, on the one side, and of biochemical and immunochemical tools for protein analysis, on the other side, a rapidly increasing number of research groups is investigating microencapsulated macromolecules, including antigens (Eldridge et al., 1992; Morris et al., 1994; Mestecky et al., 1994).

Microencapsulation of antigens and subsequent controlled release offer a number of evident, potential advantages, as briefly highlighted in the following. (i) An ideal parenteral vaccine formulation providing sustained or pulsed antigen release may obviate the necessity of booster injections. In some vaccination campaigns in developing countries a significant number of vaccinees does not return for a booster. For illustration, up to 20 % of the children vaccinated against diphtheria-tetanus-pertussis drop out at each booster round. Therefore, single shot vaccinations would improve tremendously the efficacy of vaccines in areas with insufficient infrastructural facilities. In addition, single immunization is often preferred to multiple for socio-economic and cultural reasons. (ii) Progress in immunology, molecular biology and biotechnology has increased the possibility to identify and produce the essential antigenic determinants of pathogens. However, these antigens are only seldom immunogenic enough to be used without adjuvants and/or presenting vehicle. In human vaccines, only aluminum salts are being used on a large scale. Some new adjuvants like LPS/lipid A derivatives (e.g. Fries et al., 1992) and saponins like QS21 (Livingston et al., 1994) have reached the stage of clinical testing and may be introduced eventually on the market.Vehicles presently being studied in great detail encompass liposomes and ISCOMs (Kersten and Crommelin, 1995). All these formulations do show promise and drawbacks. The non-particulate form of common adjuvants does not always suffice to stimulate immune responses, particularly with low molecular weight antigens like peptides. A multimeric presentation form is often desirable. Liposomes may possess low stability or lack intrinsic adjuvant activity, whereas ISCOMs are not designed to present soluble antigens to the immune system. Evidently, there is a niche for microspheres as adjuvating vehicle. (iii) Oral administration of vaccines is preferred to parenteral, when ease of application and compliance are considered. Efficacious non-replicating oral or otherwise locally applicable vaccines for human use still do not exist. This is not surprising, since such vaccines have to satisfy extreme criteria, i.e. the antigen must not be processed in an environment designed for processing (gut) or removal of foreign particles (respiratory tract). In the case of the oral route, the antigen must reach the cells of the mucosal immune system specifically and efficiently in the presence of excess food antigens. Microspheres may play a role in the development of vaccines for local application, since they can protect the antigen against degradation and may even serve as targeting device.

In this chapter the current state of the art on the use of biodegradable polymers and microspheres (fig. 3.2.1) in vaccine development is discussed.

3.2.2. Biodegradable polymers

The first reports on polymeric particulates as vehicles for antigens date from the early seventies (Birrenbach, 1973; Birrenbach and Speiser, 1976; Kreuter, 1974; Kreuter and Speiser, 1976). In these studies, tetanus toxoid, human IgG and Influenza virus were incorporated into nanoparticles ($< 1\ \mu m$) made of poly(acrylamide) and poly(methylmethacrylate) [PMMA], giving rise to a pronounced adjuvant effect in mice and guinea pigs. The concept of sustained antigen release systems inducing prolonged antibody formation was first proposed by Preis and Langer (1979), using model antigens incorporated into poly(ethylene-co-vinyl acetate) [PEVAc] pellets of 1 mm in diameter. These pioneering studies undoubtedly paved the way for the steadily increasing number of investigations on polymeric devices for antigen delivery. Nowadays, it is generally recognized that microspheres based on biodegradable polymers have the highest potential for single-step or single-dose immunization (Aguado and Lambert, 1991; Aguado, 1992). Although a huge number of biodegradable hydrophilic and hydrophobic polymers have been described for controlled parenteral and mucosal drug delivery (Kamath and Park, 1993; Chasin and Langer, 1990a; Leong, 1991; Scholsky et al., 1991; Heller, 1993), only relatively few of them have been associated with antigens. Clearly, the homo- and copolymers of lactic and glycolic acid [PLA, PLGA] are currently attracting the highest interest as vehicles for vaccine delivery systems for the parenteral and mucosal (oral, nasal, vaginal) routes. Significant advantages of these polymers are their long safety history, proven bio-

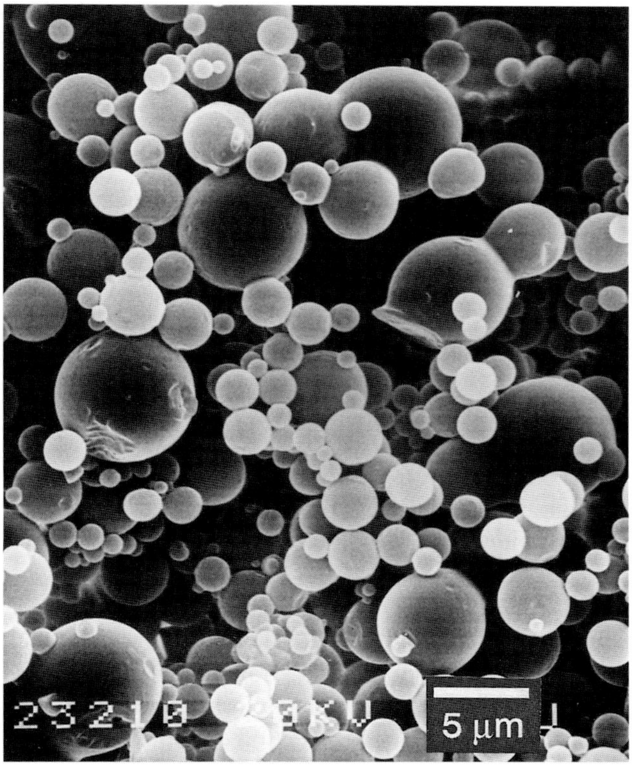

Fig. 3.2.1: Scanning electron micrograph of poly(D,L-lactide) microspheres prepared by spray-drying, using ethyl acetate as polymer solvent. The microspheres are loaded with 2,9 % bovine serum albumin.

compatibility and their property to control the time and rate at which they degrade and release incorporated active material such as antigens.

Nonetheless, alternative polymers have been proposed and deserve attention (tab. 3.2.1). The poly(acrylates) were not only the first polymers studied as adjuvants, but continue to attract great interest. Due to their very slow biodegradability, their use in vaccines is however restricted to nanoparticles, which are eliminated from the body much more easily than particles in the micrometer range. While nanoparticles of PMMA, Poly(alkyl cyanoacrylate) or of other type are very powerful adjuvants for numerous antigens and suitable for targeting of drugs and antigens, their usefulness for prolonged antigen release and elicitation of a long lasting immune response is less evident. On the other side, a great variety of hydrophilic microspheres based on cross-linked or derivatized starch, BSA and dextran or on ionic polymeric complexes, have also shown marked adjuvant effect for particular antigens. The main problem which might arise from this type of hydrophilic modified polymers is their toxicity due to residual reagents and the potential immunogenicity of the polymers themselves. Indeed, Naim and Van Oss (1992) have shown that the hydrophilicity/hydrophobicity and solubility of a polymer can be related, to some extent, to its immunogenicity. As an example, dextran and polyvinylpyrrolidone are classified as type 2 T cell independent antigens (Van Buskirk and Braley-Mullen, 1987; Brunswick et al., 1988), with dextran becoming more immunogenic when coupled to a protein. A similar effect was observed for non-immunogenic poly(acryl)-starch microspheres when they were combined with human serum albumin (Artursson et al., 1986). Therefore, the study of modified polymers, particularly if they are of natural origin, for antigen delivery microspheres should focus carefully on the immunogenicity of the polymeric carrier itself.

As pointed out above, the main polymers used as vaccine vehicles are the aliphatic polyesters based on lactic and glycolic acid, also called lactide and glycolide if considered from the mechanism of synthesis. These polymers are commercially available in medical grade and have been approved by FDA for parenteral drug delivery systems. They have been studied and used as suture material since the early seventies and received considerable attention also as materials for the controlled delivery of conventional low molecular weight drugs since about 1973 (Boswell and Scribner, 1973; Yolles et al., 1974; Wise et al., 1976; Beck et al., 1979). In 1984, a first report appeared on PLGA microspheres for the delivery of a bioactive macromolecule, i.e. luteinizing hormone-releasing hormone (Sanders et al., 1984). Finally, in 1988, Beck proposed these polymers for the delivery of bacterial and viral antigens and bovine HCG (Beck et al., 1988). Since then, a broad spectrum of antigens have been microencapsulated into PLAs and PLGAs. Table 3.2.2 gives a selection of some of the studies conducted over the past five years. Numerous investigations were published with so-called model antigens, such as BSA or ovalbumin [OVA], but their significance is very limited as results obtained with model proteins obviously cannot be extrapolated to antigens with relevance for vaccination. This is particularly true for protein compounds which can differ so greatly in their solubility, hydrophilicity/hydrophobicity, conformation, chemical and physical stability and interaction capacity with and presentation by polymers. There exists an obvious increasing interest in synthetic and recombinant, low molecular weight antigens with well defined structure. It has been reported that PLA/PLGA microspheres can stimulate the immune response not only induced by natural proteins, but also by weakly immunogenic synthetic peptides (Men et al., 1994). In addition to the various types of PLA/PLGA-microspheres summarized in table 3.2.2, some more sophisticated systems have been described which combine microspheres with phospholipids (Amselem et al., 1992) or with liposomes (Cohen et al., 1991a). However, these more complex preparations did not exhibit any marked advantages over the more simple polymer microspheres.

Proper selection of PLAs and PLGAs for antigen delivery necessitates some basic knowledge of the polymers' main physical properties. The polyesters generally used for antigen and drug delivery are either linear homopolymers of D,L-lactic acid or linear copolymers of D,L-lactic and glycolic acid with variable molar ratios, e.g. 90:10, 85:15, 75:25, 65:35 and 50:50 (fig. 3.2.2). These polymers are soluble in a

Table 3.2.1: Selected examples of biodegradable polymers, except PLAs and PLGAs, used for particulate vehicles for antigen delivery.

Polymer[1]	Antigen[2]	Delivery System[3]	Route of administration	Reference
PMMA	Influenza V.	NP	i.m.	Kreuter et al., 1986, 1988
PMMA	HIV-1, HIV-2	NP	s.c.	Stieneker et al., 1993, 1995
Poly(butyl-2-cyanoacrylate)	OVA	NP	p.o.	O'Hagan et al., 1989
Crystallized dextran, crystallized starch	Bee Venom, OVA, Schistosoma antigen	NP	s.c.	Schröder et al., 1984
Poly(acryl)-starch	HSA	MS	i.v., i.m.	Arturson et al., 1986
Poly(acryl)-starch	HSA	MS	i.v.,i.m.,i.p.	Degling et al., 1995
X-Gelatin-Chondriotin-4-sulfate	γ-IFN, GM-CSF	MS	s.c.	Golumbek et al., 1993
Proteinoid-Gum arabic	Influenza V. antigen	MS	p.o.	Santiago et al., 1993
Alginate-Spermine, Chondriotin-Spermine	Rotavirus	MS	p.o.	Offit et al., 1994
Degradable starch	Glycoprotein fragment from Influenza V.-HA	MS	intravaginal	O'Hagan et al., 1993b
Various: PS, PMMA, PHBA PLAs, PLGAs, CAP, Cellulose triacetate, EC	SEB	MS	p.o.	Eldridge et al., 1990
Gelatine	Human γ-Globulin	MS	s.c.	Nakaoka et al., 1995
X-RSA	Nodamura Virus and capsid protein	Beads	i.m.	Martin et al., 1988
X-RSA	C. botulinum, K. pneumoniae	Beads	i.m.	Langhein et al., 1987

[1] X: cross-linked; BSA: Bovine Serum Albumin; PS: Poly(styrene); PMMA: Poly(methylmethacrylate); PHBA: Poly(hydroxybutyric acid); PLA: Poly(lactic acid); PLGA: Poly(lactide-co-glycolide); CAP: Cellulose acetate phthalate; EC: Ethyl cellulose; RSA: rabbit serum albumin

[2] OVA: Ovalbumin; SEB: Staphylococcal Enterotoxin B; HSA: human serum albumin; IFN: interferon; GM-CSF: granulocyte-macrophage colony-stimulating factor

[3] NP: Nanoparticles; MS: Microspheres; Beads: approx. 100–500 μm

great number of organic solvents having some degree of polarity (dipol-dipol) and only moderate hydrogen bonding capacity. On contrast, PLAs and PLGAs are insoluble in aqueous media, in which however they take up water and show swelling to some extent, depending on the molecular weight and polymer composition. Clearly, PLGAs swell more than PLAs, due to the presence of the more hydrophilic glycolic acid (Gilding and Reed, 1979). The degree of polymerization, n and m, define the molecular weight of these macromolecules. Commonly used polymer molecular

Table 3.2.2: Selected examples of investigations using PLA and PLGA microspheres for antigen delivery.

Polymer[1]	Antigen[2]	Animal model	Route of administration[3]	Reference
PLGA 50:50, 84 kDa PLGA 75:25, 83 kDa	OVA	Mouse	i.g.	O'Hagan et al., 1994a
PLGA 50:50	OVA	Rat	s.c.	O'Hagan et al., 1991a
PLGA 50:50	OVA	Mouse	i.p., i.g.	O'Hagan et al., 1993a
PLGA 50:50	OVA	Mouse	oral	Maloy et al., 1994
PLGA 50:50, 53 kDa	OVA	Mouse	s.c.	Uchida et al., 1994
L-PLA, 2kDa	TT	Guinea-pig, rabbit, rat	i.n.	Almeida et al., 1993
PLGA 45:55-GLU	TT	Mouse	s.c.	Esparza et al., 1992
L-PLA, 3 kDa PLGA 50:50, 3 kDa, 100 k	TT	Mouse	s.c.	Gupta et al., 1993; Alonso et al., 1993, 1994
L-PLA, 2 kDa	TT	Guinea-pig	i.n.	Alpar et al., 1994
PLGA 65:35, 75 kDa	TT	Rat	s.c.	Raghuvanshi et al., 1993
D,L-PLA, 130 kDa PLGA 50:50, 12 kDa PLGA 75:25, 17 kDA	TT	Mouse	s.c.	Men et al., 1995
D,L-PLA, 49 kDa	DT	Mouse	s.c.	Singh et al., 1991, 1992
PLGA 50:50 PLGA 85:15	SEB	Mouse	i.p., i.g.	Eldridge et al., 1989, 1990, 1991a,b
L-PLA, 94 kDa D,L-PLA, 88 kDa PLGA 50:50, 75 & 75 kDA PLGA 75:25, 105 kDa	Ricin toxoid	Mouse	i.m., s.c.	Yan et al., 1995
PLGA 50:50	E. coli CFA/I and II from ETEC	Rabbit	i.g., i.m., i.d.	Edelman et al., 1993; Reid et al., 1993
PLGA 50:50	E. coli pilus protein	Rabbit	i.d.	McQueen, 1993
PLGA 50:50	B. pertussis fimbriae	Mouse	i.p.	Jones et al., 1995
PLGA?	B. pertussis antigens	Mouse	s.c.	Shahin et al., 1995
PLGA	B. pertussis filamentous haemagglutinin	Mouse	i.n., i.p.	Cahill et al., 1995
PLGA 50:50, 108 kDa	V. cholerae antigen	Rabbit	i.i., s.c., p.o.	Chandrasekhar et al., 1994
PLGA	Influenza A antigen	Mouse	i.p., p.o.	Moldoveanu et al., 1993
PLGA	HIV-1 subunit	Guinea-pig	s.c.?	Cleland et al., 1994
D,L-PLA, 12 kDa PLGA 50:50, 12 kDa	RSV synthetic antigen	Mouse	i.p.	Partidos et al., 1994
D,L-PLA, 130 kDa PLGA 50:50, 12 kDa PLGA 75:25, 17 kDa	Malaria antigen, synthetic	Mouse	s.c.	Men et al., 1994
**L-PLA, 2 kDa **PCL, 2 kDa	Malaria antigen, recombinant	Rabbit	i.m.	Amselem et al., 1992

1 Polymer molecular weights have only indicative value (as given by the authors); **PLA and **PCL (polycaprolactone) are so-called lipospheres consisting of an inner polymeric core and an outer phospholipid layer.

2 OVA: ovalbumin; TT: tetanus toxoid; DT: diphtheria toxoid; SEB: Staphylococcal Enterotoxin B; RSV: respiratory syncytial virus; CFA: Colonization factor antigen; ETEC: Enterotoxigenic E. coli.

3 i.g.: intragastric intubation; i.n.: intranasal; i.d.: intraduodenal; i.i.: intraintestinal; p.o.: per oral.

weights, M_w, for antigen delivery range from about 12,000 to 200,000. Lower M_w-PLA, e.g. 2,000, is available but its properties in terms of water uptake, glass transition temperature, antigen release and biodegradation are not adequate for controlled release microspheres. Higher M_w-polyesters cause difficulties in microsphere preparation and are not expected to provide enhanced release characteristics or immune response. Another important property of PLA is its configuration. PLA contains an optically active repeating unit, and exists therefore as isomeric L-PLA or as racemic D,L-PLA. While D,L-PLA is entirely non-crystalline, L-PLA can form a crystalline phase (semi-crystalline polymer), a property in common with PGA. Crystallinity renders the polymers less soluble in organic solvents, e.g. pure PGA dissolves only in very few fluorinated solvents, and also reduces substantially the water uptake and rate of biodegradation. The copolymers PLGA generally contain the racemic form of lactic acid, although copolymers with L-lactic acid are also available. The fact that polymerized glycolide forms a crystalline phase can lead to solubility problems of PLGA 50:50, containing 50% glycolide. A non-random distribution of the glycolide units in the polymer chain results in crystalline microdomains in PLGA 50:50 (Bendix, 1990), which in turn can cause variations in polymer degradation and solubility problems relevant for microsphere preparation (Dunn et al., 1988). In addition to polymer composition, molecular weight and crystallinity, the glassy-to-rubbery state transition temperature, so-called glass temperature T_g, of the amorphous PLAs and PLGAs is of great importance for microsphere preparation, morphological stability of microspheres, diffusion properties and biodegradation rate. In their native dry state, these aliphatic polyesters have a glass temperature ranging from approximately 40 °C for low molecular weight PLGA 50:50 to 65 °C for high molecular weight D,L-PLA. Residual organic solvents, e.g. from microsphere preparation, or humidity can lower T_g by a several degrees centigrade. Particularly for the low molecular weight PLGA 50:50 such a lowering may bring T_g down to body and *in vitro* release temperature and alter the initial phase of both antigen release and biodegradation. If in the microsphere preparation process the organic solvent is not properly eliminated and water content not reduced to a level of about 1 %, T_g might even be below storage temperature of the preparation and cause morphological deformation and agglomeration of the particles.

Biodegradation of PLA/PLGA occurs by bulk hydrolysis producing lactic and glycolic acid which can be eliminated from the body through the Krebs cycle (Lewis, 1990). Although numerous studies have focused on the degradation of these polyesters, data obtained from native polymer powders or from processed samples such as films, implants or surgical sutures are not necessarily representative for microsphere erosion. Indeed, processing conditions and sample dimensions have been shown to greatly affect degradation mechanisms and time (Vert and Garreau, 1991; Grizzi et al., 1995). In this context, Thomasin et al. (1996a) have conducted very recently a degradation study on microspheres prepared by coacervation using a broad spectrum of PLAs and PLGAs 75:25 and 50:50, varying in M_w from about 15,000 to 150,000. The results showed that the more hydrophilic low molecular weight PLGA

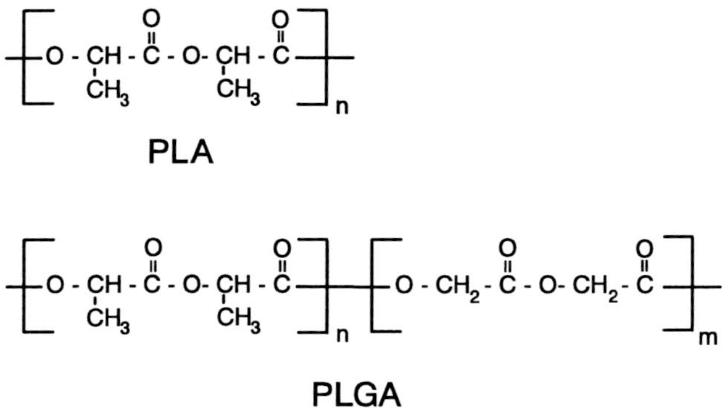

Fig. 3.2.2: Chemical composition of [PLA] and poly(D,L-lactide-co-glycolide) [PLGA].

50:50 microspheres degrade within about 1 month *in vitro*, whereas those made with more hydrophobic higher molecular weight D,L-PLA require 4 to 6 months for hydrolysis. Various studies have also been conducted to clarify the *in vitro* and *in vivo* degradation mechanism of PLA/PLGA polymers in more detail as it became evident that drug release kinetics and pharmacological responses greatly depend on the polymer characteristics and biodegradation behaviour (Makino et al., 1986, 1987; Kenley et al., 1987; Spenlehauer, 1989; Wang et al., 1990; Cohen et al., 1991b; Shah et al., 1992; Rafler and Jobmann, 1994; Park, 1994; O'Hagan et al., 1994a; Thomasin et al., 1996b). These investigations revealed that in particular the release of macromolecules is related to the polymer degradability and, hence, to the copolymer composition and molecular weight. Moreover, it has been found that the degradation pattern of polymers from different suppliers is also affected by the polymer quality, i.e. the presence of low molecular weight fractions and monomers (Schmitt et al., 1993). Special attention has further been paid to the role of water involved in polymer degradation and its influence on T_g (Vert and Garreau, 1991; Pitt, 1992). On the other side, limited knowledge is available at present of the influence of polymer degradation on the release of entrapped protein or peptide (Lawter et al., 1987; Bodmer et al., 1992; Park et al., 1995). It has also been recognized for some time, and reemphasized more recently, that during polymer degradation, the hydrolysis products are likely to create an acidic environment inside the microspheres and in their approximate surroundings, which may severely compromise protein stability in *in vitro* studies and also in an *in vivo* situation (Kenley et al., 1987; Park et al., 1995; Lu and Park, 1995). However, general conclusions about the importance of the acidic environment on protein stability cannot been drawn as every protein will be affected differently.

Besides biodegradability, biocompatibility is the crucial property of biomaterials intended to be used as vehicle for antigen delivery after parenteral or mucosal administration. The excellent biocompatibility of PLAs/PLGAs as suture material and as implants for i.m.- or s.c.-administration has been demonstrated abundantly in terms of tissue response and non-toxicity (Visscher et al., 1985, 1986, 1987; Yamaguchi and Anderson, 1993). For injectable microspheres, but not necessarily for any other types of implants (Tegnander et al., 1994), it is recognized that after i.m. injection there is an initial minimal, sharply localized, acute inflammatory response followed, at later time points, by macrophage, foreign body giant cell [FBGC], and connective tissue response, along with actual microsphere degradation. Macrophages are reported to represent the predominant cells at early time points, i.e. 15 days post injection, and at the later time points, i.e. towards the end of polymer degradation (Yamaguchi and Anderson, 1993). No fibrous capsule formation was seen in the case of these small size particles (Visscher et al., 1987). Most importantly, for vaccine vehicles, no polymorphonuclear leukocytes or lymphocytes were observed during the entire period of observation (Yamaguchi and Anderson, 1993). New results from an ongoing study also show that no significant augmentation of total amount of serum IgE could be detected at any time point when microspheres containing synthetic antigenic peptides were administered to mice (Men et al., 1996).

The promising results obtained with PLA/PLGA-based drug delivery systems have encouraged the search for and development of alternative biodegradable synthetic polymers. New classes of biocompatible and biodegradable polymers have been studied extensively and proposed for drug delivery. Amongst them, we consider as most important and best studied the polyanhydrides (Chasin et al., 1990b), polyorthoesters (Heller et al., 1990), poly(ε-caprolactones) (Pitt, 1990), polyorganophosphazenes (Allcock, 1990) and new copolymers of lactic or glycolic acid with either a more hydrophilic moiety such as polyols or ethylene oxide (Bodmer et al., 1992; Youxin et al., 1994), or a more hydrophobic moiety, such as an aliphatic rest from fatty acids or alcohols (Jobmann and Rafler, 1995). To our knowledge, however, only very few and still preliminary reports exist on the use of these polymers for antigen delivery (Esparza and Kissel, 1992; Koneberg and Kissel, 1995). On the other hand, the use of poly(amino acids) as drug delivery material has been severely hampered in the past because of the strong antigenicity and immunogenicity of this class of compounds (Kohn et al., 1990; Vermeersch and Remon, 1994).

3.2.3. Preparation Techniques of Microspheres

Antigens, like any other biologically active compounds, can be entrapped into microparticulates, in principle, by four different approaches:

(1) Physical entrapment of antigen during polymerization of a monomer dispersion (emulsion or suspension polymerization), often used for nanoparticles;
(2) Physical entrapment of antigen during particle formation with a preexisting polymer by desolvatation of the polymer solution, occasionally used for nanoparticles and frequently used for microspheres;

(3) Physical adsorption of antigen on preexisting polymeric nano- or microparticles at a defined time before administration, occasionally used for nano- and microparticles;
(4) Chemical attachment of antigens on the surface or inside preexisting nano- or microparticles.

For product development, approaches (1) and (4) are the most problematic because of the presence of reactive agents, residual monomers and initiators in the particles, which cannot be easily eliminated or neutralized. In this regard, approach (3) is the most preferable as both constituents, the microspheres (or nanospheres) and the antigen solution, can be purified and, if necessary, sterilized separately. However, unless there are mechanisms other than physical adsorption involved, the antigen will be released relatively quickly without the possibility to control the release over a prolonged period of time. Therefore, no single-dose vaccine delivery system should be expected from adsorbed antigens although the antigen presentation on the particle surface may confer sufficient adjuvancy to elicit a strong immune response (Almeida et al., 1993; Alpar and Almeida, 1994; Kreuter, 1994).

In this contribution, we shall focus on the approach (2) for biodegradable microspheres, but not for nanoparticles. For preparation techniques of the latter, excellent reviews have appeared recently (Allémann et al., 1993; Kreuter, 1994). The most common techniques for antigen microencapsulation into biodegradable polymers are the so-called solvent evaporation/extraction, coacervation and spray-drying. Excellent reviews on some of these techniques are available (Fong, 1988; Jalil and Nixon, 1990; Arshady, 1991; Newton, 1991; Aftabrouchad and Doelker, 1992), with one of them focusing particularly on encapsulation technologies for water soluble compounds (Aftabrouchad and Doelker, 1992). In addition, alternative processes may be envisaged such as cryogenic grinding of antigen loaded polymer films or extrudates, and spraying processes with or in supercritical gases, called GAS (gas antisolvent precipitation) (Randolph et al., 1993), ASES (aerosol solvent extraction system) (Bleich et al., 1993), or RESS (rapid expansion of supercritical solution) (Debenedetti et al., 1993). While solvent evaporation/extraction is undeniably the most frequently used technique in academia, particularly also for antigen microencapsulation, spray-drying has great advantages for industrial scaling-up and production. Being aware that the microencapsulation method will affect the quality of the final product, the three principle methods, discussed hereafter, can yield microspheres with the essential properties as required for particulate vaccine formulations, i.e.:

- Particle size: < 150 μm for injection with conventional syringes and needles, or < 10-20 μm for macrophage uptake, with the upper limit remaining controversial;
- Particle morphology and localization of the active compound in the microsphere: high surface regularity and low porosity are considered essential quality criteria; the microspheres may represent either a reservoir system with the active compound being localized in a central core or in multicompartments (this type is sometimes

called microcapsule), or a monolithic system with the active molecules being homogeneously distributed in the matrix as solid crystals or solid solution;
- Yield and encapsulation efficiency: should be highest possible, because antigens and biodegradable synthetic polymers are very expensive materials and not easily recoverable;
- Release kinetics: the 'burst' release during the first 24 h should be below 20 to 30% of the total dose; the release pattern can show a more constant release or a pulsatile release; for low antigen loadings, the duration of release depends essentially on the microspheres morphology, polymer hydrophobicity and degradation.

As far as microencapsulation efficiency and burst release are concerned, they may be optimized more efficiently by intimate knowledge of polymer-solvent-antigen interactions rather than by trial and error experiments (Gander et al., 1995, 1996). The three main microencapsulation techniques will be discussed in the following with respect to PLA/PLGA microspheres, although they all may apply also to other polymers. Moreover, antigens are microencapsulated preferably and more efficiently in liquid form, i.e. as aqueous solution in water or in a buffer, although lyophilized and micronized antigens can also be entrapped (Alonso et al., 1993; Thomasin et al., 1996a).

Common to the three major methods is the dispersion of the aqueous antigen solution in the organic polymer solution, non-miscible with the aqueous phase. This initial W/O-emulsion is obtained by means of standard homogenizers such as rotor-stator and ultrasound generators and high pressure homogenizers. In rare occasions, the low molecular weight synthetic antigens may be directly solubilized in the polymer solution by addition of cosolvents such as dimethylsulfoxide, acetic acid or dimethylacetamide. The polymer solvent should be sufficiently volatile to be eliminated during microsphere formation. Typical PLA/PLGA solvents include the most frequently used dichloromethane [DCM], ethyl acetate, ethyl formate, acetone, chloroform or mixtures of these solvents with alcohols or acetone. This primary W/O-emulsion is then processed by one of the methods described in the following (fig. 3.2.3).

3.2.3.1 Solvent Evaporation/Extraction

In this technique, the primary W_1/O-emulsion is transferred under vigorous stirring into a large volume of an aqueous phase (W_2) (=continuous phase) containing a surfactant or hydrocolloid as stabilizer, whereby a W_1/O/W_2-double emulsion is formed. Typically, poly(vinyl alcohol) has proven to be one of the most effective stabilizers in W_2, and is therefore most frequently used. Under continuous stirring, the organic solvent partitions from the polymer phase droplets (O) into W_2, wherefrom it evaporates ('solvent evaporation') leaving behind, after a few hours, solidified microspheres, which can be collected by filtration or centrifugation. This method has been proposed first by Takeda Industries in the mid-eighties (Yamamoto et al., 1986). Depending on the solubility of the organic solvent in W_2, the organic solvent evaporation may or

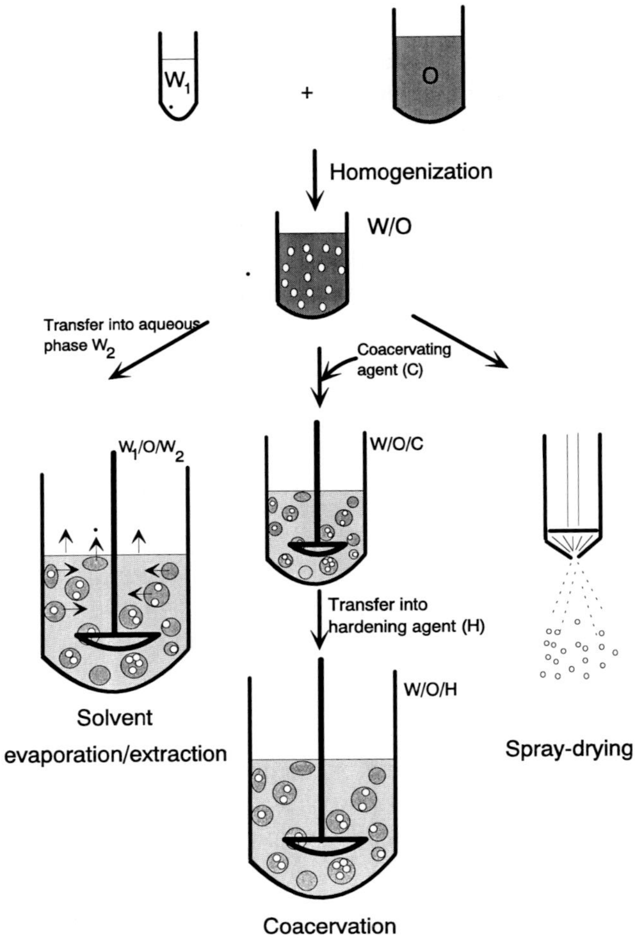

Fig. 3.2.3: Schematic illustration of the three principal methods for microencapsulation of antigens into PLA/PLGA microspheres. In the first common step, the aqueous antigen solution (W_1) is homogenized with the organic polymer solution (O) to form a W/O-emulsion. The successive steps are specific for each method (see text).

may not be a necessary step for the solidification of the polymer phase. If W_2 has a high enough capacity to extract most of the organic solvent from (O), the process is called 'solvent extraction'.

One of the major problems of the solvent evaporation/extraction method is the partition of water soluble active compounds, such as most antigens, from W_1 to W_2 resulting in poor encapsulation efficiency. Various parameters have been determined which reduce this loss of active material from W_1/O, namely a low solubility of the active compound in aqueous solutions, a high viscosity of W_1/O, small volumes of W_1,O and W_2, a low concentration of active compound in W_1, a short duration for polymer solidification and a medium size of O-droplets (Alex and Bodmeier, 1990;

Ogawa et al., 1988). A decreased water solubility of the active compound, achievable in principle through pH-adjustment in W_1 or chemical modification of the antigen, greatly improves the microencapsulation efficiency. A very efficient way to increase the viscosity of W_1/O consists of using a highly concentrated polymer solution (O) and of adding a gellable agent, such as gelatin, in W_1 (Yamamoto et al., 1986; Ogawa et al., 1988). On the other hand, reducing the time necessary for polymer solidification has been recognized as an essential parameter to achieve acceptable encapsulation efficiency (Schugens et al., 1994). This has led to the technique of solvent extraction, also called by some authors 'solvent removal' or 'in-liquid drying'. For a system consisting of an aqueous antigen solution W_1, a PLA/PLGA solution in DCM (O) and a continuous phase W_2, the addition of a defined amount of an alcohol or acetone into the W_2 phase increases greatly the solubility of DCM in W_2, inducing the hardening of the polymer droplets within a few minutes (Alonso et al., 1993).

One of the main features of the solvent evaporation/extraction method is its versatility to produce microspheres of virtually any size and also nanoparticles (Scholes et al., 1993; Le Ray et al., 1994). On the other hand, microspheres prepared by this method tend to be relatively porous (Nihant et al., 1994). Moreover, this method bears the inherent risk of losing substantial amounts of antigen, which often are highly water soluble and not available abundantly, in the W_2 phase.

3.2.3.2 Coacervation (or Organic Phase Separation)

In this method, the polymer is desolvated in a very controlled, two step process by adding slowly a so-called coacervating agent to the W_1/O emulsion, followed by transferring this mixture into a hardening agent. In the first stage, phase separation is induced, under continuous stirring, leading to a polymer rich phase, called coacervate, containing also the antigen solution, and an equilibrium phase. In the second stage, the highly viscous coacervate droplets are hardened by a more pronounced desolvatation in the hardening agent. Both coacervating and hardening agents are non-solvents for the polymer, miscible with the polymer solvent at least to some extent, but immiscible with water. A typical coacervating agent for PLAs/PLGAs is silicone oil, and examples of hardening agents encompass heptane, petroleum ether and octamethylcyclotetrasiloxane [OMCTS] (Aftabrouchad and Doelker, 1992).

Coacervation is a relatively complex microencapsulation technique which requires for every new type of polymer precise adjustment of the process conditions, such as temperature and optimal amount of coacervating agent (Ruiz et al., 1989, 1990; Thomasin et al., 1993). The most obvious drawbacks represent the amount of residual solvents in the final product, i.e. mainly coacervating and hardening agents (Thomasin et al., 1996b), and the fact that particle sizes below 10 μm are not easily produced. Despite the complexity of the process and the use of non-aqueous media, coacervation represents a most appropriate method for encapsulation of highly water soluble compounds, because water solubility does not represent a critical parameter in this process. This method has also been selected by some authors for the microencapsulation of antigens (Esparza and Kissel, 1992; McGhee et al., 1994; Men et al., 1995, 1996; Thomasin et al., 1996b).

3.2.3.3 Spray-Drying

A third and very attractive method for antigen microencapsulation is spray-drying. In this process, the W_1/O emulsion (fig. 3.2.3) is directly conducted through a nozzle and, with the aid of pressurized air, sprayed into a stream of warm air in which the solvent evaporates. The solidified particles are separated in a cyclone. This method is very simple, fast and appears particularly suitable for process scaling-up and production. Moreover, high entrapment efficiencies are generally achieved. Despite these important advantages, the literature on peptide and protein microencapsulation into PLA/PLGA is scarce (Schmiedel and Sandow, 1989; Wang et al., 1990; Gander et al., 1995; Men et al., 1995). In laboratory spray-dryers, the exposure of the product to slightly elevated temperature of about 30 to 40 °C may represent a certain drawback. Thus, polymers with a T_g below 30 °C or temperature sensitive proteins should not be used in this technique. Moreover, high molecular weight polymers (M_w above 100,000) usually yield irregularly shaped particles, and the polymer solution must be rather diluted to maintain the viscosity low for spraying. For high molecular weight polymers, this process is therefore not economic.

In summary, the three techniques of solvent evaporation/extraction, coacervation and spray-drying represent, for the time being, the main processes used in industries to produce biodegradable drug delivery systems. They have proven their suitability for peptide encapsulation into PLA/PLGA in terms of safety, reproducibility and GMP-conformity. A major obstacle for antigen containing microspheres remains the enormous difficulty of conducting and validating any of these processes under perfect aseptic conditions to avoid the necessity for terminal sterilization. For this type of heat-sensitive microspheres, γ-irradiation is the only possible sterilization method. However, it has been shown that γ-irradiation not only affects the microspheres properties (Spenlehauer et al., 1989) but also the antigenicity of microencapsulated antigens (Esparza and Kissel, 1992). Hence, new microencapsulation methods should be developed which take into consideration the special requirements of antigens and, most importantly, would allow the manufacture of such vaccine delivery systems under aseptic conditions at an industrial level.

3.2.4 Mode of Action

3.2.4.1 *In vitro* Release Kinetics

Although little is understood of what constitutes an optimal release profile for the delivery of antigen, immunologists are traditionally more interested in a pulsatile rather than a continuous release profile (Eldridge et al., 1993). There is a consensus though that a single-dose vaccine delivery system should release the antigen over a period of at least 3 to 6 months, preferably even up to 9 months. The particular aspect of antigen delivery from biodegradable microspheres, and its relevance for the immune response has been investigated by several groups. Most of these studies have considered conventional antigens such as tetanus toxoid [TT] (Alsonso et al., 1993, 1994; Almeida et al., 1993; Esparza and Kissel, 1992; Raghuvanshi et al., 1993; Thomasin et al., 1996a) and diphtheria toxoid [DT] (Singh et al., 1991), and only very few have focused on synthetic antigens (Men et al., 1994, 1996a). All of these investigations confirm the very pronounced immunostimulating properties of microspheres, without exhibiting an unambiguous boosting effect *in vivo*. At present, we have to face the open question why neither the sustained nor pulsatile *in vitro* release profiles observed result in a marked booster effect in the immunological response. Several reasons for this lack of correlation between *in vitro* release kinetics and immune response are conceivable, i.e. (i) difference between *in vitro* and *in vivo* release profiles, (ii) immunogenic instability of the antigen *in vivo*, (iii) inappropriate release kinetics. To approach these open questions, we might need to improve our knowledge on the release behaviour of antigens from PLA/PLGA microspheres.

The main body of information on the release of peptidic or proteinaceous compounds from PLA/PLGA microspheres does not originate from studies with antigens, but from those using peptide drugs and model proteins. Probably the first time pulsatile peptide/protein release from PLGA microspheres was suggested is found in a report on microencapsulated nafarelin acetate, a decapeptide, using data of the *in vivo* release form PLGA 50:50 and of its pharmacological activity (Sanders et al., 1984). In this case, the three release phases observed were assigned to diffusion of peptide located near the particle surface (first phase), to a dormant period during polymer hydrolysis (second phase) and to polymer erosion and dissolution of low molecular weight PLGA fragments (third phase). Later studies with a similar decapeptide (Lawter et al., 1987), with the octapeptide octreotide (Bodmer et al., 1992), with human serum albumin (Hora et al., 1990) or with BSA (Sah and Chien, 1993) confirmed the earlier observed triphasic release pattern, at least for drug loadings in the microspheres below 10%. At higher loadings, the second and third phases were less evident. Similarly, pronounced triphasic *in vitro* release patterns were also determined for tetanus toxoid (Gander et al., 1993) and for a synthetic malaria antigen (fig. 3.2.4) (Men et al., 1994, 1996a). It is generally agreed upon that the initial release, also called 'burst' release, and the third release phase can be explained respectively by diffusion of active compound located near the microspheres' surface during polymer hydration and by complete erosion of the microspheres due to advanced polymer hydrolysis.

On contrast, the assumed mechanisms involved in the second latent phase are more speculative. In this second phase, free diffusion is inhibited, most probably due to some type of interactions between the polymer and the peptides or proteins. Specifically, an ion exchange mechanism between the negatively charged polymeric car-

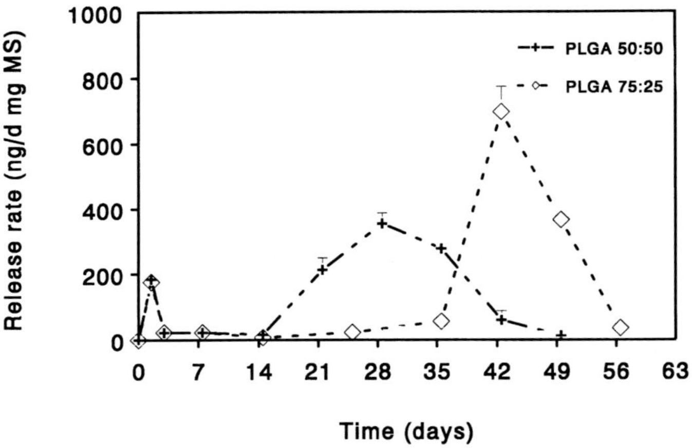

Fig. 3.2.4: *In vitro* release profiles of a synthetic malarial peptide from spray-dried PLGA 50:50 and PLGA 75:25 microspheres (from Men et al., 1994).

boxyl groups, produced through polymer hydrolysis, and positively charged peptide groups has been suggested (Bodmer et al., 1992). Such an ionic complex is of course expected only within a pH-range close to or below the pI of the protein and above the pKa of the glycolic (±3.8) and lactic acids (±3.1) (Thomasin et al., 1996a). On the other hand, charge separation on the protein may also account for this type of ionic interaction at a pH slightly above the pI. The fact that the numerous reported peptide/ protein release profiles from PLA/PLGA microspheres exhibit a relatively regular, sometimes even constant, release rate (Yamakawa et al., 1992) rather than a pulsatile pattern may still be considered in the light of an ion exchange mechanism. Indeed, ionic complexes between polymeric chains carrying carboxylate groups and positively charged peptides may form only under very restrictive conditions. These conditions should be specific for the protein itself (pI), the composition of the release medium (ionic strength, pH, buffer capacity, presence of ionic surfactants and preservatives) and a possible micro-pH inside the microspheres developing during polymer hydrolysis. If ionic complex formation is indeed the principal mechanism responsible for inhibiting protein diffusion from PLA/PLGA microspheres, the occurrence of a pulsatile release may depend very much on the actual systems used for the study. It then appears quite obvious, that *in vivo* release conditions will differ greatly from those *in vitro*, as far as the above mentioned parameters are concerned. Although the *in vivo* release data from the work of Sanders et al. (1984) indicate a triphasic pattern, the extent of the release pulse between days 7 and 21 is not very pronounced and the duration rather long. Therefore, this type of release pattern may be classified continuous rather than pulsatile. Similarly, data on pharmacological activity suggesting pulsatile release of peptides often show a lengthy interval during which the measured and sometimes only moderate effect is observed (Sanders et al., 1984; Bodmer et al., 1992; Ruiz et al., 1991). Such release patterns might not be adequate for boosting immunological responses.

The above interpretation of pulsatile release profiles possibly requires a pH-microenvironment inside the polymer particles which is lower than the pH of the release medium, generally kept at a pH of approximately 7. Such an acidic environment has also been suggested by different authors (Kenley et al., 1987; Lu and Park, 1995). The generation of high amounts of acidic degradation products inside the microspheres or in the incubation medium has caused well founded concerns about the stability of unreleased and released proteins (Gander et al., 1993; Kenley et al., 1987; Park et al., 1995; Lu and Park, 1995). Studies on the stability of tetanus toxoid in aqueous solutions and during a freeze-drying process have indeed shown that TT loses its antigenicity under certain humidity and pH conditions (Gander et al., 1993; Schwendeman et al., 1994). Aggregation was postulated to represent a major mechanism of antigen inactivation (Schwendeman et al., 1994; Alonso et al., 1994). Whether or not antigen instability is responsible for the lack of boosting effect *in vivo* should eventually depend on the particular antigen.

3.2.4.2 *In vivo* Processing

The fate of microspheres on the supracellular level is fairly well established for some routes of administration. As will be shown, particle size is an important parameter for most routes.

Although an irrelevant entry for vaccines, the central compartment is considered here for completeness, and the fate of microspheres entering the vascular system, is briefly discussed. Small microspheres and nanospheres are cleared quickly from the blood vessels and accumulate preferentially in the liver (le Ray et al., 1994). In this particular study performed with radiolabeled microspheres in mice, 30 % of the microspheres were found in the liver at thirty minutes after injection, a maximum of 83 % after 4 h and almost 60 % still after one week. Microspheres were also found in other organs belonging to the reticuloendothelial system [RES], i.e. in the spleen. After one week, radioactivity of more than 10 % was also found in faeces and urine, indicating microsphere degradation. This study did not confirm the sometimes suggested relationship between particle size and liver uptake, i.e. large particles are cleared more quickly. Presence or absence of "stealth" properties of the surface of microspheres may be a critical parameter for RES uptake. Following the fate of a radiolabeled particle does not allow the discrimination between microspheres and free or degraded label, which introduces uncertainties in the kinetics of microsphere processing.

Subcutaneous and intramuscular administration of PLGA microspheres can lead to the formation of a fibrous capsule surrounding the injection site, and to collagen deposits (Yamaguchi and Anderson, 1993). The injection site is also infiltrated by macrophages which phagocytose a fraction of the microspheres in a size dependent process. *In vitro*, PLGA microspheres smaller than 12 μm are phagocytosed by macrophages (Kanke et al., 1986, 1988), whereas *in vivo*, microspheres of sizes between 1 and 20 μm are surrounded by macrophages (Yamaguchi and Anderson, 1993). As microsphere degradation proceeds, foreign body giant cells [FBGC] are also observed. Although the role of FBGCs in the immune response, and their capacity to phagocytose are not clear, FBGCs are attracted to sites containing persistent irritants, i.e. non-phagocytosable material. At injection sites only few lymphocytes are generally observed, although B-cells have antigen processing and presenting capacity. This is in accordance with the observation that B-cells do not act as antigen presenting cells for particulate antigens like microspheres (Galelli et al., 1993). It is thought that the immunestimulating action of microspheres is mainly caused by the fraction that is phagocytosed. After macrophage uptake, this fraction is transported to the draining lymph nodes and, after intracellular processing, antigen fragments are presented to T-lymphocytes. This may explain why, in some experiments, large microspheres were found less immunogenic than small particles (Eldridge et al., 1991b; O'Hagan et al., 1993c). This observation may implicate that the controlled release kinetics of antigen at the injection site is less critical for eliciting an immune response than sometimes assumed. However, the relative importance of particle size and slow release may be an antigen related issue. The contribution of sustained release to the adjuvating effect may be of higher importance for immunogens which are immunogenic even in their free form. These considerations are speculative and should not disregard the contribution of antigen release kinetics to the immune response. For instance, two bolus immunizations with HIV gp120 resulted in antibody levels which decreased faster than those obtained with one bolus plus a booster delivered by an implanted pump over a period of two weeks (Cleland et al., 1994).

Immunization with thymus dependent antigens generally lead to the formation of memory B-cells. It is thought that these memory cells differentiate into antibody forming cells after a second contact with the antigen. It is not yet clear, if continuous stimulation results in a different response than pulsatile antigen delivery. Continuous antigenic stimulation of draining lymph nodes induced the formation of antibody forming cells (Delemarre et al., 1991). In this respect continuous and pulsatile stimulations give rise to the same events. As a matter of fact, prolonged presence of antigen mimics closer natural infections than short appearance of antigen, although the natural event is not necessarily superior in this particular context. Antigen release can proceed, in theory, for many months, provided the antigen is stable and its release is determined by the degradation of the microspheres. Typically, large 85:15 PLGA microspheres (size of 45-250 μm) can degrade over a period of 180 days, and the degradation rate of this non-phagocytosable material becomes significant only 60 days after i.m. of s.c. administration (Eldridge et al.,1990). For highly immunogenic, large and stable antigens, such formulations may induce potent booster responses.

Uptake of microspheres by mucosal tissues is well studied, at least in the case of intestinal absorption by Peyer's patches. Peyer's patches are lymphoepithelial structures playing a role in the uptake and processing of antigens. They belong to the gut-associated lymphoid tissue [GALT]. GALT and other mucosal lymphoid tissues like the bronchus-associated lymphoid tissue [BALT] form the so called common mucosal immune system. This virtual system forms an early barrier against pathogens. The current hypothesis on the induction of the mucosal humoral response is as follows (Mestecky et al., 1994). B-cells originating from the bone marrow enter Peyer's patches and probably other lymphoid tissue. They express surface IgA induced by contact with local T-cells and accessory cells. Exogenous antigens are taken up through phagocytosis and pinocytosis by specialized antigen sampling cells in Peyer's patches, namely by the M-cells. Upon interaction with antigen, T-cells, B-cells and accessory cells, sensitized B-cells migrate via regional lymph nodes to other mucosal sites and exocrine glands. There, B-cells differentiate into antibody producing plasma cells. T_{help} -cells in Peyer's patches belong to the T_{h2} subset, at least for the local anti-cholera toxin response (Xu-Amano et al., 1994). T_{h2} provide help for B-cell responses and are recognized in Peyer's patches by the release of specific cytokines: IL-5 and IL-6.

In general, particles in the lower μm size range are taken up relatively easily by Peyer's patches and mesenteric lymph nodes. Peak concentrations in these tissues are reached in 3-4 days after administration (Eldridge et al., 1991a; Offit et al., 1994). A gradual decrease to zero concentration occurs in about one month. In the spleen, the maximum microsphere concentration, which is lower than in Peyer's patches and lymph nodes, is reached after 2 weeks. Translocation of microspheres from the lumen to processing tissues is size dependent (Eldridge et al., 1990; Jani et al. 1992; Jenkins et al. 1994). Microspheres larger than 10 μm are not absorbed by Peyer's patches (Eldridge et al., 1990). A semiquantitative study in rats with fluorescent latex particles of 50 nm, 100 nm and 1000 nm demonstrated that the smallest spheres were taken up by Peyer's patches and mesenteric lymph nodes quicker than the 100 nm or 1μm particles (Jani et al., 1992). Measuring points laid between 6 and 36 hours. Particles were also detected in liver and spleen but only after 18 h or later. These results seem to contradict with other findings were optimum uptake in the Peyer's patches and lymph nodes was found for 500 nm polystyrene particles out of a tested size range of 150, 500, 1000 and 3000 nm (Jenkins et al., 1994). The mesenteric lymph nodes, however, contained mainly 3000 nm particles. Different quantification methods and time points of measurements make a comparison difficult. Beside that, the chemical nature of microspheres can influence substantially the absorption by Peyer's patches (Eldridge et al., 1990). Polystyrene microspheres are absorbed the most readily, followed by PLGA microspheres, whereas non-degradable cellulose based particles are not absorbed at all. However, only a very small portion of the administered microspheres is actually taken up by lymphoid tissue, even for 'highly' absorbable particles. A fraction of about 10^{-2} to 10^{-6} % of the initial dose could be recovered from the lymph, Peyer's patches and mesenteric lymph nodes by flow cytometric analysis (Jenkins et al., 1994). In another study, 0.5 % of radioactive microspheres with a size of 133 nm crossed the intestinal barrier within 1 hour after administration (le Ray et al., 1994). In the presence of concentrated milk, this could be increased to $\pm$ 2 %. It is questionable whether these percentages can be increased further in the gut. Nasopharyngeal delivery by aerosol immunization may be a more suitable route to optimize particle delivery to the mucosa. Intranasally administered substances are transported to the pharynx at an average speed of 6-10 mm/min due to mucociliary clearance (Vidgren et al., 1992). Contact time of microspheres with the mucosa can be extended by using mucoadhesive polymers. These polymers absorb water, swell and, as a result, adhere to mucosal glycoproteins. Whether adhesive particles can elicit a higher immune response remains to be demonstrated.

Local delivery of microencapsulated antigen often does not only lead to a local sIgA response but also to a systemic reaction (Challacombe et al., 1992). There is some evidence that the release kinetics of microencapsulated antigen determines, to some extent, the ratio local sIgA to systemic antibody level (O'Hagan et al., 1994b). Oral immunization with a fast degrading PLGA preparation resulted in a higher IgA response against entrapped OVA than slowly degrading microspheres, which in turn induced higher serum IgG. This indicates that the time of antigen release (early, inside Peyer's patches or later, in mesenteric lymph nodes) determines the type of response, i.e. local or systemic.

The mechanism of the induction of immunological memory after local immunization is far from elucidated. Sustained release and/or prolonged presence of antigen in Peyer's patches seem to stimulate the induction of memory (Challacombe et al., 1992). Repeated intragastrical immunizations (priming on 3 consecutive days, and booster doses on 3 consecutive days after 4 weeks) with OVA containing microspheres induced much higher salivary sIgA responses than administration of free antigen. On contrary, no differences were observed between free and encapsulated antigen after priming.

3.2.5 Parenteral Immunization

The number of studies investigating parenteral immunization with microspheres is limited. Results with OVA, and antigens of bacterial, viral and protozoal origin are available. In many reports on mucosal immunization, the parenteral route is used for comparison as well as for examining the immunogenicity after encapsulation.

In general, single immunization with a microencapsulated antigen results in an immune response superior to that obtained with the plain, soluble antigen and sometimes equal to the response induced with well known adjuvants, like alum salts or Freund's adjuvant, or with booster injections. Microspheres generally possess priming capacity, i.e. induce immunological memory, as can be illustrated with booster injections (Eldridge et al., 1991a; Partidos et al., 1994; Esparza and Kissel 1992).

Studies with OVA (100 μg, s.c.) in PLGA 50:50 microspheres (molecular weight of 9 kD, diameter of 5.3 μm) resulted in a priming response in rats intermediate between that obtained with plain OVA and OVA in Freund's complete adjuvant [FCA] (O'Hagan et al., 1991a). A comparable study in mice with the same batch induced a humoral response superior to that measured with FCA (O'Hagan et al., 1991b).

The authors do not provide an explanation for the different behaviour of an identical preparation in the two species. A mixture of 3 PLGA microsphere preparations, differing in polymer molecular weight and composition, and single type PLGA 50:50 microspheres induced antibody titers comparable to OVA adsorbed to aluminum hydroxide OVA (O'Hagan et al., 1993c; McGee et al., 1994). In the latter study, a high OVA dose of 300 μg protein was used. Performing dose-response experiments would reveal whether OVA in these microspheres is indeed equally immunogenic. OVA microspheres administered i.p. to mice were able to induce cellular immune responses *in vitro*, i.e. CTL response and T-cell proliferation (O'Hagan et al., 1993a) as well as *in vivo* (Maloy et al., 1994). *In vitro* T-cell proliferation was comparable to the response to OVA-alum.

A well studied bacterial antigen in parenteral controlled release systems is tetanus toxoid. In the early 1990ies, the WHO/UNDP Program for Vaccine Development initiated sponsoring and coordination of research on the development of a controlled release TT vaccine based on biodegradable polyesters (Aguado and Lambert, 1992; Galazka, 1994). Studies have shown that immunization with TT in microspheres may be competitive with alum adsorbed TT with respect to priming capacity (Esparza and Kissel, 1992). The potential of microencapsulated TT as a substitute for alum adsorbed TT is not clear. It has been claimed that single immunization with encapsulated TT is as immunogenic as two immunizations with TT-alum (Raghuvanshi et al., 1993). Other investigators report that encapsulated TT is more immunogenic than plain TT (Alonso et al., 1993) but less than alum-TT (Alonso et al., 1994). Independent testing of microsphere preparations from four manufacturers demonstrated that many preparations were able to protect mice against a toxin challenge one year after single immunization (Kersten, to be published). Some manufacturers made preparations with an overall (i.e. several types of microspheres) protection rate of 97 %. Only 42 % of mice receiving a single dose of plain TT were protected. Compared to booster immunizations with plain or alum adsorbed TT, the humoral response was considerably lower, though. Stability problems reduced the antigenicity and immunogenicity of TT in microspheres. There is evidence that this instability is caused by aggregation induced by freeze-drying, the use of organic solvents (Alonso et al., 1994) or low pH during hydrolysis of the microspheres (Gander et al., 1993, Kersten, to be published). Talwar and coworkers obtained very promising results with TT and DT in microspheres. The humoral response in mice after single administration of microspheres was comparable to that obtained with three immunizations with alum-adsorbed DT (Singh et al., 1992) or with two immunizations with TT-alum (Raghuvanshi et al., 1993). *In vitro*, TT release occurred for more than 60 days and only a burst of less than 10 % was observed. TT-microspheres of modest immunogenicity exhibit large bursts and release of antigenically active TT of less than a week (Kersten, to be published). In general research groups reporting excellent results immunize with higher doses. Apparently, microencapsulated antigens should be administered at higher doses than in conventional adjuvant formulations. An important factor contributing to the (cellular) immune response against encapsulated TT seems to be the release pattern (Men et al., 1995). With a mixture of 3 preparations good results were obtained with respect to antibody and proliferative T-cell responses. The mixture of 3 individual microsphere preparations, varying in size, polymer molecular weight and lactide/glycolide ratio, was designed to release TT in a semi-pulsatile manner (tab. 3.2.3). Although a mixture of two fast degrading preparations also resulted in high antibody levels, only the "trivalent" mixture induced T-cells in mice which proliferated on repeated contact with antigen 45 weeks after a single immunization.

Table 3.2.3: Characteristics of the mixture of TT-microspheres used to induce a long-term cellular immune response (Men et al., 1995).

Preparation*	Composition	Molecular weight	Size (μm)	Release pattern
SD502	PLGA 50/50	12,000	1-15	burst + pulse (week 3-5)
CO752	PLGA 75/25	17,000	15-80	burst + pulse (week 8-12)
CO206	PLA	129,700	32-70	no burst, continuous (months)

* SD = spray dried; CO = coacervation

Another synergistic effect between different types of microspheres, i.e. a mixture of small and large particles, was observed for inactivated staphylococcal enterotoxin B [SEB] (Eldridge et al., 1991a). A single i.p. injection of a toxoid containing PLGA microsphere mixture in mice resulted in a primary and secondary responses. After 60 days IgG anti-toxin titers of 800,000 were obtained with the mixture. The response after immunization with either small or large microspheres was 100,000 or less. If the small microspheres are given via the s.c. route, titers of almost 10^6 were reached (Eldridge et al., 1991b). This is higher than the values obtained with alum adsorbed toxoid. The soluble SEB did not induce any response after priming (Eldridge et al., 1990), but a booster dose with plain toxoid induced IgM antibodies.

In mice, microencapsulated pertussis toxin or filamentous hemagglutinin [FHA] from *Bordetella pertussis* induced IgG responses of comparable kinetics and magnitude as those observed with the alum adsorbed antigens (Shahin et al., 1995). Similar findings were reported with PLGA encapsulated fimbriae after i.p. immunization (Jones et al., 1995). Interestingly, almost complete protection was measured after intra-nasal challenge: recovery of viable bacteria from lung and trachea was 5 % or less relative to non-immunized mice. This protection level was measured throughout the 6 months of the duration of the study.

In the field of viral antigens synthetic peptides (Partidos et al., 1994), recombinant proteins (Cleland et al., 1994) and complete viruses (Offit et al., 1994; Moldoveanu et al., 1993) have been microencapsulated and used for parenteral delivery. With a peptide from the respiratory syncytial virus F and SH proteins, a mixture of PLA and PLGA microspheres was used (Partidos et al., 1994), with the more slowly degrading

PLA being expected to deliver the booster dose. Peptide loaded PLA microspheres alone induced a response after four weeks, whereas the response against peptide in PLGA microspheres occurred at least one week earlier. In a study on recombinant gp120 from HIV, a pulsatile delivery was achieved with PLGA microspheres (Cleland et al., 1994). Co-incorporation of the saponin adjuvant QS21 resulted in a vaccine whose immunogenicity in guinea pigs was comparable to three immunizations with QS21 plus alum. The use of water soluble anionic polymers allowed to encapsulate infectious rotavirus (Offit et al., 1994). I.p. immunization of mice with live virus in free form did not result in an IgG response, whereas the microspheres, containing the virus, induced IgG titers up to 1,000. For a formalin inactivated influenza virus, preservation of immunogenicity was shown after microencapsulation into PLGA (Moldoveanu et al., 1993). After systemic immunization, the microspheres induced a protective immune response in mice comparable to that achieved with free antigen. Protection was determined by measuring virus recovery in the respiratory organs 72 h after challenge.

3.2.6 Local Immunization

Many *in vivo* studies with microencapsulated antigens focus on or include local application. Apparently, expectations are high with respect to the use of microspheres as vehicles for mucosal delivery of antigens. These hopes are mainly based on the concept that antigens must reside as long as possible on and in mucosal tissue to increase the chance to become processed by local lymphoid cells. Microspheres are more stable under these *in vivo* conditions than other delivery systems like liposomes. Liposome preparations may be solubilized upon contact with bile salts (Rowland and Woodley, 1980). ISCOMs, on the other side, although immune stimulating via local routes and quite stable, even in bile salts and at low pH (Kersten, 1990), are unable to present soluble antigens unless chemically modified (Reid, 1992). If the antigen of interest is not extensively denatured during microencapsulation, microspheres offer it optimal protection. PLGA encapsulated influenza virus, for instance, is completely shielded in a simulated gastric medium (Mestecky et al., 1994). The fate of microspheres after local (oral) administration has been discussed in section 4.2 of this chapter. Here, the immunogenicity of microspheres delivered via mucosae is discussed.

The model system with OVA in PLGA microspheres, mentioned in section 3.2.5, was also used to investigate the potential of microsphere based vaccines for oral delivery (Maloy et al., 1994; O'Hagan et al., 1994b). OVA in free form induces tolerance after oral administration (Challacombe and Tomasi, 1980). Mice receiving microspheres by gastric intubation produced salivary sIgA as well as serum IgG after booster administrations (O'Hagan et al., 1994b). The secretory immune response was maximum at week 3 after boosting and faded to almost zero 10 weeks after boosting. Free OVA, as expected, was not immunogenic. The ratio sIgA/systemic IgG depended on the type of polymer. The faster degrading microspheres induced high sIgA levels as compared to the more slowly degrading particles. The time and kinetics of antigen release, as pointed out in section 3.2.4, is probably critical to the type of response. OVA microspheres also induced intestinal sIgA after oral administration (Maloy et al., 1994). Moreover, an antigen specific CTL response was detected in this latter study, although this response was much lower than after feeding with ISCOMs containing palmitoylated OVA.

Mice receiving orally PLGA entrapped toxin B subunit from *Vibrio cholerae* produced specific antibodies in serum and IgA secreting cells in mesenteric lymph nodes and spleen after 7 days (O'Hagan et al., 1993a). Free subunit B did not induce antibody secreting cells in the lymph nodes, although the serum response was comparable to the one induced by microspheres. Microspheres injected i.p. generated the highest number of antibody secreting cells in the spleen as well as highest serum response. On the contrary, PLGA microspheres containing antigenic protein fractions of a cell free lysate from *V. cholerae* were not immunogenic when administered via the oral or intestinal route in rabbits (Chandrasekhar et al., 1994). The authors suggest that a significant portion of the particles may have been too large for uptake by Peyer's patches.

Fimbrial adhesins from enterotoxigenic *Escherichia coli* [ETEC] are considered important antigens for anti-diarrhoea vaccines. A sIgA response against these colonization factors may prevent adherence to the intestinal epithelium. In a rabbit model, protection against challenge with *E. coli* was provided after three intraduodenal immunizations with 100 μg adhesin in PLGA microspheres (McQueen et al., 1993). The rabbits produced sIgA in their bile and bacterial attachment was reduced, preventing diarrhoea and weight loss. The duration of the response was not determined. Gastric intubation with one dose of 200 μg Colonization Factor Antigen I [CFA/I] in PLGA microspheres induced sIgA in only one of three rabbits (Edelman et al., 1993). All rabbits developed high systemic IgG responses. This is surprising , considering that the microspheres released their antigen within 48 h *in vitro*. As mentioned above, fast degrading OVA microspheres induced a relatively high sIgA response (O'Hagan et al., 1994b). Microencapsulated CFA/II has been tested in a phase 1 trial (Tacket et al., 1994). The vaccine, shown to be save and immunogenic in rabbits (Reid et al., 1993), was well tolerated by the 10 adult vaccinees. They received 4 immunizations of 1 mg CFA/II at week 0, 1, 2 and 4 by intestinal intubation. Seroconversion, detected by sIgA in jejunal fluid and circulating IgA secreting cells, was 50 %. Challenge with ETEC at week 8 resulted in 30 % protection whereas all ten unvaccinated controls developed diarrhoea.

Whooping cough is a disease that may be prevented by the induction of a local respiratory immune response against *Bordetella pertussis*. Currently, parenteral vaccines are used, but they do not induce substantial local responses nor long-lived immunity, a response that is observed after natural infection. Encapsulation of pertussis antigens in PLGA microspheres and subsequent intranasal immunization of mice resulted in IgG and IgA responses in bronchoalveolar lavage fluids and serum which were considerably higher than after administration of the free antigens (Shahin et al., 1995). Protection against an aerosol challenge was measured four weeks after a booster by recovery of bacteria from lung and trachea. All antigens tested, i.e. pertussis toxin, FHA and pertactin, were immunogenic in microspheres at a dose as low as 1 μg protein. In a similar study, in which FHA was used, no difference between free and encapsulated antigen was found, with respect to the humoral response and bacterial clearance rate (Cahill et al., 1995). The reason for the different findings is not clear. Antigen doses were in the same order of magnitude but no further specifications of the microspheres were given in either of the two studies.

Investigations on SEB revealed that three oral immunizations with PLGA microspheres containing inactivated SEB induced sIgA responses in saliva and in the gut, whereas the soluble form of the toxoid was not immunogenic (Eldridge et al., 1990). In a subsequent study, pulmonary and systemic responses were additionally detected (Eldridge et al., 1991a).

Amongst viral antigens, the influenza virus is relatively well studied with respect to local delivery by polymeric devices. Proteinoid microspheres, prepared by polycondensation of aminoacids, were used to present hemagglutinin-neuraminidase and M1 antigen orally in rats (Santiago et al., 1993). Encapsulated antigen induced a substantial systemic antibody response as compared to the antigens in their free form. However, compared to FCA, administered *systemically*, the microspheres elicited only a moderate response. The animals were not screened for local responses. A fragment from hemagglutinin was used as a model antigen to immunize sheep intravaginally (O'Hagan et al., 1993b). The preparation consisted of degradable starch microspheres, on which the antigen was adsorbed, and which were mixed with lysophophatidylcholine. The microspheres adhere to the mucosal epithelium and a subsequent osmotic effect opens the epithelial tight junctions. The systemic IgG and vaginal sIgA response were significantly higher than after administering the antigen in solution, but the highest systemic and local responses were obtained after i.m. immunization with alum adsorbed protein. The authors conclude that vaginal delivery may not be a suitable route because the absence of clearly defined lymphoid tissue, as opposed to the situation in other mucosal sites. There is some indirect evidence, however, that local immunization alone is generally less effective than a combination with parenteral priming. In a study in which formaline inactivated influenza virus encapsulated in PLGA microspheres, the combined immunization scheme 'systemic-oral' induced slightly higher antibody titers and better protection (virus clearance in mice) than the opposite scheme 'oral-systemic'. 'Systemic-oral' immunization as well as two systemic doses protected mice (Mestecky et al., 1994). Oral administration of influenza virus containing microspheres to primed mice induced anti-influenza antibody levels in saliva that were higher than those observed after systemic immunization. Serum responses following these schemes were comparable. Oral boosting provided complete protection after challenge (Moldoveanu et al., 1993). This is more or less confirmed by a study with inactivated simian immunodeficiency virus [SIV] incorporated in PLGA microspheres (Marx et al., 1993). Macaques could be protected against a vaginal challenge, after systemic priming followed by local (oral or intratracheal) boosting and not by oral immunization alone.

As mentioned in section 3.2.5 and table 3.2.1, live rotavirus can be microencapsulated successfully into water soluble polymers, which obviates the need for organic solvents (Offit et al., 1994). Alginate-spermine and chondroitin sulphate-spermine microspheres were both able to induce serum IgG as well as intestinal sIgA in 7 day old mice. Here again, free virus was not immunogenic. Moreover, the sIgA response was measured two months after oral immunization.

3.2.7 Safety Issues and Quality Control

Vaccines based on controlled release microspheres are accompanied by various new requirements related to quality control. Microsphere vaccines are presently in the R&D phase, and no established minimal requirements or testing procedures exist. In this paragraph, some challenging issues and possible obstacles that may hamper the introduction of such vaccine formulations on the market will be discussed.

Because PLGA is used in man on a regular basis, its safety is well established. For other members in the range of potentially useful polymers this is not yet the case. PLGA devices are considered completely biocompatible (Visscher et al., 1985, 1986, 1987; Yamaguchi and Anderson, 1993). Injection of 65/35 PLGA microspheres (25-150 µm) in rats (250 mg , s.c.) results in mild inflammatory reactions. After 150 days the microspheres are degraded completely, leaving a collagen deposit. No adverse tissue reactions were observed.

To our knowledge, the only report that exists up to now on the use of a microsphere based vaccine in man is an oral ETEC vaccine (Tacket et al., 1994). The vaccine was given to ten volunteers and was well tolerated. For preclinical and clinical evaluation, possible adverse effects have to be taken into account, which may originate not only from the polymers or their degradation products, but also from the presence of residual catalysts, used for polymerization, and organic solvents. Contamination with catalysts is not considered as a problem, because the commercially supplied medical grade polymers are generally free of detectable amounts. On the other hand, removal of organic solvents used for microsphere preparation is a more critical task. Depending on the preparation method, several % solvent residues have been reported (Lewis and Sherman, 1989; Koemen and Groenendaal, 1991; Lawter and Lanzilotti, 1992; Thomasin et al., 1996b). A variety of reliable techniques exist to determine residual solvents in microspheres (Gangrade and Price, 1992; Thomasin et al., 1996b). Whether a certain amount of residual solvents in a parenteral drug delivery system is acceptable would probably depend on the therapeutic importance. No limits have been defined so far by pharmacopoeias for most of the solvents generally used for PLA/PLGA microspheres, except for dichloromethane [DCM], for which the USP has defined a limit of 500 ppm. In this respect, a general strategy has been proposed which defines limits of residual solvents according to three toxicological solvent categories and to the dose and duration of medication (Miller, 1990). For many of the so-called organic volatile impurities, where toxicity is not a problem, the limitation can be up to 2000 ppm (de Smet et al., 1995).

A more demanding task is the necessity to produce sterile microspheres for parenteral administration. At present, no satisfactory procedure for preparing sterile microspheres and detection of non-sterile particles exist. Small amounts of encapsulated bioburden may only be found during or after an *in vitro* release study, under sterile conditions, and subsequent testing of the release medium. Microspheres, being heat-sensitive, are sterilized preferably by γ-irradiation (Spenlehauer et al., 1989), but the necessary irradiation dose, e.g. 25 kGy, may deteriorate both the polymer and the antigen (Esparza and Kissel, 1992). It has been shown indeed that γ-irradiation diminishes M_w by 30 to 40 % and shortens substantially the shelf life of microspheres, as during storage, M_w decreases further. Moreover, the release rate of irradiated samples can be considerably modified (Spenlehauer et al., 1989). The alternative approach of preparing sterile microspheres is by an aseptic process, which will require thorough, costly and time consuming validation.

Besides sterility and residual solvents, the quality parameters size, size distribution, content of antigen ('loading'), *in vitro* release and potency have to be determined on a regular basis. Particle size and size distributions can be measured readily by laser light scattering and/or scanning electron microscopy. The microscopic method also provides information on particle shape and surface morphology. Accurate determination of the antigen content in microspheres is more problematic than generally reported, as the polymeric particles must be destroyed to extract the antigen without damaging the antigen itself. Three procedures are generally used, i.e. (i) dissolution of the microspheres in an appropriate solvent, such as DCM, and extraction of the antigen from the organic phase into an aqueous phase, (ii) dissolution of the microspheres in an appropriate solvent, collecting the insoluble protein by centrifugation or filtration and dissolution of the antigen pellet or filter residue in an aqueous buffer, or (iii) accelerated hydrolysis of the polymeric particles in 0.1 N sodium hydroxide and followed by neutralization of the aqueous solution. In the first method, substantial amounts of antigen can accumulate at the interface between the aqueous phase and the organic phase and on the glass wall; this phenomenon is particularly pronounced in the case of surface active antigens, such as larger proteins. Quantitative recovery will be virtually impossible and reproducibility poor. The method of complete polymer hydrolysis is most deteriorating for the antigen, which must lose its antigenicity. The preferred method is the recovery of antigen from dissolved microspheres by centrifugation or filtration. In any case the organic solvent may damage the antigen, or irreversible precipitation may occur in the solvent. Validation of all these methods is difficult because no reference formulations are generally available. Physical mixtures of empty microspheres and free antigen, in lyophilized solid or liquid form, should not be regarded as reference for microencapsulated antigen. In the two systems, i.e. the physical mixture and the antigen containing microspheres, the antigen does not exist in identical physical form with respect to size, thermodynamic state and surface characteristics of the solid antigen particles. Finally, measurement of recovered total protein content by protein assay or spectroscopically may not be representative for antigenicity. A further and very crucial parameter for vaccine delivery systems is of course the antigen release. For slowly degrading polymers, testing time may become very long, e.g. several months, because release studies have to be conducted in real time. If reliable models would be developed for release of particular antigens from PLGA microspheres, measurement time might be reduced to a few days. Comparison with a standard or mean release profile, preferably in combination with accelerated polymer hydrolysis and antigen stability data, should allow extrapolation of release profiles. Such a procedure should suffice for routine controls.

As mentioned before, stability of protein antigens during microencapsulation, storage and release is not always satisfactory. The reported problems with tetanus toxoid (Alonso et al., 1994; Gander et al., 1993; Koneberg et al., 1994) are not unique. Since polyesters, such as PLA/PLGA, hydrolyze and produce acidic compounds, their usefulness as delivery systems for acid sensitive antigens is compromised, unless appropriate stabilizers become available. Although the pH inside degrading PLGA microspheres is not known, it can be reasonably assumed that it is lower than in the surrounding release medium. Indeed, degradation studies have shown that polymer hydrolysis is faster in the center of microspheres than at the surface, which remains intact for quite some time. Also, mass loss lags greatly behind the loss in molecular weight. In release studies in isotonic buffers, the pH can drop to 3 or lower towards the end of polymer degradation, even if the medium is replaced daily by fresh buffer. This can be fatal for the immunogenicity of antigens such as tetanus toxoid (Kersten, to be published), but also for enzyme activity as in the case of carbonic anhydrase (Lu and Park, 1995). Problems may also arise upon contact with organic solvents during microencapsulation (Alonso et al., 1994; Lu and Park., 1995) or upon freeze drying of antigen containing formulations, which is sometimes a procedure for drying microspheres. Covalent as well as non-covalent aggregation of proteins can occur (Costantino et al., 1994; Lu and Park., 1995; Schwendemann et al., 1994). Future development of PLA/PLGA based vaccines should focus strongly on antigen stability and possible stabilizers for preventing acidity induced damage to antigens.

3.2.8 Conclusions and Prospects

Despite various limitations, biodegradable microspheres based on PLA/PLGA still represent the most promising antigen delivery system for the purpose of eliciting long lasting immune response after single administration. Besides their excellent safety and biocompatibility, their potential to stimulate and prolong the immune response has been demonstrated with many antigens of natural, recombinant and synthetic origin over the past five years. Moreover, these particulate carriers may be useful for both parenteral and local immunization. Main issues which have to be addressed in future activities are the development of procedures for aseptic manufacturing of microspheres and the stability and stabilization of antigens during microencapsulation and release. Single dose vaccine formulations can be expected only for antigens whose immunogenicity remains intact over a period of several weeks to a few months under *in vivo* conditions. Ideally, the antigen should be protected entirely from body fluids until its release pulse materializes. Such a protection should be achievable by using appropriate coatings or new types of polymers which do not swell in water, but degrade heterogeneously at the particle surface without producing reactive degradation products. It remains a great challenge for the end of this century to work on such an ideal antigen delivery system for single dose vaccines.

References

Aftabrouchad, C. and Doelker, E. (1992) Méthodes de préparation des microparticules biodégradables chargées en principes actifs hydrosolubles, S.T.P.Pharma Sciences *2*, 365-380.

Aguado, M. T. (1993) Future approaches to vaccine development: Single-dose vaccines using controlled-release delivery systems, Vaccine *11*, 596-597.

Aguado, M. T. and Lambert, P.-H. (1992) Controlled release vaccines – Biodegradable polylactide/polyglycolide (PL/PG) microspheres as antigen vehicles, Immunobiology *184*, 113-125.

Alex, R. and Bodmeier, R. (1990) Encapsulation of water soluble drugs by modified solvent evaporation method. I. Effect of process and formulation variables on drug entrapment, J. Microencapsulation *7*, 347-355.

Allcock, H. R. (1990) Polyphosphazenes as new biomedical and bioactive materials, in: Biodegradable polymers as drug delivery systems, Chasin, M. and Langer, R. (Eds), Marcel Dekker, New York, U.S.A. 163-193.

Allémann, E., Gurny, R., and Doelker, E. (1993) Drug-loaded nanoparticles – Preparation methods and drug targeting issues, Europ. J. Pharm. Biopharm. *39*, 173-191.

Almeida, A. J., Alpar, H. O., and Brown, M. R. W. (1993) Immune response to nasal delivery of antigenically intact tetanus toxoid associated with poly(L-lactic acid) microspheres in rats, rabbits and guinea-pigs, J. Pharm. Pharmacol. *45*, 198-203.

Alonso, M. J., Cohen, S., Park, T. G., Gupta, R. K., Siber, G. R., and Langer, R. (1993) Determinants of release rate of tetanus vaccine from polyester microspheres, Pharm. Res. *10*, 945-953.

Alonso, M. J., Gupta, R. K., Min, C., Siber, G. R., and Langer, R. (1994) Biodegradable microspheres as controlled-release tetanus toxoid delivery systems, Vaccine *12*, 299-306.

Alpar, H. O. and Almeida, A. J. (1994) Identification of some of the physico-chemical characteristics of microspheres which influence the induction of the immune response following mucosal delivery, Eur. J. Pharm. Biopharm. *40*, 198-202.

Amselem, S., Alving, C. R., and Domb, A. J. (1992) Polymeric biodegradable Lipospheres™ as vaccine delivery systems, Polym. Adv. Technol. *3*, 351-357.

Arshady, R. (1991) Preparation of biodegradable microspheres and microcapsules: 2. Polylactides and related polyesters, J. Control. Rel. *17*, 1-22.

Artursson, P., Mårtenson, I.-L., and Sjöholm, I. (1986) Biodegradable microspheres III: Some immunological properties of polyacryl starach microparticles, J. Pharm. Sci. *75*, 697-701.

Beck, L. R., Cowsar, D. R., Lewis, D. H., Gibson, J. W., and Flowers, C. E., Jr. (1979) New long-acting injectable microcapsule contraceptive system, Am. J. Obstet. Gynecol. *135*, 419-426.

Beck, L. R., Flowers, C. F., Jr., Cowsar, D. T., and Tanquary, A. C. (1988) U.S. Patent 4,756,907.

Bendix, D. (1990) Analytical studies on the solubility problem of poly(D,L-latide-co-glycolide) 50:50, Proceed. Inter. Symp. Control. Release Bioact. Mater. *17*, 248-249.

Birrenbach, G. (1973) Ueber Mizellpolymerisate, mögliche Einschlussverbindungen (Nanokapseln) und deren Eignung als Adjuvantien, Ph.D.-thesis, No. 5071, ETH, Zürich

Birrenbach, G. and Speiser, P. P. (1976) Polymerized micelles and their use as adjuvants in immunology, J. Pharm. Sci. *65*, 1763-1766.

Bleich, J., Müller, B. W., Nagels, K. H., and Wolff, H. M. (1993) Influence of two production parameters on microparticle size in aerosol solvent extraction system (ASES), Proceed. Intern. Symp. Control. Rel. Bioact. Mater. *20*, 456-457.

Bodmer, D., Kissel, T., and Traechslin, E. (1992) Factors influencing the release of peptides and proteins from biodegradable parenteral depot systems, J. Control. Release 21, 129-138.

Boswell, G. A. and Scribner, R. M. (1973) U.S. Patent 3,773,919.

Brunswick, M., Finkelman, F. D., Highet, P. A., Inman, J. K., Dintzis, H. M., and Mond, J. J. (1988) Picogram quantities of anti-Ig antibodies coupled to dextran induce B cell proliferation, J. Immunol. *140*, 3364-3372.

Cahill, E. S., O'Hagan, D. T., Illum, L., Barnard, A., Mills, K. H. G., and Redhead, K. (1995) Immune responses and protection against *Bordetella pertussis* infection after intranasal immunization of mice with filamentous haemagglutinin in solution or incroporated in biodegradable microparticles, Vaccine *13*, 455-462.

Challacombe, S. J. and Tomasi, B. (1980) Systemic tolerance and secretory immunity after oral immunization, J. Exp. Med. *152*, 1459-1472.

Challacombe, S. J., Rahman D., Jeffery H., Davis S. S., and O'Hagan, D. T. (1992) Enhanced secretory IgA and systemic IgG antibody responses after oral immunization with biodegradable microparticles containing antigen, Immunology *76*, 164-168

Chandrasekhar, U., Sinha, S., Bhagat, H. R., Sinha, V. B., and Srivastava, B. S. (1994) Comparative efficacy of biodegradable liposomes and microspheres as carriers for delivery of Vibrio cholerae antigens in the intestine, Vaccine *12*, 1384-1388.

Chasin, M. and Langer, R. (Eds) (1990a) Biodegradable polymers as drug delivery systems, Marcel Dekker, New York, U.S.A.

Chasin, M., Domb, A., Ron, E., Mathiowitz, E., Leong, K., Laurencin, C., Brem, H., Grossman, S., and Langer R. (1990b) Polyanhydrides as drug delivery systems, in: Biodegradable polymers as drug delivery systems, Chasin, M. and Langer, R. (Eds), Marcel Dekker, New York, U.S.A. 43-70.

Cleland, J. L., Powell, M. F., Lim, A., Barrón, L., Berman, P. W., Eastman, D. J., Nunberg, J. H., Wrin, T., and Vennari, J. C. (1994) Development of a single-shot subunit vaccine for HIV-1, AIDS Res. Hum. Retroviruses *10* Suppl. 2, S21-S26.

Cohen, S., Bernstein, H., Hewes, C., Chow, M., and Langer, R.(1991a) The pharmacokinetics of, and humoral response to antigen delivered by microencapsulated liposomes, Proc. Natl. Acad. Sci. U.S.A. *88*, 10440-10444.

Cohen, S., Yoshioka, T., Lucarelli, M., Hwang, L. H., and Langer, R. (1991b) Controlled delivery systems for proteins based on poly(lactic/glycolic acid) microspheres, Pharm. Res. *8*, 713-720.

Costantino, H. R., Langer R., and Klibanov A. M. (1994) Solid-phase aggregation of proteins under pharmaceutical relevant conditions, J.Pharm. Sci. *83*, 1662-1669.

De Smet, M., Roels, K., Vanhoof, L., and Lauwers, W. (1995) Automated headspace method for organic volatile impurities in drug substances dissolved in nonaqueous medium, Pharm. Forum *21*, 501-514.

Debenedetti, P. G., Tom, J. W., Yeo, S.-D., and Lim, G.-B. (1993) Application of supercritical fluids for the production of sustained delivery devices, J. Control. Release *24*, 27-44.

Degling, L. and Stjärnkvist, P. (1995) Biodegradable microspheres XVIII: the adjuvant effect of polyacryl starch microparticles with conjugated human serum albumin, Vaccine *13*, 629-636.

Delemarre, F. G. A., Kors N., and van Rooijen N. (1991) Development of antibody-forming cells in follicles of lymph nodes of mice after continuous antigenic stimulation, Immunobiology *184*, 75-82.

Dunn, R. L., English, J. P., Strobel, J. D., Cowsar, D. R., and Tice, T. R. (1988) Preparation and evaluation of lactide/glycolide copolymers for drug delivery, in: C. Migliaresi (Ed.), Polymers in medicine III, Elsevier, Amsterdam, NL, 149-160.

Edelman, R., Russell, R. G., Losonsky, G., Tall, B. D., Tacket, C. O., Levine, M. M., and Lewis, D. H. (1993) Immunization of rabbits with enterotoxigenic E. coli colonization factor antigen (CFA/I) encapsulated in biodegradable microspheres of poly(lactide-co-glycolide), Vaccine *11*, 155-158.

Eldridge, J. H., Meulbroek, J. A., Staas, J. K., Tice, T. R., and Gilley, R. M. (1989) Vaccine-containing biodegradable microspheres specifically enter the gut-associated lymphoid tissue following oral administration and induce a disseminated mucosal immune response, Adv. Exp. Med. Biol. *251*, 192-202.

Eldridge, J. H., Hammond, C. J., Meulbroek, J. A., Staas, J. K., Gilley, R. M., and Tice, T. R. (1990) Controlled vaccine release in the gut-associated lympoid tissues. I. Orally administered biodegradable microspheres target the Peyer's patches, J. Control. Release *11*, 205-214.

Eldridge, J. H., Staas, J. K. Meulbroek, J. A., McGhee, J. R., Tice, T. R., and Gilley, R. M. (1991a) Biodegradable microspheres as a vaccine delivery system, Mol. Immunol. *28*, 287-294.

Eldridge, J. H., Staas, J. K. Meulbroek, J. A., Tice, T. R., and Gilley, R. M. (1991b) Biodegradable and biocompatible poly(DL-lactide-co-glycolide) microspheres as an adjuvant for staphylococcal enterotoxin B toxoid which enhances the level of toxin-neutralizing antibodies, Infect. Immunity. *59*, 2978-2986.

Eldridge, J. H., Staas, J. K., Tice, T. R., and Gilley, R. M. (1992) Biodegradable poly(DL-lactide-co-glycolide) microspheres. 44th Forum in Immunol. Res. Immunol. *143*, 557-563.

Eldridge, J. H., Staas, J. K., Chen, D., Marx, P. A., Tice, T. R., and Gilley, R. M. (1993) New advances in vaccine delivery systems, Seminars Hematol. *30*, Suppl. 4, 16-25.

Esparza, I. and Kissel, T. (1992) Parameters affecting the immunogenicity of microencapsulated tetanus toxoid, Vaccine *10*, 714-720.

Fong, J. W. (1988) Microencapsulation by solvent evaporation and organic phase separation processes, in Controlled release systems: fabrication technology, Hsie, D. (Ed.), CRC Press, Boca Raton, Fl, 81-108.

Fries, L. F., Gordon, D. M., Richards, R. L., Egan, J. E., Hollingdale, M. R., Gross, M., Silverman, C., and Alving, C. R.. (1992) Liposomal malaria vaccine in humans: a safe and potent adjuvant strategy, Proc. Natl. Acad. Sci. U.S.A. *89*, 358-362.

Galazka, A. M. (1994) Achievements, problems and perspectives of the expanded programme on immunization, Zentralblatt für Bakteriologie *281*, 353-364.

Galelli, A., Charlot, B., Deriaud, E., and Leclerc C. (1993) B-Cells do not present antigen covalently linked to microspheres, Immunology *79*, 69-76.

Gander, B., Thomasin, C., Merkle, H. P., Men, Y., and Corradin, G. (1993) Pulsed tetanus toxoid release from PLGA-microspheres and its relevance for immunogenicity in mice. Proceed. Intern. Symp. Control. Rel. Bioact. Mater. *20*, 65-66.

Gander, B., Wehrli, E., Alder, R., and Merkle, H. P. (1995) Quality improvement of spray-dried, protein-loaded D,L-PLA microspheres by appropriate polymer solvent selection, J. Microencapsulation *12*, 83-97.

Gander, B., Johansen, P., Hô Nam-Trân, M., and Merkle, H. P. (1996a) Thermodynamic approach to protein microencapsulation into poly(D,L-lactide) by spray drying, Int. J. Pharm., in press.

Gangrade, N. and Price, J. C. (1990) Simple gas chromatographic headspace analysis of residual organic solvent in microspheres, J. Pharm. Sci. *81*, 201-202.

Gilding, D. K. and Reed, A. M. (1979) Biodegradable polymers for use in surgery – poly(glycolic)/poly(lactic acid) homo- and copolymers: 1, Polymer *20*, 1459-1464.

Golumbek, P. T., Azhari, R., Jaffee, E. M., Levitsky, H. I., Lazenby, A., Leong, K., and Pardoll, D. M. (1993) Controlled release, biodegradable cytokine depots: A new approach in cancer vaccine design, Cancer Res. *53*, 5841-5844.

Grizzi, I., Garreau, H., Li, S., and Vert, M. (1995) Hydrolytic degradation of devices based on poly(DL-lactic acid) size-dependence, Biomaterials *16*, 305-311.

Gupta, R. K., Siber, G. R., Alonso, M. J., and Langer, R. (1993) Development of a single-dose tetanus toxoid based on controlled release from biodegradable and biocompatible polyester microspheres, in: Vaccines 93 – Modern approaches to new vaccines including prevention of AIDS, Ginsberg, H.S., Brown, F., Chanock, R.M. and Lerner, R.A. (Eds.) Cold Spring Harbor Laboratory Press, 391-396.

Heller, J., Sparer, R. V., and Zentner, G. M. (1990) Poly(ortho esters), in: Biodegradable polymers as drug delivery systems, Chasin, M. and Langer, R. (Eds), Marcel Dekker, New York, U.S.A. 121-161.

Heller, J. (1993) Polymers for controlled parenteral delivery of peptides and proteins, Adv. Drug Delivery Rev. *10*, 163-204.

Hora, M. S., Rana, R. K., Nunberg, J. H., Tice, T. R., Gilley, R. M., and Hudson, M. E. (1990) Release of human serum albumin from poly(lactide-co-glycolide) microspheres, Pharm. Res. *7*, 1190-1194.

Jalil, R. and Nixon, J. R. (1990) Biodegradable poly(lactic acid) and poly(lactide-co-glycolide) microcapsules: Problems associated with preparative techniques and release properties, J. Microencapsulation *7*, 297-325.

Jani, P. U., McCarthy, D. E., and Florence, A. T. (1992) Nanosphere and microsphere uptake via Peyer's patches. Observation of the rate of uptake in the rat after a single ral dose, Int. J. Pharm. *86*, 239-246.

Jenkins, P. G., Howard, K. A., Blackhall, N. W., Thomas, N. W., Davis, S. S., and O'Hagan, D. T. (1994) Microparticulate absorption from the rat intestine, J. Control. Release *29*, 339-350.

Jobmann, M. and Rafler. G. (1995) End-group functionalyzed polylactides: a potential protein carrier, Proceed. 5th Intern. Symp. Delivery Targeting Peptides, Proteins, Genes, Leiden, P8.

Jones, D. H., McBride, B. W., Jeffery, H., O'Hagan, D. T., Robinson, A., and Farrar, G. H. (1995) Protection of mice from *Bordetella pertussis* respiratory infection using microencapsulated pertussis fimbriae, Vaccine *13*, 675-681.

Kamath, K. R. and Park, K. (1993) Biodegradable hydrogels in drug delivery, Adv. Drug Delivery Rev. *11*, 59-84.

Kanke, M., Morlier, E., Geissler, R., Powell, D., Kaplan, A., and DeLuca, P. (1986) Interaction of microspheres with blood consituents. II. Uptake of biodegradable particles by macrophages, J Parent. Sci. *40*, 114-118.

Kanke, M., Geissler, R., Powell, D., Kaplan, A., and DeLuca, P. (1988) Interaction of microspheres with blood consituents. III. Macophage phagocytosis of various types of of polymeric drug carriers, J. Parent. Sci. *42*, 157-165.
Kenley, R. A., Lee, M. O., Mahoney II, T. R., and Sanders, L. M. (1987) Poly(lactide-co-glycolide) decomposition kinetics in vivo and in vitro, Macromolecules *20*, 2398-2403.
Kersten, G. (1990) Aspects of ISCOMs. Structural, pharmaceutical and adjuvant properties. PhD thesis University of Utrecht
Kersten, G. F. A. and Crommelin, D. J. A. (1995) Liposomes and ISCOMs as vaccine formulations, Biochim. Biophys. Acta *1241*, 117-138.
Koemen, J. and Groenendaal, J. W. (1991) Process for microencapsulation, U.S.Patent 5,066,436.
Kohn, J. (1990) Pseudopoly(amino acids), in: Biodegradable polymers as drug delivey systems, Chasin, M. and Langer, R. (Eds), Marcel Dekker, New York, US, 195-229.
Koneberg, R., Hilbert A., and Kissel, T. (1994) Parameters affecting the stability of tetanus toxoid, Eur. J. Pharm. Biopharm. *40*, S54.
Koneberg, R. and Kissel, T. (1995) Microencapsulation of tetanus toxoid using poly(lactide-co-glycolide) polymers and novel ABA-triblock-copolymers for single-shot vaccines, Proceed. 1st World Meeting APGI/APV, Budapest, 399-400.
Kreuter, J. (1974) Neue Adjuvantien auf Poly(methylmethacrylat)-Basis, Ph.D. thesis, No. 5417, ETH, Zürich.
Kreuter, J. and Speiser, P. P. (1976) New adjuvants on a poly(methylmethacrylate) base, Infect. Immunity *13*, 204-210.
Kreuter, J., Berg, U., Liehl, E., Soliva, M., and Speiser, P. P. (1986) Influence of particle size on the adjuvant effect of particulate polymeric adjuvants, Vaccine *4*, 125-129.
Kreuter, J., Liehl, E., Berg, U., Soliva, M., and Speiser, P. P. (1988) Influence of hydrophobicity on the adjuvant effect of particulate polymeric adjuvants, Vaccine *6*, 253-256.
Kreuter, J. (1988) Possibilities of using nanoparticles as carriers for drugs and vaccines, J. Microencapsulation *5*, 115-127.
Kreuter, J. (1994) Nanoparticles, in: Colloidal drug delivery systems, Kreuter, J. (Ed.) Marcel Dekker, New York, pp. 219-342.
Langer, R. (1990) New methods of drug delivery, Science *249*, 1527-1533
Langhein, C. and Newman, J. F. E. (1987) Antibody response to bacterial antigens covalently bound to biodegradable polymerized serum albumin beads, J. Appl. Bacteriol. *63*, 443-448.
Lawter, J. R., Brizzolara, N. S., Lanzilotti, M. G., and Morton, G. O. (1987) Drug release from poly(glycolid-co-lactide) microcapsules, Proceed. Intern. Symp. Control. Rel. Bioact. Mater. *14*, 99-100.
Lawter, J. R. and Lanzilotti, M. G. (1992) Hardening agent for phase separation microencapsulation, Europ. Patent Appl. 0 292 710 B1.
Le Ray, A. M., Vert, M., Gautier, J. C., and Benoît J. P. (1994) Fate of [^{14}C]poly(DL-lactide-co-glycolide) nanoparticles after intravenous and oral administration to mice, Int. J. Pharm. *106*, 201-211.
Leong, K. W. (1991) Synthetic bioerodible polymer drug systems, in: Polymers for controlled drug delivery, Tarcha, P.J. (Ed.) CRC Press, Boca Raton, Fl USA, 127-148.
Lewis, D. H. and Sherman, J. D. (1989) Low residual solvent microspheres and microencapsulation process, WO 89/03578.
Lewis, D. (1990) Controlled release of bioactive agents from lactide/glyoclide polymers, in: Biodegradable polymers as drug delivey systems, Chasin, M. and Langer, R. (Eds), Marcel Dekker, New York, USA, 1-41.
Livingston, P. O., Adluri, S., Helling, F., Yao, T., Kensil, C. R., Newman, M. J., and Marciani, D. (1994) Phase 1 trial of immunological adjuvant QS-21 with a GM2 ganglioside-keyhole

limpet haemocyanin conjugate vaccine in patients with malignant melanoma, Vaccine *12*, 1275-1280.

Lu, W. and Park, T. G. (1995) Protein release from poly(lactic-co-glycolic acid) microspheres: Protein stability problems, PDA J. Pharm. Sci. Technol. *49*, 13-19.

Makino, K., Ohshima, H., and Kondo, T. (1986) Mechanism of hydrolytic degradation of poly(L-lactide) microcapsules: effects of pH, ionic strength and buffer concentration, J. Microencapsulation *3*, 203-212.

Makino, K., Ohshima, H., and Kondo, T. (1987) Effects of plasma proteins on degradation properties of poly(L-lactide) microcapsules, Pharm. Res. *4*, 62-65.

Maloy, K. J., Donachie, A. M. O'Hagan, D. T., and Mowat, McI. (1994) Induction of mucosal and systemic immune responses by immunization with ovalbumin entrapped in poly(lactide-co-gycolide) microparticles, Immunology *81*, 661-667.

Martin, M. E. D., Dewar, J. B., and Newman, J. F. E. (1988) Polymerized serum albumin beads posessing slow release properties for use in vaccines, Vaccine *6*, 33-38.

Marx, P. A., Compans, R. W., Gettie, A., Staas, J. K., Gilley, R. M., Mulligan, M. J., Yamshchikov, G. V., Chen, D. X., and Eldridge, J. H. (1993) Protection against vaginal SIV transmission with microencapsulated vaccine, Science *260*, 1323-1327.

McGhee, J. P., Davis, S. S., and O'Hagan, D. T. (1994) The immunogenicity of a model protein entrapped in poly(lactide-co-glycolide) microparticles prepared by novel phase separation technique, J. Control. Release *31*, 55-60.

McQueen, E. C., Boedeker, E. C., Reid, R., Jarboe, D., Wolf, M., Le, M., and Brown, W. R. (1993) Pili in microspheres protect rabbits from diarrhoea induced by E. coli strain RDEC-1, Vaccine *11*, 201-206.

Men, Y., Corradin, G., Thomasin, C., Merkle, H. P., and Gander, B. (1994) Immunopotentiation of a synthetic antigen by incorporation into biodegradable microspheres, Proceed. Intern. Symp. Control. Rel. Bioact. Mater. *21*, 50-51.

Men, Y., Thomasin, C., Merkle, H. P., Gander, B., and Corradin, G. (1995) A single administration of tetanus toxoid in biodegradable microspheres elicits T cell and antibody responses similar or superior to those obtained with aluminum hydroxide, Vaccine *13*, 683-689.

Men, Y., Gander, B. Merkle, H. P., and Corradin, H. P. (1996a) Induction of strong immune responses to weakly immunogenic synthetic malarial peptides encapsulated in biodegradable polymer microspheres, Vaccine (in press).

Mestecky, J., Moldoveanu, Z., Novak, M., Huang, W. Q., Gilley, R. M., Staas, J. K., Schafer, D., and Compans, R. W. (1994) Biodegradable microspheres for the delivery of oral vaccines, J. Control. Release *28*, 131-141.

Miller, J. H. McB. (1990) Solvants résiduels, Pharmeuropa *7*, 928-934.

Moldoveanu, Z., Novak, M., Huang, W.-Q., Gilley, R. M., Staas, J. K., Schafer, D., Compans, R. W., and Mestecky, J. (1993) Oral immunization with influenza virus in biodegradable microspheres, J. Infect. Dis. *167*, 84-90.

Morris, W., Steinhoff, M. C., and Russell, P. K. (1994) Potential of polymer microencapsulation technology for vaccine innovation, Vaccine *12*, 5-11.

Naim, J. O. and van Oss, C. J. (1992) The effect of hydrophilicity-hydrophobicity and solubility on the immunogenicity of some natural and synthetic polymers, Immunol. Investig. *21*, 649-662.

Nakaoka, R., Tabata, Y., and Ikada, Y. (1995) Potentiality of gelatin microspheres as immunological adjuvant, Vaccine *13*, 653-661.

Newton, D. W. (1991) Coacervation: Principles and applications, in: Polymers for controlled drug delivery, Tarcha, P.J. (Ed.), CRC Press, Boca Raton, Fl., 67-81.

Nihant, N., Schugens, C., Grandfils, C., Jérôme, R., and Teyssié, P. (1994) Polylactide microspheres prepared by double emulsion/evaporation techique. I. Effect of primary emulsion stability, Pharm. Res. *11*, 1479-1485.

O'Hagan, D. T., Palin, K. J., and Davis, S. S. (1989) Poly(butyl-2-cyanoacrylate) particles as adjuvants for oral immunization, Vaccine *7*, 213-216.
O'Hagan, D. T., Jeffery, H., Roberts, M. J., Jr., McGee, J. P., and Davis, S. S. (1991a) Controlled release microparticles for vaccine development, Vaccine *9*, 768-771.
O'Hagan, D. T., Rahman, D., McGee, J. P., Jeffery, H., Davies, M. C., Williams, P., Davis, S., and Challacombe, S. J. (1991b) Biodegradable microparticles as controlled release antigen delivery systems, Immunology *73*, 239-242
O'Hagan, D. T., McGhee, J. P., Holmgren, J., Mowat, A. M. C., Donachie, A. M., Mills, K. H. G., Gaisford, W., Rahman, D., and Challacombe, S. J. (1993a) Biodegradable microparticles for oral immunization, Vaccine *11*, 149-154.
O'Hagan, D. T., Rafferty, D., Wharton, S., and Illum, L. (1993b) Intravaginal immunization in sheep using a bioadhesive micosphere antigen delivery system, Vaccine 11, 660-664.
O'Hagan, D. T., Jeffery, H., and Davis, S. S. (1993c) Long-Term Antibody Responses in Mice Following Subcutaneous Immunization with Ovalbumin Entrapped in Biodegradable Microparticles, Vaccine *11*, 965-969.
O'Hagan, D. T., Jeffery, H., and Davis, S. S. (1994a) The preparation and characterization of poly(lactide-co-glycolide) microparticles: III Microparticle/polymer degradation rates and the in vitro release of a model protein, Int. J. Pharm. *103*, 37-45.
O'Hagan, D. T., Rahman, D., Jeffery, H., Sharif, S., and Challacombe, S. J. (1994b) Controlled release microparticles for oral immunization, Int. J. Pharm. *108*, 133-139.
Offit, P. A., Khoury, C. A., Moser, C. A., Clark, H. F., Kim, J. E., and Speaker, T. J. (1994) Enhancement of rotavirus immunogenicity by microencapsulation, Virology *203*, 134-143.
Ogawa, Y., Yamamoto, M., Okada, H. Yashika, T., and Shimamoto, T. (1988) A new technique to efficiently entrap leuprolide acetate into microspheres of polylactic acid or copoly(lactic/glycolid) acid, Chem. Pharm. Bull. *36*, 1095-1103.
Park, T. G. (1994) Degradation of poly(D,L-lactic acid) microspheres: effect of molecular weight, J. Control. Release *30*, 161-173.
Park, T. G., Lu, W., and Crotts, G. (1995) Importance of in vitro experimental conditions on protein release kinetics, stability and polymer degradation in protein encapsulated poly(D,L-lactic acid-co-glyoclic acid) microspheres, J. Control. Release *33*, 211-222.
Partidos, C. D., Shaw, D. M., Gander, B., Merkle, H. P., Howard, C. R., and Steward, M. W. (1994) Induction of antibody responses to a synthetic peptide entrapped in biodegradable microparticles, Vaccine Res. *3*, 203-209.
Pitt, C. G. (1990) Poly-ε-caprolactone and its copolymers, in: Biodegradable polymers as drug delivery systems, Chasin, M. and Langer, R. (Eds), Marcel Dekker, New York, U.S.A. pp. 71-120.
Pitt, C. G. (1992) Non-microbial degradation of polyesters: Mechanisms and modifications, in: Vert, M., Feijen, J., Albertsson, A., Scott, G. and Chiellini, E. (Eds.), Biodegradable polymers and plastics, The Royal Society of Chemistry, Cambridge, UK, 7-17.
Preis, I. and Langer, R. S. (1979) A single-step immunization by sustained antigen release, J. Immunol. Methods *28*, 193-197.
Rafler, G. and Jobmann, M. (1994) Controlled release systems of biodegradable polymers. 1st communication: Hydrolytic degradation of biodegradable aliphatic polyesters, Pharm. Ind. *56*, 565-570.
Raghuvanshi, R. S., Singh, M., and Talwar, G. P. (1993) Biodegradable delivery system for single step immunization with tetanus toxoid, Int. J. Pharm. *93*, R1-R5.
Randolph, T. W., Randolph, A. D., Mebes, M., and Yeung, S. (1993) Sub-micrometer-sized biodegradable particles of poly(L-lactic acid) *via* the gas antisolvent spray precipitation process, Biotechnol. Prog *9*, 429-435.

Ray, A. M. le, Vert, M., Gautier, J. C., and Benoit, J. P. (1994) Fate of C-14]poly(DL- lactide-co-glycolide) nanoparticles after intravenous and oral administration to mice, Int. J. Pharm. *106*, 201-211.

Reid, G. (1992). Soluble proteins incorporate into ISCOMs after covalent attachment of fatty acid, Vaccine *10*, 597-602.

Reid, R. H., Boedeker, E. C., McQueen, C. E., Davis, D., Tseng, L. Y., Kodak, J., Sau, K., Wilhelmsen, C. L., Nellore, R., Dalal, P., and Bhagat, H. R. (1993) Preclinical evaluation of microencapsulated CFA/II oral vaccine against enterotoxigenic E. coli, Vaccine *11*, 159-167.

Rowland, R. N. and Woodley, J. F. (1980) The stability of liposomes in vitro to pH, bile salts and pancreatic lipase, Biochim. Biophys. Acta *620*, 400-409.

Ruiz, J. M., Tissier, B., and Benoit, J. P. (1989) Microencapsulation of peptide: A study of the phase separation of poly(D,L-lactic acid-co-glycolic acid) copolymers 50/50 by silicone oil, Int. J. Pharm. *49*, 69-77.

Ruiz, J. M., Busnel, J. P., and Benoit, J. P. (1990) Influence of average molecular weights of poly(D,L-lactic acid-co-glycolic acid) copolymers 50/50 on phase separation and *in vitro* drug release from microspheres, Pharm. Res. *7*, 928-934.

Sah, H. K. and Chien, Y. W. (1993) Evaluation of a microreservoir-type biodegradable microcapsule for controlled release of proteins, Drug Dev. Ind. Pharm. *19*, 1243-1263.

Sanders, L. M., Kent, J. S., McRae, G. I., Vickery, B. H., Tice, T. R., and Lewis, D. H. (1984) Controlled release of luteinizing hormone-releasing hormone analogue from poly(d,l-lactide-co-glycolide) microspheres, J. Pharm. Sci. *73*, 1294-1297.

Santiago, N., Milstein, S., Rivera, T., Garcia, E., Zaidi, T., Hong, H., and Bucher, D. (1993) Oral immunization of rats with proteinoid microspheres encapsulating influenza virus antigens, Pharm. Res. *10*, 1243-1247.

Schmiedel, R. and Sandow, J. K. (1989) Verfahren zur Herstellung von bioabbaubaren Mikrokapseln wasserlöslicher Peptide und Proteine sowie anch diesem Verfahren erhaltene Mikrokapseln, EP Patent Application 315 875.

Schmitt, E. A., Flanagan, D. R., and Linhardt, R. J. (1993) Degradation and release properties of pellets fabricated from three commercial poly(D,L-lactide-co-glycolide) biodegradable polymers, J. Pharm. Sci. *82*, 326-329.

Scholes, P. D., Coombes, A. G. A., Illum, L., Davis, S. S., Vert, M., and Davies, M. C. (1993) The preparation of sub-200 nm poly(lactide-co-glycolide) microspheres for site-specific drug delivery, J. Control. Release *25*, 145-153.

Scholsky, K. M., Tarcha, P. J., and Hsu, A. (1991) Acrylic latex systems, in: Polymers for controlled drug delivery, Tarcha, P.J. (Ed.) CRC Press, Boca Raton, Fl USA, 127-148.

Schröder, U. and Ståhl, A. (1984) Crystallized dextran nanospheres with entrapped antigen and their use as adjuvants, J. Immunol. Meth. *70*, 127-132.

Schwendeman, S. P., Lee, J. H., Gupta, R. K., Costantino, H. R., Siber, G. R., and Langer, R. (1994) Inhibition of moisture-induced aggregation of tetanus toxoid by protecting thiol groups, Proceed. Intern. Symp. Control. Rel. Bioact. Mater. 21, 54-55.

Shah, S. S., Cha, Y., and Pitt, C. G. (1992) Poly(glycolic acid-co-DL-lactic acid): Diffusion or degradation controlled drug delivery? J. Control. Release *18*, 261-270.

Shahin, R., Leef, M., Eldridge, J., Hudson, M., and Gilley, R. (1995) Adjuvanticity and protective immunity elicited by *Bordetella pertussis* antigens encapsulated in poly(D,L-lactide-co-glycolide) microspheres, Infect. Immun. *63*, 1195-1200.

Singh, M., Singh, A. and Talwar, G.P. (1991) Controlled delivery of diphtheria toxoid using biodegradable poly(D,L-lactide) microcapsules, Pharm. Res. *8*, 958-961.

Singh, M., Singh, O., Singh, A., and Talwar, G. P. (1992) Immunogenicity studies on diphtheria toxoid loaded biodegradable microspheres, Int. J. Pharm. *85*, R5-R8.

Spenlehauer, G., Vert, M., Benoit, J.-P., and Boddaert, A. (1989) In vitro and in vivo degradation of poly(D,L lactide/glycolide) type microspheres made by solvent evaporation method, Biomaterials *10*, 557-563.

Stieneker, F., Löwer, J., and Kreuter, J. (1993) Different kinetics of the humoral immune response to inactivated HIV-1 and HIV-2 in mice: Modulation by PMMA nanoparticule adjuvant, Vaccine Res. *2*, 111-118.

Stieneker, F., Kersten, G., van Bloois, L., Crommelin, D. J. A., Hem, S. L., Löwer, J., and Kreuter, J. (1995) Comparison of 24 different adjuvants for inactivated HIV-2 split whole virus as antigen in mice. Induction of titres of binding antibodies and toxicity of the formulations, Vaccine *13*, 45-53.

Tacket, C. O., Reid, R. H., Boedeker, E. C., Losonsky, G., Nataro, J. P., Bhagat, H., and Edelman, R. (1994) Enteral immunization and challenge of volunteers given enterotoxigenic E. coli CFA/II encapsulated in biodegradable microspheres, Vaccine *12*, 1270-1274.

Tegnander, A., Engebretsen, L., Bergh, K., Eide, E., Holen, K. J., and Iversen, O. J. (1994) Activation of the complement system and adverse effects of biodegradable pins of polylactic acid (Biofix®) in osteochnodritis dissecans, Acta Orthop. Scand. *65*, 472-475.

Thomasin, C., Gander, B., and Merkle, H. P. (1993) Coacervation of biodegradable polyesters for microencapsulation: A physico-chemical approach, Proceed. Intern. Symp. Contr. Rel. Bioact. Mater. *20*, 358-359.

Thomasin, C., Corradin, G., Men, Y., Merkle, H. P., and Gander B. (1996a) Tetanus toxoid and a synthetic malaria antigen in PLA/PLGA microspheres: Importance of polymer degradation and antigen release for immunological response, J. Control. Release (in press).

Thomasin, C., Johansen, P., Alder, R., Bemsel, R., Hottinger, G., Altorfer, H., Wright, A. D., Wehrli, E., Merkle, H. P., and Gander, B. (1996b) Approaching the problem of residual solvents in biodegradable microspheres prepared by coacervation, Europ. J. Pharm. Biopharm., *42*, 16-24.

Uchida, T., Martin, S., Foster, T. P., Wardley, R. C., and Grimm, S. (1994) Dose and load studies for subcutaneous and oral delivery of poly(lactide-co-glycolide) microspheres containing ovalbumin, Pharm. Res. *11*, 1009-1015.

Van Buskirk, A. M. and Braley-Mullen, H. (1987) In vitro activation of specific helper and suppressor T cells by the type 2 antigen polyvinylpyrrolidone (PVP), J. Immunol. *139*, 1400-1405.

Vermeersch, H. and Remon, J.-P. (1994) Immunogenicity of poly-D-lysine, a potential polymeric drug carrier, J. Control. Release *32*, 225-229.

Vert, M., Li, S., and Garreau, H. (1991) More about the degradation of LA/GA-derived matrices in aqueous media, J. Control. Release *16*, 15-26.

Visscher, G. E., Robison, R. L., Maulding, H. V., Fong, J. W., Pearson, J. E., and Argentieri, G. J. (1985) Biodegradation of and tissue reaction to 50:50 poly(DL-lactide-co-glycolide) microcapsules, J. Biomed. Mater. Res. *19*, 349-365.

Visscher, G. E., Robison, R. L., Maulding, H. V., Fong, J. W., Pearson, J. E., and Argentieri, G. J. (1986) Biodegradation of and tissue reaction to poly(DL-lactide) microcapsules, J. Biomed. Mater. Res. *20*, 667-676.

Visscher, G. E., Robison, R. L., and Argentieri, G. J. (1987) Tissue response to biodegradable injectable microcapsules, J. Biomater. Applic. *2*, 118-131.

Wang, H. T., Palmer, H., Linhardt, R. J., Flanagan, D. R., and Schmitt, E. (1990) Degradation of poly(ester) microspheres, Biomaterials *11*, 679-685.

Wise, D. L., McCormick, G. F., Willet, G. P., and Anderson, L. C. (1976) Sustained release of an antimalarial drug using a copolymer of glycolic/lactic acid, Life Sci. *19*, 867-874.

Xu-Amano, J., Jackson, R. J., Fujihashi, K., Kiyono, H., Staats, H. F., and McGhee, J. R. (1994) Helper Th1 and Th2 responses following mucosal or systemic immunization with cholera toxin, Vaccine *12*, 903-911.

Yamaguchi, K. and Anderson, J. M. (1993) In vivo biocompatibility studies of medisorb® 65/35 D,L-lactide/glycolide copolymer microspheres, J. Control. Release *24*, 81-93.

Yamakawa, I., Tsushima, Y., Machida, R., and Watanabe, S. (1992) *In vitro* and *in vivo* release of poly(D,L-lactic acid) microspheres containing neurotensin analogue prepared by novel oil-in-water solvent evaporation method, J. Pharm. Sci. *81*, 808-811.

Yamamoto, Y., Takada, S., and Ogawa, Y. (1986) Method of producing microcapsules, EP 0 190 833.

Yan, C., Rill, W. L., Malli, R., Hewetson, J., Tammariello, R., and Kende, M. (1995) Dependence of ricin toxoid vaccine efficacy on the structure of poly(lactide-co-glycolide) microparticle carriers, Vaccine *13*, 645-651.

Yolles, S., Eldridge, J., Leafe, T., Woodland, J. H. R., Blake, D. R., and Meyer, F. (1974) Long-acting delivery systems for narcotic antagonists, in: Controlled release of biologically active agents, Tanquary, A. C. and Lacey, R. E. (Eds), Plenum Press, New York, (Series: Adv. Exp. Med. Biol. *47*), 177-193.

Youxin, L., Volland, C., and Kissel, T. (1994) In-vitro degradation and bovine serum albumin release of the ABA triblock copolymers consisting of poly(L(+)lactic acid), or poly(L(+)lactic-co-glycolic acid) A-blocks attached to central poly(oxyethylene) B-blocks, J. Control. Release *32*, 121-128.

3.3 Peptide Based Vaccines

Hansjörg Schild and Hans-Georg Rammensee

3.3.1 Introduction

The age of peptide based vaccines started after the discoveries that antibodies generated by immunizations with synthetic peptides are sometimes able to recognize the native protein from which their sequence is derived and, much later, that cytotoxic T cells (CTL) recognize proteolytic fragments of proteins presented at the cell surface by major histocompatibility complex (MHC) molecules. By many investigators, synthetic peptide vaccines were regarded as the ultimate goal in vaccine technology because they are safe, cheap, easy to store and handle and they offered the ability to target the immune response to specific antigens expressed by the pathogens.

But up to now, no synthetic peptide vaccine has become commercially available. This situation, several years after the above mentioned discoveries, indicates that the development of peptide vaccines is not as straight forward as initially expected. Many explanations for this failure are possible, including the need for selection of peptides with higher immunogenicity according to the functional polymorphism of MHC molecules, the need for specific T helper cell epitopes for both B cell and CTL responses and the need for selection of adjuvants that allow the activation of B cells and CTL.

An improved understanding of the parameters that determine if an immune response is dominated by B cells or cell mediated immunity, a detailed knowledge of the requirements for peptides to be presented by MHC class I molecules and recognized by CTL and steady developments in the identification of rules governing MHC class II molecule/peptide interactions suggest that the outlook of synthetic peptide vaccines may not be that bad after all. In this chapter, we try to give an overview on the latest developments in the identification of T cell epitopes, progress in the administration of peptides to increase their immunogenicity and possibilities to deal with the problem of MHC polymorphism when synthetic peptides are to be used as vaccines.

3.3.2 Induction of Immune Responses Using Synthetic Peptides

3.3.2.1 B Cell Responses Induced by Synthetic Peptides

The initial efforts to generate peptide based vaccines concentrated on the induction of B cell responses. This had historical reasons because the discovery that peptides can be used to generate antibodies against native proteins precedes the characterization of peptides as the antigens recognized by T cells by a number of years (Lerner et al., 1981). Another reason is that for many pathogens, e.g. malaria, a passive transfer of monoclonal antibodies resulted in complete protection against lethal pathogen challenges in mice (Potocnjak et al., 1980) or other laboratory animals. In addition to that, a successful induction of a B cells response is very easy to monitor.

The method most commonly used to identify B cell epitopes within proteins consists of identifying which peptide fragments of the molecule of interest can be recognized by antibodies raised against the intact protein. With few exceptions, the epitopes identified by this approach are continuous epitopes representing a linear fragment of the protein as opposed to the discontinuous epitopes formed by distant parts of the protein sequence and brought together by the folding of the polypeptide chain. The degree of cross-reactivity of protein specific antibodies for peptide antigens is usually the highest for peptides covering the N- and C-termini of the protein. This is probably due to the fact that the protein termini are more frequently oriented to the surface of the protein and are more mobile than internal parts (Tainer et al., 1985). Using longer peptides does not necessarily increase the chances for recognition by protein specific antibodies as those peptides might adopt a conformation different from that present in the native protein. Therefore, large numbers of peptides are required for the identification of continuous epitopes and several methods are used for their generation . Among the most frequently used are the pepscan method (Geysen et al., 1987) and the construction of peptide libraries in bacteriophages (Scott and Smith, 1990; Cwirla et al., 1990). It is also possible to identify B cell epitopes within a given protein by using synthetic peptides for immunizations and than test the generated antibodies for recognition of the native protein.

Both approaches bear the disadvantage that a large number of peptides needs to be screened. Therefore, numerous studies were aimed at predicting potential epitopes in proteins from certain features of their amino acid sequence. The most common algorithms calculate either the hydrophilicity and the accessibility of the protein residues of interest or they try to predict epitopes based on their location in turns of the protein. All algorithms in use have the disadvantage that their prediction rate hardly exceeds 60% (Pellequer et al., 1991).

The induction of a B cell response against synthetic peptides requires still more than the selection of the correct epitopes. Synthetic peptides are poorly immunogenic and their successful application requires the use of adjuvants, and for peptides smaller than 15 amino acids the use of carrier molecules is required. Several experimental adjuvant formulations have been developed, but currently the only adjuvants approved by the Federal Drug Administration for the use in humans are aluminum salts (alum). Unfortunately, alum is a relatively weak adjuvant for the induction of B cell responses and hardly able to induce cell-mediated immunity (Schirmbeck et al., 1994c), and induces a Th2-biased response leading to an induction of antibody responses limited to certain Ig classes. Therefore, there is a definite need for new adjuvant formulations.

In addition to the classical complete Freund's adjuvant (CFA), experimental systems describe the use of incomplete Freund's adjuvant (IFA), liposomes (Frisch et al., 1991; White et al., 1995), lipopeptides (Bessler et al., 1985), phosphatidylethanolamine (PE), multiple antigen peptide (MAP) (Tam, 1988; Tam and Lu, 1989) and many others. The MAP system or the related branched lysine oligopeptide (BLO) method (Okuda et al., 1993) have the advantage that no other carrier has to be used since the polymerized peptides result in an increased molecular size. If small peptides are to be used for immunizations, carriers like proteins, synthetic polyamino acids, erythrocytes, inert particulate beads, liposomes, etc. can be successfully applied to induce a strong B cell response in combination with adjuvants.

3.3.2.2 T_H Cell Responses Induced by Synthetic Peptides

About 30 years ago it was already demonstrated that the induction of an immune response required synergism between thymus dependent lymphocytes and others. The thymus-dependent T cells were considered as helper cells that are necessary to activate the antibody-producing B cells. Since this "helper effect" could be induced using a few standard carrier proteins like KLH many studies initially concentrated on peptide recognition by B cells rather than T cells.

Several years later, after it was discovered that Th cells do not recognize the native protein but only the processed or partially degraded molecule presented by MHC class II molecules (reviewed in Unanue, 1984), the recognition of peptides by Th cells was investigated in detail, starting with the pioneering work of Berzofsky and coworkers (Berkower et al., 1982; Berkower et al., 1984). Numerous studies in recent years have identified epitopes recognized by Th cells using synthetic peptides. Inspired by the identification of ligand motifs within peptides presented by MHC class I molecules (Falk et al., 1991), many ligand and binding motifs for MHC class II molecules have been identified as well (Hammer et al., 1992; Falk et al., 1994; Chicz et al., 1992; Sette et al., 1992) (for a detailed review see Rammensee et al., 1995), allowing a more precise prediction of putative Th cell epitopes within a given protein and therefore an even faster identification. This knowledge allows to combine both B- and Th cell epitopes into entirely synthetic and chemically defined constructs with the goal to induce higher antibody titers and to ensure a specific response. Using synthetic peptides with pathogen specific Th cell epitopes in combination with synthetic B cell epitopes, instead of unrelated carrier proteins, like KLH, has also the advantage that in a addition to a specific B cell response a pathogen specific Th cell response is induced as well. Therefore, a memory response of both B- and T cells is initiated after contact with the real pathogen. Promising results have been obtained in mice immunized with a MAP containing several copies of the immunodominant B cell epitope and a selected Th epitope from the Plasmodium berghei circumsporozoite protein (de Oliveira et al., 1994).

3.3.2.3 CTL Responses Induced by Synthetic Peptides

Townsend and coworkers demonstrated in 1986 that the epitope recognized by CTL specific for the influenza nucleoprotein can be defined by short synthetic peptides presented by MHC class I molecules (Townsend et al., 1986). This finding and the increasing evidence suggesting that CTL are required for the protection from persistent viral infections (Oldstone, 1994), started to extend the aim of vaccines, especially antiviral ones, from the induction of antibody responses towards the activation of CTL responses in addition. CTL have the ability to detect very low levels of viral antigen on the surfaces of infected cells and to recognize nonglycosylated immediate-early or early proteins of a virus that are transcribed many hours before structural viral proteins are made and, most importantly, prior to the assembly of the virus into an infectious unit (Oldstone, 1991).

The isolation and identification of peptides presented by MHC molecules have provided valuable information about the rules governing peptide/MHC class I molecule interactions facilitating the identification of putative CTL epitopes (Rötzschke et al., 1990; van Bleek and Nathenson, 1990; Falk et al., 1991). This information should help to raise CTL responses against cells expressing any given protein. A detailed overview about the current knowledge about MHC class I ligand motifs and rules for the selection of CTL epitopes will be given in the next chapter.

Unfortunately, synthetic peptides containing CTL epitopes are poorly immunogenic, as mentioned earlier, and hardly induce a CTL response when injected by themselves. However, the immunogenicity can be enhanced by modifications of the peptide directly, by injecting the peptide in combination with adjuvants, by mixing the peptide with liposomes or ISCOMS or by applying the peptide complexed with carriers. Chapter 3.3.3 will give an overview about the current adjuvant technology and will discuss the use of denatured proteins, the Hepatitis B surface antigen and heat shock proteins as protein carriers for synthetic CTL epitopes.

Besides focusing on the CTL response it is nevertheless important to remember the synergistic role of other participants in the immune response. This becomes evident from experiments where CTL induced against epitopes within the M2 protein of RSV (Kulkarni et al., 1993) and LCMV NP and glycoprotein GP protect against viral challenges in mice (Jonjic et al., 1988; Hany et al., 1989; del Val et al., 1991), but CTL induced against influenza NP epitopes were not protective (Lawson et al., 1994). For a successful peptide vaccine, it appears therefore to be important to combine epitopes recognized by CTL with those inducing protective Th or B cell responses.

3.3.3 Identification and Prediction of T Cell Epitopes

3.3.3.1 Identification of the Protein

The identification of proteins carrying the epitopes recognized by protective T cells is a central issue in vaccine development because no peptide based vaccines can be developed if the antigen recognized during the immune response is not known. Several strategies, depending on the nature of the pathogen, are used. For the identification of bacterial or viral proteins, T cells generated by immunizations with the complete pathogen are tested for their ability to recognize cells expressing subgenomic fragments or individual genes. For the identification of proteins from tumor cells that are recognized by T cells similar approaches have been pioneered by Boon and coworkers, resulting in the discovery of the first gene coding for a tumor specific antigen in 1988 (De Plaen et al., 1988). Although this approach is labor intensive , a number of additional proteins recognized by tumor destroying CTL have been identified, most notably in spontaneous human tumors. Tumor rejection antigens identified by various approaches possibly include the E6 and E7 proteins of HPV16, the p53 protein (including its mutations), the aberrant fusion protein BCR-ABL, the mutated Ras proteins (reviewed in Toes et al., 1994) and the oncogene product of HER2/neu (Peoples et al., 1995).

New methods, allowing the identification of genes coding for proteins that are recognized by tumor specific T cells, have been developed. One combines the existing technique of testing peptide fractions extracted from MHC molecules with specific CTL lines with the ability to identify the peptides present in positive fractions using either automated Edman sequencing or mass spectroscopy. Very recently, this technology allowed the identification of tumor antigens (Mandelboim et al., 1994; Cox et al., 1994) and several minor H antigens in both mice and men, including H-Y (Scott et al., 1995; Wang et al., 1995). Minor H antigens share many aspects with tumor antigens. This new approach seems to be promising, at least in those cases where sufficient amounts of tumor material is available for the initial peptide extraction.

The second approach, which is a modification of a method termed autologous typing (Old, 1981), uses expression libraries of tumor cells which are screened with the IgG fraction of autologous sera (Sahin et al., 1995). The identification of proteins with antibodies of the IgG subtype implies the participation of Th cells. This approach might therefore be useful to identify, in addition to antigens recognized by B cells, new tumor specific antigens recognized by Th cells. The validity of this approach is evident from its success to independently identify the proteins MAGE-1 and tyrosinase from melanoma cells which are known to be recognized by melanoma specific CTL. In addition to that, genes coding for antigens expressed in Hodgkin's disease, renal cell carcinoma and brain tumors could be identified (Sahin et al., 1995). The sequence of those antigens now needs to be inspected for potential class I and class II restricted T cell epitopes (as described in the next chapters) which then can be tested for recognition by tumor specific T cells.

3.3.3.2 Identification of MHC Class I Restricted T cell Epitopes

The possibilities for the identification of CTL-epitopes within known proteins have been greatly enhanced by the use of ligand motifs. These motifs are derived from the information obtained by the study of biochemically isolated peptides from MHC molecules (Falk et al., 1991) and explained by the information obtained by X-ray analysis of MHC crystals (Madden et al., 1993; Zhang et al., 1992; Fremont et al., 1992), and by binding assays with synthetic peptides (Ruppert et al., 1993; Kast et al., 1994; Parker et al., 1994). Numerous MHC class I ligand motifs have been identified to date. These motifs provide information about the preferred length and amino acid composition at certain positions of peptides associated with the different MHC molecules. Some examples are given in table 3.1.1 and a summary is provided in a review by Rammensee et al. (1995).

For the prediction of epitopes it is important not only to consider the basic motif (main anchor positions and length) but also the amino acid composition at the non-anchor residues, since they can contribute considerably to the interaction of the peptides with the MHC molecules. This is evident from the preferences seen for certain residues at non-anchor positions in pool sequencing data, from the interaction of such residues with MHC sites as observed in MHC/peptide crystals, and from binding studies showing that certain residues at a given position within the peptide can be detrimental for binding.

All this should be kept in mind, when a protein sequence is searched for potential epitopes able to be presented by MHC class I molecules. The first step is to inspect the sequence for stretches fitting to the basic anchor motif (two anchors in most cases), whereby variations in peptide length as well as in anchor position occupancy should be allowed. If a motif, for example, calls for 9mers with I or L at position 9, 10mers with a fit at the C-terminus should be considered as well. In addition, the list of potential epitopes should also include peptides with other aliphatic residues at the C-terminus, for example V or M. This register of peptides is now inspected for the presence of as many non-anchor residues as possible in common with known ligands. Every decent ligand motif contains these residues, they are listed under 'preferred residues' and 'others' (tab. 3.3.1). Such preferences at non anchor residues are explained by the individual amino acids contributing to respective contact sites within the MHC groove, for example at the P1 position in many motifs. Based on the motifs derived from the analysis of natural ligands and peptide binding studies, computer programms have been developed that assist in the prediction of CTL epitopes (D'Amaro et al., 1995; Adams and Kozoil, 1995)

Table 3.3.1: HLA class I motifs

HLA-A1

	Position									Source
	1	2	**3**	4	5	6	7	8	**9**	
Anchor or auxiliary anchor residues		T	**D**	P			L		**Y**	
		S	**E**							
Other preferred residues		L		G	G	G				
				I	N	V				
					Y	I				
Examples for ligands	A	T	**D**	F	K	F	A	M	**Y**	Cyclin-like protein 135-143
	I	A	**D**	M	G	H	L	K	**Y**	Proliferation cell nuclear antigen 241-249
	M	I	**E**	P	R	T	L	Q	**Y**	Ribosomal protein S16 40-48
	Y	T	S	D	Y	F	I	S	**Y**	Ets-1 154-162
T-cell epitopes	E	A	**D**	P	T	G	H	S	**Y**	MAGE-1 161-169
	V	S	**D**	G	G	P	N	L	**Y**	Influenza A PB1 591-599
	C	T	**E**	L	K	L	S	D	**Y**	Influenza A NP 44-52
	E	V	**D**	P	I	G	H	L	**Y**	MAGE-3

HLA-A*0201

	Position										Source
	1	**2**	3	4	5	6	7	8	**9**		
Anchor or auxiliary anchor residues		**L**				V			**V**		
		M							L		
Preferred residues				E				K			
				K							
Other residues	I		A	G	I	I	A	E			
	L		Y	P	K	L	Y	S			
	F		F	D	Y	T	H				
	K		P	T	N						
	M		M		G						
	Y		S		F						
	V		R		V						
					H						
Examples for ligands	S	**L**	L	P	A	I	V	E	**L**		Protein phosphatase 2A 389-397
	Y	**L**	L	P	A	I	V	H	**I**		ATP-dependent RNA Helicase 148-156
	T	**L**	W	V	D	P	Y	E	**V**		B cell transloc. gene 1 protein 103-111
	S	X	P	S	G	G	X	G	**V**		Unknown
T-cell epitopes	I	**L**	K	E	P	V	H	G	**V**		HIV-1 RT 476-484
	I	**L**	G	F	V	F	T	L	T	**V**	Influenza matrix protein 59-68
	L	**L**	F	G	Y	P	V	Y	**V**		HTLV-1 tax 11-19

HLA-A3

	Position											Source
	1	**2**	3	4	5	6	7	8	**9**	**10**		
Anchor or auxiliary anchor residues		**L**	F			I	I		**K**	K		
		V	Y			M	L		**Y**			
		M				F	M		**F**			
						V	F					
						L						
Other preferred residues	I				I	T		Q				
					P			S				
					V			T				
					K			K				
Examples for ligands	K	X	E	K	M	I	L	R	**K**			Unknown
	K	**L**	E	K	N	I	L	Y	**K**			Unknown
	Y	**L**	X	V	R	X	A	X	i	V		Unknown
	K	**L**	H	K	Q	R	A	K	S			Unknown
T-cell epitopes	R	**L**	R	D	L	L	L	I	V	T	**R**	HIV-1 env gp41 768-778
	Q	**V**	P	L	R	P	M	T	Y	**K**		HIV-1 nef 73-82
	T	**V**	Y	Y	G	V	P	V	W	**K**		HIV-1 env gp120 36-45
	R	**L**	R	P	G	G	K	K	K			HIV-1 gag p17 20-29

HLA-B7

	Position										Source
	1	**2**	3	4	5	6	7	8	**9**		
Anchor or auxiliary anchor residues		**P**	R						**L**		
									F		
Other preferred residues				D	D	F	L				
				G	P	T					
Also detected	A		D	E	I	R	V				
	H		E	H	V	L					
	S		Q	L		I					
			K	K							
			Y	S							
			F	T							
			M	P							
			N								
			A								
Examples for ligands	A	**P**	R	T	V	A	L	T	A		HLA-DP signal sequence 9-17
	A	**P**	R	T	V	A	L	T	A	**L**	HLA-DP signal sequence 9-18
	A	**P**	R	A	X	X	X	X	X		Unknown
	A	**P**	R	X	P	X	T	G	X		Unknown
T-cell epitope	T	**P**	G	P	G	V	R	Y	P	**L**	HIV-1 nef 128-137

(for references see review by Rammensee et al., 1995)

Peptide binding studies can provide additional information for non-anchor residues increasing or decreasing peptide binding. For the identification of natural ligands it is however important to keep in mind that the information derived from peptide binding studies is biased for peptide binding only and does not account for influences of antigen processing, such as the specificity of enzymes, peptide transporters and chaperones involved.

After the initial list of putative ligands has been trimmed down to those peptides fulfilling the theoretical considerations, a peptide binding assay should be performed, since for some MHC alleles only 20 – 40 % of the predicted epitopes are able to bind. This is exemplified in table 3.3.2 showing several ligands predicted to bind to HLA-A3 molecules. All eight peptides carry the basic A3 motif (L, V, M) x 6 (K, Y, F), but only two of them actually bind to the A3 molecule. The peptide with the highest affinity is the only one that displays, in addition to the basic anchors optimal amino acid selections at two auxiliary anchor positions. Thus, this example illustrates that adherence to a motif is required but by no means sufficient for a peptide to be a good binder. In addition, this example underlines the importance of auxiliary anchor residues. The availability of HLA-transgenic mice makes it possible to test predicted human CTL epitopes in vivo, as performed for peptides derived from hepatitis C proteins (Shirai et al., 1995). It should be kept in mind that even though most natural ligands will fit to the motifs, exceptions are possible as observed by Mandelboim and coworkers (Mandelboim et al., 1994) who identified a H-2K^b restricted tumor antigen that did not contain the ligand motif predicted for H-2K^b.

Table 3.3.2: Binding of predicted HLA-A3 epitopes

										Protein	relative binding
L	**V**	G	F	L	**L**	**L**	K	**Y**		Mage 1	3,0
M	**L**	E	S	V	**I**	K	N	**Y**		Mage 1	1,5
L	**L**	D	L	S	K	E	E	**K**		TyrR	1,0
F	**V**	R	A	L	D	**M**	A	**K**		TyrR	1,0
D	**L**	S	A	P	E	K	D	**K**		TyrO	1,0
L	**L**	D	I	L	D	T	A	G	**K**	N-Ras	1,0
I	**L**	D	T	A	G	K	E	E	**Y**	N-Ras	1,0
I	**L**	D	T	A	G	**L**	E	E	**Y**	N-Ras	1,0

3.3.3.3 Identification of MHC Class II Restricted T Cell Epitopes

The analysis of peptides associated with MHC class I molecules revealed clear ligand motifs, typically characterized by a certain peptide length and two dominant anchor positions that are occupied by closely related amino acids. However, such motifs are less apparent with ligands eluted from class II MHC molecules because these peptides vary extensively in their length, which complicates the alignment of isolated natural ligands. Inspite of that, the combined efforts of natural ligand analysis, individual and as pools, peptide binding assays, and structural analysis led to the elucidation of clear-cut motifs for a number of class II molecules as well (see below). Before these motifs were available, class II restricted epitopes were identified by using sets of overlapping peptides of a given protein in competition assays with known ligands, either labeled with biotin or radioactivity or detected by specific T cells (Sette et al., 1989; Leighton et al., 1991; Reay et al., 1994). As already mentioned for class I MHC ligands, the identification of epitopes according to these parameters can be misleading, since the influences of antigen processing are not accounted for. Time and money also allow the study of only a limited set of peptides. Therefore, complementary effects between different residues might stay undetected, since not all combinatorial possibilities can be studied.

Aided by the X-ray analysis of class II molecules which has given invaluable information about peptide/class II MHC interactions (Brown et al., 1993; Stern et al., 1994), several class II ligand motifs have been obtained recently, using the approaches of peptide binding (Reay et al., 1994; Hammer et al., 1993; Hammer et al., 1995) and pool sequencing and alignment of natural ligands as previously performed for class I MHC ligands (Falk et al., 1994; Kropshofer et al., 1992; Kropshofer et al., 1993; Vogt et al., 1994; Schild et al., 1995). These motifs have the following characteristics. A stretch of nine amino acids corresponding to that part of the peptide embedded in the MHC groove contains the allele-specific motif. On average, the motif starts at absolute positions 3 to 5 of the natural ligands. The first anchor position, called P1, represents hydrophobic amino acids for most of the class II MHC motifs known so far. The interaction of the MHC molecules with the first anchor position of the peptide seems to be particularly important as indicated by the strong clustering of hydrophobic amino acids in cycles 3 to 5 of sequencing analysis of self-peptide pools, the striking influence of P1 in peptide binding assays and probably most impressively illustrated by the deep pocket seen in the crystal of the HLA-DR1 molecule complexed with the influenza hemagglutinin peptide 306-318. Besides P1, additional anchor residues can be observed up to position 9. For the DR1 molecule the anchors are located at positions P4, P6, P7 and P9. A very similar anchor spacing is observed for other members of the DR family, like DR2, DR3, DR4 molecules and also for the highly homologous family of the mouse H-2E molecules (tab. 3.3.3).

Less is known about the motifs of HLA-DP and DQ molecules (Falk et al., 1994) and the murine homolog of the latter (H-2A) but the information available so far suggests a different anchor spacing and possibly no prominent anchor at P1 which complicates the alignment of known natural ligands.

The prediction of epitopes for class II ligands within a protein is complicated by the fact that, unlike for class I ligands, some of the anchor positions are very degenerate in their specificity. Therefore, the first step after the selection of P1 (at least for DR and H-2E molecules) is to pick out the most allele-specific anchor residue which is for

Table 3.3.3: HLA class II motifs

HLA-DRB1*0101

		Relative position										Source
		1	2	3	**4**	5	**6**	7	8	**9**		
Anchor residues		**Y,V,**			**L,A**		**A,G**			L,A		
		L,F			I,V		S,T			I,V		
		I,A			M,N		P			N,F		
		M,W			Q					Y		
Examples for ligands	VGSD	**W**	R	F	**L**	R	**G**	Y	H	Q	YA	HLA-A2 103-117
	VGSD	**W**	R	F	**L**	R	**G**	Y	H	Q	YAYDG	HLA-A2 103-120
	LPKPPKPVSK	**M**	R	M	**A**	T	**P**	L	L	M	QALPM	Invariant chain 97-120
	IPAD	**L**	R	I	**I**	S	**A**	N	G	C	K	Na^+-K^+-ATPase 199-216
	RVE	**Y**	H	F	**L**	S	**P**	Y	V	S	PKESP	Transferrin receptor 680-696
	YKHT	**L**	N	Q	**I**	D	**S**	V	K	**V**	WPRRPT	Cattle fetuin 56-74

HLA-DRB1*0301 (DR17)

		Relative position										Source
		1	2	3	**4**	5	**6**	7	8	**9**		
Anchor or auxiliary anchor residues		**L,I**			**D**		K, R			Y,L		
		F,M					E,Q			F		
		V					**N**					
Examples for ligands	ISNQ	**L**	T	L	**D**	S	**N**	T	K	**Y**	FHKLN	Apolipoprotein B 2877-2894
	ISNQ	**L**	T	L	**D**	S	**N**	T	K	**Y**	FHKL	Apolipoprotein B 2877-2893
	VDT	**F**	L	E	**D**	V	**K**	N	L	**Y**	HSEA	α1-Antitrypsin 149-164
	KPRA	**I**	V	V	**D**	P	V	H	G	**F**	MY	LDL-Receptor 518-532
	KQT	**I**	S	P	**D**	Y	**R**	N	M	I		IgG2a, Membrane domain
	YPD	**F**	I	M	**D**	P	**K**	E	K	D	KV	Unknown

example P4 for DRB1*0301, P6 for DRB1*0101 and P9 H-2E and DRB1*0405. After that, the nonamer stretch should by inspected for the presence of additional anchor residues, keeping in mind that optimal fits at some of the anchors can compensate a mismatch at others residues, with the exception of P1 for DR and H-2E molecules.

An approach to calculate the influences of all amino acids at the different positions within the nonamer has been worked out by Hammer and coworkers and allowed a very accurate prediction of peptides binding to DRB1*0401 (Hammer et al., 1995). This method is based on the idea that the binding of peptides to DR4 correlates with the net result of all side chain interactions of the peptide with the MHC molecule, and that most side chain effects depend on their relative position to the P1 anchor. These interactions can either enhance or decrease the binding or can have a neutral effect. Peptide libraries were constructed consisting of short, P1-anchored and alanine-based peptides where all amino acids had been substituted individually at positions 2 to 9 and tested in a peptide binding assay to the DR4 molecule. The values, obtained by

HLA-DRB1*0401 (DR4Dw4)

		Relative position										Source
		1	2	3	**4**	5	**6**	**7**	8	**9**		
Anchor or preferred residues		**F,Y**			F,W		N,S	pol.*		pol.*		
		W,I,			I,L		T,Q	chg.*		ali.*		
		L,V			V,A		H,R	ali.*		K		
		M			D,E							
					no							
					R,K							
Examples for ligands		**F**	V	R	**F**	D	**S**	**D**	A	**A**	SQRMEP	HLA-A2 33-47
	VDDTQ	**F**	V	R	**F**	D	**S**	**D**	A	**A**	SQRM	HLA-A2 28-45
	DGKD	**Y**	I	A	**L**	N	E	**D**	L	**S**	S	HLA-B44 143-156
	DVA	**F**	V	K	**D**	Q	**T**	**V**	I	**Q**	NTD	Cattle transferrin 68-82
	YDHN	**F**	V	K	**A**	I	**N**	**A**	I	**Q**	KSW	Cathepsin C 170-185
	KHKV	**Y**	A	C	**E**	V	**T**	**H**	Q	**G**	...	Igκ chain C region 80-?

* pol.: Polar; chg.: charged; ali.: aliphatic

HLA-DRB1*0405 (DR4Dw15)

		Relative position										Source
		1	2	3	**4**	5	**6**	**7**	8	**9**		
Anchor or preferred residues		**F,Y**			V,I		N,S	pol.*,		**D**,E		
		W,V			L,M		T,Q	chg.*		Q		
		I,L			D,E		K,D	ali.*				
		M										
Examples for ligands	YPTQRAR	**Y**	Q	W	**V**	R	C	**N**	P	**D**	SNS	PGSG 1-19
	QRAR	**Y**	Q	W	**V**	R	C	**N**	P	**D**	SNS	PGSG 4-19
	EPDH	**Y**	V	V	**V**	G	A	**Q**	R	**D**	A	Transferrin receptor 398-411
	THY	**Y**	A	V	A	V	V	**K**	K	**D**	TDFK	Transferrin 92-107
	KELK	**I**	D	I	**I**	P	**N**	**P**	Q	**E**	R	Hsp 90-beta 68-81
	YLL	**Y**	Y	T	**E**	F	**T**	**P**	T	**E**	KD	β_2-microglobulin 83-96

* pol.: Polar; chg.: charged; ali.: aliphatic

(for references see review by Rammensee et al., 1995)

this so called 'side chain scanning', were than assigned to the individual amino acids from positions 2 to 9, and it was found that the sum of these values correlated indeed with the MHC-binding affinity of selected peptides. This method now allows a computer-based identification of DR4 ligands from any protein based on their calculated scores derived from side chain scanning. Similar approaches should be possible for a detailed analysis of other class II ligand motifs with a dominant anchor at P1.

A related, although not quite mature method is to analyze the results of class II ligand pool sequencing for attributing quantitative scores for each amino acid at every position (Davenport et al., 1995). This approach, if it can be perfected, has the advantage of integrating the imprinting of antigen processing on the motif, in addition to binding.

3.3.4 Administration of Peptides

3.3.4.1 Adjuvants

Many subunit vaccines require the use of adjuvants as immunostimulators for the successful induction of immune responses. Emulsions of mineral oil are often the adjuvant of choice for farm and laboratory animals since they evoke high immune responses, including protective CTL responses in mice (Schulz et al., 1991). Unfortunately, these adjuvants, especially those of the water-in-oil type, can provoke undesirable side-effects, like tissue damage at the site of injection, granulomatous reactions or arthritis and are therefore not allowed for the use in humans. Numerous attempts to develop strong and safe adjuvants that could replace alum, have been undertaken, some of which will be listed below. Most of them have been tested to induce strong B cell responses but their ability to induce specific and more important, protective CTL responses, needs to be determined for many of them.

One type of adjuvants frequently used contain the metabolizable oil squalene and surfactants like Tween 80 and Span 85. Several components can be added, like the active ingredient of mycobacteria, N-acetyl-muramyl-L-alanyl-D-isoglutamine (MDP) and derivatives or synthetic sulfolipopolysaccharides (Hilgers et al., 1994). Another adjuvant formulation used is the Syntex adjuvant formulation (SAF) either alone or mixed with MDP. Additional adjuvants that should be mentioned are the Quil A saponins which were able to induce both B cell and CTL responses, monophosphoryl lipid A (MPL), and the Ribi formulation, containing MPL and mycobacterial cell walls. The adjuvant activity of MPL could be enhanced using squalene or liposomes (Richards et al., 1989). Recently, Dyall et al. (1995) showed that, quite surprisingly, a commercial adjuvant (Titermax), developed for the induction of a strong B cell response in mice, was also the best to induce peptide-specific CTL in mice, among a number of adjuvants tested.

3.3.4.2 Delivery Systems

Antigen-containing liposomes and microspheres have been shown to deliver antigens in a highly immunogenic form. A wide array of delivery systems is used, ranging from liposomes over biodegradable microspheres to immune stimulating complexes (ISCOMS). Liposomes carry the antigen either trapped inside a lipid bilayer or anchored to the surface and the appropriate constructs may have distinct enhancing effects which can be augmented by including MDP in the liposome preparation (Phillips and Chedid, 1988). The antigens delivered by liposomes can either be processed through endosomal mechanisms and presented by class II molecules or through cytosolic processes leading to a class I restricted presentation (Alving and Richards, 1990; Harding et al., 1991). In addition, liposomes have been shown to induce a B cell as well as a CTL response against a synthetic peptide (White et al., 1995)

Biodegradable microspheres formulated from poly (D,L) lactic co-glycolic acid have been used to induce immune responses against recombinant proteins or inactivated pathogens. They are able to provoke significant IgA responses after oral or intratestinal administration. Their use in the induction of immune responses against synthetic peptides still needs to be investigated in more detail, as currently performed for an antifertility vaccine consisting of the C-terminal peptide of beta-hCG (Jones, 1994).

ISCOMS are symmetrically arranged repeated units of a single antigen. They are prepared by mixing the antigen with detergent and a glycoside, called Quil A. This mixture forms micelles that interact with the antigen expressed on the surface of the micelle as a multivalent complex. After the initial report by Osterhaus and coworkers in 1984 that proteins incorporated into ISCOMS are highly immunogenic in vivo (Morein et al., 1984) and induce humoral and cellular responses including MHC class I restricted CTL (Jones et al., 1988; Takahashi et al., 1990), many protein- and peptide antigens have been tested in immunization experiments using ISCOMS.

3.3.4.3 Lipopeptides

After the discovery that the N-terminal part of the outer membrane lipoprotein of E. coli is mitogenic for B cells and macrophages (Bessler et al., 1985; Hoffmann et al., 1989), it was conjugated to synthetic peptides to increase their immunogenicity. These lipopeptides, carrying N-terminally the lipo-amino acid tripalmitoyl-S-glycerylcysteinyl-seryl-serine (Pam3CSS) were able to induce potent antibody responses (Bessler et al., 1985), stimulated virus specific CTL in vivo (Deres et al., 1989) and induced T cell memory (Schild et al., 1991) without the need for any additional adjuvants. Later on, different lipid modifications of peptides, like α-aminohexadecanoic acid (Hda) were used, and it could be demonstrated that a lipopeptide consisting of 34 amino acids and the Hda modification is able to activate CTL and helper T cells and induces a B cell response (Martinon et al., 1992; Defoort et al., 1992). These experiments suggest that it should be possible to synthezise lipopeptides carrying multiple epitopes that allow the simultaneous stimulation of either B- and T cells or that activate T cells restricted by different MHC alleles. The latter would be an important prerequisite for the use of lipopeptide vaccines in outbred populations.

The exact mechanism of T cell activation is not known but the mitogenic capacities of the P3CSS anchor are probably not important for the immunogenicity if these lipopeptides, since non-mitogenic lipid anchors also stimulate B- and T cell responses. The data available so far suggest that the ability of lipopeptides to cross cell membranes (Metzger et al., 1993) and deliver the peptide to the cytosol where they might enter the class I restricted antigen presentation pathway, seem to be a central feature. The lipid part could allow either a passive diffusion across the cell membrane or be responsible for an active transport through a receptor specific for the lipid moiety, as suggested for the uptake of DNA oligonucleotides modified with cholesterol (Krieg et al., 1993). Another possibility is, however, the uptake of lipopeptide aggregates by phagocytosis into macrophages and their feeding into one of the alternative class I loading pathways for exogenous antigens.

3.3.4.4 Protein Carriers

Protein carriers like KLH are commonly used to increase the immunogenicity of peptides to induce B- and helper T cell mediated immune responses. The dogma that peptides originating from proteins in the cytosol are presented by class I MHC molecules, and peptides from exogenous proteins have only access to class II MHC molecules, has excluded proteins from the use as CTL activating vaccines. Recent experiments have challenged this view. It was shown that peptides from soluble proteins can be presented in vitro by APCs (Rock et al., 1990), most likely macrophages (Kovacsovics Bankowski et al., 1993; Kovacsovics Bankowski and Rock, 1994; Bachmann et al., 1995) and that the efficiency of the presentation can be enhanced if the antigen is provided in a particulate form, for example coupled to Latex-beads (Kovacsovics Bankowski et al., 1993; Harding and Song, 1994; Kovacsovics Bankowski and Rock, 1995). In some experiments performed with macrophages from TAP k.o. mice the results suggest that the peptides are loaded onto class I MHC molecules in a TAP-independent manner (Harding and Song, 1994; Bachmann et al., 1995), in other experiments this peptide loading is TAP-dependent (Kovacsovics Bankowski and Rock, 1995), indicating that this issue is not resolved yet.

For the development of new vaccine strategies it was even more interesting to observe that immunizations with proteins that either form particles by themselves, like the hepatitis B virus surface antigen (Schirmbeck et al., 1994b), or that are denatured by detergents (Schirmbeck et al., 1994a; Weidt et al., 1994; Speidel et al., unpublished) are able to induce specific CTL responses without the need for additional adjuvants. These denatured proteins can now be used for immunizations directly or can act as a carrier for synthetic peptides containing CTL epitopes. That peptides coupled to proteins exogenously added to cells can be recognized by class I restricted CTL has been shown for the influenza matrix peptide 57-68 coupled to transferrin (Brander et al., 1993).

The use of a new group of proteins as potential peptide carriers has emerged from the pioneering work of Srivastava and coworkers. They were able to demonstrate that tumor specific, protective immunity could be induced by immunizations with heat shock proteins purified from tumor cells (reviewed in Srivastava and Maki, 1991). The heat shock proteins able to induce immunity include the Hsp70 and Hsp90 family and the ER-resident heat shock protein gp96 (Udono and Srivastava, 1994). No adjuvants were required for the immunizations. The tumor rejection is mediated by

CD8+ T cells and the presence of macrophages is essential (Udono et al., 1994). It was further suggested that peptides chaperoned by the heat shock proteins are responsible for the specificity of the immune response (Udono and Srivastava, 1994). By now, heat shock proteins have also been shown to induce CTL responses against viral- (Blachere et al., 1993) and minor H antigens (Arnold et al., 1995) in vivo, indicating that their potential is not limited to tumor antigens. What makes heat shock proteins interesting candidates as a carrier for peptides is their ability to be loaded in vitro with synthetic peptides and to channel these peptides into the class I MHC restricted antigen presentation pathway (Suto and Srivastava, 1995). The mechanisms behind this phenomenon are not clear, but receptor-mediated uptake of the HSP as well as a TAP-dependent transport of the peptides into the ER seems to be involved (Arnold et al., unpublished).

3.3.5 Dealing with MHC Polymorphism

The fact that MHC molecules are very polymorphic and select different peptides of the same protein for the presentation by MHC molecules complicates the use of peptide vaccines in outbred populations. HLA typing before vaccination would allow to use the correct peptide mixture for each individual but would boost the costs dramatically, excluding this approach to be used on a routine basis, although it seems feasible to do this as a therapeutic measure in individual cancer patients. A vaccination with a mixture of T cell epitopes, able to be presented by a representative selection of MHC molecules occurring widespread in the population, would solve this problem. But it should be made sure that peptides present in the mixture do not influence the binding of others, as possible for peptides that are HLA-B27 restricted which might also compete for the binding to HLA-B8 molecules (Tussey et al., 1995). The mixture of peptides containing T cell epitopes could be generated by different methods, for example: i) a mixture of lipopeptides containing single T cell epitopes; ii) lipopeptides containing multiple T cell epitopes in a linear arrangement; iii) multiple T cell epitopes polymerized according to the BLO method; iv) T cell epitope mixtures, covalently attached to aggregated proteins or non-covalently to HSPs, v) synthetic peptides mixed in a suitable adjuvant formulation, if found.

3.3.6 Controlling the Efficacy

Along with the evolution of vaccination strategies, concepts to monitor specific immune responses after immunizations need to be developed. For many years now, serum levels of antigen specific antibodies are tested, mainly in ELISA based assays, to control for the induction of a specific B cell response on a routine basis. No comparable tests are available to monitor the induction of a specific CTL response. CTL assays with in vitro cultivated PBLs of immunized individuals can not be performed on a routine basis. Recently, a new method to examine a CTL response in situ was developed (Puccetti et al., 1994). Based on the local injection of class I binding peptides, a skin test analogous to the conventional DTH assay allowed to monitor the activation status of CD8+ T cells in mice that were immunized with tumor cells, antigenic peptides or peptide-pulsed dendritic cells. Antigen specific DTH responses, mediated by CD8+, class I MHC restricted T cells were detectable for up to six months after priming. Specific DTH responses could also be induced in stage IV melanoma patients (Mukherji et al., 1995) after the injection of autologous APCs incubated with the HLA-A1 restricted peptide of the MAGE-1 antigen. This suggests that peptide-based skin tests are a useful tool to monitor in vivo CTL responses to document and optimize immunization strategies.

3.3.7 Conclusion

Synthetic peptide vaccines appear attractive due to their safety and economical features but their successful application in general vaccination schemes requires additional information. A detailed knowledge is crucial especially for the following: i) what are the cellular mechanisms that are responsible for the protection against the pathogen; ii) which epitopes are recognized during the immune response; iii) how can these epitopes be administered in a form that is immunogenic for all the relevant cellular partners and induces immunologic memory of the correct immune branch; iv) how can the ongoing immune response be monitored on a routine basis after the vaccination? As summarized in the previous chapters, promising results have been obtained in the last years addressing many of the abovementioned questions, including better strategies for the identification of B-and T cell epitopes, an improved understanding of the balance between a B-cell or CTL dominated immune response and the development of new adjuvant formulations. If this knowledge is combined, peptide-based vaccines can become reality after all.

References

Adams, H.-P. and Kozoil, J. A. (1995) Prediction of binding to MHC class I molecules. J. Immunol. Meth. *185*, 181-190.

Alving, C. R. and Richards, R. L. (1990) Liposomes containing lipid A: a potent nontoxic adjuvant for a human malaria sporozoite vaccine. Immunol. Lett. *25*, 275-279.

Arnold, D., Faath, S., Rammensee, H.-G., and Schild, H. (1995) Cross-priming of minor histocompatibility antigen-specific cytotoxic T cells upon immunization with the heat shock protein gp96. J. Exp. Med. *182*, 885-890.

Bachmann, M. F., Oxenius, A., Pircher, H., Hengartner, H., Ashton-Richardt, P. A., Tonegawa, S., and Zinkernagel, R. M. (1995) TAP1-independent loading of class I molecules by exogenous viral proteins. Eur. J. Immunol. *25*, 1739-1743.

Berkower, I., Buckenmeyer, G. K., Gurd, F. R., and Berzofsky, J. A. (1982) A possible immunodominant epitope recognized by murine T lymphocytes immune to different myoglobins. Proc. Natl. Acad. Sci. USA *79*, 4723-4727.

Berkower, I., Matis, L. A., Buckenmeyer, G. K., Gurd, F. R., Longo, D. L., and Berzofsky, J. A. (1984) Identification of distinct predominant epitopes recognized by myoglobin-specific T cells under the control of different Ir genes and characterization of representative T cell clones. J. Immunol. *132*, 1370-1378.

Bessler, W. G., Cox, M., Lex, A., Suhr, B., Wiesmuller, K. H., and Jung, G. (1985) Synthetic lipopeptide analogs of bacterial lipoprotein are potent polyclonal activators for murine B lymphocytes. J. Immunol. *135*, 1900-1905.

Blachere, N. E., Udono, H., Janetzki, S., Li, Z., Heike, M., and Srivastava, P. K. (1993) Heat shock protein vaccines against cancer. J. Immunother. *14*, 352-356.

Brander, C., Wyss Coray, T., Mauri, D., Bettens, F., and Pichler, W. J. (1993) Carrier-mediated uptake and presentation of a major histocompatibility complex class I-restricted peptide. Eur. J. Immunol. *23*, 3217-3223.

Brown, J. H., Jardetzky, T. S., Gorga, J. C., Stern, L. J., Urban, R. G., Strominger, J. L., and Wiley, D. C. (1993) Three-dimensional structure of the human class II histocompatibility antigen HLA-DR1 [see comments]. Nature *364*, 33-39.

Chicz, R. M., Urban, R. G., Lane, W. S., Gorga, J. C., Stern, L. J., Vignali, D. A., and Strominger, J. L. (1992) Predominant naturally processed peptides bound to HLA-DR1 are derived from MHC-related molecules and are heterogeneous in size. Nature *358*, 764-768.

Cox, A. L., Skipper, J., Chen, Y., Henderson, R. A., Darrow, T. L., Shabanowitz, J., Engelhard, V. H., Hunt, D. F., and Slingluff, C. L. (1994) Identification of a single peptide recognized by five melanoma-specific human cytotoxic T cel lines. Science *264*, 716-719

Cwirla, S. E., Peters, E. A., Barrett, R. W., and Dower, W. J. (1990) Peptides on phage: a vast library of peptides for identifying ligands. Proc. Natl. Acad. Sci. USA *87*, 6378-6382.

D'Amaro, J., Houbiers, J. G. A., Drijfhout, J. W., Brandt, R. M. P., Schipper, R., Bouves Bavinck, J. N., Melief, C. J. M., and Kast, W. M. (1995) A computer programm for predicting possible cytotoxic T lymphocyte epitopes based on HLA class I peptide-binding motifs. Human Immunology *43*, 13-18

Davenport, M. P., Ho Shon, I. A. P., and Hill, A. V. S. (1995) An empirical method for the prediction of T-cell epitopes. Immunogenetics *42*, 392-397.

de Oliveira, G. A., Clavijo, P., Nussenzweig, R. S., and Nardin, E. H. (1994) Immunogenicity of an alum-adsorbed synthetic multiple-antigen peptide based on B- and T-cell epitopes of the Plasmodium falciparum CS protein: possible vaccine application. Vaccine *12*, 1012-1017.

De Plaen, E., Lurquin, C., Van Pel, A., Mariame, B., Szikora, J. P., Wolfel, T., Sibille, C., Chomez, P., and Boon, T. (1988) Immunogenic (tum-) variants of mouse tumor P815: cloning of the gene of tum- antigen P91A and identification of the tum- mutation. Proc. Natl. Acad. Sci. USA *85*, 2274-2278.

Defoort, J. P., Nardelli, B., Huang, W., and Tam, J. P. (1992) A rational design of synthetic peptide vaccine with a built-in adjuvant. A modular approach for unambiguity. Int. J. Pept. Protein Res. *40*, 214-221.

del Val, M., Schlicht, H. J., Volkmer, H., Messerle, M., Reddehase, M. J., and Koszinowski, U. H. (1991) Protection against lethal cytomegalovirus infection by a recombinant vaccine containing a single nonameric T-cell epitope. J. Virol. *65*, 3641-3646.

Deres, K., Schild, H., Wiesmuller, K. H., Jung, G., and Rammensee, H. G. (1989) In vivo priming of virus-specific cytotoxic T lymphocytes with synthetic lipopeptide vaccine [see comments]. Nature *342*, 561-564.

Dyall, R., Vasovic, L. V., Molano, A., and Nicoliczugiz, J. (1995) CD4-independent in vivo priming of murine CTL by optimal MHC class I-restricted peptides derived from intracellular pathogens. Int. Immunol. *7*, 1205-1212

Falk, K., Rötzschke, O., Stevanovic, S., Jung, G., and Rammensee, H. G. (1991) Allele-specific motifs revealed by sequencing of self-peptides eluted from MHC molecules. Nature *351*, 290-296.

Falk, K., Rötzschke, O., Stevanovic, S., Jung, G., and Rammensee, H. G. (1994) Pool sequencing of natural HLA-DR, DQ, and DP ligands reveals detailed peptide motifs, constraints of processing, and general rules. Immunogenetics *39*, 230-242.

Fremont, D. H., Matsumura, M., Stura, E. A., Peterson, P. A., and Wilson, I. A. (1992) Crystal structures of two viral peptides in complex with murine MHC class I H-2Kb [see comments]. Science *257*, 919-927.

Frisch, B., Muller, S., Briand, J. P., Van Regenmortel, M. H., and Schuber, F. (1991) Parameters affecting the immunogenicity of a liposome-associated synthetic hexapeptide antigen. Eur. J. Immunol. *21*, 185-193.

Geysen, H. M., Rodda, S. J., Mason, T. J., Tribbick, G., and Schoofs, P. G. (1987) Strategies for epitope analysis using peptide synthesis. J. Immunol. Methods *102*, 259-274.

Hammer, J., Takacs, B., and Sinigaglia, F. (1992) Identification of a motif for HLA-DR1 binding peptides using M13 display libraries. J. Exp. Med. *176*, 1007-1013.

Hammer, J., Valsasnini, P., Tolba, K., Bolin, D., Higelin, J., Takacs, B., and Sinigaglia, F. (1993) Promiscuous and allele-specific anchors in HLA-DR-binding peptides. Cell *74*, 197-203.

Hammer, J., Gallazzi, F., Bono, E., Karr, R. W., Guenot, J., Valsasnini, P., Nagy, Z. A., and Sinigaglia, F. (1995) Peptide binding specificity of HLA-DR4 molecules: correlation with rheumatoid arthritis association [see comments]. J. Exp. Med. *181*, 1847-1855.

Hany, M., Oehen, S., Schulz, M., Hengartner, H., Mackett, M., Bishop, D. H., Overton, H., and Zinkernagel, R. M. (1989) Anti-viral protection and prevention of lymphocytic choriomeningitis or of the local footpad swelling reaction in mice by immunization with vaccinia-recombinant virus expressing LCMV-WE nucleoprotein or glycoprotein. Eur. J. Immunol. *19*, 417-424.

Harding, C. V., Collins, D. S., Kanagawa, O., and Unanue, E. R. (1991) Liposome-encapsulated antigens engender lysosomal processing for class II MHC presentation and cytosolic processing for class I presentation. J. Immunol. *147*, 2860-2863.

Harding, C. V. and Song, R. (1994) Phagocytic processing of exogenous particulate antigens by macrophages for presentation by class I MHC molecules. J. Immunol. *153*, 4925-4933.

Hilgers, L. A., Platenburg, P. L., Luitjens, A., Groenveld, B., Dazelle, T., and Weststrate, M. W. (1994) A novel non-mineral oil-based adjuvant. II. Efficacy of a synthetic sulfolipopolysaccharide in a squalane-in-water emulsion in pigs. Vaccine *12*, 661-665.

Hoffmann, P., Wiesmuller, K. H., Metzger, J., Jung, G., and Bessler, W. G. (1989) Induction of tumor cytotoxicity in murine bone marrow-derived macrophages by two synthetic lipopeptide analogues. Biol. Chem. Hoppe-Seyler *370*, 575-582.

Jones, W. R. (1994) Vaccination for contraception. Aust. N.-Z. Obstet. Gynaecol. *34*, 320-329.

Jones, P. D., Tha Hla, R., Morein, B., Lovgren, K., and Ada, G. L. (1988) Cellular immune responses in the murine lung to local immunization with influenza A virus glycoproteins in micelles and immunostimulatory complexes (iscoms). Scand. J. Immunol. *27*, 645-652.

Jonjic, S., del Val, M., Keil, G. M., Reddehase, M. J., and Koszinowski, U. H. (1988) A nonstructural viral protein expressed by a recombinant vaccinia virus protects against lethal cytomegalovirus infection. J. Virol. *62*, 1653-1658.

Kast, W. M., Brandt, R. M., Sidney, J., Drijfhout, J. W., Kubo, R. T., Grey, H. M., Melief, C. J., and Sette, A. (1994) Role of HLA-A motifs in identification of potential CTL epitopes in human papillomavirus type 16 E6 and E7 proteins. J. Immunol. *152*, 3904-3912.

Kovacsovics Bankowski, M., Clark, K., Benacerraf, B., and Rock, K. L. (1993) Efficient major histocompatibility complex class I presentation of exogenous antigen upon phagocytosis by macrophages. Proc. Natl. Acad. Sci. USA *90*, 4942-4946.

Kovacsovics Bankowski, M. and Rock, K. L. (1994) Presentation of exogenous antigens by macrophages: analysis of major histocompatibility complex class I and II presentation and regulation by cytokines. Eur. J. Immunol. *24*, 2421-2428.

Kovacsovics Bankowski, M. and Rock, K. L. (1995) A phagosome-to-cytosol pathway for exogenous antigens presented on MHC class I molecules. Science *267*, 243-246.

Krieg, A. M., Tonkinson, J., Matson, S., Zhao, Q., Saxon, M., Zhang, L. M., Bhanja, U., Yakubov, L., and Stein, C. A. (1993) Modification of antisense phosphodiester oligodeoxynucleotides by a 5' cholesteryl moiety increases cellular association and improves efficacy. Proc. Natl. Acad. Sci. USA *90*, 1048-1052.

Kropshofer, H., Max, H., Muller, C. A., Hesse, F., Stevanovic, S., Jung, G., and Kalbacher, H. (1992) Self-peptide released from class II HLA-DR1 exhibits a hydrophobic two-residue contact motif. J. Exp. Med. *175*, 1799-1803.

Kropshofer, H., Max, H., Halder, T., Kalbus, M., Muller, C. A., and Kalbacher, H. (1993) Self-peptides from four HLA-DR alleles share hydrophobic anchor residues near the NH2-terminal including proline as a stop signal for trimming. J. Immunol. *151*, 4732-4742.

Kulkarni, A. B., Connors, M., Firestone, C. Y., Morse, H. C., and Murphy, B. R. (1993) The cytolytic activity of pulmonary CD8+ lymphocytes, induced by infection with a vaccinia virus recombinant expressing the M2 protein of respiratory syncytial virus (RSV), correlates with resistance to RSV infection in mice. J. Virol. *67*, 1044-1049.

Lawson, C. M., Bennink, J. R., Restifo, N. P., Yewdell, J. W., and Murphy, B. R. (1994) Primary pulmonary cytotoxic T lymphocytes induced by immunization with a vaccinia virus recombinant expressing influenza A virus nucleoprotein peptide do not protect mice against challenge. J. Virol. *68*, 3505-3511.

Leighton, J., Sette, A., Sidney, J., Appella, E., Ehrhardt, C., Fuchs, S., and Adorini, L. (1991) Comparison of structural requirements for interaction of the same peptide with I-Ek and I-Ed molecules in the activation of MHC class II-restricted T cells. J. Immunol. *147*, 198-204.

Lerner, R. A., Green, N., Alexander, H., Liu, F. T., Sutcliffe, J. G., and Shinnick, T. M. (1981) Chemically synthesized peptides predicted from the nucleotide sequence of the hepatitis B virus genome elicit antibodies reactive with the native envelope protein of Dane particles. Proc. Natl. Acad. Sci. USA *78*, 3403-3407.

Madden, D. R., Garboczi, D. N., and Wiley, D. C. (1993) The antigenic identity of peptide-MHC complexes: a comparison of the conformations of five viral peptides presented by HLA-A2 [published erratum appears in Cell 1994 Jan 28;76(2):following 410]. Cell *75*, 693-708.

Mandelboim, O., Berke, G., Fridkin, M., Feldman, M., Eisenstein, M., and Eisenbach, L. (1994) CTL induction by a tumour-associated antigen octapeptide derived from a murine lung carcinoma [see comments]. Nature *369*, 67-71.

Martinon, F., Gras Masse, H., Boutillon, C., Chirat, F., Deprez, B., Guillet, J. G., Gomard, E., Tartar, A., and Levy, J. P. (1992) Immunization of mice with lipopeptides bypasses the prerequisite for adjuvant. Immune response of BALB/c mice to human immunodeficiency virus envelope glycoprotein. J. Immunol. *149*, 3416-3422.

Metzger, J. W., Sawyer, W. H., Wille, B., Biesert, L., Bessler, W. G., and Jung, G. (1993) Interaction of immunologically-active lipopeptides with membranes. Biochim. Biophys. Acta *1149*, 29-39.

Morein, B., Sundquist, B., Hoglund, S., Dalsgaard, K., and Osterhaus, A. (1984) Iscom, a novel structure for antigenic presentation of membrane proteins from enveloped viruses. Nature ***308***, 457-460.

Mukherji, B., Chakraborty, N. G., Yamasaki, S., Okino, T., Yamase, H., Sporn, J. R., Kurtzman, S. K., Ergin, M. T., Ozols, J., Meehan, J., and Mauri, F. (1995) Induction of antigen-specific cytolytic T cells in situ in human melanoma by immunization with synthetic peptide-pulsed autologous antigen presenting cells. Proc. Natl. Acad. Sci. USA *92*, 8078-8082.

Okuda, K., Kaneko, T., Yamakawa, T., Tanaka, S., Hamajima, K., Shigematsu, T., Yamamato, A., and Kawamoto, S. (1993) Strong synergistic effects of multicomponent vaccine for human immunodeficiency virus infection. J. Clin. Lab. Immunol. *40*, 97-113.

Old, L. J. (1981) Cancer immunology: the search for specificity. G. H. A. Clowes Memorial lecture. Cancer Res. *41*, 361-375.

Oldstone, M. B. A. (1991) Molecular anatomy of viral persistence. J. Virol. *65*, 6381-6386.

Oldstone, M. B. A. (1994) The role of cytotoxic T lymphocytes in infectious disease: History, Criteria, and state of the Art. Curr. Top. Microbiol. Immunol. *189*, 1-9.

Parker, K. C., Bednarek, M. A., and Coligan, J. E. (1994) Scheme for ranking potential HLA-A2 binding peptides based on independent binding of individual peptide side-chains. J. Immunol. *152*, 163-175.

Pellequer, J. L., Westhof, E., and Van Regenmortel, M. H. (1991) Predicting location of continuous epitopes in proteins from their primary structures. Methods Enzymol. *203*, 176-201.

Peoples, G. E., Goedegebuure, P. S., Smith, R., Linehan, D. C., Yoshino, I., and Eberlein, T. J. (1995) Breast and ovarian cancer-specific cytotoxic T lymphocytes recognize the same HER2/neu-derived peptide. Proc. Natl. Acad. Sci. USA *92*, 432-436.

Phillips, N. C. and Chedid, L. (1988) MDP and liposomes. In: Liposomes as Drug Carriers. Gregoriadis, G. (ed.), John Wiley, New York, 243-259.

Potocnjak, P., Yoshida, N., Nussenzweig, R. S., and Nussenzweig, V. (1980) Monovalent fragments (Fab) of monoclonal antibodies to a sporozoite surface antigen (Pb44) protect mice against malarial infection. J. Exp. Med. *151*, 1504-1513.

Puccetti, P., Bianchi, R., Fioretti, M. C., Ayroldi, E., Uyttenhove, C., Van Pel, A., Boon, T., and Grohmann, U. (1994) Use of a skin test assay to determine tumor-specific CD8+ T cell reactivity. Eur. J. Immunol. *24*, 1446-1452.

Rammensee, H. G., Friede, T., and Stevanovic, S. (1995) MHC ligands and peptide motifs: first listing. Immunogenetics *41*, 178-228.

Reay, P. A., Kantor, R. M., and Davis, M. M. (1994) Use of global amino acid replacements to define the requirements for MHC binding and T cell recognition of moth cytochrome c (93-103). J. Immunol. *152*, 3946-3957.

Richards, R. L., Swartz, G. M., Jr., Schultz, C., Hayre, M. D., Ward, G. S., Ballou, W. R., Chulay, J. D., Hockmeyer, W. T., Berman, S. L., and Alving, C. R. (1989) Immunogenicity of liposomal malaria sporozoite antigen in monkeys: adjuvant effects of aluminium hydroxide and non-pyrogenic liposomal lipid A. Vaccine *7*, 506-512.

Rock, K. L., Gamble, S., and Rothstein, L. (1990) Presentation of exogenous antigen with class I major histocompatibility complex molecules. Science *249*, 918-921.

Rötzschke, O., Falk, K., Wallny, H. J., Faath, S., and Rammensee, H. G. (1990) Characterization of naturally occurring minor histocompatibility peptides including H-4 and H-Y. Science *249*, 283-287.

Ruppert, J., Sidney, J., Celis, E., Kubo, R. T., Grey, H. M., and Sette, A. (1993) Prominent role of secondary anchor residues in peptide binding to HLA-A2.1 molecules. Cell *74*, 929-937.

Sahin, U., Türeci, Ö., Schmitt, H., Cochlovius, B., Johannes, T., Schmits, R., and Pfreundschuh, M. (1995) Proc. Natl. Acad. Sci. USA, in press.

Schild, H., Deres, K., Wiesmuller, K. H., Jung, G., and Rammensee, H. G. (1991) Efficiency of peptides and lipopeptides for in vivo priming of virus-specific cytotoxic T cells. Eur. J. Immunol. *21*, 2649-2654.

Schild, H., Grüneberg, U., Pougialis, G., Wallny, H.-J., Keilholz, W., Stevanivic, S., and Rammensee, H.-G. (1995) Natural ligand motifs of H-2E molecules are allele specific and illustrate homology to HLA-DR molecules. Int. Immunol., in press.

Schirmbeck, R., Bohm, W., and Reimann, J. (1994a) Injection of detergent-denatured ovalbumin primes murine class I-restricted cytotoxic T cells in vivo. Eur. J. Immunol. *24*, 2068-2072.

Schirmbeck, R., Melber, K., Kuhrober, A., Janowicz, Z. A., and Reimann, J. (1994b) Immunization with soluble hepatitis B virus surface protein elicits murine H-2 class I-restricted CD8+ cytotoxic T lymphocyte responses in vivo. J. Immunol. *152*, 1110-1119.

Schirmbeck, R., Melber, K., Mertens, T., and Reimann, J. (1994c) Selective stimulation of murine cytotoxic T cell and antibody responses by particulate or monomeric hepatitis B virus surface (S) antigen. Eur. J. Immunol. *24*, 1088-1096.

Schulz, M., Zinkernagel, R. M., and Hengartner, H. (1991) Peptide-induced antiviral protection by cytotoxic T cells. Proc. Natl. Acad. Sci. USA *88*, 991-993.

Scott, D. M., Ehrmann, I. E., Ellis, P. S., Bishop, C. E., Agulnik, A. I., Simpson, E., and Mitchell, M. J. (1995) Identification of a mouse male-specific transplantation antigen, H-Y. Nature *376*, 695-698.

Scott, J. K. and Smith, G. P. (1990) Searching for peptide ligands with an epitope library. Science *249*, 386-390.

Sette, A., Adorini, L., Appella, E., Colon, S. M., Miles, C., Tanaka, S., Ehrhardt, C., Doria, G., Nagy, Z. A., Buus, S. et al. (1989) Structural requirements for the interaction between peptide antigens and I-Ed molecules. J. Immunol. *143*, 3289-3294.

Sette, A., Ceman, S., Kubo, R. T., Sakaguchi, K., Appella, E., Hunt, D. F., Davis, T. A., Michel, H., Shabanowitz, J., Rudersdorf, R. et al. (1992) Invariant chain peptides in most HLA-DR molecules of an antigen-processing mutant. Science *258*, 1801-1804.

Shirai, M., Arichi, T., Nishioka, M., Nomura, T., Ikeda, K., Kawanishi, K., Engelhard, V. H., Feinstone, S. M., and Berzofsky, J. A. (1995) CTL responses of HLA-A2.1-transgenic mice specific for hepatitis C viral peptides predict epitopes for CTL of humans carrying HLA-A2.1. J. Immunol. *154*, 2733-2742.

Srivastava, P. K. and Maki, R. G. (1991) Stress-induced proteins in immune response to cancer. Curr. Top. Microbiol. Immunol. *167*, 109-123.

Stern, L. J., Brown, J. H., Jardetzky, T. S., Gorga, J. C., Urban, R. G., Strominger, J. L., and Wiley, D. C. (1994) Crystal structure of the human class II MHC protein HLA-DR1 complexed with an influenza virus peptide. Nature *368*, 215-221.

Suto, R. and Srivastava, P. K. (1995) A mechanism for the specific immunogenicity of heat shock protein-chaperoned peptides. Science *269*, 1585-1588.

Tainer, J. A., Getzoff, E. D., Paterson, Y., Olson, A. J., and Lerner, R. A. (1985) The atomic mobility component of protein antigenicity. Annu. Rev. Immunol. *3*, 501-535.

Takahashi, H., Takeshita, T., Morein, B., Putney, S., Germain, R. N., and Berzofsky, J. A. (1990) Induction of CD8+ cytotoxic T cells by immunization with purified HIV-1 envelope protein in ISCOMs [see comments]. Nature *344*, 873-875.

Tam, J. P. (1988) Synthetic peptide vaccine design: synthesis and properties of a high-density multiple antigenic peptide system. Proc. Natl. Acad. Sci. USA *85*, 5409-5413.

Tam, J. P. and Lu, Y. A. (1989) Vaccine engineering: enhancement of immunogenicity of synthetic peptide vaccines related to hepatitis in chemically defined models consisting of T- and B-cell epitopes. Proc. Natl. Acad. Sci. USA *86*, 9084-9088.

Toes, R. E. M., Offringa, R., Feltkamp, M. C. W., Visseren, M. J. W., Schoenberger, S. P., Melief, C. J. M., and Kast, W. M. (1994) Tumor rejection antigens and tumor specific cytotoxic T lymphocytes. Behring Inst. Mitt. *94*, 72-86.

Townsend, A. R., Rothbard, J., Gotch, F. M., Bahadur, G., Wraith, D., and McMichael, A. J. (1986) The epitopes of influenza nucleoprotein recognized by cytotoxic T lymphocytes can be defined with short synthetic peptides. Cell *44*, 959-968.

Tussey, L. G., Rowland-Jones, S., Zheng, T. S., Androlewicz, M. J., Cresswell, P., Frelinger, J. A., and McMichael, A. J. (1995) Different MHC class I alleles compete for presentation of overlapping viral epitopes. Immunity *3*, 65-77.

Udono, H., Levey, D. L., and Srivastava, P. K. (1994) Cellular requirements for tumor-specific immunity elicited by heat shock proteins: tumor rejection antigen gp96 primes CD8+ T cells in vivo. Proc. Natl. Acad. Sci. USA *91*, 3077-3081.

Udono, H. and Srivastava, P. K. (1994) Comparison of tumor-specific immunogenicities of stress-induced proteins gp96, hsp90, and hsp70. J. Immunol. *152*, 5398-5403.

Unanue, E. R. (1984) Antigen-presenting function of the macrophage. Annu. Rev. Immunol. *2*, 395-428.

van Bleek, G. M. and Nathenson, S. G. (1990) Isolation of an endogenously processed immunodominant viral peptide from the class I H-2Kb molecule [see comments]. Nature *348*, 213-216.

Vogt, A. B., Kropshofer, H., Kalbacher, H., Kalbus, M., Rammensee, H. G., Coligan, J. E., and Martin, R. (1994) Ligand motifs of HLA-DRB5*0101 and DRB1*1501 molecules delineated from self-peptides. J. Immunol. *153*, 1665-1673.

Wang, W., Meadows, L. R., den Haan, J. M. M., Sherman, N. E., Chen, Y., Blokland, E., Shabanowitz, J., Agulnik, A. I., Hendrickson, R. C., Bishop, C. E., Hunt, D. F., Goulmy, E., and Engelhard, V. H. (1995) Human H-Y: A male-specific histocompatibility antigen derived from the SMCY protein. Science *269*, 1588-1590.

Weidt, G., Utermohlen, O., Zerrahn, J., Reimann, J., Deppert, W., and Lehmann Grube, F. (1994) CD8+ T lymphocyte-mediated antiviral immunity in mice as a result of injection of recombinant viral proteins. J. Immunol. *153*, 2554-2561.

White, W. I., Cassatt, D. R., Madsen, J., Burke, S. J., Woods, R. M., Wassef, N. M., Alving, C. R., and Koenig, S. (1995) Antibody and cytotoxic T-lymphocyte responses to a single liposome-associated peptide antigen. Vaccine *13*, 1111-1122.

Zhang, W., Young, A. C., Imarai, M., Nathenson, S. G., and Sacchettini, J. C. (1992) Crystal structure of the major histocompatibility complex class I H-2Kb molecule containing a single viral peptide: implications for peptide binding and T-cell receptor recognition. Proc. Natl. Acad. Sci. USA *89*, 8403-8407.

3.4 Recombinant Bacteria as Vaccine Carriers of Heterologous Antigens

Carlos E. Hormaeche and C. M. Anjam Khan

3.4.1 Use of Live Attenuated Bacteria as Antigen Delivery Systems

There is a continuing need for improved vaccines for man and animals. Many of the existing vaccines have proved remarkably successful, e.g. smallpox was eradicated over a decade ago, toxoids have controlled tetanus and diphtheria, and the new capsular vaccines for the agents of bacterial meningitis are causing a pronounced decrease in disease.

However, problems remain in several areas, either due to poor immunogenicity of available preparations, lack of suitable immunogens, or as in the case of diseases caused by eukaryotic parasites, difficulties in the preparations of vaccines. It is noteworthy that five of the six diseases identified by the WHO special programme – malaria, schistosomiasis, leishmaniasis, trypanosomiasis, filariasis – are parasitic diseases. The causative agent of the sixth – leprosy – is non-cultivable *in vitro*. As growth of large quantities of parasite material for mass immunisation is often impractical, vaccine development must proceed by other avenues such as chemical synthesis of defined antigens or their preparation by recombinant DNA technology.

Several possible delivery systems for synthetic or recombinant antigens have been considered, such as microparticles, liposomes, ISCOMs, etc (reviewed in Woodrow and Levine, 1990). An alternative approach which is currently receiving much attention is the use of live attenuated vectors. Vaccinia was the first live vector employed, and it has proved very successful for vaccination of feral foxes against rabies (Pastoret et al., 1995). However, with the eradication of smallpox mass immunisation is no longer practised, and this may prompt the use of vectors other than vaccinia in humans. Other viral live vectors have also been considered (Morin et al., 1990).

There has been considerable work on the development of live bacterial vectors as delivery systems for recombinant antigens. Attenuated *Yersinia* strains have been used to deliver the binding subunits of the enterotoxins from *E. coli* (O'Gaora et al., 1990) and *Vibrio cholerae* (Sory, 1990). Oral administration of *Lactococcus* expressing tetanus toxin fragment C protects mice against lethal challenge with tetanus toxin (Wells et al., 1993). *Listeria* vaccines expressing recombinant antigens can elicit CTLs to beta galactosidase (Schafer et al., 1992), protect mice from lethal tumour challenge (Pan et al., 1995) and from infection with LCMV (Shen et al., 1995).

The two most widely used bacterial carriers have been *Mycobacterium bovis* BCG, the current tuberculosis vaccine, and live attenuated salmonellae, also in current use as live vaccines for typhoid fever. Both of these systems have advantages and drawbacks.

BCG is in widespread use and it has been given to more subjects than any other vaccine. It has a good safety record, and the resurgence of tuberculosis will require the use of BCG or the like for years to come. Possible disadvantages are its frequent use by injection rather than oral administration, and the controversy regarding the efficacy of BCG, which has ranged from better than 80 % to negative protection (Bloom and Fine, 1994). The reasons for these wide discrepancies in the efficacy of BCG remain to be clarified; it may be that cross-immunity to environmental mycobacteria in some test areas has interfered with the results of BCG trials (Bloom and Fine, 1994).

There have also been problems with expression of recombinant antigens in mycobacteria. However, many of these have now been resolved, and there are examples of antigens from HIV, SIV, *Borrelia,* pneumococci, *M. tuberculosis* and IL-2 expressed in BCG. These have variously elicited humoral and cell mediated responses to the recombinant antigens and protection from streptococci and *Borrelia* (see tab. 3.4.1). The location of the antigen in the mycobacterial cell was found to be important for the elicitation of the immune response, with expression on the mycobacterial membrane eliciting a stronger response than when the antigen was expressed intracellularly or secreted from the cell (Stover et al., 1993). The use of BCG as a delivery system for recombinant antigens has been reviewed recently by Stover (Stover, 1994).

3.4.2 Live Salmonella Vaccines

Conventional whole-cell killed salmonella vaccines provide only partial protection, and their excessive reactogenicity has prompted the search for alternative preparations. A new typhoid vaccine consists of purified Vi antigen, the capsular polysaccharide of *Salmonella typhi*, the agent of human typhoid fever. Given by injection, it elicits antibody and confers significant although incomplete protection (Keitel et al., 1994). The live attenuated Ty21a typhoid vaccine, a Vi^- *galE* mutant, has been in use for some years (Tacket and Levine, 1994). Ty21a was obtained by nitrosoguanidine mutagenesis, and contains many lesions. More recently, a Vi^- strain of *S. typhi* harbouring a precise deletion in *galE* was found to be still capable of causing typhoid fever in volunteers, suggesting that the *galE* mutation is not fully attenuating in *S. typhi* (Hone et al., 1988). Therefore, the precise reason for the attenuation of Ty21a is unclear. Ty21a is however very safe; it is given orally as three doses, elicits cell me-

Table 3.4.1: Recombinant molecules expressed in BCG

Antigen	Organism	Ab	CMI	protection	refs
gag,pol, env reverse transcriptase, gp20, gp41	HIV1	+	+	ND	Stover,1994; Stover et al., 1992; Winter et al., 1991
env V3 loop	HIV1		+	ND	Kameoka et al., 1994
gag, nef	SIVmac		+	ND	Yasutomi et al. 1993; Winter et al., 1995
beta galactosidase	*E. coli*	+	+	NA	Murray et al., 1992; Lagranderie et al., 1993
OspA	*Borrelia burgdorferi*	+		+	Stover et al., 1993; Langermann et al., 1994b
PspA	*Streptococcus pneumoniae*	+		+	Langermann et al., 1994a
MPT70	*Mycobacterium tuberculosis*	ND	ND	ND	Matsumoto et al., 1995
IL-2	mouse	stimulates IFNγ			O'Donnell et al., 1994

diated immunity (CMI) and also provides significant but incomplete protection from typhoid fever (Tacket and Levine, 1994). However, attempts to use Ty21a as a vehicle for delivering recombinant antigens have not enjoyed great success (Herrington et al., 1990), perhaps because Ty21a is very attenuated in humans.

Live vaccines are nevertheless generally considered to confer better protection against salmonellosis than killed vaccines, perhaps due to their ability to elicit cell mediated immunity as well as antibody (Hormaeche et al., 1993). There has been much work on a search for improved live oral salmonella vaccines which would confer protection following a single oral dose, thus avoiding the need for multiple doses and parenteral injection, with the attendant problems following possible re-use of needles and transmission of HIV and hepatitis viruses. Several new vaccine candidates, the "new generation" of live salmonella vaccines with defined, non-reverting genetic lesions in known attenuating genes, have been described. Salmonellae harbouring mutations in genes encoding adenylate cyclase and cyclic AMP receptor protein (*cya crp* mutants) are attenuated and have been used as live vaccines (Chatfield et al., 1995). Strains with lesions in regulatory genes such as *ompR* and *phoP* mutants have also been investigated as vaccine candidates (reviewed in Chatfield et al., 1995).

Probably the most widely used of the current live attenuated salmonella vaccines are the Aro mutants described by Stocker's group (Stocker, 1990). Aro mutants harbour mutations in genes of the aromatic pathway, and appear to owe their avirulence to their requirement for *p*-amino benzoic acid (PABA) which is lacking in mammalian tissues; the result is an arrest in protein synthesis due to a lack of fMet-tRNA$^{met}{}_{f}$ (reviewed in Stocker, 1990). The growth of Aro salmonellae in the reticulo-endothelial system (RES) is very slow as compared to the wild type, perhaps "coasting" on the residual intracellular pool of PABA, thus apparently allowing time for the development of the protective cellular and humoral response which leads to clearance of the vaccine organisms from the RES leaving immunity to rechallenge. Mutations in genes *aroA, aroC* and *aroD* have all been found to be attenuating. Aro salmonella vaccines are attenuated and effective in mice (Stocker, 1990), cattle (Smith et al., 1984a; Smith et al., 1984b; Jones et al., 1991; Segall and Lindberg, 1991; Robertsson et al., 1983), sheep (Mukkur et al., 1987) and chickens (Cooper et al., 1992; Cooper et al., 1993; Cooper et al., 1990; Cooper et al., 1994a; Cooper et al., 1994b), and Aro mutants are currently showing great promise in human volunteer trials (Tacket and Levine, 1994). The requirement for PABA may explain their lack of invasiveness in immunosuppressed hosts (sublethal irradiation, agammagobulinaemia, cyclophosphamide, the *scid* phenotype, or administration of silica – reviewed in Stocker, 1990). Much of the work referred to below concerns mice immunised with Aro strains of *S. typhimurium* or *S. enteritidis*.

We have also reported that salmonella *htrA* mutants are effective vaccines (Johnson et al., 1991; Chatfield et al., 1991; Strahan et al., 1992). HtrA is a member of the s^{E} family of stress proteins, the deduced amino acid sequence of the salmonella HtrA protein showing 88 % homology with its *Escherichia coli* counterpart. Salmonella *htrA* mutants are likewise more susceptible to oxidative stress, and are attenuated in animals; they are effective vaccines, and are not more invasive in immunosuppressed mice (Strahan et al., 1992). In *E. coli*, the HtrA phenotype confers high temperature resistance, but we did not find this to be the case with the salmonella *htrA* mutants. *E. coli htrA* mutants also show increased susceptibility to oxidative stress, and we found that the same was true for the salmonella *htrA* mutants. Although the precise reason for the reduced virulence of the salmonella *htrA* mutants remains to be determined, one possibility could be impaired survival within phagocytic cells due to increased susceptibility to oxidative killing mechanisms. The impaired ability of *htrA* mutants to degrade foreign proteins could be useful when contemplating their use as antigen delivery systems for recombinant antigens (see below).

3.4.3 Live Salmonella Vaccines as Antigen Delivery Systems

There is a steadily increasing body of evidence to show that live attenuated salmonella vaccines constitute an effective delivery system for recombinant antigens from a wide variety of pathogens, thereby increasing the potential of vaccines which are in themselves a necessary requirement (Hormaeche et al., 1994). Live attenuated salmonella vaccines are effective when given orally; the new generation vaccines require only a single dose for seroconversion and protection. This is perhaps due to the persistence and limited multiplication of the vaccine with *de novo* synthesis of antigen within the RES itself for a period, which together with the adjuvanticity of the organism itself, provides a potent stimulus to the immune response. We have observed that mice immunised with Aro salmonellae develop an IgG response to salmonella LPS, suggesting the existence of bystander T cell help to a T cell independent antigen (J. Harrison et al., to be published).

Notably, live salmonella vaccines have been shown to elicit humoral, secretory and cell mediated immunity, including in some cases cytotoxic T cells, to the recombinant antigens, thereby conceivably providing the capability to elicit the immune responses necessary for protection against pathogens with varied and complex pathogenesis (Hormaeche et al., 1994). This could be especially useful when considering their use for delivery of recombinant antigens from multicellular parasites which undergo complex development within the host, and for which protective immunity may require the participation and combined effects of more than one arm of the immune system. The wide variety of pathogens for which the system is being successfully employed bears witness to the potential of salmonellae as antigen delivery systems. The area has recently been extensively reviewed (Chatfield et al., 1995; Hormaeche et al., 1994; Roberts, 1994), and this chapter will deal mainly with new developments.

3.4.4 Problems Encountered in the Expression of Recombinant Antigens in Salmonella Vaccines

One of the main problems in the use of salmonellae as antigen delivery systems is the expression of the recombinant antigen in the salmonella vaccine strain in a way that will elicit an immune response. There are many similarities between *Salmonella* and *E. coli,* the commonly used expression host used in recombinant DNA technology. Furthermore, many of the expression plasmids developed for use in *E. coli* can be used in salmonellae, and moving the vector from *E. coli* to the salmonella vaccine can be easily done. The simple expression of the antigen to levels detectable by e.g. Western blotting is often readily achievable.

Intuitively, it may be expected that the magnitude of an immune response to a guest antigen depends on, among other factors, the antigen dose. The expression level of a cloned gene is affected by a number of factors which include promoter strength, translational initiation sequences, transcriptional termination sequences, plasmid copy number and stability. Unfortunately, high levels of expression of some recombinant antigens, e.g. some eukaryotic antigens, from strong constitutive promoters can be very unstable in salmonella vaccines (see below). It is perhaps not surprising that some of the effective constructs prepared to date have been those which expressed the antigen to appreciable levels (1 – 2 % of cell protein) in the salmonella vaccine (Brown et al., 1987).

However, it is by now very clear from the experience of several different research groups that the expression of a recombinant antigen in a salmonella vaccine strain is by itself no guarantee that the construct will elicit a response to the guest antigen. Sadly, the requirements for a recombinant antigen to be immunogenic when expressed in salmonella vaccines are far from clear, and many empirical solutions have been attempted; some are described below.

3.4.5 Degradation of the Recombinant Antigen by Salmonella Proteases

Proteolytic degradation of the recombinant antigens in salmonella vaccines can pose a difficult problem. Different approaches to avoid proteolytic degradation in *E. coli* have been described (Miller, 1987), but the protease systems in salmonellae are not as yet characterised to the point where they can be manipulated to increase stability of recombinant antigens in salmonella vaccines. We observed that a recombinant antigen from *Schistosoma mansoni*, P50, was well expressed in *E. coli*, but extensively degraded to small fragments (still recognised by antisera on Western blots) when expressed in salmonella vaccine strains (our unpublished observations; collaboration with J. Havercroft).

We also observed that a construct encoding multiple copies of an epitope from foot and mouth disease virus as a fusion to beta galactosidase was well expressed (as detected by a monoclonal antibody to the peptide) in *E. coli*. However, no peptide – only beta galactosidase – could be detected when the plasmid was transferred from the *E. coli* host to salmonellae. Recovery of the expression plasmid from *Salmonella* and transfer back to *E. coli* again showed expression of the peptide, suggesting that the failure to detect the peptide in *Salmonella* was not due to corruption of the DNA encoding the peptide, but perhaps due to degradation of the antigen (our unpublished observations).

An additional characteristic of the HtrA phenotype is an impaired ability to degrade aberrant proteins (the *degP* phenotype in *E. coli*). We suggested that, if the impaired ability to degrade extraneous proteins also applied to salmonella HtrA strains, this could perhaps prove useful for expression of some recombinant antigens in salmonella vaccines (Johnson et al., 1991). We have indeed observed that the herpes simplex virus glycoprotein D appears to be expressed better in salmonella *htrA* mutants than in Aro mutants (see below; Chabalgoity et al., 1996). If this applies to other recombinant antigens, then *htrA* mutants may deserve consideration when planning expression of recombinants antigens in salmonella vaccines.

However, it must be emphasised that HtrA is one of several proteins forming part of a complex stress response. The normal substrate and enzymatic cleavage site for HtrA are unknown, and it is not immediately apparent why a mutation in a periplasmic serine protease such as HtrA should improve the survival of a recombinant protein expressed in the salmonella cytoplasm.

3.4.6 Plasmid Stability and Level of Expression

Expression of recombinant antigens – particularly from eukaryotic pathogens – in salmonellae can cause problems, especially when using expression vectors designed for production of maximal amounts of protein in *E. coli*. Commonly used *E. coli* expression vectors often rely on a strong promoter such as *tac*, regulated in *E. coli* by the *lacIq* repressor. The latter is normally absent in *Salmonella*, so that *tac* and other similarly regulated promoters are expressed constitutively to a high level. This can be highly desirable if the guest antigen is tolerated well by the salmonella host, but this is often not the case.

3.4.6.1 Plasmid Loss

High level expression of recombinant proteins in salmonellae from an expression plasmid can often be achieved, but all too frequently the plasmid can only be maintained in the salmonella vaccine by growth in antibiotic-containing media. Viable counts on organs of mice inoculated with cultures in which the plasmid itself is unstable can show almost total loss of the expression plasmid in a few days, with a dramatic reduction in antibiotic-resistant bacteria in the RES and consequently no immune responses to the guest antigen (Maskell et al., 1987). Such a strain is obviously of little use, as a vaccine must be capable of retaining the expression plasmid in the tissues and producing the recombinant protein in the absence of the antibiotic.

It is therefore essential to check the ability of the recombinant vaccine to retain the plasmid in the tissues by performing viable counts on organ homogenates in parallel on media with and without antibiotic.

3.4.6.2 Impaired Persistence of the Vaccine

In other cases the salmonella vaccine does express the recombinant protein and retains the expression plasmid in the tissues, but the growth of the salmonella is so impaired that it fails to proliferate in the tissues and is cleared, again without inducing an immune response. For reasons that remain unclear, the most effective live vaccines appear to be those which are capable of colonising and persisting in the tissues for a period before being cleared (Hormaeche et al., 1993).

It is therefore essential to check the ability of a recombinant vaccine to persist in the tissues as compared with the persistence of the parent strain, especially when making comparisons between the ability of different vaccines to elicit responses to guest antigens.

3.4.6.3 Stable *in vitro,* Unstable *in vivo*

A puzzling situation we observed was that of a salmonella vaccine expressing an antigen from herpes simplex virus which expressed the antigen stably in culture in the absence of antibiotic, but which lost the plasmid very rapidly when injected into mice (M. Izhar, Ph.D. thesis, Cambridge). This unexplained result is a reminder that the conditions and stresses under which the salmonella vaccine must proliferate and express the recombinant antigen in the tissues – perhaps intracellularly (Buchmeier and Heffron, 1990) – are probably very different to those in culture. It may prove useful to attempt to mimic the conditions of intracellular growth *in vitro* as an additional screening test for vaccine candidates.

3.4.6.4 Repeated Inoculations

Unstable constructs such as those described above can sometimes elicit immune responses to the guest antigens if they are administered repeatedly by parenteral injection. However, the same result can probably be obtained using a killed preparation of the vaccine. Recombinant live vaccines which require multiple parenteral doses should be tested in parallel with a killed preparation of the same to ensure that the immune response to the guest antigen is indeed due to the carrier effect of the live vaccine, and not merely to repeated inoculations of recombinant protein. One of the attractive features of the salmonella delivery system is precisely its single-dose capability, and this is more difficult to achieve by killed preparations.

3.4.6.5 Vaccine Strain Background

We have observed that the parent strain background from which the salmonella vaccine is derived can make the difference between failure and success in expressing recombinant antigens in salmonella vaccines (our unpublished observations and Chabalgoity et al., 1996). We routinely test new antigens in a panel of different salmonella vaccine strains, and in some cases we have found marked differences in their ability to express the recombinant antigens. Even derivatives of the same parental strain carrying different attenuating mutations in the aromatic pathway have varied in their ability to express a given antigen. We have no explanation for these effects.

3.4.6.6 Repeated Passage

Repeated passage is an empirical approach to stabilising a culture which has yielded positive results in at least two cases. Coulson et al. (1994) found that the stability of a plasmid expressing a clone of the protective antigen of *Bacillus anthracis* driven by the *lac* promoter in an *aroA S. typhimurium* vaccine increased following passage in mice treated with ampicillin; a clone recovered from mice showed increased stability, which was accompanied by a reduction in plasmid copy number. We found that a clone expressing an antigen from herpes simplex glycoprotein D as a fusion to tetanus toxin fragment C (see below) was unstable *in vitro*, yielding 10 % or less of plasmid-containing organisms following overnight growth without ampicillin. However, after several passages in antibiotic containing medium, this increased to over 90 % and the construct was immunogenic in mice (Chabalgoity et al., 1996).

3.4.6.7 Removal of Toxic Subregions of the Antigen

In some cases the gene encoding the recombinant protein can be modified to remove subregions causing problems in expression, *e.g.* removal of signal peptides or anchor sequences from normally secreted proteins can reduce their toxicity for salmonellae, allowing expression intracellularly at a higher level and increasing stability of the vaccine. Cohen et al. (1990) found that a clone of dengue virus envelope antigen which expressed 93 % of the gene was toxic for salmonella vaccine strains, but another clone which expressed 73 % of the gene and excluded hydrophobic sequences at either end of the protein was well tolerated and immunogenic in mice, eliciting antibodies which recognised the native virus.

3.4.6.8 Codon Optimization

The codon bias of a gene can have profound effects on its level of expression (Grosjean and Friers, 1982). The codon usage of highly expressed genes reflects a selection against codons specifying tRNA molecules of low abundance and codons whose binding energy for interactions with the anti-codon loops of the cognate tRNA is suboptimal. As codon choice also varies from organism to organism, heterologous genes are often expressed at lower levels when there is a poor fit between the codon bias of the gene and that of the host organism. Expression of tetanus toxin fragment C (TetC) was improved by codon optimisation. By substituting rare codons with others more frequently used in *E. coli*, Makoff *et al.* improved the expression level of TetC in *E. coli* from less than 1 % to up to 20% of total cell protein in *E. coli* (Oxer et al., 1991). This increase in level of expression as a result of codon optimisation may have contributed to the success of the salmonella- tetanus vaccine (see below), although this has not been formally proven.

3.4.6.9 Reducing the Expression Level with a Weaker Promoter

This is an obvious approach, but if stabilisation can only be obtained with a very low expression level, it runs the risk that the construct will not elicit an immune response to the guest antigen. However, a promoter which causes a moderate reduction in expression level is worth testing, especially with proteins that are normally exported to the periplasm. We have reported that expression of the binding subunit of the *E. coli* heat-labile enterotoxin (LT-B) in an Aro salmonella vaccine from the strong constitutive *tac* promoter was very unstable, the plasmid being rapidly lost in the tissues after inoculation. However, expression of LT-B from the weaker P_l promoter yielded a stable vaccine which elicited antibodies which neutralised the holotoxin (Maskell et al., 1987).

3.4.6.10 Incorporation of the Foreign Gene into the Salmonella Chromosome

This is an attractive solution which has several advantages. It eliminates the problems of an unstable plasmid; moreover, the antibiotic resistance marker can also be removed with in a second step (Chatfield et al., 1992), which is especially desirable when contemplating the construction of vaccines to be used in the field. The foreign gene is introduced into the salmonella vaccine on a suicide plasmid with flanking regions which allow incorporation into the salmonella chromosome by homologous recombination into a predetermined site.

Two systems have been described, one using the *his* gene (Hone et al., 1988) and another in which the foreign gene is inserted into the *aroC* gene (Strugnell et al., 1990). The latter technique has the additional advantage that insertion of the foreign gene into the chromosome in itself introduces an attenuating mutation into the vaccine strain. However, this technique effectively reduces gene copy number to one, and in some cases it may be the attendant decrease in expression level which is responsible for the increase in stability. Expression level of the guest antigen can be quite low, less than 1 % of total cell protein; this was however sufficient for eliciting a response to the *Plasmodium falciparum* circumsporozoite antigen in an *aroC aroD S. typhi* vaccine in human volunteers, although not all subjects gave a positive response (González et al., 1994).

3.4.6.11 Stabilisation by Incorporation of Essential Genes into Expression Plasmids

Curtiss's group have developed a system (Nakayama et al., 1988) making use of expression plasmids which carry an essential gene (*e.g. asd*) which complements in *trans* a corresponding chromosomal lesion in the salmonella vaccine, which also carries additional attenuating lesions, e.g. a *cya crp* mutant. Cells which lose the expression plasmid will be overly attenuated and unable to survive in the tissues, and will be lost *in vivo*. This technique will not necessarily make the plasmid better tolerated by the vaccine strain, but will ensure that cells which have lost the plasmid will not outrun the vaccine strain in the tissues

3.4.6.12 Incompatible Plasmids

One possible approach to the expression of toxic proteins is to construct a vaccine in which expression of the guest antigen is normally repressed and which persists in the tissues, but which constantly generates cells producing the protein. Ervin et al. (1993) have described a system in which control of expression is achieved by means of two incompatible plasmids, one carrying the gene for the guest antigen driven by the *trc* promoter, and another plasmid carrying the LacI repressor. The system is designed so that during growth cells which lose the repressor plasmid by segregation will express the antigen. The system was successful in obtaining a response to beta galactosidase in mice. This system could prove useful for toxic proteins.

3.4.6.13 "On-Off" Promoters

A different approach to the expression of toxic proteins was described by Tijhaar et al. (1994). They have made use of an "on-off" promoter controlled by a randomly invertible sequence derived from the bacteriophage Mu Gin invertase. This generates a non-expressing bacterial population which is continually yielding producing bacteria. The system was applied to the expression of CT-B in an *aroA* salmonella vaccine, which was immunogenic in mice.

3.4.6.14 *In vivo* Inducible Promoters

The salmonella vaccine has to be capable of producing the antigen in the tissues, and promoters which respond to environmental signals when the organism is growing *in vivo*, including within host cells, could prove particularly useful. Some salmonella proteins are upregulated during growth inside macrophages (Buchmeier and Heffron, 1990); salmonella *phoP* activated genes are transcriptionally activated within antigen-processing macrophages. Hohmann et al. (1995) found that an *aroA S. typhimurium* vaccine elicited a response to a model antigen, alkaline phosphatase, when the antigen was expressed as a genetic fusion to the PagC protein. The *pagC* gene is *phoP*-activated, and the gene encoding the fusion protein was incorporated as a single copy in the salmonella chromosome. The authors noted that two other fusions of alkaline phosphatase to genes not controlled by *phoP* were not immunogenic, even though the amounts of alkaline phosphatase produced during growth *in vitro* were similar to or greater than that produced by the *pagC* fusion. Assuming that all strains colonised and persisted equally in the tissues, the result suggests that the ability of the vaccine to produce the antigen *in vivo* can be as important as, if not more so than, the total amount produced *in vitro*.

Everest et al. (1995) have recently studied the ability of three different promoters to express a model antigen in salmonellae. The expression of *lacZ* from the *htrA, groE* and the anaerobically inducible *E. coli* nitrite reductase *nirB* (Oxer et al., 1991) promoters in Aro salmonellae was studied. Expression of beta galactosidase increased with temperature in all cases. However, when expression was studied in salmonellae growing inside eukaryotic cells, it was found that the *htrA* and *nirB* promoters increased expression inside all cells, whereas the *groE* promoter increased expression in macrophages, but not in epithelial cells. Newton *et al.* have found that the level of expression of MalE from *nirB* was influenced by the distance between the promoter and the Shine-Dalgarno sequence, with the best yields being up to 20 % of total cell protein (Newton et al., 1995). The use of the *nirB* promoter to drive the expression of recombinant antigen fusions in salmonellae is described in section 3.4.9 below.

3.4.7 Location of the Recombinant Antigen on the Bacterial Cell and Antigen Presentation

There are many examples of salmonella vaccines which have elicited responses to guest antigens expressed in the salmonella cytoplasm; we have shown that sera from mice immunised with attenuated salmonella vaccines recognise salmonella proteins such as RNA polymerase (Brown and Hormaeche, 1989) or intracellular guest antigens like beta galactosidase (Brown et al., 1987). However, location of the recombinant antigen on the bacterial cell can be important for certain antigens, as it may affect assembly and expression of conformational epitopes. In some cases this may require export to the periplasm, and systems for exporting the guest antigen from the cell to the periplasm or to the bacterial surface have been described utilising MalE and OmpA export signals or as fusions to LT-B (Lipscombe et al., 1991; Schodel and Will, 1990).

Several outer membrane proteins are themselves potent immunogens, and systems have been described for expression of epitopes in flagellin, fimbriae, LamB, OmpA and PhoE (O'Callaghan et al., 1990; Agteberg et al., 1990; Leclerc et al., 1989; Su et al., 1992; Schnorr et al., 1991; Newton et al., 1989; Stockner, 1994; Klemm and Hedegaard, 1990; Reeves et al., 1991). However, we have recently found that a conformational determinant (which requires formation of a disulphide bridge) on a malarial protein is recognised by the corresponding monoclonal antibody when the protein is expressed in the salmonella cytoplasm (Khan et al., to be published).

The combined effects on immunogenicity of either secretion of the antigen from the bacterial cell or intracellular expression, and escape of the salmonella vaccine into the cytosol or growth within the phagosome, have been recently investigated by Hess et al. (1996). Variants of an *aroA S. typhimurium* vaccine were constructed which expressed two immunogenic proteins from *Listeria monocytogenes* (Hess et al., 1995), the haemolysin (Hly; listeriolysin) or the P60 protein either in secreted form or intracellularly (Hess et al., 1996). Only the vaccines expressing the secreted proteins conferred protection from infection with *Listeria*; protection was better with the vaccine expressing Hly, which allows translocation of the vaccine into the cytosol, suggesting that growth of the vaccine and expression of the antigen in the cytosol had increased the induction of MHCI dependent T-cell immunity. Further experiments with MHCI and MHCII deficient gene disruption mice suggested that both vaccines were capable of generating CD4+ and CD8+ protective T-lymphocytes.

Aro *S. dublin* expressing class I and class II MHC-restricted epitopes of listeriolysin from *Listeria monocytogenes* as inserts in the salmonella flagellin expressed from a single copy in the chromosome were immunogenic in mice; the amino acids constituting the flanking regions on either side of the MHC I restricted epitope were important in influencing the magnitude of the T cell response, perhaps by facilitating excision of the epitope in the endosomal compartment of antigen-presenting cells. The recombinant vaccines protected mice from challenge with *Listeria* (Verma et al., 1995a). Studies on Aro salmonellae expressing an epitope from moth cytochrome C showed that antigen processing and presentation of class II-restricted epitopes in the salmonella flagellin was facilitated by the presence of flanking regions containing cathepsin B sites which can be cleaved in the endosome (Verma et al., 1995b).

The ability of salmonella vaccines expressing antigens intracellularly to stimulate T cells may also depend on the type of cell in which the organism is growing. It is now clear that salmonellae can grow both within phagocytes and also within epithelial cells e. g. hepatocytes (Hsu, 1993) Salmonellae expressing recombinant antigens elicit cytotoxic T cells against ovalbumin (Turner et al., 1993), the malaria circumsporozoite antigen (González et al., 1994; Aggarwal et al., 1990; Flynn et al., 1990) and influenza nucleoprotein (Gao et al., 1992; Tite et al., 1990). Macrophages infected with salmonellae expressing an ovalbumin (OVA) peptide expressed as fusions to the cytoplasmic protein Crl will present the peptide to T cells in association with MHC I molecules by a novel pathway which does not require *de novo* synthesis of class I molecules (Pfeifer et al., 1993). Macrophages process salmonellae with mutations in *phoP*, a regulatory gene, more efficiently than wild type organisms, as measured by their ability to present the OVA peptide to T cells (Wick et al., 1995). In contrast, Chinese hamster ovary (CHO) cells infected with salmonellae expressing recombinant influenza A virus nucleoprotein (NP) did not present the antigen to T cells (Gao et al., 1992).

3.4.8 Examples of Multivalent Vaccine Strains

Table 3.4.2 lists antigens from viruses, bacteria, parasites and model antigens, as well as cytokines, which have been expressed in salmonella vaccines. The immune responses – and in some cases protection – obtained are indicated. Adhesins such as *E. coli* K88 have been expressed in *S. typhimurium*, eliciting systemic and local immunity and also protection from salmonellosis (Stevenson and Manning, 1985; Dougan et al., 1986). The CFA/I and CS3 colonisation factor antigens of enterotoxigenic *E. coli* have been expressed in the *aroC, aroD S. typhi* CVD908 *Salmonella typhi* vaccine strain as a potential combined vaccine for typhoid fever and traveller's diarrhoea (Girón et al., 1995). The filamentous haemagglutinin (Molina and Parker, 1990) and the surface P.69 protein, pertactin (Strugnell et al., 1992) from *Bordetella pertussis* have been expressed in salmonellae. High level expression of pertactin proved deleterious to salmonellae; this was overcome by integration of the pertactin gene into the salmonella chromosome, and the recombinant protein was expressed on the cell surface of the *Salmonella.* The vaccine failed to elicit antibody to P.69 in mice, although it did evoke a positive cellular response. The vaccine conferred protection from aerogenic challenge with *B. pertussis* and oral challenge with virulent salmonellae (Strugnell et al., 1992).

Several groups have expressed the binding subunit of *E. coli* heat labile enterotoxin, LT-B, in attenuated salmonellae (Maskell et al., 1987; Clements and Cárdenas, 1990; Clements et al., 1988). LT-B has also been exploited as a carrier for epitopes expressed as C-terminal genetic fusions. LT-B-epitope fusions are immunogenic as the purified proteins, eliciting a response to a *B. pertussis* P.69 epitope when given intranasally to mice (Lipscombe et al., 1991). Epitopes from hepatitis B virus antigens have been similarly fused to LT-B and delivered with salmonellae (Schodel and Will, 1990). The nature of the linker joining the epitope to LT-B was important for the immunogenicity of *Streptococcus sobrinus* antigens fusions (Jagusztyn-Krynicka et al., 1993). Prior immunisation with the salmonella vaccine alone did not prevent the development of an immune response to LT-B, but rather boosted the anti-LT-B response (Bao and Clements, 1991).

Mice given large amounts of dead attenuated salmonellae expressing LT-B orally made serum and mucosal responses to LT-B similar to those elicited by the live vaccine (Cárdenas et al., 1994); the authors suggested that it was the initial antigen dose rather than persistence in tissues which was important for immunogenicity. However, the very potency of LT-B as a mucosal immunogen may have influenced this result; it had been previously shown that salmonellae expressing tetanus toxin fragment C (TetC) were only effective as a living oral vaccine, with large oral doses of dead organisms failing to elicit a response to TetC (Fairweather et al., 1990).

Salmonellae expressing the *Streptococcus pyogenes* M protein protected mice from challenge with virulent streptococci, protection being M-protein specific (Poirier et al., 1988). An *aroA* salmonella vaccine expressing the pneumococcal pneu-

Table 3.4.2: Examples of the use of live attenuated *S.typhimurium* and *S.typhi* strains as carriers of heterologous antigens

Antigen	Organism	Salmonella *strain*	Immune response[a]			Pro-tection[b]	Ref.
			Ab	CMI	CTL		
	Bacterial						
C-fragment	*C. tetani*	*S. typhimurium aroA, aroA aroD, aroA aroC.*	+	ND[c]	ND	+	Chatfield et al., 1992a,b; Fairweather et al., 1990
P.69	*B. pertussis*	*S. typhimurium aroA aroC aroD*	–	+	ND	+	Strugnell et al., 1992
FHA	*B. pertussis*	*S. typhimurium aroA*	+	ND	ND	ND	Guzmán et al., 1991; Molina & Parker, 1990
PTX-S1	*B. pertussis*	*S. typhimurium aroA*	+	ND	ND	ND	Walker et al., 1992
M protein	*S. pyogenes*	*S. typhimurium aroA*	+	ND	ND	+	Poirier et al., 1988
LTB	*E. coli*	*S. typhimurium aroA,*	+	ND	ND	+	Maskell et al., 1987
LTB	*E. coli*	*S. typhimurium aroA*	+	ND	ND	ND	Cárdenas et al., 1994
K1 capsule	*E. coli*	*S. typhimurium aroA*	–	ND	ND	ND	O'Callaghan et al., 1988
K88	*E. coli*	*S. typhimurium aroA, galE*	+	ND	ND	+	Stevenson & Manning, 1985; Dougan et al., 1986; Attridge et al., 1988; Morona et al., 1994
β–galactosidase	*E. coli*	*S. typhimurium aroA, galE*	+	+	ND	ND	Brown et al., 1987; Ervin et al., 1993; Everest et al. 1995
Haemagglutinin	*P. gingivalis*	*S. typhimurium aroA*	+	ND	ND	ND	Dusek et al. 1993
28 kD OMP	*N. meningitidis*	*S. typhimurium aroA*	+	ND	ND	ND	Tarkka et al. 1989
Various	*T. pallidum*	*S. typhimurium aroA, aroA aroC*	+/-	ND	ND	–[d]	Strugnell et al. 1989, 1990
31kD	*B. abortus*	*S. choleraesuis cya crp cdt, S. typhimurium cya crp*	+	–	–	ND	Stabel et al. 1990, 1991, 1993
17kD	*F. tularensis*	*S. typhimurium cya crp asd*	+	ND	ND	+	Sjösted et al. 1990, 1992
SpaA	*S. mutans*	*S. typhimurium cya crp*	ND	ND	ND	ND	Curtiss et al. 1986, 1989; Katz et al. 1987
SpaA	*S. sobrinus*	*S. typhimurium cya crp*	+	ND	ND	ND	Nakayama et al. 1988; Redman et al., 1995; Curtiss et al., 1988
SpaA Dex	*S. sobrinus*	*S. typhimurium cya crp*	ND	ND	ND	ND	Jagusztyn-Krynicka, 1993

Antigen	Organism	Salmonella *strain*	Immune response[a]			Protection[b]	Ref.
			Ab	CMI	CTL		
pneumolysin toxoid	*S. pneumoniae*	*S. typhimurium aroA*	+	ND	ND	ND	Paton et al., 1993
CT-B	*V. cholerae*	*S. dublin, S. typhimurium*	+	ND	ND	ND	Tijhaar et al., 1994; Su et al., 1992
invasin	*Y. pseudo-tuberculosis*	*S. typhimurium aroA*	+	ND	ND	+, trans-location	Simonet et al. 1994
Vi antigen	*S. typhi*	*S. typhimurium cya crp*	+	+	ND	ND	Cao et al. 1992
F1 capsular antigen	*Y. pestis*	*S. typhimurium aroA*	+	+	ND	+	Oyston et al. 1995
listeriolysin	*Listeria monocytogenes*	*S. dublin aroA*		+	+	+	Verma et al. 1995 a,b
listeriolysin P60	*Listeria monocytogenes*	*S. typhimurium aroA*	ND	+	ND	+	Hess et al. 1995a,b; Gentschev et al. 1992
Various	*M. leprae*	*S. typhimurium cya crp*	ND	ND	ND	ND	Clark-Curtiss et al. 1990
Protective antigen	*B. anthracis*	*S. typhimurium aroA*	–	ND	ND	(+)	Coulson et al. 1994
alkaline phosphatase (PagC fusion)	*E. coli*	*S. typhimurium aroA*	+	ND	ND		Hohmann et al. 1995
Viral							
Nucleoprotein	Influenza A	*S. typhimurium aroA*	+	+	+[e]	+[f]	Gao et al. 1992; Tite et al. 1990
Surface antigen	Hepatitis B	*S. dublin*	+	ND	ND	ND	Wu et al. 1989
Core protein	Hepatitis B	*S. typhimurium aroA, aroA aroD, cya crp, Phopc*	+	+	ND	ND	Schödel 1992; Schödel & Will 1990; Schödel et al. 1990, 1991 Hopkins et al. 1995
gpD	Herpes simplex	*S. typhimurium aroA*	ND	ND	ND	ND	Bowen et al. 1990
gpD	Herpes simplex	*S. typhimurium htrA*	+	ND	ND	+	Chabalgoity et al. 1995
Envelope antigen	Dengue 4	*S. typhimurium aroA*	ND	ND	ND	ND	Cohen et al. 1990
Core protein	Wood chuck	*S. typhimurium cya crp*	+	ND	ND	ND	Schödel & Will 1990
VP7	Rotavirus SA11	*S. typhimurium cya crp*	–	ND	ND	ND	Sala et al. 1990
env V3 loop	HIV1	*S. typhimurium aroA*	+	+	ND	ND	Charbit et al. 1993; Newton et al. 1995
p27gag	SIV	*S. typhimurium aroD, htrA, S. dublin aroA*	–	ND	ND	ND	Strahan et al. 1992
gag, gp120	HIV2	*S. typhimurium Aro*	–		(+)	– suppression	Franchini et al. 1995

Antigen	Organism	Salmonella *strain*	Immune response[a]			Pro-tection[b]	Ref.
			Ab	CMI	CTL		
Parasites							
CSP	*P.berghei*	*S. typhimurium WR4017,WR4024*	–	+	+	+	Aggarwal et al. 1990; Sadoff et al. 1988
CSP	*P.falciparum*	*S. typhimurium WR4024*	–	ND	+	ND	Aggarwal et al. 1990
SERP, HRPII	*P. falci-parum*	*S. typhimurium cya crp*	+	ND	ND	ND	Leclerc et al. 1989
CSP	*P. yoelii*	*S. typhimurium aroA*	–	ND	+	–	Flynn et al. 1990
gp63	*L. major*	*S. typhimurium aroA, aroA aroD*	+	+ IL-2, IFN	ND	+	Yanf et al. 1990; Xu et al. 1995
Species specific Ag	*E. multilocularis*	*S. typhimurium galE*	+	+	ND	ND	Gottstein et al. 1990
Surface antigen	*Entamoeba histolytica*	*S. typhimurium cya crp*	ND	ND	ND	ND	Cieslak et al. 1993
Glutathione S-transferase	*Schistosoma mansoni*	*S. typhimurium aroA*	+	+	ND	+++[g]	Khan et al. 1994a,b
Cytokines							
IL-1β	Human	*S. typhimurium aroA*	+	ND	ND	+	Carrier et al. 1992
IL-4	Mouse	*S. typhimurium aroA*					Denich et al. 1993
TGF-β	Mouse	*S. typhimurium aroA aroD*					Ianaro et al. 1995
Other antigens							
ovalbumin	NA	*S. typhimurium aroA aroD*	–	–	+	NA	Turner et al. 1993
MalE	*E. coli*	*S. typhimurium aroA*	+	ND	ND	NA	Newton et al. 1995; Fayolle et al., 1994
Bet vI, Bet v II	tree pollen antigens	*S. typhimurium aroA*	Th1 respons eIgG2a /b, no IgG1 or IgE	ND	ND	NA	Vitala et al. 1995
S. typhi carriers							
O antigen	*S. sonnei*	*S. typhi Ty21a*	+	ND	ND	+/–	Formal et al. 1981; Tramont et al. 1984; Black et al. 1987; Hartman et al. 1991; Seid et al. 1984; van de Verg et al. 1990
O antigen	*S. flexneri*	*S. typhi Ty21a*	+	ND	ND	+	Baron et al. 1987

Antigen	Organism	Salmonella *strain*	Immune response[a]			Pro-tection[b]	Ref.
			Ab	CMI	CTL		
O antigen	*V. cholerae*	*S. typhi Ty21a*	+	ND	ND	+/–	Attridge, 1991; Attridge et al. 1990; Dearlove et al. 1992; Forrest et al. 1991; Tacket et al. 1990
LTB	*E. coli*	*S. typhi Ty21a*	+	ND	ND	+	Maskell et al. 1987
CSP	*P. falciparum*	*S. typhi aroA aroD*	+	ND	+	ND	González et al. 1994
Fragment C	Tetanus toxin	*S. typhi aroC aroD*	ND	ND	ND	ND	Chatfield et al. 1992
CFA/I, CS3	Enterotoxigenic *E. coli*	*S. typhi aroC aroD*	ND	ND	ND	ND	Girón et al. 1995

[a] Ab, antibody response; CMI, cell mediated response, (macrophage activation, T-cell proliferation, cytokine release or delayed hypersensitivity); CTL, T cytotoxic cells

[b] Protection in an animals, volunteer trialss or in vittro neutralisation.

[c] ND, not done.

[d] Immunisation enhanced the development of syphilitic lesions in rabbits

[e] CD4+ class II-restricted T cytotoxic cells

[f] Protection required boosting intranasally using the purified nucleoprotein

[g] Protection from tetanus, salmonella and schistisimiasis

molysin elicited serum IgG and IgA responses to pneumolysin in mice (Paton et al., 1993). A *cya crp S. typhimurium* vaccine expressing the *Streptococcus sobrinus* surface protein A antigen induced protective humoral immune responses in rats (Redman et al., 1995).

An *aroA* salmonella expressing the *Yersinia pseudotuberculosis* invasin showed increased translocation from the gut to the mesenteric lymph nodes. The mice made circulating and secretory antibody to the invasin, and showed a decrease of translocation (but not of systemic dissemination) following oral challenge with *Yersinia* (Simonet et al., 1994). An Aro *S. typhimurium* expressing the *Borrelia burgdorferi* OspA antigen protected mice from challenge with *Borrelia* (Dunne et al., 1995). An Aro *S. typhimurium* expressing the protective antigen of *Bacillus anthracis* likewise conferred protection from challenge with anthrax spore challenge (Coulson et al., 1994). An *aroA S. typhimurium* expressing the *Yersinia pestis* F1 antigen protected mice against plague following i/v or oral immunisation (Oyston et al., 1995).

Several antigens from viruses have been successfully delivered using salmonella vaccines, including hepatitis B virus hybrid nucleocapsid antigen/pre-S2 particles (Schodel, 1992); antigens from simian immunodeficiency virus P27 *gag* antigen (Strahan et al., 1992) and HIV (Charbit et al., 1993) have also been expressed in salmonellae, and the HIV-1 gp120 has been expressed in an attenuated strain of *Salmonella typhi* (Fouts et al., 1995). A salmonella expressing influenza virus A nucleoprotein protected from challenge with live virus (Tite et al., 1990). Aro salmonellae expressing an 18-residue epitope of HIV1 gp42 inserted in the flagellar protein elicited antibodies in mice which recognised the peptide and in some cases the recombinant gp160 (Newton et al., 1995). A Phop^{c} *S. typhimurium* strain expressing a recombinant form of the hepatitis B virus core antigen administered by the oral, nasal, rectal and vaginal routes elicited systemic and also secretory immune responses to the recombinant antigen at sites proximal and also distal to the site of immunisation (Hopkins et al., 1995) after a single immunising dose.

A recent report (Franchini et al., 1995) describes highly attenuated HIV type 2 recombinant poxviruses which were able to confer protection from challenge with 100 infectious doses of HIV2 in macaques. Immunisation with an *aroA S. typhimurium* expressing the HIV-2 gag or the gp120 portion of the envelope did not elicit strong immune responses and failed to confer protection.

Salmonella carriers have also proved effective for antigens from parasites, eliciting CD8+ cytotoxic T cells to the circumsporozoite (CSP) antigen from *Plasmodium spp.* and protection from challenge with sporozoites (Aggarwal et al, 1990; Flynn et al., 1990; Sadoff et al., 1988). An *aroA, aroD S. typhimurium* vaccine expressing the *Leishmania major* gp63 protein protects mice from infection following oral immunisation with the vaccine (Xu et al., 1995). Studies in human volunteers have been conducted using strain CVD908 *S. typhi aroC aroD*, the human candidate live oral typhoid vaccine. The strain was engineered to express the *P. falciparum* CSP from the *tac* promoter as a single copy in the salmonella chromosome (González et al., 1994). Volunteers given two doses made antibody to the CSP, and one subject showed cytotoxic CD8+ T cells to the CSP (González et al., 1994).

3.4.9 Expression of Recombinant Antigens as Fusions to Tetanus Toxin Fragment C (TetC) Driven from the Anaerobically Inducible *nirB* Promoter

The 50 kDa C-terminal C fragment of tetanus toxin (TetC) is immunogenic and confers protection against experimental tetanus. Expression of TetC in salmonella vaccines produced an effective combined salmonella-tetanus vaccine which was protective in mice (Fairweather et al., 1990). The vaccines were greatly improved by driving expression of TetC from the anaerobically inducible *E. coli* nitrite reductase *nirB* promoter. The latter has been used in biotechnology for expression of TetC in *E. coli* (Chatfield et al., 1992). The *nirB*-TetC Aro salmonella vaccines were more immunogenic in mice than constructs using other promoters, conferring full protection from challenge with tetanus toxin after a single oral dose of the *Salmonella-nirB* – TetC vaccine (Chatfield et al., 1992). A *nirB*-TetC *S. typhi* vaccine is a candidate human typhoid-tetanus vaccine (Chatfield et al., 1992).

Tetanus toxoid has been frequently used as a carrier for other antigens coupled chemically to it making use of its potent immunogenicity. We have developed the *nirB*-TetC system to allow expression of guest antigens in salmonellae as C-terminal fusions to TetC via a short "hinge" domain, designed to provide spatial and temporal separation between the two proteins to promote correct folding (Khan et al., 1994). The 28 kDa glutathione *S*-transferase from *Schistosoma mansoni* (P28) is a candidate immunogen which confers protection from schistosomiasis in animal models. We expressed P28 as a full length fusion to TetC in Aro *S. typhimurium* from the *nirB* promoter (Khan et al., 1994 in collaboration with A. Capron's group, Lille). The resulting fusion was soluble, of the size expected for a full length fusion, and bound to glutathione, suggesting that P28 had folded correctly. The level of expression of the TetC-P28 fusion in *Salmonella* was higher than that for P28 expressed from the *nirB* promoter on its own, which was not immunogenic for mice.

A single dose of the salmonella-tetanus-P28 construct was immunogenic in mice, eliciting antibody to P28 and TetC. The mice were protected from tetanus and salmonellosis (Khan et al., 1994). A single oral dose of the trivalent vaccine elicited secretory IgA to P28, and more significantly conferred protection from challenge with schistosomula as measured by a reduction in worm burden in the vaccinated mice (A. Capron pers. comm; to be published). The result suggests that a trivalent oral vaccine for typhoid, tetanus and schistosomiasis may be feasible.

3.4.10 Fusions of Repeating Epitopes to TetC

We further developed the *nirB*-TetC system for the expression of defined epitopes as multimeric tandem repeats (repitopes) fused to TetC. Antigens with repeating copies of an epitope can be more immunogenic than single copies; the immunogenicity of a peptide from foot and mouth disease virus increased with increasing copies of the peptide (Broekhuijsen et al., 1987). The 115-131 epitope of P28 confers protection in animal models, especially when constructed as a synthetic octavalent branched "octopus" (Wolowczuk et al., 1991). We expressed the 115-131 epitope as genetic fusions to TetC of tandem repeats of increasing length (1, 2, 4 and 8 copies); their immunogenicity for mice increased with the number of copies of the construct, the monomer eliciting the lowest response and the octamer the highest following a single dose (Khan et al., 1994).

3.4.11 Preimmunisation with Tetanus Toxoid Did Not Decrease Vaccine Efficacy

Preimmunisation with tetanus toxoid has been shown to decrease the response to antigens coupled to it in humans, which could pose a problem with the use of TetC-guest antigen fusions in salmonella vaccines (Schütze et al., 1985; DiJohn et al, 1989). We investigated this possibility in mice immunised with tetanus toxoid in alum and later immunised with the Aro salmonella vaccines expressing the TetC-full length P28 fusion, and also the TetC-octameric peptide fusion described above (Chabalgoity et al., 1995). The immune response to the guest antigens was not suppressed by preimmunisation. The response to the full length P28 was the same in preimmunised and control mice. Preimmunisation with toxoid actually *increased* the response to the peptide (Chabalgoity et al., 1995).

3.4.12 Fusions of TetC and Antigens from Herpes Simplex Virus

We have expressed antigens from herpes simplex glycoprotein D (gD) using the *nirB*-TetC system (Chabalgoity et al., 1996). We found that expression of full length gD was greater in *htrA* salmonellae. The level of expression of fusions of full length gD to TetC is not high, and the fusions were poorly immunogenic and not protective. We have now used this system to express a peptide from gD containing a protective epitope as multiple tandem copies to TetC. These were expressed more effectively in HtrA than in Aro salmonellae. A single dose of the different constructs elicited antibodies to the peptide, with immunogenicity again improving with the fusions consisting of higher numbers of the peptide; these elicited a good antibody response which neutralised the virus in *in vitro* assays. Most significantly, animals were protected from challenge with herpes simplex virus (Chabalgoity et al., 1996).

3.4.13 Cytokines and Immunomodulation

Salmonella vaccines may also be useful for delivering biologically active molecules other than antigens to the host. An *aroA S. typhimurium* expressing human IL-1β conferred protection from radiation in mice (Carrier et al., 1992). (The mice also developed antibodies to the human IL-1). An *aroA S. typhimurium* expressing recombinant murine IL-4 (Denich et al., 1993) persisted longer in the tissues and was more resistant to killing by macrophages than the control strain. Mice administered derivatives of an *aroA aroD S. typhimurium* expressing either murine IL-2, IFNγ or TNFα displayed increased non-specific resistance to infection with *Leishmania major*; simultaneous vaccination with a strain expressing the *L. major* surface gp63 (Yang et al., 1990) conferred complete protection against *L. major* (F. Y. Liew, pers. comm., to be published). An Aro *S. typhimurium* expressing transforming growth factor-beta reduced the inflammation in carrageenin-induced oedema in mice (Ianaro et al., 1995).

An Aro *S. typhimurium* expressing tree pollen antigens was found to elicit a TH1 type of response to these antigens without an IgE response; the authors suggest that it may be possible to decrease or modulate specific IgE responses in vivo (Vrtala et al., 1995).

3.4.14 Influence of the Host Background on the Response to Recombinant Antigens

The possible influences of the host's genetic background on the immune response to guest antigens delivered in live salmonella vaccines remain to be fully investigated. Some hosts can be high or low responders to the target antigens (Tite et al., 1990). The *E. coli* MalE protein was expressed in an *aroA S. typhimurium*; studies on different mouse strains showed that the innate resistance gene *Ity* (Vidal et al., 1993) may modulate the response to the guest antigen, but the major effect on responsiveness was effected by genes linked to the major histocompatibility complex. However, the low responsiveness of some mouse strains could be overridden by increasing the inoculum dose (Fayolle et al., 1994).

3.4.15 Summary and Conclusions

Live attenuated salmonella vaccines have proved themselves as effective antigen delivery systems in experimental models. It has been repeatedly shown by various groups that they can elicit humoral, secretory and cell mediated immune responses to recombinant antigens, and in several cases protection from infection with viruses, bacteria and parasite. Several different methods are now available for the expression of different antigens in salmonella vaccines. As expression is one of the major problems in vaccine construction, it is therefore possible to contemplate the use of this system for a wide variety of antigens. It is possible to express full length proteins or small peptides or epitopes in salmonellae in immunogenic form. The multivaccine potential of the system is amplified by the possibility of simultaneously expressing more than one recombinant antigen in a given salmonella vaccine. A single dose of an experimental combined vaccine protects mice from *Salmonella*, tetanus and schistosomiasis. Human volunteer trials with recombinant *S. typhi* vaccines are currently under way.

References

Aggarwal, A., Kumar, S., Jaffe, R., Hone, D., Gross, M., and Sadoff, J. (1990) Oral *Salmonella*: malaria recombinants induce specific cytotoxic CD8+ cytotoxic T cells. J Exp Med *172*, 1083-1090.

Agteberg, M., Adriaanse, H., Lankhof, H., Meloen, R., and Tommassen, J. (1990) Outer membrane protein PhoE protein of *Escherichia coli* as a carrier for foreign antigenic determinants: immunogenicity of epitopes of foot-and-mouth disease virus. Vaccine *8*, 85-91.

Aldovini, A. and Young, R. A. (1991) Humoral and cell mediated immune responses to live recombinant BCG-HIV vaccines. Nature *351*, 479-482.

Attridge, S. (1991) Oral immunization with *Salmonella typhi* Ty21a-based clones expressing *Vibrio cholerae* O-antigen: serum bactericidal antibody responses in man in relation to pre-immunization antibody levels. Vaccine *9*, 877-882.

Attridge, S. R., Daniels, D., Morona, J. K., and Morona, R. (1990) Surface co-expression of *Vibrio cholerae* and *Salmonella typhi* O-antigens on Ty21a clone EX210. Microb Pathogen *8*, 177-188.

Attridge, S., Hackett, J., Morona, R., and Whyte, P. (1988) Towards a live oral vaccine against enterotoxigenic *Escherichia coli*. Vaccine *6*, 387-390.

Bao, J. X. and Clements, J. D. (1991) Prior immunologic experience potentiates the subsequent antibody response when *Salmonella* strains are used as vaccine carriers. Infect Immun *59*, 3841-3845.

Baron, L. S., Kopecko, D. J., Formal, S. J., Seid, R., Guerry, P., and Powell, C. (1987) Introduction of *Shigella flexneri* 2a type and group antigens genes into oral typhoid vaccine strain *Salmonella typhi* Ty21a. Infect Immun *55*, 2797-2801.

Black, R. E., Levine, M. M., Clements, M. L., Losonsky, G., Herrington, D., Berman, S., and Formal, S. B. (1987) Prevention of shigellosis by a *Salmonella typhi – Shigella sonnei* bivalent vaccine. J Infect Dis *155*, 1260-1265.

Bloom, B. R. and Fine, P. E. M. (1994) The BCG experience: implications for future vaccines against tuberculosis. In: Tuberculosis: pathogenesis, prevention and control. Bloom, B. R (ed.), American Society for Microbiology, Washington, 531-557.

Bowen, J. C., Alpar, O., Philpotts, O., Roberts, I. S., and Brown, M. R. W. (1990) Preliminary studies on infection by attenuated salmonella in guinea pigs and on expression of herpes simplex virus. Res Microbiol *141*, 873-877.

Broekhuijsen, M. P., van Rijn, J. M. M., Blom, A. J. M., Pouwels, P. H., Enger-Valk, B. E., Brown, F., and Francis, M. (1987) Fusions proteins with multiple copies of the major antigenic determinant of foot-and-mouth disease protect both the natural host and laboratory animals. J Gen Virol *68*, 3137-3143.

Brown, A. and Hormaeche, C. E. (1989) The immune response to salmonellae in mice and humans studied by immunoblots and ELISA. Microb Pathogen *6*, 445-454.

Brown, A., Hormaeche, C. E., Demarco de Hormaeche, R. A., Dougan, G., Winther, M., Maskell, D., and Stocker, B. A. D. (1987) An attenuated *aroA S.typhimurium* vaccine elicits humoral and cellular immunity to cloned beta galactosidase in mice. J Infect Dis *155*, 86-92.

Buchmeier, N. A. and Heffron, F. (1990) Induction of salmonella stress proteins upon infection of macrophages. Science *248*, 730-732.

Cao, Y., Wen, Z., and Lu, D. (1992) Construction of a recombinant oral vaccine against *Salmonella typhi* and *Salmonella typhimurium*. Infect Immun *60*, 2823-2827.

Cárdenas, L., Dasgupta, U., and Clements, J. D. (1994) Influence of strain viability and antigen dose on the use of attenuated mutants of *Salmonella* as vaccine carriers. Vaccine *12*, 833-840.

Carrier, M. J., Chatfield, S. N., Dougan, G., Nowicka, U. T., O'Callaghan, D., Beesley, J. E., Milano, S., Cillari, E., and Liew, F. Y. (1992) Expression of human IL-1 beta in *Salmonella typhimurium*: A model system for the delivery of recombinant therapeutic proteins *in vivo*. J. Immunol. *148*, 1176-1181.

Chabalgoity, J. A., Khan, C. M. A., Nash, A. A., and Hormaeche, C. E. (1996) A *Salmonella typhimurium htrA* live vaccine expressing multiple copies of a peptide comprising amino acids 8-23 of herpes simplex virus glycoprotein D as a genetic fusion to tetanus toxin fragment C protects mice from herpes simplex virus infection. Mol Microbiol *19*, 791-801.

Chabalgoity, J. A., Villarreal-Ramos, B., Khan, C. M. A., Chatfield, S. N., Demarco de Hormaeche, R. A., and Hormaeche, C. E. (1995) Influence of preimmunisation with tetanus toxoid on immune responses to tetanus toxin fragment C-guest antigen fusions in a *Salmonella* vaccine carrier. Infect Immun *63*, 2564-2569.

Charbit, A., Martineau, P., Ronco, J., Leclerc, C., Lo-Man, R., Michel, V., O'Callaghan, D., and Hofnung, M. (1993) Expression and immunogenicity of the V3 loop from the envelope of human immunodeficiency virus type 1 in an attenuated *aroA* strain of *Salmonella typhimurium* upon genetic coupling to *Escherichia coli* carrier proteins. Vaccine *12*, 1221-1228.

Chatfield, S. N., Charles, I. G., Makoff, A. J., Oxer, M. D., Dougan, G., Pickard, D., Slater, D., and Fairweather, N. F. (1992) Use of the *nir*B promoter to direct the stable expression of heterologous antigens in *Salmonella* oral vaccine strains: Development of a single-dose oral tetanus vaccine. BioTechnology *10*, 888-892.

Chatfield, S. N., Fairweather, N. F., Charles, I., Pickard, D., Levine, M., Hone, D., Posada, M., Strugnell, R. A., and Dougan, G. (1992) Construction of a genetically defined *Salmonella typhi* Ty2 *aro*A, *aro*C mutant for the engineering of a candidate oral typhoid-tetanus vaccine. Infect Immun *10*,53-60.

Chatfield, S. N., Roberts, M., Dougan, G., Hormaeche, C., and Khan, C. M. A. (1995) The development of oral vaccines against parasitic diseases utilising live attenuated *Salmonella*. Parasitology *10*, S17-S24.

Chatfield, S. N., Strahan, K., Pickard, D., Charles, I. G., Hormaeche, C. E., and Dougan, G. (1991c) Evaluation of *Salmonella typhimurium* strains harboring defined mutations in *htrA* and *aroA* in the murine salmonellosis model. Microb Pathogen *12*, 145-151.

Cieslak, P. R, Tonghai, Z., and Stanley, Jr. S. L. (1993) Expression of a recombinant *Entamoeba histolytica* antigen in a *Salmonella typhimurium* vaccine strain. Vaccine *11*, 773-776.

Clark-Curtiss, J. E., Thole, J. E. R., Sathish, M., Bosecker, B. A., Sela, S., de Carvalho, E. F., and Esser, R. E. (1990) Protein antigens of *Mycobacterium leprae*. Res Microbiol *141*, 859-871.

Clements, J. D. and Cárdenas, L. (1990) Vaccines against enterotoxigenic bacterial pathogens based on hybrid *Salmonella* that express heterologous antigens. Res Microbiol *141*, 981-993.

Clements, J. D., Lyon, F. L., Lowe, K. L., Farrand, A. L., and El-Morshidy, S. (1988) Oral immunization of mice with attenuated *Salmonella enteritidis* containing a recombinant plasmid which codes for production of the B subunit of heat labile enterotoxin of *Escherichia coli*. Infect Immun *53*, 685-692.

Cohen, S., Powell, C. J., Dubois, D. R., Hartman, A., Summers, P. L., and Eckels, K. H. (1990) Expression of the envelope antigen of dengue virus in vaccine strains of *Salmonella*. Res Microbiol *141*, 855-858.

Cooper, G. A., Nicholas, R. A. J., Cullen, G. A., and Hormaeche, C. E. (1990) Vaccination of chickens with an *S.enteritidis aroA* live oral salmonella vaccine. Microb Pathogen *9*, 255-265.

Cooper, G. L., Venables, L. M., Nicholas, R. A. J., Cullen, G. A., and Hormaeche, C. E. (1992) Vaccination of chickens with chicken-derived *Salmonella enteritidis* phage type 4 live oral salmonella vaccines. Vaccine *10*, 247-254.

Cooper, G. L., Venables, L. M., Nicholas, R. A. J., Cullen, G. A., and Hormaeche, C. E. (1993) Further studies of the application of liver *Salmonella enteritidis aroA* vaccines in chickens. Vet Record *133*, 31-36.

Cooper, G. L., Venables, L. M., Woodward, M. J., and Hormaeche, C. E. (1994) The invasiveness and persistence of *S. enteritidis, S. typhimurium* and a genetically-defined *S. enteritidis aroA* strain in young chickens. Infect Immun *62*, 4739-4746.

Cooper, G. L., Venables, L. M., Woodward, M. J., and Hormaeche, C. E. (1994) Vaccination of chickens with CVL30, a genetically defined *Salmonella enteritidis aroA* live oral vaccine. Infect Immun *62*, 4747-4754.

Coulson, N. M., Fulop, M., and Titball, R. W. (1994) *Bacillus anthracis* protective antigen, expressed in *Salmonella typhimurium* SL3261, affords protection against anthrax spore challenge. Vaccine *12*, 1395-1401.

Curtiss III, R., Goldschmidt, R. M., Fletchall, N. B., and Kelly, S. M. (1988) Avirulent *Salmonella typhimurium Dcya Dcrp* oral vaccine strains expressing a streptococcal colonization and virulence antigen. Vaccine *6*, 155-160.

Curtiss III, R., Kelly, S. M., Gulig, P. A., and Nakayama, K. (1989) Selective delivery of antigens by recombinant bacteria. Curr Top Microbiol Immunol *146*, 35-49.

Curtiss, R. I. (1986) Genetic analysis of *Streptococcus mutans* virulence and prospects for an anticaries vaccine. J Dent Res *65*, 1034-1045.

Dearlove, C. E., Forrest, B. D., van den Bosch, L., and LaBrooy, J. T. (1992) The antibody response to an oral Ty21a-based typhoid-cholera hybrid is unaffected by prior oral vaccination with Ty21a (letter). J Infect Dis *165*, 182-183.

Denich, K., Börlin, P., O'Hanley, P., Howard, M., and Heath, A. W. (1993) Expression of the mouse interleukin-4 gene in an attenuated *aroA* strain of *Salmonella typhimurium*: persistence and immune response in BALB/c mice and susceptibility to macrophage killing. Infect Immun *61*, 4818-4827.

DiJohn, D., Wassermann, S. S., Torres, J. R., Cortesía, M. J., Murillo, J., Losonsky, G. A., Herrington, D. A., Stürcher, D., Levine, M. M. (1989) Effect of priming with carrier on response to conjugate vaccine. Lancet 1415-1418.

Dougan, G., Sellwood, R., Maskell, D., Sweeney, K., Beesley, J., Hormaeche, C. E. (1986) *In vivo* properties of a cloned K88 adherence antigen determinant. Infect Immun *52*, 344-347.

Dunne, M., Al-Ramadi, B. K., Barthold, S. W., Flavell, R. A., and Fikrig, E. (1995) Oral vaccination with an attenuated *Salmonella typhimurium* strain expressing *Borrelia burgdorferi* OspA prevent murine Lyme borreliosis. Infect Immun. *63*, 1611-1614.

Dusek, D. M., Progulske-Fox, A., Whitlock, J., and Brown, T. A. (1993) Isolation and characterisation of a cloned *Porphyromonas gingivalis* hemagglutinin from an avirulent strain of *Salmonella typhimurium*. Infect Immun *61*, 940-946.

Ervin, S. E., Small, P. A. Jr., and Gulig, P. A. (1993) Use of incompatible plasmids to control expression of antigen by *Salmonella typhimurium* and analysis of immunogenicity in mice. Microb Pathogen *15*, 93-101.

Everest, P., Frankel, G., Li, J., Lund, P., Chatfield, S., and Dougan, G. (1995) Expression of *lacZ* from the *htrA, nirB* and *groE* promoters in a salmonella vaccine strain: influence of growth in mammalian cells. FEMS Microbiol Letts *125*, 97-101.

Fairweather, N. F., Chatfield, S. N., Makoff, A., Strugnell, R., Bester, J., Maskell, D. J., and Dougan, G. (1990) Oral vaccination of mice against tetanus by use of a live attenuated *Salmonella* carrier. Infect Immun *58*, 1323-1326.

Fayolle, C., O'Callaghan, D., Martineau, P., Charbit, A., Clément, J. M., Hofnung, M., and Leclerc, C. (1994) Genetic control of antibody responses induced against an antigen delivered by recombinant attenuated *Salmonella typhimurium*. Infect Immun *62*, 4310-4319.

Flynn, J. L., Weiss, W. R., Norris, K. A., Seifert, H. S., Kumar, S., and So, M. (1990) Generation of a cytotoxic T-lymphocyte response using a *Salmonella* antigen delivery system. Mol Microbiol *4*, 2111-2118.

Formal, S. B., Baron, L. S., Kopecko, D. J., Washington, O., Powell, C., and Life, C. A. (1981) Construction of a potential bivalent vaccine strain: introduction of *Shigella sonnei* form I antigen genes into into the *galE Salmonella typhi* Ty21a typhoid vaccine strain. Infect Immun *34*, 746-750.

Forrest, B. D., LaBrooy, J. T., Robinson, P., Dearlove, C. E., and Shearman, D. J. (1991) Specific immune response in the human respiratory tract following oral immunization with live typhoid vaccine. Infect Immun *59*, 1206-1209.

Fouts, T. R., Lewis, G. K., and Hone, D. M. (1995) Construction and characterisation of a *Salmonella typhi*-based human immunodeficiency virus type 1 vector vaccine. Vaccine *13*, 561-569.

Franchini, G., Robert-Guroff, M., Tartaglia, J., Aggarwal, A., Abimiku, A., Benson, J., Markham, P., Limbach, K., Hurteau, G., Fullen, J., Aldrick, K., Miller, N., Sadoff, J., Paoletti, E., and Gallo, R. C. (1995) Highly attenuated HIV type 2 recombinant poxviruses, but mot HIV-2 recombinant *Salmonella* vacines, induce long-lasting protection in rhesus macaques. AIDS Research and Human Retroviruses, *8*, 909-920.

Gao, X.-M., Tite, J., Lipscombe, M., Rowland-Jones, S., Ferguson, D. J. P., and McMichael, A. J. (1992) Recombinant *Salmonella typhimurium* strains that invade nonphagocytic cells are resistant to recognition by antigen-specific T-lymphocytes. Infect Immun *60*, 3780-3789.

Gentschev, I., Sokolovic, Z., Kohler, S., Krohne, G. F., Hof, H., Wagner, J., and Goebel, W. (1992) Identification of p60 antibodies in human sera and presentation of this listerial antigen on the surface of attenuated salmonellae by the HlyB-HlyD secretion system. Infect. Immun. *60*, 5091-5098.

Girón, J. A., Xu, J.-G., González, C. R., Hone, D., Kaper, J. B., and Levine, M. M. (1995) Simultaneous expression of CFA/I and CS3 colonisation factor antigens of enterotoxigenic *Escherichia coli* by *DaroC, DaroD Salmonella typhi* vaccine strain CVD 908. Vaccine *13*, 939-946.

González, C., Hone, D., Noriega, F. R., Tacket, C. O., Davis, J. R., Losonsky, G., Nataro, J. P., Hoffman, S., Malik, A., Nardin, E., Sztein, M. B., Heppner, G., Fouts, T. R., Isibasi, A., and Levine, M. M. (1994) *Salmonella typhi* vaccine strain CVD908 expressing the circumsporozoite protein of *Plasmodium falciparum*: strain construction and safety and immunogenicity in humans. J Infect Dis *169*, 927-931.

Gottstein, B., Muller, N., Cryz, S. J. J., Vogel, M., Tanner, I., and Seebeck, T. (1990) Humoral and cellular immune response in mice and dogs induced by a recombinant *Echinococcus multilocularis* antigen produced by a *Salmonella typhimurium* vaccine strain. Paras Immunol *12*, 163-174.

Grosjean, J. and Friers, W. (1982) Preferential codon usage in prokaryotic genes: the optimal codon – anticodon interaction energy and the selective codon usage in efficiently expressed genes. Gene *18*, 199-209.

Guzmán, C. A., Brownlie, R. M., Kadurugamuwa, J., Walker, M. J., and Timmis, K. N. (1991) Antibody responses in the lungs of mice following oral immunization with *Salmonella typhimurium aro*A and invasive *Escherichia coli* strains expressing the filamentous hemagglutinin of *Bordetella pertussis*. Infect Immun *59*, 4391-4397.

Hartman, A. B., Ruiz, M. M., and Shultz, C. L. (1991) Molecular analysis of variant plasmid forms of a bivalent *Salmonella typhi-Shigella sonnei* vaccine strain. J Clin Microbiol *29*, 27-32.

Herrington, D. A., van de Verg, L., Formal, S. B., Hale, T. L., Tall, B. D., Crysz, S. J., Tramont, E. C., and Levine, M. M. (1990) Studies in volunteers to evaluate candidate *Shigella* vaccines: further experience with a bivalent *Salmonella typhi – Shigella sonnei* vaccine and protection conferred by previous *Shigella sonnei* disease. Vaccine *8*, 353-357.

Hess, J., Gentschev, I., Miko, D., Welzel, M., Ladel, Ch. H., Goebel, W., and Kaufmann, S. H. E. (1996) Superior efficacy of secreted over somatic antigen display in vaccine induced protection against an intracellular bacterial pathogen. Proc. Natl. Acad. Sci. USA *93*, 1458-1463.

Hess, J., Gentschev, I., Szalay, G., Ladel, C., Bubert, A., Goebel, W., and Kaufmann, S. H. E. (1995) *Listeria monocytogenes* P60 supports host cell invasion by and *in vivo* survival of attenuated *Salmonella typhimurium*. Infect. Immun. *63*, 2047-2053.

Hohmann, E. L., Oletta, C. A., Loomis, W. P., and Miller, S. I. (1995) Macrophage-inducible expression of a model antigen in *Salmonella typhimurium* enhances immunogenicity. Proc Natl Acad Sci USA *92*, 2904-2908.

Hone, D. M., Attridge, S. R., Forrest, B., Morona, R., Daniels, D., LaBrooy, J. T., Bartholomeusz, R. C. A., Shearman, D. J. C., and Hackett, J. (1988) A *galE* (Vi antigen negative) mutant of *Salmonella typhimurium* Ty2 retains virulence in humans.Infect. Immun. *56*, 1326-1333.

Hone, D., Attridge, S., van den Bosch, L., and Hackett, J. (1988) A chromosomal integration system for stabilization of heterologous genes in *Salmonella* based vaccine strains. Microb Pathogen *5*, 407-418.

Hopkins, S., Kraehenbul, J.-P., Schödel, F., Potts, A., Peterson, D., de Grandi, P., and Nardelli-Haefliger, D. (1995) A recombinant *Salmonella typhimurium* vaccine induces local immunity by four different routes of immunization. Infect Immun *63*, 3279-3286.

Hormaeche, C. E., Khan, C. M. A., Mastroeni, P., Villarreal, B., Dougan, G., Roberts, M., and Chatfield, S. N. (1994) (1995) Salmonella vaccines: mechanisms of immunity and their use as carriers of recombinant antigens. In Ala'Aldeen, D. and Hormaeche, C. E. (eds) Molecular and Clinical Aspects of Vaccine Development. Chichester, John Wiley, 119-153.

Hormaeche, C. E., Villarreal, B., Mastroeni, P., Dougan, G., and Chatfield, S. N. (1993) Immunity mechanisms in experimental salmonellosis. In: Biology of *Salmonella*, Cabello, F., Hormaeche, C. E., Mastroeni, P., and Bonina, L. (eds.) New York, Plenum Press, 223-235.

Hsu, H. S. (1993) Is *Salmonella* an intracellular pathogen? In: Biology of *Salmonella*, Cabello, F., Hormaeche, C. E., Mastroeni, P., Bonina, L. (eds.) New York, Plenum Press, 121-129.

Ianaro, A., Xu, D., O'Donnell, M., Di Rosa, M., and Liew, F. Y. (1995) Expression of TGF-b in attenuated *Salmonella typhimurium*: oral administration leads to the reduction of inflammation, IL-2 and IFN-g, but enhancement of IL-10, in carrageenin-induced oedema in mice. Immunology *84*, 8-15.

Jagusztyn-Krynicka, E. K., Clark-Curtiss, J., and Curtiss III, R. (1993) *Escherichia coli* heat-labile toxin subunit B fusions with *Streptococcus sobrinus* antigens expressed by *Salmonella typhimurium* oral vaccine strains: importance of the linker for antigenicity and biological activities of the hybrid proteins. Infect Immun *61*, 1004-1015.

Johnson, K. S., Charles, I. G., Dougan, G., Miller, I. A., Pickard, D., O'Goara, P., Costa, G., Ali, T., and Hormaeche, C. E. (1991) The role of a stress-response protein in bacterial virulence. Mol Microbiol *5*, 401-407.

Jones, P. W., Dougan, G., Hayward, C., Mackensie, N., Collins, P., and Chatfield, S. N. (1991) Oral vaccination of calves against experimental salmonellosis using a double *aro* mutant of *Salmonella typhimurium*. Vaccine *9*, 29-34.

Kameoka, M., Nishino, Y., Matsuo, K., Ohara, N., Kimura, T., Yamazaki, A., Yamada, T., and Ikuta, K. (1994) Cytotoxic T lymphocyte response in mice induced by a recombinant BCG vaccination which produces an extracellular alpha antigen that fused with the human immunodeficiency virus type 1 envelope immunodominant domain in the V3 loop. Vaccine *12*, 153-158.

Katz, J., Michalek, S. M., Curtiss, R. I., Harmon, C., Richardson, G., and Mestecky, J. (1987) Novel oral vaccines: the effectiveness of cloned gene products on inducing secretory immune responses. Adv Exp Med Biol *216B*, 1741-1747.

Keitel, W. A., Bond, N. L., Zahradnik, J. M., Cramton, T. A., and Robbins, J. B. (1994) Clinical and serologic responses following primary and booster immunisation with *Salmonella typhi* Vi capsular polysacharide vaccines. Vaccine *12*, 195-199.

Khan, C. M. A., Villarreal-Ramos, B., Pierce, R. J., Riveau, G., Demarco de Hormaeche, R., McNeill, H., Ali, T., Chatfield, S., Capron, A., Dougan, G., and Hormaeche, C. E. (1994) The construction, expression and immunogenicity of the *Schistosoma mansoni* P28 glutathione *S*-transferase as a genetic fusion to tetanus toxin fragment C in a live Aro attenuated vaccine strain of *Salmonella*. Proc Natl Acad Sci USA *91*, 11261-11265.

Khan, C. M. A., Villarreal-Ramos, B., Pierce, R. J., Riveau, G., Demarco de Hormaeche, R., McNeill, H., Ali, T., Chatfield, S., Capron, A., Dougan, G., and Hormaeche, C. E. (1994) The construction, expression and immunogenicity of multiple tandem copies of the *Schistosoma mansoni* peptide comprising amino acids 115-121 of the P28 glutathione *S*-transferase, expressed as C-terminal fusions to tetanus toxin fragment C in a live Aro attenuated vaccine strain of *Salmonella*. J Immunol *153*, 5634-5642.

Klemm, P. and Hedegaard, L. (1990) Fimbriae of *Escherichia coli* as carriers of heterologous antigenic sequences. Res Microbiol *141*, 1013-1017.

Lagranderie, M., Murray, A., Gicquel, B., Leclerc, C., and Gheorghiu, M. (1993) Oral immunization with recombinant BCG induces cellular and humoral immune responses against the foreign antigen. Vaccine *11*, 1283-1290.

Langermann, S., Palaszynski, S. R., Burlein, J. E., Koenig, S., Hanson, M. S., Briles, D. E., and Stover, C. K. (1994) Protective humoral response against pneumococcal infection in mice elicited by recombinant Bacille Calmette-Guerin vaccines expressing pneumococcal surface protein A. J Exp Med *180*, 2277-2286.

Langermann, S., Palaszynski, S., Sadziene, A., Stover, C. K., and Koenig, S. (1994) Systemic and mucosal immunity induced by BCG vector expressing outer- surface protein A of *Borrelia burgdorferi*. Nature *372*, 552-555.

Leclerc, C., Charbit, A., Molla, A., and Hofnung, M. (1989) Antibody response to a foreign epitope expressed at the surface of recombinant bacteria: importance of the route of immunization. Vaccine *7*, 242-248.

Lipscombe, M., Charles, I. G., Roberts, M., Dougan, G., Tite, J., and Fairweather, N. (1991) Intranasal immunisation using the B subunit of the *Escherichia coli* heat-labile toxin fused to an epitope of the *Bordetella pertussis* P.69 antigen. Mol Microbiol *5*, 1385-1392.

Maskell, D. J., Sweeney, K. J., O'Callaghan, D., Hormaeche, C. E., Liew, F. Y., and Dougan, G. (1987) *Salmonella typhimurium aroA* mutants as carriers of the *Escherichia coli* heat-labile enterotoxin B subunit to the murine secretory and systemic immune systems. Microb Pathogen *2*, 211-221.

Matsumoto, S., Matsuo, T., Ohara, N., Hotokezaka, H., Naito, M., Minami, J., and Yamada, T. (1995) Cloning and sequencing of a unique antigen MPT70 from *Mycobacterium tuberculosis* H37Rv and expression in BCG using *E. coli*-mycobacteria shuttle vector. Scand J Immunol *41*, 281-287.

Miller, C. G. (1987) Protein degradation and protelolytic modification. In Niedhardt, F. C., Ingraham, J. L., Low, K. B., Magasinik, B., Schaechter, M., and Umbarger, H. E. (eds.) *Escherichia coli* and *Salmonella typhimurium*: cellular and molecular biology. Washington, American Society for Microbiology, 680-691.

Molina, N. C. and Parker, C. D. (1990) Murine antibody response to oral infection with live *aroA* recombinant *Salmonella dublin* vaccine strains expressing filamentous haemagglutinin antigen from *Bordetella pertussis*. Infect Immun *58*, 2523-2528.

Morin, J. E., Lubeck, M. D., Mason, B. B., Molnar-Kimber, K. L., Dheer, S. K., Bhat, B. M., Chanda, P. K., Natuk, R. J., Chengalvala, M. V. R., Mizutani, S., Davis, A. R., and Hung, P. P. (1990) Recombinant adenovirus vaccines for hepatitis B virus. In: New generation vaccines, Woodrow, J. C., Levine, M. M. (eds.), Marcel Dekker, New York, 448-457.

Morona, R., Morona, J. K., Considine, A., Hackett, J. A., van den Bosch, L., Beyer, L., and Attridge, S. R. (1994) Construction of K88- and K99-expressing clones of *Salmonella ty-*

phimurium G30: immunogenicity following oral administration to pigs. Vaccine *12*, 513-517.

Mukkur, T. K. S., McDowell, G. H., Stocker, B. A. D, and Lascelles, A. K. (1987) Protection in salmonellosis in mice and sheep by immunization with aromatic dependent *Salmonella typhimurium*. J Med Microbiol *24*, 11-19.

Murray, A., Winter, N., Lagranderie, M., Hill, D. F., Rauzier, J., Timm, J., Leclerc, C., Moriarty, K. M., Gheorghiu, M., and Gicquel, B. (1992) Expression of *Escherichia coli* beta-galactosidase in *Mycobacterium bovis* BCG using an expression system isolated from Mycobacterium paratuberculosis which induced humoral and cellular immune responses. Mol. Microbiol. *22*, 3331-3342

Nakayama, K., Kelly, S. M., and Curtiss III, R. (1988) Construction of an Asd^+ expression – cloning vector: stable maintenance and high level expression of cloned genes in a salmonella vaccine strain. Bio/Technology *6*, 693-697.

Newton, S. M. C., Jacob, C. O., and Stocker, B. A. D. (1989) Immune response to cholera toxin epitope inserted in *Salmonella* flagellin. Science *244*, 70-72.

Newton, S. M. C., Joys, T. M., Anderson, S. A., Kennedy, R. C., Hovi, M. E., and Stocker, B. A. D. (1995) Expression and immunogenicity of an 18-residue epitope of HIV gp41 inserted in the flagellar protein of a *Salmonella* live vaccine. Res Microbiol *146*, 203-216.

Newton, S. M. C., Klebba, P. E., Hofnung, M., and Charbit, A. (1995) Studies of the anaerobically inducible promoter p*nirB* and the improved expression of bacterial antigens. Res Microbiol *146*, 193-202.

O'Callaghan, D., Charbit, A., Martineau, P., Leclerc, C., van der Werf, S., Nauciel, C., and Hofnung, M. (1990) Immunogenicity of foreign peptide epitopes expressed in bacterial envelope proteins. Res Microbiol *141*, 963-969.

O'Callaghan, D., Maskell, D. J., Beesley, J. E., Lifely, M. R., Roberts, M., Boulnois, G., and Dougan, G. (1988) Characterisation and *in vivo* behaviour of a *Salmonella typhimurium aro*A strain expressing *Escherichia coli* K1 polysaccharide. FEMS Microbiol Letts *52*, 269-274.

O'Donnell, M. A., Aldovini, A., Duda, R. B., Yang, H., Szilvasi, A., Young, R. A., and DeWolf, W. C. (1994) Recombinant *Mycobacterium bovis* BCG secreting functional interleukin- 2 enhances gamma interferon production by splenocytes. Infect Immun *62*, 2508-2514.

O'Gaora, P., Roberts, M., Bowe, F., Hormaeche, C. E., Demarco de Hormaeche, R. A., Cafferkey, M., Tite, J., and Dougan, G. (1990) *Yersinia enterocolitica aroA* mutants as carriers of the B subunit of the *Escherichia coli* heat-labile enterotoxin to the murine immune system. Microb Pathogen, *9*, 105-116.

Oxer, M. D., Bentley, C. M., Doyle, J. G., Peakman, T. C., Charles, I. G., and Makoff, A. J. (1991) High level heterologous expression in *E. coli* using the anaerobically activated *nirB* promoter. Nucl Acids Res *19*, 2889-2892.

Oyston, P. C. F., Williamson, E. D., Leary, S. E. C., Eley, S. M., Griffin, K. F., and Titball, R. W. (1995) Immunisation with live recombinant *Salmonella typhimurium aroA* producing F1 antigen protects against plague. Infect Immun *63*, 563-568.

Pan, Z. K., Ikonomidis, G., Lazenby, A., Pardoll, D., and Paterson, Y. (1995) A recombinant *Listeria monocytogenes* vaccine expressing a model tumour antigen protects mice against lethal tumour challenge and causes regression of established tumours. Nature Medicine *1*, 471-480.

Pastoret, P. P., Boulanger, D., and Brochier, B. (1995) Field trials of a recombinant rabies vaccines. Parasitology, *110*, S37-S42.

Paton, J. C, Morona, J. K, Harrer, S., Hansman, D., and Morona, R. (1993) Immunisation of mice with *Salmonella typhimurium aroA* expressing a genetically toxoided derivative of the pneumococcal toxin pneumolysin. Microb Pathogen *14*, 95-102.

Pfeifer, J. D., Wick, M. J., Roberts, R. L., Findlay, K., Normark, S. J, and Harding, C. (1993) Phagocytic processing of bacterial antigens for presentation to T cells. Nature *361*, 359-362.

Poirier, T. P., Kehoe, M. A., and Beachey, E. H. (1988) Protective immunity evoked by oral administration of attenuated *aroA Salmonella typhimurium* expressing cloned streptococcal M protein. J Exp Med *168*, 25-32.

Redman, T. K., Harmon, C. C., Lallone, R., and Michalek, S. M. (1995) Oral immunisation with recombinant *Salmonela typhimurium* expressing surface protein antigen A of *Streptococcu sobrinus*: dose response and induction of protective humoral responses in rats. Infect Immun *63*, 2004-2011.

Reeves, P. R., Johnson, M. A., Holmes, J. H., and Dyall-Smith, M. L. (1991) Expression of rotavirus VP7 antigens in fusions with bacterial proteins. Res Microbiol *141*, 1019-1025.

Roberts, M., Chatfield, S. N., and Dougan, G. (1994) Salmonella as carriers of heterologous antigens. In O'Hagan, D. T. (ed.) Novel delivery systems for oral vaccines. London, CRC Press, 27-58.

Robertsson, J. A., Lindberg, A. A., Hoiseth, S., and Stocker, B. A. D. (1983) *Salmonella typhimurium* infection in calves: protection and survival of virulent challenge bacteria after immunisation with live or inactivated vaccines. Infect Immun *41*, 742-757.

Sadoff, J. C., Balloul, W. R., Baron, L. S., Majarian, W. R., Brey, R. N., Hockmeyer, W. T., Young, J., Crysz, J. J., Ou, J., Lowell, G. J., and Chulay, J. D. (1988) Oral *Salmonella typhimurium* vaccine expressing circumsporozoite protein protects against malaria. Science *240*, 336-338.

Salas, V. E., Plebanski, M., Castro, S., Perales, G., Mata, E., López, S., Arias, C. F. (1990) Synthesis of the surface glycoprotein of rotavirus SAL1 in the *aro*A strain of *Salmonella typhimurium* SL3261. Res Microbiol *141*, 883-886.

Schafer, R., Portnoy, D. R., Brassell, S. A., and Paterson, Y. (1992) Induction of a cellular immune response to a foreign antigen by a recombinant *Listeria monocytogenes* vaccine. J Immunol *149*, 53-59.

Schodel, F. (1992) Prospects for oral vaccination using recombinant bacteria expressing viral epitopes. Adv Virus Res *41*, 409-460.

Schodel, F. and Will, H. (1990) Expression of hepatitis B virus antigens in attenuated salmonellae for oral immunisation. Res Microbiol *141*, 831-837.

Schodel, F., Milich, D. R., and Will, H. (1990) Hepatitis B virus nucleocapsid / pre-S2 fusion proteins expressed in attenuated salmonellae for oral vaccination. J Immunol *145*, 4317-4321.

Schödel, F., Will, H., and Milich, D. R. (1991) Hybrid hepatitis-B virus core/pre-S particles expressed in live attenuated salmonellae for oral immunization In: Vaccines 91: modern approaches to new vaccines including prevention of AIDS. Chanock, R. M., Ginsberg, H. S., Brown, F., and Lerner, A. (eds), New York: Cold Spring Harbor Laboratory Press, 319-325.

Schorr, J., Knapp, B., Hundt, E., Küpper, H. A., and Amann, E. (1991) Surface expression of malaria antigens in *Salmonella typhimurium*: induction of serum antibody response upon oral vaccination of mice. Vaccine *9*, 675-681.

Schütze, M.-P., Leclerc, C., Jolivet, M., Audibert, F., and Chedid, L. (1985) Carrier-induced suppression, a major issue for future synthetic vaccines. J Immunol *135*, 1354-1423.

Segall, T. and Lindberg, A. A. (1991) *Salmonella dublin* experimetal infection in calves: protection after oral immunisation with an auxotrophic *aroA* live vaccine. J Vet Med *38*, 142-160.

Seid, R. C. J., Kopecko, D. J., Sadoff, J. C., Schneider, H., Baron, L. S., and Formal, S. B. (1984) Unusual lipopolysaccharide antigens of a *Salmonella typhi* oral vaccine strain expressing the *Shigella sonnei* form I antigen. J Biol Chem *259*, 9028-9034.

Shen, H., Slifka, M. K., Matloubian, M., Jensen, E. R., Ahmed, R., and Miller, J. F. (1995) Recombinant *Listeria monocytogenes* as a live vaccine vehicle for the induction of protective anti-viral cell-mediated immunity. Proc Natl Acad Sci USA *92*, 3987-3991.

Simonet, M., Fortineau, N., Beretti, J. L., and Berche, P. (1994) Immunisation with live *aroA* recombinant *Salmonella typhimurium* producing invasin inhibits intestinal translocation of *Yersinia pseudotuberculosis*. Infect Immun *62*, 863-867.

Sjösted, A., Sandström, G., and Tärnvik, A. (1990) Immunization of mice with an attenuated *Salmonella typhimurium* strain expressing a membrane protein of *Francisella tularensis*: a model for identification of bacterial determinants relevant to the host defence against tularemia. Res Microbiol *141*, 887-891.

Sjöstedt, A., Sandström G., and Tärnvik, A. (1992) Humoral and cell-mediated immunity in mice to a 17-kilodalton lipoprotein of *Francisella tularensis* expressed by *Salmonella typhimurium*. Infect Immun *60*, 2855-2862.

Smith, B. P., Reina-Guerra, M., Hoiseth, S. K., Stocker, B. A. D., Habasha, F., Johnson, E., and Merritt, F. (1984) Aromatic-dependent *Salmonella typhimurium* as modified live vaccines for calves. Am J Vet Res *45*, 59-66.

Smith, B. P., Reina-Guerra, M., Stocker, B. A. D., Hoiseth, S. K., and Johnson, E. (1984) Aromatic-dependent *Salmonella dublin* as a parenteral modified live vaccine for calves. Am J Vet Res *45*, 2231-2235.

Sory, M.-P. and Cornelis, G. R. (1990) Delivery of cholera toxin B subunit by using a recombinant *Yersinia enterocolitica* strain as a live oral vector. Res Microbiol, *141*, 921-931.

Stabel, T. J., Mayfield, J. E., Morfitt, D. C., and Wannemuehler, M. J. (1993) Oral immunization of mice and swine with an attenuated *Salmonella choleraesuis* [*cya*-12 (*crp-cdt*)19] mutant containing a recombinant plasmid. Infect Immun *61*, 610-618.

Stabel, T. J., Mayfield, J. E., Tabatabai, L. B, and Wannemuehler, M. J. (1991) Swine immunity to an attenuated *Salmonella typhimurium* mutant containing a recombinant plasmid which codes for production of a 31 kilodalton protein of *Brucella abortus*. Infect Immun. *59*, 2941-2947.

Stabel, T. J., Mayfield, J. E., Tabatabai, L. B., and Wannemuehler, M. J. (1990) Oral immunization of mice with attenuated *Salmonella typhimurium* containing a recombinant plasmid which codes for production of a 31-kilodalton protein of *Brucella abortus*. Infect Immun *58*, 2048-2055.

Stevenson, G. and Manning, P. (1985) Galactose epimeraseless (*galE*) mutant G30 of *Salmonella typhimurium* is a good potential live oral vaccine carrier for fimbrial antigens. FEMS Microbiology Letters *28*, 317-321.

Stocker, B. A. D. (1990) Aromatic-dependent *Salmonella* as live vaccine presenters of foreign inserts in flagellin. Res Microbiol *141*, 787-796.

Stocker, B. A. D. and Newton, S. M. (1994) Immune responses to epitopes inserted in salmonella flagellin. Intern Rev Immunol *11*, 167-178.

Stover, C. K. (1994) Recombinant vaccine delivery systems and encoded vaccines. Curr Opin Immunol *6*, 568-571.

Stover, C. K., Bansal, G. P., Hanson, M. S., Burlein, J. E., Palaszynski, S. R., Young, J. F., Koenig, S., Young, D. B., Sadziene, A., and Barbour, A. G. (1993) Protective immunity elicited by recombinant bacille Calmette-Guerin (BCG) expressing outer surface protein A (OspA) lipoprotein: A candidate Lyme disease vaccine. J Exp Med *178*, 197-209.

Stover, C. K., De la Cruz, V. F., Bansal, G. P., Hanson, M. S., Fuerst, T. R., Jacobs, W. R. Jr., and Bloom, B. R. (1992) Use of recombinant BCG as a vaccine delivery vehicle. Adv Exp Med Biol *327*, 175-182.

Strahan, K. M., Kitchin, P., and Hormaeche, C. E. (1992) SIV-*Salmonella* constructs and their potential as vaccine candidates. Vaccine Res *1*, 257-263.

Strahan, K., Chatfield, S. N., Tite, J., Dougan, G., and Hormaeche, C. E. (1992) Impaired resistance to infection does not increase the virulence of *Salmonella htrA* live vaccines. Microb Pathogen *12*, 311-317.

Strugnell, R. A., Maskell, D., Fairweather, N. F., Pickard, D., Cockayne, A., Penn, C., and Dougan, G. (1990) Stable expression of foreign antigens from the chromosome of *Salmonella typhimurium* vaccine strains. Gene *88*, 57-63.

Strugnell, R., Dougan, G., Chatfield, S., Charles, I., Fairweather, N., Tite, J., Li, J. L., Beesley, J., and Roberts, M. (1992) Characterization of a *Salmonella typhimurium aro* vaccine strain expressing the P.69 antigen of *Bordetella pertussis*. Infect Immun *60*, 3994-4002.

Strugnell, R., Schouls, L., Cockayne, A., Bailey, M., van Embden, J., and Penn, C. (1989) Experimental syphilis vaccines: use of *aro*A *Salmonella typhimurium* to deliver recombinant *Treponema pallidum* antigens. In: Meheus, A. and Spier, R. E. (eds.), Vaccines for Sexually Transmitted Diseases. Oxford, Butterworths, 107-113.

Su, G. F., Brahmbhatt, H. N., Wehland, J., Rohde, M., and Timmis, K. N. (1992) Construction of stable LamB-Shiga toxin B subunit hybrids: analysis of expression in *Salmonella typhimurium aro*A strains and stimulation of B subunit-specific mucosal and serum antibody responses. Infect Immun *60*, 3345-3359.

Tacket, C. O. and Levine, M. M. (1994) Typhoid vaccines – old and new. In Ala'Aldeen, D. and Hormaeche, C. E. (eds) Molecular and Clinical Aspects of Vaccine Development. Chichester: John Wiley, 156-178.

Tacket, C. O., Forrest, B., Morona, R., Attridge, S. R., LaBrooy, J., Tall, B. D., Reymann, M., Rowley, D., and Levine, M. M. (1990) Safety, immunogenicity and efficacy against cholera challenge in humans of a typhoid – cholera hybrid vaccine derived from *Salmonella typhi* Ty21a. Infect Immun *58*, 1620-1627.

Tarkka, E., Muotiala, A., Karvonen, M., Saukkonen-Laitinen, K., and Sarvas, M. (1989) Antibody production to a meningococcal outer membrane protein cloned into a live *Salmonella typhimurium aroA* vaccine strain. Microb Pathogen *6*, 327-335.

Tijhaar, E. J., Zheng-Xin, Y., Karlas, J. A., Meyer, T. F., Stukart, M., Osterhaus, A. D. M. E., and Mooi, F. R. (1994) Construction and evaluation of an expression vector allowing the stable expression of foreign antigens in a *Salmonella typhimurium* vaccine strain. Vaccine *12*, 1004-1011.

Tite, J. P., Gao, X.-M., Hughes-Jenkins, C. M., Lipscombe, M., O'Callaghan, D., Dougan, G., and Liew, F.-Y. (1990) Anti-viral immunity induced by recombinant nucleoprotein of influenza A virus. III. Delivery of recombinant nucleoprotein to the immune system using attenuated *Salmonella typhimurium* as a live carrier. Immunology *70*, 540-546.

Tramont, E. C., Chung, R., Berman, S., Keren, D., Kapfer, C., and Formal, S. B. (1984) Safety and antigenicity of typhoid-*Shigella sonnei* vaccine (strain 5076-1C). J Infect Dis *149*, 133-136.

Turner, S. J., Carbone, F. R., and Strugnell, R. A. (1993) *Salmonella typhimurium* Δ*aroA* Δ*aroD* mutants expressing a foreign recombinant protein induce specific major histocompatibility complex class I-restricted cytotoxic T-lymphocytes in mice. Infect Immun *61*, 5374-5380.

Van de Verg, L., Herrington, D. A., Murphy, J. R., Wasserman, S. S., Formal, S. B., and Levine, M. M. (1990) Specific immunoglobulin A-secreting cells in peripheral blood of humans following oral immunization with a bivalent *Salmonella typhi-Shigella sonnei* vaccine or infection by pathogenic *S.sonnei*. Infect Immun *58*, 2002-2004.

Verma, N. K., Zeigler, H. K., Stocker, B. A. D., and Schoolnik, G. K. (1995) Induction of a cellular immune response to a defined T cell epitope as an insert in the flagellin of a live vaccine strain of *Salmonella*. Vaccine *13*, 235-244.

Verma, N. K., Zeigler, H. K., Wilson, M., Khan, M., Safley, S., Stocker, B. A. D., and Schoolnikm G. K. (1995) Delivery of a class I and class II MHC restricted T cell epitopes of listeriolysin of *Listeria monocytogenes* by attenuated *Salmonella*. Vaccine *13*, 142-150.

Vidal, S. M., Malo, D., Vogan, K., Skamene, E., and Gros, P. (1993) Natural resistance to infection by intracellular parasites: isolation of a candidate for *Bcg*. Cell *73*, 469-485.

Vrtala, S., Grote, M., Ferreira, F., Susani, M., Stocker, B., Kraft, D., and Valenta, R. (1995) Humoral immune responss to recombinant tree pollen allergens (*Bet v I* and *Bet v II*) in mice: construction of a live oral allergy vaccine. Int Arc Allergy Immunol *107*, 290-294.

Walker, M. J., Rohde, M., Timmis, K. N., and Guzman, C. A. (1992) Specific lung mucosal and systemic immune responses after oral immunization of mice with *Salmonella typhimurium aro*A, *Salmonella typhi* Ty21a, and invasive *Escherichia coli* expressing recombinant pertussis toxin S1 subunit. Infect Immun *60*, 4260-4268.

Wells, J. M., Wilson, P. W., Norton, P. M., Casson, M. J., and Le Page, R. W. F. (1993) *Lactococcus lactis*: high level expression of tetanus toxin fragment C and protection against lethal challenge. Mol Microbiol *8*, 1155-1162.

Wick, M. J., Harding, C. V., Twesten, N. J., Normark, S. J., and Pfeifer, J. D. (1995) The *phoP* locus influences processing and presentation of *Salmonella typhimurium* antigens by activated macrophages. Mol Microbiol *16*, 465-476.

Winter, N., Lagranderie, M., Gangloff, S., Leclerc, C., Gheorghiu, M., and Gicquel, B. (1995) Recombinant BCG strains expressing the SIV(mac251)nef gene induces proliferative and CTL responses against nef synthetic peptides in mice. Vaccine *13*, 471-478.

Winter, N., Lagranderie, M., Rauzier, J., Timm, J., Leclerc, C., Guy, B. et al. (1991) Expression of heterologous genes in *Mycobacterium tuberculosis* BCG: induction of a cellular response against HIV-1 Nef protein. Gene *109*, 47-54

Wolowczuk, I., Auriault, C., Bossus, M., Boulanger, D., Gras-Masse, H., Mazingue, C., Pierce, R. J., Grezel, D., Reid, G. D., Tartar, A., and Capron, M. (1991) Antigenicity and immunogenicity of a multiple peptidic construction of the *Schistosoma mansoni* SM28 GST antigen in rat, mouse and monkey: 1: partial protection of Fischer rats after active immunisation. J Immunol *146*, 1987-1995.

Woodrow, G. C. and Levine, M. M. (eds.) (1990) New generation vaccines. Marcel Dekker, New York.

Wu, J. Y., Newton, S. M. C., Judd, A., Stocker, B. A. D., and Robinson, W. S. (1989) Expression of immunogenic epitopes of hepatitis B surface antigen with hybrid flagellin proteins by a vaccine strain of *Salmonella*. Proc Natl Acad Sci USA *86*, 4726-4730.

Xu, D., McSorley, S. J., Chatfield, S. N., Dougan, G., and Liew, F. Y. (1995) Protection against *Leishmania major* infection in genetically susceptible BALB/c mice by GP63 delivered orally in attenuated *Salmonella typhimurium* ($AroA^-$ $AroD^-$) Immunology *85*, 1-7.

Yang, D. M., Fairweather, N., Button, L. L., McMaster, W. R., Kahl, L. P., and Liew, F. Y. (1990) Oral *Salmonella typhimurium* (AroA-) vaccine expressing a major leishmanial surface protein (gp63) preferentially induces T helper 1 cells and protective immunity against leishmaniasis. J. Immunol. *145*, 2281-2285.

Yasutomi, Y., Koenig, S., Haun, S. S., Stover, C. K., Jackson, R. K., Conard, P., Conley, A. J., Emini, E. A., Fuerst, T. R., and Letvin, N. L. (1993) Immunization with recombinant BCG-SIV elicits SIV-specific cytotoxic T lymphocytes in rhesus monkeys. J Immunol *150*, 3101-3107.

3.5 Genetic Detoxification of Bacterial Toxins

Rino Rappuoli and Mariagrazia Pizza

3.5.1 Introduction

Serendipity often provokes discoveries and technological breakthroughs. A milestone in vaccine development was the discovery by Glenny and Ramon, that formalin treatment of a culture supernatant of *Corynebacterium diphtheriae*, renders it unable to kill guinea pigs, that became resistant to subsequent challenges with diphtheria toxin (Ramon, 1924; Glenny and Hopkins, 1923). This observation made feasible the large scale detoxification of diphtheria and tetanus toxins and the introduction of mass vaccination against these diseases. Subsequently, the use of formaldehyde or other chemical agents to inactivate bacteria, viruses and other toxins has allowed their use in vaccine development. Today, many vaccines are still produced using exactly the same procedure of formaldehyde detoxification described by Ramon and Glenny in the early twenties.

Widely used vaccines based on chemically detoxified components are diphtheria, tetanus, whole cell pertussis, inactivated polio, influenza, and many others containing killed bacteria or viruses. Even vaccines developed in the last decade have been obtained using chemical treatment to eliminate the toxicity of bacterial toxins. The best examples of this are the acellular vaccines against pertussis that have been developed during the period spanning 1975-1990. In this case, formalin and many other chemical agents have been used to inactivate pure, or partially purified, pertussis toxin (Rappuoli et al., 1992; Moxon and Rappuoli, 1990).

A new way to inactivate bacterial toxins for vaccine development was discovered in 1971 by Uchida and Pappenheimer who, following random mutagenesis of the bacteriophage coding for diphtheria toxin, isolated a phage that coded for a non toxic protein that was immunologically indistinguishable from diphtheria toxin (Cross reacting Material or CRM). This protein was later found to differ from the wild type toxin only for one aminoacid located in the active site of the toxin (Glycine 52 had been replaced by Glutamic acid) (Uchida et al., 1971; Giannini et al., 1984). The advent of new technologies for gene manipulation and site-directed mutagenesis have made possible the rational construction of non toxic derivatives of bacterial toxins. The first example of this novel approach has been the construction of a non toxic derivative of pertussis toxin, used to develop a third generation acellular pertussis vaccine, that is already in use for infant immunization (Pizza et al., 1989). In this chapter we will describe the genetic construction of non toxic derivatives of pertussis, heat-labile and cholera toxins, highlighting the advantages of the genetic detoxification versus the chemical detoxification.

3.5.2 Genetic Detoxification of Pertussis Toxin, Cholera Toxin and Heat-Labile Toxin

3.5.2.1 Structure of the Proteins

Pertussis toxin (PT), cholera toxin (CT), and heat-labile toxin (LT) are proteins formed by two functionally distinct subunits, named A and B, that are held together by non covalent bonds (Rappuoli and Pizza, 1991). A is responsible for the toxicity, while B is a non toxic oligomer with receptor binding activity. A schematic structure of the three proteins is shown in figure 3.5.1.

In the case of PT, the A protomer is formed by the subunit S1, an enzymatic protein of 26.220 daltons with ADP-ribosyltransferase activity. The S1 subunit binds NAD and transfers the ADP-ribose group to a family of GTP-binding proteins, such as G_i and G_o, that are involved in signal transduction in eukaryotic cells (Linder and Gilman, 1992; Hepler and Gilman, 1992), thus altering their response to extracellular stimuli (Bokoch et al., 1983; Katada and Ui, 1982; Rappuoli and Pizza, 1991). B is a non toxic oligomer formed by four distinct subunits, named S2, S3, S4, and S5, of 21.920, 21.860, 12.060 and 11.770 daltons respectively, that are present in a 1:1:2:1 ratio (Tamura et al., 1982). The B oligomer binds the receptor on the surface of eukaryotic cells and facilitates the translocation of the enzymatically active subunit across the cell membrane so that it can reach the target G proteins. The genes coding for the five subunits of pertussis toxin are clustered in a 3.2 Kilobase fragment of the chromosomal DNA and have the typical organization of a bacterial operon, that was sequenced by two different research groups in 1986 (Nicosia et al., 1986; Locht and Keith, 1986). Downstream of the PT operon there is another operon coding for the secretion apparatus of PT (Covacci and Rappuoli, 1993; Weiss et al., 1993). The crystal structure of the molecule has also been recently solved (Stein et al., 1994).

In CT and LT, the A subunit is formed by a 239 polypeptide chain bearing ADP-ribosyltransferase activity that, as in the case of pertussis toxin, binds NAD and transfers the ADP-ribose group to GTP-binding proteins involved in transmembrane signaling. The main target of CT and LT is G_s, a protein that activates the adenylate cyclase thus inducing the synthesis of the cAMP second messanger (Linder and Gilman, 1992; Hepler and Gilman, 1992). The B oligomer is formed by five identical sub-

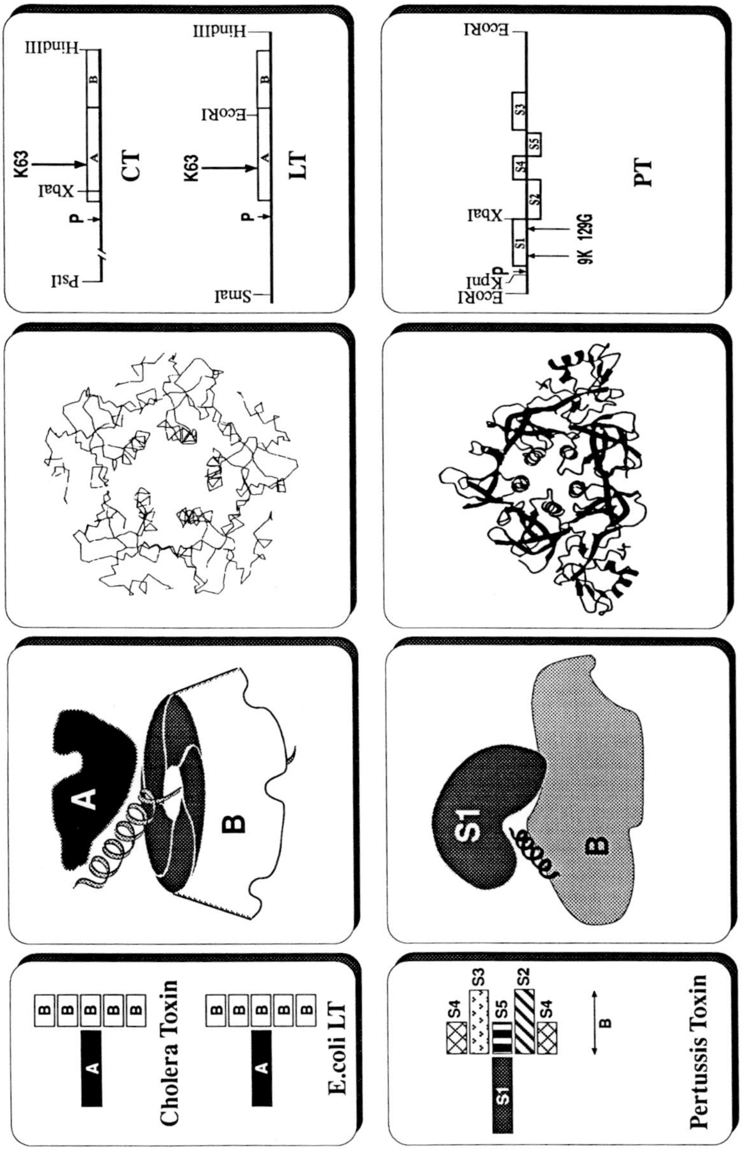

Fig. 3.5.1: Schematic structure of CT, LT and PT. From left to right: schematic representation of subunits composition, three-dimensional structure, alpha-carbon trace of the B subunits and structure of the toxin genes, indicating the aminoacids that have been mutagenized to obtain non toxic derivatives.

units of 103 aminoacids that assemble into a pentameric structure; this structure binds the GM1 receptor ganglioside. LT, in addition to GM1, binds also other receptors containing a terminal galactosyl moiety. The genes coding for CT and LT are highly homologous and are organized in a 1.6 Kb operon, located on the chromosome of *Vibrio cholerae* and on a plasmid of *Escherichia coli*. The aminoacid sequences of both toxins and the nucleotide sequences of their genes are known (Dallas and Falkow, 1980; Spicer et al., 1981; Mekalanos et al., 1983). A review of the different versions of the primary aminoacid sequences of these toxins has been recently published (Domenighini et al., 1995). The crystal structure of LT, with and without the bound galactose, has been solved (Sixma et al., 1993; Sixma et al., 1992; Sixma et al., 1991).

3.5.2.2 The Active Site of ADP-Ribosylating Toxins

The crystal structure of ADP-ribosylating toxins (Sixma et al., 1991; Stein et al., 1994; Allured et al., 1986; Choe et al., 1992) has shown that, in spite of the absence of obvious similarities in the primary structure, the active site of these toxins is remarkably conserved (Domenighini et al., 1994; Domenighini et al., 1991). PT, CT, LT, diphtheria toxin (DT) and pseudomonas exotoxin A (PAETA), share a common structure of the NAD-binding and catalytic site. This can be described as a cavity formed by an alpha-helix bent over a beta-strand that form the ceiling and the floor of the NAD-binding cavity, respectively (fig. 3.5.2). Two aminoacids that are essential for catalysis are conserved in all toxins and are located in the same position, at the two sides of the cavity. These aminoacids are a Glutamic acid that is common to all toxins and that had been shown to be essential for catalysis by biochemical and genetic studies before the structure was known (Carroll and Collier, 1984; Douglas and Collier, 1987; Pizza et al., 1988; Barbieri et al., 1990); and a second residue that can be either an histidine (Papini et al., 1989) in DT and PAETA, or an arginine in PT, CT and LT (Burnette et al., 1988; Burnette et al., 1991; Lobet et al., 1991; Pizza et al., 1994a).

3.5.3 Construction of the PT-9K/129G Non Toxic Derivative of PT

The mutagenesis of the pertussis toxin S1 gene started immediately after the sequence of the gene became available, long before the structure of the active site was known. Several groups started to mutagenize the S1 gene, to express the mutant genes in *E. coli* or *Bordetella pertussis*, and to check the enzymatic activity of the mutant proteins. After extensive mutagenesis studies, three groups identified aminoacid substitutions that reduced considerably the enzymatic activity without changing the structure of the molecules. The aminoacid mutations identified were Arg9→Lys (Burnette et al., 1988), Asp11→Ser, Asp13→Ser (Barbieri and Cortina, 1988) and Glu129→Gly (Pizza et al., 1988). The above mutations were then introduced by allelic exchange (Stibitz et al., 1986) into the chromosome of *B. pertussis* containing a deletion of the wild type PT gene. A schematic representation of the process used to replace the wild type PT gene with the mutated PT gene is shown in figure 3.5.3. The mutant *B. pertussis* strains obtained by this process produced proteins that had 10-, 100- or even 1000-fold reduced enzymatic activity and toxicity. Howev-

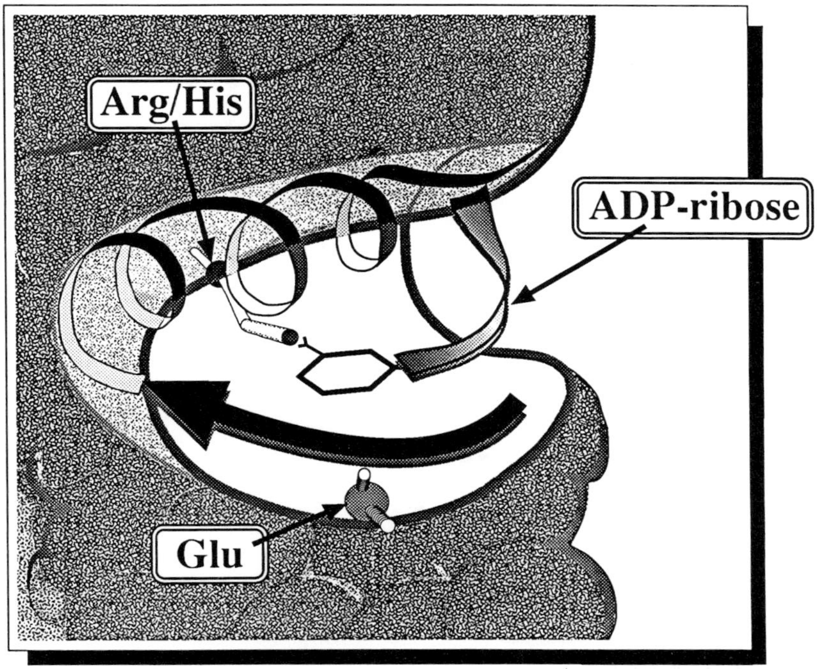

Fig. 3.5.2: Schematic drawing of the active site indicating the aminoacids important for catalysis and the supposed position of the NAD molecule.

er, none of the mutant PT molecules containing a single aminoacid mutation had a toxicity which was low enough to be considered safe for inclusion in vaccines. To further reduce the toxicity, two or more of the above mutations were combined and the mutated genes were then transferred to *B. pertussis* to obtain strains producing proteins containing two or three aminoacid substitutions. Some of the double mutants showed complete loss of the enzymatic activity and a complete conservation of the molecule structure. Among the mutants obtained, the best one turned out to be PT-9K/129G, a mutant containing two mutations, one in position 9 (Arg9→Lys) and the other in position 129 (Glu129→Gly) (Pizza et al., 1989). Remarkably, when the 3D structure of PT became available, the two mutated aminoacids in the PT-9K/129G mutant were found to be located in the NAD binding site and to be the two aminoacids important for catalysis. These residues are shown in figure 3.5.2.

3.5.3.1 Clinical Development Shows the Superior Immunogenicity of PT-9K/129G

The complete characterization of the PT double mutant showed that it was at least a million times less toxic than wild type PT, and that it had maintained intact all the other activities of the pertussis toxin, such as receptor binding, hemagglutination,

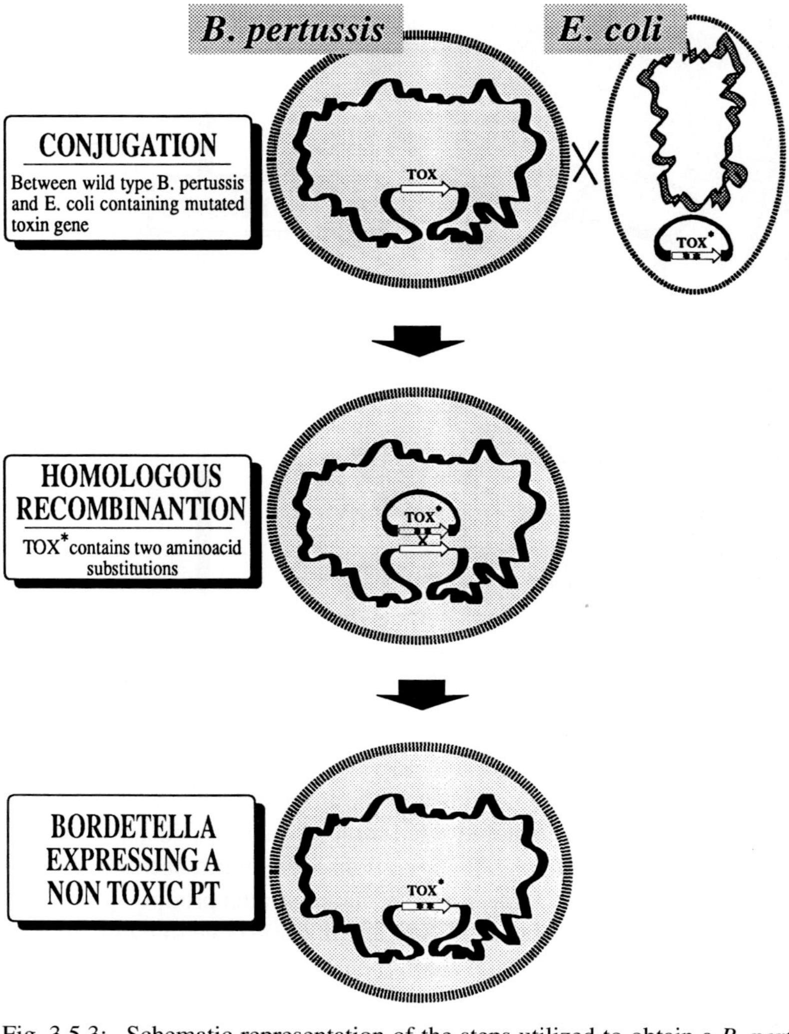

Fig. 3.5.3: Schematic representation of the steps utilized to obtain a *B. pertussis* strain producing a non toxic PT.

binding to protective polyclonal and monoclonal antibodies and T cell mitogenicity. Moreover, no toxic activity was detectable both *in vitro* and *in vivo* (tab. 3.5.1).

After showing the complete absence of toxicity, PT-9K/129G was tested for its immunogenicity in animal models. The mutant PT molecule when used to immunize mice and guinea pigs, induced high titers of anti PT antibodies with a high toxin-neutralizing activity evaluated on CHO cells. Finally, the ability of PT-9K/129G to protect mice from death in the intracerebral challenge with virulent *B. pertussis* was tested. This model had been shown to correlate with vaccine efficacy in infants. As

Table 3.5.1: Toxic and non toxic properties of PT and PT-9K/129G mutant

Property	Native PT	PT-9K/129G Mutant	Reference
Toxic properties of PT			
CHO cell-clustered growth (ng/ml)	0.005	>5000[a]	Pizza et al., 1989
Histamine-sensitivization (µg/mouse)	0.1–0.5	>50[a]	Nencioni et al., 1990
Leukocytosis stimulation (µg/mouse)	0.02	>50[a]	Nencioni et al., 1990
Anaphylaxis potentiation (µg/mouse)	0.04	>7.5[a]	Nencioni et al., 1990
Enhanced insulin secretion (µg/mouse)	<1	>25[a]	Nencioni et al., 1990
IgE induction (*in vitro*) (ng/ml)	0.8	>100[a]	van der Pouw-Kraan et al., 1995
IgE induction (*in vitro*) (ng/rat)	10	>200[a]	Kosecka et al., 1994
Long-lasting enhancement of nerve-mediated intestinal permeabilization of antigen uptake (ng/rat)	1	>200[a]	Kosecka et al., 1994
Inhibition of IL1-induced IL2 release in EL4 6.1 cells (µg/ml)	0.1	>100[a]	Zumbihl et al., 1995
Lethal dose (µg/kg)	15	>1500	Rappuoli, unpublished data
ADP-ribosylation (ng)	1	>20.000	Pizza et al., 1989
Non toxic properties of PT			
T cell mitogenicity (µg/ml)	0.1–0.3	0.1–0.3	Nencioni et al., 1991
Hemagglutination (µg/well)	0.1–0.5	0.1–0.5	Nencioni et al., 1991
Mitogenicity for PT-specific T cells (µg/ml)	3	3	Pizza et al., 1989
Platelet activation (µg/ml)	5	5	Sindt et al., 1994
Mucosal adjuvanicity (µg/mouse)	3	3	Roberts et al., 1995
Affinity constant (monoclonal 1B7, anti-S1) [K([b]L/mol)]	2.4×10^8	6.1×10^8	Nencioni et al., 1990
Affinity constant (polyclonal anti-PT) [K([b]L/mol)]	2.0×10^{10}	9.8×10^9	Nencioni et al., 1991

[a] Means that no effect was observed at the highest dose reported that was used in the assay
[b] L = liters

shown in table 3.5.2, the protection was dose-dependent (Pizza et al., 1989; Nencioni et al., 1990; Nencioni et al., 1991).

The results obtained in animal models encouraged the clinical development of a vaccine containing this molecule. Phase I clinical trials using the purified PT-9K/129G molecule alone, or in combination with other pertussis antigens, such as Filamentous Hemagglutinin (FHA) and Pertactin (69K), were performed. The results

Table 3.5.2: Protection of mice from intracerebral challenge with virulent *B. pertussis* strain, using a standard whole cell pertussis vaccine provided by NIH and purified PT-9K/129G.

NIH standard cellular vaccine		Acellular vaccine containing PT-9K/129G	
Dose (ml)	Survivors	Dose (μg)	Survivors
0.04	15/16	30	16/16
0.008	13/16	12	16/16
0.0010	9/16	4.8	12/16
0.00032	2/16	1.92	10/16
		0.77	7/16
		0.25	3/16

confirmed what had been previously shown in animal models: PT-9K/129G was safe and very immunogenic (Podda et al., 1990; Podda et al., 1991; Podda et al., 1993; Podda et al., 1994). The high immunogenicity allowed the formulation of a DTP vaccine which contained a very low amount of antigens: 5 μg of PT-9K/129G, 2.5 μg of FHA and 2.5 μg of 69K, plus Diphtheria and Tetanus toxoids. The new vaccine was tested in many phase II studies in Italy and the USA, giving consistently very good immunogenicity. One of the trials performed in the United States by the National Institute of Allergy and Infectious Diseases (NIAID), offered the unique opportunity to compare the safety and the immunogenicity of two vaccine formulations containing the genetically detoxified PT (one formulation with 10 μg of mutant PT alone, the other with 5 μg of mutant PT, 2.5 μg of FHA and 2.5 μg of 69K), with eleven vaccines containing chemically detoxified PT (most of these vaccines contained at least 25 μg of chemically detoxified PT and 25 μg of FHA) (Edwards, 1993). Three doses of each vaccine were given to 120 infants. The results showed that all acellular vaccines were remarkably less reactogenic than the whole cell vaccine. Nevertheless, a trend towards an increase in the reactogenicity was observed after the second and third dose, with many of the vaccines containing an high amount of PT and FHA (Decker et al., 1995) (fig. 3.5.4). The vaccines containing the genetically detoxified PT did not show this trend, suggesting that they would show a safer profile, especially during booster immunizations. This phenomenon is very likely due to the low antigen content of the pertussis vaccine.

Moreover, the immunogenicity of the vaccines showed dramatic differences. In spite of the very low antigen content, the vaccines containing genetically detoxified PT induced the highest anti-PT antibodies in ELISA and CHO toxin neutralization titers (Edwards et al., 1995). When immunogenicity was expressed as antibody units induced per μg of protein, the genetically detoxified molecule induced titers that were from five to 20 times higher than those induced by the chemically detoxified molecules. This impressive result, shown in figure 3.5.5, confirmed in infants what

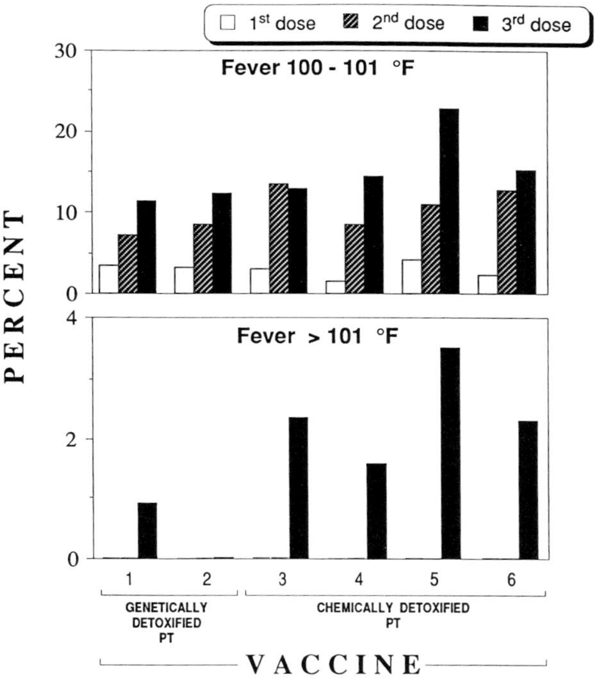

Fig. 3.5.4: Safety profile (fever) of the most representative vaccines of the phase II NIAID trial.

had been consistently seen in animal models and in the previous clinical trials. Furthermore, it allowed to definitely conclude that the genetically detoxified PT is an antigen with superior immunogenicity compared to the chemically detoxified PT.

We are currently trying to understand this phenomenon and to verify if, in addition to a superior antibody titer, the genetically detoxified PT induces also an immune response that is qualitively different. In an attempt to understand the difference between chemically and genetically detoxified PT molecules, we have analyzed them under the electron microscope, using the deep-freeze etching technique. Preliminary experiments suggest that the formalin treatment of PT changes remarkably the shape of the molecule that will be seen by the immune system as a ghost, barely resembling the native molecule, while the genetically detoxified one is identical in shape to the wild type PT. A similar effect of formalin treatment has been described for vacuolating cytotoxin A (VacA) of *Helicobacter pylori* (Rappuoli et al., 1995).

Another important factor to consider is that PT-9K/129G has a functional receptor binding site, that allows the molecule to bind to the surface of antigen presenting cells and interact directly with them. This may result in a different path of antigen presentation, and in an immune response of different quality.

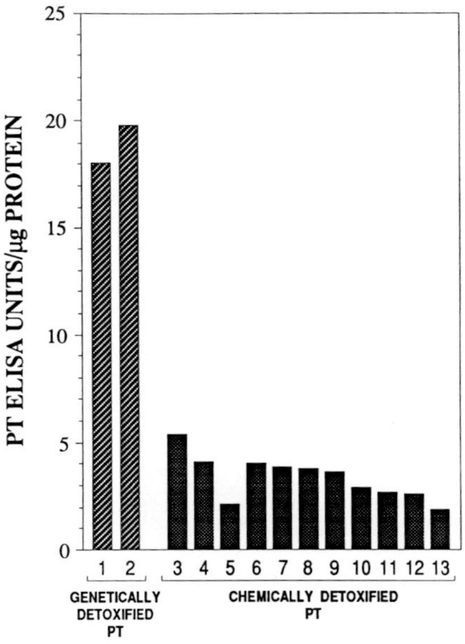

Fig. 3.5.5: Comparison of immunogenicity expressed as ELISA units per µg of protein in vaccines containing genetically or chemically detoxified PT.

3.5.3.2 Genetically Detoxified PT is Effective in Protecting Against the Disease

Following the phase II clinical studies described above, a vaccine containing 5µg of PT-9K/129G, 2.5 µg of FHA and 2.5 µg of 69K was tested for efficacy in a large scale trial sponsored by NIAID and performed by Donato Greco in Italy during the period 1992-1995. In this trial, approximately 15.000 infants divided into four groups were immunized. The first group received the vaccine containing the genetically detoxified PT, the second group received the vaccine containing a chemically detoxified PT, the third group received the whole cell vaccine, and the fourth group received the placebo. The efficacy was determined starting from 30 days to 18 months after the third immunization. During this period of active surveillance, the results showed that the vaccine containing genetically detoxified PT with only 5 µg of PT-9K/129G, 2.5 µg of FHA and 2.5 µg of 69K, was as protective as the vaccine containing 25 µg of chemically detoxified PT, 25 µg of FHA and 8 µg of 69K (84 % protection against the disease) (Greco et al., 1995). However, during the period which followed the first immunization up to 30 days post third immunization, the vaccine containing genetically detoxified PT showed higher protection. Similarly, during the period 18-27 months after the third immunization the vaccine containing genetically detoxified PT showed greater protection. These results suggest that 5 µg of genetically detoxified PT induces an immunity which is more effective in preventing disease than 25 µg of chemically detoxified PT. In conclusion, in this phase III study the vaccine containing genetically detoxified PT, in spite of the lower content of protein, showed superior immunogenicity, efficacy, and safety profile.

3.5.4. Genetic Detoxification of LT and CT

Encouraged by the results obtained with the PT and on the basis of this experience, we have started the mutagenesis work on heat-labile and cholera toxins. Initially, the same aminoacids found to be important for PT were mutagenized. Arg7 (equivalent to Arg9 of PT) and Glu112 (equivalent to Glu129 of PT) were changed into Lys and Ala, respectively. In contrast to the finding observed with pertussis toxin, where single mutation reduced but did not eliminate the toxicity, in LT and CT the single mutations were able to eliminate completely the enzymatic activity of the A subunit and the *in vitro* toxicity of the molecule. However, the aminoacid mutations that were optimal in PT, turned out not to be very useful in LT and CT. In fact, the mutants containing the above mutations were very sensitive to protease digestion and very unstable to storage and manipulation (Burnette et al., 1991; Hase et al., 1994). Therefore, new aminoacid mutations were designed using computer modelling of the LT structure that had become available in the meantime (Pizza et al., 1994a).

Of the many mutants tested, LT-K63 (containing a Ser63→Lys substitution, designed to fill the active site with the bulky side chain of the Lys residue), was found to be devoid of enzymatic activity, non toxic both *in vitro* and *in vivo*, and very stable to protease treatment. This molecule was then purified and tested in immunogenicity and adjuvanticity studies. A CT molecule containing the same mutation was also developed and shown to have similar properties.

3.5.4.1 Non Toxic Mutants of CT and LT Induce Neutralizing Antibodies Against the A Subunit

One of the things that was well established in the literature on CT and LT, is the fact that neutralizing antibodies are induced only by the B oligomer, and that the A subunit has no role in protective immunity. Based on this observation, only the B oligomer has been used for the development of vaccines against cholera and enterotoxinogenic *E. coli*. Because all the previous results had been obtained using either the fully active toxins that may bias the immune response, or molecules that had been detoxified using harsh treatment with chemicals or denaturing agents, we decided to tackle again the question using the genetically detoxified LT and CT mutants. Indeed, we found that immunization with the native non toxic derivatives of LT and CT, not only induced strong neutralizing antibody response against the B subunit, but neutralizing antibodies could be detected also against the A subunit (Pizza et al., 1994b; Fontana et al., 1995). This observation suggested that non toxic derivatives of LT and CT may be superior immunogens for the development of vaccines, confirming that the observations made with pertussis toxin may be true also for LT and CT, and possibly have a general value.

3.5.4.2 Non Toxic Derivatives of LT are Mucosal Adjuvants

Mucosal delivery of vaccines would induce an immune response at the mucosal sites in addition to the systemic response. This type of vaccination would therefore have the advantage to block pathogens at their portal of entry, before they invade the organism. In spite of the obvious advantages that mucosal vaccines would provide, so far most of the vaccines are delivered systemically by injection. Mucosal vaccines have not been developed mostly because conventional vaccines are not effective if delivered mucosally. Moreover, there are no appropriate adjuvants to enhance the immune response to mucosally delivered antigens. CT and LT are known to induce an immune response to mucosally delivered antigens that are co-administered; however, because of their toxicity, they cannot be used in humans. The availability of non toxic derivatives of LT and CT allowed us to test whether the mucosal adjuvanticity was linked to the enzymatic activity of these molecules, or whether also the non toxic mutants had a similar activity.

In spite of published observations that linked the enzymatic activity to mucosal adjuvanticity (Lycke et al., 1992), in a study performed in collaboration with Gill Douce and Gordon Dougan of the Imperial College for Science and Technology in London, we found that 1 µg of non toxic mutants of LT induced a strong systemic and mucosal immunity to antigens that were co-delivered intranasally (Douce et al., 1995). This observation has now been extended to many different antigens such as ovalbumin, KLH, tetanus toxin fragment C (tab. 3.5.3), the hemagglutinin of the influenza virus, HIV gp120 and to several antigens of *H. pylori*. In the case of *H. pylori*, oral immunization using as vaccine the vacuolating cytotoxin and LT-K63 as adjuvant, induced protection from bacterial challenge. The above results obtained mostly in mice and guinea pigs show that the non toxic derivatives of LT may provide the way to promote the development of mucosal vaccines in humans. Clinical trials using LT-K63 as mucosal adjuvants are planned and will be performed during the next few months.

Table 3.5.3: Protection of mice from challenge with tetanus toxin.

Immunogen	Serum IgG response	Mucosal IgA response	Mice surviving tetanus challenge
Tetanus fragment C i.n.	–	–	0/10
Tetanus fragment C + non toxic LT mutant i.n.	+	+	10/10
Tetanus fragment C + LT wild type i.n.	+	+	10/10
Tetanus fragment C s.c.	+	–	10/10

3.5.5 Conclusions

Genetic detoxification of pertussis toxin and the development of a vaccine containing this molecule, has shown that this technology offers vaccines that have tremendous benefits compared to those containing toxoids derived by chemical detoxification. The most outstanding advantages are the followings:

1) Increased immunogenicity. The non toxic derivative of PT was found to be 5-20 times more immunogenic than any of the chemically detoxified PTs (both in animal models and in infants).

2) Increased safety. The low dose of genetically detoxified PT necessary for inducing an appropriate immune response allows the use of a 5-10 fold less protein in the vaccine. This resulted in a reduced frequency of side effects such as fever, especially after the second, third and fourth dose of vaccine.

3) Absence of enzymatic activity and no risk of reversion. Many of the chemically detoxified vaccines contain a residual activity of pertussis toxin, that in some cases is close to that present in whole cell pertussis vaccine (Storsaeter et al., 1990; Miller et al., 1995). Other vaccines showed reversion to toxicity of the chemically detoxified toxin during storage (Storsaeter et al., 1990). Because the presence of active toxin in the whole cell vaccine has been linked to the neurological side effects associated to it (Munoz et al., 1985; Steinman et al., 1985), the residual toxicity present in the chemically detoxified vaccines makes them ethically unacceptable.

   Recent observations indicate that the effects induced by the active pertussis toxin are very long-lasting up to 8 months in rats (Kosecka et al., 1994), and therefore the presence of active PT may have unpredictable consequences even for a long time after vaccination. Thus, it is unknown what consequences may be encountered with an active toxin in acellular pertussis vaccines.

All the above problems are absent in the vaccine containing the genetically detoxified PT-9K/129G, that is already produced by the bacterium in a non toxic form.

4. In addition to these advantages, genetic detoxification provides the opportunity to design totally novel vaccines that were not feasible with conventional technologies. In fact, genetically detoxified pertussis toxin has been shown to be an excellent mucosal vaccine when given intranasally to mice (Shahin et al., 1995). Consequently, PT-9K/129G could be used as the basis for the development of a mucosal vaccine against pertussis. This would not be feasible with conventional, chemically detoxified molecules that are inactive in this system.

5. Finally, the finding that LT-K63 is an adjuvant for antigens that are co-administered at the mucosal level, opens a new world for vaccine development.

In conclusion, genetic detoxification of bacterial toxins allowed the improvement of conventional vaccines and the development of vaccines that so far were not feasible, thus opening a new era in vaccinology.

References

Allured, V. S., Collier, R. J., Carroll, S. F., and McKay, D. B. (1986) Structure of exotoxin A of Pseudomonas aeruginosa at 3.0-Angstrom resolution. Proc. Natl. Acad. Sci. U. S. A. *83*, 1320-1324.

Barbieri, J. T., Pizza, M., Cortina, G., and Rappuoli, R. (1990) Biochemical and biological activities of recombinant S1 subunit of pertussis toxin. Infect. Immun. *58*, 999-1003.

Barbieri, J. T. and Cortina, G. (1988) ADP-ribosyltransferase mutations in the catalytic S-1 subunit of pertussis toxin. Infect. Immun. *56*, 1934-1941.

Bokoch, G. M., Katada, T., Northup, J. K., Hewlett, E. L., and Gilman, A. G. (1983) Identification of the predominant substrate for ADP-ribosylation by islet activating protein. J. Biol. Chem. *258*, 2072-2075.

Burnette, W. N., Cieplak, W., Mar, V. L., Kaljot, K. T., Sato, H., and Keith, J. M. (1988) Pertussis toxin S1 mutant with reduced enzyme activity and a conserved protective epitope. Science *242*, 72-74.

Burnette, W. N., Mar, V. L., Platler, B. W., Schlotterbeck, J. D., McGinley, M. D., Stoney, K. S., Rohde, M. F., and Kaslow, H. R. (1991) Site-specific mutagenesis of the catalytic subunit of cholera toxin: substituting lysine for arginine 7 causes loss of activity. Infect. Immun. *59*, 4266-4270.

Carroll, S. F. and Collier, R. J. (1984) NAD binding site of diphtheria toxin: identification of a residue within the nicotinamide subsite by photochemical modification with NAD. Proc. Natl. Acad. Sci. U. S. A. *81*, 3307-3311.

Choe, S., Bennett, M. J., Fujii, G., Curmi, P. M., Kantardjieff, K. A., Collier, R. J., and Eisenberg, D. (1992) The crystal structure of diphtheria toxin. Nature *357*, 216- 222.

Covacci, A. and Rappuoli, R. (1993) Pertussis Toxin Export Requires Accessory Genes Located Downstream from the Pertussis Toxin Operon. Mol. Microbiol. *8*, 429-434.

Dallas, W. S. and Falkow, S. (1980) Amino acid homology between cholera toxin and Escherichia Coli heat labile toxin. Nature *288*, 499-501.

Decker, M. D., Edwards, K. M., Steinhoff, M. C., Rennels, M. B., Pichichero, M. E., Englund, J. A., Anderson, E. L., Deloria, M. A., and Reed, G. F. (1995) Comparison of thirteen acellular pertussis vaccines: adverse reactions. Pediatrics. *96 (3 PT.2)*, 557-566.

Domenighini, M., Montecucco, C., Ripka, W. C., and Rappuoli, R. (1991) Computer modelling of the NAD binding site of ADP- ribosylating toxins: active-site structure and mechanism of NAD binding. Mol. Microbiol. *5*, 23-31.

Domenighini, M., Magagnoli, C., Pizza, M., and Rappuoli, R. (1994) Common features of the NAD-binding and catalytic site of ADP-ribosylating toxins. Mol. Microbiol. *14* (1), 41-50.

Domenighini, M., Pizza, M., Jobling, M. G., Holmes, R. K., and Rappuoli, R. (1995) Identification of errors among database sequence entries and comparison of correct amino acid sequences for the heat-labile enterotoxins of Escherichia coli and Vibrio cholerae. Mol. Microbiol. *15* (6), 1165-1167.

Douce, G., Turcotte, C., Cropley, I., Roberts, M., Pizza, M., Domenghini, M., Rappuoli, R., and Dougan, G. (1995) Mutants of Escherichia coli heat-labile toxin lacking ADP- ribosyltransferase activity act as nontoxic, mucosal adjuvants. Proc. Natl. Acad. Sci. USA *92*, 1644-1648.

Douglas, C. M. and Collier, R. J. (1987) Exotoxin A of Pseudomonas aeruginosa: substitution of glutamic acid 553 with aspartic acid drastically reduces toxicity and enzymatic activity. J. Bacteriol. *169*, 4967-4971.

Edwards, K. M., Meade, B. D., Decker, M. D., Reed, G. F., Rennels, M. B., Steinhoff, M. C., Anderson, E. L., Englund, J. A., Pichichero, M. E., Deloria, M. A., and Deforest, A. (1995) Comparison of thirteen acellular pertussis vaccines: overview and serologic response. Pediatrics. *96 (3 PT.2)*, 548-557.

Edwards, K. M. (1993) Acellular Pertussis Vaccines – A Solution to the Pertussis Problem. J. Infect. Dis. *168*, 15-20.

Fontana, M. R., Manetti, R., Giannelli, V., Magagnoli, C., Marchini, A., Domenighini, M., Rappuoli, R., and Pizza, M. (1995) Construction of non toxic derivatives of cholera toxin and characterization of the immunological response against the A subunit. Infect. Immun. *63*, 2356-2360.

Giannini, G., Rappuoli, R., and Ratti, G. (1984) The amino-acid sequence of two non-toxic mutants of diphtheria toxin: CRM45 and CRM197. Nucleic. Acids. Res. *12*, 4063-4069.

Glenny, A. T. and Hopkins, B. E. (1923) Diphtheria toxoid as an immunizing agent. Br. J. Exp. Pathol. *4*, 283-288.

Greco, D., Salmaso, S., Mastrantonio, P., Giuliano, M., Tozzi, A. E., Ciofi, M. L., Giammanco, A., Panei, P., Blackwelder, W. C., Klein, D. L., Wassilack, S. G. F., and The Progetto Pertosse Working Group (1996) Clinical efficacy, immunogenicity and safety of two acellular and one whole-cell pertussis vaccines: results from the Italian trial. New Engl. J. Med. 334 (6), 341-348.

Hase, C. C., Thai, L. S., Boesmanfinkelstein, M., Mar, V. L., Burnette, W. N., Kaslow, H. R., Stevens, L. A., Moss, J., and Finkelstein, R. A. (1994) Construction and characterization of recombinant Vibrio cholerae strains producing inactive cholera toxin analogs. Infect. Immun. *62*, 3051-3057.

Hepler, J. R. and Gilman, A. G. (1992) G Proteins. TIBS *17*, 383-387.

Katada, T. and Ui, M. (1982) ADP ribosylation of the specific membrane protein of C6 cells by islet-activating protein associated with modification of adenylate cyclase activity. J. Biol. Chem. *257*, 7210-7-16.

Kosecka, U., Marshall, J. S., Crowe, S. E., Bienenstock, J., and Perdue, M. H. (1994) Pertussis toxin stimulates hypersensitivity and enhances nerve-mediated antigen uptake in rat intestine. Amer. J. Physiol-Gastrointest. L. *30*, G745-G752.

Linder, M. E. and Gilman, A. G. (1992) G proteins. Sci. Am. *267* (1), 56-61, 64-65.

Lobet, Y., Cluff, C. W., and Cieplak, W., Jr. (1991) Effect of site-directed mutagenic alterations on ADP-ribosyltransferase activity of the A subunit of Escherichia coli heat-labile enterotoxin. Infect. Immun. *59*, 2870-2879.

Locht, C. and Keith, J. M. (1986) Pertussis toxin gene: nucleotide sequence and genetic organization. Science *232*, 1258-1264.

Lycke, N., Tsuji, T., and Holmgren, J. (1992) The adjuvant effect of Vibrio cholerae and Escherichia coli heat-labile enterotoxins is linked to their ADP-ribosyltransferase activity. Eur. J. Immunol. *22*, 2277-2281.

Mekalanos, J. J., Swartz, D. J., Pearson, G. D., Harford, N., Groyne, F., and de Wilde, M. (1983) Cholera toxin genes: nucleotide sequence, deletion analysis and vaccine development. Nature *306*, 551-557.

Miller, E., Waight, P., Ashworth, E., Thornton, C., and Redhead, K. (1995) Summary of clinical and laboratory data for acellular and whole cell pertussis vaccines from UK studies. In: Compatibility of acellular pertussis with other vaccines in UK primary immunization and boosting schedules, 1-16. Edited by Salisbury, M., London, PHLS, CAMR and NIBSC.

Moxon, E. R. and Rappuoli, R. (1990) Haemophilus influenzae infections and whooping cough. Lancet *335*, 1324-1329.

Munoz, J. J. (1985). Biological activity of pertussigen (Pertussis Toxin). In Pertussis Toxin. Sekura, R. D., Moss, J., and Vaughan, M. (eds.), Academic Press, Orlando, pp. 1-18.

Nencioni, L., Pizza, M., Bugnoli, M., De Magistris, T., Di Tommaso, A., Giovannoni, F., Manetti, R., Marsili, I., Matteucci, G., Nucci, D. et al. (1990) Characterization of genetically inactivated pertussis toxin mutants: candidates for a new vaccine against whooping cough. Infect. Immun. *58*, 1308-1315.

Nencioni, L., Volpini, G., Peppoloni, S., de Magistris, M. T., Marsili, I., and Rappuoli, R. (1991) Properties of the pertussis toxin mutant PT-9K/129G after formaldeyde treatment. Infect. Immun. *59*, 625-630.

Nicosia, A., Perugini, M., Franzini, C., Casagli, M. C., Borri, M. G., Antoni, G., Almoni, M., Neri, P., Ratti, G., and Rappuoli, R. (1986) Cloning and sequencing of the pertussis toxin genes: operon structure and gene duplication. Proc. Natl. Acad. Sci. U. S. A. *83*, 4631-4635.

Papini, E., Schiavo, G., Sandona, D., Rappuoli, R., and Montecucco, C. (1989) Histidine 21 is at the NAD+ binding site of diphtheria toxin. J. Biol. Chem. *264*, 12385-12388.

Pizza, M., Bartoloni, A., Prugnola, A., Silvestri, S., and Rappuoli, R. (1988) Subunit S1 of pertussis toxin: mapping of the regions essential for ADP- ribosyltransferase activity. Proc. Natl. Acad. Sci. U. S. A. *85*, 7521-7525.

Pizza, M., Covacci, A., Bartoloni, A., Perugini, M., Nencioni, L., de Magistris, M. T., Villa, L., Nucci, D., Manetti, R., Bugnoli, M., Giovannoni, F., Olivieri, R., Barbieri, J. T., Sato, H., and Rappuoli, R. (1989) Mutants of pertussis toxin suitable for vaccine development. Science *246*, 497-500.

Pizza, M., Domenighini, M., Hol, W., Giannelli, V., Fontana, M. R., Giuliani, M. M., Magagnoli, C., Peppoloni, S., Manetti, R., and Rappuoli, R. (1994a) Probing the structure-activity relationship of Escherichia coli LT-A by site-directed mutagenesis. Mol. Microbiol. *14*, 51-60.

Pizza, M., Fontana, M. R., Giuliani, M. M., Domenighini, M., Magagnoli, C., Giannelli, V., Nucci, D., Hol, W., Manetti, R., and Rappuoli, R. (1994b) A genetically detoxified derivative of heat-labile E. coli enterotoxin induces neutralizing antibodies against the A subunit. J. Exp. Med. *6*, 2147-2153.

Podda, A., Nencioni, L., de Magistris, M. T., Di Tommaso, A., Bossu, P., Nuti, S., Pileri, P., Peppoloni, S., Bugnoli, M., Ruggiero, P. et al. (1990) Metabolic, humoral, and cellular responses in adult volunteers immunized with the genetically inactivated pertussis toxin mutant PT-9K/129G. J. Exp. Med. *172*, 861-868.

Podda, A., Nencioni, L., Marsili, I., Peppoloni, S., Volpini, G., Donati, D., Di Tommaso, A., de Magistris, M. T., and Rappuoli, R. (1991) Phase I clinical trial of an acellular pertussis vaccine composed of genetically detoxified pertussis toxin combined with FHA and 69 kDa. Vaccine *9*, 741- 745.

Podda, A., Deluca, E. C., Titone, L., Casadei, A. M., Cascio, A., Bartalini, M., Volpini, G., Peppoloni, S., Marsili, I., Nencioni, L., and Rappuoli, R. (1993) Immunogenicity of an Acellular Pertussis Vaccine Composed of Genetically Inactivated Pertussis Toxin Combined with Filamentous Hemagglutinin and Pertactin in Infants and Children. J. Pediatr. *123*, 81-84.

Podda, A., Deluca, E. C., Contu, B., Furlan, R., Maida, A., Moiraghi, A., Stramare, D., Titone, L., Uxa, F., Dipisa, F., Peppoloni, S., Nencioni, L., Rappuoli, R., Bartalini, M., Bona, G., Budroni, M., Pistilli, A. M. C., Cascio, A., Cascio, G., Cossu, M., Dallorto, P., Dileo, G., Furlan, A., Macagno, F., Marsili, I., Meloni, T., Regoli, D., Rigo, G., Trappan, A., Vargiu, G., and Volpini, G. (1994) Comparative study of a whole-cell pertussis vaccine and a recombinant acellular pertussis vaccine. J. Pediatr. *124*, 921-926.

Ramon, G. (1924) Sur la toxine et sur l'anatoxine diphtheriques. Ann. Inst. Pasteur *38*, 1-10.

Rappuoli, R., Podda, A., Pizza, M., Covacci, A., Bartoloni, A., de Magistris, M. T., and Nencioni, L. (1992) Progress towards the development of new vaccines against whooping cough. Vaccine *10*, 1027-1032.

Rappuoli, R. and Pizza, M. (1991) Structure and evolutionary aspects of ADP-ribosylating toxins. In: Sourcebook of bacterial protein toxins, 1-20. Edited by Alouf, J. and Freer, J., Academic press.

Rappuoli, R., Douce, G., Dougan, G., and Pizza, M. (1995). Genetic detoxification of bacterial toxins: A new approach to vaccine development. Int. Arch. Allergy. Immunol. *108*, 327-333.

Roberts, M., Bacon, A., Rappuoli, R., Pizza, M., Cropley, I., Douce, G., Dougan, G., Marinaro, M., McGhee, J. and Chatfield, S. (1995). A mutant pertussis toxin molecule that lacks ADP-ribosyltransferase activity, PT-9K/129G, is an effective mucosal adjuvant for intranasally delivered proteins. Infect. Immun. *63(6)*, 2100-2108.

Shahin, R., Leef, M., Eldridge, J., Hudson, M., and Gilley, R. (1995) Adjuvanicity and protective immunity elicited by Bordetella pertussis antigens encapsulated in poly(DL-lactide-co-glycolide) microspheres. Infect. Immun. 63 (4), 1195-1200.

Sindt, K. A., Hewlett, E. L., Redpath, G. T., Rappuoli, R., Gray, L. S. and Vandenberg, S. R. (1994). Pertussis toxin activates platelets through an interaction with platelet glycoprotein Ib. Infect. Immun. *62*, 3108-3114.

Sixma, T. K., Pronk, S. E., Kalk, K. H., Wartna, E. S., van Zanten, B. A., Witholt, B., and Hol, W. G. (1991) Crystal structure of a cholera toxin-related heat-labile enterotoxin from E. coli. Nature *351*, 371-377.

Sixma, T. K., Pronk, S. E., Kalk, K. H., van Zanten, B. A., Berghuis, A. M., and Hol, W. G. (1992) Lactose binding to heat-labile enterotoxin revealed by X-ray crystallography. Nature *355*, 561-564.

Sixma, T. K., Kalk, K. H., Vanzanten, B. A. M., Dauter, Z., Kingma, J., Witholt, B., and Hol, W. G. J. (1993) Refined Structure of Escherichia-Coli Heat-Labile Enterotoxin, a Close Relative of Cholera Toxin. J. Mol. Biol. *230*, 890-918.

Spicer, E. K., Kavanaugh, W. M., Dallas, W. S., Falkow, S., Konigsberg, W. H., and Shafer, D. (1981) Sequence homologies between A subunits of E. coli and V. cholerae enterotoxin. Proc. Natl. Acad. Sci. USA. *78*, 50-54.

Stein, P. E., Boodhoo, A., Armstrong, G. D., Cockle, S. A., Klein, M. H., and Read, R. J. (1994) The crystal structure of pertussis toxin. Structure *2*, 45-57.

Steinman, L., Weiss, A., Adelman, N., Lim, M., Zuniga, R., Ochlert, J., Hewlett, E., and Falkow, S. (1985). Pertussis toxin is required for pertussis vaccine encephalopathy. Proc. Natl. Acad. Sci. U. S. A *82*, 8733-8736.

Stibitz, S., Black, W., and Falkow, S. (1986) The construction of a cloning vector designed for gene replacement in Bordetella pertussis. Gene *50*, 133-140.

Storsaeter, J., Hallander, H., Farrington, C. P., Olin, P., Mollby, R., and Miller, E. (1990) Secondary analyses of the efficacy of two acellular pertussis vaccines evaluated in a Swedish phase III trial. Vaccine *8*, 457-461.

Tamura, M., Nogimori, K., Katada, T., Ui, M., and Ishii, S. (1982) Subunit structure of islet-activating protein, pertussis toxin, in conformity with the A-B model. Biochemistry *21*, 5516-5520.

Uchida, T., Gill, D. M., and Pappenheimer, A. M. J. (1971) Mutation in the structural gene for diphtheria toxin, carried by the temperate phage beta. Nature *233*, 8-11.

Weiss, A. A., Johnson, F. D., and Burns, D. L. (1993) Molecular Characterization of an Operon Required for Pertussis Toxin Secretion. Proc. Natl. Acad. Sci. USA *90*, 2970-2974.

van der Pouw-Kraan, T., Rensink, I., Rappuoli, R. and Aarden, L. (1995). Co-stimulation of T cells via CD28 inhibits human IgE production. Reversal by pertussis toxin. Clin. Exp. Immunol. *99*, 473-478.

Zumbihl, R., Dornand, J., Fischer, T., Cabane, S., Rappuoli, R., Bouaboula, M., Casellas, P. and Rouot, B. (1995). IL-1 stimulates a diverging signaling pathway in EL4 6.1 thymoma cells. J. Immunol. *155*, 181-189.

3.6 Recombinant Virus as Vaccination Carrier of Heterologous Antigens

Marion E. Perkus and Enzo Paoletti

3.6.1 Introduction

The recombinant DNA technology of the 1970's quickly led to the development of viral vectors for the insertion and expression of foreign genes. By the early 1980's, techniques were developed for expression of heterologous genes in a variety of DNA viruses, including vaccinia (Panicali and Paoletti, 1982; Mackett et al., 1982), herpes simplex virus (Post et al., 1982), adenovirus (Berkner and Sharp, 1982), the papovaviruses SV40 (Southern and Berg, 1982) and polyoma virus (Fried and Ruley, 1982), as well as RNA retroviruses (Wei et al., 1981; Gilboa et al., 1982).

Currently, a wide variety of viral vectors have been utilized for the construction of recombinants for use as vaccines against human and veterinary diseases, in cancer therapy, and in gene therapy for congenital diseases. This review will provide an overview of the field, including innovative approaches. It is not intended as an exhaustive listing of recombinant viruses expressing heterologous antigens. It will describe the biology of the various viral vectors only as this applies to vaccine design and use.

3.6.1.1 Advantages of Live Recombinant Viruses as Vaccines

Live recombinant viruses offer several advantages as vaccines, compared to conventional live or killed vaccines, or subunit approaches. First, a recombinant vaccine contains only the antigen(s) of interest in a heterologous background. Since the vaccinee is not exposed to the entire pathogen, there is no danger of disease through loss of attenuation or improper inactivation of the vaccine. Second, since viruses express antigens intracellularly, the antigens are generally presented to the immune system in a manner analogous to infection by the pathogen. This results in proper presentation of peptides in conjunction with MHC class I molecules, which is essential for eliciting cellular responses.

Replication competent live recombinant viruses have the advantage of increasing the amount of antigen delivered through replication of the vector. In contrast, vectors which are deficient for replication may have other advantages, including attenuation, persistence, and inability to lyse the host cell. Clearly, the "ideal vector" will vary with different applications. In some cases, a combination of vectors may be beneficial.

Vaccine vectors must strike the proper balance between immunogenicity and innocuity. Development of a vector often entails deletion or inactivation of genes involved in pathogenicity, for example the thymidine kinase (TK) gene in herpesviruses. A highly attenuated vaccinia virus vector, NYVAC, was developed by deletion of multiple genes implicated in pathogenicity (Tartaglia et al., 1992). Though NYVAC is restricted for growth on mammalian cells from several mammalian species, including humans, it retains immunogenicity.

Another approach to attenuation of viral vectors is exemplified in the adenovirus system. Recombinant adenovirus are typically deleted for the E1 region, which contains genes required for viral growth. E1 functions are supplied by growing the recombinant virus in the 293 complementing tissue culture cell line which contains an integrated copy of the E1 genes, and expresses the E1 functions. Similarly, in the herpesvirus system, HSV amplicons, which do not encode the functions necessary for packaging into virions, are propagated in the presence of HSV helper virus.

Natural attenuation of vectors can also be achieved by use of viruses which are restricted for growth in the target species. Thus canarypoxvirus, like all avipoxviruses, is capable of productive infection only in avian cells. However, antigens expressed in the ALVAC canarypox vector elicit protective immune responses in a variety of mammalian species (Taylor et al., 1995). An ALVAC recombinant expressing the rabies glycoprotein has been shown to elicit immune responses in humans in a Phase I study (Cadoz et al., 1992).

3.6.2 Double Stranded DNA Viruses: Poxviruses

3.6.2.1 Poxviruses

3.6.2.1.1 Orthopoxvirus Vectors

Poxviruses are large, enveloped, double stranded DNA viruses which replicate in the cytoplasm of infected cells. Because of their cytoplasmic location, messages are unspliced. Poxviruses carry within their virions the enzymes necessary for the initiation of infection. They also encode many of the functions necessary for DNA replication and gene expression, with minimal involvement of the cell's nuclear machinery. Not surprisingly, poxvirus DNA is not infectious.

Vaccinia, the prototypic poxvirus, is an orthopoxvirus with a very wide host range. Vaccinia, which has been used as an immunizing agent for two hundred years, was responsible for the eradication of smallpox in 1980 .

Vaccinia virus (VV) was first utilized as a cloning vector in 1982 (Panicali and Paoletti, 1982; Mackett et al., 1982). Recombination occurs by infecting tissue culture cells with the virus and transfecting with plasmids containing a heterologous gene inserted in a nonessential site flanked by vaccinia arms. The heterologous gene must be inserted downstream from an endogenous or translocated poxvirus promoter, since poxvirus DNA polymerase does not recognize cellular or other nuclear promoters. Wild type vaccinia, with a genomic size of ca. 190 kb, can accept inserts of at least 25 kb, enough to accommodate multiple genes. Many non-essential regions have been identified where heterologous genes can be inserted (Perkus et al., 1986, Tartaglia et al., 1992). Large regions on both termini totalling over 47 kb are dispensable for growth in tissue culture (Perkus et al., 1991). The total amount of heterologous DNA which can be accommodated by a vaccinia genome with extensive deletions is unknown. A VV-based recombinant virus expressing 7 antigens from 4 stages of the life cycle of the human malaria parasite P. falciparum was constructed using the attenuated NYVAC vector (Perkus et al. 1995a, Tine et al., submitted). This recombinant, NYVAC-Pf7, is currently in Phase I clinical trials.

Vaccinia has been used as a vector for a vast number of heterologous genes (for recent reviews see Perkus et al., 1995b; Cox et al., 1995; Cox et al., 1992). Originally, most work with recombinant vaccinia utilized the neurotropic WR laboratory strain, which is not suitable for vaccine development. Table 3.6.1 contains a partial listing of vaccinia vaccine strain-based recombinants which have shown protective efficacy in animals.

As seen in table 3.6.1, the Copenhagen vaccine strain of VV, which was used extensively in the smallpox eradication program, has been used successfully as a vaccine vector to protect animals against a number of pathogens. A Copenhagen based rabies vaccine bait, RABORAL, has been used successfully in the field to eliminate fox rabies in Belgium (Brochier et al., 1994). Recombinants based on NYVAC, the attenuated, host restricted vaccine vector derived from the Copenhagen vaccine strain, have been shown to protect mice from rabies (Tartaglia et al., 1992), swine from Japanese Encephalitis virus (Konishi et al., 1992) and horses from Equine influenza virus (Taylor et al. 1992a). Some recombinants based on the Wyeth vaccine strain have shown protective efficacy in animals (tab. 3.6.1). However, the usefulness of the Wyeth strain as an immunization vehicle has been questioned (Lee et al., 1992). Morgan et al. (1988) reported that cottontop tamarins inoculated with a Wyeth-based recombinant expressing the gp340 envelope antigen of Epstein-Barr virus were not protected against Epstein-Barr virus-induced lymphoma, but animals inoculated with the analogous WR-based recombinant were protected.

Several mammalian species-specific poxviruses have been used to engineer recombinant virus which protect the target species from disease. A raccoonpox recombinant protects raccoons from rabies challenge (Esposito et al., 1988), a capripox re-

Table 3.6.1: Protective Vaccinia Recombinants

Pathogen	Vaccine Strain	Protection
Rabies	Copenhagen	mice, raccoons, foxes
	NYCBH	mice, dogs
	NYVAC	mice
Measles	Copenhagen	mice, rats, dogs (from CDV)
Mouse Cytomegalous Virus (MCMV)	Copenhagen	mice
Yellow Fever	Copenhagen	mice
Japanese Encephalitis Virus (JEV)	Copenhagen	mice
	NYVAC	swine
Human Papilloma Virus (HPV)	Copenhagen	mice, rats
Lassa Fever	Wyeth	guinea pigs
	Lister	guinea pigs
Rinderpest	Wyeth	cattle
Peste des petits ruminants	Wyeth	goats
Equine Herpes Virus (EHV)	Copenhagen	hamsters
Pseudorabies Virus (PRV)	Copenhagen	mice, swine
Equine Influenza Virus (EIV)	NYVAC	horses
Bovine Leukemia Virus (BLV)	Copenhagen	sheep
Bovine Papilloma Virus (BPV)	Copenhagen	rats
Polyoma Virus (PV)	Copenhagen	rats
Venezuelan Equine Encephalomyelitis (VEE)	Wyeth	mice, monkeys, horses
Avian Influenza	Wyeth	chickens

combinant protects cattle from both rinderpest and lumpy skin disease (Romero et al., 1993) and protects goats against peste des petits ruminants (Romero et al., 1995), and a swinepox recombinant partially protects pigs from pseudorabies virus challenge (van der Leek et al., 1994).

3.6.2.1.2 Avipoxvirus Vectors

Attenuated avipoxviruses, especially fowlpoxvirus (FPV), have long been used as vaccines in the field. In addition to conferring protection against FPV disease (Ogawa et al., 1990), FPV-based recombinant viruses have been used to demonstrate protection of chickens against avian influenza (Taylor et al., 1988a), Newcastle disease virus (Taylor et al., 1990), and Marek's disease (Nazerian et al., 1992). FPV-based recombinant virus has also been shown to partially protect turkeys against turkey rhinotracheitis pneumovirus challenge (Quinzhong et al., 1994) .

Avipoxviruses are incapable of replicating in mammalian cells. Thus it is somewhat surprising that recombinant avipoxviruses can function as effective immunizing agents for mammalian species. In 1988 Taylor et al. (1988b), using a FPV-based recombinant virus expressing the rabies glycoprotein G, demonstrated protection of mice, cats and dogs from lethal rabies challenge. Wild et al. (1990) reported that a FPV-based recombinant virus expressing the measles fusion (F) protein protected mice from fatal measles encephalitis, but that protection required a considerably higher dose of the FPV recombinant, compared to a replication competent vaccinia-based recombinant.

Canarypoxvirus (CPV), another avipoxvirus, is a much more potent vector for mammalian species than is FPV. A recombinant CPV-based recombinant expressing the rabies glycoprotein G was approximately 100 times as efficacious in protecting mice against rabies as the corresponding FPV recombinant, and is equivalent in efficacy to a replication competent VV-rabies recombinant (Taylor et al., 1991a; Tartaglia et al., 1992). Similarly, a CPV-recombinant expressing measles virus hemagglutinin and F antigens protected dogs from canine distemper challenge with the same level of protection as was seen with a replication competent VV recombinant expressing the same antigens (Taylor et al., 1991b; 1992b).

ALVAC, a plaque-cloned isolate derived from a canarypox vaccine strain (Tartaglia et al., 1992), is being developed as a vaccine vector for human and veterinary use. Like NYVAC, ALVAC is highly attenuated compared to existing vaccinia vaccine strains (Tartaglia et al., 1992). A single dose of 6.7 $\log_{10}$ $TCID_{50}$ of vCP65, an ALVAC-based recombinant expressing the rabies glycoprotein G, is sufficient to protect dogs against a lethal rabies challenge 36 months post-inoculation (Taylor et al., 1995). Safety and immunogenicity of vCP65 was demonstrated in humans (Cadoz et al., 1992) and other primate species (Taylor et al., 1995).

An ALVAC-based measles recombinant expressing hemagglutinin and F is also being evaluated in human clinical trials. An ALVAC-based recombinant expressing genes from Japanese encephalitis virus (JEV) protects mice from JEV challenge (Konishi et al., 1994). This ALVAC-JEV recombinant is in human Phase I clinical trials, along with an analogous NYVAC-based JEV recombinant.

Recombinant virus ALVAC-EIV expresses hemagglutinin genes from types A1 and A2 Equine Influenza. Following exposure to type A2 EIV, control horses succumbed to infection, while horses inoculated with ALVAC-EIV displayed markedly attenuated symptoms, accompanied by increased levels of A2-type specific antibody (Taylor et al., 1992a). Another ALVAC-based recombinant, ALVAC-FL, expresses the envelope and gag-specific antigens of the retrovirus feline leukemia virus (FeLV). ALVAC-FL was shown to protect cats from FeLV challenge, although vaccinated cats did not demonstrate measurable levels of FeLV neutralizing antibody at any time before challenge (Tartaglia et al., 1993b).

3.6.2.1.3 Poxvirus-Based Vaccine Candidates Against Human Immunodeficiency Virus (HIV) and Related Viruses

Poxvirus-based recombinants show promise as vaccine candidates against HIV and related retroviruses. VV-based recombinants expressing the envelope glycoprotein from simian immunodeficiency virus (SIV) protect macaques from SIV challenge (Hu et al., 1992). In this system, as with HIV, better immune responses are generated by priming with a recombinant poxvirus and boosting with purified subunit protein, compared to inoculation by either immunogen alone. Specifically, priming with HIVac1e, a Wyeth-based HIV recombinant expressing the HIV-1_{IIIB} envelope gp160, and boosting with baculovirus-derived gp160 subunit resulted in both $CD4^+$ and $CD8^+$ responses, whereas HIVac1e alone elicited only $CD4^+$ responses. As might be expected, gp160 subunit did not elicit cellular responses (Cooney et al., 1993, Graham et al., 1993). Another VV-based HIV recombinant which is scheduled for phase I trial, TBC-3B, expresses the HIV-1_{IIIB} gag/pol proteins, as well as the envelope. In tissue culture, TBC-3B has been shown to produce non-infectious HIV-like particles (Dru et al., 1994). Such particles may be expected to enhance immunogenicity, either through direct administration as a protein subunit boost, or by in vivo production within the vaccinee following inoculation of TBC-3B.

HIV infection is accompanied by immunosuppresion. Since vaccination with VV can cause severe side effects in an immunosuppressed host (Fenner et al., 1988), it is unlikely that a recombinant based on a traditional vaccinia vaccine strain such as Wyeth would be used as a vaccine candidate in HIV high risk populations. NYVAC and ALVAC, which are highly attenuated compared to Wyeth (Tartaglia et al., 1992), are currently being used as bases for the development of vaccine candidates against HIV. There is special interest in the avipoxvirus vector ALVAC, which is unaffected by pre-existing immunity to vaccinia. ALVAC- and NYVAC-based recombinants expressing the HIV envelope alone, or in combination with gag, elicit both humoral and cell mediated responses against HIV antigens in mice (Cox et al., 1993) and macaques (Abimiku et al., 1995).

Two ALVAC-based HIV-1 recombinants are in phase I clinical studies. The first is a prime-boost protocol consisting of two inoculations with a live non-replicating ALVAC-HIV recombinant (vCP125) expressing the HIV-1 gp160, followed by booster injections of recombinant envelope subunit. This prime-boost regimen was well tolerated, and elicited both humoral and cell-mediated immune responses (Pialoux et al., 1995). More recently, a second prime-boost clinical trial has been initiated using vCP205, an ALVAC-based recombinant expressing HIV-1 gag/protease as well as an HIV-1 envelope gene.

NYVAC and ALVAC based recombinants expressing antigens from HIV-2 have been assessed in prime/boost protocols using the HIV-2/rhesus macaque challenge system (Tartaglia et al., 1993a). Seven of eight immunized animals resisted initial challenge with 100 infectious doses of HIV-2 virus, and five of seven resisted rechallenge six months later. Significantly, HIV-1 NYVAC and ALVAC recombinants also elicited cross-protection against HIV-2 challenge in this model system (Tartaglia et al., 1993a; Abimiku et al., 1995).

3.6.2.1.4 Poxvirus Recombinants and Tumor Therapy

VV recombinants expressing viral or cellular tumor associated antigens (TAAs) have been shown to confer protection from tumor challenge in a number of mouse and rat experimental models (reviewed in Perkus et al., 1995b). For example, a VV-based recombinant expressing the carcinoembryonic antigen (CEA), which is expressed in a number of human cancers, elicits humoral and cell-mediated responses in mice and rhesus monkeys, and protects mice from tumor challenge (Kantor et al., 1992a,b). VV recombinants expressing the human papilloma virus tumor antigens E6 or E7, which are linked to cervical cancer, yield partial protection to mice from tumor challenge. Protection was shown to be mediated by $CD8^+$ T lymphocytes (Chen et al., 1991, 1992).

Since tumor formation requires evasion of the host immune surveillance system, human tumor associated antigens (TAAs) are most likely weakly immunogenic. The potential utility of poxvirus vectors to increase immunogenicty of TAAs was demonstrated in a mouse model system. Weakly immunogenic ML38 murine colonic adenocarcinoma tumor cell lines were infected with VV recombinants expressing the B7.1 or B7.2 costimulatory molecules, leading to tumor rejection in immunocompetent syngeneic mice (Hodge et al., 1994). VV recombinants have also been used to manipulate the immune response by expression of a variety of cytokines (reviewed in Ruby et al., 1992; Perkus et al., 1995b).

ALVAC and NYVAC have potential as vectors in many aspects of antitumor therapy. Like all poxviruses, both ALVAC and NYVAC are potent inducers of cell-mediated immune responses (Cox et al., 1993; Tartaglia et al., 1993c; Egan et al., 1995) which are important in combating tumors. ALVAC and NYVAC based recombinants expressing TAAs and/or cytokines or other biological response modifiers could be administered parenterally or intratumorally. Such recombinants could also be useful in both adoptive and active cell-based immunotherapy (see Perkus et al., 1995b).

3.6.2.2 Herpesviruses

Of the approximately 100 members of the Herpesviridae family, seven types have been described which infect humans. Herpesviruses are large, enveloped viruses, containing a linear double stranded DNA genome ranging in size from 125 kb for varicella-zoster virus to 229 kb for cytomegalovirus. Unlike poxvirus DNA, herpesvirus DNA is infectious. After infection, the genome migrates to the cell nucleus, where it can either circularize and be maintained during a latent phase as an episome, or proceed with a lytic replication cycle, resulting in virion production and cell death. The target cell for latency varies with the virus; the neurotropic herpes simplex 1 virus

(HSV-1) establishes latency in neurons, while the lymphotropic Epstein Barr Virus (EBV) establishes latency in B lymphocytes.

The use of herpesviruses as genetic vectors in human gene therapy was reviewed recently (Vos 1995). HSV-1 was one of the first viruses developed as a vector for the expression of heterologous genes (Post et al., 1982). Herpesvirus vectors have been developed in two distinct forms (1) as "helper free" virus vectors analogous to the poxvirus vectors described above and (2) as helper dependent mini virus vectors (amplicons). In the first case, the recombinant virus encodes all necessary viral functions for replication and packaging. Heterologous antigens are inserted into nonessential regions of the genome. The exact amounts of heterologous DNA which can be accommodated within the genome have not been determined. It is estimated that at least 30 kb of non-essential DNA in the HSV genome can be replaced by foreign DNA, and that at least 9 kb of additional DNA can be packaged into the herpesviral capsid.

In the second case, the recombinant "virus" vector is reduced to a small plasmid, or amplicon, containing the minimal viral elements required for replication and packaging into infectious virions. These amplicons are defective for viral production, and require coinfection with a helper virus to supply necessary viral proteins in trans. Amplicons replicate by a "rolling circle" mechanism, and are packaged as concatemers within the herpesviral capsid. In theory, the amount of heterologous DNA which could be accommodated in an amplicon has been estimated at 65kb (Vos, 1995). In practice, large inserts above 15 kb appear to be unstable in HSV-1 amplicons. Inserts of up to 29 kb are stable in EBV amplicons.

3.6.2.2.1 Herpesvirus Vectors

The human herpesvirus, varicella zoster virus (VZV), is the causative agent of chickenpox. Oka, the attenuated vaccine strain of VZV, is approved for use in immunocompromised as well as normal individuals. Since infection with Epstein Barr virus, another herpesvirus, may lead to life-threatening complications in immunocompromised individuals, an EBV vaccine would be beneficial. Lowe et al. (1987) engineered Oka as a vaccine vector by inserting an expression cassette containing the EBV gp350/220 coding sequence into the TK gene. Since VZV replicates in the nucleus, normal splicing of the transcript was observed, resulting in production of both the gp350 and the 220 glycoproteins.

Several animal herpesviruses which are currently used as vaccines are being engineered as vectors. In 1991, van Ziji et al. reported that a live attenuated Pseudorabies (PRV) herpesvirus vaccine strain expressing the envelope glycoprotein E1 of hog cholera virus protects swine against both pseudorabies virus and hog cholera. More recently, Mulder et al. (1994) reported that, although PRV and hog cholera virus exhibit very different tropism and pathogenesis, expression of the foreign antigen did not change tropism or virulence of the PRV vaccine.

Bovine herpesvirus-1 (infectious bovine rhinotracheitis virus; IBRV) was used as a vector for the expression of epitopes from the major immunogenic regions of the

VP1 capsid protein of foot and mouth disease virus (FMDV). In this recombinant, the FMDV epitopes were expressed in a fusion protein, by engineering them onto the amino terminus of the nonessential IBRV gIII glycoprotein (Kit et al., 1991). Vaccination with the IBRV-FMDV recombinant induced protective levels of anti-FMDV antibodies in calves, and protected them from challenge with virulent IBRV.

HVT (herpesvirus of turkeys) is used extensively as a vaccine to protect chickens from Marek's disease, which is caused by another avian herpesvirus, Marek's disease virus (MDV). In an effort to broaden the immune response, Ross et al. (1993) expressed the MDV glycoprotein gB in an HVT vector. Another group expressed the fusion protein from Newcastle disease virus (NDV) in a HVT vector. This recombinant protected chickens from both Newcastle disease and Marek's disease (Morgan et al., 1992)

The non oncogenic primate herpesvirus saimiri (HVS) has been considered favorably as a possible vector for future gene therapy. Recombinant HVS virus have been shown to be capable of persistence as episomes in a variety of human haematopoietic and epithelial cell lines (Simmer et al., 1991).

3.6.2.2.1.1 Herpes Simplex Virus-1 as a Vector

In a normal HSV-1 infection, virus is taken up by axonal terminals and transported to sensory ganglia. The virus can persist in a latent state for an extended period of time without damage to the neuron. Thus HSV-1 is an obvious candidate for development as a neurotropic viral vector for gene therapy. Since neurovirulent HSV-1 can lead to encephalitis and death, attenuation of the vector is of highest importance before it can be used in clinical trials. Investigators have used E. coli b-galactosidase as a reporter gene to study the attenuation and establishment of latency of various replication-compromised HIV-1 mutants in a rat brain model system (Chiocca et al., 1990; Fink et al., 1992). For discussion of safety issues and the relative advantages of replicative vs non-replicative vectors etc., see reviews by Breakefield and DeLuca (1991) and Vos (1995).

One potential use for HSV-1 mediated gene therapy is in the treatment of mucopolysaccharidosis (MPS) VII (Sly disease), a human genetic disease of the CNS which is caused by a deficiency of b-glucuronidase. Wolfe et al. (1992) used MPS VII mice as a model system. They inoculated an HSV-1 vector expressing b-glucuronidase into the cornea of the mice. This resulted in latent infection of the trigeminal ganglion, and expression of the enzyme for up to four months post-inoculation. In this recombinant virus, b-glucuronidase was expressed under the control of the LAT promoter, the only HSV promoter which functions during latency.

HSV-1 has also been investigated as a vector for gene transfer to the liver. Miyanohara et al. (1992) infected mice with recombinant virus injected either directly into the liver or through the portal vein. Using a number of heterologous genes, including factor IX, they found that genes placed under the control of the human cytomegalous virus IE promoter were expressed transiently, while the same genes under the control of the LAT promoter were expressed for an extended period of time.

HSV-1 amplicons have also been used as vectors for transfer of heterologous genes to neurons in rats. Unlike other HSV vectors, amplicons do not establish latency. Ho et al. (1993) injected an HSV amplicon overexpressing the rat brain glucose transporter gene into adult rat hippocampus, resulting in a small but significant increase in glucose transport. Federoff et al. (1992) reported that the injection into the superior cervical ganglion of adult rats of an amplicon expressing nerve growth factor can prevent the decline of tyrosine hydroxylase levels that typically follows axotomy. Recently, Lu and Federoff (1995) reported the development of glucocorticoid-induceable HSV-1 amplicon vectors, which could be potentially useful in applications where inducible gene expression is required.

3.6.2.3 Adenoviruses

Adenoviruses are widespread in nature, infecting mammalian and avian species. Currently, 47 different serotypes have been isolated from humans. Adenoviruses are non-enveloped DNA viruses containing a linear, double stranded genome ranging in size from about 30 to 40 kbp. During infection, the adenovirus virion enters the cell through receptor-mediated endocytosis. Viral DNA is also infectious. Replication occurs in the nucleus, generally without integration of the viral DNA into the host genome. The transcriptional pattern, which is very complex and includes extensive splicing of mRNA transcripts, will not be discussed here (see Horwitz, 1990). Basically, early genes in six transcription units, E1A, E1B, E2A/B, E3A/B, E4 and L1, are expressed before DNA replication. After DNA replication commences, five groups of late genes are expressed from transcription initiating at the major late promoter (MLP) and subjected to differential mRNA splicing. Late transcripts possess a common 5'-untranslated 200 bp tripartite leader (TPL) which enhances translation.

3.6.2.3.1 Adenovirus Vectors

Interest in recombinant adenoviruses (rAds) as potential vaccines is based in part on the successful use of types 4 and 7 (Ad4, Ad7) as oral vaccines since 1969 for prevention of respiratory disease in military recruits. Ad5, which has been studied extensively, and which is less pathogenic compared to Ad4 and Ad7, is often the strain of choice for the development of recombinant vaccines. Another advantage of adenoviruses as potential vectors is the ease of growing them to high titers in tissue culture.

Techniques for preparation of rAds using plasmids of viral DNA were well developed by the early 1980s (Berkner and Sharp, 1982). A major consideration in the construction of rAds is the amount of DNA which can be packaged into the virion. Bett et al. (1993) determined that Ad5 has a packaging capacity of 105 % of the genome, about 1.2kb more than the wild type genome, with larger recombinants showing instability. Additional space for heterologous DNA can be gained by deletions within the genome. The E3 region is dispensable for growth in tissue culture. Also, E1 functions can be supplied in trans by growing the virus in 293 cells, a human embryonic kidney cell line which contains the left 11 % of the Ad5 genome, including the E1 region, integrated into its chromosome. Through expanded deletions, Bett et al. (1994) have constructed an Ad5 vector designed to accommodate 8.3 kb of heterologous DNA. In rAds, heterologous genes are commonly inserted into the E1 or E3 region, or in an intergenic region between the right terminus and the E4 promoter. Foreign genes are commonly placed under the transcriptional control of the E3 promoter or the MLP.

Since E1 functions are essential for viral growth, deletion of E1 produces virus which is replication defective. This feature is often desirable, especially for gene therapy applications. However, care must be taken, since Lochmuller et al. (1994) recently reported that multiple passages of E1-deleted adenovirus in 293 cells can yield some $E1^+$, replication competent virus through recombination with E1 sequences in the host chromosome.

Work with cotton rats has questioned the inocuity of deleting the E3 region. Ginsberg et al. (1989) inoculated cotton rats intranasally with an adenovirus deleted of the open reading frame in the E3 region encoding the 19 kDa protein. They found that inoculation with the E3 deleted virus resulted in an increased inflammatory response in the lungs, compared to inoculation with wild-type virus. However, recombinant adenoviruses deleted for E3 may be safe by the oral route for chimpanzees (Natuk et al., 1993) and humans (Tacket et al., 1992). The relative merits of retaining or deleting E3 from recombinant adenovirus vaccines are still being debated (Natuk et al., 1993; Tacket et al., 1992; Chengalvala et al., 1994; Randrianarison-Jewtoukoff and Perricaudet, 1995).

Recently, there has been an effort toward the development of a minimal adenovirus cloning vector, analogous to the HSV amplicon described above. Mitani et al. (1995) are working toward engineering a defective Ad5-based vector which does not encode any viral proteins. Such a vector which could, in theory, accommodate up to 37 kb of foreign DNA, would be propagated in tissue culture by coinfection with a helper virus. Thus far the system has not been perfected, but the authors are hopeful that obstacles can be overcome. In another development aimed at simplifying the construction of adenovirus-based vectors, Ketner et al. (1994) report the construction of a yeast artificial chromosome containing a complete copy of the Ad2 genome.

3.6.2.3.2 Recombinant Adenoviruses as Vaccines

Recombinant human adenoviruses have been used to express antigens from a wide variety of heterologous viruses (reviewed in Randrianarison-Jewtoukoff and Perricaudet, 1995), beginning with expression of the surface antigen from Hepatitis B Virus (HBsAg) in 1985 (Saito et al., 1985). Antibody responses to HBsAg vectored by adenoviruses were initially demonstrated in hamsters (Morin et al., 1987) and rabbits (Levrero et al., 1988).

Dogs are a semipermissive model system for Ad4 and Ad7 infection. Chengalvala et al. (1991, 1994) inoculated dogs by the intratrachael route with Ad4 or Ad7 based recombinant virus expressing HBsAg. They demonstrated antibody responses which were boosted following reinoculation with the heterotypic recombinant virus. Previously, Lubeck et al. (1989) had immunized chimpanzees using enteric coated capsules containing an Ad7 recombinant expressing HBsAg, followed by a heterotypic boost using an analogous Ad4-based recombinant. They reported antibody response to HBsAg in two of three chimpanzees, and protection from subsequent HBV challenge in one of three animals. In an initial safety and immunogenicity study, Tacket et al. (1992) reported that one administration of an enteric coated capsule containing an $E3^-$ Ad7 recombinant expressing the HBsAg was not sufficient to elicit an HBsAg-specific antibody response in any of three volunteers tested.

Studies with Ad5 recombinants expressing herpesvirus antigens emphasize the importance of route of administration in the elicitation of protective immune responses. Gallichan et al. (1993) evaluated an Ad5 recombinant expressing the glycoprotein B (gB) of HSV in intranasal (i.n.) and intraperitoneal (i.p.) inoculation into mice. They found that i.n. inoculation was superior, eliciting both serum IgG and secretory IgA, compared to only serum IgG following i.p. inoculation. Similarly, anti-HSV cytotoxic lymphocytes (CTLs) were observed in the spleen following both routes of inoculation, but in the mediastinial lymph nodes only after i.n. inoculation. Furthermore, mice immunized by the i.n. route showed superior protection against heterologous i.n challenge with HSV-2, compared to mice inoculated by the i.p. route.

Ad5 vectors have also been used to express antigens from other herpesviruses. Eloit et al. (1990) inoculated mice (i.p. route) and rabbits (i.p., intramuscular (i.m.), or intravenous (i.v.) route) with an Ad5 recombinant expressing PRV gp50. Neutralizing antibodies were elicited in at least some animals by all routes tested, and some animals survived virulent PRV challenge. Ragot et al. (1993a) were able to protect cotton top tamarins from EBV induced lymphoma by three i.m. inoculations of a Ad5 recombinant expressing the EBV gp340/220 glycoprotein. Protection was in the absence of neutralizing antibodies, suggesting a cellular response. Finally, CBA mice immunized by the i.p. route with an Ad5 recombinant expressing the CMV gB responded with gB-specific CTLs (Berencsi et al., 1993).

Dogs and mice inoculated either oronasally or parenterally with an Ad5 vector expressing the rabies glycoprotein responded with rabies-specific neutralizing antibody. Mice were subsequently protected from intracerebral challenge, though neutralizing antibody was undetectable at the time of challenge (Prevec et al., 1990). Similarly, a recombinant Ad5 vector expressing the glycoprotein from another rhabdovirus, Vesicular stomatitis virus (VSV), elicited VSV-specific neutralizing antibodies in calves, piglets and dogs when inoculated by the subcutaneous (s.c.) or oral route. Mice inoculated with this recombinant by the i.p. route were protected against i.v. challenge with a lethal dose of VSV (Prevec et al., 1989).

Like adenovirus, the paromyxovirus respiratory syncitial virus (RSV) infects the lungs. Although RSV is the major cause of severe lower respiratory disease in infants, a suitable vaccine for RSV has not yet been developed. The formalin-inactivated RSV vaccine tested in the 1960s not only failed to protect, but actually enhanced disease. Hsu and co-workers constructed Ad4, Ad5, and Ad7-based recombinants expressing the RSV F and/or G antigens. Following various combinations of sequential, heterotypic inoculation of dogs by the intratracheal route, they detected boosting of RSV antibody, and protection of dogs from subsequent challenge. In contrast, a single chimpanzee responded with low-level RSV antibody following an initial oral immunization with the Ad5-based recombinant, but the response was not increased by heterologous boosting (Hsu et al., 1992). Using a ferret model, the same group demonstrated boosting of RSV antibody responses following sequential heterotypic i.n. inoculations. Ferrets were protected from RSV challenge in a dose dependent manner (Hsu et al., 1994).

To test for potential enhanced pathology, Connors et al (1992) inoculated cotton rats with various potential RSV vaccine candidates, including formalin inactivated RSV, purified F protein, a chimeric FG protein produced from a baculovirus vector, and Ad5 and Vaccinia vectors expressing the F protein. Following RSV challenge, they found that animals which had been inoculated with purified F protein, baculovirus-FG or formalin inactivated RSV demonstrated both bronchial and alveolar enhanced histopathology relative to unimmunized animals. Ad-F immunized animals displayed control levels of alveolar histopathology and only inconsistent levels of bronchiolar histopathology. VV-F immunized animals demonstrated no enhanced histopathology.

Various other investigators have demonstrated the efficacy of Ad5-based recombinants expressing heterologous viral antigens to protect mice from challenge by the corresponding pathogen. Thus immunization by an Ad5 recombinant expressing the spike and nucleocapsid antigen from mouse hepatitis virus (MHV), a coronavirus, protects mice against a lethal MHV challenge (Wesseling et al., 1993). Similarly, immunization by an Ad5 recombinant expressing the non-structural NS1 glycoprotein from tick-borne encephalitis virus (TBEV), a flavivirus, protects mice against both viraemic and encephalitic infections following TBEV challenge (Jacobs et al., 1994). Furthermore, inoculation of mice with an Ad5 recombinant expressing an antigen from rotavirus, a reovirus, protects not only the immunized mice, but also their sucking offspring from rotavirus-induced diarrhea (Both et al., 1993). Similar passive protection was observed following inoculation of mice with a recombinant VV expressing the same antigen (Andrew et al., 1992).

Natuk et al. inoculated dogs (1992) and chimpanzees (1993) with adenovirus recombinants expressing antigens from the HIV retrovirus. Animals received booster inoculations first by heterotypic Ad inoculation, then by subunit. Dogs inoculated sequentially by the intratracheal route with Ad4-env, Ad5-env and Ad-7-env recombinants showed enhanced immune responses following the first, but not the second heterotypic boost. Three chimpanzees were immunized with a series of three sequential, heterotypic administrations of Ad-HIV recombinants by the oral route, followed by a subunit boost, then followed by administration of an Ad7-HIV recombinant by the i.n. route. This regimen elicited secretory anti-HIV antibodies, stimulated cell-mediated immune responses, and resulted in boosted anti-HIV serum responses. The experiment was terminated shortly after a final inoculation of an Ad4-HIV recombinant by the i.n. route, since all animals showed evidence of infection by Streptococcus pneumoniae, resulting in the death of one animal.

3.6.2.3.3 Use of Recombinant Adenoviruses in Cancer Therapy

Investigators have also turned their attention to exploring the potential utility of adenovirus recombinants in combating cancer. Haddada et al. (1993) report that infection of P815 tumor cells with an Ad5 recombinant expressing murine interleukin-2 (mIL-2) abrogates their tumorigenicity and induces antitumoral immunity in mice. Members of the same group report that direct intratumoral delivery of the Ad5/mIL-2 recombinant leads to recovery of the mice and rejection of further tumor challenge (Cordier et al., 1994). Using a mouse model system, other workers have shown that a rAds expressing the human p53 anti-oncogene can suppress tumor growth (Wills et al., 1994).

3.6.2.3.4 Use of Recombinant Adenoviruses in Gene Therapy

Adenoviruses have generated considerable interest as a vector for gene therapy (reviewed in Perricaudet and Stratford-Perricaudet, 1995). Nonreplicating Ads deleted for E1A and E1B genes are especially useful in gene therapy, since there is no spread of virus. Also, since host cells do not supply the missing E1 functions, infection with E1 deleted viruses does not result in expression of Ad proteins and cell death. Unlike retroviruses, which require cell division for integration and gene expression, Ads can be used as vectors for gene therapy in non-dividing as well as dividing cells. Although Ads generally do not integrate into the cell genome, persistence of Ad vector and gene expression is extremely stable in non-dividing cells such as muscle cells and neurons. More transient expression is observed in replicating cells, due to cell turnover.

Terminally differentiated cells of the respiratory epithelium are infected by Ad during a natural infection. Thus Ad is a natural vector to consider for gene therapy for cystic fibrosis (CF), a common lethal hereditary disease. CF is caused by a mutation of the cystic fibrosis conductance regulator (CFTR) gene, which results in defective cAMP-mediated chloride ion transport in the respiratory epithelium of affected individuals. In a preliminary experiment, Rosenfeld et al. (1992) inoculated cotton rats intratracheally with a defective Ad5 recombinant expressing the human CFTR gene. Human CFTR protein was detected in epithelial cells two weeks after inoculation, and CFTR mRNA transcripts were present for up to six weeks. In the first human trial, Zabner et al. (1993) administered an Ad2/CFTR recombinant to a small area of the nasal epithelium of three CF patients. They detected evidence of correction of the CFTR gene defect for at least three weeks. The investigators propose a larger clinical study to investigate the safety of repetitive administration of their Ad2/CFTR recombinant at increased doses (Welsh et al., 1995). This study will also evaluate clinical efficacy. An Ad5-based CFTR recombinant is also being evaluated in the clinic (Genetic Therapy, Inc., 1994).

For purposes of gene therapy, the ideal vector would not elicit an immune response against itself. Engelhardt et al. (1994) report a second generation E1 deleted adenovirus vector which has a ts mutation in the E2a gene. Compared to an E1 deleted adenovirus vector which is wild type for E2a, the new vector elicits less inflammation and $CD8^+$ T cell infiltration in the airway epithelium of inoculated cotton rats, accompanied by longer expression of a b-galactosidase reporter gene. Eissa et al. (1994) sound a cautionary note on the use of adenovirus-based vectors for use in gene therapy for CF. They report the presence of Ad2 or Ad5 E1a sequences in 21 % of normal and 13 % of CF individuals tested. They raise the possibility that preexisting E1a sequences in the host cells could complement or recombine with the vector, supporting replication of the recombinant virus.

Adenoviruses are also being considered for gene therapy for the common hereditary disease a1-antitrypsin (a1AT) deficiency. Like CF, a1AT deficiency manifests in the lung, where lack of the functional a1-antitrypsin enzyme allows destruction of lung tissue by the proteolytic enzyme NE. Rosenfeld et al. (1991) inoculated cotton rats by the intratracheal route with a defective Ad5 recombinant carrying the human a1AT gene. They detect the presence of human a1AT mRNA in the respiratory epithelium, and secretion of human a1AT into the epithelial lining fluid for at least a week.

The a1-antitrypsin secretory glycoprotein is normally produced in the liver. Jaffe et al. (1992) reported that in vivo intraportal administration into rats of an Ad5 recombinant expressing human a1AT produced detectable serum levels of human a1AT for 4 weeks. Earlier, Stratford-Perricaudet et al. (1990) had demonstrated the utility of adenovirus-based gene therapy in mice to replace a defective liver enzyme (ornithine transcarbamylase) for over a year following i.v. inoculation of the recombinant virus.

Li et al. (1993) quantitated the efficiency of gene transfer to mouse hepatocytes following intraportal infusion of 10^{10} recombinant adenovirus carrying the E. coli b-galactosidase reporter gene. They report initial expression in over 95 % of hepatocytes, which declines to less than 10 % after four months. Recently, Lieber et al. (1995), made use of the respective advantages of adenoviruses and retroviruses to increase the level of permanent gene transfer to mouse hepatocytes in vivo. First they infused a recombinant adenovirus expressing the human urokinase gene into mice through the portal vein. This caused high rates of asynchronous liver regeneration, which dramatically increased the transduction of a retroviral vector carrying the gene of interest.

Adenovirus vectors are being considered for gene therapy in a variety of cells. Lemarchand et al. (1992), demonstrated Ad5-vectored expression in the endothelial cells of human blood vessels, using the human a1AT gene as a prototype gene. More recently, Rome et al. (1994) used a double balloon catheter to deliver an adenovirus vector expressing the B-galactosidase reporter gene to luminal endothelial cells in sheep arteries. Expression of B-galactosidase has also been demonstrated in neurons and other cells of the nervous system following stereotactic inoculations of a replication deficient adenovirus recombinant into the rat hippocampus (Le Gal La Salle et al., 1993). In other applications, biologically active human factor VIII (Connelly et al., 1995) and erythropoietin (Descamps et al., 1994) were produced in mice following inoculation of adenovirus recombinants by the i.v. route.

Gene therapy to muscle cells is being investigated as a treatment for muscular dystrophy, a lethal and common genetic disease caused by the absence of dystrophin. As part of this investigation, Ragot et al. (1993b) were able to demonstrate intramuscular expression of an abbreviated form of the gene following i.m. injection of mice with a defective Ad5 recombinant encoding this "minidystrophin". Previously, using b-galactosidase as a marker gene, Stratford-Perricaudet et al. (1992) had reported that i.v. inoculation of infant mice with an Ad5 recombinant gives stable b-galactosidase expression in many tissues, including heart and skeletal muscle, while i.m. inoculation of adult mice results in stable, localized expression. Quantin et al. (1992) inoculated mice with a similar recombinant virus, but using a muscle-specific promoter. They reported expression of b-galactosidase in muscle cells for 75 days, with no expression observed in nonmuscle tissues surrounding the injection site.

3.6.2.3.5 Use of Recombinant Adenoviruses as Suicide Vectors

One of the most intriguing potential uses of rAd is as a "suicide vector". This technique involves expression of a conditionally lethal function, with the purpose of using this function to destroy unwanted cells. Unlike cellular kinases, the herpes thymidine kinase (HSV tk), is capable of phosphorylating the nucleoside analog ganciclovir. Thus, expression of HSV tk by a rAd renders the infected cell sensitive to killing by ganciclovir treatment. Ohno et al. (1994) used a rAd to introduce the HSV tk into porcine arteries immediately after balloon injury. They report that intimal hyperplasia, which is generally induced following such arterial injury, decreased after ganciclovir treatment.

Previously, Venkatesh et al. (1990) had demonstrated the principle of a "suicide vector" in vitro using a replication defective Ad vector which expresses the HSV-tk under the control of the HIV-1 long terminal repeat (LTR). When human cells expressing the HIV-1 transactivator Tat are infected with the rAd, the Tat activates the LTR promoter, resulting in high levels of HSV-tk expression. Treatment with ganciclovir resulted in cytotoxicity of infected cells. The ultimate aim of this approach is to accomplish selective killing of HIV-infected cells in vivo. Since, in this system, Tat is required for HSV-tk synthesis following rAd infection, only HIV-infected cells would be killed by ganciclovir treatment.

3.6.2.3.6 Adenovirus Capsids as Delivery Vehicles

In addition to serving as a molecular vector for expression of heterologous genes, adenovirus has another, quite different potential use a vaccination carrier. Adenoviruses enter cells through receptor-mediated endocytosis. Once within the endosome, the virus escapes from the vesicle by disrupting the membrane. Free in the cytoplasm, the virus targets the nucleus through the nuclear pores. This trafficking can be accomplished by Ad empty capsids in the absence of Ad DNA. Investigators have taken advantage of this to utilize Ad, either as intact virions or as empty capsids, as a delivery vehicle for targeting heterologous DNA to cells (reviewed in Curiel, 1995). Basically, Ad capsids are linked to polylysine/DNA conjugates containing the heterologous gene of interest. These complexes enter the cell through the Ad receptor, and the DNA is delivered to the nucleus. Since DNA is carried on the outside of the capsid, packaging constraints do not apply, and there is no theoretical limit to the size of the DNA that can be carried.

As might be expected, the efficiency of the Ad capsid/DNA delivery system depends on the abundance of adenovirus receptors on the cell. The host range can be broadened by addition of another ligand to the complex. Wagner et al. (1992) demonstrated that Ad/DNA complexes coupled to transferrin/polylysine could enter cells by interaction with either adenovirus or transferrin receptors. For targeting of the complex to specific cells, the adenovirus fiber protein can be neutralized by specific monoclonal antibody. Non-specific entry of the complex can be prevented by neutralizing the positive charge on the polylysine-condensed DNA by the use of a negatively charged polynucleotide, such as yeast tRNA.

3.6.2.3.7 Animal Adenoviruses as Vectors

Relatively little work has been reported on the use of animal Ads as vectors. Mittal et al. (1995) are developing bovine adenovirus type 3 (BAd3) as an expression vector. Thus far, they report the expression of the firefly luciferase reporter gene in bovine MDBK and human 293 cells. Besides its potential use as a vaccination vector for immunization of cattle, the authors suggest BAd3 as vector for gene transfer into human cells.

Cotten et al. (1993) are investigating the potential of chicken adenovirus CELO (chicken embryo lethal orphan) virus as a carrier for the introduction of polylysine-complexed DNA into cells. They demonstrate that, like human Ad5, CELO virus displays endosomolytic activity. Also, as in the case of Ad5, CELO virus/polylysine/DNA complexes can deliver DNA efficiently to mammalian cells. The authors suggest CELO virus, complexed to an appropriate cell-specific ligand, as an innocuous vector for targeting of DNA to selected human cell types.

3.6.2.4 Bovine Papilloma Virus as a Vector for Gene Therapy

Bovine papilloma virus (BPV) is a small bovine virus containing a double stranded DNA circular genome which remains as a multicopy episome in the nucleus of the infected cell. Cassileth et al. (1995) propose a phase I clinical trial using a defective BPV expressing human IL-2. In preclinical studies, the authors demonstrated that nonimmunogenic Lewis lung carcinoma (LLC) cells could be made immunogenic by expression of IL-2, among other lymphokines. IL-2 cDNA was introduced into the LLC cell by a BPV vector, which also expressed the Neo^r gene for purposes of selection. In contrast to non-transfected LLC cells, which cause lethal tumors in mice, s.c. injection of BPV-containing LLC clones expressing IL-2 led to rejection of the IL-2 producing tumor cells and survival of the mice. In the clinical trial, the authors propose to establish tumor cell lines from patients with limited disease (LD) small cell lung cancer (SCLS), then to transfect the cells with BPV vector carrying the Neo^r and IL-2 genes. After amplification of clones, the cells will be irradiated and inoculated back into the patient by the s.c. route. The BPV vector used in these studies is deleted for BPV late genes L1 and L2, preventing the formation of intact, infectious virus particles. Work is also underway to improve the vector by also deleting the E5, E6, and E7 transforming early genes (Ohe et al., 1995).

3.6.3 Single Stranded DNA Viruses as Vectors for Gene Therapy

Adeno-associated virus 2 (AAV) is a nonpathogenic human parvovirus. Approximately 80 % of the population shows evidence of exposure to AAV. The virus contains a single stranded linear DNA genome of 4680 nt. Message for both structural and non-structural genes is spliced. AAV is a defective virus, requiring coinfection with adenovirus or herpes virus for a productive, lytic infection. In the absence of helper virus, AAV integrates into the host genome, generally into a defined location on chromosome 19, where it persists in a latent phase until it is rescued by Ad infection. Like Ad, AAV infects a broad range of cultured cells. Also like Ad, AAV can infect non-dividing cells. AAV has been proposed as a vector for human gene therapy (Samulski, 1995; Rolling and Samulski, 1995; Kotin, 1994). The genome can be manipulated in plasmid form in vitro. Up to 96 % of the genome can be replaced with foreign genes, with up to about 5 kb of heterologous sequences.

Flotte et al. (1993) inoculated a rAAV expressing the cystic fibrosis transmembrane conductance regulator (CFTR) gene into rabbit lungs. They report CFTR mRNA and protein in the airway epithelium for up to six months after inoculation.

Current problems which must be overcome before AAV can be used for gene therapy include inferior packaging systems, difficulty of separating rAAV from Ad helper viruses, and the inability to grow rAAV to as high titers as wild type AAV. Another concern is the observation that rAAV do not appear to display the same specificity of chromosomal integration as wild type AAV.

Like AAV, LuIII is a human parvovirus with a single stranded genome of about 5 kb. Unlike AAV, LuIII does not integrate into the host genome, and does not require a helper virus for productive infection. LuIII has been suggested as a vector for gene therapy for certain applications, such as toxin gene targeting, where transient expression resulting from a non-integrating vector would be preferable. Maxwell et al. (1993) are developing defective LuIII as expression vectors which can be grown in tissue culture by complementation with wild type LuIII virus. They report that expression from a rLuIII virus containing the luciferase marker gene could be detected in HeLa cells for up to seven days following transduction.

3.6.4 RNA Viruses: Retroviruses

3.6.4.1 Retroviruses

Retroviruses are widespread among mammals and many other species. Each virion contains two copies of a positive strand RNA genome. Upon entry into the cell, the viral core travels to the nucleus. Reverse transcriptase present in the core copies the genome to a double stranded DNA form. This cDNA copy is integrated, as a provirus, at random locations into the genome of the host cell. Duplicated at each end of the provirus are long terminal repeats (LTRs). Transcription, which results in the production of spliced mRNA, is initiated from promoters in the LTRs. Retroviruses were first used as vectors for the expression of heterologous genes in the early 1980s (Wei et al., 1981). Since that time, a variety of forms of elegant retroviral vectors and packaging cell lines have been developed, which will not be discussed here (for review, see Morgan, 1995).

3.6.4.1.1 Retroviruses as Vaccine Vectors

Recombinant retroviruses have been used to a very limited extent as immunizing agents against infectious disease. Hunt et al. (1988) used a vector derived from Rous sarcoma virus (RSV), an avian leukosis virus, to express the HA from avian influenza. When inoculated into chickens, this recombinant virus elicited very low levels of hemagglutination-inhibiting and neutralizing antibody. However, immunized chickens were completely protected from pathogenic avian influenza challenge. In contrast, the same group (Brown et al., 1992) report that immunization of chickens with a retroviral vector expressing the avian influenza NP elicits antibody but does not protect chickens from challenge.

3.6.4.1.2 Use of Retroviral Vectors for Gene Transfer

Retroviruses are the family of viruses which is used most commonly in gene transfer. There is great interest in the potential of retroviruses for gene therapy to correct genetic defects, and for gene transfer, especially in the treatment of cancer and HIV. The Moloney murine leukemia virus (Mo-MLV) is the most widely utilized retroviral vector.

From the viewpoint of gene therapy, retroviruses have the advantage of stability of expression, since the recombinant virus is integrated into the host genome and is replicated along with host DNA prior to division into daughter cells. Conversely, since the retroviral genome can not integrate into the DNA of non-dividing cells, retroviruses are not useful vectors for gene transfer into post-mitotic target cells.

Familial hypercholesterolemia (FH) in humans is a genetic disorder caused by a deficiency of low density lipoprotein receptors (LDLRs). The Watanabe Heritable Hyperlipidemic (WWHL) rabbit is a useful animal model for FH. Chowdhury et al. (1991) performed lobectomies on WWHL rabbit livers, then transfected primary cultures of the liver cells ex vivo with a recombinant retrovirus expressing the LDLR. The transfected, autologous liver cells were then transplanted back into the rabbits, resulting in long term decrease in total serum cholesterol. Wilson et al. (1992) published a similar clinical protocol for treatment of patients with FH. Recently, Pages et al. (1995) reported improvement in retroviral-transduced LDL gene expression in hepatocytes through the use of a tissue-specific promoter. They also increased transduction efficiency through the use of hepatocyte growth factor. In another approach, Lieber et al. (1995) transduced mouse hepatocytes in vivo with a rAd expressing urokinase. This resulted in high rates of asynchronous liver regeneration, which increased by 5 to 10 fold the rate of subsequent retroviral-mediated gene transfer, compared to the rates observed following traditional partial hepatectomy.

Morgan et al. (1987) transduced cultured human keratinocytes with a retroviral vector containing the human growth hormone gene. They reported that the cells continued to produce the hormone, following grafting as an epithelial sheet onto athymic mice. Similarly, Geary et al. (1994) transduced baboon smooth muscle cells obtained by biopsy with a retroviral vector expressing the B-galactosidase marker gene, then reintroduced the cells in vascular grafts. The grafts were well tolerated, with 10% of the cells expressing B-galactosidase.

Congenital erythropoietic porphyria (CEP), a human genetic disease with a defect in heme biosynthesis, is caused by a deficiency of uroporphyrinogen III synthase (UROIIIS) activity. Moreau-Gaudry et al. (1995) transduced human fibroblasts deficient for UROIIIS with retroviral vectors carrying the gene, resulting in clones that express the enzyme to at least wild type values. They also report high rates of gene transfer into peripheral blood progenitor cells.

3.6.4.1.3 Use of Retroviral Vectors in Cancer Therapy

In broad terms, the main use of retroviral vectors in cancer therapy is to target the host immune response to the malignancy. Tumor infiltrating lymphocytes can be isolated from a human melanoma, expanded in vitro, and then reintroduced into the patient by adoptive transfer as a therapeutic measure. Rosenberg (1991) has used a Mo-MLV vector to introduce heterologous genes into TILs. After initial experiments using the Neo^r gene as a marker, he inserted various cytokine genes into TILs. In contrast to results observed in tumor cells, Rosenberg reported difficulty in achieving high level expression of retrovirally-transduced cytokines in TILs. TILs expressing the retrovirally-transduced cytokine, tumor necrosis factor (TNF), are being tested in clinical trials in patients with advanced cancer.

Retrovirally transduced IL-2 cytokine is being investigated as a potential tool for anti-tumor therapy. Arienti et al. (1994) reported that two human melanoma cell lines transduced with a retroviral vector expressing hIL-2 lose their tumorigenicity in nude mice. They suggest that these IL-2 secreting cell lines will be of value in the vaccination of melanoma patients. Fakhrai et al. (1995) evaluated the effect of mouse fibroblasts retrovirally transduced with IL-2 in the treatment of colorectal carcinoma in the CT-26 murine tumor model. They reported that immunization with a mixture of irradiated tumor cells and IL-2-transduced fibroblasts gave greater protection against tumor challenge than did irradiated tumor cells alone. In the light of these results, a phase I clinical protocol is planned in which colon carcinoma patients will be injected with autologous irradiated tumor cells in combination with autologous fibroblasts which have been transduced with an IL-2 recombinant retrovirus (Sobol et al., 1995).

The interleukin 4 (IL-4) cytokine is known to produce tumor regression and prolonged survival of mice with intracerebral gliomas. Wei et al. (1995) report that implantation of fibroblasts producing a retrovirus vector which contains the IL-4 gene promotes tumor regression and host survival better than mere implantation of fibroblasts expressing a comparable level of IL-4 without the retrovirus.

In another approach to combating cancer, Zhang et al. (1993) used a retroviral vector which directs the synthesis of K-ras antisense RNA to transduce human lung cancer cells expressing a mutated, tumorigenic form of K-ras. They report that the intracellularly synthesized antisense RNA inhibited translation of the mutated K-ras protein, reduced cell proliferation, and reduced tumorigenicity of the cells in a mouse model system.

3.6.4.1.4 Retroviruses as Suicide Vectors for Treatment of Brain Tumors

Like adenoviral vectors, retroviruses have been used successfully as suicide vectors. In 1992 (Culver et al.), rats with cerebral glioma were injected intratumorally with murine fibroblasts that were producing a retroviral vector expressing the HSV-TK. After five days, rats were treated with ganciclovir (GCV), resulting in complete regression of the tumor. This protocol takes advantage of the ability of retroviruses to spread to nearby dividing cells, i.e., tumor cells, but not normal neural tissue, which is not undergoing cell division.

The HSV-TK/GCV technique is being used in clinical trials of human patients with brain tumors, both adult (Oldfield et al., 1993) and pediatric (Raffel et al., 1994). In these protocols, mouse NIH 3T3 cells producing the retroviral vector expressing HSV-TK are inoculated intratumorally, followed after one week by a fourteen day regimen of treatment with GCV.

3.6.4.1.5 Recombinant Retroviruses in Treatment of HIV

Recombinant retroviruses are being investigated for potential utility in combating infection by the HIV retrovirus using several different approaches. Laube et al. (1994) inoculated rhesus monkeys with four doses of autologous fibroblasts which had been transduced with a non-replicating retroviral vector expressing the HIV-1_{IIIB} env/rev protein. The authors report that immunized monkeys developed CTL responses specific for HIV-1 env/rev. They suggest a similar protocol for genetic immunization in HIV-infected humans.

In another approach, Chuah et al. (1994) used a retroviral vector to express RNA antisense to the HIV-1 TAR region. They found that a human T cell line transduced with this vector and challenged with HIV-1_{MN} virus was inhibited for HIV-1 viral production. In yet another approach, Leavitt et al. (1994) used a retroviral vector to transfer an anti-HIV-1 ribozyme to human peripheral blood lymphocytes (PBLs). They found that transduced PBL clones from multiple donors expressed the ribozyme and resisted HIV-1 infection. The authors report that a human clinical trial is in development to test the safety and efficacy of treatment of HIV patients with autologous PBLs expressing the ribozyme.

3.6.4.1.6 Issues Related to the Use of Retroviral Vectors

To be effective as a vehicle for gene therapy, a vector must survive in the host long enough to fulfill its function. Russell et al. (1995) investigated the effects of human serum and cerebrospinal fluid on retroviral vectors. They found that human serum inactivates the murine retroviral vectors commonly used, as well as mouse NIH-3T3 cells which are often used to deliver such vectors. In contrast, human cerebrospinal fluid did not inactivate retroviral vectors or lyse murine retrovirus packaging cells. On the basis of these results, the authors suggest that gene transfer by retroviral vectors may be more successful within the central nervous system than elsewhere. This is borne out by the successful use of retroviral vectors as suicide vectors for brain tumors, described above. The problem of serum inactivation of retroviral vectors was also addressed by Rother et al. (1995). They report that inactivation of retroviral vectors is dependent on membrane attack complex formation, and that inactivation of vector particles in human serum or blood can be abrogated with the use of soluble complement inhibitors.

Mo-MLV, the most widely used retroviral vector, was previously thought to be completely non pathogenic in primates. A note of caution was sounded by Donahue and co-workers in 1992. They used very high concentrations of retroviral vector particles contaminated with smaller amounts of replication-competent virus for gene transduction into stem cells from rhesus monkeys. Following reconstitution of the monkeys' bone marrow using the transduced cells, formation of lymphomas was observed in several of the monkeys. The authors stress the importance of avoiding replication-competent virus contamination in all retroviral vectors used for human gene transfer studies.

Currently, a number of viruses, containing either a positive or negative RNA genome, are being developed as vectors for expression of heterologous antigens (reviewed in Bredenbeek and Rice, 1992, Schlesinger, 1995). In general, since replication of RNA is inherently more error-prone than DNA, RNA viral vectors are considered more appropriate for uses requiring transient, rather than long term, expression of heterologous genes.

3.6.4.2 Positive Strand RNA Viruses

The genomic RNA of positive stranded RNA virus is infectious. Since RNA amplification and transcription occurs in the cytoplasm, transcripts are not spliced. Positive stranded RNA viruses include some viruses, like the togaviruses, which utilize subgenomic mRNAs as part of their life cycle, as well as other viruses, like the picornaviruses and flaviviruses, which synthesize only full length message.

3.6.4.2.1 Alphaviruses

Alphaviruses are members of the togavirus family of enveloped RNA viruses. Alphaviruses are transmitted by mosquitoes to vertebrate hosts. Infection of vertebrate cells in culture results in a lytic infection. The alphavirus genome consists of a single stranded 12 kb RNA molecule of positive polarity. Genomic replication occurs through the synthesis of an intermediate (-) stranded genome length DNA molecule. Subgenomic (+) strand RNA serves as mRNA.

Two members of the alphavirus genus, Sinbis virus and Semliki Forest virus (SFV), have been used for expression of heterologous genes. Recombinant Sinbis virus vectors have been of two types (1) defective interfering (DI) RNAs that require co-infection with Sinbis virus to provide necessary viral functions, and (2) replication competent, but packaging defective virus. The size of heterologous genetic material that can be maintained in recombinant genomes is limited by packaging constraints. Bredenbeek and Rice (1992) report that the amount of heterologous genetic material which can be stably maintained in double subgenomic (i.e., containing two promoters for mRNA synthesis) vectors is limited to under 2 kb. A packaging system has been developed for SFV recombinants which should, in theory, allow stable insertion of 5 kb of heterologous genetic material into that vector (Liljestrom and Garoff, 1991).

Two studies have demonstrated immunogenicity in mice of Sinbis-based recombinants. Hahn et al. (1992) observed priming of influenza-specific CTL responses in mice following intraperitoneal injection of Sinbis vectors expressing either a truncated form of the Influenza HA or mini-genes encoding two CTL epitopes of the HA. Using another approach, London et al. (1992) inserted a neutralization epitope from Rift Valley fever virus into nonessential regions in the Sinbis virus E2 or E3 glycoprotein. They found that, in the E2 chimera, the epitope was expressed on the virion surface. Immunization of mice with this recombinant resulted in partial protection against Rift Valley fever challenge.

Since RNA from (+) stranded RNA viruses is infectious, Johanning et al. (1995) investigated non-replicative vs replication competent Sinbis recombinant RNA as a means of introducing genetic information directly into cells. Mice were injected in the tongue or quadriceps muscle with Sinbis-based RNA expressing the firefly luciferase reporter gene. The authors report that replication competent Sinbis recombinant RNA was vastly superior to non-replicative RNA in the levels and duration of luciferase expression that were induced.

Zhou et al. (1994) inoculated mice with two forms of a SFV recombinant expressing the Influenza nucleoprotein (NP). Mice were inoculated either with recombinant viral particles by the i.p. and i.v. routes, or with self-replicative recombinant RNA in the quadriceps muscle. Both regimens resulted in humoral responses. In addition, inoculation with viral particles elicited a CTL response.

3.6.4.2.2 Picornaviruses

Picornaviruses (literally, small rna viruses) are small, non-enveloped viruses containing a single stranded genome of (+) polarity. Pathogenic picornaviruses include the human enteroviruses poliovirus and hepatitis A, the cardioviruses encephalomyocarditis (EMCV), Theiler's virus, Coxsackievirus and Mengo virus, and the rhinoviruses, as well as the highly pathogenic foot and mouth disease virus of cattle (FMDV). Poliovirus, the picornavirus member which has been used most extensively as an expression vector, has a genome of about 7.5 kb. Unlike alphaviruses, the picoranviruses synthesize only full length mRNA. For a recent review on the use of picornavairuses as vectors, see Girard et al. (1995).

3.6.4.2.2.1 Poliovirus Vectors

The first recombinant polioviruses were intertypic chimeras with an antigenic site from one poliovirus type inserted in the same location into the capsid protein of another type (Burke et al., 1988; Murray et al., 1988). Murray et al. (1988) reported the construction of a type1/type 3 hybrid that elicited both type 1 and type 3 neutralizing antibodies in rabbits and monkeys. Burke et al. (1989) engineered a Sabin cDNA vector for insertion of epitopes into the antigenic 1 site. They constructed recombinant polioviruses with epitopes from hepatitis A (HAV), rhinovirus and Coxsackie picornaviruses. Two poliovirus/HAV chimeras elicited low levels of HAV neutralizing antibodies in some immunized rabbits and mice (Lemon et al., 1992). Poliovirus chimeras containing neutralizing epitopes from the FMDV picornavirus were found to induce neutralizing antibodies in guinea pigs, and s.c. immunization with one poliovirus/FMDV chimera protected some guinea pigs against FMDV challenge (Kitson et al., 1991). Jenkins et al. (1990) and Dedieu et al. (1992) made similar poliovirus chimera containing epitopes from the human papilloma virus (HPV)-16 capsid protein and the HIV envelope. In both cases, rabbits inoculated with the chimera produced antibodies to the heterologous epitope.

3.6.4.2.2.2 Poliovirus Vectors Expressing HIV Antigens

More recently, investigators have used other approaches to express larger heterologous antigens in poliovirus vectors. Poliovirus is first translated as a polyprotein, which is processed at specific cleavage sites into the mature poliovirus proteins. Andino et al. (1994) engineered recombinant poliovirus genomes such that a heterologous protein would be expressed at the amino terminus of the polyprotein, or at junctions between normal poliovirus cleavage products. They were able to achieve cleavage of the heterologous protein from the mature poliovirus proteins, which assembled normally into a capsid. Using this method, they report that poliovirus can stably carry insertions equivalent in size to 15 % of the viral genome, or about 1 kb. Two of the heterologous proteins expressed by this method were HIV gag and Nef. Inoculation of mice by the i.p. route with a mixture of poliovirus recombinants expressing gag or nef elicited antibody response to both HIV proteins, as well as to poliovirus. Rectal inoculation of a monkey with the poliovirus/Nef recombinant elicited Nef-specific serum IgG and IgA, and secretory IgA.

Porter et al. (1993) constructed poliovirus-based "minireplicons" in which the VP2 and VP3 capsid genes in the P1 capsid precursor region of the poliovirus polyprotein were replaced by the gag or pol genes of HIV-1, resulting in HIV-1-P1 fusion proteins. The recombinant genome was encapsidated by transfection into cells previously infected with a recombinant vaccinia virus which expresses the authentic poliovirus capsid precursor protein, P1. The authors report that the chimeric genomes are stable, and suggest the utility of the encapsidates minireplicons as a potential vaccine vector.

3.6.4.2.2.3 Other Picornaviruses Vectors

Mengo virus is a cardiovirus with a broad host range. Altmeyer et al. (1994, 1995) have engineered an attenuated Mengo virus as an expression vector. They insert heterologous sequences into the leader (L) polypeptide, resulting in fusion proteins. A Mengo virus recombinant expressing HIV-1 epitopes elicits HIV-1 specific antibody and CTLs in mice following two i.p. inoculations. HIV-1 specific antibody responses were also observed in monkeys inoculated once with this recombinant virus by the i.m. route (Altmeyer et al., 1994). Mice inoculated once by the i.p. route with a Mengo virus recombinant expressing an immunodominant CTL epitope of lymphocytic choriomeningitis virus (LCMV) respond with LCMV-specific CTLs, and are protected against intracranial LCMV challenge (Altmeyer et al., 1995). Recently Zhang et al. (1995) demonstrated that heterologous genetic material can also be inserted into the L polypeptide of the murine cardiovirus, Theiler's virus, resulting in stable recombinant viruses.

Recently, intertypic chimeras were engineered for the picornavirus FMDV, similar to the poliovirus intertypic chimeras described previously. Using FMDV serotype A as a backbone, Rieder et al. (1994) replaced the G-H loop of the VP1 capsid protein with homologous sequences from serotypes C or O. They reported that swine inoculated with their A/C chimera were protected from challenge with FMDV serotype A, and partially protected from challenge with FMDV serotype C. Reimann et al. (1991) constructed intertypic chimeras of coxsackieviruses B3 and B4 by inserting a B4 epitope into its homologous region in VP1 of B3. When inoculated into rabbits, the chimera elicited antibodies capable of neutralizing both serotypes.

Finally, Arnold et al. (1994) demonstrated the feasibility of making viable chimeras of human rhinoviruses displaying heterologous immunogens. The same group (Smith et al., 1994) used random systematic mutagenesis to generate a library of human rhinovirus 14 chimeras expressing possible variations of the HIV-1 immunodominant V3 loop, inserted into an exposed portion of the rhinovirus VP2 capsid protein.

3.6.4.3 Negative Strand RNA Viruses

The genomic RNA of negative strand RNA viruses is not infectious, since transcription of the genome to positive strand mRNA is necessary for completion of the viral life cycle and translation of encoded polypeptides. Through the use of reverse genetics, negative stranded viruses including the influenza orthomyxovirus and various paramyxoviruses are being engineered as expression systems for heterologous antigens (reviewed in Schlesinger, 1995; Bredenbeek and Rice, 1992; Garcia-Sastre and Palese, 1995).

3.6.4.3.1 Paramyxoviruses as Vectors

Paramyxoviruses are enveloped viruses which replicate in the cytoplasm of infected cells. Members include Sendai virus and respiratory syncitial virus (RSV), as well as the causative agents of measles, mumps, and Newcastle disease. The paramyxovirus genome consists of a single strand of RNA of negative polarity, surrounded by a nucleocapsid which contains the viral proteins necessary for primary transcription. Thus far, two paramyxoviruses, Sendai, and RSV have been engineered as expression vectors.

Sendai virus causes respiratory tract infection in mice and rats. Sendai, a prototypic paramyxovirus, contains a negative sense RNA genome of 15.3 kb. In 1991, Park et al. transcribed a chimeric Sendai negative sense RNA genome from cDNA. In this chimera, the (antisense) coding region was replaced by antisense RNA specifying the chloramphenicol acetyltransferase (CAT) gene. Replication and encapsulation of this chimeric genome required coinfection with wild type Sendai helper virus.

Human RSV is the leading cause of severe pediatric respiratory tract disease. Collins et al. (1991) using a very similar technique to that employed with Sendai virus, replaced the coding region of RSV with the CAT reporter gene. Analogous to the Sendai virus chimera, the chimeric RSV genomic RNA was amplified, expressed and incorporated into virions when transfected into RSV infected cells.

3.6.4.3.2 Influenza Virus Vectors

Influenza viruses, members of the orthomyxoviridae family, are a major cause of respiratory disease, both in humans and animals. Orthomyxoviruses are enveloped viruses which, unlike the paramyxoviruses, replicate in the nucleus of infected cells. The genome of human influenza virus consists of eight different negative stranded RNA segments. Most messenger RNA contains transcripts of full length segments, with the exception of M2 mRNA, which is spliced. The current status of Influenza virus vectors was reviewed very recently (Garcia-Sastre and Palese, 1995).

The first reverse genetics system for manipulating the genome of a negative strand RNA virus was described for human influenza A in 1989, when Luytjes and coworkers replaced the NS gene of human influenza A with the CAT reporter gene. Recombinant, chimeric, RNA was expressed in vitro by transcription from a DNA plasmid. When this chimeric NS/CAT RNA was transfected into cells in the presence of wild type helper virus and purified influenza A polymerase proteins, it was amplified, expressed, and incorporated into influenza virus particles. Reverse genetics has also been demonstrated for human influenza B virus (Barclay and Palese, 1995) and avian influenza A virus (Horimoto and Kawaoka, 1994). The reverse genetics system has been used extensively to study the genetics and biology of human influenza A virus, through in vitro manipulation of influenza genes followed by reintroduction back into infectious virions (reviewed by Garcia-Sastre and Palese, 1995). Recombinants created by this technique include infectious human influenza A viruses containing chimeric hemagglutinin (HA) molecules expressing neutralizing epitopes from different subtypes (Li et al., 1992).

3.6.4.3.2.1 Influenza Virus Vectors for Expression of Foreign Genes

Human influenza A virus has been used as a vector in two ways; for the expression of immunogenic epitopes on influenza proteins, and for the expression of heterologous proteins. Both B cell and T cell immunogenic epitopes have been successfully presented on chimeric influenza proteins. As might be expected, B cell epitopes are most immunogenic when presented in the exposed, hypervariable antigenic sites of the HA protein. Heterologous T cell epitopes, which require cleavage to peptides before presentation to the immune system, can be inserted into various locations in a number of influenza proteins.

A CTL epitope from lethal lymphocytic choriomenengitis virus (LCMV) nucleoprotein was expressed in the stalk region of the influenza neuraminidase (NA). Mice inoculated with this recombinant by the i.n. or i.p. route were protected against i.c. challenge by LCMV (Castrucci et al., 1994). Li et al. (1993b) inserted a CTL epitope from the circumsporozoite protein of mouse malaria Plasmodium yoelii into the antigenic E site of the influenza HA protein. They found that mice immunized with this influenza-based recombinant, then boosted by a vaccinia-based recombinant expressing the entire circumsporite protein, were protected against sporozoite challenge. Interestingly, the opposite order of immunization with the two recombinant viruses did not confer protection from challenge. Rodrigues et al. (1994) also reported the utility of immunization of mice with an influenza-based P.yoelii recombinant, followed by boosters with a recombinant vaccinia virus.

Muster et al. (1994) inserted a B cell epitope from HIV gp41 into the HA of Influenza A. Mice immunized with this chimeric virus responded with serum neutralizing antibodies, and with IgA antibodies in nasal and intestinal secretions. Another influenza/HIV chimera, expressing the immunodominant V3 loop inserted in the influenza HA, elicited both neutralizing antibodies and CTLs in immunized mice (Li et al., 1993a).

Heterologous sequences of under about 100 nucleotides have been expressed in influenza vectors as insertions into influenza proteins as described above. To express longer heterologous genes in an influenza vector, two strategies have been developed. In one strategy, Percy et al. (1994) increased the length of the NA gene segment by inserting the entire coding sequence for the CAT reporter gene in frame upstream from NA coding sequences. The polyprotein translated from this chimeric RNA is processed by cleavage by a 16 amino acid self-cleaving 2A protease from FMDV which was inserted between CAT and NA sequences. The result is authentic, biologically active, NS and CAT polypeptides. Using another strategy, Garcia-Sastre et al. (1994) changed the NA message into a bicistronic mRNA by the introduction of an internal ribosomal entry site (IRES) between the coding sequences derived from the gp41 of HIV-1 and the influenza NA. By these means, expression of heterolgous genetic material of about 1 kb can be achieved using influenza-based vectors.

3.6.4.3.3 Hepatitis Delta Virus as a Vector

Hepatitis Delta virus (HDV) is a defective virus which requires coinfection with Hepatitis B virus (HBV) for replication. It is estimated that HDV accounts for 20 to 40 % of the chronic HBV-associated liver disease. The single stranded, negative polarity, circular RNA HDV genome, of about 1700 nt, does not encode any envelope proteins. Instead, HDV utilizes HBV surface antigens as its envelope. Like HBV, HDV replicates in the nucleus of infected cells. The extremely high copy number of genomic HDV RNA present in the infected cell has motivated some investigators to consider the HDV as a potential vector for the expression of heterologous genetic information (summarized in Netter et al., 1995).

The insertion site for foreign genetic material is at the end of a rod structure formed by the self annealing of the HDV genome. Viable viruses have been constructed containing inserts of up to 60 nt of heterologous genetic information. Potentially useful biologically active RNAs which could be presented in chimeric HDV include ribozymes, antisense RNAs, and sense RNA decoy molecules. Thus far, a chimeric HDV containing a ribozyme with activity against CAT mRNA has been constructed. Netter et al. (1995) describe the construction of a chimeric HDV containing the TAR sequence of HIV-1, to act as a decoy for Tat protein.

The most likely use for recombinant HDV is likely to be in the amelioration of chronic hepatitis in humans. Netter et al. (1995) report that they are constructing a HDV chimera containing, as a decoy, a sequence which is important in hepadnavirus replication. They will test this chimera first, in vitro, for its ability to interfere with the replication of woodchuck hepatitis virus (WHV) in transfected cells, and second, in vivo, by intrahepatic injection into a woodchuck which is a chronic carrier for WHV. The degree of cytopathogenicity of various naturally occurring and genetically modified form of HDV is still under investigation.

3.6.5 Summary

As described above, a wide variety of recombinant virus vector systems, based on DNA viruses, retroviruses and (+) and (-) stranded RNA viruses, have been used for the expression of heterologous antigens and biologically relevant nucleic acids in humans and animals. Uses include immunization for the prevention or amelioration of disease caused by a variety of pathogens, treatment of cancer, and gene transfer to correct genetic deficiencies. Some vectors, such as poxviruses, have been used extensively for development of vaccines for immunization against heterologous pathogens. Other viruses, most notably retroviruses, have been used mainly for gene transfer.

The choice of an optimal viral vector is very dependent on the specific application for which it will be used. Important factors to be considered in vaccination against heterologous pathogens include the immunogenicity and safety of the vector, and the relative advantages of replicative vs. nonreplicative vectors. Much work has gone into improving the safety of various viral vectors, for example the development of the ALVAC and NYVAC attenuated poxvirus vectors, and the use of replication deficient viruses in other systems. For many pathogens, the ability of the immunizing vector to elicit mucosal as well as humoral immunity is important. This may also be a function of the route of inoculation, as shown for vaccinia virus (Small et al., 1985; Kanesaki et al., 1991; Meitin et al., 1994).

For gene transfer, the tropism of the vector is important, as evidenced by the use of HSV vectors for gene transfer to neurons, and adenovirus vectors for gene transfer to lung epithelial cells. Persistence of the vector is also a consideration. Retrovirus vectors, which integrate into the DNA of dividing cells, provide a source for persistent expression of antigen. Unlike retroviruses, adenoviruses are useful vectors for persistent expression of heterologous genes in terminally differentiated cells, while providing more transient expression in dividing cells.

Choice of vector will also depend on the amount of heterologous genetic material to be expressed. For certain applications, expression of B cell or T cell epitopes, perhaps in the context of a viral structural protein in a picornavirus or influenza virus, is sufficient. In other cases, expression of multiple foreign genes will be required. Some viruses, notably poxviruses, can accept large inserts, encoding several genes, while many other viruses are severely limited by packaging constraints. Work is ongoing in several systems to expand the amount of foreign genetic material which can be incorporated through the development of appropriate helper viruses and packaging cells. The ingenious use of adenoviral capsids as carriers for extra-virion, chemically linked DNA, overcomes the limitation of packaging constraints while maintaining the advantages of adenovirus targeting for delivery of heterologous DNA.

Often, there may be advantages to sequential use of different vectors. Pre-existing immunity to a vector can be overcome through use of a vector of a different serotype, for example sequential use of recombinant human adenovirus types 4, 5, and 7. Similarly, recombinant canarypox vector is not affected by preexisting immunity to vaccinia. Influenza virus, with several serotypes of HA, and human rhinoviruses, which contain over 100 serotypes, have been suggested as appropriate vectors for use in repeated immunizations. Sequential use of viral vectors of different classes can sometimes accomplish more than use of a single vector. As one example, recombinant adenovirus expressing urokinase has been used in vivo to make liver cells more susceptible for gene transfer by retroviral vectors. In another example, inoculation of mice with an influenza-based malaria recombinant, followed by boosting with a VV-based malaria recombinant, conferred protection against challenge, while immunization with either recombinant alone did not.

References

Abimiku, A. G., Franchini, G., Tartaglia, J., Aldrich, K., Myagkikh, M., Markam, P. D., Chong, P., Klein, M., Kieny, M.-P., Paoletti, E., Gallo, R. C., and Robert-Guroff, M. (1995) HIV-1 recombinant poxvirus vaccine induces cross-protection against HIV-2 challenge in rhesus macaques. Nature Med. *1*, 321-329.

Altmeyer, R., Escriou, N., Girard, M., Palmenberg, A., and van der Werf, S. (1994) Attenuated Mengo virus as a vector for immunogenic human immunodeficiency virus type 1 glycoprotein 120. Proc. Natl. Acad. Sci. USA *91*, 9775-9779.

Altmeyer, R., Girard, M., van der Werf, S., Mimic, V., Seigneur, L., and Saron, M.-F. (1995) Attenuated Mengo virus: a new vector for live recombinant vaccines. J. Virol. *69*, 3193-3196.

Andino, R., Silvera, D., Suggett, S. D., Achacoso, P. L., Miller, C. J., Baltimore, D., and Feinberg, M. B. (1994) Engineering poliovirus as a vaccine vector for the expression of diverse antigens. Science *265*, 1448-1451.

Andrew, M. E., Boyle, D. B., Coupar, B. E. H., Reddy, D., Bellamy, A. R., and Both, G. W. (1992) Vaccinia-rotavirus VP7 recombinants protect mice against rotavirus-induced diarrhoea. Vaccine *10*, 185-191.

Arienti, F., Sule-Suso, J., Melani, C., Maccalli, C., Belli, F., Illeni, M. T., Anichini, A., Cascinelli, N., Colombo, M. P., and Parmiani, G. (1994) Interleukin-2 gene-transduced human melanoma cells efficiently stimulate MHC-unrestricted and MHC-restricted autologous lymphocytes. Hum. Gene Ther. *5*, 1139-1150.

Arnold, G. F., Resnick, D. A., Li, Y., Zhang, A., Smith, A. D., Geisler, S. C., Jacobo-Molina, A., Lee, W.-M., Webster, R. G., and Arnold, E. (1994) Design and construction of rhinovirus chimeras incorporating immunogens from polio, influenza and human immunodeficiency viruses. Vir. *198*, 703-708.

Barclay, W. S. and Palese, P. (1995) Influenza B viruses with site-specific mutations introduced into the HA gene. J. Virol. *69*, 1275-1279.

Berencsi, K., Rando, R. F., de Taisne, C., Paoletti, E., Plotkin, S. A., and Gonzol, E. (1993) Murine cytotoxic T cell response specific for human cytomegalovirus glycoprotein B (gB) induced by adenovirus and vaccinia virus recombinants expressing gB. J. Gen. Vir. *74*, 2507-2512.

Berkner, K. L. and Sharp, P.A. (1982) Preparation of adenovirus recombinants using plasmids of viral DNA. In: Gluzman, Y. (ed.) Eukaryotic Viral Vectors, Cold Spring Harbor Laboratory, Cold Spring Harbor, New York. 193-198.

Bett, A. J., Haddara, W., Prevec, L., and Graham, F. L. (1994) An efficient and flexible system for construction of adenovirus vectors with insertions or deletions in early regions 1 and 3. Proc. Natl. Acad. Sci. USA *91*, 8802-8806.

Bett, A. J., Prevec, L., and Graham, F. L. (1993) Packaging capacity and stability of human adenovirus type 5 vectors. J. Virol. *67*, 5911-5921.

Both, G. W., Lockett, L. J., Janardhana, V., Edwards, S. J., Bellamy, A. R., Graham, F. L., Prevec, L. and Andrew, M. E. (1993) Protective immunity to rotavirus-induced diarrhoea is passively transferred to newborn mice from naive dams vaccinated with a single dose of a recombinant adenovirus expressing rotavirus VP7sc Vir. *193*, 940-950.

Breakefield, X. O. and DeLuca, N. A. (1991) Herpes simplex virus for gene delivery to neurons. New Biologist. *3*, 203-218.

Bredenbeek, P. J. and Rice, C. M. (1992) Animal RNA virus expression systems. Seminars in Virology *3*, 297-310.

Brochier, B., Boulanger, D., Costy, F., and Pastoret, P.-P. (1994) Towards rabies elimination in Belgium by fox vaccination using a vaccinia-rabies glycoprotein recombinant virus. Vaccine *12*, 1368-1371.

Brown, D. W., Kawaoka, Y., Webster, R. G., and Robinson, H. L. (1992) Assessment of retrovirus-expressed nucleoprotein as a vaccine against lethal influenza virus infections of chickens. Av. Dis. *36*, 515-520.

Burke, K. L, Dunn, G., Ferguson, M., Minor, P. D. and Almond, J. W. (1988) Antigen chimaeras of poliovirus as potential new vaccines. Nature *322*, 81-82.

Burke, K. L., Evans, D. J., Jenkins, O., Meredith, J., D'Souza, E. D. A., and Almond, J. W. (1989) A cassette vector for the construction of antigenic chimaeras of poliovirus. J. Gen. Virol. *70*, 2475-2479.

Cadoz, M., Strady, M., Meignier, B., Taylor, J., Tartaglia, J., Paoletti, E., and Plotkin, S. (1992) Immunization of humans using a novel vector: Canarypox expressing rabies glycoprotein. Lancet *339*, 1429-1432.

Cassileth, P. A., Podack, E., Sridhar, K., and Savaraj, N. (1995) Clinical protocol. Phase I study of transfected cancer cells expressing the interleukin-2 gene product in limited stage small cell lung cancer. Hum. Gene Ther. *6*, 369-383.

Castrucci, M. R., Hou, S., Doherty, P. C., and Kawaoka, Y. (1994) Protection against lethal lymphocytic choriomeningitis virus (LCMV) infection by immunization of mice with an influenza virus containing an LCMV epitope recognized by cytotoxic T lymphocytes. J. Virol. *68*, 3486-3490.

Chen, L., Thomas, E. K., Hu, S.-L., Hellstrom, I., and Hellstrom, K. E. (1991) Human papillomavirus type 16 nucleoprotein E7 is a tumor rejection antigen. Proc. Natl. Acad. Sci. USA *88*, 110-114.

Chen, L., Mizuno, M. T., Singhal, M. C., Hu, S.-L., Galloway, D. A., Hellstrom, I., and Hellstrom, E. K. (1992) Induction of cytotoxic T lymphocytes specific for a syngeneic tumor expressing the E6 oncoprotein of human papillomavirus type 16. J. Immunol. *148*, 2617-2621.

Chengalvala, M., Lubeck, M. D., Davis, A. R., Mizutani, S., Molnar-Kimber, K., Morin, J., and Hung, P. P. (1991) Evaluation of adenovirus type 4 and type 7 recombinant hepatitis B vaccines in dogs. Vaccine *9*, 485-490.

Chengalvala, M., Bhat, B. M., Bhat, R., Lubeck, M. D., Mizutani, S., Davis, A. R., and Hung, P. P. (1994) Immunogenicity of high expression adenovirus-hepatitis B virus recombinant vaccines in dogs. J. Gen. Vir. *75*, 125-131.

Chiocca, E. A., Choi, B. B., Cai, W., DeLuca, N. A., Schaffer, P. A., DiFiglia, M., Breakefield, X. O., and Martuza, R. L. (1990) Transfer and expression of the lacZ gene in rat brain neurons mediated by herpes simplex virus mutants. New Biologist. *2*, 739-746.

Chowdhury, J. R., Grossman, M., Gupta, S., Chowdhury, N. R., Baker, J. R., Jr., and Wilson, J. M. (1991) Long-term improvement in hypercholesterolemia after ex vivo gene therapy in LDLR-deficient rabbits. Science *254*, 1802-1805.

Chuah, M. K. L., Vandendriessche, T., Chang, H.-K., Ensoli, B., and Morgan, R. A. (1994) Inhibition of human immunodeficiency virus type-1 by retroviral vectors expressing antisense-TAR. Hum. Gene Ther. *5*, 1467-1475.

Collins, P. L., Mink, M. A., and Stec, D. S. (1991) Rescue of synthetic analogs of respiratory syncytial virus genomic RNA and effect of truncations and mutations on the expression of a foreign reporter gene. Proc. Natl. Acad. Sci. USA *88*, 9663-9667.

Connelly, S., Smith, T. A. G., Dhir, G., Gardner, J. M., Mehaffey, M. G., Zaret, K. S., McClelland, A., and Kaleko, M. (1995) *In vivo* gene delivery and expression of physiological levels of functional human factor VIII in mice. Hum. Gene Ther. *6*, 185-193.

Connors, M., Collins, P. L., Firestone, C.-Y., Sotnikov, A. V., Waitze, A., Davis, A. R., Hung, P. P., Chanock, R. M., and Murphy, B. R. (1992) Cotton rats previously immunized with a chimeric RSV FG glycoprotein develop enhanced pulmonary pathology when infected with RSV, a phenomenon not encountered following immunization with vaccinia-RSV recombinants or RSV. Vaccine *10*, 475-484.

Cooney, E. L., McElrath, M. J., Corey, L., Hu, S.-L., Collier, A. C., Arditti, D., Hoffman, M., Coombs, R. W., and Smith, G. E. (1993) Enhanced immunity to human immunodeficiency virus (HIV) envelope elicited by a combined vaccine regimen consisting of priming with a vaccinia recombinant expressing HIV envelope and boosting with gp160 protein. Proc. Natl. Acad. Sci. USA *90*, 1882-1886.

Cordier, L., Duffour, M. T., Sabourin, J. C., Lee, M. G., Cabannes, J., Ragot, T., Perricaudet, M., and Haddada, H. (1994) Complete recovery of mice from a pre-established tumour by direct intratumoral delivery of an adenovirus vector harbouring the murine IL-2 gene. Gene Ther. *1*, 1-6.

Cotten, M., Wagner, E., Zatloukal, K., and Birnstiel, M. L. (1993) Chicken adenovirus (CELO virus) particles augment receptor-mediated DNA delivery to mammalian cells and yield exceptional levels of stable transformants. J. Virol. *67*, 3777-3785.

Cox, W. I., Tartaglia, J., and Paoletti, E. (1992) Poxvirus recombinants as live vaccines. Recombinant Poxviruses. Binns, M. and Smith, G. (eds.). CRC Press. 123-162.

Cox, W. I., Gettig, R. R., and Paoletti, E. (1995) Poxviruses as genetic vectors. In: Viruses in human gene therapy. Vos, J.-M. H. (ed.), Chapman & Hall Carolina Academic Press, Durham NC. 141-178.

Cox, W. I., Tartaglia, J., and Paoletti, E. (1993) Induction of cytotoxic T lymphocytes by recombinant canarypox (ALVAC) and attenuated vaccinia (NYVAC) viruses expressing the HIV-1 envelope glycoprotein. Virology *195*, 845-850.

Culver, K. W., Ram, Z., Wallbridge, S., Ishii, H., Oldfield, E. H., and Blaese, R. M. (1992) In vivo gene transfer with retroviral vector-producer cells for treatment of experimental brain tumors. Science *256*, 1550-1552.

Curiel, D. T. (1995) Gene transfer mediated by adenovirus-polylysine-DNA complexes. In: Viruses in human gene therapy. Vos, J. M. H. (ed.), Chapman & Hall Carolina Academic Press, Durham NC. 179-212.

Dedieu, J.-F., Ronco, J., van der Werf, S., Hogle, J. M., Henin, Y., and Girard, M. (1992) Poliovirus chimeras expressing sequences from the principal neutralization domain of human immunodeficiency virus type 1. J. Virol. *66*, 3161-3167.

Descamps, V., Blumenfeld, N., Villeval, J.-L., Vainchenker, W., Perricaudet, M., and Beuzard, Y. (1994) Erythropoietin gene transfer and expression in adult normal mice: use of an adenovirus vector. Hum. Gene Ther. *5*, 979-985.

Donahue, R. E., Kessler, S. W., Bodine, D., McDonagh, K., Dunbar, C., Goodman, S., Agricola, B., Byrne, E., Raffeld, M., Moen, R., Bacher, J., Zsebo, K. M., and Nienhuis, A. W. (1992) Helper virus induced T cell lymphoma in nonhuman primates after retroviral mediated gene transfer. J. Exp. Med. *176*, 1125-1135.

Dru, A., Ludosky, M-.A., Cartaud, J., and Beaud, G. (1994) Cell line-dependent release of HIV-like gag particles after infection of mammalian cells with recombinant vaccinia virus. AIDS Res. Hum. Retrovir. *10*, 383-390.

Egan, M. A., Pavlat, W. A., Tartaglia, J., Paoletti, E., Weinhold, K. J., Clements, M. L., and Siliciano, R. F. (1995) Induction of human immunodeficiency virus type 1 (HIV-1)-specific cytolytic T lymphocyte responses in seronegative adults by a nonreplicating, host-range-restricted canarypox vector (ALVAC) carrying the HIV-1_{MN} env gene. J. Infec. Dis. *171*, 1623-1627.

Eissa, N. T., Chu, C.-S., Danel, C., and Crystal, R. G. (1994) Evaluation of the respiratory epithelium of normals and individuals with cystic fibrosis for the presence of adenovirus E1a sequences relevant to the use of $E1a^-$ adenovirus vectors for gene therapy for the respiratory manifestations of cystic fibrosis. Hum. Gene Ther. *5*, 1105-1114.

Eloit, M., Gilardi-Hebenstreit, P., Toma, B., and Perricaudet, M. (1990) Construction of a defective adenovirus vector expressing the pseudorabies virus glycoprotein gp50 and its use as a live vaccine. J. Gen. Vir. *71*, 2425-2431.

Engelhardt, J. F., Litzky, L., and Wilson, J. A. (1994) Prolonged transgene expression in cotton rat lung with recombinant adenoviruses defective in E2a. Hum. Gene Ther. *5*, 1217-1229.

Esposito, J. J., Knight, J. C., Shaddock, J. H., Novembre, F. J., and Bauer, G. M. (1988) Successful oral rabies vaccination of raccoons with raccoon poxvirus recombinants expressing rabies virus glycoprotein. Virology *167*, 313-316.

Fakharai, H., Shawler, D. L., Gjerset, R., Naviaux, R. K., Kozoil, J., Royston, I., and Sobol, R. E. (1995) Cytokine gene therapy with interleukin-2-transduced fibroblasts: effects of IL-2 dose on anti-tumor immunity. Hum. Gene Ther. *6*, 591-601.

Federoff, H. J., Geschwind, M. D., Geller, A. I., and Kessler, J. A. (1992) Expression of nerve growth factor in vivo from a defective herpes simplex virus 1 vector prevents effects on axotomy on sympathetic ganglia. Proc. Natl. Acad. Sci. USA *89*, 1636-1640.

Fenner, F., Henderson, D. A., Arita, I., Jezek, J., and Ladnyi, I. D. (1988) Smallpox and Its Eradication. Geneva, World Health Organization.

Fink, D. J., Sternberg, L. R., Weber, P. C., Mata, M., Goins, W. F., and Glorioso, J. C. (1992) In vivo expression of b-galactosidase in hippocampal neurons by HSV-mediated gene transfer. Hum Gene Ther. *3*, 11-19.

Flotte, T. R., Afione, S. A., Conrad, C., McGrath, S. A., Solow, R., Oka, H., Zeitlin, P. L., Guggino, W. B., and Carter, B. J. (1993) Stable in vivo expression of the cystic fibrosis transmembrane conductance regulator with an adeno-associated virus vector. Proc Natl. Acad. Sci. USA *90*, 10613-10617.

Fried, M. and Ruley, E. (1982) Use of polyoma virus vector. In: Eukaryotic Viral Vectors. Gluzman, Y. (ed.), Cold Spring Harbor Laboratory, Cold Spring Harbor, New York. 67-70.

Gallichan, W. S., Johnson, D. C., Graham, F. L., and Rosenthal, K. L. (1993) Mucosal immunity and protection after intranasal immunization with recombinant adenovirus expressing herpes simplex virus glycoprotein B. J. Infec. Dis. *168*, 622-629.

Garcia-Sastre, A., Muster, T., Barclay, W. S., Percy, N., and Palese, P. (1994) Use of a mammalian internal ribosomal entry site element for expression of a foreign protein by a transfectant influenza virus. J. Virol. *68*, 6254-6261.

Garcia-Sastre, A. and Palese, P. (1995) Influenza virus vectors. Biologicals *23*, 171-178.

Geary, R. L., Clowes, A. W., Lau, S., Vergel, S., Dale, D. C., and Osborne, W. R. A. (1994) Gene transfer in baboons using prosthetic vascular grafts seeded with retrovirally transduced smooth muscle cells: a model for local and systemic gene therapy. Hum. Gene Ther. *5*, 1211-1216.

Genetic Therapy, Inc.(1994) Clinical Protocol.Gene therapy for cyctic fibrosis utilizing a replication deficient recombinant adenovirus vector to deliver the human cystic fibrosis transmembrane conductance regulator cDNA to the airways. A phase I study. Hum. Gene Ther. *5*, 1019-1057.

Gilboa, E., Park, J., Kolbe, M., Hwang, S., Kucherlapati, R., Noonan K., and Freeman, H. (1982) Transduction and expression of nonselectable genes using retrovirus-derived vectors. In: Eukaryotic Viral Vectors. Gluzman, Y. (ed.), Cold Spring Harbor Laboratory, Cold Spring Harbor, New York. 145-151.

Ginsberg, H. S., Lundholm-Beauchamp, U., Horswood, R. L., Pernis, B., Wold, W. S. M., Chanock, R. M., and Prince, G. A. (1989) Role of early region 3 (E3) in pathogenesis of adenovirus disease. Proc. Natl. Acad. Sci. USA *86*, 3823-3827.

Girard, M., Altmeyer, R., van der Werf, S.,Wychowski, C., and Martin, A. (1995) The use of picornaviruses as vectors for the engineering of live recombinant vaccines. Biologicals *23*, 165-169.

Graham, B. S., Matthews, T. J., Belshe, R. B., Clements, M. L., Dolin, R., Wright, P. F., Gorse, G. J., Schwartz, D. H., Keefer, M. C., Bolognesi, D. P., Corey, L., Stablein, D. M., Esterlitz, J. R., Hu, S.-L., Smith, G. E., Fast, P. E., Koff, W. C., and the NIAID AIDS Vaccine Clinical Trials Network (1993) Augmentation of Human Immunodeficiency Virus Type 1 neutralizing antibody by priming with gp160 recombinant vaccinia and boosting with rgp160 in vaccinia-naive adults. J. Infec. Dis. *167*, 533-537.

Haddada, H., Ragot, T., Cordier, L., Duffour, M. T., and Perricaudet, M. (1993) Adenoviral interleukin-2 gene transfer into P815 tumour cells abrogates tumorigenicity and induces antitumoral immunity in mice. Hum. Gene Ther *4*, 299-306.

Hahn, C. S., Hahn, Y. S., Braciale, T. J., and Rice, C. M. (1992) Infectious Sinbis virus transient expression vectors for studying antigen processing and presentation. Proc. Natl. Acad. Sci. USA *89*, 2679-2683.

Ho, D. Y., Mocarski, E. S., and Sapolsky, R. M. (1993) Altering central nervous system physiology with a defective herpes simplex virus vector expressing the glucose transporter gene. Proc. Natl. Acad. Sci USA *90*, 3655-3659.

Hodge, J. W., Abrams, S., Schlom, J., and Kantor, J. (1994) Induction of antitumor immunity by recombinant vaccinia viruses expressing B7-1 or B7-2 costimulatory molecules. Cancer Res. *54*, 5552-5555.

Horimoto, T. and Kawaoka, Y. (1994) Reverse genetics provides direct evidence for a correlation of hemagglutinin cleavability and virulence of an avian influenza A virus. J. Virol. *68*, 3120-3128.

Horwitz, M. S. (1990) Adenoviridae and their replication. In: Virology 2nd Edition. Fields, B. N. and Knipe, D. N. et al. (eds.), Raven Press Ltd. New York, N.Y.

Hsu, K.-H. L., Lubeck, M. D., Davis, A. R., Bhat, R. A., Selling, B. H., Bhat, B. M., Mizutani, S., Murphy, B. R., Collins, P. L., Chanock, R. M., and Hung, P. P. (1992) Immunogenicity of recombinant adenovirus-respiratory syncytial virus vaccines with adenovirus types 4, 5, and 7 vectors in dogs and a chimpanzee. J. Infec. Dis. *166*, 769-775.

Hsu, K.-H. L., Lubeck, M. D., Bhat, B. M., Bhat, R. A., Kostek, B., Selling, B. H., Mizutani, S., Davis, A. R., and Hung, P. P. (1994) Efficacy of adenovirus-vectored respiratory syncytial virus vaccines in a new ferret model. Vaccine *12*, 607-612.

Hu, S.-L., Abrams, K., Barber, G. N., Moran, P., Zarling, J. M., Langlois, A. J., Kuller, L., Morton, W. R., and Benveniste, R. E. (1992) Protection of macaques against SIV infection by subunit vaccines of SIV envelope glycoprotein gp160. Science *255*, 456-459.

Hunt, L. A., Brown, D. W., Robinson, H. L., Naeve, C. W., and Webster, R. G. (1988) Retrovirus-expressed hemagglutinin protects against lethal influenza virus infections. J. Virol. *62*, 3014-3019.

Jacobs, S. C., Stephenson, J. R., and Wilkinson, G. W. G. (1994) Protection elicited by a replication-defective adenovirus vector expressing the tick-borne encephalitis virus non-structural glycoprotein NS1. J. Gen. Vir. *75*, 2399-2402.

Jaffe, H. A., Danel, C., Longnecker, G., Metzger, M., Setoguchi, Y., Rosenfeld, M. A., Gant, T. W., Thorgeirsson, S. S., Stratford-Perricaudet, L. D., Perricaudet, M., Pavirani, A., Lecocq, J.-P., and Crystal, R. G. (1992) Adenovirus-mediated in vivo gene transfer and expression in normal rat liver. Nature Gen. *1*, 372-378.

Jenkins, O., Cason, J., Burke, K. L., Lunney, D., Gillen, A., Patel, D., McCance, D. J., and Almond, J. W. (1990) An antigen chimera of poliovirus induces antibodies against human papillomavirus type 16. J. Virol. *64*, 1201-1206.

Johanning, F. W., Conry, R. M., LoBuglio, A. F., Wright, M., Sumerel, L. A., Pike, M. J., and Curiel, D. T. (1995) A sinbis virus mRNA polynucleotide vector achieves prolonged and high level heterologous gene expression in vivo. Nuc. Acids Res. *9*, 1495-1501.

Kanesaki, T., Murphy, B. R., Collins, P. L., and Ogra, P. L. (1991) Effectiveness of enteric immunization in the development of secretory immunoglobin A response and the outcome of infection with respiratory syncytial virus. J. Virol. *65*, 657-663.

Kantor, J., Irvine, K., Abrams, S., Kaufman, H., DiPietro, J., and Schlom, J. (1992a) Antitumor activity and immune responses induced by a recombinant carcinoembryonic antigen-vaccinia virus vaccine. J. Natl. Cancer Ins. *84*, 1084-1091.

Kantor, J., Irvine, K., Abrams, S., Snoy, P., Olsen, R., Greiner, J., Kaufman, H., Eggensperger, D., and Schlom, J. (1992b) Immunogenicity and safety of a recombinant vaccinia virus vaccine expressing the carcinoembryonic antigen gene in a nonhuman primate. Cancer Res. *52*, 6917-6925.

Ketner, G., Spencer, F., Tugendreich, S., Connelly, C., and Hieter, P. (1994) Efficient manipulation of the human adenovirus genome as an infectious yeast artificial chromosome clone. Proc. Natl. Acad. Sci. USA *91*, 6186-6190.

Kit, M., Kit, S., Little, S. P., Di Marchi, R. D., and Gale, C. (1991) Bovine herpesvirus-1 (infectious bovine rhinotracheitis virus)-based viral vector which expresses foot-and-mouth disease epitopes. Vaccine *9*, 564-572.

Kitson, J. D. A., Burke, K. L., Pullen, L. A., Belsham, G. J., and Almond, J. W. (1991) Chimeric polioviruses that include sequences derived from two independent antigenic sites of foot and mouth disease virus (FMDV) induce neutralizing antibodies against FMDV in guinea pigs. J. Virol. *65*, 3068.

Konishi, E., Pincus, S., Paoletti, E., Laegreid, W. W., Shope, R. E., and Mason, P. W. (1992) A highly attenuated host range-restricted vaccinia virus strain, NYVAC, encoding the prM, E, and NS1 genes of Japanese encephalitis virus prevents JEV viremia in swine. Virology *190*, 454-458.

Konishi, E., Pincus, S., Paoletti, E., Shope, R. E., and Mason, P. W. (1994) Avipox virus-vectored Japanese encephalitis virus vaccines: use as vaccine candidates in combination with purified subunit immunogens. Vaccine *12*, 633-638.

Kotin, R. M. (1994) Prospects for the use of adeno-associated virus as a vector for human gene therapy. Hum. Gene Ther. *5*, 793-801.

Laube, L. S., Burrascano, M., Dejusus, C. E., Howard, B. D., Johnson, M. A., Lee, W. T. L., Lynn, A. E., Peters, G., Ronlov, G. S., Townsend, K. S., Eason, R. L., Jolly, D. J., Merchant, B., and Warner, J. F. (1994) Cytotoxic T lymphocyte and antibody responses generated in Rhesus monkeys immunized with retroviral vector-transduced fibroblasts expressing Human Immunodeficiency virus type-1 IIIB ENV/REV proteins. Hum. Gene Ther. *5*, 853-862.

Leavitt, M. C., Yu, M., Yamada, O., Kraus, G., Looney, D., Poeschla, E., and Wong-Staal, F. (1994) Transfer of an anti-HIV-1 ribozyme gene into primary human lymphocytes. Hum. Gene Ther. *5*, 1115-1120.

Lee, M. S., Roos, M., McGuigan, L. C., Smith, K. A., Cormier, N., Cohen, L. K., Roberts, B. E., and Payne, L. G. (1992) Molecular attenuation of vaccinia virus: mutant generation and animal characterization. J. Virol. *66*, 2617-2630.

Le Gal La Salle, G., Robert, J. J., Berrard, S., Ridoux, V., Stratford-Perricaudet, L. D., Perricaudet, M., and Mallet, J. (1993) An adenovirus vector for gene transfer into neurons and glia in the brain. Science *259*, 988-990.

Lemarchand, P., Ari Jaffe, H., Danel, C., Cid, M. C., Kleinman, H. K., Stratford-Perricaudet, L. D., Perricaudet, M., Pavirani, A., Lecocq, J.-P., and Crystal, R. G. (1992) Adenovirus-mediated transfer of a recombinant human a_1 -antitrypsin cDNA to human endothelial cells. Proc. Natl. Acad. Sci. USA *89*, 6482-6486.

Lemon, S. M., Barclay, W., Ferguson, M., Murphy, P., Jing, L., Burke, K., Wood, D., Katrak, K., Sangar, D., Minor, P. D., and Almond, J. W. (1992) Immunogenicity and antigenicity of chimeric picornaviruses which express hepatitis A virus (HAV) peptide sequences: evidence for a neutralization domain near the amino terminus of VP1 of HAV. Vir. *188*, 285-295.

Levrero, M., Ballay, A., Skellekens, H., Tiollais, P., and Perricaudet, M. (1988) Hepatitis B adenovirus recombinant as a potential live vaccine. In: Proceedings of the 8th International Biotechnology Symposium. Durand., G., Bobichon, L., and Florent, J. (eds.), Paris, 702-712.

Li, Q., Kay, M. A., Finegold, M., Stratford-Perricaudet, L. D., and Woo, A. L. C. (1993) Assessment of recombinant adenoviral vectors for hepatic gene therapy. Hum. Gen. Ther. *4*, 403-409.

Li, S., Schulman, J. L., Moran, T., Bona, C., and Palese, P. (1992) Influenza A virus transfectants with chimeric hemagglutinins containing epitopes from different subtypes. J. Virol. *66*, 399-404.

Li, S., Polonis, V., Isobe, H., Zaghouani, H., Guinea, R., Moran, T., Bona, C., and Palese, P. (1993a) Chimeric influenza virus induces neutralizing antibodies and cytotoxic T cells against human immunodeficiency virus type 1. J. Virol. *67*, 6659-6666.

Li, S., Rodrigues, M., Rodriguez, D., Rodriguez, J. R., Esteban, M., Palese, P., Nusssenzweig, R. S., and Zavala, F. (1993b) Priming with recombinant influenza virus followed by administration of recombinant vaccinia virus induces $CD8^+$ T-cell-mediated protective immunity against malaria. Proc. Natl. Acad. Sci. USA *90*, 5214-5218.

Lieber, A., Vrancken Peeters, M.-J. T. F. D., Meuse, L., Fausto, N., Perkins, J., and Kay, M. A. (1995) Adenovirus-mediated urokinase gene transfer induces liver regeneration and allows for efficient retrovirus transduction of hepatocytes in vivo. Proc. Natl. Acad. Sci. USA *92*, 6210-6214.

Liljestrom, P. and Garoff, H. (1991) A new generation of animal cell expression vectors based on the Semliki Forest virus replicon. Biotechnology *9*, 1356-1361.

Lochmuller, H., Jani, A., Huard, J., Prescott, S., Simoneau, M., Massie, B., Karpati, G., and Acsadi, G. (1994) Emergence of early region 1-containing replication-competent adenovirus in stocks of replication-deficient adenovirus recombinants (DE1+DE3) during multiple passages in 293 cells. Hum. Gen. Ther. *5*, 1485-1491.

London, S. D., Schmaljohn, A. L., Dalrymple, J. M., and Rice, C. M. (1992) Infectious enveloped RNA virus antigenic chimeras. Proc. Natl. Acad. Sci. USA *89*, 207-211.

Lowe, R. S., Keller, P. M., Keech, B. J., Davison, A. J., Whang, Y., Morgan, A. J., Kieff, E., and Ellis, R. W. (1987) Varicella-zoster virus as a live vector for the expression of foreign genes. Proc. Natl. Acad. Sci. USA *84*, 3896-3900.

Lu, B. and Federoff, H. J. (1995) Herpes simplex virus type 1 amplicon vectors with glucocorticoid-inducible gene expression. Human Gene Ther. *6*, 419-428.

Lubeck, M. D., Davis, A. R., Chengalvala, M., Natuk, R. J., Morin, J. E., Molnar-Kimber, K., Mason, B. B., Bhat, B. M., Mizutani, S., Hung, P. P., and Purcell, R. H. (1989) Immunogenicity and efficacy testing in chimpanzees of an oral hepatitis B vaccine based on live recombinant adenovirus. Proc. Natl. Acad. Sci. USA *86*, 6763-6767.

Luytjes, W., Krystal, M., Enami, M., Parvin, J. D., and Palese, P. (1989) Amplification, expression, and packaging of a foreign gene by influenza virus. Cell *59*, 1107-1113.

Mackett, M., Smith, G. L., and Moss, B. (1982) Vaccinia virus: a selectable eukaryotic cloning and expression vector. Proc Natl Acad Sci USA *79*, 7415-7419.

Maxwell, I. H., Maxwell, F., Rhode, S. L. III, Corsini, J., and Carlson, J. O. (1993) Recombinant LuIII autonomoous parvovirus as a transient transducing vector for human cells. Hum. Gen. Ther. *4*, 411-450.

Meitin, C. A., Bender, B. S., and Small, P. A., Jr. (1994) Enteric immunization of mice against influenza with recombinant vaccinia. Proc. Natl. Acad. Sci. USA *91*, 11187-11191.

Mitani, K., Graham, F. L., Caskey, C. T., and Kochanek, S. (1995) Rescue, propagation, and partial purification of a helper virus-dependent adenovirus vector. Proc. Natl. Acad. Sci. USA *92*, 3854-3858.

Mittal, S. K., Prevec, L., Graham, F. L., and Babiuk, L. A. (1995) Development of a bovine adenovirus type-3 based expression vector. J. Gen. Vir. *76*, 93-102.

Miyanohara, A., Johnson, P. A., Elam, R. L., Dai, Y., Witztum, J. L., Verma, I. M., and Friedmann, T. (1992) Direct gene transfer to the liver with herpes simplex virus type 1 vectors: transient production of physiologically relevant levels of circulating factor IX. New Biologist *4*, 238-246.

Moreau-Gaudry, F., Ged, C., Barbot, C., Mazurier, F., Boiron, J. M., Bensidhoum, M., Reiffers, J., and de Verneuil, H. (1995) Correction of the enzyme defect in cultured congenital erythropoietic porphyria disease cells by retrovirus-mediated gene transfer. Hum. Gene Ther. *6*, 13-20.

Morgan, A. J., Mackett, M., Finerty, S., Arrand, J. R., Scullion, F. T., and Epstein, M. A. (1988) Recombinant vaccinia virus expressing Epstein-Barr virus glycoprotein gp340 protects cottontop tamarins against EB virus-induced malignant lymphomas. J Med Virol *25*, 189-195.

Morgan, J. R., Barrandon, Y., Green, H., and Mulligan, R. C. (1987) Expression of an exogenous growth hormone gene by transplantable human epidermal cells. Science *237*, 1476-1479.

Morgan, R. A. (1995) Retroviral vectors in human gene therapy. In: Viruses in human gene therapy. Vos, J.-M. H. (ed.), Chapman & Hall Carolina Academic Press, Durham NC, 77-107.

Morgan, R. W., Gelb, J. Jr., Scheurs, C. S., Lutticken, D., Rosenberger, J. K., and Sondermeijer, P. J. A. (1992) Protection of chickens from Newcastle and Marek's diseases with a recombinant herpesvirus of turkeys vaccine expressing the Newcastle disease virus fusion protein. Avian Dis. *36*, 858-870.

Morin, J. E., Lubeck, M. D., Barton, J. E., Conley, A. J., Davis, A. R., and Hung, P. P. (1987) Recombinant adenovirus induces antibody response to hepatitis B virus surface antigen in hamsters. Proc. Natl. Acad. Sci. USA *84*, 4626-4630.

Mulder, W. A. M., Priem, J., Glazenburg, K. L., Wagenaar, F., Gruys, E., Gielkens, A. L. J., Pol, J. M. A., and Kimmaan, T. G. (1994) Virulence and pathogenesis of non-virulent and virulent strains of pseudorabies virus expressing glycoprotein E1 of hog cholera virus. J. Gen. Virol. *75*, 117-124.

Murray, M. G., Kuhn, R. J., Arita, M., Kawamura, N., Nomoto, A., and Wimmer, E. (1988) Poliovirus type1/type3 antigenic hybrid virus constructed in vitro elicits type 1 and type 3 neutralizing antibodies in rabbits and monkeys. Proc. Natl. Acad. Sci. USA *85*, 3202-3207.

Muster, T., Guinea, R., Trkola, A., Purtscher, M., Klima, A., Steindl, F., Palese, P., and Katinger, H. (1994) Cross-neutralizing activity against divergent human immunodeficiency virus type 1 isolates induced by the gp41 sequence ELDKWAS. J. Virol. *68*, 4031-4034.

Natuk, R. J., Chanda, P. K., Lubeck, M. D., Davis, A. R., Wilhelm, J., Hjorth, R., Wade, M. S., Bhat, B. M., Mizutani, S., Lee, S., Eichberg, J., Gallo, R. C., Hung, P. P., and Robert-Guroff, M. (1992) Adenovirus-human immunodeficiency virus (HIV) envelope recombinant vaccines elicit high-titered HIV-neutralizing antibodies in the dog model. Proc. Natl. Acad. Sci. USA *89*, 7777-7781.

Natuk, R. J., Lubeck, M. D., Chanda, P. K., Chengalvala, M., Wade, M. S., Murthy, S. C. S., Wilhelm, J., Vernon, S. K., Dheer, S. K., Mizutani, S., Lee, S.-G., Murthy, K. K., Eichberg, J. W., Davis, A. R., and Hung, P. P. (1993) Immunogenicity of recombinant human adenovirus-human immunodeficiency virus vaccines in chimpanzees. Aids Res. Hum. Ret. *9*, 395-404.

Nazerian, K., Lee, L. F., Yanagida, N., and Ogawa, R. (1992) Protection against Marek's Disease by a fowlpox virus recombinant expressing the glycoprotein B of Marek's Disease Virus. J. Virol. *66*, 1409-1413.

Netter, H. J., Lazinski, D. W., and Taylor, J. M. (1995) Hepatitis delta viruses. In: Viruses in human gene therapy. Vos, J.-M. H. (ed.), Chapman & Hall Carolina Academic Press, Durham NC, 33-52.

Ogawa, R., Yanagida, N., Saeki, S., Saito, S., Ohkawa, S., Gotoh, H., Kodama, K., Kamogawa, K., Sawaguchi, K., and Iritani, Y. (1990) Recombinant fowlpox viruses inducing protective immunity against Newcastle disease and fowlpox viruses. Vaccine *8*, 486-490.

Ohe, Y., Zhao, D., Saijo, N., and Podack, E. R. (1995) Construction of a novel bovine papillomavirus vector without detectable transforming activity suitable for gene transfer. Hum. Gene Ther. *6*, 325-333.

Ohno, T., Gordon, D., San, H., Pompili, V. J., Imperiale, M. J., Nabel, G. J., and Nabel, E. G. (1994) Gene therapy for vascular smooth muscle cell proliferation after arterial injury. Science *265*, 781-784.

Oldfield, E. H., Ram, Z., Culver, K. W., Blaese, R. M., DeVroom, H. L., and Anderson, W. F. (1993) Clinical protocol. Gene therapy for the treatment of brain tumors using intratumoral transduction with the thymidine kinase gene and intravenous ganciclovir.Hum. Gene Ther. *4*, 39-69.

Pages, J.-C., Andreoletti, M., Bennoun, M., Vons, C., Elcheroth, J., Lehn, P., Houssin, D., Chapman, J., Briand, P., Benarous, R., Franco, D., and Weber, A. (1995) Efficient retroviral-mediated gene transfer into primary culture of murine and human hepatocytes: expression of the LDL receptor. Hum. Gen. Ther. *6*, 21-30.

Panicali, D. and Paoletti, E. (1982) Construction of poxviruses as cloning vectors: Insertion of the thymidine kinase gene from herpes simplex virus into the DNA of infectious vaccinia virus. Proc Natl Acad Sci USA *79*, 4927-4931.

Park, K. H., Huang, T., Correia, F. F., and Krystal, M. (1991) Rescue of a foreign gene by Sendai virus. Proc. Natl. Acad. Sci. USA *88*, 5537-5541.

Percy, N., Barclay, W. S., Garcia-Sastre, A., and Palese, P. (1994) Expression of a foreign protein by influenza A virus. J. Virol. *68*, 4486-4492.

Perkus, M. E., Panicali, D., Mercer, S., and Paoletti, E. (1986) Insertion and deletion mutants of vaccinia virus. Virology *152*, 285-297.

Perkus, M. E., Goebel, S. J., Davis, S. W., Johnson, G. P., Norton, E. K., and Paoletti, E. (1991) Deletion of 55 open reading frames from the termini of vaccinia virus. Virology *180*, 406-410.

Perkus, M. E., Taylor, J., Tartaglia, J., Pincus, S., Kauffman, E. B., Tine, J. A., and Paoletti, E. (1995a) Live Attenuated Vaccinia and Other Poxviruses as Delivery Systems: Public Health Issues. Annals of the NY Acad. of Sci. *754*, 222-233.

Perkus, M. E., Tartaglia, J., and Paoletti, E. (1995b) Poxvirus-Based Vaccine Candidates for Cancer, AIDS, and Infectious Diseases. J. Leukocyte Biology *58*, 1-13.

Perricaudet, M. and Stratford-Perricaudet, L. D. (1995) Adenovirus-mediated in vivo gene therapy. In: Viruses in human gene therapy. Vos, J.-M. H. (ed.), Chapman & Hall Carolina Academic Press, Durham NC., 1-32.

Pialoux, G., Excler, J. L., Riviere, Y., Gonzalez-Canali, G., Feuillie, V., Coulaud, P., Gluckman, J.-C., Matthews, T. J., Meignier, B., Kieny, M.-P., Gonnet, P., Diaz, I., Meric, C., Paoletti, E., Tartaglia, J., Salomon, H., Plotkin, S., The AGIS group, and l'Agence Nationale de Recherche sur le SIDA (1995) A prime-boost approach to HIV preventive vaccine using a recombinant canarypox virus expressing glycoprotein 160 (MN) followed by a recombinant glycoprotein 160 (MN/LAI). AIDS Res. and Hum. Retrovir. *11*, 373-381.

Porter, D. C., Ansardi, D. C., Choi, W. S., and Morrow, C. D. (1993) Encapsidation of genetically engineered poliovirus minireplicons which express human immunodeficiency virus type 1 gag and pol proteins upon infection. J. Virol. *67*, 3712-3719.

Post, L. E., Norrild, B., Simpson, T., and Roizman, B. (1982) Chicken ovalbumin gene fused to a herpes simplex virus a promoter and linked to a thymidine kinase gene is regulated like a viral gene. Mol. Cell. Biol. *2*, 233-240.

Prevec, L., Schneider, M., Rosenthal, K. L., Belbeck, L. W., Derbyshire, J. B., and Graham, F. L. (1989) Use of human adenovirus-based vectors for antigen expression in animals. J. Gen. Vir. *70*, 429-434.

Prevec, L., Campbell, J. B., Christie, B. S., Belbeck, L., and Graham, F. L. (1990) A recombinant human adenovirus vaccine against rabies. J. Infec. Dis. *161*, 27-30.

Qingzhong, Y., Barrett, T., Brown, T. D. K., Cook, J. K. A., Green, P., Skinner, M. A., and Cavanagh, D. (1994) Protection against turkey rhinotracheitis pneumovirus (TRTV) induced by a fowlpox virus recombinant expressing the TRTV fusion glycoprotein (F). Vaccine *12*, 569-573.

Quantin., B., Perricaudet, L. D., Tajbakhsh, S., and Mandel, J.-L. (1992) Adenovirus as an expression vector in muscle cells in vivo. Proc. Natl. Acad. Sci. USA *89*, 2581-2584.

Raffel, C., Culver, K., Kohn, D., Nelson, M., Siegel, S., Gillis, F., Link, C. J., Villablanca, J. G., and Anderson, W. F. (1994) Clinical protocol. Gene therapy for the treatment of recurrent pediatric malignant astrocytomas with in vivo tumor transduction with the Herpes Simplex thymidine kinase gene/ganciclovir system. Hum. Gene Ther. *5*, 863-890.

Ragot, T., Finerty, S., Watkins, P. E., Perricaudet, M., and Morgan, A. J. (1993a) Replication-defective recombinant adenovirus expressing the Epstein-Barr virus (EBV) envelope glycoprotein gp340/220 induces protective immunity against EBV-induced lymphomas in the cottontop tamarin. J. Gen. Vir. *74*, 501-507.

Ragot, T., Vincent, N., Chafey, P., Vigne, E., Gilgenkrantz, H., Couton, D., Cartaud, J., Briand, P., Kaplan, J.-C., Perricaudet, M., and Kahn, A. (1993b) Efficient adenovirus-mediated transfer of a human minidystrophin gene to skeletal muscle of *mdx* mice Nature *361*, 647-650.

Randrianarison-Jewtoukoff, V. and Perricaudet, M. (1995) Recombinant adenoviruses as vaccines. Biologicals *23*, 145-157.

Reider, E., Baxt, B., Lubroth, J., and Mason, P. W. (1994) Vaccines prepared from chimeras of foot-and-mouth disease virus (FMDV) induce neutralizing antibodies and protective immunity to multiple serotypes of FMDV. J. Virol. *68*, 7092-7098.

Reimann, B.-Y, Zell, R., and Kandolf, R. (1991) Mapping of a neutralizing antigenic site of coxsackievirus B4 by construction of an antigen chimera. J. Virol. *65*, 3475-3480.

Rodrigues, M., Li, S., Murata, K., Rodriguez, D., Rodriguez, J. R., Bacik, I., Bennink, J. R., Yewdell, J. W., Garcia-Sastre, A., Nussenzweig, R. S., Esteban, M., Palese, P., and Zavala, F. (1994) Influenza and vaccinia viruses expressing malaria $CD8^+$ T and B cell epitopes. J. Immunol. *153*, 4636-4648.

Rolling, F. and Samulski, R. J. (1995) AAV as a viral vector for human gene therapy. Generation of recombinant virus. Mol. Biotech *3*, 9-15.

Rome, J. J., Shayani, V., Newman, K. D., Farrell, S., Lee, S. W., Virmani, R., and Dichek, D. A. (1994) Adenoviral vector-mediated gene transfer into sheep arteries using a double-balloon catheter. Hum. Gene Ther *5*, 1249-1258.

Romero, C. H., Barrett, T., Evans, S. A., Kitching, R. P., Gershon, P. D., Bostock, C., and Black, D. N. (1993) Single capripoxvirus recombinant vaccine for the protection of cattle against rinderpest and lumpy skin disease. Vaccine *11*, 737-742.

Romero, C. H., Barrett, T., Kitching, R. P., Bostock, C., and Black, D. N. (1995) Protection of goats against peste des petits ruminants with recombinant capripoxviruses expressing the fusion and haemagglutinin protein genes of rinderpest virus. Vaccine *13*, 36-40.

Rosenberg, S. A. (1991) Immunotherapy and gene therapy of cancer. Cancer Res. (Suppl.) *51*, 5074s-5079s.

Rosenfeld, M. A., Siegfried, W., Yoshimura, K., Yoneyama, K., Fukayama, M., Stier, L. E., Paakko, P. K., Gilardi, P., Stratford-Perricaudet, L. D., Perricaudet, M., Jallat, S., Pavirani, A., Lecocq, J.-P., and Crystal, R. G. (1991) Adenovirus-mediated transfer of a recombinant a1-antitrypsin gene to the lung epithelium in vivo. Science *252*, 431-434.

Rosenfeld, M. A., Yoshimura, K., Trapnell, B. C., Yoneyama, K., Rosenthal, E. R., Dalemans, W., Fukayama, M., Bargon, J., Stier, L. E., Stratford-Perricaudet, L., Perricaudet, M., Guggino, W. B., Pavirani, A., Lecocq, J.-P., and Crystal, R. G. (1992) In vivo transfer of the human cystic fibrosis transmembrane conductance regulator gene to the airway epithelium. Cell *68*, 143-155.

Ross, L. J. N., Binns, M. M., Tyers, P., Pastorek, J., Zelnik, V., and Scott, S. (1993) Construction and properties of a turkey herpesvirus recombinant expressing the Marek's disease virus homologue of glycoprotein B of herpes simplex virus. J. Gen. Vir. *74*, 371-377.

Rother, R. P., Squinto, S. P., Mason, J. M., and Rollins, S. A. (1995) Protection of retroviral vector particles in human blood through complement inhibition. Hum. Gene Ther. *6*, 429-435.

Ruby, J., Ramshaw, A., Karupiah, G., and Ramshaw, I. (1992) Recombinant virus vectors that coexpress cytokines – a new vaccine strategy. Vaccine Res. *1*, 347-356.

Russell, D. W., Berger, M. S., and Miller, A. D. (1995) The effects of human serum and cerebrospinal fluid on retroviral vectors and packaging cell lines. Hum. Gene Ther. *6*, 635-641.

Saito, I., Oya, Y., Yamamoto, K., Yuasa, T., and Shimojo, H. (1985) Construction of nondefective adenovirus type 5 bearing a 2.8-kilobase hepatitis B virus DNA near the right end of its genome. J.Virol. *54*, 711-719.

Samulski, R. J. (1995) Adeno-associated viral vectors. In: Viruses in human gene therapy. Vos, J.-M. H. (ed.), Chapman & Hall Carolina Academic Press, Durham NC, 53-76.

Schlesinger, S. (1995) RNA viruses as vectors for the expression of heterologous proteins. Mol. Biotech. *3*, 155-165.

Simmer, B., Alt, M., Buckreus, I., Berthold, S., Fleckenstein, B., Platzer, E., and Grassmann, R. (1991) Persistence of selectable herpesvirus saimiri in various human haematopoietic and epithelial cell lines. J. Gen. Vir. *72*, 1953-1958.

Small, P. A. Jr., Smith, G. L., and Moss, B. (1985) Intranasal vaccination with a recombinant vaccinia virus containing influenza hemagglutinin prevents both influenza virus pneumonia and nasal infection: intradermal vaccination prevents only viral pneumonia. In: Vaccines

1985. Lerner, R. A., Chanock, R. M., and Brown, F. (eds.), Cold Spring Harbor Laboratory, Cold Spring Harbor, NY, 175-176.

Smith, A. D., Resnick, D. A., Zhang, A., Geisler, S. C., Arnold, E., and Arnold, G. F. (1994) Use of random systematic mutagenesis to generate viable human rhinovirus 14 chimeras displaying human immunodeficiency virus type 1 V3 loop sequences. J. Virol. *68*, 575-579.

Sobol, R. E., Royston, I., Fakhrai, H., Shawler, D. L., Carson, C., Dorigo, O., Gjerset, R., Gold, D. P., Kozoil, J., Mercola, D., Van Beveren, C., and Wilson, D. (1995) Clinical protocol. Injection of colon carcinoma patients with autologous irradiated tumor cells and fibroblasts genetically modified to secrete Interleukin-2 (IL-2): a phase I study. Hum. Gene Ther. *6*, 195-204.

Southern, P. and Berg, P. (1982) Mammalian cell transformation with SV40 vector. In: Eukaryotic Viral Vectors. Gluzman, Y. (ed.), Cold Spring Harbor Laboratory, Cold Spring Harbor, New York, 41-45.

Stratford-Perricaudet, L. D., Levrero, M., Chasse J.-F., Perricaudet, M., and Briand, P. (1990) Evaluation of the transfer and expression in mice of an enzyme-encoding gene using a human adenovirus vector. Hum. Gene Ther. *1*, 241-256.

Stratford-Perricaudet, L. D., Makeh, I., Perricaudet, M., and Briand, P. (1992) Widespread long-term gene transfer to mouse skeletal muscles and heart. J. Clin. Invest. *90*, 626-630.

Tacket, C. O., Losonsky, G., Lubeck, M. D., Davis, A. R., Mizutani, S., Horwith, G., Hung, P., Edelman, R., and Levine, M. M. (1992) Initial safety and immunogenicity studies of an oral recombinant adenohepatitis B vaccine. Vaccine *10*, 673-676.

Tartaglia, J., Perkus, M. E., Taylor, J., Norton, E. K., Audonnet, J.-C., Cox, W. I., Davis, S. W., VanderHoeven, J., Meignier, B., Riviere, M., Languet, B., and Paoletti, E. (1992) NYVAC: a highly attenuated strain of vaccinia virus. Virology *188*, 217-232.

Tartaglia, J., Franchini, G., Robert-Guroff, M., Abimuku, A., Benson, J., Limbach, K., Wills, M., Gallo, R. C., and Paoletti, E. (1993a) Highly attenuated poxvirus vector strains, NYVAC and ALVAC, in retrovirus vaccine development. In: Retroviruses of Human A.I.D.S. and Related Animal Diseases. Girard, M. and Valette, L. (eds.), Huitieme Colloque des Cent Gardes Marnes-La-Coquette, Paris France, Pasteur Merieux Serums et Vaccins/ANRS, Paris, 293-298.

Tartaglia, J., Jarrett, O., Neil, J. C., Desmettre, P., and Paoletti, E. (1993b) Protection of cats against feline leukemia virus by vaccination with a canarypox virus recombinant, ALVAC-FL. J. Virol. *67*, 2370-2375.

Tartaglia, J., Taylor, J., Cox, W. I., Audonnet, J.-C., Perkus, M. E., Radaelli, A., de Giuli Morghen, C., Meignier, B., Riviere, M., Weinhold, K. J., and Paoletti, E. (1993c) Novel poxvirus strains as research tools and vaccine vectors. In: AIDS Research Reviews (Vol 3). Koff, W. C., Wong-Staal, F., and Kennedy, R. C. (eds.), Marcel Dekker, New York, 361-378.

Taylor, J., Meignier, B., Tartaglia, J., Languet, B., VanderHoeven, J., Franchini, G., Trimarchi, C., and Paoletti, E. (1995) Biological and immunogenic properties of a canarypox-rabies recombinant, ALVAC-RG (vCP65) in non-avian species. Vaccine *13*, 539-549.

Taylor, J., Tartaglia, J., Moran, T., Webster, R. G., Bouquet, J.-F., Quimby, F. W., Holmes, D., Laplace, E., Mickle, T., and Paoletti, E. (1992a) The role of poxvirus vectors in influenza vaccine development. In: Proceedings of the Third International Symposium on Avian Influenza. University of Wisconsin-Madison Extension Duplicating Services, 311-335.

Taylor, J., Weinberg, R., Kawaoka, Y., Webster, R., and Paoletti, E. (1988a) Protective immunity against avian influenza induced by a fowlpox virus recombinant. Vaccine *6*, 504-508.

Taylor, J., Weinberg, R., Languet, B., Desmettre, P., and Paoletti, E. (1988b) Recombinant fowlpox virus inducing protective immunity in non-avian species. Vaccine *6*, 497-503.

Taylor, J., Edbauer, C., Rey-Senelonge, A., Bouquet, J.-F., Norton, E., Goebel, S., Desmettre, P., and Paoletti, E. (1990) Newcastle disease virus fusion protein expressed in a fowlpox virus recombinant confers protection in chickens. J. Virol. *64*, 1441-1450.

Taylor, J., Trimarchi, C., Weinberg, R., Languet, B., Guillemin, F., Desmettre, P., and Paoletti, E. (1991a) Efficacy studies on a canarypox-rabies recombinant virus. Vaccine *9*, 190-193.

Taylor, J., Pincus, S., Tartaglia, J., Richardson, C., Alkhatib, G., Briedis, D., Appel, M., Norton, E., and Paoletti, E. (1991b) Vaccinia virus recombinants expressing either the measles virus fusion or hemagglutinin glycoprotein protect dogs against canine distemper virus challenge. J. Virol. *65*, 4263-4274.

Taylor, J., Weinberg, R., Tartaglia, J., Richardson, C., Alkhatib, G., Briedis, D., Appel, M., Norton, E., and Paoletti, E. (1992b) Nonreplicating viral vectors as potential vaccines: Recombinant canarypox virus expressing measles virus fusion (F) and hemagglutinin (HA) glycoproteins. Virology *187*, 321-328.

van der Leek, M. L., Feller, J. A., Sorensen, G., Isaacson, W., Adams, C. L., Borde, D. J., Pfeiffer, N., Tran, T., Moyer, R. W., and Gibbs, E. P. J. (1994) Evaluation of swinepox virus as a vaccine vector in pigs using an Aujeszky's disease (pseudorabies) virus gene insert coding for glycoproteins gp50 and gp63. Vet. Record *134*, 13-18.

van Zijl, M., Wensvoort, G., de Kluyver, E., Hulst, M., van der Gulden, H., Gielkens, A, Berns, A., and Moormann, R. (1991) Live attenuated pseudorabies virus expressing envelope glycoprotein E1 of hog cholera virus protects swine against both pseudorabies and hog cholera. J. Virol. *65*, 2761-2765.

Venkatesh, L. K., Arens, M. Q., Subramanian, T., and Chinnadurai, G. (1990) Selective induction of toxicity to human cells expressing human immunodeficiency virus type 1 Tat by a conditionally cytotoxic adenovirus vector. Proc. Natl. Acad. Sci. USA *87*, 8746-8750.

Vos, J.-M. H. (1995) Herpesviruses as genetic vectors. In: Viruses in human gene therapy. Vos, J.-M. H. (ed.) Chapman & Hall Carolina Academic Press, Durham NC, 109-140.

Wagner, E., Zatloukal, K., Cotten, M., Kirlappos, H., Mechtler, K., Curiel, D. T., and Birnstiel, M. L. (1992) Coupling of adenovirus to transferrin-polylysine/DNA complexes greatly enhances receptor-mediated gene delivery and expression of transfected genes. Proc. Natl. Acad. Sci. USA *89*, 6099-6103.

Wei, C.-M., Gibson, M., Spear, P. G., and Scolnick, E. M. (1981) Construction and isolation of a transmissable retrovirus containing the src gene of Harvey murine sarcoma virus and the thymidine kinase gene of herpes simplex virus type 1. J. Virol. *39*, 935-944.

Wei, M. X., Tamiya, T., Hurford, R. K., Jr., Boviatsis, E. J., Tepper, R. I., and Chiocca, E. A. (1995) Enhancement of interleukin-4-mediated tumor regression in athymic mice by in situ retroviral gene transfer. Hum. Gene Ther. *6*, 437-443.

Welsh, M. J., Zabner, J., Graham, S. M., Smith, A. E., Moscicki, R., and Wadsworth, S. (1995) Clinical Protocol. Adenovirus-mediated gene transfer for cystic fibrosis: Part A. Safety of dose and repeat administration in the nasal epithelium. Part B: Clinical efficacy in the maxillary sinus. Hum. Gen. Ther. *6*, 205-218.

Wesseling, J. G., Godeke, G.-J., Schijns, V. E. C. J., Prevec, L., Graham, F. L., Horzinek, M. C., and Rottier, P. J. M. (1993) Mouse hepatitis virus spike and nucleocapsid proteins expressed by adenovirus vectors protect mice against lethal infection. J. Gen. Vir. *74*, 2061-2069.

Wild, F., Giraudon, P., Spehner, D., Drillien, R., and Lecocq, J.-P. (1990) Fowlpox virus recombinant encoding the measles virus fusion protein: protection of mice against fatal measles encephalitis. Vaccine *8*, 441-442.

Wills, K. N., Maneval, D. C., Menzel, P., Harris, M. P., Sutjipto, S., Vaillancourt, M.-T., Huang, W.-M., Johnson, D. E., Anderson, S. C., Wen, S. F., Bookstein, R., Shepard, H. M., and Gregory, R. J. (1994) Development and characterization of recombinant adenoviruses encoding human p53 for gene therapy of cancer. Hum. Gene Ther. *5*, 1079-1088.

Wilson, J. M., Grossman, M., Raper, S. E., Baker, J. R., Newton, R. S., and Thoene, J. G. (1992) Clinical protocol. Ex vivo gene therapy of familial hypercholesterolemia. Hum Gen. Ther. *3*, 179-222.

Wolfe, J. H., Deshmane, S. L., and Fraser, N. W. (1992) Herpesvirus vector gene transfer and expression of b-glucuronidase in the central nervous system of MPS VII mice. Nature gen. *1*, 379-384.

Zabner, J., Couture, L. A., Gregory, R. J., Graham, S. M., Smith, A. E., and Welsh, M. J. (1993) Adenovirus-mediated gene transfer transiently corrects the chloride transport defect in nasal epithelia of patients with cystic fibrosis. Cell *75*, 207-216.

Zhang, L., Sato, S., Kim, J. I., and Roos, R. P. (1995) Theiler's virus as a vector for foreign gene delivery. J. Virol. *69*, 3171-3175.

Zhang, Y., Mukhopadhyay, T., Donehower, L. A., Georges, R. N., and Roth, J. A. (1993) Retroviral vector-mediated transduction of K-ras antisense RNA into human lung cancer cells inhibits expression of the malignant phenotype. Hum. Gen. Ther. *4*, 451-460.

Zhou, X., Berglund, P., Rhodes, G., Parker, S. E., Jondal, M., and Liljestrom, P. (1994) Self-replicating Semliki Forest virus RNA as a recombinant vaccine. Vaccine *12*, 1510-1514.

3.7 Naked DNA Vaccination

John W. Shiver, Jeffrey B. Ulmer, John J. Donnelly, and Margaret A. Liu

3.7.1 Introduction

There has been a great deal of interest recently in the use of plasmid DNA encoding antigens for immunization (Waine, 1994). This interest began with an observation made, during experiments to test potential approaches to gene therapy, that direct injection into mouse muscle tissue of "naked" plasmid DNA or RNA encoding either ß-galactosidase, luciferase, or chloramphenicol acetyltransferase reporter genes resulted in functional gene expression by myocytes (Wolff et al., 1990). These investigators also reported similar results following DNA vaccination of Rhesus monkeys demonstrating that this phenomenon was not limited to rodents (Jiao et al., 1992). These results were particularly surprising because higher levels of gene expression resulted from using DNA alone compared to vaccination with DNA formulated with cationic lipid, hence the use of the qualifier "naked".

The potential significance these results held for vaccine development became apparent when Ulmer et al. (1993) demonstrated that immunization of mice with a plasmid encoding a full-length influenza nucleoprotein (NP) gene conferred protection to vaccined mice from lethal challenge with a flu strain that was heterosubtypic to the one from which the NP gene was derived (H3 vs. H1, which were separated by 34 years of virus evolution). Protection in this challenge model was probably mediated by antigen-specific T lymphocyte immunity (anti-NP antibodies did not confer protection) and, significantly, vaccination produced potent cytotoxic T lymphocyte (CTL) responses directed against an NP peptide epitope. DNA vaccination of mice using plasmid constructs encoding the hemagglutinin (HA) gene resulted in strong neutralizing antibodies against influenza and complete protection from homologous viral challenge (Montgomery et al., 1993, Ulmer et al., 1993, 1994). Other laboratories reported having achieved partial protection from influenza challenge with HA DNA-vaccinated chickens, although neutralizing antibodies were not detected (Robinson et al., 1993; Webster et al., 1994), and HIV gp160-specific antibody and

T cell responses in mice using a gp160/rev-expressing plasmid (Wang et al., 1993; Coney et al., 1994).

A sampling of the recent literature reveals that DNA vaccines have been used to elicit specific immune responses against a variety of antigens and animal species in addition to those noted above, such as hepatitis B surface antigen in mice (Davis et al., 1993, 1994), herpes simplex virus-1 glycoprotein B in mice (Manickan et al., 1995), bovine herpesvirus 1 glycoprotein IV in cattle (Cox et al., 1993), rabies virus glycoprotein in mice (Xiang et al., 1994, 1995), malaria circumsporozoite protein in mice (Sedegah et al., 1994; Hoffman et al., 1994), *Leishmania* gp63 in mice (Xu and Liew, 1995), lymphocytic choriomeningitis virus (LCMV) NP in mice (Pedroz Martins et al., 1995; Yokoyama et al., 1995), carcinoembryonic antigen in mice (Conry et al., 1994), MHC Class I antigen in rats (Geissler et al., 1994), cottontail rabbit papillomavirus (CRPV) L1 in rabbits (Donnelly et al., 1996b), and *M. tuberculosis* antigen 85 complex proteins in mice (Huygen et al., 1996) comprising a partial list. A common feature of most of these reports has been the development of antibody, CTL, and, to a lesser extent, helper T cell responses in vaccinated animals as was described above for influenza and HIV antigens.

Thus, DNA vaccines have shown a remarkable ability to elicit both cellular and humoral immunities. Generally, both types of these immune responses have been obtained only with live, attenuated viral vaccines or live, recombinant vectors (e.g., vaccinia, adenovirus, etc.) rather than killed organisms or recombinant subunit vaccines that produce humoral immunity predominantly. Whether a vaccine can elicit strong T lymphocyte and CTL activities appears to correlate with its ability to produce protein synthesis *in vivo* following vaccination. *In vivo* synthesis of foreign protein allows processing and presentation of vaccine-derived peptide epitopes in association with MHC Class I molecules, which is required for the generation of CD8+ CTL responses, as well as MHC Class II. Furthermore, because DNA vaccine-derived antigens are synthesized *in vivo*, native protein structure and conformation, including the potential for oligomer formation and appropriate post-translational modifications, should be obtained. Thus, it could be expected that more potent neutralizing antibody responses would be generated using DNA vaccines than may be obtained when using vaccines based on chemically modified proteins.

3.7.2 Vaccine Plasmid Design

The essential components of plasmid DNA designed for vaccination are basically the same as for mammalian expression vectors: a promoter/enhancer and polyadenylation signal/transcript termination sequences, which may be selected from a variety of mammalian or viral sources; and, a bacterial-derived plasmid backbone containing both an origin of replication for *E. coli* and a selectable marker. It may also be advantageous to include intron-containing sequences that have been shown to greatly improve expression within transfected cell lines for many constructs. Obviously, there are many permutations possible within this framework. The promoter/enhancers that have been most widely used for DNA vaccination are the cytomegalovirus immediate/early promoter (pCMVIE) enhancer (Boshart et al., 1985) and the Rous sarcoma virus (RSV) LTR (Gorman et al., 1982). Not surprisingly, perhaps, these promoters (especially pCMVIE) are well-known for mediating the highest levels of protein expression by *in vitro* transfection of cell lines compared to most other promoters. While it is difficult to make precise correlations between the abilities of a plasmid to express heterologous protein *in vitro* and to elicit immune responses *in vivo*, it appears that, in general, plasmids providing better expression *in vitro* elicit more potent immune responses *in vivo*.

Our experiments to date have indicated that pCMVIE provides the greatest utility for DNA vaccine use. Figure 3.7.1 shows a schematic of a basic vaccine vector, V1J. This vector, which was described previously (Montgomery et al., 1993), is comprised of pCMVIE, intron A derived from CMV, bovine growth hormone (BGH) polyadenylation/transcript termination sequence, and ampicillin resistance (amp^r) gene, contained within a pUC plasmid backbone from which the lac operon and multicloning site have been deleted. Other vectors using a kan^r gene in place of amp^r have also been used for vaccination. Genes may be inserted at either the BglII (preferred) or PstI restriction enzyme sites. V1J provides a potent and versatile expression vector which does not replicate in mammalian cells, does not contain any sequences known to promote plasmid integration into host genomic DNA, and can be produced in large quantities by growth in *E. coli*. These properties should help ensure the safety of DNA vaccination, by minimizing the probability for cell-transforming integration events. In addition, preparation of large quantities of plasmid is achievable. While it is probable that no single construct may be the best vaccine vehicle for all possible genes and that additional improvements may still be possible, V1J has proven to work very well with a variety of viral and non-viral genes and animal species (mice, African green and Rhesus monkeys, ferrets, and rabbits).

3.7.3 Formulation and Delivery of DNA

It is clear from data obtained from many animal models that “naked” DNA vaccines can elicit diverse and potent immune responses. However, the mechanism of DNA uptake as well as the identity of the antigen presenting cells (APC) to the immune system remain unknown. Without this knowledge it is probable that the best formulation/delivery system is also unknown.

A variety of conditions have been used for vaccinations by different groups. While our best results have been obtained using normal saline solutions of plasmid, other groups have reported achieving improved immune responses with other formulations. Davis et al., (1993, 1994) used 25 % sucrose in PBS either as a pre-treatment of

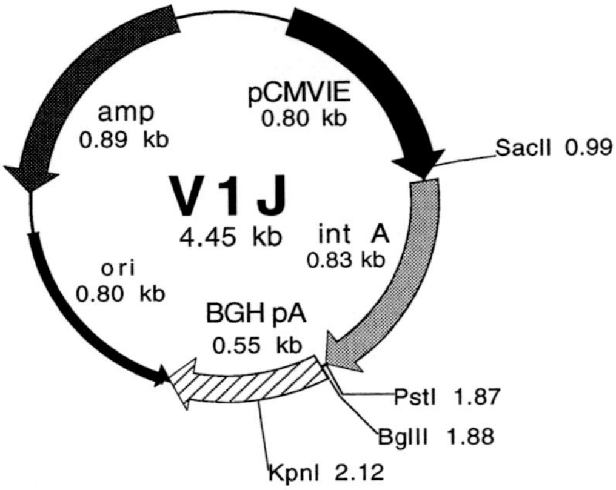

Fig. 3.7.1: Schematic diagram of vaccination vector, V1J. DNA segments are drawn to scale with useful restriction enzyme sites labeled.

muscle prior to injection of DNA or co-delivered with DNA. Weiner and co-workers (Wang et al., 1993; Coney et al., 1994) used solutions of the lidocaine-like anaesthetic, bupivicaine, to pre-treat muscle sites 1-2 days prior to DNA inoculation as well as co-delivery with plasmid. In experiments focused on gene transfer rather than vaccination, DNA was formulated with liposomes for optimal delivery (Zhu et al., 1993). The latter approach has been used especially for intravenous or intraperitoneal injection routes rather than intramuscular. We have not found sucrose or bupivicaine formulations to give increased immune responses compared to DNA dissolved in saline solutions in our experiments to date using a high-expressing vector (Shiver et al., unpublished data). While only a small percentage of myotubules appear to take up and express DNA following intramuscular injection using saline formulation, this has been sufficient for obtaining significant immune responses (although other cell types may play important roles for eliciting immune responses). However, optimizing the formulation and delivery of DNA vaccines should remain an active area of research.

"Gene guns" have also been used as an alternative means for delivering DNA. This technique uses an instrument designed to propel DNA-coated gold particles directly into cells within the epidermis and dermis. Particle bombardment delivery is a distinctly different approach to DNA vaccination by direct injection but some of the results obtained by this method are worth noting here. The earliest report for eliciting immune responses by DNA delivery was obtained with mice using a "gene gun" and a plasmid expressing human growth hormone (hGH) (Tang et al., 1992). These researchers observed detectable levels of hGH protein and anti-hGH antibodies in the serum of vaccined mice. Other investigators subsequently reported development of antibody and T cell responses for influenza, HIV, and hGH by particle bombardment of DNA (Eisenbraun et al., 1993; Haynes et al., 1993; Fuller and Haynes, 1994). A careful comparison of immune responses elicited by vaccination using "naked" DNA vs. "gene gun" delivery has yet to be made although it is clear that small quantities of DNA ($\leq$1 μg) can successfully generate immune responses by either method (figs. 3.7.3 and 3.7.4; Ulmer et al., 1994; Haynes et al., 1993; Fynan et al., 1993).

3.7.4 DNA Vaccine-Mediated Immunity

DNA vaccines are capable of eliciting both humoral and cellular immunity as described previously. For most antigens, this immunity is both potent and long-lived in certain animal models. Because DNA vaccines are novel it is of particular interest to determine in some detail the characteristics of the immune responses generated. This discussion will focus on results obtained using vaccine plasmids encoding influenza- and HIV-derived genes.

3.7.4.1 Humoral Immunity

Initial experiments utilizing plasmids containing the influenza NP gene focused on determining whether anti-NP CTL responses and cross-strain protection could be induced in mice (Ulmer et al., 1993). Over the course of these experiments, however, it became apparent that anti-NP antibodies also were generated following DNA vaccination, with ELISA endpoint titers often exceeding 10^6 (although anti-NP antibodies do not provide protection to influenza infection). Subsequently, these antibody titers were found to persist throughout the lifetime of the mouse and could be induced with 1 μg or less of DNA (Yankauckas et al., 1993; Ulmer et al., 1994). These were surprising results considering that NP may be expected to localize within the nucleus, based upon its viral function and because the NP protein sequence contains apparent nuclear localization motifs. However, it is possible that some of the NP can be extruded from cells, albeit by a nonclassical secretory pathway, and fluorescent microscopy has shown NP at the plasma membrane surface when expressed by influenza virus-infected cell lines (Stitz et al., 1990).

Influenza HA DNA vaccination also induced high levels of antibodies including those with hemagglutination-inhibiting (HI) activitity (which represents a key subpopulation of viral neutralizing antibodies). As with NP-specific antibodies, long-lived HI antibody titers that were comparable to those obtained from mice surviving live influenza challenges were elicited by a single low dose of DNA. Similar results were found with HA DNA vaccination of African green monkeys although somewhat higher doses of DNA (10 μg) with boosting were generally required to achieve maximal responses (Donnelly et al., 1995). Significantly, HI antibody titers obtained in monkeys after two immunizations were comparable to or better than those obtained using licensed influenza vaccines and exceeded levels expected to confer protective humoral immunity in humans.

We have also tested DNA vaccines using genes derived from HIV-1. One of these constructs produces a secreted form of gp120 by using a chimeric gene comprised of the signal peptide sequence of the tissue-specific plasminogen activator gene (tPA) and the mature HIV-1 gp120-encoding sequence. This construct has been shown to mediate rev-independent expression of secreted gp120 (Chapman et al., 1991). Mice and monkeys vaccinated with gp120 DNA developed gp120-specific antibody responses that depended on both dose and number of vaccinations (Shiver et al., 1995a, 1995b; Liu et al., 1994). For mice, maximum antibody titers were obtained using 2-3 vaccinations with $\geq$ 10 μg of DNA (see fig. 3.7.2). Interestingly, high levels of antibody responses to gp120 were found even though the inoculating gp120 DNA was derived from the MN isolate of HIV-1 while the assay antigen was derived from II-IB. The primary structure of these genes diverge by $\sim$ 25 % showing that antibody responses against conserved domains were generated by DNA vaccination. Similar titers were obtained using either intramuscular (i.m.) or intradermal (i.d.) vaccination routes although the i.d. route generally gave greater variation of response as well as lower magnitudes of T lymphocyte reactivities (Shiver et al., unpublished observations). Based upon our overall results, i.m. vaccination appears to be the preferred route.

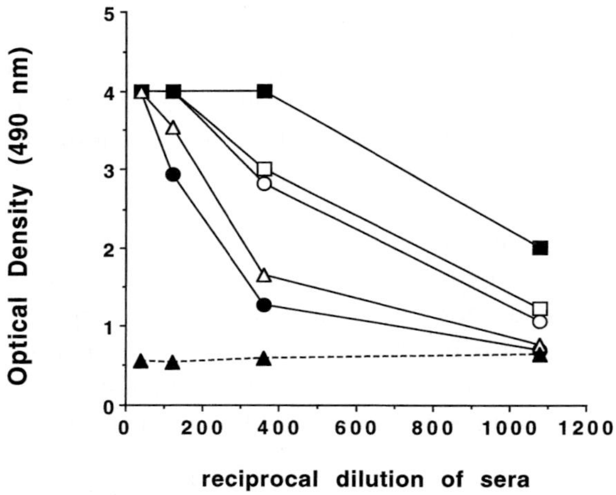

Fig. 3.7.2: Measurement of anti-gp120 serum antibody responses for mice receiving a tPA-gp120 (MN) DNA vaccine. Five Balb/c mice (solid lines) were vaccinated 3 times with 100 μg of DNA at 0, 4, and 8 weeks. Sera were obtained 3 weeks following the final vaccination and antibody responses were tested by ELISA in comparison with a pre-vaccination serum pool. rgp120 (IIIB) (Repligen, Cambridge, MA, USA) was used for assay antigen.

gp120 DNA vaccination of African green monkeys elicited similar antibody titers as found with mice although larger doses of DNA were required (0.2-2 mg). Sera from these monkeys contained low titers (<1/100) of HIV-1 (homologous) neutralizing antibodies which could not be increased with additional vaccinations. Recent reports in the literature suggest that oligomeric, membrane-bound *env*, similar to that produced by viral infection, may provide superior immunogen for generating neutralizing antibodies compared with secreted, monomeric gp120 (Moore et al., 1993, 1994). We are currently testing this possibility with DNA vaccines that express oligomeric envelope protein.

3.7.4.2 Cellular Immunity

3.7.4.2.1 Generation of Anti-Viral CTL

DNA vaccination using influenza NP-expressing plasmid clearly established that potent anti-NP CTL responses could be induced in vaccinated mice (Ulmer et al., 1993, 1994). These responses were detected in splenocytes from immunized animals following *in vitro* restimulation of splenocytes using either influenza virus infection or treatment with an MHC Class I-restricted NP peptide epitope (a.a. 147-155). Anti-NP CTL showed significant levels of cytotoxicity with P815 mastocytoma target cells sensitized with antigen by either means (see fig. 3.7.3). Significantly, primary CTL could also be detected with these target cells using the less sensitive method of restimulation with ConA and IL-2 treatment in the absence of influenza antigen. Furthermore, CTL responses were long-lived and could be elicited by small amounts of DNA (see fig. 3.7.3).

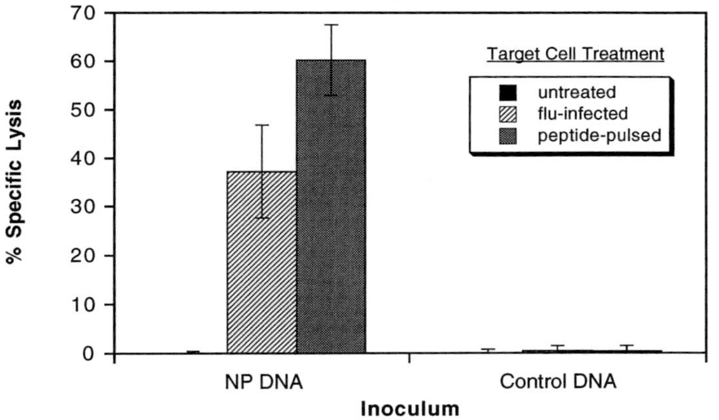

Fig. 3.7.3: Induction of anti-NP CTL responses in mice by injection of NP DNA. Female BALB/c mice (4- to 6-weeks) were injected i.m. with NP DNA (A/PR/8/34) (1 μg) 3 times at 3-week intervals. Spleen cells were necropsied 2 years later and restimulated *in vitro* with influenza virus-infected, irradiated syngeneic spleen cells for 7 days in the presence of IL-2. Anti-NP CTL (effector:target = 25) were detected by lysis of P815 target cells infected with influenza virus or pulsed with the H-2^d-restricted CTL epitope (amino acids 147-155). Untreated target cells were not lysed. Data represent % specific lysis ± sd, where n=3.

gp120 DNA vaccination also induced CTL activities in mice (Shiver et al., manuscript in preparation). These responses were detected using a previously defined peptide epitope for BALB/c mice corresponding to the V3 loop domain of gp120

(Takahashi et al., 1988) for both *in vitro* restimulation of T cells and sensitization of P815 target cells. We also tested for CTL activities in Rhesus monkeys that had been vaccinated using both a gp120 and a dicistronic gp160/rev plasmids. Strong anti-*env* CD8+ CTL were induced in 4/4 monkeys following two vaccinations with these vectors (Yasutomi et al., manuscript in preparation). These responses were boosted to higher levels by additional vaccination and lasted at least seven months following the final inoculation date (the longest time point tested). Thus, these experiments prove that DNA vaccines' ability to induce MHC I-restricted CTL responses is not limited to a single antigen and is effective in a nonhuman primate species as well as rodents.

CTL have also been induced by DNA vaccines encoding rabies virus gp (Xiang et al., 1993), LCMV NP (Pedroz Martins et al., 1995; Yokoyama et al., 1995; Zarozinski et al., 1995), malaria circumsporozoite protein (Sedegah et al., 1994), and MHC class I (Geissler et al., 1994).

3.7.4.2.2 Helper T Lymphocyte Responses

Lymphocyte proliferation following treatment with recall antigen suggests the presence of antigen-specific memory T cells. Additionally, the effector functions of T cells may largely result from the types of cytokines that activated T cells secrete following antigenic stimulation. T cells may be grouped into functional subsets by the cytokine patterns they generate (Street and Mosmann, 1991). Type 1-like helper T cell (T_H1) cytokines (IL-2, γ-interferon) have been shown to promote cellular immune responses, including CTL activities, and the IgG2a immunoglobulin subtype while type two (T_H2) cytokines (IL-4, IL-5, IL-6, IL-10) play important roles for development of antibody responses. IL-10 and γ-interferon can also cross-regulate between T cell subtypes. The type of T cell help generated has been shown to have an important role in the development of protective immunity, or susceptibility, to disease in murine models (Romagnani, 1991; Heinzel, 1995). Evidence is increasing that similar helper T cell subsets exist in humans demonstrating the relevance of the murine data (Romagnani, 1995). For HIV infections, T_H1-like responses have been linked to maintenance of disease-free infection (Clerici and Shearer, 1993). These observations suggest that the type of helper T function immunization elicits may play an important role in vaccine efficacy.

Immunization with DNA encoding HIV, influenza, and tuberculosis genes resulted in the generation of lymphocytes capable of proliferation and cytokine secretion during culture with protein antigen. Mice developed long-term T cell memory (at least seven months following the final immunization) responses after two vaccinations with plasmids encoding either HIV gp120 or *rev* (Shiver et al., 1995a,b). Strong proliferation was observed at all DNA doses tested, down to 1 μg, even in the absence of measureable antigen-specific serum antibody responses, indicating that less antigen may be required to elicit T cell responses by DNA vaccination than for antibody generation. Supernatants from these lymphocyte cultures showed high levels of IL-2 and γ-interferon for lymphocytes from all vaccinees with little or no IL-4 indicating that DNA vaccination produced T_H1-like responses. For gp120 DNA, similar lymphocyte responses were found for all lymph sites tested including the spleen, peripheral blood mononuclear cells, and the inguinal, iliac, and mesenteric lymph nodes. Separation of lymphocyte populations from spleens for these experiments showed that CD4+ T cells were the primary responders although CD8+ T cells also showed some proliferation and cytokine secretion (Shiver et al., manuscript in preparation).

Influenza NP and HA DNA mouse vaccinees also showed *in vitro* antigen-specific lymphocyte memory responses using either live or inactivated virus as well as purified antigen (Ulmer et al., unpublished results). Vaccination with DNA encoding *M. tuberculosis* antigen 85 similarly resulted in strong proliferative responses (Huygen et al., 1996). The profile of cytokines secreted in responses to purified antigen 85 also was found to be indicative of a T_H1-like T cell phenotype as found for HIV *env* vaccines. In addition, GM-CSF was secreted in significant amounts. Both γ-interferon and GM-CSF are thought to be important cytokines for immunity to tuberculosis.

Interestingly, Fuller and Haynes (1994) reported that gene gun immunization with an HIV gp120 gene induced T_H1-like cytokine responses which switched to a T_H2-type of cytokine profile following subsequent immunizations. We have not observed this phenomenon with any of our vaccine constructs. To date, i.m. DNA injection helper T cell profiles have been stable and durable regardless of either the number of vaccinations or type of gene used. Other investigators have also reported T_H1-like cytokine responses following DNA vaccination (Xu and Liew, 1995; Manickan et al., 1995).

3.7.5 Protection in Animal Challenge Models

As described previously, vaccination of mice with either NP or HA DNA provided protection from lethal influenza challenge. While HA DNA elicited protective immunity for homologous viral challenge (Montgomery et al., 1993; Ulmer et al., 1994), NP DNA vaccination protected against a heterosubtypic virus that arose 34 years after the strain from which the NP gene had been cloned (Ulmer et al., 1993). Not only were mice protected from death in these experiments, but significant reduction in morbidity, determined by weight loss following infectious challenge, was also obtained. Figure 3.7.4 shows the results of a homologous (influenza A/PR8/34) challenge of BALB/c mice following three vaccinations using a 1 μg dose of DNA. All HA DNA vaccinees survived a lethal challenge that killed nearly all control mice which had received parental plasmid that did not contain an influenza gene. Moreover, the HA DNA vaccinated mice showed no detectable signs of morbidity (e.g., weight loss).

Donnelly et al. (1995) extended these observations with the DNA influenza vaccine to a ferret challenge model. Ferrets are exquisitely susceptible to influenza infection within the upper respiratory tract and are considered the best available model for

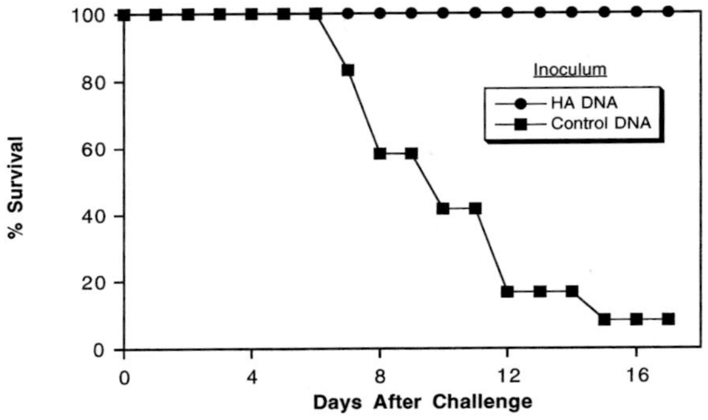

Fig. 3.7.4: Generation of protective immunity by injection of influenza HA DNA. Female BALB/c mice (4- to 6-weeks) were injected i.m. with control DNA not containing a gene insert or with HA DNA (A/PR/8/34) (1 μg) 3 times at 3-week intervals, then challenged 3 weeks later with a lethal dose of A/PR/8/34 influenza virus. Data are plotted as % survival versus days after challenge.

evaluating antigenic drift of influenza field isolates. Although some strains of influenza virus can induce febrile responses in ferrets, the principal disease manifestations in our studies of recent human influenza isolates were sneezing and thickened nasal mucus. The progress of influenza infection in the ferrets was monitored by recovery of virus from nasal washes. This method also is used for evaluation of the efficacy of influenza vaccines in humans. Immunization with DNA encoding the HA, NP and matrix (M1) protein of an H3N2 influenza virus was more effective than immunization with whole inactivated influenza virus at reducing nasal viral shedding when the ferrets were challenged with an antigenic drift variant also of the H3N2 subtype.

The protective efficacy of DNA vaccination also has been demonstrated in other animal challenge models including bovine herpesvirus (Cox et al., 1993), rabies virus (Xiang et al., 1994, 1995), LCMV (Pedroz-Martins et al., 1995; Yokoyama et al., 1995; Zarozinski et al., 1995), CRPV (Donnelly et al., 1996), and *Plasmodium yoelii* malaria (Sedegah et al., 1994; Hoffman et al., 1994). Cumulatively, these studies argue strongly for a potential role for DNA vaccines in the protection of human and animal populations from infectious diseases.

3.7.6 Conclusions and Future of DNA Vaccines

The experiments described above demonstrate the remarkable success for DNA vaccines obtained thus far in animals as well as hint of their potential for clinical application. Although experimental data is not available for humans at this time, clinical trials are planned, or recently begun, by several groups. Thus, within the near future data should become available to evaluate the utility of DNA vaccination in humans. Besides the question of whether DNA can elicit similar immune responses in humans as observed in animal models, important safety considerations remain to be addressed as well. These include the possibilities of creating mitotically transformed cells by insertion of plasmid DNA into the host genome as well as the generation of anti-DNA antibodies. To date there has been no observation for either of these potential events (Nichols et al., 1995; Ulmer et al., unpublished observations). However, more work remains to be done in these areas before these issues may be considered resolved.

References

Borrow, P., Lewicki, H., Hahn, B. H., Shaw, G. M., and Oldstone, M. B. A. (1994) Virus-specific $CD8^+$ cytotoxic T-lymphocyte activity associated with control of viremia in primary human immunodeficiency virus type 1 infection, J. Virol. *68*, 5103-6110.

Boshart, M., Weber, F., Jahn, G., Dorsch-Hasler, K., Fleckenstein, B., and Schaffner, W. (1985) A very strong enhancer is located upstream of an immediate early gene of human cytomegalovirus, Cell *41*, 521-530.

Broder, C. C., Earl, P. L., Long, D., Abedon, S. T., Moss, B., and Doms, R. W. (1994) Antigenic implications of human immunodeficiency virus type 1 envelope quaternary structure: Oligomer-specific and sensitive monoclonal antibodies, Proc. Natl. Acad. Sci. USA. *91*, 11699-11703.

Chapman, B. S., Thayer, R. M., Vincent, K. A., and Haigwood, N. L. (1991) Effect of intron A from human cytomegalovirus (Towne) immediate-early gene on heterologous expression in mammalian cells, Nucleic Acids Res. *19*, 3979-3986.

Clerici, M. and Shearer, G. M. (1993) A Th1 > Th2 switch is a critical step in the etiology of HIV infection, Immunol. Today, *14*, 107-110.

Coney, L., Wang, B., Ugen, K. E., Boyer, J., McCallus, D., Srikantan, V., Agadjanyan, M., Pachuk, C. J., Herold, K., Merva, M., Gilbert, L., Deng, K., Moelling, K., Newman, M., Williams, W. V., and Weiner, D. B. (1994) Facilitated DNA inoculation induces anti-HIV-1 immunity *in vivo*, Vaccine *12*, 1545-1550.

Conry, R. M., LoBuglio, L. F., Kantor, J., Schlom, J., Loechel, F., Moore, S. E., Sumerel, L. A., Barlow, D. L., Abrams, S., and Curiel, D. T. (1994) Immune response to a carcinoembryonic antigen polynucleotide vaccine, Cancer Res. *54*, 1164-1168.

Cox, G. J. M., Zamb, T. J., and Babiuk, L. A. (1993) Bovine herpesvirus 1: Immune responses in mice and cattle injected with plasmid DNA, J. Virol. *67*, 5664-5667.

Danko, I. and Wolff, J. A. (1994) Direct gene transfer into muscle, Vaccine *12*, 1499-1502.

Davis, H. L., Michel, M.-L., and Whalen, R. G. (1993) DNA-based immunization induces continuous secretion of hepatitis B surface antigen and high levels of circulating antibody, Hum. Mol. Genet. *2*, 1847-1851.

Davis, H. L., Michel, M.-L., Mancini, M., Schleef, M., and Whalen, R. G. (1994) Direct gene transfer in skeletal muscle: plasmid DNA-based immunization against hepatitis B virus surface antigen, Vaccine *12*, 1503-1509.

Donnelly, J. J., Friedman, A., Martinez, D., Montgomery, D. L., Shiver, J. W., Motzel, S. L., Ulmer, J. B., and Liu, M. A. (1995) Preclinical efficacy of a prototype DNA vaccine: Enhanced protection against antigenic drift in influenza virus, Nature Medicine *1*, 583-587.

Donnelly, J. J., Martinez, D., Jansen, K., Montgomery, D. L., Ellis, R. W., and Liu, M. A. (1996) Protection against papilloma virus with a polynucleotide vaccine, J. Infect. Dis. 713, 314-320.

Eisenbraun, M. D., Fuller, D. H., and Haynes, J. R. (1993) Examination of parameters affecting the elicitation of humoral immune responses by particle bombardment-mediated genetic immunization, DNA and Cell Biol. *12*, 791-797.

Fuller, D. H. and Haynes, J. R. (1994) A qualitative progression in HIV type 1 glycoprotein 120-specific cytotoxic cellular and humoral immune responses in mice receiving a DNA-based glycoprotein 120 vaccine, AIDS Res. Hum. Retrovir. *10*, 1433-1441.

Fynan, E. F., Webster, R. G., Fuller, D. H., Haynes, J. R., Santoro, J. C., and Robinson, H. L. (1993) DNA vaccines: Protective immunizations by parenteral, mucosal, and gene-gun inoculations, Proc. Natl. Acad. Sci. U.S.A. *90*, 11478-11482.

Geissler, E. K., Wang, J., Fechner, J. H., Jr., Burlingham, W. J., and Knechtle, S. J. (1994) Immunity to MHC Class I antigen after direct DNA transfer into skeletal muscle, J. Immunol. *152*, 413-421.

Gorman, C. M., Merlino, G. T., Willingham, M. C., Pastan, I., and Howard, B. H. (1982) The Rous sarcoma virus long terminal repeat is a strong promoter when introduced into a variety of eukaryotic cells by DNA-mediated transfection, Proc. Natl. Acad. Sci. U.S.A. *79*, 6777-6781.

Haynes, J. R., Eisenbraun, M. D., Fuller, D. H., Fynan, E. F., and Robinson, H. L. (1994) Gene gun-mediated DNA immunization elicits humoral, cytotoxic, and protective immune responses, Vaccines 94, Cold Spring Harbor Laboratory Press, 65-70.

Heinzel, F. P. (1995) Th1 and Th2 cells in the cure and pathogenesis of infectious diseases, Curr. Opin. Infect. Dis. *8*, 151-155.

Hoffman, S. L., Sedegah, M., and Hedstrom, R. C. (1994) Protection against malaria by immunization with a *Plasmodium yoelii* circumsporozoite protein nucleic acid vaccine, Vaccine *12*, 1529-1533.

Huygen, K., Content, J., Denis, O., Montgomery, D. L., Yawman, A., Drowart, A., Lozes, E., Vandenbussche, P., van Vooren, J,. Liu, M. A., and Ulmer, J. (1996) submitted for publication.

Jiao, S., Williams, P., Berg, R. K., Hodgeman, B. A., Liu, L., Repetto, G., and Wolff, J. A. (1992) Direct gene transfer to nonhuman primate myofibers in vivo, Hum. Gene Ther. *3*, 21-33.

Koup, R. A., Safrit, J. T., Cao, Y., Andrews, C. A., McLeod, G., Borkowsky, W., Farthing, C., and Ho, D. D. (1994) Temporal association of cellular immune responses with the initial control of viremia in primary human immunodeficiency virus type 1 syndrome, J. Virol. *68*, 4650-4655.

Liu, M. A., Davies, M.-E., Yasutomi, Y., Perry, H. C., Letvin, N. L., and Shiver, J. W. (1994) Immune responses to HIV generated by DNA vaccines, *in* Retroviruses of Human AIDS and Related Animal Diseases, Fondation Marcel Mérieux, M. Girard and B. Dodet, eds., 197-200.

Lowrie, D. B., Tascon, R. E., Colston, M. J., and Silva, C. L. (1994) Towards a DNA vaccine against tuberculosis, Vaccine *12*, 1537-1540.

Manickan, E., Rouse, R. J. D., Yu, Z., Wire, W. S., and Rouse, B. T. (1995) Genetic immunization against herpes simplex virus: Protection is mediated by CD4+ T lymphocytes, J. Immunol. *155*, 259-265.

Montgomery, D. L., Shiver, J. W., Leander, K. R., Perry, H. C., Friedman, A. Martinez, D., Ulmer, J. B., Donnelly, J. J., and Liu, M. A. (1993) Heterologous and homologous protection against influenza A by DNA vaccination: Optimization of DNA vectors, DNA and Cell Biol. *12*, 777-783.

Moore, J. P., Cao, Y., Ho, D. D., and Koup, R. A. (1994) Development of anti-gp120 antibody response during seroconversion to human immunodeficiency virus type 1, J. Virol. *68*, 5142-5155.

Moore, J. P. and Ho, D. D. (1993) Antibodies to discontinuous or conformationally sensitive epitopes on the gp120 glycoprotein of human immunodeficiency virus type 1 are highly prevalent in sera of infected humans, J. Virol. *67*, 863-875.

Nichols, W. W., Manam, S., and Ledwith, B. (1995) Ann. NY Acad. Sci. in press.

Pedroz Martins, L., Lau, L. L., Asano, M. S., and Ahmed, R. (1995) DNA vaccination against persistent viral infection, J. Virol. *69*, 2574-2582.

Raz, E., Carson, D. A., Parker, S. E., Parr, T. B., Abai, A. M., Aichinger, G., Gromkowski, S. H., Singh, M., Lew, D., Yaankauckas, M. A., Baird, S. M., and Rhodes, G. H. (1994) Proc. Natl. Acad. Sci. U.S.A. *91*, 9519-9523.

Robinson, H. L., Hunt, L. A., and Webster, R. G. (1993) Protection against a lethal influenza virus challenge by immunization with a haemagglutinin-expressing plasmid DNA, Vaccine *9*, 957-960.

Romagnani, S. (1991) Type-1 T helper and type-2 T helper cells: Functions, regulation and role in protection and disease, Int. J. Clin. Lab. Res. *21*, 152-158.

Romagnani, S. (1995) Biology of human T_H1 and T_H2 cells, J. Clin. Immunol. *15*, 121-129.

Sedegah, M., Hedstrom, R., Hobart, P., and Hoffman, S. L. (1994) Protection against malaria by immunization with plasmid DNA encoding circumsporozoite protein, Proc. Natl. Acad. Sci. USA. *91*, 9866-9870.

Shiver, J. W., Perry, H. C., Davies, M.-E., and Liu, M. A. (1995a) Immune responses to HIV gp120 elicited by DNA vaccination, Vaccines 95, Chanock, R. M., Brown, F., Ginsberg, H. S., and Norrby, E., eds., Cold Spring Harbor Laboratory Press, 95-98.

Shiver, J. W., Perry, H. C., Davies, M.-E., Freed, D. L., and Liu, M. A. (1995b) Cytotoxic T lymphocyte and helper T cell responses following HIV polynucleotide vaccination, Ann. NY Acad. Sci., in press.

Stitz, L., Schmitz, C., Binder, D., Zinkernagel, R., Paoletti, E., and Becht, H. (1990) Characterization and immunological properties of influenza A virus nucleoprotein (NP): Cell-associated NP isolated from infected cells or viral NP expressed by vaccinia recombinant virus do not confer protection, J. Gen. Virol. *71*, 1169-1179.

Street, N. E. and Mosmann, T. R. (1991) Functional diversity of T lymphocytes due to secretion of different cytokine patterns, Faseb J. *5*, 171-177.

Takahashi, H., Cohen, J., Hosmalin, A., Cease, K. B., Houghten, R., Cornette, J. L., DeLisi, C., Moss, B., Germain, R. N., and Berzofsky, J. A. (1988) An immunodominant epitope of the human immunodeficiency virus envelope glycoprotein gp160 recognized by class I major histocompatibility complex molecule-restricted murine cytotoxic T lymphocytes, Proc. Natl. Acad. Sci. USA. *85*, 3105-3109.

Tang, D.-C., DeVit, M., and Johnston, S. A. (1992) Genetic immunization is a simple method for eliciting an immune response, Nature, *356*, 152-154.

Ulmer, J. B., Deck, R. R., DeWitt, C. M., Friedman, A., Donnelly, J. J., and Liu, M. A. (1994) Protective immunity by intramuscular injection of low doses of influenza virus DNA vaccines, Vaccine *12*, 1541-1544.

Ulmer, J. B., Donnelly, J. J., Parker, S. E., Rhodes, G. H., Felgner, P. L., Dwarki, V. J., Gromkowski, S. H., Deck, R. R., DeWitt, C. M., Friedman, A., Hawe, L. A., Leander, K. R., Martinez, D., Perry, H. C., Shiver, J. W., Montgomery, D. L., and Liu, M. A. (1993) Heterologous protection against influenza by injection of DNA encoding a viral protein, Science *259*, 1745-1749.

Waine, G. J. (1994) Nucleic Acid Vaccines, Parasitology Today *10*, 453-454.

Wang, B., Ugen, K. E., Srikantan, V., Sato, A. I., Boyer, J., Williams, W. V., and Weiner, D. B. (1993) Gene inoculation generates immune responses against human immunodeficiency virus type 1, Proc. Natl. Acad. Sci. USA. *90*, 4156-4160.

Webster, R. G., Fynan, E. F., Santoro, J. C., and Robinson, H. (1994) Protection of ferrets against influenza challenge with a DNA vaccine to the haemagglutinin, Vaccine *12*, 1495-1498.

Wolff, J. A., Malone, R. W., Williams, P., Chong, W., Acsadi, G., Jani, A., and Felgner, P. L. (1990) Direct gene transfer into mouse muscle in vivo, Science *247*, 1465-1468.

Xiang, Z. G., Spitalnik, S. L., Cheng, J., Erikson, J., Wojczyk, B., and Ertl, H. C. (1995) Immune responses to nucleic acid vaccines to rabies virus, Virology, *209*, 569-579.

Xiang, Z. G., Spitalnik, S. L., Tran, M., Wunner, W. H., Cheng, J., and Ertl, H. C. (1994) Vaccination with a plasmid vector carrying the rabies virus glycoprotein gene induces protective immunity against rabies virus, Virology, *199*, 132-140.

Xu, D. and Liew, F. Y. (1995) Protection against leishmaniasis by injection of DNA encoding a major surface glycoprotein, gp63, of *L. major*, Immunology, *84*, 173-176.

Yankauckas, M. A., Morrow, J. E., Parker, S. E., Abai, A., Rhodes, G. H., Dwarki, V. J., and Gromkowski, S. H. (1993) Long-term anti-nucleoprotein cellular and humoral immunity is induced by intramuscular injection of plasmid DNA containing NP gene, DNA and Cell Biol. *12*, 771-776.

Yokohama, M., Zhang, J., and Whitton, J. L. (1995) DNA immunization confers protection against lethal lymphocytic choriomeningitis virus infection, J. Virol. *69*, 2684-2688.

Zarozinski, C. C., Fynan, E. F., Selin, L. K., Robinson, H. L., and Welsh, R. M. (1995) "Protective CTL-dependent immunity and enhanced immunopathology in mice immunized by particle bombardment with DNA encoding and internal virion protein". J. Immunol. 154, 4010-4017.

Zhu, N., Ligitt, D., Liu, Y., and Debs, R. (1993) Systemic gene-expression after intravenous DNA delivery into adult mice, Science, *2261*, 209-211.

3.8 Oral Vaccination, Mucosal Immunity and Oral Tolerance with Special Reference to Cholera Toxin

Jan Holmgren, Cecil Czerkinsky, Jia-Bin Sun and Ann-Mari Svennerholm

3.8.1 Introduction

There are several good reasons why mucosal, and especially oral vaccination has got rapidly increasing attention. Firstly, the vast majority of infections, whether caused by viruses or bacteria, occur at or take their departure from a mucosal surface and topical application of vaccine is usually needed for inducing a protective immune response at mucosal surfaces (Mestecky, 1987; McDermott and Elson, 1991, 1992). Secondly, it has been found that immune cells stimulated by vaccination at one mucosal surface, especially in the gut or the nose, may disseminate to and give rise to an immune response also at certain distant mucosae which gives promise for orally or nasally administered vaccines to be used against a broad spectrum of mucosal infections (McDermott and Bienenstock, 1979; Mestecky, 1987). Thirdly, although the selective immunological advantage of using a mucosal rather than a systemic route of immunization mainly resides in the superiority of mucosal immunization for stimulating localized immune responses in mucosal and glandular tissues, a mucosal immunization route may also be used to stimulate high-titred serum antibody responses and in some reported cases also cytotoxic lymphocytes (CTL) (Ogra et al., 1976; Mestecky, 1987). Since both the manufacturing (including product quality control) and especially the administration of many oral vaccines may be much simpler, safer and cheaper than with injectable vaccines there would also be substantial logistic advantages in being able to replace some of the currently used injectable vaccines with corresponding oral ones. Finally, an aspect which has only recently been high-lighted in the area of mucosal vaccinology is that the oral route of immunization also has the potential of inducing systemic tolerance – "oral tolerance" – against especially T cell mediated immune reactions associated with delayed type hypersensitivity (DTH) inflammatory reactions (Chase, 1946; Challacombe and Tomasi, 1980; Weiner et al., 1993a). This phenomenon may open up novel prospects for the development of oral anti-inflammatory immunizing agents in addition to the more conventional anti-infectious vaccines, and this in its turn brings the field of mucosal immunology and mucosal vaccines to have significant bearing also on the prevention and therapy of e.g. certain autoimmune and allergic diseases where DTH immunoreactions is important in pathogenesis and/or pathophysiology.

However, despite these attractive features it has in practice often proved to be rather difficult to stimulate strong mucosal IgA immune responses by oral-mucosal administration of antigens and experience with soluble protein antigens has usually been disappointing. A striking exception in this regard is cholera toxin (CT) and also, in humans better than in other species, its non-toxic B subunit pentamer moiety (CTB). Both CT and CTB have proved to be very potent mucosal immunogens (Pierce and Gowans, 1975; Holmgren et al., 1977; Svennerholm et al., 1982; Holmgren et al., 1993) and also to be a very efficient oral (or nasal, rectal or vaginal) carrier-delivery system for bringing various other chemically or genetically fused antigens to induce much enhanced IgA responses at mucosal surfaces (McKenzie and Halsey, 1984; Czerkinsky et al., 1989; Holmgren et al., 1994). Furthermore, whilst both CT and CTB have been shown to be strong enhancers with regard to immunogenicity and mucosal IgA response carrier-delivery properties, it has become apparent from our recent work that these proteins have quite opposite effects with regard to the induction of systemic oral tolerance (Sun et al., 1994, Czerkinsky and Holmgren, 1995). Thus, it has been well established that CT very efficiently inhibits oral tolerance (Elson and Ealding, 1984) whereas it was recently found that CTB is a very strong inducer of oral tolerance for antigens coupled to it (Sun et al., 1994; 1996). Indeed, the very potent efficiency of CTB-based conjugates to induce oral tolerance to T cell mediated inflammatory reactions have given rise to optimism that a whole new class of immunotherapeutic agents and vaccines may be developed based on this novel mucosal tolerization principle (Czerkinsky and Holmgren, 1995). In this article we describe the effects of these two proteins as oral-mucosal immunogens and transmucosal carrier and immunomodulating systems.

3.8.2 Oral Vaccines Against Cholera and *E. coli* Diarrhea

Enteric infections caused mainly by different bacteria but also by rotaviruses and certain parasites represent one of the leading global health problems; together these enteric infections cause millions of deaths each year, especially in children below the age of five years (World Health Organization, 1994).

Foremost among these infections are those caused by bacteria evoking disease by the production of one or more enterotoxins, which together account for more than

50 % of all diarrheal disease cases. Cholera is the most severe of these so-called enterotoxic enteropathies, whereas diarrhea caused by enterotoxigenic E. coli bacteria (ETEC) by far accounts for the largest number of cases (Farthing and Keusch, 1989).

The pathogenesis and mechanisms of immunity in cholera and ETEC diarrhea have many similarities (Holmgren, 1981). The disease, which is characterized by watery diarrhea in the absence of any signs of invasion or inflammation, is preceded by the adherence, colonization and multiplication of the bacteria on the epithelial surface of the small intestine and the release and cytotonic action on the gut epithelium of closely similar and immunologically cross-reactive protein enterotoxins (in addition, or alternatively, ETEC may produce another enterotoxin which does not resemble cholera toxin, see below). Protective immunity is mainly, if not exclusively mediated by locally produced IgA antibodies (and immunologic memory) which can interfere with the bacterial colonization on one hand and with the enterotoxin attachment and action on the other hand; indeed it was shown in animal studies that for both ETEC and cholera disease these different types of immunity co-operate synergistically in protecting from disease (Holmgren and Svennerholm, 1983).

Based on these similarities oral vaccines have recently been developed against both cholera and ETEC diarrhea in which the oral administration route is used to optimally stimulate the local gut immune system and in which one important vaccine component consists of CTB to induce protective antitoxic IgA immunity and another component consists of killed whole-cell cholera or ETEC organisms prepared so as to be able to induce mucosal IgA antibacterial immunity interfering with bacterial colonisation and adherence.

3.8.2.1 Cholera Vaccines

An oral cholera vaccine consisting of the non-toxic, highly immunogenic CTB protein in combination with heat-and formalin-killed *V. cholerae* O1 classical and El Tor vibrios (tab. 3.8.1) has been developed and is now a licensed vaccine (Holmgren et al., 1992). This CTB-whole cell (B-WC) vaccine, which is given together with a bicarbonate buffer to preserve the CTB pentameric structure, has in extensive clinical trials, including large field trials, proved to be completely safe and to provide good protection against cholera and also partial protection against diarrhea caused by LT-producing ETEC.

The B-WC vaccine was designed to evoke antitoxic as well as antibacterial intestinal immunity, since in animal studies these types of immunity have been shown to provide synergistic co-operative protection (Holmgren et al., 1977; Holmgren and Svennerholm, 1983). Phase 1 and phase 2 clinical studies established that the vaccine does not cause any detectable side effects and that, after either two or three doses, it stimulates a gut mucosal IgA antitoxic and antibacterial immune response (including memory) comparable to that induced by cholera disease itself (Svennerholm et al., 1984a,b; Jertborn et al., 1988; Quiding et al., 1991). Furthermore, immunization with either the complete B-WC vaccine or the WC component alone was found to protect

Table 3.8.1: Composition of lincensed oral B subunit-whole cell cholera vaccine*

	Content per dose
B subunit	1 mg recombinant CTB
Whole-cell component	1 x 1011 inactivated V. cholerae O1 bacteria: – heat-killed Inaba (strain Cairo 48) – heat-killed Ogawa (strain Cairo 50) – formalin-killed classical (strain Cairo 50) – formalin-killed El Tor (strain Phil 6973) (2.5 x 1010 bacteria of each preparation)

* Produced by SBL Vaccin, Stockholm, Sweden

American volunteers against challenge with a dose of live cholera vibrios (biotype El Tor) that caused disease in 100 % of concurrently tested unvaccinated controls (Black et al., 1987). On this basis, a large, double-blind placebo-controlled field trial with more than 90,000 participants was undertaken in rural Bangladesh. The results established that both the B-WC vaccine and the WC component alone confer long-lasting protection against cholera. The B-WC vaccine had a higher initial efficacy level than the WC vaccine alone (85 % *versus* 58 % for the initial 4- to 6-month period) with the protective efficacy of B-WC in comparison with WC alone being 73 % in support of the significant protective immunogenicity of the CTB component (Clemens et al., 1986). The B-WC continued to be significantly more protective than the WC alone vaccine for the first eight months after vaccination; thereafter, however, the efficacy was similar, approximately 60 % for both vaccines, if calculated for the whole population above age 2 years, for a 3-year follow-up period (Clemens et al., 1990). Protection was of similar magnitude after two or three doses of vaccine (Clemens et al., 1990). Still higher (70 %) long-term protective efficacy was seen in those over age 5 when vaccinated. It is likely that the age group below 5 years, in which immunity rapidly waned after the initial high-level protection for the first 6-9 months, could also be provided with long-lasting high-level protection by a booster immunization after one year.

Two additional recent field trials have confirmed the strong protective efficacy of the oral cholera vaccine in two different settings. In Peru, Sanchez et al. (1994) found that vaccination conferred 86% protection against cholera in Peruvian military recruits. It is especially noteworthy that this high level of protection was (i) seen with two doses of vaccine given only 1-2 weeks apart, (ii) was obtained with vaccine based on recombinantly CTB, (iii) was directed against severe cholera of exclusively the O1 El Tor biotype, and (iv) was achieved in a population being almost exclusively of blood group O; these were factors that earlier were thought to possibly reduce the efficacy of the vaccine. In Vietnam, a locally produced vaccine being closely similar to the Swedish version except for lacking the CTB component has also been found to give ca 80% protection against O1 El Tor cholera (Trach et al., to be published).

Through its B subunit component, the B-WC vaccine also provided substantial, approximately 70%, short-term protection against diarrhea caused by ETEC producing LT (which cross-reacts immunologically with cholera toxin) either alone or together with the non-cross-reactive heat-stable toxin (ST) (Clemens et al., 1988a). Furthermore, both the B-WC and the WC vaccines substantially reduced the overall diarrhea morbidity among those vaccinated, such that there was a 50% reduction in admissions for life-threatening diarrhea in the vaccinated as compared with the placebo group over a 3-year follow-up period (Clemens et al., 1988b). A substantial protective efficacy of the oral B-WC cholera vaccine against LT ETEC diarrhea was also demonstrated by Peltola et al. (1991), who studied Finnish tourists travelling to Morocco: in comparison with a placebo vaccine two doses of the B-WC vaccine given shortly before the travel conferred 60-82% protection against diarrhea associated with LT-producing ETEC (60% against ETEC being isolated from the stool as the only pathogen and 82% against ETEC being isolated together with another pathogen, mainly Salmonella).

3.8.2.2 ETEC Vaccine

Infection with ETEC is the most frequent cause of diarrhea both in developing countries and among travellers. Estimates indicate that infections with different types of ETEC account for more than one billion diarrheal episodes and one million deaths annually among children in developing countries (Holme et al., 1981; Farthing and Keusch, 1989). Yet, no vaccine for use in humans is available. New knowledge about virulence factors and protective antigens of ETEC, however, has suggested practical approaches towards development of a useful vaccine against ETEC diarrhea, and an oral ETEC vaccine has recently been developed based on these principles.

Adhesion of ETEC to intestinal mucosa is mediated by antigenically distinct fimbriae. In strains pathogenic for humans, three main adhesins have been identified, i.e. the colonization factor antigens CFA/I, CFA/II, and CFA/IV (Evans and Evans, 1989) and all of these antigens should be included in an oral ETEC vaccine. However, purified CFAs have proved to be sensitive to proteolytic degradation in the human gastrointestinal tract, complicating the use of such a vaccine preparation. Likewise, it would be difficult to construct a vaccine (e.g. a live vaccine) based on the concomitant expression of all these antigens by a single organism, since despite considerable efforts the different CFAs could not be expressed on the same host strain. A more practical way to construct a vaccine may instead be to prepare killed ETEC bacteria of different strains that express the most important CFAs on their surface and combine these organisms with an appropriate toxoid component. Such a vaccine should be given orally in order to ideally evoke both anti-colonization and antitoxic IgA local immune responses in the gut.

We (A-M S and J H) have now in collaboration with SBL Vaccin, Sweden, developed such a prototype ETEC based on a mixture of *E. coli* strains that express the key CFA in immunogenic form and CTB (tab. 3.8.2). The inactivation of bacteria, which includes mild formalin treatment, causes complete killing of bacteria without significant loss in antigenicity of the different CFA. Furthermore, the CFA of these inactivated organisms have been stable during storage for years and during incubation in gastric juice. As in the oral cholera vaccine, the WC component is combined with CTB to induce antitoxic mucosal immunity, which is known to co-operate synergistically with anti-CFA immunity against experimental ETEC infections in animals (Åhrén et al., 1982). As mentioned, the results from the cholera vaccine field trial in Bangladesh (Clemens et al., 1988) as well as those from a prospective placebo-controlled study in Finnish travellers to Morocco (Peltola et al., 1991) demonstrated that CTB, which cross-reacts immunologically with the B subunit of *E. coli* heat-labile toxin, LTB, induced significant protection against *E. coli* LT as well as LT/ST diarrhea. Our studies in rabbits confirm that immunization with purified CTB affords protective immunity against challenge with LT-producing organisms comparable to that of immunization with corresponding doses of purified LTB (Svennerholm et al., 1991). These findings support the view that CTB, which can be produced inexpensively and on a large scale through the use of a recombinant production system (Sanchez and Holmgren, 1989) can replace LTB as "toxoid" in an ETEC vaccine.

Table 3.8.2: Composition of oral ETEC vaccine*

	Content per dose
B subunit	1 mg recombinant CTB
Whole-cell component	Five strains of formalin-killed ETEC (2 x 10^{10} bacteria each) expressing: – CFA/I – CFA/II (CS1) – CFA/II (CS2+CS3) – CFA/IV (CS4+CS6) – CFA/IV(CS5+CS6)

* Produced by SBL Vaccin, Stockholm, Sweden

The results of clinical trials in Swedish volunteers indicate that such an ETEC vaccine is safe and gives rise to substantial intestinal anti-CFA as well as anti-LT antibody production in nearly 100% of those vaccinated. This was first shown with a prototype vaccine only containing CFA/I and CFA/II strains (Åhrén et al., 1993) and was recently confirmed with the more complex, definitive vaccine shown in Table 3.8.2 (Jertborn et al., 1996). Indeed, both the frequency and the magnitude of these responses in Swedish volunteers, as measured in intestinal lavage fluid, have been comparable to those previously seen in response to severe clinical ETEC disease in convalescent patients in Bangladesh (Stoll et al., 1986). Studies of the protective efficacy of the ETEC vaccine are now planned both in travellers and in children in endemic areas.

3.8.3 Use of CTB as Carrier for Unrelated Vaccine Antigens

The concept of a common mucosal immune system, through which a fraction of activated lymphocytes from the gut can disseminate immunity to other mucosal and glandular tissues (McDermott and Bienenstock, 1979), has generated much interest in the possibility of developing oral vaccines also against infections in e.g. the respiratory and urogenital tracts (Mestecky and McGhee, 1989).

However, as mentioned, it has often proved difficult to generate strong mucosal responses to especially non-replicating antigens, which largely can be explained by a combination of the poor stability of many antigens in the mucosal milieu and their poor uptake into the mucosal immune system compartment, in e.g. the intestine mainly the Peyer's patches. Recent studies in animal systems have demonstrated an important finding: oral administration of small amounts of protein antigens (that are not immunogenic per se when given alone by this route) covalently coupled to CTB can elicit vigorous mucosal as well as extramucosal antibody responses. McKenzie and Halsey (1984) were the first ones to report that covalent coupling of an antigen (horseradish peroxidase) to CTB markedly enhanced gut immune responses to the peroxidase antigen after oral administration. More recently, we and others have confirmed and extended those findings to also demonstrate an effective oral delivery potential of CTB for coupled foreign antigens with regard to stimulation of IgA mucosal responses also outside the intestine, e.g. in salivary glands and in the respiratory and urogenital tracts, as well as for eliciting circulating antibodies (Liang et al., 1988; Czerkinsky et al., 1989; Drew et al., 1992). For instance, Czerkinsky et al. (1989) demonstrated that oral administration of small amounts of a streptococcal protein antigen covalently linked to CTB (and given with a low dose of free CT as adjuvant) induced both mucosal and extramucosal IgA and IgG antistreptococcal antibody responses in mice. The mucosal response was associated with the development of high frequencies of specific antibody-secreting cells in mesenteric lymph nodes, submandibular salivary glands, and the spleen and with high levels of circulating IgA and IgG antibodies to the streptococcal antigen in serum. In contrast, equivalent or 10-fold higher doses of streptococcal antigen given alone or conjugated to a non-binding protein (bovine serum albumin) were ineffective or, at best, poor in eliciting antibody responses to the streptococcal antigen. Analogous findings were reported by Liang et al. (1988), who found that oral administration of Sendai virus chemically coupled to CT potentiated the IgA immune responses in the upper respiratory tract against Sendai virus in comparison with immunizations with other formulations of this antigen. Likewise, Drew et al. (1992) showed that chemical conjugates between CT and peptide antigens derived from herpes simplex virus (HSV)-2 glycoprotein D when administered to mice by intraperitoneal (i.p.) priming followed by either i.p. or intragastric boosting gave rise to IgA anti-HSV-2 antibodies in vaginal washings and to protection against a lethal intravaginal challenge with HSV-2.

These findings lend support to the notion that peroral immunization with antigens linked (chemically or genetically) with CTB may be a useful strategy for vaccinating against pathogens encountered not only at enteromucosal surfaces but also at extraintestinal mucosal and non-mucosal sites. With the aid of recombinant DNA engineering, foreign antigens have been linked to either the amino or carboxy ends of the CTB subunit, and a gene overexpression system has been developed to permit production of the hybrid proteins in substantial quantities (Sanchez and Holmgren, 1989; Sanchez et al., 1990). Several vaccine candidates prepared along these lines have been described but none of them have yet reached into human testing.

The promising features of CTB as a mucosal carrier system (and analogous findings have been reported for *E. coli* LTB) clearly do not exclude the potentials of other, competing technologies including both live vectors and inert particulate delivery systems (Michalek et al., 1994). It is known that replicating antigens, such as bacteria and viruses having special ability to penetrate into and replicate in the Peyer's patches or corresponding immune response inductive sites in other mucosae, often give rise to much stronger immune response than mucosally administered non-replicating antigens. Thus, it has been attractive to try to develop genetically engineered attenuated live viruses or bacteria as mucosal vectors for expressing various foreign antigens. Promising bacterial vectors include e.g. Salmonella spp. and BCG, and among viral vectors, most attention has been placed on vaccinia, adeno, polio and canary pox viruses. Studies have revealed promising as well as problematic features and to date no entirely satisfactory system has been identified.

From the observation that particulate antigens given orally are more effective in inducing local and systemic immune responses than soluble antigens, recent research has also focused on the development of particulate mucosal delivery systems. These systems may enhance uptake by M cells and will also protect the antigen from acidic and proteolytic degradation. Available particulate systems include e.g. microspheres, liposomes and iscoms (Michalek et al., 1994). A system which has received much attention recently for its use in mucosal delivery is biodegradable microspheres based on co-polymers formed by poly DL-lactide-co-glycolide (DL-PLG). Various antigens have been encapsulated in DL-PLG microspheres and administrated orally in animal models where they were shown to induce mucosal and systemic immunity. Induction of mucosal and systemic responses by the oral route where shown to be influenced by the size of the microspheres: microspheres <5 μm in diameter are not retained in the Peyer's patches but found in the spleen and in lymph nodes where they induce systemic immunity, while those with the size range 5-10 μm remain in the Peyer's patches and induce mucosal immunity, and those with a larger size still are not taken up at all and do not produce an immune response. Therefore, it may be possible to exploit differential trafficking of orally absorbed microspheres to induce concomitantly or selectively mucosal and systemic immune responses. Recent reports have also demonstrated the efficacy of intranasal administration of antigen entrapped in microspheres. However, several problems will have to be addressed for the use of microspheres in vaccines. In general, their preparations requires large amounts of antigens, and are often prepared under harsh conditions which will denature many of the candidate antigens. Moreover, the uptake of orally administered microspheres into Peyer's patches is low and >99% of the material is eliminated without uptake.

3.8.4 Oral Tolerance and Anti-Inflammatory Active Immunization

Mucosal uptake of antigen not only can induce secretory IgA antibody responses in various mucosal tissues but also, and often, systemic tolerance. Teleologically, the induction of either or both of mucosal immunity and systemic tolerance could be advantageous to protect the host from infection and disease by mucosal pathogens and from pathogenic systemic responses that may develop against absorbed food antigens or inhaled environmental matters or against antigens produced by infectious agents.

3.8.4.1 The Oral Tolerance Phenomenon

Mucosal administration of antigens is in fact a long-recognized method of inducing peripheral tolerance (Wells, 1911; Chase, 1946; Mowat, 1987). The phenomenon, often referred to as "oral tolerance" (because initially documented by the effect of oral administration of antigen), is characterized by the fact that animals fed or having inhaled an antigen become refractory or have diminished capability to develop an immune response when re-exposed to that very same antigen introduced by a systemic route, e. g. by injection. This phenomenon is an important natural physiological mechanism whereby we avoid to develop DTH reactions to many ingested food proteins and other antigens (Mowat, 1987). Depending upon the dose of antigen administered, clonal deletion (excessive doses) or anergy (high doses) of antigen-specific T cells (Whitacre et al., 1991; Weiner et al., 1993a) and/or expansion of regulatory cells producing immunosuppressive cytokines (IL-4, IL-10 and TGF-b) (Weiner et al., 1993a) (induced by lower antigen doses) may result in decreased T cell immune responsiveness. It is interesting to note that the latter scenario involves cytokines (Weiner et al., 1993a) that are also known to up-regulate IgA production (Brandtzaeg, 1995) and is thus compatible with the observation that secretory IgA immune responses and systemic T cell tolerance may develop concomitantly (Challacombe and Tomasi, 1980). Because tolerance can be transferred by both serum and cells from tolerized animals, it is possible that humoral antibodies (IgA?), circulating undegraded antigens or tolerogenic fragments and cytokines may act synergistically to confer T cell unresponsiveness. Since oral tolerance is exquisitely specific for the antigen initially ingested or inhaled and thus does not influence the development of systemic immune responses against other antigens, its manipulation has become an increasingly attractive strategy for preventing and possibly treating illnesses associated with or resulting from the development of untoward immunological reactions against specific antigens encountered or expressed in non-mucosal tissues including both foreign antigens and auto-antigens.

3.8.4.2 Medical Potential and Limitations

Mucosally induced immunological tolerance has earlier been proposed as a strategy to prevent or to reduce the intensity of allergic reactions to chemical drugs (Chase, 1946), soluble protein antigens and particulate antigens (Wells, 1911; Mowat, 1987), and to reduce or suppress immune responses against self antigens (Whitacre et al., 1991; Trentham et al., 1993; Weiner et al., 1993a,b). It has been possible to delay the onset and/or to decrease the intensity of experimentally induced autoimmune diseases in a variety of animal systems by mucosal deposition of auto-antigens onto the intestinal (by feeding) or the respiratory mucosa (by aerosolization or intranasal instillation of antigens). Furthermore, pilot clinical trials of oral tolerance have recently been conducted in patients with autoimmune diseases and promising clinical effects have been reported (Trentham et al., 1993; Weiner et al., 1993). Much in the same way, oral administration of antigens had earlier been proposed to prevent and/or treat allergic reactions to common allergens such as house dust components or substances present in grass pollen (Wortmann, 1977; Rebien et al., 1982).

Although the above examples indicate that oral tolerance offers good promise for inducing specific immunologic tolerance, its therapeutic potential has remained limited due to practical problems. To be clinically broadly applicable, mucosally-induced immunological tolerance must be effective in patients in whom the disease process is already established and/or in whom potentially tissue-damaging immune cells exist. This is especially important when considering strategies of tolerance induction in patients suffering from an autoimmune disease, an allergic condition, or a chronic inflammatory reaction to a persistent microorganism. However, current protocols of mucosally induced tolerance have had limited success in suppressing the expression of an already established state of systemic immunological sensitization (Challacombe and Tomasi, 1980; Hansson et al., 1979). Most importantly and in analogy with mucosal vaccines aimed at inducing immune responses to infectious pathogens, induction of systemic immunological tolerance by mucosal application of most antigens often requires considerable amounts (mg to grams) of tolerogen/antigen and, unless the tolerogen/antigen is administered repeatedly over long periods of time the immunosuppressive effect is of relatively short duration. Therefore, oral tolerance has largely continued to be regarded as a physiological homeostatic mechanism to suppress the development of harmful immune reactions to food proteins ingested in large quantities on a more or less daily basis, whilst its practical medical utility for treatment of immunological disorders has remained elusive.

3.8.4.3 Cholera Toxin and Oral Tolerance

It has been widely assumed that oral tolerance leading to clonal deletion or anergy is induced by antigen fragments presented to T cells in the absence of the co-stimulatory signals needed for induction of a productive immune response, a situation that has been presented to prevail in the intestine and other mucosae in relation to their handling of most foreign matters. Thus, most antigens are extensively degraded before entering a mucosal tissue, and it has also been reported that antigen presentation by mucosal epithelial cells provides tolerogenic rather than immunogenic signals (Mowat, 1987). It has thus been widely assumed that only protease-resistant molecules with known mucosa-binding properties leading to facilitated up-take into Peyer's patches in the gut or corresponding lymphoid aggregate inductive sites in other mucosae can induce local and systemic immune responses when administered by a mucosal route without inducing systemic immunological tolerance (Walker et al., 1975; de Aizpurua et al., 1988; Michalek et al., 1994). A notable example is cholera toxin, one of the most potent mucosal immunogens known so far (Pierce and Gowans, 1975; Lange et al., 1980; Elson and Ealding, 1984a; Lycke and Holmgren, 1986; Holmgren et al., 1993) and which when administered simultaneously with an unrelated antigen by the oral route can also concomitantly dramatically adjuvant the mucosal IgA immune response (Lycke and Holmgren, 1986; Dertzbaugh and Elson, 1991; Holmgren et al., 1993) and prevent the induction of systemic immunological tolerance to said antigen (Elson and Ealding, 1984b). Also, the cholera toxin binding subunit (CTB) was reported to have similar albeit weaker adjuvant activity and tolerance inhibition effect (Elson and Ealding, 1984b; Hirabayashi et al., 1991; Wilson et al., 1990), although Lycke and Holmgren (1986) did not see any adjuvant activity of highly purified CTB. Based on these observations and as discussed above, mucosal administration of antigens coupled to mucosa-binding molecules such as cholera toxin or its mucosa-binding fragment cholera toxin B subunit, has been proposed as a strategy to induce local and systemic immune responses rather than systemic tolerance (McKenzie and Halsey, 1984; Nedrud et al., 1987; Liang et al., 1988; Czerkinsky et al., 1989; Lehner et al., 1992; Holmgren et al., 1993).

3.8.4.4 CTB and the Induction of Tolerance

We suspected that the tolerance-breaking properties attributed to both cholera toxin and CTB (Elson and Ealding, 1984b) might be selective for cholera toxin and thus, as concerns CTB, be explained by low yet significant levels of contamination by cholera toxin of commercial CTB preparations used in previous studies. Consistent with this hypothesis, we have observed that physical coupling of an antigen to recombinantly produced and thus inherently uncontaminated CTB led to effects contrary to those reported previously: when given by various mucosal (oral, intranasal, vaginal, rectal) routes and in the absence of CT adjuvant, CTB instead of abrogating profoundly enhanced systemic tolerance to the conjugated antigens it (Sun et al., 1994). Based on this unexpected finding and on the results of additional experiments with several soluble protein antigens (gamma-globulins, myelin basic protein, collagen II, insulin), haptens, and particulate antigens (red blood cells, allogeneic thymocytes), all giving similar results, we have good reasons to believe that such a mucosal delivery system based on coupling antigens to CTB and possibly also to other mucosa-binding non-toxic carrier molecules, may be quite useful and effective for inducing peripheral tolerance (Sun et al., 1994; Czerkinsky and Holmgren, 1995; unpublished data).

This concept was, as mentioned, first documented by the use of recombinant CTB as mucosa-binding molecule, and sheep red blood cells (SRBC) as antigen/tolerogen in a murine system. SRBC was chosen as a model antigen since it is one of the best characterized oral tolerogens with regard to both antibody formation and cell-mediated immune reactions, the latter reactions being typified by the classical delayed type hypersensitivity (DTH) reaction. Both types of immune reactions have been implicated in the development of autoimmune diseases, allergic reactions, acute graft rejection and in a number of chronic inflammatory conditions (Auchincloss and Sachs, 1993; Gallin, 1993; Schwartz, 1993). CTB was conjugated to the surface of SRBC by first attaching the CTB receptor ganglioside GM1 to the SRBC surface, and then reacting the CTB to the GM1-coated SRBC under conditions (excess CTB) leaving part of the receptor-binding sites of CTB free to react with GM1 receptors on other cells; the properties of the CTB-SRBC conjugate were ascertained by receptor-binding *in vitro* assays (Sun et al., 1994). When the CTB-SRBC conjugate was fed to mice, it was found that a single dose could suppress *in vitro* antigen-induced proliferative responses of T cells, and in vivo DTH reactivity to SRBC, and although to a lesser extent also serum antibody responses (Sun et al., 1994). This much more limited effect on antibody formation than on DTH reactions is very much in line with the appreciation that in the murine system, the primary effector cell in DTH belongs to the Th1 cell subset, whereas antibody formation depends on B cells and their cognate interaction with mainly Th2 cells. It is well recognized that after both systemic and mucosal tolerization regimens Th1 cells are more readily tolerized than Th2 cells, and B cells are even more difficult to tolerize in an antigen-specific manner (Chiller et al., 1971; Schwartz, 1993; Husby et al., 1994). In the case of DTH reactivity, both early (2-4 hrs) and late (24-48 hrs) responses were either abrogated or considerably reduced. In contrast, it required daily consecutive administration of unconjugated SRBC to suppress antibody responses and DTH reactivity to levels comparable to those obtained after feeding a single dose of CTB-conjugated antigen, and with respect to DTH reactivity, only the late (24-48 hrs) reactions were suppressed with no apparent effect on the early component. The latter observation is especially important inasmuch as it suggests that the suppressive effects of oral administration of antigens coupled to CTB involve mechanisms that appear to be distinct from those implicated in conventional regimens of oral tolerance induction with feeding of multiple doses of free antigen (tab. 3.8.3). Most importantly, this new strategy could be employed to suppress DTH responses even in animals previously sensitized at systemic sites (tab. 3.8.3). Thus, when mice were first systemically sensitized (by footpad injection) with SRBC and then fed a single dose of SRBC-CTB conjugate, they failed to develop early as well as late DTH reactivity to a subsequent systemic challenge with SRBC. In contrast, animals fed the same dose of unconjugated SRBC displayed practically normal skin DTH reactivity to SRBC.

The role of cholera toxin on oral tolerance was tested in the same system, and the results confirmed that cholera toxin both prevents and breaks oral tolerance, i.e. has the opposite effect as compared to CTB. Furthermore, it was shown that physical conjugation of CTB to SRBC was imperative for the strong tolerogenic effect. Thus, adding as little as 100 ng to 1 microgram of cholera toxin to the oral SRBC-CTB conjugate (tab. 3.8.3) or feeding SRBC coupled to cholera toxin rather than to CTB (not shown) abrogated the tolerogenic effects of the conjugate and simply mixing SRBC with free SRBC had no effect beyond that of feeding SRBC alone (not shown).

As mentioned, the above initial finding have now been extended to several other antigens, including allogeneic thymocytes, and a number of soluble protein antigens including selected autoantigens, and also TNP as model of a contact irritant haptenic compound. In all instances, single or double mucosal (oral or intranasal) administration of CTB-conjugated antigens were effectively tolerogenic at doses 100- to 1000-fold lower than those of corresponding unconjugated antigens (Sun et al., to be published).

3.8.4.5 Mechanisms of Tolerization

As mentioned, the more conventional schedules of oral tolerization with multiple feedings of relatively large doses of free antigen may involve one or more of Th1 clonal deletion, Th1 clonal anergy and the generation of regulatory Th2 cells producing TGF-β and other anti-inflammatory cytokines (Weiner et al., 1993).

Several findings suggest that the dramatically more effective tolerization achieved with conjugates between antigens and CTB involves at least partly different mechanisms. We described earlier the strong effect on the early (initiation) stage of the DTH reaction achieved with CTB-antigen conjugates but never seen with the conventional regimens feeding free antigen. In more recent studies, we have been able to transfer suppression of the early and the late components of DTH reactions to SRBC independently from each other with cells from animals orally tolerized with SRBC-CTB (Sun et al., to be published). The implications of the latter finding may be considerable inasmuch as they demonstrate that this novel strategy of tolerance induction acts on the very early stages of a T-cell mediated inflammatory reaction. It has also been possible with CTB-SRBC conjugate feeding to induce tolerance against the early stage of DTH in the "nude" mouse model, as well as to induce tolerance against both early and late DTH in CD8 $^{--}$ transgenic "knock-out" mice (Sun et al., to be published).

Table 3.8.3: Inhibition of early and late delayed-type hypersensitivity (DTH) reactions by oral administration to non-sensitized ("Prevention") or immune mice ("Treatment")of sheep red blood cells (SRBC) given alone or coupled to the B subunit of cholera toxin (CTB)

	Specific thickness increment x 10^{-3} cm	
Oral tolerogen	2 hr	24 hr
"Prevention"		
CTB-SRBC x 1	0 ± 2.8**	2 + 0.6**
CTB-HRBC x 1	9 ± 3.5	36 ± 6.1
SRBC x 1	13 ± 8.0	32 ± 5.0
SRBC x 5	11 ± 4.2	38 ± 6.0
SRBC x 10	11 ± 4.8	24 ± 8.5
SRBC x 15	10 ± 2.0	12 ± 2.2*
SRBC x 20	10 ± 1.6	1.8 ± 0.4**
Saline	8 ± 3.0	38 ± 4.5
"Treatment"		
CTB-SRBC	0 ± 3.3**	0 ± 7.1 **
CTB-SRBC + CT	20 ± 3.9	57 ± 7.8
SRBC	19 ± 2.6	22 ± 4.4
Saline	24 ± 2.3	33 ± 3.6

Prevention experiment.
BALB/c mice were fed single or daily consecutive doses of SRBC-CTB or SRBC. One week after the last oral administration, animals were primed and challenged by systemic injections of SRBC in the left footpad followed 5 days later by the right footpad. The DTH reaction was measured as specific footpad swelling at various times after the second infection and showed a liphasic pattern with an early peak at 2 hours and a later stronger peak at 24 hours. It was found that the daily oral administration of SRBC for 3-4 weeks was required to suppress the 24 hr DTH reactions to a level comparable to that achieved by a single administration of SRBC conjugated to CTB. As many as 20 consecutive feedings with SRBC over a 4 week period had no effect on the development of the early phase (2-4 hours) of the DTH response, in contrast to the situation seen with animals fed a single dose of SRBC conjugated to CTB who failed to develop an early DTH response (* P<0.01; **, P<0.001; Mann-Whitney U test). The tolerance induction by CTB-SRBC was antigen specific as evident from feeding CTB-HRBC (human RBC) conjugate, which was ineffective.

Treatment experiment.
SRBC were first injected in the left rear footpad of BALB/c mice to induce a state of primary systemic immunity. Four days later, animals were fed a single oral dose of SRBC conjugated to CTB with or without free cholera toxin (CT 100 ng), SRBC alone, or saline. Two days after the latter feeding, animals were given a second injection of SRBC in the right footpad to elicit DTH reactions. Whereas mice fed SRBC alone developed DTH responses not significant different from those seen in control animals fed only saline, mice fed SRBC conjugated to CTB had significantly reduced early and late DTH responses to SRBC (**, P<0.001; Wilcoxon rank test).

3.8.4.6 CTB-Tolerogen Conjugates for Immunotherapy

For obvious reasons, we have been especially interested in seeing whether the new oral tolerization principle using CTB-antigen conjugates would also work in immunotherapeutic animal models of autoimmune diseases associated with DTH inflammatory reactions. The disease models of particular interest in this regard include myelinbasic protein (MBP)-induced experimental autoimmune encephalomyelitis (EAE), autoimmune diabetes and collagen II-induced arthritis.

Experimental Autoimmune Encephalomyelitis (EAE). Of special interest because of the similarity of this animal model with human multiple sclerosis, we have found that a single dose of oral cholera B subunit conjugated to MBP both prevents and can "reverse" EAE development in Lewis rats (Table 4, Sun et al, 1996). Animals were injected in the hind footpad with MBP plus Freund's complete adjuvant after being pretreated with different antigens by the oral route. Animals who were fed saline, or an irrelevant antigen coupled to cholera B subunit, or repeated moderate doses of MBP all developed EAE disease with severe paralysis. In contrast, rats fed repeatedly with a high dose of MBP antigen (5 x 1 mg) or with a single low (25 μg) dose of cholera B subunit-coupled MBP remained healthy. The clinical picture was in complete accordance with the presence or absence of inflammation in the central nervous system as judged by histopathology (Sun et al., 1996). It thus appears that by using CTB-coupled MBP one can both reduce the number of doses and dramatically decrease the amount of antigen needed for preventing this autoimmune disease which has many similarities with multiple sclerosis in humans. Most importantly, feeding animals with CTB-MBP six days after disease induction could also abrogate clinical EAE (tab. 3.8.4).

Analyses of cytokine production after in vitro MBP challenge of lymph node cells from CTB-MBP treated animals have disclosed that oral administration of CTB-conjugated MBP leads to profound down-regulation of IL-2 production and concurrent up-regulation of interferon-g secretion. The latter finding is in sharp contrast with the effect of feeding repeated large doses (5 x 1mg) of unconjugated MBP which resulted in suppression of both IL-2 and IFN-γ production (not shown), suggesting again that CTB exerts unique immunomodulating properties on peripheral T cell responses to co-fed antigens (Sun et al., 1996).

Autoimmune diabetes. More recently, using the NOD mouse model of spontaneous autoimmune diabetes, we have also found that a single oral dose of insulin conjugated to CTB could protect animals against diabetes. In adoptive transfer experiments, a low number of T cell-enriched spleen cells from animals fed CTB-insulin were shown to effectively suppress autoimmune diabetes when co-transferred with syngeneic diabetogenic T cells (Thivolet et al., to be published).

Collagen II-induced arthritis. We have also been able to demonstrate the efficiency of this strategy of tolerance induction in a murine model of antigen-induced arthritis. DBA mice having inhaled as little as 25 micrograms of collagen type II chemically coupled to CTB given one week after disease induction (by intracutaneous injection

Table 3.8.4: Oral administration of CTB-conjugated myelin basic protein (MBP) prevents and reverses induction of experimental autoimmune encephalomyelitis (EAE) in Lewis rats

Oral tolerogen (admin. day)	EAE Clinical score	Incidence of paralysis
"Prevention"		
CTB-MBP 25 μg x 1 (–4)	0.4 ± 0.2*	0/5
MBP 100 μg x 5 (–2, –4, –6, –8, –10)	3.0 ± 0.5	5/5
MBP 1000 μg x 5 (–2, –4, –6, –8, –10)	0.5 ± 0.6*	0/5
Saline (–4 or +6)	4.2 ± 0.5	8/8
"Treatment"		
CTB-MBP 50 μg x 1 (+6)	0.2 ± 0.3*	0/4
MBP 50 μg x 1 (+6)	3.3 ± 0.5	4/4
MBP 1000 μg x 5 (+2, +4, +6, +8, +10)	0.2 ± 0.2*	0/4

Female Lewis rats were injected in the rear footpads with myelin basic protein (MBP) emulsified in Freund's complete adjuvant so as to induce EAE. MBP alone or conjugated to CTB "oral tolerogen" was administered by ingastric instillation together with bicarbonate at the indicated doses and times before or after disease induction. The intensity of EAE was monitored by standard clinical examination of animals for a period of 20 days, using a clinical scoring system (0, no symptoms; 1, mild limp tail paresis; 1.5, moderate; 2, complete tail paralysis with rear leg weakness; 3, rear leg paralysis; 4, front and rear leg paralysis; 5, death). The incidence of paralysis refers to the number out of total tested exhibiting a clinical score of ≥ 1.5.

* $P<0.05$; Wilcoxon rank test.

of collagen type II in Freund's adjuvant) showed no or considerably delayed (by more than 30 days) onset of arthritis and had decreased disease severity (joint swelling and erosiveness) as compared to animals given comparable doses of unconjugated collagen type II or CTB alone (Tarkowski et al., to be published).

Allograft rejection. Furthermore, by coupling thymocytes to CTB and feeding this conjugate to mice we have also been able to significantly prolong the survival of transplanted hearts in allogeneic mouse recipients. Again, the effect was superior to that obtained by feeding the cells alone (Sun et al., to be published).

3.8.5 Conclusions

As discussed in the above there is a great need for the design of vaccines aimed at promoting SIgA immune responses against the numerous infectious pathogens that enter the host through mucosal membranes, as well as protecting the host from developing potentially harmful cell-mediated immune responses against the same agents or other environmental antigens. The relative inefficiency of injectable vaccines to evoke secretory IgA immune responses in mucosal tissues and the fact that they can induce DTH reactivity and thereby by-stander tissue damage upon subsequent encounter with the corresponding pathogen constitute two major reasons to encourage the development of alternative strategies to stimulate appropriate immune responses in MALT against infectious agents. Mucosal administration of antigens may theoretically result in the concomitant expression of SIgA antibody responses in various mucosal tissues and secretions, and under appropriate conditions, in the simultaneous down-regulation of cell-mediated immune reactivity at local and systemic sites. Genital and eye infections with *Chlamydia trachomatis*, respiratory infection with respiratory syncytical virus, gastrointestinal infection with *Helicobacter pylori*, and periodontal infection with *Bacteroides spp* are just a few examples where this kind of vaccine development strategy may be important.

It is evident from this article that both cholera toxin and its non-toxic CTB moiety are potent protective oral-mucosal immunogens, and that based on this recombinantly produced CTB is now an important component of licensed and currently developed oral vaccines against cholera and ETEC diarrhea, respectively. CTB has also in animal models been shown to be a very promising carrier protein for various foreign antigens, chemically linked or genetically fused to the CTB molecule, thus significantly enhancing the mucosal IgA immunogenicity of the attached antigens. More surprising and even more impressive with regard to its potency, our recent findings as discussed here have shown that CTB is also a very efficient transmucosal carrier-delivery system for induction of peripheral T cell tolerance. Based upon these new findings, it may be possible to use cholera B subunit as a most effective mucosal carrier delivery system for inducing both mucosal surface immunity and/or specific systemic T cell tolerization. Although still at the early stages of animal experimentation, this new principle may lead to the development of novel anti-infectious and anti-inflammatory vaccines against infectious diseases as well as potentially of medically useful immunotherapeutic agents in selected autoimmune and other DTH-type diseases. Thus, the findings discussed in this article and such perspectives may broaden our perception of oral and other mucosal vaccines to include not only prophylaxis against various infections but also prevention and/or therapy against tissue-damaging inflammatory sequences derived from both infections, autoimmune and allergic disorders.

Acknowledgements

We gratefully acknowledge the active contributions of Marianne Jertborn, Marianne Lindblad, Michael Lebens and many other colleagues and coworkers in the vaccine development work, of M. Lindblad, M. Lebens, Carola Rask, B-L Li and Margaretha Fredriksson in the basic oral tolerance studies and of Thomas Olsson, Charles Thivolet, and Andrej Tarkowski in the autoimmunity disease model studies. The studies summarized here were supported in part by the Swedish Medical and Technical Research Councils, SAREC and the Institut National de la Santé et de la Recherche Médicale (France).

References

Auchincloss, Jr., H. and Sachs, D. H. (1993) Transplantation and graft rejection. In: Paul, W. E. (ed.), Fundmental Immunology, 3 ed., Raven Press, New York, 1099-1142.

Black, R. E., Levine, M. M., Clements, M. L., Young, C. R., Svennerholm, A.-M., and Holmgren, J. (1987) Protective efficacy in man of killed whole vibrio oral cholera vaccine with and without the B subunit of cholera toxin. Infect. Immun. *55*, 1116-1129.

Brandtzaeg, P. (1995) Basic mechanisms of mucosal immunity – A major adaptive defense system. The Immunologist *3*, 89-96.

Challacombe, S. J. and Tomasi, T. B. Jr. (1980) Systemic tolerance and secretory immunity after oral immunization. J. Exp. Med. *152*, 1459-1472.

Chase, *M. W.* (1946) Inhibition of experimental drug allergy by prior feeding of the sensitizing agent. Proc. Soc. Exp. Biol. *61*, 257-259.

Chiller, J. M., Habicht, G. S., and Weigle, W. O. (1991) Kinetic differences in unresponsivenss of thymus and bone narrow cells. Science *171*, 813-815.

Clemens, J., Sack, D. A., Harris, J. R., Chakraborty, J., Khan, M. R., Stanton, B. F., Kay, B. A., Khan, M. U., Yunus, M. D., Svennerholm, A.-M., and Holmgren, J. (1986) Field trial of oral cholera vaccines in Bangladesh. Lancet i, 124-127.

Clemens, J., Sack, D., Harris, J. R., Chakraborty, J., Neogy, P. K., Stanton, B., Huda, N., Khan, M. U., Kay, B. A., Khan, M. R., Ansaruzzaman, M., Yunus, M., Rao, M. R., Svennerholm, A.-M., and Holmgren, J. (1988a) Cross-protection by B subunit-whole cell cholera vaccine against diarrhea associated with heat-labile toxin-producing enterotoxigenic *Escherichia coli*: Results of a large-scale field trial. J. Infect. Dis. *158*, 372-377.

Clemens, J. D., Sack, D. A., Harris, J. R., Chakraborty, J., Khan, M. R., Stanton, B. F., Ali, M., Ahmed, F., Yunus, M., Kay, B. A., Khan, M. U., Rao, M. R., Svennerholm, A.-M., and Holmgren, J. (1988b) Impact of B subunit killed whole-cell and killed whole-cell-only oral vaccines against cholera upon treated diarrhoeal illness and mortality in an area endemic for cholera. Lancet i, 1375-1379.

Clemens, J. D., Sack, D. A., Harris, J. R., Van Loon, F., Chakraborty, J., Ahmed, F., Rao, M. R., Khan, M. R., Yunus, M. D., Huda, N., Stanton, B. F., Kay, B. A., Walter, S., Eeckels, R., Svennerholm, A.-M., and Holmgren, J. (1990) Field trial of oral cholera vaccines in Bangladesh: results from three-year follow-up. Lancet i, 270-273.

Czerkinsky, C. and Holmgren, J. (1995) The mucosal immune system and prospects for anti-infectious and anti-inflammatory vaccines. The Immunologist *3*, 97-103.

Czerkinsky, C., Russell, M. W., Lycke, N., Lindblad, M., and Holmgren, J. (1989) Oral administration of a streptococcal antigen coupled to cholera toxin B subunit evokes strong antibody responses in salivary glands and extramucosal tissues, Infect. Immun. *57*, 1072-1077.

de Aizpurua, H. J. and Russell-Jones, G. J. (1988) Oral vaccination. Identification of classes of proteins that provoke an immune response upon oral feeding. J. Exp. Med. *167*, 440-451.

Dertzbaugh, M. T. and Elson, C. O. (1991) Cholera toxin as a mucosal adjuvant. In: Spriggs, D. R. and Koff, W. C. (eds.), Topics in Vaccine Adjuvant Research, CRC Press, Boca Raton, 119-132.

Drew, M. D., Estrada-Correa, A., Underdown, B. J., and McDermott, M. R. (1992) Vaccination by cholera toxin conjugated to a herpes simplex virus type 2 glycoprotein D peptide. J. Gen. Virol. *73*, 2357-2366.

Elson, C. O. and Ealding, W. (1984a) Generalized systemic and mucosal immunity in mice after mucosal stimulation with cholera toxin. J. Immunol. *132*, 27-36.

Elson, C. O. and Ealding, W. (1984b) Cholera toxin did not induce oral tolerance in mice and abrogated oral tolerance to an unrelated antigen. J. Immunol. *133*, 2892.

Evans, D. J. Jr. and Evans, D. G. (1989) Determinants of microbial attachment and their genetic control. In: Farthing, M. J. G. and Keusch, G. T. (eds.), Enteric infection. Mechanisms, manifestations and management. Chapman and Hall, London, 15.

Farthing, M. J. G. and Keusch, G. T. (eds.) (1989) Enteric infection. Mechanisms, manifestations and management. Chapman and Hall, London.

Gallin, J. I. (1993) Inflammation. In: Paul, W. E. (ed.), Fundmental Immunology, 3 ed., Raven Press, New York, 1015-1032.

Hansson, D. G., Vaz, N. M., Rawlings, L. A., and Lynch, J. M. (1979) Inhibition of specific immune responses by feeding protein antigens. II. Effects of prior passive and active immunization. J. Immunol. *122*, 2261-2266.

Hirabayashi, Y., Tamamura, S.-I., Suzuki, Y., Nagamine, T., Aizawa, C., Shimada, K., and Kurata, T. (1991) H-2-unrestricted adjuvant effect of cholera toxin B subunit on murine antibody responses to influenza virus haemagglutinin. Immunology *72*, 329-335.

Holme, T., Holmgren, J., Merson, M. H. et al. (eds.) (1981) Acute enteric infections in children: New prospects for treatment and prevention. Elsevier/North-Holland, Amsterdam.

Holmgren, J. (1981) Actions of cholera toxin and the prevention and treatment of cholera. Nature *292*, 413-417.

Holmgren, J. and Svennerholm, A.-M. (1983) Cholera and the immune response. Progr. Allergy *33*, 106-119.

Holmgren, J. and Lindholm, L. (1976) Cholera toxin, ganglioside receptors and the immune response. Immunol. Commun. *5*, 737-756.

Holmgren, J., Svennerholm, A.-M., Lönnroth, I., Fall-Persson, M., Markman, B., and Lundbäck, H. (1977) Development of improved cholera vaccine based on subunit toxoid. Nature (Lond) *269*, 602-604.

Holmgren, J., Svennerholm, A.-M., Jertborn, M., Clemens, J., Sack, D. A., Salenstedt, R., and Wigzell, H. (1992) An oral B subunit: whole cell vaccine against cholera. Vaccine *10*, 911-914.

Holmgren, J., Lycke, N., and Czerkinsky, C. (1993) Cholera toxin and cholera B subunit as oral-mucosal adjuvant and antigen vector systems. Vaccine *11*, 1179-1184.

Holmgren, J., Czerkinsky, C., Lycke, N., and Svennerholm, A.-M. (1994) Strategies for the induction of immune responses at mucosal surfaces making use of cholera toxin B subunit as immunogen, carrier and adjuvant. Am. J. Trop. Med. *50*, 42-54.

Husby, S., Mestecky, J., Moldoveanu, Z., Holland, S., and Elson, C. O. (1994) Oral tolerance in humans. T cell but not B cell tolerance after antigen feeding. J. Immunol. *152*, 4663-4670.

Jertborn, M., Svennerholm, A.-M., and Holmgren, J. (1988) Five-year immunological memory in Swedish volunteers after oral cholera vaccination. J. Infect. Dis. *157*, 374-377.

Jertborn, M., Åhrén, C., Holmgren, J., and Svennerholm, A.-M. (1996) Clinical trials of an oral inactivated enterotoxigenic *Escherichia coli* vaccine. In press.

Lange, S., Nygren, H., Svennerholm, A.-M., and Holmgren, J. (1980) Antitoxic cholera immunity in mice: Influence of antigen deposition on antitoxin-containing cells and protective immunity in different parts of intestine. Infect. Immun. *28*, 17-23.

Lehner, T., Bergmeyer, L. A., Panagiotidi, C., Tao, L., Brookes, R., Klavinskis, L. S., Walker, P., Walker, J., Ward, R. G. et al. (1992) Induction of mucosal and systemic immunity to a recombinant simian immunodeficiency viral protein. Science *258*, 1365-1369.

Liang, X., Lamm, M. E., and Nedrud, J. G. (1988) Oral administration of cholera toxin Sendai virus conjugate potentiates gut and respiratory immunity against Sendai virus. J. Immunol. *141*, 3781-3787.

Lycke, N. and Holmgren, J. (1986) Strong adjuvant properties of cholera toxin on gut mucosal immune responses to orally presented antigens. Immunology *59*, 301-308.

McDermott, M. R. and Bienenstock, J. (1979) Evidence for a common mucosal immunologic system: I. Migration of B immunoblasts into intestinal, respiratory, and genital tissues. J. Immunol. 122, 1892-1898.

McDermott, R. P. and Elson, C. O. (eds.) (1991) Mucosal Immunology I: Basic Principles. Gastroenterology Clinics of North America, WB Saunders, Philadelphia 1991 Mucosal Immunology II: Clinical Applications.

McDermott, R. P. and Elson, C. O. (eds.) (1992) Mucosal Immunology II: Clinical Applications. Gastroenterology Clinics of North America, WB Saunders, Philadelphia 1992.

McKenzie, S. J. and Halsey, J. F. (1984) Cholera toxin B subunit as a carrier protein to simulate a mucosal immune response. J. Immunol. *133*, 1818-1824.

Mestecky, J. (1987) The common mucosal immune system and current strategies for induction of immune responses in external secretions. J. Clin. Immunol. *7*, 265-276.

Mestecky, J. and McGhee, J. R. (1989) Oral immunization: past and present. Curr. Top. Microbiol. Immunol. *146*, 1-11.

Michalek, S. M., Eldridge, J. H., Curtiss III, R., and Rosenthal, K. L. (1994) Antigen delivery systems: new approaches to mucosal immunization. Handbook of Mucosal Immunology, Academic Press, 373-390.

Mowat, A. M. (1987) The regulation of immune response to dietary protein antigens. Immunol. Today *8*, 93-98.

Nedrud, J. G., Liang, X., Hague, N., and Lamm, M. E. (1987) Combined oral/nasal immunization protects mice from Sendai virus infection. J. Immunol. *139*, 3484-3492.

Ogra, P. L., Chiba, Y., Beutner, K. R., and Morag, A. (1976) Vaccination by non-parenteral routes: characterization of immune response. Dev. Biol. Stand. *33*, 19-26.

Peltola, H., Siitonen, A., Kyrönseppä, H., Simula, I., Mattila, L., Oksanen, P. et al. (1991) Prevention of travellers' diarrhoea by oral B-subunit/whole cell cholera vaccine. Lancet *338*, 1285-1289.

Pierce, N. S. and Gowans, J. L. (1975) Cellular kinetics of the intestinal immune response to cholera toxoid in rats. J. Exp. Med. *142*, 1550-1563.

Quiding, M., Nordström, I., Kilander, A., Andersson, G., Hanson, L. Å., Holmgren, J., and Czerkinsky, C. (1991) Intestinal immune responses in humans. Oral cholera vaccination induces strong intestinal antibody responses, gamma-interferon production, and evokes local immunological memory. J. Clin. Invest. *88*, 143-148.

Rebien, W., Puttonen, E., Maasch, H. J., Stix, E., and Wahn, U. *(1982)* Clinical and immunological response to oral and subcutaneous immunotherapy with grass pollen extracts. A prospective study. Eur. J. Pediatry *138*, 341-344.

Sanchez, J. and Holmgren, J. (1989) Recombinant system for overexpression of cholera toxin B subunit in *Vibrio cholerae* as a basis for vaccine development. Proc. Natl. Acad. Sci. USA *86*, 481-485.

Sanchez, J., Johansson, S., Löwenadler, B., Svennerholm, A.-M., and Holmgren, J. (1990) Recombinant cholera toxin B-subunit and gene fusion proteins for oral vaccination. Res. Microbiol. *141*, 971-979.

Sanchez, J. L., Vasques, B., Begue, R. R., Meza, R., Castellares, G., Cabezas, C., Watts, D. M., Svennerholm, A.-M., Sadoff, J. C., and Taylor, D. N. (1994) Protective efficacy of the oral, whole cell/recombinant B subunit cholera vaccine in Peruvian military recruits. Lancet *344*, 1273-1276.

Schwartz, R. H. (1993) Immunological tolerance. In: Paul, W. E. (ed.) Fundamental Immunology, 3 ed., Raven Press, New York, 677-732.

Schwartz, R. S. (1993) Autoimmunity and autoimmune diseases. In: Paul, W. E. (ed.) Fundmental Immunology, 3 ed., Raven Press, New York, 1033-1098.

Stoll, B. J., Svennerholm, A.-M., Gothefors, L., Barua, D., Huda, S., and Holmgren, J. (1986) Local and systemic antibody responses to naturally acquired enterotoxigenic *Escherichia coli* diarrhea in endemic areas. J. Infect. Dis. *153*, 527-534.

Sun, J. B., Holmgren, J., and Czerkinsky, C. (1994) Cholera toxin B subunit: An efficient transmucosal delivery system for induction of peripheral tolerance. Proc. Natl. Acad. Sci. USA *91*, 10795-10799.

Sun, J. B., Rask, C., Olsson, T., Holmgren, J., and Czerkinsky, C. (1996) Prevention and reversal of experimental autoimmune encephalomyelitis by oral tolerization with myelin basic protein conjugated to cholera toxin B subunit. Proc. Natl. Acad. Sci. USA, in press.

Svennerholm, A.-M., Sack, D. A., Holmgren. J., and Bardhan, P. K. (1982) Intestinal antibody responses after immunization with cholera B subunit. Lancet i, 305-308.

Svennerholm, A.-M., Jertborn, M., Gothefors, L., Karim, A. M. M. M., Sack, D. A., and Holmgren, J. (1984a) Mucosal antitoxic and antibacterial immunity after cholera disease and after immunization with a combined B subunit-whole cell vaccine. J. Infect. Dis. *149*, 884-893.

Svennerholm, A.-M., Gothefors, L., Sack, D. A., Bardhan, P. K., and Holmgren, J. (1984b) Local and systemic antibody responses and immunological memory in humans after immunization with cholera B subunit by different routes. Bull WHO *62*, 909-918.

Svennerholm, A.-M., Åhren, C., Wennerås, C., and Holmgren, J. (1991) Development of an oral vaccine against enterotoxigenic *Escherichia coli* diarrhea. In: Molecular pathogenesis of gastrointestinal infections. Wadström, T. et al. (eds.) PLENUM, London, 287-294.

Trentham, D. E., Dynesius-Trentham, R. A., Orav, E. J., Combitchi, D., Lorenzo, C., Sewett, K. L., Hafler, D. A., and Weiner, H. L. (1993) Effects of oral administration of collagen on rheumatorid arthritis. Science *261*, 1727-1730.

Walker, W. A., Wu, M., Isselbacher, K. J., and Block, K. J. (1975) Intestinal uptake of micromolecules. IV. The effect of pancreatic duct ligation on the break down of antigen-antibody complexes on the intestinal surface. Gastroenterology *69*, 1223-1229.

Weiner, H. L., Friedman, A., Miller, A., Khoury, S. J., Al-Sabbagh, A., Santos, L., Sayegh, M., Nussenblatt, R. B., Trentham, D. E., and Hafler, D. A. (1993a) Oral tolerance: Immunological mechanisms and treatment of murine and human organ specific autoimmune diseases by oral administration of autoantigens. Ann. Rev. Immunol. *12*, 809-837.

Weiner, H. L., Mackin, G. A., Matsui, M., Orav, E. J., Khoury, S. J., Dawson, D. M., and Hafler, D. A. (1993) Double-blind pilot trial of oral tolerization with myelin antigens in multiple schlerosis. Science *259*, 1321-2324.

Wells, H. (1911) Studies on the chemistry of anaphylaxis III. Experiments with isolated proteins, especially those of hen's egg. J. Infect. Dis. *9*, 147-171.

Whitacre, *C. C.,* Gienapp, C., Orosz, I. E., and Bitar, D. (1991) Oral tolerance in experimental autoimmune encephalomyelitis. III. Evidence for clonal anergy. J. Immunol. *147*, 2155-2163.

Wilson, A. D., Clarke, C. J., and Stokes, C. R. (1990) Whole cholera toxin and B subunit act synergistically as an adjuvant for the mucosal immune response of mice to keyhole limpet haemocyanin. Scand. J. Immunol. *31*, 433-451.

World Health Organization (1994) Diarrhoeal and acute respiratory disease control. Interim report, Geneva.

Wortmann, *F.* (1977) Oral hyposensitization of children with pollinosis or house dust asthma. Allergol et Immunopathol. *5*, 15-26.

Åhrén, C, Gothefors, L, Stoll, B., and Salek, M. A. (1982) Synergistic protective effect of antibodies against *Escherichia coli* enterotoxin and colonization factor antigens. Infect. Immun. *38*, 74-79.

Åhrén, C., Wennerås, C., Holmgren, J., and Svennerholm, A.-M. (1993) Intestinal antibody response after oral immunization with a prototype cholera B subunit-colonization factor antigen enterotoxigenic *Escherichia coli* vaccine. Vaccine *11*, 929-934.

4. Specific Vaccination Strategies

4.1 *Helicobacter pylori*: Pathogenic Determinants and Strategies for Vaccine Design

Paolo Ghiara, Antonello Covacci, John L. Telford and Rino Rappuoli

4.1.1 Introduction

The presence of 'spiral bacteria' in mammalian stomach was reported for the first time in 1893 by the Italian pathologist Giulio Bizzozzero, mentor of Camillo Golgi, while working at the University of Turin (Bizzozzero, 1893). He had observed a member of the genus *Helicobacter* colonizing the gastric mucosa of apparently healthy dogs. About 100 years later the curved, flagellated, Gram-negative bacillus *Helicobacter pylori*, has been officially recognized to be responsible of causing chronic gastritis, peptic ulcer and gastric tumors in humans (Parsonnet et al., 1991; Parsonnet et al., 1994; Blaser and Parsonnet, 1994).

The re-discovery started in 1982 when the routine use of the endoscope had allowed a more in-depth examination of the gastrointestinal tract of patients. A gastroenterologist, Barry Marshall, and a pathologist, Robin Warren, both working at the Royal Perth Hospital in Western Australia, isolated a 'new' spiral bacterium from patients with peptic ulcer and chronic gastritis. They called it '*Campylobacter pyloridis*' and proposed that it was the causative agent of the diseases (Marshall and Warren, 1983). In 1988 the bacterium received the name *Helicobacter pylori*. At those times only a handful of bacteriologists and gastroenterologists were familiar with this organism. Since then the interest for *H.pylori* has increased in a terrific way. New bacteria related to *Helicobacter* have been isolated from many hosts and the number is increasing every year. The genus *Helicobacter* now includes many species (tab. 4.1.1) that have the capacity of colonizing the gastro-intestinal mucosae of many vertebrate species. *H. pylori* is a component of this family, that has evolved to adapt to the stomach of humans and non-human primates.

Table 4.1.1: The *Helicobacter* genus 1995

Species	natural hosts	localization	reference
Helicobacter pylori	humans, monkeys, swine and cat	stomach	Marshall and Warren, 1983
Helicobacter felis	cat, dog and monkeys	stomach	Lee *et al.* 1988
Helicobacter mustelae	ferret	stomach	Fox *et al.* 1986
Helicobacter nemestrinae	primates	stomach	Brondson *et al.* 1988
Helicobacter acinonyx	cheetas	stomach	Eaton *et al.* 1993
Helicobacter heilmannii	humans and monkeys and dog	stomach	Solnick *et al.* 1993
Helicobacter sp. 'Bird-C'	birds	intestine	Dewhirst *et al.* 1994
Helicobacter sp. 'Bird-B'	birds	intestine	Dewhirst *et al.* 1994
Helicobacter canis	dog	intestine	Stanley *et al.* 1993
Helicobacter cinaedi	humans, dog	intestine	Orlicek *et al.* 1993
Helicobacter fennelliae	humans, monkeys, dog	intestine	Kemper *et al.* 1993
Helicobacter muridarum	rodents	intestine*	Lee *et al.* 1992
Helicobacter sp. 'CLO-3'	humans, swine and dog	intestine	Dewhirst *et al.* 1994
Helicobacter pametensis	birds, swine	intestine	Dewhirst *et al.* 1994
Helicobacter pullorum	poultry, humans	intestine	Stanley *et al.* 1994
Helicobacter 'rappini'	humans, cows, rodents	intestine	Schauer *et al.* 1993
Helicobacter bilis	rodents	bile, liver, intestine	Fox *et al.* 1995
Helicobacter hepaticus	rodents	liver, intestine	Fox *et al.* 1994

* in aged mice (≥12 months old) this species can be isolated also from stomach (Lee *et al.* 1992)

4.1.1.1 The Second Decade of *H.pylori*: Relevance of Infection and Epidemiology Defined

Following the original report of Marshall and Warren, the concept that peptic ulcer, a disease usually associated with bad dietary habits and stress, was caused by a bacterial infection, took about 12 years to be officially accepted. In 1994 a Consensus Conference held at the NIH recognized *H.pylori* as the causative agent of all duodenal ulcers, of nearly all gastric ulcers and of chronic active gastritis (NIH Consensus Development Panel on Helicobacter pylori in Peptic Ulcer Disease, 1994). In the same year the WHO International Agency for Research on Cancer has included *H.pylori* in the 'group 1' (definite) carcinogens (Logan, 1994), thus ranking it along with other microbial pathogens, like hepatitis B and C viruses, that are known to play an important role in the occurrence of malignancies of the hepatobiliary system.

In developed countries the infection increases with age and more than 60 % of the elderly are infected. More than 80 % of population in developing areas is infected as early as childhood, or even infancy. The prevalence of infection is often associated with low socio-economical status and poor hygiene conditions. The most conceivable theory to explain the epidemiological data is that the infection is probably acquired during childhood, but in developed countries, due to improvement of hygiene and socio-economical conditions, there has been a progressive decrease in the risk of infection along the generations (cohort phenomenon) during the last 50 years (Banatvala et al., 1993a; Megraud, 1993; Megraud, 1994). This hypothesis may allow to predict that the overall prevalence of *H.pylori* infection will progressively diminish in developed countries.

The mechanism of transmission is unknown, but it may probably occur through the oro-oral or the oro-faecal routes (Mendall and Northfield, 1995). Although traces of *H.pylori* can be detected in dental plaque (Shames et al., 1989; Desai et al., 1991; Lambert et al., 1993; Olsson et al., 1993; Banatvala et al., 1993b) or faeces (Luzzi et al., 1993; Leverstein-van Hall et al., 1993; Mapstone et al., 1993; Vanzwet et al., 1994; Kelly et al., 1994; Namavar et al., 1995), there has however not been a formal demonstration that any one of these routes is important. Some authors consider the transmission of the infection via faeces quite unlikely (Lee et al., 1991; Hazell et al., 1994). Of interest is a theory of transmission recently proposed by Dr Axon, that has been defined as the 'gastro-oral' route (Axon, 1995). This theory is based on the fact that the infection is probably acquired during childhood and on the observation that children vomit quite frequently; *H.pylori* could thus be transmitted to another child by accidental oral contamination with vomit. Transmission of infection from natural animal hosts to humans cannot be excluded. In this light, interesting is the recent report concerning a natural infection detected in cats, a new natural host for *H.pylori,* that therefore could represent a potential reservoir (Handt et al., 1994).

About 10 % of the infected population in developed countries is diagnosed to have peptic ulcer disease. Patients with duodenal ulcers rarely develop gastric cancer (Eidt et al., 1995; Blaser et al., 1995), a disease that is slowly decreasing all over the world.

However, the adenocarcinoma of the antrum is still an important cause of death in many developed countries, and it is probably the commonest type of cancer in the male population of China (Forman et al., 1990; Mitchell et al., 1992; Armstrong, 1980) and Japan (Blaser et al., 1994; Tsugane et al., 1993; Blaser et al., 1993a).

4.1.1.2 The Fight Against Infection: Chemotherapy Versus Vaccination

H.pylori infection is currently treated with an association of antisecretory drugs and antibiotics (Tytgat, 1994). There is some variability in the therapeutic regimens applied. 'Dual' therapy, usually poorly successful, consists of combination of a proton pump inhibitor (omeprazole) with either amoxycillin, or metronidazole. The 'triple' therapy comprises omeprazole, a tetracycline, or metronidazole, and amoxycillin. All these therapeutic approaches force the patient to swallow an high number of pills every day for 1-2 weeks. Depending on the therapeutic regimens applied, the success of eradication varies between 50 to 90 % of cases. About 3-5 % of patients undergoing eradication regimens are forced to withdraw the therapy due to serious side effects. A major drawback of the antibiotic intervention in developing countries is the ratio between infected individuals and the extremely poor economical status. Furthermore, this therapeutic regimen will be progressively made less effective by the occurrence of antibiotic resistance (Rautelin et al., 1992; Westblom et al., 1992; Malfertheiner, 1993; Malanoski et al., 1993; Rautelin et al., 1994).

Vaccination appears to be a more promising approach to combat this relevant human pathogen. Here we will review the state-of-the-art in the development of vaccines against *H.pylori*.

4.1.1.3 Host-Parasite Interactions

A common histopathological hallmark of the gastric biopsies of chronically infected patients is the marked infiltration of the lamina propria by inflammatory cells (Blaser, 1993b; Solnick and Tompkins 1993; Blaser et al., 1994; Blaser et al., 1995; Telford et al., 1994a). In chronic active gastritis polymorphonuclear leukocytes are the prominent infiltrating cells. This may be an asymptomatic condition, but it is thought to precede the occurrence of peptic ulcer. In old chronically infected patients histopathology reveals either a progressive metaplastic intestinalization of the gastric mucosa, or a marked loss of parietal cell function (gastric atrophy), this picture usually correlates with an increased risk of gastric adenocarcinoma (Fiocca et al., 1989; Solcia et al., 1990; Caselli et al., 1993).

4.1.1.3.1 *H.pylori* Adaptation to a Hostile Environment

The stomach mucosa is an acid environment that usually does not allow bacteria to grow or survive. Histological sections of biopsies from stomach mucosae of patients infected with *H.pylori* (Caselli et al., 1989) shows that: a) a dominant bacterial population swimming free in the closely neutral pH of the viscous mucus layer lining the gastric epithelium, b) a small fraction of the organisms adhere tightly to the luminal surface of gastric cells, which, at the adhesion interface organize their membrane in 'pedestal' structures strongly resembling those found when enteropathogenic *Escherichia coli* adhere to target cells (Smoot et al., 1993; Noach et al., 1994), c) sparse bacteria may be found associated to the intercellular spaces between adjacent epithelial cells in correspondence of ongoing active damage to tight junctions .

To survive in this environment *H.pylori* uses a rather unique mixture of metabolic pathways. *H.pylori* has been shown to catabolize carbohydrates like glucose (Mendz et al., 1993). Although growing in a microaerophylic atmosphere, it has pyruvate and fumarate metabolic pathways that are characteristic of anaerobiotic bacteria (Mendz et al., 1994a). Biosynthetic pathways of purine nucleotides resemble those of an intracellular organism (Mendz et al., 1994b), but *H.pylori* is not a clearly invasive pathogen. There is some evidence that the bacterium processes excess nitrogen from the cytoplasm using a 'urea cycle' (Mendz et al., 1994c), a characteristic which is rather uncommon for a prokaryotic organism. *H.pylori* has been also reported to express a 54 kDa protein factor, termed AIF, with the capacity to reduce the acid output from the gastric epithelium (Huang et al., 1995), this feature could help the bacterium to survive in its hostile niche. A striking feature of this bacterium is its capacity to assume *in vitro*, upon prolonged or non-optimal culture condition, a spherical shape that is usually called 'coccoid form' (Bode et al., 1993; Nilius et al., 1994). The significance of this transformation into a 'quiescent' form is not well understood and its relevance to pathogenesis awaits to be assessed.

4.1.1.3.2 Immune Response Against *H.pylori*

A good understanding of immune response to the pathogen is required to evaluate the feasibility of immune intervention against *H.pylori*. Despite the great interest to the immune mechanisms at other mucosal sites, the study of immunology of the stomach is almost completely neglected. This is possibly due to the fact that, before the evidence of gastric colonization by *H.pylori*, this organ was not included among the sites where immune competence had to be accomplished, because it was thought to be 'sterile' due to its acidic environment.

Activation of immune system upon *H.pylori* colonization of the gastric mucosa is apparently inefficient in eliminating the infection. Although a clear antibody response may be elicited, both systematically and locally (Sobala et al., 1991; Crabtree et al., 1991a), there is no clear relationship between cellular and humoral responses in infected people (Karttunen, 1991; Sharma et al., 1994). Moreover, *H.pylori*-specific CD4+ T cell clones may be isolated from both seronegative and seropositive patients (Di Tommaso et al., 1995). The cellular infiltration that occurs in chronic infection is sustained by the expression of soluble mediators, such as TNFα, IL6 and IL8 (Crabtree et al., 1991b; Crabtree et al., 1993). It can be hypothesized that the only outcome of these responses is to exacerbate the pathology rather than efficiently fight off the infection. This seems to be a major drawback for the success of active immunotherapy against *H.pylori*. However, the vaccination experiments using animal models have suggested that an active immunization with strong mucosal adjuvant like Choleratoxin (CT) or heatlabile enterotoxin of *E. coli* (LT) can prevent, and even eradicate, the infection (see below). Thus the development of good vaccine formulations for efficient presentation of protective epitopes to the host is a major challenge.

4.1.2 *H.pylori* Antigens and Their Potential as Vaccine Candidates

Proliferation of molecular studies on *H.pylori* is causing continuous increase of the number of genes cloned and sequenced. Most of the genes have been characterized because of the interest in assessing their putative roles in pathogenesis, but a major result of these investigations is that a significant number of molecules can now be considered as protective antigens for vaccination studies in animal models.

Before entering into the details of the known relevant genes and factors, it is necessary to point out that only a small percentage of individuals carrying the infection develop severe pathology. Host factors have been proposed to be important for the development of more severe diseases (Lee, 1993). However, a more convincing hypothesis to explain this phenomenon is that the patients with the more severe diseases are infected by *H.pylori* strains, that may express specific virulence factors associated with the disease (Telford et al., 1994a).

A cytotoxin causing vacuolation of epithelial cells in culture was described by Leunk et al in the broth supernatants of *H.pylori* strains obtained from ulcer patients (Leunk et al., 1988). This cytotoxic activity was significantly more frequently observed in strains isolated from patients with peptic ulcer (Figura et al., 1989).The gene coding for the toxin has been cloned, sequenced and named *vacA* (Vacuolating cytotoxin gene A) (Cover et al., 1994; Schmitt et al., 1994; Telford et al., 1994b; Phadnis et al., 1994). More than 90 % of duodenal ulcer patients have circulating IgG, as well as mucosal IgA, specific for a 128kDa antigen, CagA (cytotoxin associated gene A) (Crabtree et al., 1991c), that has been cloned and sequenced (Covacci et al., 1993; Tummuru et al., 1993).

Genetic analysis of the strains obtained from infected patients from different geographical areas has allowed to partition the isolates in at least two families, named Type I and Type II (Xiang et al., 1995). Type I bacteria contain the *cagA* gene, express the CagA antigen and an active form of the vacuolating cytotoxin. Type II bacteria do not contain the *cagA* gene, do not express the CagA antigen, and do not produce an active form of the vacuolating cytotoxin. Patients with more severe gastric pathologies are significantly more frequently infected by the Type I bacteria. Serological studies have shown that infection by this kind of bacteria is also associated with the occurrence of duodenititis and duodenal ulcerations. Type I strains represent about 84 % of the isolates, about 28 % of the Type I strains may either express CagA molecule, in the absence of an active VacA, or they may express the active vacuolating toxin without apparent transcription of *cagA* gene, or with a deletion interesting the *cagA* gene. The remaining 16 % of isolates are Type II bacteria that do not have the *cagA* gene and do not express the VacA cytotoxin.

H.pylori antigens and cloned genes can be thus distinguished into two groups: 1) those common to both Type I and Type II strains; and 2) those that are expressed only in Type I strains.

The catalogue of factors and genes is continuously growing (tab. 4.1.2). Of course, the major challenge would be now to demonstrate their importance in pathogenicity by testing the virulence of specific isogenic mutants in appropriate *in vitro* and *in vivo* models.

4.1.2.1 Antigens Common to Both Type I and Type II Strains

4.1.2.1.1 Urease

Urease (urea amidohydrolase; E.C. 3.5.1.5) is a nickel-dependent enzyme that catalyzes the hydrolysis of urea to form ammonia and carbon dioxide. Its activity gives thus rise to two potent acid-buffering chemical entities, one of which (ammonia) may be used as a metabolic source of nitrogen and is also highly toxic (Collins et al., 1993). Several ureases have been purified from various bacteria, fungi and several plant species (Jabri et al., 1995). Although being different in their quaternary structure, amino acid sequence comparison of the known ureases shows an high degree of sequence identity (more than 50 %). Their primary role in bacteria is probably to allow the producing organism to utilize urea as nitrogen source for the metabolism. In plants they are thought to be involved in both nitrogen metabolic pathways and in defense mechanisms. Bacterial ureases are usually considered relevant virulence factors. For example, the urease activity of *Proteus mirabilis* is involved in the formation of urinary stones following infection of the urinary tract (Collins et al., 1993).

H.pylori urease is produced in a large amount by the bacterium (Mobley et al., 1991) and constitutes about 6 % of the water-extractable proteins (Monath, 1995). It has the peculiar property to be surface-exposed, while most of the other bacterial ureases are mainly cytoplasmic (Bode et al., 1989). At variance with other homologous ureases, the *H.pylori* enzyme is able to catalyze the hydrolysis of urea at the low pH values, that are typical of the gastric environment. The native enzyme is composed of two subunits, UreA (27kDa) and UreB (62kDa), that are stably associated in a 1:1 heterodimer. The heterodimers are then assembled together to form a macromolecular $(A{:}B)_6$ ring-like structure of high molecular mass (Dunn et al., 1990; Austin et al., 1991). The two urease subunits are encoded by *ureA* and *ureB* genes, that belong to a

Table 4.1.2: The 1995 Catalogue of *H.pylori* characterized genes

Gene name	Product name	Function	Reference
ure *(CDABEFGHI)* operon	urease (UreA + UreB)$_6$	Hydrolysis of urea, essential for colonization	Labigne *et al.* 1991; Hu *et al.* 1993
vacA	VacA*	Epithelial cell vacuolization, major role in the genesis of peptic ulcer	Cover *et al.* 1994; Schmitt *et al.* 1994; Telford *et al.* 1994b; Phadnis *et al.* 1994
cagA	CagA*	Unknown	Covacci *et al.* 1993; Tummuru *et al.* 1993
hspA	HspA	Heat shock protein, Ni^{++}-binding domain	Suerbaum *et al.* 1994
hspB	HspB	Heat shock protein, Chaperonin	Macchia *et al.* 1993; Suerbaum *et al.* 1994
flaA	FlaA	Flagellin (major), motility essential for colonization	Leying *et al.* 1992
flaB	FlaB	Flagellin (minor), motility	Suerbaum *et al.* 1993
flgE	FlgE	Flagellar hook component	O'Toole *et al.* 1994
sod	SOD homodimer	Superoxide dismutase	Spiegelhalder *et al.* 1993
kat	Kat	Catalase	Bukanov *et al.* 1994
hpaA	HpaA	Adhesin	Janzon *et al.* 1993
lpp20	Lpp20	Envelope lipoprotein	Kostrzynska *et al.* 1994
alpA	AlpA	Adhesin	Odenbriet *et al.* 1995
hly	Hly	Hemolysin	Segal and Tompkins 1993
hap	Hap	Protease, Mucinase	Smith *et al.* 1994
hpcopA	HpcopA	P-type ATPase	Ge *et al.* 1995
hpcopP	HpcopP	Cu^{++}-transporter protein	Ge *et al.* 1995
nixA	NixA	Ni^{++}-binding protein, essential for urease activity	Mobley *et al.* 1995
hpn	Hpn	Ni^{++}-binding protein, not essential for urease activity	Gilbert *et al.* 1995
recA	RecA	DNA recombinase	Thompson and Blaser, 1995
napA	HP-NAP	Neutrophil proadhesive activity	Evans *et al.* 1995

* Expressed only by Type I strains

large operon composed of nine genes (named A trough I) (Labigne et al., 1991; Hu et al., 1993). The precise role of the other seven *H.pylori* urease genes has to be determined. The expression is probably regulated by the availability of nitrogen (Cussac et al., 1992) and by incorporation of nickel, that is controlled by a recently cloned nickel-binding protein named NixA (Mobley et al., 1995).

A possible role of urease in *H.pylori* pathogenesis could be to form a neutral pH cloud around the bacterium by producing the acid buffering ammonia and carbon dioxide. Isogenic mutants carrying disrupted urease genes have been reported to loose their ability to colonize experimental animals (Eaton et al., 1991; Tsuda et al., 1994). However, the precise role of urease in pathogenesis is quite controversial since the same isogenic strains were still unable to colonize animals, in which gastric acidity had been abolished by treatment with the proton pump blocker omeprazole (Eaton et al., 1991). Consequently, it is interesting that *H.pylori* has been reported to survive better in an acid environment when exposed to urea, probably because at neutral pH values the urease is poorly active and urea may have toxic activity on the bacterium (Clyne et al., 1995). An urease negative mutant has been reported to be more susceptible to killing by polymorphonuclear leukocytes (Kist, 1995). It is possible that urease is primarily a metabolic enzyme, and its essential role for colonization resides in the capacity to quickly make available ammonia as a nitrogen source.

Urease has been reported to have pro-inflammatory activities *in vitro* and it is present in deep mucosal tissue (lamina propria), whereas *H.pylori* cells are absent (Mai et al., 1991; Mai et al., 1992). Furthermore, there have been several reports suggesting its direct or indirect toxic activity on gastric mucosa (Murakami et al., 1990; Smoot et al., 1990; Xu et al., 1990; Cover et al., 1991; Desai et al., 1993; Takeuchi et al., 1995). However, a recent study that employed a mouse model of acute gastric injury failed to sustain these hypotheses (Ghiara et al., 1995a).

Urease has been reported by many groups to be a protective antigen in animal models of infection (Pallen et al., 1990; Ferrero et al., 1994; Michetti et al., 1994; Marchetti et al., 1995; Lee et al., 1995). Recently, a recombinant UreB subunit has been also successfully used in therapeutic immunizations to eradicate an otherwise chronic infection of *H.felis* in the mouse (Corthésy-Theulaz et al., 1995).

4.1.2.1.2 Flagellins

Motility is often an important phenotypic trait of bacteria that colonize the viscous mucous lining of the gastrointestinal tract. *H.pylori* usually bears 4-6 sheathed polar flagella. These flagella are composed of a basal portion, that contains the motor and the hook structure, and a central filament that is enveloped in a sheath composed of a bilayered membrane with a composition closely resembling that of the outer membrane of the cell wall (Geis et al., 1989; Geis et al., 1993). Motility is essential for gastric colonization by *H.pylori*, thus it is considered a major virulence factor (Eaton et al., 1992a). The flagellin filament is composed by a major 53kDa flagellin molecule (FlaA) (Leying et al., 1992). Sequence alignments with other flagellin genes demonstrates an high degree of homology (about 60 %) at the C- and N- termini with the related genus *Campylobacter*, while the central domain of the molecule is much less conserved. Further investigations have shown the presence of a second flagellin molecule with a slightly higher molecular mass (54kDa). This second flagellin has been named FlaB (Suerbaum et al., 1993). Targeted disruption of *flaA* gene results in complete loss of flagellum, while the *flaB* mutants are less motile but flagellated (Suerbaum et al. 1993). In a more recent study, in which *flaAflaB* double mutants were also used, FlaB expression was shown to be essential for optimal motility of *H.pylori* (Josenhans et al., 1995). The reasons for flagellin gene duplication in *H. pylori* is unclear, however, similar duplications of flagellin genes occur in other gastrointestinal pathogens and may represent a functional advantage (Alm et al., 1993).

The flagellar hook has been recently the object of great interest (O'Toole et al., 1994). It consists of a 70x16 nm structure composed of a major 78kDa subunits called FlgE. Mutants defective in hook production generated by allelic replacement were both non-motile and non flagellated. At variance with *fla* gene regulation mechanisms found in other *Enterobacteriaceae*, the genetic disruption of *flgE* gene did not affect the production of both FlaA and FlaB proteins that, however, are not assembled in flagellar filaments and remain cytoplasmic.

There is a lack of extensive studies on the relevance of the above proteins for the pathogenesis *in vivo*. A recent study has shown that any of them are involved in the stimulation of IL-8 synthesis by *H.pylori in vitro* (Huang et al., 1995). There has not been any study aimed at assessing the feasibility of these antigens as vaccine candidates. However, a major limit in using flagellins as antigens is that their sequence contains epitopes that are quite conserved in other bacterial species.

4.1.2.1.3 Heat Shock Proteins

Two *H.pylori* proteins, named HspA and HspB, that are functionally equivalent to *E.coli* GroES and GroEL, respectively, have been described (Macchia et al., 1993; Kansau et al., 1994; Suerbaum et al., 1994). These proteins are encoded by a bicistronic operon. The HspA protein bears at its C-terminus a cysteine- and histidine-rich stretch resembling a nickel-binding domain, which is absent from the other known GroES molecules. Coexpression in *E.coli* of *hspB* gene with Urease gene cluster (*ureABEFGHI*) has shown a significant increase in the secretion of active urease enzyme, suggesting that also in *H.pylori* these kind of proteins may function as chaperonins (Suerbaum et al., 1994).

The recombinant HspA and HspB proteins retain their antigenicity and are recognized by *H.pylori*-infected patients. Nevertheless, it has not been possible to associate this seropositivity to a particular gastric pathology (Macchia et al., 1993). HspB has an high sequence homology with other members of heat shock protein family in both bacterial and eukaryotic organisms; in particular, it has about 54 % sequence identity with the human homolog (Macchia et al., 1993). Thus the antibody response induced by this antigen may be potentially cross-reacting with the host molecule. A role of autoimmunity during *H.pylori* infection has still, however, to be convincingly demonstrated. Finally, heat shock proteins could contribute to the pathogenesis of gastric disease via the chronic activation of antigen-specific T-cells bearing the γ/δ T cell receptor (Engstrand et al., 1991; Trejdosiewicz et al., 1991), that may contribute to the exacerbation of pathology (Kaufmann et al., 1990).

Although these antigens have been recently reported to induce protection in the *H.felis* model (Ferrero, 1994), and to potentiate the protection induced by recombinant UreB (Ferrero et al., 1995), their use as vaccine candidates in humans is probably not recommendable due to the possible presence of common epitopes in the bacterial molecule and in the human counterpart, and to the possibility to induce autoimmunity.

4.1.2.1.4 Superoxide Dismutase

H.pylori can be actively phagocytosed by macrophages and neutrophils. However, in the absence of opsonins, the bacterium is able to survive within the phagocyte. A major killing mechanism of phagocytes is represented by the activation of the oxidative burst that leads to the formation of highly reactive superoxide anions. To counteract this toxic activity *H.pylori*, like many other pathogens, has evolved to express a Superoxide dismutase (SOD) that catalyzes the dismutation of superoxide anions. The enzyme is composed of two identical subunits of 24 kDa *H.pylori* SOD has been cloned and sequenced. Sequence alignment studies have revealed significant homology with many known SODs of other facultative intracellular pathogens (Spiegelhalder et al., 1993). Despite the lack of a clear leader sequence, like the urease, the protein is actively secreted to the surface of the bacterium. The enzyme is associated with the outer membrane of the cell wall as well as with the external surface of flagellar sheath. After its cloning and sequencing, there has not been any subsequent study to assess its relevance in pathogenesis. The relevance of this molecule as a vaccine candidate is probably poor due to the high sequence homology with other bacterial SODs as well as to the presence of similar antigens in the host.

4.1.2.1.5 Adhesins

H.pylori cells are usually found swimming in the mucous gel lining the gastric mucosa. A small percentage of bacteria can also tightly adhere to the luminal surface of gastric cells and in the spaces between adjacent cells (Caselli et al., 1989). The adherence is totally absent in areas of intestinal metaplasia (Solcia et al., 1990; Fiocca et al., 1989). This clear-cut specificity indicates that adhesion is mediated by specific recognition between structures present on the bacterium and on the host gastric cells. Many putative adhesins have been characterized and active or passive immunization against these molecules could be relevant to prevent or combat the *H.pylori* infection.

The gene *hpaA* encodes for the 20kDa receptor-binding subunit of the N-acetylneuraminyllactose-binding fibrillar hemagglutinin (NLBH) (Evans et al., 1993). The protein contains the sequence KRTIQK that represents the functionally homologous sialic acid-binding moiety found in some *E.coli* sialic acid-specific adhesins. Polyclonal antibodies raised against this hexapeptide inhibited the hemagglutination of human red blood cells induced by *H.pylori*. Construction of deletion *H.pylori* mutants in the *hpaA* gene has been reported (Janzon et al., 1993), but they have not yet been tested in any infection model. A recent report has shown that this antigen is not important for induction of IL-8 by epithelial cell lines (Huang et al., 1995). The use in some animal models of isogenic mutants lacking HpaA expression is eagerly needed to assess the role of this molecule in colonization and pathogenesis.

Haas and colleagues have recently described a gene, called *alpA*, encoding for a putative new adhesin of 53kDa. This gene is present in all *H. pylori* strains examined, but absent in *H.mustelae* or *H.felis*, suggesting that the putative adhesin might be specific for the interaction of *H. pylori* with human gastric cells (Odenbriet et al., 1995).

Recent elegant studies by Boren et al have shown that blood group O antigen Le^b, expressed by the human gastric mucosa, is also involved in *H.pylori* adhesion (Boren et al., 1993; Falk et al., 1993; Falk et al., 1995). These observations account for the known higher susceptibility of people carrying these antigens to develop peptic ulcer disease. Research is actively ongoing for the identification on *H.pylori* surface of the binding moiety responsible for this recognition.

Active immunization against structures involved in *H.pylori* adhesion to epithelial cells could be important to protect the host from colonization.

4.1.2.1.6 LPS

H.pylori lipopolysaccharide (LPS), like other gram negative LPS, consists of a lipid component, Lipid A, that ensures incorporation into the outer layer of the outer membrane, and an outer carbohydrate domain that spans into the outer environment. Lipid A structure is quite conserved among different strains (Moran et al., 1992) and its biological activity has been reported to be low (Mattsby-Baltzer et al., 1992) as compared to *E.coli* LPS. The carbohydrate chain consists of an oligosaccharide core, that is conserved among strains, and a polysaccharide side chain that has been found to be more variable in antigenicity (Mills et al., 1992).

Of outstanding interest are recent data (Aspinall *et al.* 1994) indicating that the O side chain can mimic the structure of the blood group antigens Le^x and Le^y. These antigens are also expressed in the gastric mucosa (Sakamoto et al., 1989). Thus this mimicry of gastric mucosa antigens by *H.pylori* LPS may have importance in evading the immune response and may also represent a structural explanation of the finding (Negrini et al., 1991), whereby they hypothesized the generation of autoantibodies reacting to the gastric mucosa upon chronic infection.

4.1.2.1.7 Others

Many new genes and/or factors have been described recently. Their relevance as vaccine components needs to be assessed. *hap* gene encodes for a metalloprotease endowed with the capacity to degrade the mucin that may be relevant in colonization (Smith et al., 1994). *kat* gene encodes for a potent catalase that presumably works in association with SOD in eliminating the oxygen radicals (Bukanov and Berg, 1994). Other recently cloned genes include: 1) a major envelope lipoprotein (*lpp20*) gene encoding for a membrane-associated 20kDa protein (Kostrzynska *et al.* 1994); 2) hemolysin (*hly*) (Segal and Tompkins, 1993); 3) a P-type (H^+/K^+) ATPase (*hpcopA*) (Ge et al., 1995); 4) a cation binding protein involved in copper transport (*hpcopP*) (Ge et al., 1995); 5) a new histidine-rich nickel-binding protein named Hpn (Gilbert et al., 1995); 6) a group of outer membrane proteins belonging to the porin family, named HopA-D (Exner et al., 1995); 7) the gene *recA* encoding for a DNA recombinase that may be relevant to the development of attenuated live vaccines or live carriers (Thompson and Blaser, 1995) and, 8) the neutrophil activating protein, called HP-NAP, consisting of a 150 kDa homopolymer of ten identical 15kDa monomers encoded by the *napA* gene (Evans et al. 1995). There is finally a new yet uncharacterized toxic factor called 'Granulating cytotoxin', mainly found in the related organism *H.hepaticus,* but released in minimal amounts also by several strains of *H.pylori* (Taylor et al., 1995). This toxic factor might contribute to the gastric damages observed during infection.

4.1.2.2 Antigens Expressed Only in Type I Strains

4.1.2.2.1 CagA

The CagA is an immunodominant antigen, associated with the bacterial cell wall. The native molecule is synthesized in about 80 % of all *H. pylori* strains and is generally co-expressed with the active form of the vacuolating cytotoxin (Covacci et al., 1993; Xiang et al., 1995). The *cagA* gene codes for a protein of 1147 aa, with a theorethical mass 128,012.73 Da and a calculated pI of 9.72. This antigen is very hydrophilic, uses a signal peptide-independent secretion pathway, and shows a size heterogeneity that is obtained via internal duplication. The nucleotide sequence of different clinical strains reveals that, at nucleotide position 3406, a 102 bp conserved DNA motif is repeated once in the strain 84-183 (ATCC 53726) and twice in the strain G39. The amino acids encoded by this DNA motif are very hydrophilic: it is quite interesting to note that protective antigens of malaria parasites, *Candida albicans* and *Streptococcus pneumoniae* are immunodominant molecules with an high number of repeated epitopes. The CagA antigen is absent in noncytotoxic Type II strains strains due to a gene deletion (Xiang et al., 1995). The antibody levels against the CagA in patients infected by *H.pylori* show a 100 % correlation with duodenal ulcer and a 75 % correlation with gastroduodenal diseases ($p < 0.0005$), suggesting that only bacteria expressing this antigen and the active vacuolating toxin are ulcerogenic (Xiang et al., 1993a; Xiang et al., 1993b). From 44 different clinical isolates of *H. pylori* tested using probes specific for the *cagA* and the *vacA* genes we isolated two spontaneous mutants that expressed the vacuolating cytotoxin in the absence of CagA synthesis, and one mutant that expressed the CagA antigen without any detectable cytotoxic activity (Xiang et al., 1995). These clinical isolates and two isogenic mutants for the *cagA* and *vacA* genes produced by insertional mutagenesis clearly indicate that the function of CagA is dissociated from the vacuolating activity (Cover et al., 1994; Tummuru et al., 1994).

The complete nucleotide sequence of the chromosomal region that contain the *cagA* gene, and that is present only in Type I bacteria, has been tentatively named CagI (Berg, 1995; Covacci et al., 1995). This region contains genes coding for virulence factors associated with the export machinery and with the ability to induce gastric epithelial interleukin-8 secretion (Crabtree et al., 1995). Increased IL-8 synthesis induced by CagI positive strains may explain the association of these strains with active neutrophilic gastritis and gastroduodenal disease.

The purified CagA antigen may be also included in a second-generation, non invasive, diagnostic serological test (Xiang et al., 1993a) to identify patients at risk of developing duodenal ulcer disease, and to verify complete eradication after antibiotic treatment.

The potential of CagA as a vaccine candidate is currently being assessed. Preliminary results from our laboratory indicate that this antigen is able to induce protective immunity against *H.pylori* in mice (Ghiara et al., manuscript in preparation).

4.1.2.2.2 VacA

Type I strains of *H. pylori* produce a cytotoxin which causes massive vacuolation and death in a variety of target cells *in vitro*. Serological studies (Cover et al., 1992b; Cover et al., 1993) and animal models of infection and disease (Telford et al., 1994b; Ghiara et al., 1995a; Marchetti et al., 1995) have implicated the toxin in peptic ulcer disease. Recently, immunization of mice with purified VacA has been shown to confer protective immunity against infection by Type I strains of *H. pylori* (Marchetti et al., 1995). Hence, as in other infectious diseases caused by toxin producing pathogens, the VacA protein is an important candidate for inclusion in a therapeutic or prophylactic vaccine.

The vacuoles induced in target cells by treatment with the toxin are acidic and derive from the fluid phase endocytic pathway, as evidenced by the accumulation of lucifer yellow dye. Furthermore, the vacuoles can be stained with antibodies against the rab7 small GTP binding protein, which associates specifically with the late endosome compartment of the endocytic pathway, suggesting that the toxin may interfere at this stage in vesicle structure, or cycling (Papini et al., 1994). The mechanism of action of the toxin is not yet known.

Purification of cytotoxic activity from *H. pylori* culture supernatants revealed that the toxin is a protein which migrates in gel filtration at an apparent molecular mass of around 700 kDa (Cover et al., 1992a). The complex consists of several copies of a single polypeptide of 95 kDa, which is released from the bacteria after cleavage from a 140 kDa precursor. Similarities have been noted between the carboxyterminal 45 kDa of the precursor polypeptide which remains associated with the bacterial cells and the carboxyterminal region of the IgA protease family of exotoxins, which is involved in outer membrane transport, suggesting that the 95 kDa polypeptide may be exported in a similar fashion (Telford et al., 1994b; Schmitt et al., 1994). With time, the 95 kDa polypeptide undergoes specific cleavage at a short repeated sequence of eight amino acids to produce a 37 kDa aminoterminal fragment and a 58 kDa carboxyterminal fragment which copurify on gel filtration, suggesting that they remain associated after cleavage.

Many bacterial toxins which act on intracellular targets are composed of two functionally distinct subunits. These di-chain or AB toxins contain an enzymatically active subunit (A) whose translocation to the cytosol is dependent on interaction of the binding (B) subunit with the cell membrane (Domenighini et al., 1995). One common property of the isolated B subunits of these toxins is the capacity to interact with artificial membranes to form ion channels. This membrane interaction is usually associated with oligomerization of the B subunits which may form channels through which the enzymatically active A subunit is translocated. For example, the B subunit of anthrax toxin oligomerizes and inserts in the membrane on reaching the acidic environment of the endosome (Milne et al., 1994). Other toxins such as cholera toxin are assembled in the bacteria with oligomeric B subunits. An unusual variation of this theme is the heteropentameric B subunit of pertussis toxin (Nicosia et al., 1986).

Deep-freeze etch electron microscopy has revealed that the active toxin is an high molecular weight homoligomer formed from either six or seven 95 kDa monomer polypeptides (Telford et al., submitted). Within each monomer, two distinct domains can be recognized. Cleavage of an exposed loop between these two domains produces the 37 kDa amino terminal subunit and the 58 kDa carboxyterminal subunit. After cleavage, however, the subunits remain associated with the oligomeric structure. The physiological role of this cleavage is as yet unknown but by analogy with the family of AB bacterial toxins, it may be important for the activation of toxicity.

Parenteral immunization of rabbits with highly purified oligomeric VacA induced high titres of toxin neutralizing antibodies, whereas recombinant 95 kDa monomer polypeptide, which failed to assemble and was inactive, did not. Hence, as with other oligomeric bacterial toxins, native conformation is likely to be necessary for the induction of a protective immune response. In support of this, comparison of antibody titres against native and denatured VacA in neutralizing sera from rabbits immunized with native material revealed that as much as 90 % of the response was against conformational epitopes (Manetti et al., 1995). However, when the recombinant 95 kDa monomer was used as antigen in oral immunizations, we found that, also in this non correctly folded form, the VacA antigen can induce protective immunity against

H.pylori infection in mice (Ghiara et al., manuscript in preparation). This result indicates that in order to prevent infection *in vivo* opsonizing antibodies may be sufficient. Alternatively, it could be hypothesized that conformational epitopes are not directly involved in protection against infection, or that also cellular mechanisms of protective immunity may be involved.

The capability of induction of good neutralizing antibodies upon immunization may be however desirable, particularly to prevent possible toxic systemic effects of the toxin. Hence an important goal for preparation of a candidate toxoid for inclusion in a vaccine will be to detoxify the molecule by substitution of a few key amino acid residues, which do not alter the overall structure of the molecule. This approach has proven highly successful in the generation of inactive pertussis toxin for use in a new vaccine against whooping cough (Pizza et al., 1989).

4.1.2.3 Large Scale Production of Antigens

An important step in the development of a vaccine against *H.pylori* is the large-scale production of antigens for their characterization. The purification of *H.pylori* antigens is made very difficult by the low yield of biomass obtained with conventional culture systems, that require the use of horse blood or calf serum. Cyclodextrin derivatives can substitute serum in both solid and liquid media for the growth of *H.pylori* (Olivieri et al., 1993). These culture systems have been recently used to optimize the growth in large fermentors for the production of important virulence factors such as VacA (Marchini et al., 1994).

4.1.3. Animal Models

The availability of suitable animal models is essential for the understanding of pathogenesis of *H.pylori* infection and for the development of vaccination strategies. Currently, developed animal models can be distinguished in two groups a) *disease models*, in which some of the features of *H.pylori*-induced gastric disease have been reproduced by the oral administration of bacterial components to experimental animals, and b) *infection models*, in which orally administered live bacteria are found to colonize the gastric mucosa of hosts.

4.1.3.1 Animal Models of Disease

The first report describing the occurrence of some histopathological features similar to *H.pylori* chronic infections by administration of crude bacterial supernatants to experimental animals was published by Karita and colleagues (Karita et al., 1991). These authors reported that oral administration to mice of a filtered broth supernatant of *H.pylori* was able to induce gastritis (i.e. inflammatory cell migration in the lamina propria) and epithelial ulcerations strongly resembling that observed in chronically infected humans. This was the first observation suggesting that mice, a well characterized and moderately inexpensive host, would be susceptible to the virulence factors expressed by *H.pylori*. Using rats as recipient animals, Ross and colleagues (Bui et al., 1991; Ross et al., 1992) reported that chemically induced gastric ulcers can be exacerbated by the oral administration of *H.pylori* whole bacteria or bacterial extracts.

These two reports were considered extremely interesting since they described valid alternatives to overcome the reported failure to establish persistent infection by *H.pylori* in laboratory rodents.

In 1994 these results were confirmed in BALB/c mice by our group (Telford et al., 1994b). Moreover, it was further shown that: a) only bacterial extracts obtained from Type I strains were able to induce gastritis, ulcerations and severe mucin depletion, while those obtained from Type II strain only induced mucin layer depletion; b) administration of purified urease did not have any pathological effect on gastric mucosae; and c) purified cytotoxin (VacA) was able to induce remarkable gastric epithelial derangements and focal ulcerations (Ghiara et al., 1995a). Using in the same test crude extracts from strains carrying disrupted or mutated genes encoding for the major virulence factors, we have recently confirmed that VacA has an important role in the induction of gastric epithelial damage, while urease activity does not seem to play a significant role in inducing histopathological damage. Furthermore, we have shown that accumulation of inflammatory cell infiltrate is induced by factors present in the Type I strains that are distinct from CagA or VacA proteins (Ghiara et al., 1995a). Figure 4.1.1 illustrates the result of a representative experiment in which epithelial damage scores (EDS) and inflammatory scores (IS) were assessed on histological sections of mice treated with saline alone or with 500 μg of crude sonic extracts of wild type *H.pylori* , or of its isogenic mutants carrying disrupted genes for VacA (VacA:Km) or for CagA (CagA:Km). Deletion of VacA expression induced a significant reduction in only EDS with respect to the effect induced by wild type sonicate; IS remained unchanged. Deletion of CagA expression did not induce any significant change in EDS or IS induced by the sonic extract of the wild type parent strain.

These kind of models are useful to determine the *in vivo* acute toxicity of *H.pylori* components and, for example, they can be used for the initial screening of chemically or genetically detoxified preparations of relevant virulence factors (i.e. VacA) that are being considered as candidates in human vaccines.

Major limits of models reproducing disease through pulsed oral administration of high amounts of bacterial components possibly consist in a less efficient delivery of virulence factors to the gastric mucosae as compared to the slow and continuous delivery that conceivably occurs during infection by live bacteria. Moreover, these models are not suitable for studying vaccination since they do not reproduce all the aspects of *H.pylori* induced disease that consist in colonization and establishment of chronic infection of the gastric mucosae.

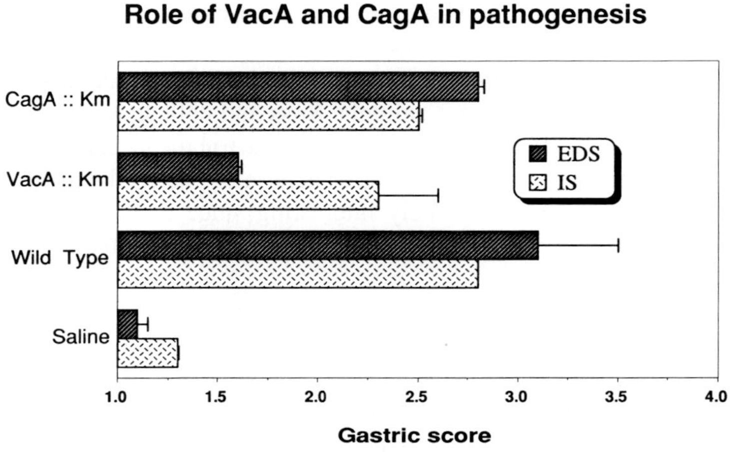

Fig. 4.1.1: Gastric damage induced in BALB/c mice by oral administration of saline alone, or containing 500 μg of sonic extracts from a wild type *H.pylori* strain, or isogenic mutant strains lacking expression of VacA (VacA:Km), or of CagA (CagA:Km). EDS = epithelial damage score. IS = inflammatory score. Results are expressed as mean ± SEM obtained from 3-6 mice pre group. See Ghiara et al. (1995a) for detailed experimental procedure.

4.1.3.2 Animal Models of Infection

Persistent infection by *H.pylori*, with consequent development of gastric disease, has been established in many animal species. Models employing monkeys have the invaluable advantage to reproduce a disease that is almost indistinguishable from that observed in humans (Baskerville et al., 1988; Euler et al., 1990; Hazell et al., 1992; Dubois et al., 1994). Major limits to the diffusion of this model are the sophisticated facilities required for animal housing and the very high cost for their maintenance. Moreover, due to the susceptibility of monkey species to natural infection (like humans) it is very difficult to select a significant number of individuals that are not naturally infected by *H.pylori*, or by related species of the genus *Helicobacter* (Bronsdon et al., 1991). This model is therefore currently used only in a few specialized laboratories. Surprisingly, no vaccination study has been yet reported with this model host.

Krakowka and his coworkers have described an extremely interesting infection model in the gnotobiotic piglets (Krakowka et al., 1987). This animal has largely been used in biomedical research because the physiology of many of its organs and systems is quite similar to that of humans. This model has been recently used to elegantly show the possibility of mixed infection by both Type I and Type II strains in the same individual (Akopyants et al., 1995). The same group has recently reported that coccoid forms of *H.pylori* are not able to establish infection in this host (Eaton et al., 1995) The pathological consequence of *H.pylori* infection in pigs has, however, some difference with that observed in humans, since it consists mainly in a lymphoplasmacytic gastritis, a condition which is more often observed in children rather than in adult patients. In fact, infected human adults usually have a chronic active gastritis in which the cell infiltrate is composed mainly of mononuclear phagocytes and neutrophils. A major limit to the broad diffusion to many laboratories of this model, like the ones that use monkeys, is its difficulty that again consists in specialized housing facilities and relatively high cost of maintenance. Thus, as far as we know, this model is presently used by only a few laboratories.

Eaton and Krakowka have used their model to assess the induction of protection after either oral or parenteral immunization with whole bacteria without any adjuvant (Eaton et al., 1992b). All the animals were killed one week after the infective oral challenge with *H.pylori*. No protection from infection was observed. On the other hand, these authors reported an exacerbation of gastritis in parenterally immunized animals after the oral challenge with live bacteria. These authors suggested that the choice of a better immunization protocol and route of immunization would probably improve the results; however, there has not yet been any report describing such improvement.

Gnotobiotic beagle dogs have been reported to be infected by *H.pylori*, but this model has not been used thereafter (Radin et al., 1990). Much interest has been attracted by recent reports describing domestic cats as suitable hosts, and possibly natural reservoir, of *H.pylori* (Handt et al., 1994). Gastritis induced by natural and artificial infection is quite similar to that induced by infection in humans. Future developments of this observation are eagerly waited to evaluate the reliability of this animal as a valuable host for studying induction of protective immunity.

The lack of suitable and less expensive animal models of *H.pylori* infection has prompted many investigators to test the relevance of infection models that utilize related members of *Helicobacter* genus.

Ferrets have been experimentally infected with their natural gastric colonizer *H.mustelae*. The infection by this bacteria causes gastritis and occasional ulcers to the host (Fox et al., 1990; Otto et al., 1990; Fox et al., 1988). Mucosal and systemic immune response can be induced in these animals after immunization with *H.pylori* lysate plus CT (Czinn et al., 1991), and interesting studies of therapeutic immunizations has been recently carried out (Monath et al., 1994).

H.felis infection of mice has been the most extensively used and characterized model so far (Lee et al., 1990; Dick-Hegedus et al., 1991; Fox et al., 1993; Danon et al., 1995). The cellular infiltrate observed in chronically infected mice shares many similarities with that observed in humans. Chronic infection induces a measurable systemic and local humoral immunity. Due to its apparent simplicity, this model has been largely diffused among several groups studying the feasibility of vaccination. Chen et al. in 1992 reported for the first time that vaccination against *Helicobacter* was feasible immunizing the mice orally with bacterial lysates plus CT as mucosal adjuvant (Chen et al., 1992). The major limit of this model is its poor relevance to the human disease because: a) it employs a bacterium that is not an human pathogen; b) *H.felis* is not able to express the pathogenic determinants (i.e. VacA and CagA) (Xiang et al., 1995) that have been demonstrated to be important in human pathology;

and , c) chronic infection does not induce ulcerations to the host. Two recent reports have used recombinant *H.pylori* urease subunits plus CT to induce protective immunity against infection (Ferrero et al., 1994; Michetti et al., 1994). Despite the high sequence homology between *H.pylori* and *H.felis* ureases, in one of the reports it was shown that *H.felis* urease B subunit induces protection in 60 % of immunized mice, while only 25 % of mice immunized with *H.pylori* urease B subunit were protected from challenge with *H.felis* (Ferrero et al., 1994). This model, although having been the first conveniently available model of infection with an *Helicobacter*, may thus not be ideal for giving insight into the pathogenesis of human infection and for the development of human vaccines.

Much effort has been devoted to the establishment of persistent infection with *H.pylori* in small laboratory rodents, i.e. mice and rats.

An initial attempt to infect small rodents with *H.pylori* gave negative results (Cantorna et al., 1990). In 1989 Henneresse et al. reported successful colonization by *H.pylori* in germ-free rats (Henneresse et al., 1989). The infection persisted for up to eight weeks and the animals developed a mild chronic inflammation of the gastric antral mucosa. Karita and colleagues in 1991 reported for the first time the establishment of infection in athymic (nude) mice and euthymic germ-free mice using bacteria isolated from patients (Karita et al., 1991). In euthymic non germ-free mice the colonization was only transient and persisted for less than two weeks, while nude mice and germ-free mice were found to develop a more persistent infection ($\geq$ 10 weeks). The infected mice developed gastritis and ulcerations strongly resembling that observed in humans. Nude or germ-free mice are not suitable for studies aimed at assessing the feasibility of vaccination because of their particular immune status and also because they need sophisticated housing facilities.

Cellini et al have reported at the end of 1994 an improvement to this mouse *H.pylori* model (Cellini et al., 1994). Using fresh isolates infection could persist for at least 4 weeks in normal BALB/c mice. Mice develop gastric pathology and mount an humoral immune response against the host, that is detectable in the serum even at times when the infection is no more detectable by culture (i.e. at 8 and 12 weeks). It is noteworthy that the detection system used by these authors is based on culture of homogenized gastric specimens within one hour after collection of the organs. These authors also obtained efficient colonization of mice using the *in vitro* non cultivable coccoid forms of *H.pylori*. This result, that is of extreme importance for the understanding of *H.pylori* pathogenesis, is however in contrast with a more recent report (Eaton et al., 1995), that describes the failure of this 'quiescent' form of *H.pylori* to establish infection in the well characterized gnotobiotic pig model.

A major improvement to the mouse model of infection has been recently reported by our laboratory (Marchetti et al., 1995). We used very fresh clinical isolates cultured in microaerophylic conditions for no more than three weeks after isolation from patients' biopsies. Specific pathogen free (SPF) CD1 mice were fasted for 18-24 hours before each treatment and gastric acidity was buffered before bacterial inoculations with Na bicarbonate. The mice were not allowed to drink or eat for three hours after each treatment. The mice received three bacterial inocula of 10^9 CFUs over a week. At weekly intervals infection was assessed by culture. Gastric samples were processed immediately after sacrifice and isolation of colonizing bacteria was accomplished by gently streaking the gastric mucosa onto the surface of blood agar plates that subsequently were rapidly cultured in microaerophylic conditions. Using this procedure we have been able to obtain stable infections detectable for at least 12 weeks. Infections were performed with phenotypically characterized *H.pylori* strains. Cytotoxic (Type I) freshly isolated strains were able to establish gastric colonization of mouse gastric mucosa more quickly than non cytotoxic (Type II) strains. A moderate gastritis was observed in all infected mice after 4-8 weeks, but in mice infected with Type I strains the gastric pathology was much more evident and consisted of both cell infiltration in the lamina propria and superficial erosions of epithelium.

Bacteria isolated from infected mice two weeks after the primary inoculum were then orally inoculated to other mice. Several cycles of isolation-reinoculation were then performed with an apparent increase in colonization efficiency. This suggests that 'good colonizers' could be selected and maintained by *in vivo* passaging the bacteria. Using these 'mouse adapted' *H.pylori* strains we can now observe infections that persist for a long time ($\geq$ 8 months) with no signs of decrease.

Interestingly, while the efficiency of colonization does not seem to vary among different mouse strains, the extent of gastric pathology is different in various mouse strains (Marchetti and Ghiara, unpublished observations). In particular, after the same time of infection (8 weeks) with a Type I strain, BALB/c mice showed only a very mild gastritis with no ulceration, while C57BL/6 mice had a more evident infiltration of inflammatory cells in the lamina propria and several ulcerations were observed in each infected mice. Figure 4.1.2 illustrates the histological appearance of gastric mucosa of a representative control C57BL/6 mouse (panel A), and of a representative C57BL/6 mouse infected for 8 weeks with a type I *H.pylori* strain (panel B). Note the induction of epithelial ulceration and the intense infiltration of neutrophils and mononuclear cells both in epithelium and in the submucosa.

The availability of this mouse model of *H.pylori* infection and disease allows to study pathogenesis and to assess the feasibility of vaccine in an animal model that is highly relevant to human disease. We have reported successful protection of mice from infection by *H.pylori,* following intragastric immunization with *H.pylori* antigens together with LT as mucosal adjuvant. Table 4.1.3 summarizes the results obtained. While immunization with urease was able to prevent infection by both Type I and Type II strains, the VacA induced a Type I specific protection.

4.1.4 Antigen Delivery and Mucosal Adjuvants

Most of human pathogens begin their interaction with the host at mucosal sites. Nonetheless, most of currently used vaccines are given parenterally. This is possibly

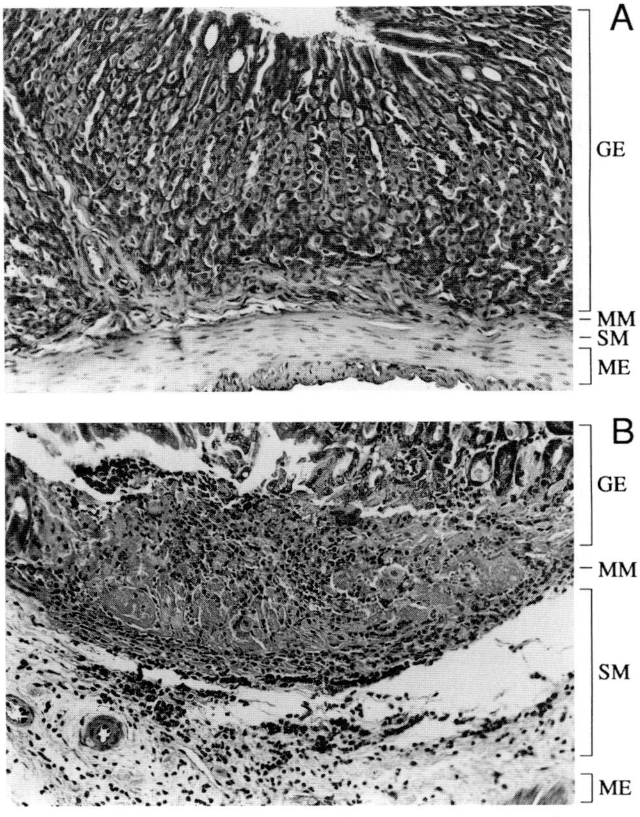

Fig. 4.1.2: Hematoxylin and Eosin staining of gastric mucosa of a control uninfected C57BL/6 mouse (panel A) and of a C57BL/6 mouse infected for 8 weeks with Type I strain SPM326 (panel B). Note in panel B the ulceration and the intense inflammatory infiltrate. GE = gastric epithelium; MM = muscolaris mucosae; SM = sub mucosa; ME = muscolaris externa. (original magnification = 200X).

due to the fact that effective mucosal immunization is rather difficult to be achieved and parenteral vaccines require a much lower amount of antigens than that required to elicit mucosal immunity. Major problems to be overcome in this area are: a) efficient delivery to mucosal immune system and b) use of efficacious adjuvants. The development of an efficacious human *H.pylori* vaccine will certainly face these issues.

Strategies to achieve efficacious antigen delivery at mucosal sites may include 1) expression of candidate molecules in living microorganisms that can colonize transiently some mucosa of the host and express sufficient amounts of antigens, 2) covalent conjugation of antigens to molecules, like CT-B, that have the property to adhere to the mucosal epithelia, although this approach has been reported also to induce tolerance (Sun et al., 1994), 3) inclusion of antigens in non-living microparticle carriers, like DL-lactide-*co*-glicolide microspheres or liposomes, that may adhere efficiently to mucosal epithelia and also allow the antigen to be released slowly (Walker, 1994).

Table 4.1.3: Protection induced by intragastric immunizations in mice*

Immunization with	Infected/total	% protected mice
	Challenge with Type I bacteria	
saline	14/14	0
LT	13/14	7.1
urease + LT	2/13	84.6
VacA + LT	3/14	78.6
Type I lysate + LT	0/20	100.0
Formalin killed bacteria +LT	0/10	100.0
	Challenge with Type II bacteria	
saline	8/8	0
LT	6/8	25
urease + LT	1/8	87.5
VacA + LT	7/7	0
Type II lysate + LT	0/10	100.0

* For the details of immunization protocol see Marchetti et al. (1995).

Adjuvant development for mucosal vaccines is moving quickly in these years. The active component of parenteral Freund's adjuvant, muramyl-dipeptide (MDP) has been found to have also adjuvant properties toward mucosally delivered antigens (Morisaki et al. 1983). Its adjuvant properties are possibly due to the capacity to activate macrophages and to induce the release of cytokines. A major drawback to the use of this adjuvant are its neuropharmacological side effects. However, MDP has been recently found to be ineffective adjuvant in oral vaccinations with *H.pylori* urease (Monath, 1995). MF59, an water-oil emulsion recently developed at Chiron Biocine as parenteral adjuvant, has proved to exert significant adjuvant activity when given intranasally with antigens to mice (Barchfeld et al., 1995). Therefore, efficacy of this new adjuvant in mucosal vaccination against *H.pylori* is currently being assessed.

Cholera toxin (CT), or the homologous heat-labile toxin of enterotoxigenic *E. coli* (LT), have received recently much attention due to their potent activity as mucosal adjuvants (Walker, 1994). In animal models CT and LT have been reported to be necessary to achieve protective immunity against *H.pylori* or *H.felis* infections in mice (Ferrero et al., 1994; Michetti et al., 1994; Marchetti et al., 1995). These toxins are, however, not useful in humans due to their high toxic activity.

The development of safe mucosal adjuvants for use in humans in the place of the highly toxic and life threatening CT is thus mandatory. Total bacterial lysates have proved to be efficacious immunogens in *H.felis* model when administered orally with commercially available purified non-toxic CT-B (Lee et al., 1994). However, the use of a recombinant form of CT-B, devoid of any contamination by CT-A subunit, did not confirm this result (Nedrud et al., 1995).

A more promising approach has been to develop genetically-detoxified mutants of LT that maintain their adjuvanticity (Douce et al., 1995). One of these non-toxic mutants of LT, called LTK63, has been tested with promising results in oral immunizations in the recently developed mouse model of *H.pylori* infection (Ghiara et al., 1995b).

4.1.5 Vaccine Feasibility

Investigations in mice have shown that protective immunity can be achieved by mucosal immunization with *H.pylori* antigens plus CT or LT. So far, the identified protective antigens are urease, VacA, CagA, heat shock proteins, bacterial lysates and formaldehyde-killed bacteria (tab. 4.1.4). A phase I trial is ongoing in Switzerland with urease (Ghiara and Michetti, 1995). To the present, vaccines based only on a single purified virulence factors did not induce total protection in animal models. The use of only one component in acellular vaccines may thus not be ideal also for possible occurrence of antigenic escape in the target microorganism.

Depending on the animal model that has been used, complete protection has been achieved only when whole bacterial cell preparations or combinations of purified antigens were used. A classical vaccine based on whole cell preparations has the advantage to be relatively easy to prepare and to be quite cheap. Contrarily, they may have higher toxicity than those based of purified proteins and, due to antigen cross-reactivity, they may also induce immunity against components of the normal bacterial flora of the gut.

Combinations of purified and recombinant virulence factors have proved to be safe and to induce strong protective immunity in recently developed acellular pertussis vaccines (Englund et al., 1994). A similar approach is probably the best also for *H.pylori* vaccination. A combination of HspA and UreB subunit has been recently reported to induce total protection against *H.felis* in mice (Ferrero et al., 1995). However, the use of heat shock proteins as immunogens in humans may be dangerous due to the potential induction of autoimmunity. The choice of good candidates will be a major challenge for the future. Antigens expressed only by Type I strains will confer a strain-specific protection (Marchetti et al., 1995). A good vaccine formulation should induce immunity against all *H.pylori* strains and also to confer protective immunity against virulence factors that are important in the induction of the disease. Preliminary experiments in our laboratory have shown that a combination of VacA and Urease can induce total protection against *H.pylori* in mice.

Of great interest is the recent finding that oral therapeutic immunization is able to eradicate an otherwise chronic infection by *H.felis* in the mouse, using either a whole cell sonicate (Doidge et al., 1994) or purified recombinant UreB subunit (Corthésy-

Table 4.1.4: Prophylactic vaccination: current status of animal studies

H.pylori antigen	Adjuvant	Route	Animal species	Infection with	% protection	reference
Sonicate	CT, LT, LTK63	IG	mouse	*H.pylori*	80–100	Marchetti et al, 1995
	CT, LT	IG	mouse	*H.felis*	≈ 70	Michetti et al., 1994
	none	IG, IP	pig	*H.pylori*	0	Eaton et al., 1992b
Formaldehyde-killed bacteria	LTK63	IG	mouse	*H.pylori*	100	Ghiara et al., manuscript in preparation
Urease*	LT, CT, LTK63	IG	mouse	*H.pylori*	≈ 80	Marchetti et al., 1995
	LT, CT	IG, P.O.	mouse	*H.felis*	≈ 70	Monath et al., 1994
	MDP	IG	mouse	*H.felis*	0	Monath et al., 1995
	CFA	IP	mouse	*H.felis*	0	Monath et al., 1995
UreB subunit	LT, CT	IG	mouse	*H.felis*	25–70	Ferrero et al.,1994 Michetti et al., 1994
HspA	CT	IG	mouse	*H.felis*	≈ 80	Ferrero et al., 1995
HspB	LTK63	IG	mouse	*H.pylori*	≈ 50	Ghiara et al.,
	CT	IG	mouse	*H.felis*	≈ 70	Ferrero et al., 1995 unpublished observations
UreB + HspA	LT	IG	mouse	*H.felis*	100	Ferrero et al., 1995
VacA	LT, CT, LTK63	IG	mouse	*H.pylori*	≈ 80	Marchetti et al., 1995
CagA	LTK63	IG	mouse	*H.pylori*	≈ 70	Ghiara et al., manuscript in preparation
VacA + Urease*	LT, LTK63	IG	mouse	*H.pylori*	80–100	Ghiara et al., unpublished observations

* holoenzyme; MDP = muramyl dipeptide; CFA= Complete Freund's adjuvant; IG = intragastric; P.O. = administration in the oral cavity; IP = intraperitoneal

Theulaz et al., 1995) plus CT as adjuvant. The latter study also showed that some of the cured mice were also protected from a second infective challenge. Similar results have been obtained by Monath et al., who reported successful elimination of *H.mustelae* infection in ferrets (Monath et al., 1994). This observation is very important because, since it has been done on naturally infected ferrets, it allows to predict that similar results could be obtained on infected humans. Preliminary results obtained in our laboratory have shown that intragastric inoculation with *H.pylori* antigens could efficiently eradicate *H.pylori* infection in mice (Ghiara et al., manuscript in preparation) (tab. 4.1.5).

The challenge of the next years will be to identify a vaccine formulation to be tested in efficacy trials and to identify the target population of a candidate *H.pylori* vaccine, as well as to test the efficacy of therapeutic immunization. Probably, the initial target of a prophylactic vaccine is the Third World. In fact, due to the high prevalence of infection early in life in this area of the planet, efficacy of preventive immunization could be easily evaluated by vaccination in the first months of life. Development of efficacious regimens of therapeutic immunization would probably be more relevant

Table 4.1.5: Therapeutic vaccination: current status of animal studies

Infection	Animal species	*H.pylori* antigen	adjuvant	eradication (%)	reference
H.felis	mouse	sonicate	CT	70–90	Doidge et al 1994
		UreB	CT	≈ 50	Corthésy-Teulaz et al. 1995
H.pylori	mouse	sonicate	LTK63	70	Ghiara et al.,
		recombinant VacA	LTK63	90	manuscript in preparation
H.mustelae (natural infection)	ferret	Urease holoenzyme	CT	≈ 50	Monath et al. 1994

in developed countries to eradicate infections that are resistant to antibiotic treatment, but it would be of great importance also in developing countries.

Stimulation of active immunity has been historically the best strategy to combat infectious diseases. *H.pylori* infection is a relevant human pathogen that infects about 50 % of the world population. Prevention or eradication of infection by active immunization with protective epitopes would probably also prevent the occurrence of gastric cancer, a malignancy that kills millions of people every year. The next decade will witness the transfer of experience accumulated on the laboratory animals to clinical trials, that will assess the feasibility of human vaccination against this important human pathogen.

Acknowledgements

We are grateful to Dr. Anthony Moran (University College, Galway, Ireland) and Dr. Natale Figura (University of Siena, Italy) for critical readings of the manuscript. We also thank Giorgio Corsi for the artwork and Catherine Mallia for the preparation of the manuscript.

References

Akopyants, N. S., Eaton, K. A., and Berg, D. E. (1995) Adaptive mutation and cocolonization during Helicobacter pylori infection of gnotobiotic piglets. Infect. Immun. *63*, 116-121.

Alm, R. A., Guerry, P., and Trust, T. J. (1993) Significance of duplicated flagellin genes in Campylobacter. J. Mol. Biol. *230*, 359-363.

Armstrong, B. (1980) The epidemiology of cancer in the Peoples Republic of China. Int. J. Epidemiol. *9*, 305-315.

Aspinall, G. O., Monteiro, M. A., Pang, H., Walsh, E. J., and Moran, A. P. (1994) O antigen in the lipopolysaccharide of Helicobacter pylori NCTC 11637. Carbohydrate Lett *1*, 151-156.

Austin, J. W., Doig, P., Stewart, M., and Trust, T. J. (1991) Macromolecular structure and aggregation states of Helicobacter pylori urease. J. Bacteriol. *173*, 5663-5667.

Axon, A. T. (1995) Personal communication. VIIIth Int. Workshop of EHPSG, Edinburgh, 7-9 July 1995.

Banatvala, N., Mayo, K., Megraud, F., Jennings, R., Deeks, J. J., and Feldman, R. A. (1993a) The cohort effect and Helicobacter pylori. J. Infect. Dis. *168*, 219-221.

Banatvala, N., Lopez, C. R., Owen, R., Abdi, Y., Davies, G., Hardie, J., and Feldman, R. (1993b) Helicobacter pylori in dental plaque. Lancet. *341*, 380.

Barchfeld, G., Hessler, A., Sharp, M., Cataldo, D., Tarquina, P., Pizza, M., De Magistris, M. T., Douce, G., Dougan, G., and Van Nest G. (1995) Intranasal influenza vaccination of mice and baboons with adjuvants MF59 and LTK63. Clin. Immunol. Immunopathol. *76*, S93.

Baskerville, A. and Newell, D. G. (1988) Naturally occurring chronic gastritis and C.Pylori infection in the Rhesus monkey: a potential for gastritis in man. Gut *29*, 465-472.

Berg, D. E. et al. (1995) Personal communication. VIIIth Int Workshop of EHPSG, Edinburgh, 7-9 July 1995.

Bizzozero, G. (1893) Über die schlauchförmingen Drüsen des Magendarmkanals und die Beziehungen ihres Epithels zu dem Oberflächenepithel der Schleimhaut. Arch. F. Mikr. Anat. *42*, 82-152.

Blaser, M. J., Kobayashi, K., Cover, T. L., Cao, P., Feurer, I. D., and Perez-Perez, G. I. (1993a) Helicobacter pylori infection in Japanese patients with adenocarcinoma of the stomach. Int. J. Cancer *55*, 799-802.

Blaser, M. J. (1993b) Helicobacter pylori: microbiology of a 'slow' bacterial infection. Trends Microbiol *1*, 255-260.

Blaser, M. J. and Parsonnet, J. (1994) Parasitism by the ''slow'' bacterium Helicobacter pylori leads to altered gastric homeostasis and neoplasia. J. Clin Invest. *94*, 4-8.

Blaser, M. J., Chyou, P. H., and Nomura, A. (1995) Age at establishment of Helicobacter pylori infection and gastric carcinoma, gastric ulcer, and duodenal ulcer risk. Cancer Res. *55*, 562-565.

Bode, G., Malfertheiner, P., Nilius, M., Lehnhardt, G., and Ditschuneit, H. (1989) Ultrastructural localisation of urease in outer membrane and periplasm of Campylobacter pylori. J. Clin. Pathol. *42*, 778-789.

Bode, G., Mauch, F., and Malfertheiner, P. (1993) The coccoid forms of Helicobacter pylori. Criteria for their viability. Epidemiol. Infect. *111*, 483-490.

Boren, T., Falk, P., Roth, K. A., Larson, G., and Normark, S. (1993) Attachment of Helicobacter pylori to Human Gastric Epithelium Mediated by Blood Group Antigens. Science *262*, 1892-1895.

Bronsdon, M. A. and Schoenknecht, F. D. (1988) Campylobacter pylori isolated from the stomach of the monkey, Macaca Nemestrina. J. Clin. Microbiol. *26*, 1725-1728.

Bronsdon, M. A., Goodwin, C. S., Sly, L. I., Chilvers, T., and Schoenknecht, F. D. (1991) Helicobacter nemestrinae sp. nov., a spiral bacterium found in the stomach of a pigtailed macaque (Macaca nemestrina). Int.. J. Syst. Bacteriol. *41*, 148-153.

Bui, H. X., del Rosario, A., Sonbati, H., Lee, C. Y., George, M., and Ross, J. S. (1991) Helicobacter pylori affects the quality of experimental gastric ulcer healing in a new animal model. Exp. Mol. Pathol. *55*, 261-268.

Bukanov, N. O. and Berg, D. E. (1994) Ordered Cosmid library and high-resolution physical-genetic map of Helicobacter pylori strain NCTC11638. Mol. Microbiol. *11*, 509-523.

Cantorna, M. T. and Balish, E. (1990) Inability of human clinical strains of Helicobacter pylori to colonize the alimentary tract of germfree rodents. Can. J. Microbiol. *36*, 237-241.

Caselli, M., Figura, N., Trevisani, L., Pazzi, P., Guglielmetti, P., Bovolenta, M. R., and Stabellini, G. (1989) Patterns of physical modes of contact between Campylobacter pylori and gastric epithelium: implications about the bacterial pathogenicity. Am. J. Gastroenterol. *84*, 511-513.

Caselli, M., Aleotti, A., Trevisani, L., and Alvisi, V. (1993) Helicobacter pylori and intestinal metaplasia: possible implications in gastric carcinogenesis. Endoscopy. *25*, 437.

Cellini, L., Allocati, N., Angelucci, D., Iezzi, T., Dicampli, E., Marzio, L., and Dainelli, B. (1994) Coccoid Helicobacter pylori not culturable in vitro reverts in mice. Microbiol. Immunol. *38*, 843-850.

Chen, M., Lee, A., and Hazell, S. (1992) Immunisation against gastric helicobacter infection in a mouse/Helicobacter felis model. Lancet *339*, 1120-1121.

Clyne, M., Labigne, A., and Drumm, B. (1995) Helicobacter pylori requires an acidic environment to survive in the presence of urea. Infect. Immun. *63*, 1669-1673.

Collins, C. M. and D'Orazio, S. E. F. (1993) Bacterial Ureases: structure, regulation of expression and role in pathogenesis. Mol. Microbiol. *9*, 907-913.

Corthésy-Theulaz, I., Porta, N., Glauser, M., Saraga, E., Vaney, A. C., Haas, R., Kraehenbuhl, J. P., Blum, A. L., and Michetti, P. (1995) Oral immunization with Helicobacter pylori urease B subunit as a treatment against Helicobacter infection in mice. Gastroenterol *109*, 115-121.

Covacci et al. (1995) Personal communication. VIIIth Int Workshop on Campylobacters, Helicobacters and Related Organisms, Winchester, 10-13 July 1995.

Covacci, A., Censini, S., Bugnoli, M., Petracca, R., Burroni, D., Macchia, G., Massone, A., Papini, E., Xiang, Z., Figura, N., and Rappuoli, R. (1993) Molecular characterization of the 128-kDa immunodominant antigen of Helicobacter pylori associated with cytotoxicity and duodenal ulcer. Proc. Natl. Acad. Sci. U. S. A. *90*, 5791-5795.

Cover, T. L., Puryear, W., Perez-Perez, G. I., and Blaser, M. J. (1991) Effect of urease on HeLa cell vacuolation induced by Helicobacter pylori cytotoxin. Infect. Immun. *59*, 1264-1270.

Cover, T. L. and Blaser, M. J. (1992a) Purification and characterization of the vacuolating toxin from Helicobacter pylori. J. Biol. Chem. *267*, 10570-10575.

Cover, T. L., Cao, P., Murthy, U. K., Sipple, M. S., and Blaser, M. J. (1992b) Serum neutralizing antibody response to the vacuolating cytotoxin of Helicobacter pylori. J.Clin. Invest. *90*, 913-918.

Cover, T. L., Cao, P., Lind, C. D., Tham, K. T., and Blaser, M. J. (1993) Correlation between vacuolating cytotoxin production by Helicobacter pylori isolates in vitro and in vivo. Infect. Immun. *61*, 5008-5012.

Cover, T. L., Tummuru, M. K. R., Cao, P., Thompson, S. A., and Blaser, M. J. (1994) Divergence of Genetic Sequences for the Vacuolating Cytotoxin Among Helicobacter Pylori Strains. J. Biol.Chem. *269*, 10566-10573.

Crabtree, J. E., Shallcross, T. M., Wyatt, J. I., Taylor, J. D., Heatley, R. V., Rathbone, B. J., and Losowsky, M. S. (1991a) Mucosal humoral immune response to Helicobacter pylori in patients with duodenitis. Dig. Dis. Sci. *36*, 1266-1273.

Crabtree, J. E., Shallcross, T. M., Heatley, R. V. and Wyatt, J. I. (1991b) Mucosal tumour necrosis factor alpha and interleukin-6 in patients with Helicobacter pylori associated gastritis. Gut. *32*, 1473-147.

Crabtree, J. E., Taylor, J. D., Wyatt, J. I., Heatley, R. V., Shallcross, T. M., Tompkins, D. S., and Rathbone, B. J. (1991c) Mucosal IgA recognition of Helicobacter pylori 120 kDa protein, peptic ulceration, and gastric pathology. Lancet. *338*, 332-335.

Crabtree, J. E., Peichl, P., Wyatt, J. I., Stachl, U., and Lindley, I. J. (1993) Gastric interleukin-8 and IgA IL-8 autoantibodies in Helicobacter pylori infection. Scand. J. Immunol. *37*, 65-70.

Crabtree, J. E., Covacci, A., Farmery, S. M., Xiang, Z., Tompkins, D. S., Perry, S., Lindley, I. J. D., and Rappuoli, R. (1995) Helicobacter pylori induced interleukin-8 expression in gastric epithelial cells is associated with CagA positive phenotype. J. Clin. Pathol. *48*, 41-45.

Cussac, V., Ferrero, R. L., and Labigne, A. (1992) Expression of Helicobacter pylori urease genes in Escherichia coli grown under nitrogen-limiting conditions. J. Bacteriol. *174*, 2466-2473.

Czinn, S. J. and Nedrud, J. G. (1991) Oral immunization against Helicobacter pylori. Infect. Immun. *59*, 2359-2363.

Danon, S. J., O'Rourke, J. L., Moss, N. D., and Lee, A. (1995) The importance of local acid production in the distribution of Helicobacter felis in the mouse stomach. Gastroenterol. *108*, 1386-1395.

Desai, H. G., Gill, H. H., Shankaran, K., Mehta, P. R., and Prabhu, S. R. (1991) Dental plaque: a permanent reservoir of Helicobacter pylori? Scand. J. Gastroenterol. *26*, 1205-1208.

Desai, M. A. and Vadgama, P. M. (1993) An in vitro study of enhanced H+ diffusion by urease action on urea. Implications for Helicobacter pylori-associated peptic ulceration. Scand. J Gastroenterol. *28*, 915-999.

Dewhirst, F. E., Seymour, C., Fraser, G. J., Paster, B. J., and Fox, J. G. (1994) Phylogeny of Helicobacter isolates from bird and swine feces and description of Helicobacter pametensis sp. nov. Int. J. Syst. Bact. *44*, 553-560.

Di Tommaso, A., Xiang, Z. Y., Bugnoli, M., Pileri, P., Figura, N., Bayeli, P. F., Rappuoli, R., Abrignani, S., and De Magistris, M. T. (1995) Helicobacter pylori-specific CD4(+) T-cell clones from peripheral blood and gastric biopsies. Infect. Immun. *63*, 1102-1106

Dick-Hegedus, E. and Lee, A. (1991) Use of a mouse model to examine anti-Helicobacter pylori agents. Scand. J. Gastroenterol. *26*, 909-915.

Doidge, C., Gust, I., Lee, A., Buck, F., Hazell, S., and Manne, U. (1994) Therapeutic Immunisation Against Helicobacter Infection. Lancet *343*, 914-915

Domenighini, M., Pizza, M., and Rappuoli, R. (1995) Bacterial ADP-ribosyltransferases. In: Handbook of natural toxins, Vol. 8: 'Bacterial toxins and virulence factors in disase', 59-80, Moss, J., Iglewski, B., Vaughan, M., and Tu A. T. (eds.). Marcel Dekker, Inc., New York.

Douce, G., Trucotte, C., Cropley, I., Roberts, M., Pizza, M., Domenighini, M., Rappuoli, R., and Dougan, G. (1995) Mutants of Escherichia coli heat-labile toxin lacking ADP-ribosyltrasferase activity act as nontoxic, mucosal adjuvants. Proc. Nat. Acad. Sci. USA *92*, 1644-1648.

Dubois, A., Fiala, N., Hemanackah, L. M., Drazek, E. S., Tarnawski, A., Fishbein, W. N., Perez-Perez, G. I., and Blaser, M. J. (1994) Natural gastric infection with Helicobacter pylori in monkeys: A model for spiral bacteria infection in humans. Gastroenterol. *106*, 1405-1417.

Dunn, B. E., Campbell, G. P., Perez-Perez, G. I., and Blaser, M. J. (1990) Purification and characterization of urease from Helicobacter pylori. J. Biol. Chem. *265*, 9464-9449.

Eaton, K. A., Brooks, C. L., Morgan, D. R., and Krakowka, S. (1991) Essential role of urease in pathogenesis of gastritis induced by Helicobacter pylori in gnotobiotic piglets. Infect. Immun. *59*, 2470-2475.

Eaton, K. A., Morgan, D. R., and Krakowka, S. (1992a) Motility as a factor in the colonisation of gnotobiotic piglets by Helicobacter pylori. J. Med. Microbiol. *37*, 123-127.

Eaton, K. A. and Krakowka, S. (1992b) Chronic active gastritis due to Helicobacter pylori in immunized gnotobiotic piglets. Gastroenterol. *103*, 1580-1586.

Eaton, K. A., Dewhirst, F. E., Radin, M. J., Fox, J. G., Paster, B. J., Krakowka, S., and Morgan, D. R. (1993) Helicobacter acinonyx sp. nov., isolated from cheetahs with gastritis. Int. J. Syst. Bacteriol. *43*, 99-106.

Eaton, K. A., Catrenich, C. E., Makin, K. M., and Krakowka, S. (1995) Virulence of coccoid and bacillary forms of Helicobacter pylori in gnotobiotic piglets. J. Infect. Dis. *171*, 459-462.

Eidt, S. and Stolte, M. (1995) The significance of Helicobacter pylori infection in relation to gastric cancer and lymphoma. Eur. J. Gastroenterol. Hepatol. *7*, 318-321.

Englund, J. A., Decker, M. D., Edwards, K. M., Pichichero, M. E, Steinhoff, M. C., and Anderson, E. L. (1994). Acellular and whole-cell pertussis vaccines as booster doses: a multicenter study. Pediatrics *93*, 37-43.

Engstrand, L., Scheynius, A., and Pahlson, C. (1991) An increased number of gamma/delta T-cells and gastric epithelial cell expression of the groEL stress-protein homologue in Helicobacter pylori-associated chronic gastritis of the antrum. Am. J. Gastroenterol. *86*, 976-980.

Euler, A. R., Zurenko, G. E., Moe, J. B., Ulrich, R. G., and Yagi, Y. (1990) Evaluation of two monkey species (Macaca mulatta and Macaca fascicularis) as possible models for human Helicobacter pylori disease. J. Clin. Microbiol. *28*, 2285-2290.

Evans, D. G., Karjalainen, T. K., Evans, D. J., Jr., Graham, D. Y., and Lee, C. H. (1993) Cloning, nucleotide sequence, and expression of a gene encoding an adhesin subunit protein of Helicobacter pylori. J. Bacteriol. *175*, 674-683.

Evans, D. J., Evans, D. G., Takemura, T., Nakano, H., Lampert, H. C., Graham, D. Y., Granger, D. N., and Kvietys, P. R. (1995) Characterization of a Helicobacter pylori neutrophil-activating protein. Infect. Immun. *63*, 2213-2220.

Exner, M. M., Doig, P., Trust, T. J., and Hancock, R. E. W. (1995) Isolation and characterization of a family of porin proteins from Helicobacter pylori. Infect. Immun. *63*, 1567-1572.

Falk, P., Roth, K. A., Boren, T., Westblom, T. U., Gordon, J. I., and Normark, S. (1993) An in vitro adherence assay reveals that Helicobacter pylori exhibits cell lineage-specific tropism in the human gastric epithelium. Proc. Natl. Acad. Sci. USA. *90*, 2035-2039.

Falk, P. G., Bry, L., Holgersson, J., and Gordon, J. I. (1995) Expression of a human alpha-1,3/4-fucosyltransferase in the pit cell lineage of FVB/N mouse stomach results in production of Le(b)-containing glycoconjugates: A potential transgenic mouse model for studying Helicobacter pylori infection. Proc. Natl. Acad. Sci. USA *92*, 1515-1519.

Ferrero, R. L. et al. (1994) Personal communication. VIIth Int Workshop of EHPSG, Houston, 1994.

Ferrero, R. L., Thiberge, J. M., Huerre, M., and Labigne, A. (1994) Recombinant antigens prepared from the urease subunits of Helicobacter spp: evidence of protection in a mouse model of gastric infection. Infect. Immun. *62*, 4981-4989.

Ferrero, R. L., Thiberge, M., Kansau, I., Wusher, N., Huerre, M., and Labigne, A. (1995) Immunisation with H.pylori heat shock protein A (HspA) and urease subunit B (UreB) affords total protection against H.felis infection in mice. Gut *37 (Suppl 1)*, A51.

Figura, N., Guglielmetti, P., Rossolini, A., Barberi, A., Cusi, G., Musmanno, R. A., Russi, M., and Quaranta, S. (1989) Cytotoxin production by Campylobacter pylori strains isolated from patients with peptic ulcers and from patients with chronic gastritis only. J. Clin. Microbiol. *27*, 225-226.

Fiocca, R., Villani, L., DeGiacomo, C., Perego, M., Trespo, E., and Solcia, E. (1989) Morphological evidence of Campylobacter pylori pathogenicity in chronic gastritis and peptic ulcer. Acta Gastroenterol Belgica *52*, 324-335.

Forman, D., Sitas, F., Newell, D. G., Stacey, A. R., Boreham, J., Peto, R., Campbell, T. C., Li, J., and Chen, J. (1990) Geographic association of Helicobacter pylori antibody prevalence and gastric cancer mortality in rural China. Int. J. Cancer *46*, 608-611.

Fox, J. G., Edrise, B. M., Cabot, E. B., Beacage, C., Murphy, J. C., and Prostak, K. (1986) Campylobacter-like organism isolated from the gastric mucosa of ferrets. Am. J. Vet.Res. *47*, 236-239

Fox, J. G., Cabot, E. B., Taylor, N. S., and Laraway, R. (1988) Gastric colonization by Campylobacter pylori subsp. mustelae in ferrets. Infect. Immun. *56*, 2994-2996.

Fox, J. G., Correa, P., Taylor, N. S., Lee, A., Otto, G., Murphy, J. C., and Rose, R. (1990) Helicobacter mustelae-associated gastritis in ferrets. An animal model of Helicobacter pylori gastritis in humans. Gastroenterol. *99*, 352-361.

Fox, J. G., Blanco, M., Murphy, J. C., Taylor, N. S., Lee, A., Kabok, Z., and Pappo, J. (1993) Local and systemic immune responses in murine Helicobacter felis active chronic gastritis. Infect. Immun. *61*, 2309-2315.

Fox, J. G., Dewhirst, F. E., Tully, J. G., Paster, B. J., Yan, L., Taylor, N. S., Collins, M. J., Gorelick, P. L., and Ward, J. M. (1994) Helicobacter Hepaticus Sp Nov, a Microaerophilic Bacterium Isolated from Livers and Intestinal Mucosal Scrapings from Mice. J. Clin. Microbiol. *32*, 1238-1245.

Fox, J. G., Yan, L. L., Dewhirst, F. E., Paster, B. J., Shames, B., Murphy, J. C., Hayward, A., Belcher, J. C., and Mendes, E. N. (1995) Helicobacter bilis sp nov, a novel Helicobacter species isolated from bile, livers, and intestines of aged, inbred mice. J. Clin. Microbiol. *33*, 445-454.

Ge, Z. M., Hiratsuka, K., and Taylor, D. E. (1995) Nucleotide sequence and mutational analysis indicate that two Helicobacter pylori genes encode a P-type ATPase and a cation-binding protein associated with copper transport. Mol. Microbiol. *15*, 97-106.

Geis, G., Leying, H., Suerbaum, S., Mai, U., and Opferkuch, W. (1989) Ultrastructure and chemical analysis of Campylobacter pylori flagella. J. Clin Microbiol *27*, 436-441.

Geis, G., Suerbaum, S., Forsthoff, B., Leying, H., and Opferkuch, W. (1993) Ultrastructure and biochemical studies of the flagellar sheath of Helicobacter pylori. J. Med. Microbiol. *38*, 371-377.

Ghiara, P., Marchetti, M., Blaser, M. J., Tummuru, M. K. R., Cover, T. L., Segal, E. D., Tompkins, L. S., and Rappuoli, R. (1995a) Role of the Helicobacter pylori virulence factors Vacuolating Cytotoxin, CagA and Urease in a mouse model of disease. Infect Immun. *63*, 4154-4160.

Ghiara, P., Marchetti, M., Di Tommaso, A., Saletti, G., Burroni, D., Figura, N., Manetti, R., Massari, P., Marchini, A., Aricò, B., Censini, S., Xiang, Z., Olivieri, R., Pizza, M., De Magistris, M. T., Abrignani, S., Covacci, A., Telford, J. L., and Rappuoli, R. (1995b) Infection by Helicobacter pylori in a mouse model that mimics human disease: protection by oral vaccination. Gut *37 (Suppl 1)*, A51.

Ghiara, P. and Michetti, P. (1995) Development of a vaccine. Current opinion in gastroenterology. 'The year in H. pylori'. *11 (Suppl 1)*, 52-56.

Gilbert, J. V., Ramakrishna, J., Sunderman Jr., F. W., Wright, A., and Plaut, A. G. (1995) Protein Hpn: cloning and characterization of a histidine-rich metal binding polypeptide in Helicobacter pylori and Helicobacter mustelae. Infect. Immun. *63*, 2682-2688.

Handt, L. K., Fox, J. G., Dewhirst, F. E., Fraser, G. J., Paster, B. J., Yan, L. L., Rozmiarek, H., Rufo, R., and Stalis, I. H. (1994) Helicobacter pylori isolated from the domestic cat: Public health implications. Infect. Immun. *62*, 2367-2374.

Hazell, S. L., Eichberg, J. W., Lee, D. R., Alpert, L., Evans, D. G., Evans, D. J. Jr., and Graham, D. Y. (1992) Selection of the chimpanzee over the baboon as a model for Helicobacter pylori infection. Gastroenterol. *103*, 848-854.

Hazell, S. L., Mitchell, H. M., Hedges, M., Shi, X., Hu, P. J., Li, Y. Y., Lee, A., and Reisslevy, E. (1994) Hepatitis A and evidence against the community dissemination of Helicobacter pylori via feces. J. Infect. Dis. *170*, 686-689.

Henneresse, P., Eyssen, H., Ectors, N., Geboes, K., Janssens, J., and Vantrappen, G. (1989) Infestation experimentale du rat exenique par le Campylobacter pylori. Gastroenterol. Clin. Bio. *96*, 167.

Hu, L. T. and Mobley, H. L. (1993) Expression of catalytically active recombinant Helicobacter pylori urease at wild-type levels in Escherichia coli. Infect. Immun. *61*, 2563-2569.

Huang, J. Z., O'Toole, P. W., Doig, P., and Trust, T. J. (1995) Stimulation of interleukin-8 production in epithelial cell lines by Helicobacter pylori. Infect. Immun. *63*, 1732-1738.

Huang, L. L., Cave, D. R., and Kane, A. V. (1995) Identification of an acid-inhibitory protein from Helicobacter pylori. FASEB J. *9*, A1441.

Jabri, E., Carr, M. B., Hausinger, R. P., and Karplus, P. A. (1995) The crystal structure of urease from Klebsiella aerogenes. Science *268*, 998-1004.

Janzon, L. and Wickman, M. (1993) Construction of deletion mutants in the hpaA gene encoding the binding subunit of fibrillar adhesin of Helicobacter pylori. Acta Gastroenterol. Belgica *56 (Suppl)*, 103.

Josenhans, C., Labigne, A., and Suerbaum, S. (1995) Comparative ultrastructural and functional studies of Helicobacter pylori and Helicobacter mustelae flagellin mutants: Both flagellin subunits, FlaA and FlaB, are necessary for full motility in Helicobacter species. J. Bacteriol. *177*, 3010-3020.

Kansau, I., Thiberge, J. M., Ferrero, R. L., Suerbaum, S., and Labigne, A. (1994) Immunological and functional properties of the HspA and HspB heat shock proteins of Helicobacter pylori. Am. J. Gastroenterol. *89*, 1326-1320.

Karita, M., Kouchiyama, T., Okita, K., and Nakazawa, T. (1991) New small animal model for human gastric Helicobacter pylori infection: success in both nude and euthymic mice. Am. J. Gastroenterol. *86*, 1596-1603.

Karttunen, R. (1991) Blood lymphocyte proliferation, cytokine secretion and appearance of T cells with activation surface markers in cultures with Helicobacter pylori. Comparison of the responses of subjects with and without antibodies to H. pylori. Clin. Exp. Immunol. *83*, 396-400.

Kaufmann, S. H. E., Schoel, E. B., Wand-Wurttenberg, A., Steinhof, U., Munk, M. E., and Koga, T. (1990). T cells, stress proteins and pathoegenesis of mycobacterial infections. Curr Topics Microbiol. Immunol. *155*, 125-141.

Kelly, S. M., Pitcher, M. C. L., Farmery, S. M., and Gibson, G. R. (1994) Isolation of Helicobacter pylori from feces of patients with dyspepsia in the United Kingdom. Gastroenterol. *107*, 1671-1674.

Kemper, C. A., Mickelsen, P., Morton, A., Walton, B., and Deresinski, S. C. (1993) Helicobacter (Campylobacter) fennelliae-like organisms as an important but occult cause of bacteraemia in a patient with AIDS. J. Infect. *26*, 97-101.

Kist, M. et al. (1995) Personal Communication. VIIIth Int Workshop on Campylobacters, Helicobacters and related organisms, Winchester, 10-13 July 1995.

Kostrzynska, M., O'Toole, P. W., Taylor, D. E., and Trust, T. J. (1994) Molecular characterization of a conserved 20-kilodalton membrane-associated lipoprotein antigen of Helicobacter pylori. J. Bacteriol. *176*, 5938-5948.

Krakowka, S., Morgan, D. R., Kraft, W. G., and Leunk, R. D. (1987) Establishment of gastric Campylobacter pylori infection in the neonatal gnotobiotic piglet. Infect. Immun. *55*, 2789-2796.

Labigne, A., Cussac, V., and Courcoux, P. (1991) Shuttle cloning and nucleotide sequences of Helicobacter pylori genes responsible for urease activity. J. Bacteriol. *173*, 1920-1931.

Lambert, I., Clyne, M., and Drumm, B. (1993) H pylori in dental plaques. Lancet. *341*, 957.

Lee, A., Hazell, S. L., O'Rourke, J., and Kouprach, S. (1988) Isolation of a spiral-shaped bacterium from the cat stomach. Infect. Immun. *56*, 2843-2850.

Lee, A., Fox, J. G., Otto, G., and Murphy, J. (1990) A small animal model of human Helicobacter pylori active chronic gastritis. Gastroenterol. *99*, 1315-1323.

Lee, A., Fox, J. G., Otto, G., Dick, E. H., and Krakowka, S. (1991) Transmission of Helicobacter spp. A challenge to the dogma of faecal-oral spread. Epidemiol. Infect. *107*, 99-109.

Lee, A., Phillips, M. W., O'Rourke J. L., Paster, B. J., Dewhirst, F. E., Fraser, G. J., Fox, J. G., Sly, L. I., Romaniuk, P. J., and Trust, T. J. (1992) Helicobacter muridarum sp. nov., a microaerophilic helical bacterium with a novel ultrastructure isolated from the intestinal mucosa of rodents. Int. J. Syst. Bacteriol. *42*, 27-36.

Lee, A. (1993) H pylori-initiated ulcerogenesis: look to the host. Lancet. *341*, 280-281.

Lee, A. and Chen, M. H. (1994) Successful immunization against gastric infection with Helicobacter species: Use of a cholera toxin B-subunit- whole-cell vaccine. Infect. Immun. *62*, 3594-3597.

Lee, C. K., Weltzin, R., Thomas, W. D., Kleanthous, H., Hermak, T. H., Soman, G., Hill, J. E., Ackerman, S. K., and Monath, T. P. (1995) Oral immunization with recombinant Helicobacter pylori urease induces secrtory IgA antibodies and protects mice from challenge with Helicobacter felis. J. Infect. Dis. *172*, 161-172.

Leunk, R. D., Johnson, P. T., David, B. C., Kraft, W. G., and Morgan, D. R. (1988) Cytotoxic activity in broth-culture filtrates of Campylobacter pylori. J. Med. Microbiol *26*, 93-99.
Leverstein-van Hall, M. A., van der Ende, A., van Milligen de Wit, M., Tytgat, G. N., and Dankert, J. (1993) Transmission of Helicobacter pylori via faeces. Lancet. *342*, 1419-1420.
Leying, H., Suerbaum, S., Geis, G., and Haas, R. (1992) Cloning and genetic characterization of a Helicobacter pylori flagellin gene. Mol. Microbiol. *6*, 2863-2874.
Logan, R. P. H. (1994) Helicobacter pylori and gastric cancer. Lancet *344*, 1078-1079.
Luzzi, I., Pezzella, C., Caprioli, A., Covacci, A., Bugnoli, M., and Censini, S. (1993) Detection of vacuolating toxin of Helicobacter pylori in human faeces. Lancet. *341*, 1348.
Macchia, G., Massone, A., Burroni, D., Covacci, A., Censini, S., and Rappuoli, R. (1993) The Hsp60 protein of Helicobacter pylori: structure and immune response in patients with gastroduodenal diseases. Mol. Microbiol. *9*, 645-652.
Mai, U. E., Perez-Perez, G. I., Wahl, L. M., Wahl, S. M., Blaser, M. J., and Smith, P. D. (1991) Soluble surface proteins from Helicobacter pylori activate monocytes/macrophages by lipopolysaccharide-independent mechanism. J. Clin. Invest. *87*, 894-900.
Mai, U. E., Perez-Perez, G. I., Allen, J. B., Wahl, S. M., Blaser, M. J., and Smith, P. D. (1992) Surface proteins from Helicobacter pylori exhibit chemotactic activity for human leukocytes and are present in gastric mucosa. J. Exp. Med. *175*, 517-525.
Malanoski, G. J., Eliopoulos, G. M., Ferraro, M. J., and Moellering, R. C., Jr. (1993) Effect of pH variation on the susceptibility of Helicobacter pylori to three macrolide antimicrobial agents and temafloxacin. Eur. J. Clin. Microbiol. Infect. Dis. *12*, 131-133.
Malfertheiner, P. (1993) Compliance, adverse events and antibiotic resistance in Helicobacter pylori treatment. Scand. J. Gastroenterol. Suppl. *196*, 34-37.
Manetti, R., Massari, P., Burroni, D., De Bernard, M., Marchini, A., Olivieri, R., Papini, E., Montecucco, C., Rappuoli, R., and Telford, J. L. (1995) Helicobacter pylori cytotoxin: importance of native conformation for induction of neutralizing antibodies. Infect. Immun. *63*, 4476-4480.
Mapstone, N. P., Lynch, D. A., Lewis, F. A., Axon, A. T., Tompkins, D. S., Dixon, M. F., and Quirke, P. (1993) PCR identification of Helicobacter pylori in faeces from gastritis patients. Lancet. *341*, 447.
Marchetti, M., Arico, B., Burroni, D., Figura, N., Rappuoli, R., and Ghiara, P. (1995) Development of a mouse model of Helicobacter pylori infection that mimics human disease. Science *267*, 1655-1658.
Marchini, A., Massari, P., Manetti, R., and Olivieri, R. (1994) Optimized conditions for the fermentation of Helicobacter pylori and production of vacuolating cytotoxin. FEMS Microbiol. Lett. *124*, 55-59.
Marshall, B. J. and Warren, J. R. (1983) Unidentified curved bacilli in the stomach of patients with gastritis and peptic ulceration. Lancet *1*, 1311-1315.
Mattsby-Baltzer, I., Mielniczuk, Z., Larsson, L., Lindgren, K., and Goodwin, S. (1992) Lipid A in Helicobacter pylori. Infect. Immun. *60*, 4383-4387.
Megraud, F. (1993) Epidemiology of Helicobacter pylori infection. Gastroenterol. Clin. North Am. *22*, 73-88.
Megraud, F. (1994) Epidemiology of Helicobacter pylori infection: where are we in 1995? Eur. J. Gastroenterol. Hepatol. *7*, 292-295.
Mendall, M. A. and Northfield, T. C. (1995) Transmission of Helicobacter pylori. Gut. *37*, 1-3.
Mendz, G. L., Hazell, S. L., and Burns, B. P. (1993) Glucose Utilization and Lactate Production by Helicobacter Pylori. J. Gen. Microbiol *139*, 3023-3028.
Mendz, G. L., Hazell, S. L., and Vangorkom, L. (1994a) Pyruvate metabolism in Helicobacter pylori. Arch. Microbiol. *162*, 187-192.
Mendz, G. L., Jimenez, B. M., Hazell, S. L., Gero, A. M., and Osullivan, W. J. (1994b) Salvage synthesis of purine nucleotides by Helicobacter pylori. J. Appl. Bacteriol. *77*, 674-681.

Mendz, G. L. and Hazell, S. H. (1994c) The urea cycle of Helicobacter pylori. Am. J. Gastroenterol. *89*, 1289.

Michetti, P., Corthésy-Theulaz, I., Davin, C., Haas, R., Vaney, A. C., Heitz, M., Bille, J., Kraehenbuhl, J. P., Saraga, E., and Blum, A. L. (1994) Immunization of BALB/c mice against Helicobacter felis infection with H. pylori urease. Gastroenterol. *107*, 1002-1011.

Mills, S. D., Kurjanczyk, L. A., and Penner, J. L. (1992) Antigenicity of Helicobacter pylori lipopolysaccharides. J. Clin. Microbiol. *30*, 3175-3180.

Milne, J. C., Furlong, D., Hanna, P. C., Wall, J. S., and Collier, R. J. (1994) Antrax protective antigen forms oligomers during intoxication of mammalian cells. J. Biol. Chem. *269*, 20607-20612.

Mitchell, H. M., Li, Y. Y., Hu, P. J., Liu, Q., Chen, M., Du, G. G., Wang, Z. J., Lee, A., and Hazell, S. L. (1992) Epidemiology of Helicobacter pylori in southern China: identification of early childhood as the critical period for acquisition. J. Infect. Dis. *166*, 149-153.

Mobley, H. L., Hu, L. T., and Foxal, P. A. (1991) Helicobacter pylori urease: properties and role in pathogenesis. Scand. J. Gastroenterol. Suppl. *187*, 39-46.

Mobley, H. L. T., Garner, R. M., and Bauerfeind, P. (1995) Helicobacter pylori nickel-transport gene nixA: Synthesis of catalytically active urease in Escherichia coli independent of growth conditions. Mol. Microbiol. *16*, 97-109.

Monath, T. P., Thomas, W., Weltzin, R. A., Soman, G., Ackerman, S. A., Pappo, J., Ermak, T., Bhagat, H., and Lee, C. K. (1994) Progress towards a vaccine against Helicobacter pylori. Am. J. Gastroenterol. *89*, 1383.

Monath, T. P. (1995). Personal communication. Int Workshop of EHPSG, Edinburgh, 7-9 July 1995.

Moran, A. P., Helander, I. M., and Kosunen, T. U. (1992) Compositional analysis of Helicobacter pylori rough-form lipopolysaccharides. J. Bacteriol. *174*, 1370-1377.

Morisaki, I., Michalek, S. M., Harmon, C. C., Torii, M., Hamada, S., and McGhee, J. R. (1983) Effective immunity to dental caries: enhancement of salivary anti-Streptococcus mutans antibody response with oral adjuvants. Infect. Immun. *40*, 577-581.

Murakami, M., Yoo, J. K., Teramura, S., Yamamoto, K., Saita, H., Matuo, K., Asada, T., and Kita, T. (1990) Generation of ammonia and mucosal lesion formation following hydrolysis of urea by urease in the rat stomach. J. Clin. Gastroenterol. *12 Suppl 1*, S104-S109.

Namavar, F., Roosendaal, R., Kuipers, E. J., Degroot, P., Vanderbijl, M. W., Pena, A. S., and Degraaff, J. (1995) Presence of Helicobacter pylori in the oral cavity, oesophagus, stomach and faeces of patients with gastritis. Eur. J. Clin. Microbiol. Infect. Dis. *14*, 234-237.

Nedrud, J., Blanchard, T., Czinn, S., and Lycke, N. (1995) Recombinant cholera toxin B subunit is not an adjuvant for oral immunization against helicobacter felis in mice. J. Cell. Biochem. *Suppl 19A*, 261.

Negrini, R., Lisato, L., Zanella, I., Cavazzini, L., Gullini, S., Villanacci, V., Poiesi, C., Albertini, A., and Ghielmi, S. (1991) Helicobacter pylori infection induces antibodies cross-reacting with human gastric mucosa. Gastroenterol. *101*, 437-445.

Nicosia, A., Perugini, M., Franzini, C., Casagli, M. C., Borri, M. G., Antoni, G., Almoni, M., Neri, P., Ratti, G., and Rappuoli, R. (1986) Cloning and sequencing of the pertussis toxin genes: operon structure and gene duplication. Proc. Natl. Acad. Sci. USA *83*, 4631-4635.

NIH Consensus Development Panel on Helicobacter pylori in Peptic Ulcer Disease (1994) Helicobacter pylori in peptic ulcer disease. JAMA *272*, 65-69.

Nilius, M., Strohle, A., Bode, G., and Malfertheiner, P. (1994) Coccoid like forms (CLF) of Helicobacter pylori. Enzyme activity and antigenicity. Int. J. Med. Microbiol. *280*, 259-272.

Noach, L. A., Rolf, T. M., and Tytgat, G. N. J. (1994) Electron microscopic study of association between Helicobacter pylori and gastric and duodenal mucosa. J. Clin. Pathol. *47*, 699-704.

Odenbriet, S., Till, M., and Haas, R. (1995) Identification of a Helicobacter pylori specific protein involved in adherence to gastric cells. Gut *37 (Suppl 1)*, A1.

Olivieri, R., Bugnoli, M., Armellini, D., Bianciardi, S., Rappuoli, R., Bayeli, P. F., Abate, L., Esposito, E., de Gregorio, L., and Aziz, J. (1993) Growth of Helicobacter pylori in media containing cyclodextrins. J. Clin. Microbiol. *31*, 160-162.

Olsson, K., Wadstrom, T., and Tyszkiewicz, T. (1993) H pylori in dental plaques. Lancet. *341*, 956-957.

Orlicek, S. L., Welch, D. F., and Kuhls, T. L. (1993) Septicemia and meningitis caused by Helicobacter cinaedi in a neonate. J. Clin. Microbiol. *31*, 569-571.

O'Toole, P. W., Kostrzynska, M., and Trust, T. J. (1994) Non-motile mutants of Helicobacter pylori and Helicobacter mustelae defective in flagellar hook production. Mol.. Microbiol. *14*, 691-703.

Otto, G., Fox, J. G., Wu, P. Y., and Taylor, N. S. (1990) Eradication of Helicobacter mustelae from the ferret stomach: an animal model of Helicobacter (Campylobacter) pylori chemotherapy. Antimicrob. Agents Chemother. *34*, 1232-1236.

Pallen, M. J. and Clayton, C. L. (1990) Vaccination against Helicobacter pylori urease. Lancet. *336*, 186-187.

Papini, E., De Bernard, M., Milia, E., Bugnoli, M., Zerial, M., Rappuoli, R., and Montecucco, C. (1994) Cellular vacuoles induced by Helicobacter pylori originate from late endosomal compartments. Proc. Natl. Acad. Sci. USA *91*, 9720-9724.

Parsonnet, J., Friedman, G. D., Vandersteen, D. P., Chang, Y., Vogelman, J. H., Orentreich, N., and Sibley, R. K. (1991) Helicobacter pylori infection and the risk of gastric carcinoma. N. Engl. J. Med. *325*, 1127-1231.

Parsonnet, J., Hansen, S., Rodriguez, L., Gelb, A. B., Warnke, R. A., Jellum, E., Orentreich, N., Vogelman, J. H., and Friedman, G. D. (1994) Helicobacter Pylori Infection and Gastric Lymphoma. N. Engl J Med *330*, 1267-1271.

Phadnis, S. H., Ilver, D., Janzon, L., Normark, S., and Westblom, T. U. (1994) Pathological Significance and Molecular Characterization of the Vacuolating Toxin Gene of Helicobacter Pylori. Infect. Immu. *62*, 1557-1565.

Pizza, M., Covacci, A., Bartoloni, A., Perugini, M., Nencioni, L., Manetti, R., Bugnoli, M., Barbieri, J. T., Sato, H., and Rappuoli, R. (1989) Mutants of pertussis toxin suitable for vaccine development. Science *246*, 497-500.

Radin, M. J., Eaton, K. A., Krakowka, S., Morgan, D. R., Lee, A., Otto, G., and Fox, J. (1990) Helicobacter pylori gastric infection in gnotobiotic beagle dogs. Infect. Immun. *58*, 2606-2612.

Rautelin, H., Seppala, K., Renkonen, O. V., Vainio, U., and Kosunen, T. U. (1992) Role of metronidazole resistance in therapy of Helicobacter pylori infections. Antimicrob. Agents Chemother. *36*, 163-166.

Rautelin, H., Tee, W., Seppala, K., and Kosunen, T. U. (1994) Ribotyping Patterns and Emergence of Metronidazole Resistance in Paired Clinical Samples of Helicobacter pylori. J. Clin. Microbiol. *32*, 1079-1082.

Ross, J. S., Bui, H. X., del Rosario, A., Sonbati, H., George, M., and Lee, C. Y. (1992) Helicobacter pylori. Its role in the pathogenesis of peptic ulcer disease in a new animal model. Am. J. Pathol. *141*, 721-727.

Sakamoto, J., Watanabe, T., Tokumaru, T., Takagi, H., Nakazato, H., and Lloyd, K. O. (1989) Expression of Lewis^{a}, Lewis^{b}, Lewis^{x}, Lewis^{y}, sialyl-Lewis^{a}, and sialyl-Lewis^{x} blood group antigens in human gastric carcinoma and in normal gastric tissue. Cancer Res. *49*, 745-752.

Schauer, D. B., Ghori, N., and Falkow, S. (1993) Isolation and characterization of "Flexispira rappini" from laboratory mice. J. Clin Microbiol *31*, 2709-2714.

Schmitt, W. and Haas, R. (1994) Genetic Analysis of the Helicobacter Pylori Vacuolating Cytotoxin – Structural Similarities with the IgA Protease Type of Exported Protein. Mol Microbiol *12*, 307-319.

Segal, E. D. and Tompkins, L. S. (1993) Identification of an RTX-like hemolysin produced by Helicobacter pylori. Acta Gastroenterol. Belgica *56 (Suppl)*, 63.

Shames, B., Krajden, S., Fuksa, M., Babida, C., and Penner, J. L. (1989) Evidence for the occurrence of the same strain of Campylobacter pylori in the stomach and dental plaque. J. Clin Microbiol *27*, 2849-2850.

Sharma, S. A., Miller, G. G., Perez-Perez, G. I., Gupta, R. S., and Blaser, M. J. (1994) Humoral and cellular immune recognition of Helicobacter pylori proteins are not concordant. Clin. Exp. Immunol. *97*, 126-132.

Smith, A. W., Chahal, B., and French, G. L. (1994) The human gastric pathogen Helicobacter pylori has a gene encoding an enzyme first classified as a mucinase in Vibrio cholerae. Mol. Microbiol. *13*, 153-160.

Smoot, D. T., Mobley, H. L., Chippendale, G. R., Lewison, J. F., and Resau, J. H. (1990) Helicobacter pylori urease activity is toxic to human gastric epithelial cells. Infect. Immun. *58*, 1992-1994.

Smoot, D. T., Resau, J. H., Naab, T., Desbordes, B. C., Gilliam, T. M., Bull-Henry, K., Curry, S. B., Nidiry, J., Sewchand, J., and Mills-Robertson, K. (1993) Adherence of Helicobacter pylori to cultured human gastric epithelial cells. Infect. Immun. *61*, 350-355.

Sobala, G. M., Crabtree, J. E., Dixon, M. F., Schorah, C. J., Taylor, J. D., Rathbone, B. J., Heatley, R. V., and Axon, A. T. (1991) Acute Helicobacter pylori infection: clinical features, local and systemic immune response, gastric mucosal histology, and gastric juice ascorbic acid concentrations. Gut. *32*, 1415-1418.

Solcia, E., Cappella, E., Fiocca, R., Cornaggia, M., Rindi, G., Villani, L., Bosi, F., and Ambrosiani, L. (1990) Exocrine and endocrine epithelial changes in types A and B chronic gastritis. In: Helicobacter pylori, gastritis and peptic ulcer, 245-258, Malfertheiner, P. and Dischuneit, H. (eds.). Springer Verlag, New York.

Solnick, J. V., O'Rourke J, Lee, A., Paster, B. J., Dewhirst, F. E., and Tompkins, L. S. (1993) An uncultured gastric spiral organism is a newly identified Helicobacter in humans. J. Infect. Dis. *168*, 379-385.

Solnick, J. V. and Tompkins, L. S. (1993) Helicobacter pylori and gastroduodenal disease: pathogenesis and host-parasite interaction. Infect. Ag. Dis. *1*, 294-309.

Spiegelhalder, C., Gerstenecker, B., Kersten, A., Schiltz, E., and Kist, M. (1993) Purification of Helicobacter pylori superoxide dismutase and cloning and sequencing of the gene. Infect. Immun. *61*, 5315-5325.

Stanley, J., Linton, D., Burnens, A. P., Dewhirst, F. E., Owen, R. J., Porter, A., On, S. L., and Costas, M. (1993) Helicobacter canis sp. nov., a new species from dogs: an integrated study of phenotype and genotype. J. Gen. Microbiol. *139*, 2495-2504.

Stanley, J., Linton, D., Burnens, A. P., Dewhirst, F. E., On, S. L. W., Porter, A., Owen, R. J., and Costas, M. (1994) Helicobacter pullorum sp nov – Genotype and Phenotype of a new species isolated from poultry and human patients with gastroenetritis. Microbiology *140*, 3441-3449.

Suerbaum, S., Josenhans, C., and Labigne, A. (1993) Cloning and genetic characterization of the Helicobacter pylori and Helicobacter mustelae flaB flagellin genes and construction of H. pylori flaA- and flaB-negative mutants by electroporation-mediated allelic exchange. J. Bacteriol. *175*, 3278-3288.

Suerbaum, S., Thiberge, J., Kansau, I., Ferrero, R. L., and Labigne, A. (1994) Helicobacter pylori hspA-hspB heat -shock gene cluster: nucleotide sequence, expression, putative function and immunogenicity. Mol. Microbiol. *14*, 959-974.

Sun, J. B., Holmgren, J., and Czerkinsky, C. (1994) Cholera toxin B subunit: an efficient transmucosal carrier-delivery system for induction of peripheral immunological tolerance. Proc. Natl. Acad. Sci. USA *91*, 10795-10799.

Takeuchi, K., Ohuchi, T., Harada, H., and Okabe, S. (1995) Irritant and protective action of urea urease ammonia in rat gastric mucosa: Different effects of ammonia and ammonium ion. Dig. Dis. Sci. *40*, 274-281.

Taylor, N. S., Fox, J. G., and Yan, L. (1995) In-vitro hepatotoxic factor in Helicobacter hepaticus, H- pylori and other Helicobacter species. J. Med. Microbiol. *42*, 48-52.

Telford, J. L., Covacci, A., Ghiara, P., Montecucco, C., and Rappuoli, R. (1994a) Unravelling the pathogenic role of Helicobacter pylori in peptic ulcer: potential for new therapies and vaccines. Trends Biotechnol. *12*, 420-426.

Telford, J. L., Ghiara, P., Dellorco, M., Comanducci, M., Burroni, D., Bugnoli, M., Tecce, M. F., Censini, S., Covacci, A., Xiang, Z. Y., Papini, E., Montecucco, C., Parente, L., and Rappuoli, R. (1994b) Gene Structure of the Helicobacter Pylori Cytotoxin and Evidence of Its Key Role in Gastric Disease. J. Exp. Med. *179*, 1653-1658.

Thompson, S. A and Blaser, M. J. (1995) Isolation of the Helicobacter pylori recA gene and involvment of the recA region in resistance to low pH. Infect. Immun. *63*, 2185-2193.

Trejdosiewicz, L. K., Calabrese, A., Smart, C. J., Oakes, D. J., Howdle, P. D., Crabtree, J. E., Losowsky, M. S., Lancaster, F., and Boylston, A. W. (1991) Gamma delta T cell receptor-positive cells of the human gastrointestinal mucosa: occurrence and V region gene expression in Heliobacter pylori-associated gastritis, coeliac disease and inflammatory bowel disease. Clin. Exp. Immunol. *84*, 440-444.

Tsuda, M., Karita, M., Morshed, M. G., Okita, K., and Nakazawa, T. (1994) A Urease-negative mutant of Helicobacter pylori constructed by allelic exchange mutagenesis lacks the ability to colonize the nude mouse stomach. Infect. Immun. *62*, 3586-3589.

Tsugane, S., Kabuto, M., Imai, H., Gey, F., Tei, Y., Hanaoka, T., Sugano, K., and Watanabe, S. (1993) Helicobacter pylori, dietary factors, and atrophic gastritis in five Japanese populations with different gastric cancer mortality. Cancer. Causes Control. *4*, 297-305.

Tummuru, M. K., Cover, T. L., and Blaser, M. J. (1993) Cloning and expression of a high-molecular-mass major antigen of Helicobacter pylori: evidence of linkage to cytotoxin production. Infect. Immun. *61*, 1799-1809.

Tummuru, M. K. R., Cover, T. L., and Blaser, M. J. (1994) Mutation of the cytotoxin-associated cagA gene does not affect the vacuolating cytotoxin activity of Helicobacter pylori. Infect. Immun. *62*, 2609-2613.

Tytgat, G. N. J. (1994) Review Article – Treatments That Impact Favourably upon the Eradication of Helicobacter pylori and Ulcer Recurrence. Aliment. Pharmacol. Therapeut. *8*, 359-368.

Vanzwet, A. A., Thijs, J. C., Kooistrasmid, A. M. D., Schirm, J., and Snijder, J. A. M. (1994) Use of PCR with Feces for Detection of Helicobacter Pylori Infections in Patients. J. Clin. Microbiol. *32*, 1346-1348.

Walker, R. I. (1994) New strategies for using mucosal vaccination to achieve more effective immunization. Vaccine *12*, 387-400.

Westblom, T. U. and Unge, P. (1992) Drug resistance of Helicobacter pylori: memorandum from a meeting at the Sixth International Workshop on Campylobacter, Helicobacter, and Related Organisms. J. Infect. Dis. *165*, 974-975.

Xiang, Z., Bugnoli, M., Ponzetto, A., Morgando, A., Figura, N., Covacci, A., Petracca, R., Pennatini, C., Censini, S., and Armellini, D. (1993a) Detection in an enzyme immunoassay of an immune response to a recombinant fragment of the 128 kilodalton protein (CagA) of Helicobacter pylori. Eur. J Clin Microbiol Infect Dis *12*, 739-745.

Xiang, Z., Bugnoli, M., Rappuoli, R., Covacci, A., Ponzetto, A., and Crabtree, J. E. (1993b) Helicobacter pylori: host responses in peptic ulceration. Lancet. *341*, 900-901.

Xiang, Z. Y., Censini, S., Bayeli, P. F., Telford, J. L., Figura, N., Rappuoli, R., and Covacci, A. (1995) Analysis of expression of CagA and VacA virulence factors in 43 strains of He-

licobacter pylori reveals that clinical isolates can be divided into two major types and that CagA is not necessary for expression of the vacuolating cytotoxin. Infect. Immun. *63*, 94-98.
Xu, J. K., Goodwin, C. S., Cooper, M., and Robinson, J. (1990) Intracellular vacuolization caused by the urease of Helicobacter pylori. J. Infect. Dis. *161*, 1302-1304.

4.2 Malaria Vaccination

Manuel E. Patarroyo and Roberto Amador

4.2.1 Introduction

Development of a malaria vaccine against *Plasmodium falciparum* is a priority in developing countries. The disease is spread world-wide in more than 100 tropical countries, and represents around 300 million clinical cases a year and as many as 2.7 million deaths (WHO, 1992). Child mortality (1-5 years old) in Sub-Saharan Africa is 175 per 1000 live births, and it is estimated that one million African children die each year due to *P. falciparum* malaria (Greenwood et al., 1987). In addition, malaria in pregnant women is the major cause of abortion and low birth weight.

Even though vaccines are among the most cost-effective medical interventions, they are unfortunately not big money makers, especially if they are against diseases that do not afflict developed countries.

The malaria infection commences with the inoculation of sporozoites by infected mosquitoes. The sporozoites infect hepatocytes and develop into merozoites which thereafter infect erythrocytes, where a huge amplification of the parasite population occurs and produces the clinical signs and symptoms. Afterwards the parasite differentiates to form gametocytes, the stage which is infectious to mosquitoes. Because *P. falciparum* is essentially an intracellular parasite, it is present in its cell-free form for a period too short to ensure adequate antigen presentation of the immune system.

The first attempts at producing a vaccine against malaria were begun in the 1940's, and for the next 50 years the main focus of this research was directed towards the study of candidate antigens from the sporozoite stages. It was only with the advent of *in vitro* culture techniques in the mid 70's and early 80's that the study of antigens from the asexual blood stages was begun. More recently hepatic or exoerythrocitic stage antigens have been investigated (Romero, 1992).

Although knowledge of the parasite's biology is incomplete, research has allowed insight into some of the mechanisms that the parasite uses to evade host immunity. From this still unclear picture, certain principles have been clarified for the development of a malarial vaccine. One of these is the necessity to adopt an 'antigenic cocktail' approach to obtain a synthetic or recombinant subunit vaccine. One simple epitope or monovalent vaccines do not take into account the complexity of the parasite cycle.

The synthetic Colombian Malaria vaccine, SPf66 (Amador et al., 1992; Lopez et al., 1994; Moreno and Patarroyo, 1989), is an example of this cocktail methodology, and the one that probably endorses the success of this approach. It is the first generation of multiantigenic, multistage synthetic vaccines, and the first malaria vaccine to be safely inoculated into large groups of people.

SPf66 is a chimeric molecule consisting of four synthetic epitopes (Merrifield, 1963; Houghten, 1995), one derived from the repeat domain of the circumsporozoite protein, the PNANP sequence and three blood stage epitopes selected on amino acid sequences from three distinct blood stage proteins that elicited delayed or a protective response in *Aotus* monkeys.

4.2.2 Immunological Considerations for Malaria Vaccines

It was noted in early times that with some viral infections, those who survived a first attack rarely suffered a clinical infection of the same type. Based on this observation, attenuated live viruses have been used as biological vaccines to give long-lasting immunity, after one or two administrations (Ada, 1995). Nevertheless from the immunological point of view, the finer aspects related with molecular mechanisms of protection were not explored.

In contrast, malaria induces a partial immunity and clinical recurrence is common. Basic knowledge about natural immunity and possible functional mechanisms of interactions with the SPf66 vaccine and other vaccine candidates are becoming better understood.

Simultaneously, experimental and epidemiological evidence have provided some information about protective malaria immunity:

a) It is well known that sporozoite challenge of volunteers, previously immunised with radiation-attenuated sporozoites, does not induce any detectable blood-stage infection (Egan et al., 1995; Edelman et al., 1995) and therefore malaria immunity is stage-specific.
b) Natural immunity under conditions of permanent exposure takes a long time to develop, and is age dependent. Furthermore low level of infection remains, despite natural immunity, meaning that is not total efficacious against parasitemia.
c) Severe malaria infection no longer occurs in children over 4-5 years of age; this phenomenon had been denominated "premunition", which implies that the state of immunity is maintained by the persistence of a low level infection over a long period of time (Hommel, 1993). An antitoxin immunity, has been related with this response (Kwiatkowski, 1992).

The interaction between the malaria parasite and the host is not a straightforward one. Genetic restriction of the immune response to malaria, observed for both T-cell and B-cell epitopes (Nardin and Nussenzweig, 1993) has to be addressed and a vaccine construct probably would have to include multiple T epitopes from different target antigens, or alternatively, an universal T-cell activator epitope to ensure that the great majority of people are able to recognise these epitopes. Extensive studies on the cellular response at a clonal level (Hollingdale et al., 1990), preferential induction of T helper subsets, antibody-dependent cellular inhibition activity (ADCI) and antibody-dependent cellular cytotoxicity (ADCC), will need to be considered to block the parasite at the hepatocyte level.

Although epidemiological and experimental evidence point convincingly to the importance of various T-cell functions in the acquisition and maintenance of protective immunity, a reliable in vitro assay to test the behaviour of the clinical immunity does not exist (Hviid et al., 1995).

Based on the evidence of the limitations of the protective acquired immunity, different groups have focused on the characterisation of immunodominant epitopes expressed on the parasitized RBCs. Although adults in endemic areas develop antibodies to conserved epitopes on the RBC surface, unfortunately they are not immunodominant (Marsh and Howard, 1986).

Some believe that including either protective immunity against sequences common to different strains or a sequence from each strain, could be potentially useful in vaccine design. However, these polymorphisms are a possible adaptive response of the parasite to host immune response. T cell clones have been reported as having an extensive cross-reactivity to variant sequences (Moreno et al., 1993).

After the epitopes have been well characterized, certain mechanisms to improve the immune response, such as modifying directly the chemical structure of them, may be employed. Other possible methods include: chemical linkage via covalent bonds to a carrier (Romero, 1992), genetic fusion of epitopes, polymerisation of the molecule (as was the case with the SPf66 vaccine; see Radack et al., 1986), the use of deposit-based adjuvants or possible built-in adjuvants such as P3C used in MAPs (Huang et al., 1995).

The peptides have been shown to be good immunogens with different deposit-based adjuvants, but they are animal species specific. Several new adjuvants are currently being developed. The new compounds have been involved in different immunological activities such as enhancement of the uptake and presentation of antigens by antigen presenting cells, stimulation of lymphokine production and modulation of T helper and CTL response (Hui, 1994). However, the limitation of a new adjuvant formulation would be its side effects and high cost per dose.

The ideal adjuvant would have the efficacy of Freund's complete adjuvant but without its various side effects such as granuloma formation and carcinogenic activity. Different strategies exist to develop adjuvants such as liposomes, multimeric immunostimulant complexes (ISCOMs) as QS-21(Kensil et al., 1995) or squalane emulsions.

It is also necessary to address the question of the use of controlled release formulations for vaccine delivery, thus overcoming the need for multiple administration, and the future development of vaccines for oral administration.

4.2.3 Novel Malaria Vaccine Approaches

Given the unique basic characteristics of the immune response, it is understandable that the main goal is to try and mimic or increase the natural immunity. The best candidates for design of a malaria vaccine are the proteins required for parasite survival – with low mutation rates, conserved epitopes from multiple or alternative steps in the invasion process and they also have to be targets of protective immune response.

Only some of the thousands of proteins that are probably synthesised by the parasite have been cloned and characterized, and of these few have been tested in humans or monkey trials.

During invasion, a few host proteins, some soluble and others belonging to the merozoite membrane, interact with the erythrocyte (Hadley Terence et al., 1995).

Recombinant DNA cloning and sequencing techniques combined with serological studies have enabled epitopes to be mapped precisely to amino acid residues. Conformational and linear epitopes have been identified but it will be necessary to increase the immunogenicity in order to make them more suitable as vaccine antigens.

In order to select epitopes from proteins with demonstrated invasion properties to be included in a malaria synthetic vaccine, our approach has been to identify the ones with conserved structure which bind the RBC. This has been achieved by analysing the binding kinetics of radiolabelled synthetic peptides containing sequences derived from these proteins and their analogs. Aminoacids which serve as contact residues for a dominant number of antibodies within a polyclonal serum cannot be replaced without altering recognition (Molano et al., 1992). The immunogenicity of peptides have been tested in the Aotus monkey model. Several peptides have been identified, based on recognition by monkey sera and cross reactivity with native sequences.

Another issue to be resolved is the three-dimensional conformation of the peptide in solution. Synthetic peptides and recombinant proteins do not necessarily adopt their native conformation as in the intact molecule. The structural constraints present in the native molecule affect an amino acid sequences binding specificity and affinity with an antibody or receptor. Although a recent article reports the expression of antigens in their correct conformation (Murphy et al., 1994), this appears to be a rather isolated case, and the study of this particular expression vector merits study. Methods are currently available for conformational restricting peptides with covalent replacements for hydrogen bonds, thus locking the peptide into a configuration that mimics the epitope's native conformation (Satterthwait et al., 1989; Satterthwait et al., 1990).

The nature of the antigen may itself influence the profile of peptide regulatory factors (PRFs) such as cytokines, which are important in lymphocyte activation. The modification of antigen structure or presentation, direct incorporation of PRFs or inducers of PRF production into vaccine vehicles or incorporation of genes into live viral or bacterial vectors may all be exploited.

It is encouraging that Multiple antigen peptide system (MAPs) (Tam et al., 1990), molecular biology subunit vaccines, vaccinia constructs and DNA immunisation designs are being developed but it is still too early to be able to judge their efficacy (Ada, 1995).

4.2.4 Scenarios and Methods of Evaluation

During the last ten years, there has been dramatic progress in malaria vaccine development. In order to get to the humans trial stage one has to screen not only a range of vaccines but also a range of adjuvants and vaccine/adjuvant combinations. This demonstrates the importance of bench research and the use of animal models before proceeding to extremely expensive trials in humans.

SPf66 vaccine became the only vaccine identified from such trials and its development has involved many disciplines. Different methods for epidemiological evaluation were designed and updated in order to be useful under clinical and field conditions (Patarroyo et al., 1988; Alonso et al., 1994).

The use of randomised double-blind and placebo-controlled trials plays a central role in the evaluation of the protective efficacy and side effects of new malaria vaccines (WHO, 1989; Ballou, 1995; Alonso et al., 1994).

Before engaging in a costly phase III trials, it is necessary to conduct Phase I and II trials for evaluation of the safety, immunogenicity and trends in protective efficacy of a designated group under controlled conditions.

The most critical step in the development of new products involves the phase II trial. Safe and adequate methods for experimental challenge must be implemented. To test a vaccine with epitopes from different parasite stages, is necessary to set challenges through mosquito bites and/or intravenous infected RBCs.

When testing a Sporozoite vaccine, the availability of a *P. falciparum* strain able to grow under stringent in vitro conditions and to form gametocytes is necessary. Mosquitoes are infected by membrane feeding, and are expected to produce 3-5 ineffective bites per human volunteer with a 75 % of sporozoite charge. As an endpoint, the criteria for the stopping time is the appearance of clinical symptoms. At this moment the volunteers must be treated whatever the parasitaemia levels (Davis, 1995; Beier et al., 1991).

Merozoite vaccine candidates are tested by intravenous challenge with infected RBCs and the volunteers have to be followed very closely. The strain of *P. falciparum* has to be a low virulence native strain, able to produce moderate to high parasitaemias with a low clinical profile.

For merozoitic challenge the endpoints to measure under experimental conditions are the delay in time to infection, delay of pre-patent period and parasitaemia autocontrol. Theoretically, under natural conditions, a merozoite vaccine could prevent clinical illness without resolving the infection, or control parasite numbers after their release from the liver, thus preventing or reducing malaria related morbidity and mortality.

To date, the only reported human merozoite challenge have been performed by our group (Patarroyo et al., 1988). In this trial all immunized subjects became clinically ill, but the infection in three of five volunteers receiving one of the vaccines SPf66, was controlled without treatment, while all control subjects required treatment.

Several human trials to test safety, immunogenicity, protection and the formulation schedule of SPf66 were done with increasing number of volunteers (Amador et al., 1992a; Amador et al., 1992b; Patarroyo et al., 1992; Rocha et al., 1992).

In order to try and cover the whole clinical malaria spectrum under stable and unstable malaria epidemiological conditions, we performed the studies of SPf66 in Latin America and Tanzania. We conducted phase III trials in several Latin American locations which are characterized by low transmission conditions (Unstable malaria), few asymptomatic cases of parasite infection and composed of different ethnic groups. The results from these studies showed a protective efficacy ranging between 38.8 to 60.2 % against *P. falciparum* (Valero et al., 1993; Noya et al., 1993; Sempértegui et al., 1994).

To test the vaccine under conditions of intense malaria transmission (Stable malaria) a study in Tanzania was undertaken. The African study showed similar results as described previously (Teuscher et al., 1994).

The decrease in the frequency of episodes detected in children vaccinated with SPf66 in the Tanzanian trial, could result from a decrease in intensity of the symptoms rather than the number of attacks, thus preventing or reducing malaria related morbidity and mortality.

Infections with no symptoms occur at a high frequency in sub-Saharan Africa. For this reason, in the Tanzanian study, a case of malaria was defined as a parasitemia of 20,000 or more and axillary temperature of 37.5° or more. It is possible that the vaccine was able to reduce malaria mortality in a higher proportion than the 31 % recorded for the clinical episodes that were detected. These episodes could include most of the potentially severe cases (Trape and Rogier, 1995). Recently, a randomised trail in The Gambia with SPf66, in a group of infants aged 6-11 months was published (D'Alessandro et al., 1995). The vaccine had no significant protective effect against clinical malaria episodes, but is not yet conclusive, under the trial conditions, since the infants were followed only for three months. An analysis of the strong active case detection and a long follow-up period during the next rainy season will be necessary to increase our knowledge in this important risk group. The next step is to test the vaccine against new endpoints such as severe malaria in infants, the high risk group, under EPI programme conditions and other different epidemiological conditions as East Asia, where the drug resistance is very high.

The knowledge required in a given situation to move from controlled trials to routine evaluation of a vaccine will certainly be different from one geographical area to another considering the different strains, vectors, populations and magnitude of the problem.

Clinical trials with recombinant and synthetic vaccines against sporozoites were done on the late eighties (Pauker and Kassirer, 1987). The vaccines were safe, able to induce immune response but with low or no protection (Romero, 1992). Because of that an increase of research has been devoted to analyse relevant T and B cell epitopes and preclinicals trials with recombinant products and MAPs constructs have been shown to be promising for future developments (Tam et al., 1990).

In the sexual blood stage, the most promising antigen from the zygote surface antigen is Pfs25. The immune response raised against it, is able to partially inhibit, inside the mosquito, the development of the ookinete and the production of sporozoites impeding the transmission of the disease, a so-called transmission-blocking vaccine. Field trials are expected to begin soon.

4.2.5 Public Health Impact of the Malaria Vaccine Around the World

Parasite and insect vector control measures with insecticides and chloroquine, at first thought to be an effective remedy, later on turned to be ineffective due to the mosquito and parasite mutation and subsequent resistance development as well as social behaviour of the host. Constraints on resources have been the major obstacle for the implementation of a control program for malaria (WHO, 1993).

Vaccines against pre-erythrocitic stages are important in protection of special high risk groups as primigravid women and migratory people such as tourists and soldiers.

A vaccine to the erythrocytic stage, targeted to block the phase where the parasite acquires clinical significance, is the major goal for implementation in highly endemic areas. A cheap and effective malaria vaccine would have tremendous public health and economic consequences in these kind of situations. It would reduce childhood mortality in Africa, it would substantially reduce the costs presently spent on treatment, it would improve nutrition and could bring direct economic benefits by reducing absenteeism and by improving work efficiency.

The so-called transmission blocking vaccines have no individual prophylactic value but would aid in protecting populations living in endemic areas by preventing transmission of a malarial infection to the mosquito vector, thus breaking its life cycle. Indirectly, the other stage vaccines could decrease transmission too.

In Africa alone the estimated annual direct and indirect cost of malaria in 1987 was US $800 million. This figure is expected to rise to more than US $1,800 million by 1995. Despite financial limitations, a proposed goal for the control of malaria in

Africa under the Brazzaville-Congo plan for 1997 is a reduction of 20 % in the incidence of severe malaria and mortality in at least 32 of the 42 countries with endemic malaria, compared with 1995 baseline figures than still to be determined (World Bank, 1993).

Currently in Latin America the areas with greatest problems are the areas of transition of the economic development, such as agricultural and mining exploitations in jungles newly opened and in the zones where there is conflict, smuggling and colonisation. In Southeast Asia the main problem is the developing resistance to the most currently available drugs.

The enormous cost in lives and days of labour lost, added to the cost of treatment of the patient make malaria a considerable economic burden. The possible great socioeconomic saving associated with vaccine use underlines the need for greater emphasis on immunoprophylactics. TDR/ WHO has estimated if SPf66 were to lead 30 % reduction in general mortality as well as malaria morbidity, the cost per disability lifeyear saved (Daly) would amount around US $9. This would be a figure competitive with other control measures as bednets or chemoprophylaxis (WHO, 1995).

4.2.6 Conclusions

The basic research and the clinical / field stages of vaccine development is slow. A vaccine requires a minimum of ten years in the developmental process before being implemented in the field.

Although knowledge of the parasite's biology is incomplete, research has allowed insight into some of the mechanisms that the parasite uses to evade host immunity. These is the base to adopt an 'antigenic cocktail' approach to obtain a synthetic or recombinant subunit vaccine. The synthetic Colombian Malaria vaccine SPf66, is an example of this cocktail methodology.

A vaccine to the erythrocytic stage, targeted to block the phase where the parasite acquires clinical significance, is the major goal for implementation in highly endemic areas. A cheap and effective malaria vaccine would have tremendous public health and economic consequences in these kind of situations. With the offered donation of The SPf66 Colombian vaccine to WHO, to produce the bulk peptide at cost ensure that developing countries are able to acquire the vaccine cheap. The vaccine would reduce childhood mortality in Africa, it would substantially reduce the costs presently spent on treatment, it would improve nutrition and could bring direct economic benefits by reducing absenteeism and by improving work efficiency.

In the meantime, the stratification of areas on the basis of local microepidemiological pictures, the rationalisation of the use of different control measures and integration of Public Health Control Systems in the implementation of immunopreventive tools currently under field research will be necessary to overcome the problem of the malaria disease (Nájera et al., 1992).

Acknowledgements

Thank you to John Robinson, Ph.D. for help in preparation and translation of the manuscript.

References

Ada, G. (1995) Strategies for exploiting the immune system in the design of vaccines. Mol Immunol, *28*, 225-230.

Alonso, P. L., Tanner, M., Smith, T., Hayes, R. J., Schellenberg, J. R. M., Lopez, M. C., Bastos de Azevedo, I., Menendez, C., Lyimo, E., Weiss, N., Kilama, W. L., and Teuscher, T. (1994) A trial of the synthetic malaria vaccine SPf66 in Tanzania: rationale and design. Vaccine, *12*, 181-186.

Alonso, P. L., Teuscher, T., and Tanner, M. (1994) SPf66: Research for development or development of research? Vaccine, *12*, 99-101.

Amador, R., Moreno, A., Murillo, L. A., Sierra, O., Saavedra, D., Mora, A. L., Rocha, C. L., Alvarado, F., Falla, J. C., Orozco, M. A., Coronell, C. G., Ortega, N. M., Molano, A., Velásquez, J. F., Valero, M. V., Franco, L. C., Guzmán, F., Salazar, L. M., Espejo, F., Mora, E., Farfán, R., Zapata, N., Rosas, J. E., Calvo, J. C., Castro, J., Quiñonez, T., Nuñez, F., and Patarroyo, M. E. (1992) Safety and immunogenicity of the synthetic malaria vaccine SPf66 in a large field trial. J Infect Dis, *166*, 139-144.

Amador, R., Moreno, A., Valero, M. V., Murillo, L. A., Mora, A. L., Rojas, M., Rocha, C. L., Salcedo, M., Guzmán, F., Espejo, F., Nuñez, F., and Patarroyo, M. E. (1992) The first field trials of the chemically synthetized malaria vaccine SPf66: safety, immunogenicity and protectivity. Vaccine, *10*, 179-184.

Ballou, W. R. (1995) Clinical trials of plasmodium falciparum erythrocytic stage vaccines. Am J Trop Med Hyg, *50*, 59-65.

Beier, J. C., Davis, J. R., Vaughan, J. A., Noden, B. H., and Beier, M. S. (1991) Quantitation of *Plasmodium Falciparum* sporozoites transmitted in vitro by experimentally infected *anopheles gambie* and *anopheles stephensi*. Am J Trop Med Hyg, *44*, 564-570.

D'Alessandro, U., Leach, A., Drakely, C. J., Bennett, S., Olaleye, B. O., Fegan, G. W., Jawara, M., Lanrerock, P., George, M. O., Targett, G. A. T., and Greenwood, B. M. (1995) Efficacy trial of malaria vaccine SPf66 in Gambian infants. Lancet, *146*, 462-467.

Davis, J. R. (1995) Laboratory methods for the conduct of experimental malaria challenge of volunteers. Vaccine, *12*, 321-326.

Day, K. P. and Marsh, K. (1991) Naturally acquired immunity to *Plasmodium falciparum*. Immun Today, *12*, 68-71.

Edelman, R., Hoffman, S. L., Davis, J. R., Beier, M., Sztein, M. B., Losonsky, G., Herrington, D. A., Eddy, H. A., Hollingdale, M. R., Gordon, D. M., and Clyde, D. F. (1995) Long-term persistence of sterile immunity in a volunteer immunized with X-irradiated *Plasmodium falciparum* sporozoites. J Infect Dis, *168*, 1066-1070.

Egan, J. E., Hoffman, S. L., Haynes, D., Sadoff, J. C., Schneider Imogene, Grau, G., Hollingdale, M. R., Ballou W., Ripley, and Gordon, D. M. (1995) Humoral immune responses in volunteers immunized with irradiated *plasmodium falciparum* sporozoites. Am J Trop Med Hyg, *49*, 166-173.

Greenwood, B. M., Bradley, A. K., Greenwood, A. M., Byass, P., Jammeh, K., Marsh, K., Tulloch, S., Oldfield, F. S., and Hayes, R. (1987) Mortality and morbidity from malaria among children in a rural area of The Gambia, West Africa. Trans. R. Soc. Trop. Med. Hyg. *81*, 478-486.

Hadley Terence, J., Miller, L. H., and Haynes, J. D. (1995) Recognition of red cells by malaria parasites: the role of erythrocyte-binding proteins. Transfus Med Rev, *5*, 108-122.

Hollingdale, M. R., Aikawa, M., Atkinson, C. T., Ballou, W. R., Chen, G. X., Li, J., Meis, J. F., Sina, B., Wright, C., and Zhu, J. D. (1990) Non-CS pre-erythrocytic protective antigens. Immunology Letters, *25*, 71-76.

Hommel, M. (1993) Amplification of cytoadherence in cerebral malaria: towards a more rational explanation of disease pathophysiology. Ann Trop Med Parasitol, *87*, 627-635.

Houghten, R. (1995) General method for the rapid solid phase synthesis of large numbers of peptides: specificity of antigen antibody interaction at the level of individual aminoacids. Proc Natl Acad Sci USA, *82*, 5131-5135.

Huang, W., Nardelli, B., and Tam, J. (1995) Lipophilic multiple antigen peptide system for peptide immunogen and synthetic vaccine. Mol Immunol, *31*, 1191-1199.

Hui, G. S. N. (1994) Liposomes, muramyl dipeptide derivatives, and nontoxic lipid A derivatives as adjuvants for human malaria vaccines. Am. J. Trop. Med. Hyg. *50*, 41-51.

Hviid, L., Jakobsen, P. H., Abu-Zeid, Y. A., and Theander, T. G. (1995) T-cell response in malaria. APMIS, *100*, 95-106.

Kensil, C. R., Newman, M. J., Coughlin, R. T., Soltysik, S., Bedore, D., Recchia, J., Wu, J., and Marciani, D. J. (1995) The use of Stimulon adjuvant to boost vaccine response. Vacc Res, *2*, 273-281.

Kwiatkowski, D. (1992) Malaria:becoming more specific about non-specific immunity. Curr Opin Immunol, *4*, 425-431.

Lopez, M. C., Silva, Y., Thomas, M. C., Garcia, A., Faus, M. J., Alonso, P. L., Martínez, F., Del Real, G., and Alonso, C. (1994) Characterization of SPf(66)n: a chimeric molecule used as a malaria vaccine. Vaccine, *12*, 585-591.

Marsh, K. and Howard, R. J. (1986) Antigens induced on erythrocytes by P. falciparum: Expression of diverse and conserved determinants. Science, *231*, 150-153.

Merrifield, R. (1963) Solid-phase peptide synthesis. I. The synthesis of of a tetrapeptide. J Am Chem Soc , *85*, 2149-2154.

Molano, A., Segura, C., Guzman, F., Lozada, D., and Patarroyo, M. E. (1992) In human malaria protective antibodies are directed mainly against the Lys-Glu ion pair within the Lys-Glu-Lys motif of the synthetic vaccine SPf66. Parasite Immunol. *14*, 111-124.

Moreno, A. and Patarroyo, M. E. (1989) Development of an asexual blood stage malaria vaccine. Blood, *74*, 537-546.

Moreno, A., Clavijo, P., Edelman, R., Davis, J., Sztein, M., Sinigaglia, F., and Nardin, E. (1993) CD4+ T cell clones obtained from *Plasmodium falciparum* sporozoite-immunized volunteers recognize polymorphic sequences of the Circumsporozoite protein. J. Immunol. *151*, 489-499.

Murphy, V. F., Rowan, W. C., Page, M. J., and Holder, A. A. (1994) Expression of hybrid malaria antigens in insect cells and their engineering for correct folding and secretion. Parasitology, *100*, 177-183.

Nájera, J. A., Liese, B. H., and Hammer, J. (1992) Malaria. New patterns and Perspectives. 1-32.

Nardin, E. H. and Nussenzweig, R. S. (1993) T cell responses to pre-erythrocytic stages of malaria: role in protection and vaccine development against pre-erythrocytic stages. Annual Review of Immunology, *11*, 687-727.

Noya, O., Gabaldon, I., de Noya, B., Borges, R., Zerpa, N., Urbáez, J. D., Madonna, A., Garrido, E., Jimenez, M. A., Borges, R. E., García, P., Reyes, I., Prieto, W., Colmenares, C.,

Pabón, R., Barraez, T., de Caceres, L. G., Godoy, N., and Sifontes, R. (1993) A population-based clinical trial with the SPf66 synthetic *Plasmodium falciparum* malaria vaccine in Venezuela. J. Infect. Dis. *170*, 396-402.

Patarroyo, G., Franco, L. C., Amador, R., Murillo, L. A., Rocha, C. L., Rojas, M., and Patarroyo, M. E. (1992) Study of the safety and immunogenicity of the synthetic malaria SPf66 vaccine in children aged 1-14 years. Vaccine, *10*, 175-178.

Patarroyo, M. E., Amador, R., Clavijo, P., Moreno, A., Guzmán, F., Romero, P., Tascón, R., Franco, A., Murillo, L. A., Pontón, G., and Trujillo, G. (1988) A synthetic vaccine protects humans against challenge with asexual blood stages of Plasmodium falciparum malaria. Nature, *332*, 158-161.

Pauker, S. G. and Kassirer, J. P. (1987) Decision Analysis. N Engl J Med, *316*, 250-272.

Radack, K. L., Rouan, G., and Hedges, J. (1986) The likelihood ratio. An improved measure for reporting and evaluating diagnostic test results. Arch Pathol Lab Med, *110*, 689-693.

Rocha, C. L., Murillo, L. A., Mora, A. L., Rojas, M., Franco, L. C., Cote, J., Valero, M. V., Moreno, A., Amador, R., Nuñez, F., Coronell, C. G., and Patarroyo, M. E. (1992) Determination of the immunization schedule for field trials with the sybthetic malaria vaccine SPf66. Parasite Immunol, *14*, 95-109.

Romero, P. (1992) Malaria vaccines. Current Biol, *4*, 432-441.

Satterthwait, A. C., Arrhenius, T., Hagopian, R. A., Zavala, F., Nussenzweig, V., and Lerner, R. A. (1989) The conformational restriction of synthetic peptides, including a malaria peptide, for use as immunogens. Philosophical Transactions of the Royal Society of London – Series B: Biological Sciences, *323*, Biological Sciences 3-Biological Science 72.

Satterthwait, A. C., Chiang, L. C., Arrhenius, T., Cabezas, E., Zavala, F., Dyson, H. J., Wright, P. E., and Lerner, R. A. (1990) The conformational restriction of synthetic vaccines for malaria. Bull WHO, *68 Suppl*, 17-25.

Sempértegui, F., Estrella, B., Moscoso, J., Piedrahita C. L., Hernández, D., Gaybor, J., Naranjo, P., Mancero, O., Arias, S., Bernal, R., Córdova, M. E., Suarez, J., and Zicker, F. (1994) Safety, immunogenicity and protective effect of the SPf66 malaria synthetic vaccine against *Plasmodium falciparum* infection in randomized double-blind placebo-controlled field trial in an endemic area of Ecuador. Vaccine, *12*, 337-342.

Tam, J. P., Clavijo, P., Lu, Y. A., Nussenzweig, V., Nussenzweig, R., and Zavala, F. (1990) Incorporation of T and B epitopes of the circumsporozoite protein in a chemically defined synthetic vaccine against malaria. Journal of Experimental Medicine, *171*, 299-306.

Teuscher, T., Schellenberg, J. R. M., Bastos de Azevedo, I., Hurt, N., Smith, T., Hayes, R. J., Masanja, H., Silva, Y., Machin, D., Kitua, A., Kilama, W. L., Tanner, M., and Alonso, P. L. (1994) SPf66, a chemically synthesized subunit malaria vaccine, is safe and immunogenic in Tanzanians exposed to intense malaria transmission. Vaccine, *12*, 328-336.

Trape, J. and Rogier, C. (1995) Efficacy of SPf66 vaccine against *Plasmodium falciparum* malaria in children. Lancet, *345*, 134-135.

Valero, M. V., Amador, R., Galindo, C., Figueroa, J., Bello, M. S., Murillo, L. A., Mora, A. L., Patarroyo, G., Rocha, C. L., Rojas, M., Aponte, J. J., Sarmiento, L. E., Lozada, D. M., Coronell, C. G., Ortega, N. M., Rosas, J. E., Alonso, P. L., and Patarroyo, M. E. (1993) Vaccination with SPf66, a chemically synthesised vaccine, against *Plasmodium falciparum* malaria in Colombia. Lancet, *341*, 705-710.

WHO Division of Control of Tropical Diseases and World Bank The potential impact of research and development on malaria control activities. (1993) 1-20.

WHO Division of Control of Tropical Diseases Outline for WHO's plan of work for malaria control 1993-2000. (1992) 1-21.

WHO Malaria Action Programme Guidelines for the evaluation of *Plasmodium Falciparum* asexual blood-stage vaccines in populations exposed to natural infection. (1989) 1-34.

WHO Tropical disease research: progress 1975-94: highlights 1993-94: twelfth programme report of the UNDP/World Bank/ WHO Special Programme for Research and Training in Tropical Diseases (TDR). (1995) *1*, 57-76.

World Bank, WHO, and UNDP Strategic plan for applied field research in tropical diseases. Draft for discussion at the AFR planning meeting 26-28 May 1993, Geneva. (1993) (UnPub).

4.3 Vaccination Against Schistosomiasis: Concepts and Strategies

Gilles J. Riveau and André Capron

4.3.1 Introduction

Schistosomiasis, the second major parasite disease in the world after malaria affects at least 200 millions people, 500 millions being exposed to there risk of infection and is responsible for 300 to 500,000 deaths per year.

Morbidity observed in this chronic and debilitating disease is essentially related to the remarkable female worm fecundity, hundreds of eggs being laid every day and deposited in numerous mucous membranes and tissues. Granuloma formation around eggs, in particular in the liver, leads to the development of severe fibrotic and often irreversible lesions.

Although active drugs, such as praziquantel are available, evidence is now accumulating that, while they can reduce the overall incidence of severe forms of the disease, they do not prevent reinfection, have little effect on already developed hepatosplenic manifestations and do not significantly affect transmission.

Unlike protozoan parasites, schistosomes, as metazoans, do not replicate in their vertebrate hosts. It is agreed, on the basis of experimental and epidemiological studies, that a significant but partial reduction, estimated around 60 percent, of the worm burden following infection would considerably reduce pathology and affect parasite transmission.

Considered in terms of biological features the most relevant to immunology, the life cycle of schistosomes indeed offers a series of events that promote a close interaction between the parasite and the immune system: active cutaneous penetration of larvae and their transformation into schistosomula, complex migratory cycle in the vertebrate host, intravascular localisation and blood feeding habits of adult worms, egg deposition in mucosae, liver, and various tissues, and prolonged survival in immune vertebrate hosts. These prominent characters of a vertebrate infection have elicited numerous immunological research in three major areas: effector mechanisms of resistance; immune mechanisms of pathogenesis; mechanisms of parasite survival and regulation of host immune system. In each of these research areas, novel mechanisms have been uncovered, leading to more general concepts bearing implications far beyond the specific field of schistosomiasis. From rodent models to man, these studies have allowed promising approaches, from bench to bedside, towards the possible immunological control of one of the major human parasitic diseases, by pointing to which parasite antigen may induce protection and which component of the immune response should be elicited by the immunisation schedule.

4.3.2 Escape Mechanisms in Schistosomiasis

Survival and growth of infectious organisms in the face of the immune response mooted by their immunocompetent hosts is made possible through one of two strategies: one is to outrun rejection by fast replication and/or mutation, observed in some bacterial or viral infections; the other is to impair development or expression of immunity so as to evade the deleterious consequences of the host's counter-attack. Indeed, it is now recognised that in most infections, it is the infectious organism that determines whether or not to allow itself to be rejected by immune effectors.

This is clearly the case of parasites which in general induce chronic infections and are thus confronted for months and often years with the immune response that they have to maintain within limits compatible with both their own survival and that of their mammalian hosts. This leads in many instances to the refined elaboration in these organisms of a complex network of processes allowing them to escape the defence mechanisms of their host.

Indeed, one of the simplest ideas when considering the host-parasite relation and its stability in terms of evolution, population dynamics, and individual infection, is the likely existence of structures common to parasites and their hosts which would allow a precise adaptation of their respective metabolic requirements.

In the case of schistosomes, there are cross-reactivities between the various developmental stages of the parasite and antigenic components of its invertebrate host. Numerous parasite genes cloned so far show high levels of nucleotide sequence homology with mammalian genes, attributable to functional conservation since the corresponding proteins are potent immunogens. Besides, antigenic disguises appears to be employed by schistosomes during their life in their vertebrate host: indeed, within hours after their penetration through the skin they acquire a masking coat of host molecules, both glycolipids and glycoproteins. An interesting observation is acquisition by schistosomes of class I and class II major histocompatibility complex products (Sher et al., 1978). Although these acquired MHC products are recognisable on the surface of worms by alloantibodies or CD8+ T cells (Sher et al., 1978, Butterworth et al., 1979), they do not appear to play a role in promoting parasite recognition by T cells or in immunoregulation. It is generally assumed that these host antigens limit access of immune effectors to target antigens since their acquisition coincides with the capacity of transforming schistosomula to resist immune damage.

In fact, the contribution of antigenic disguise or antigenic mimicry in the reduction of surface recognition of parasites is difficult to appreciate. In schistosomes, the acquisition of so-called host antigens proceeds concomitantly with intrinsic changes in membrane susceptibility to antibody-dependent killing and with the loss of expression in the worm membrane of the major target antigens that are expressed on the schistosomulum surface. Sharing of epitopes between host and parasite may accordingly appear to be related more to adaptation to common metabolic environments than to a strategy of evasion from immune clearance.

Evidence has been given that invertebrates synthesise neurohormones and that functions of invertebrate immunocytes can be regulated through mechanisms very similar to the vertebrate neuroimmune interactions (Stefano et al., 1989). Considering the immunomodulatory properties of proopiomelanocortin (POMC)-derived peptides, and their phylogenetic conservation, we have postulated their existence in Schistosoma mansoni and their possible implication in the interactions with both intermediate and definitive hosts.

POMC-related peptides were identified in the main stages of the parasite life cycle (cercariae, schistosomula, adult worms and miracidia). Radio immunoassays detected molecules related to ACTH, αMSH and β endorphin in crude extracts of all stages (Duvaux-Miret et al., 1990; Duvaux-Miret and Capron, 1991). The presence of three main POMC-derived peptides led us to hypothesise the existence of a POMC gene in the parasite. Using oligonucleotide probes specific for the regions of the genes encoding POMC conserved amino-acid motifs, we could demonstrate that related sequences are present in the parasite genome (Duvaux-Miret and Capron, 1991).

The hypothesis of the implication of neuropepdide like substances was assessed by the ability of release by *S. mansoni* adult worms of ACTH and β-endorphin and physiological conditions (Duvaux-Miret et al., 1992).

The role of the parasite POMC-derived peptides immuno-suppression in the definitive host (man) and in the intermediate host (*biomphalaria glabrata* snail) has been examined. Coincubation of adult worms with human polymorphonuclear leukocytes or *B. glabrata* immunocytes led to the appearance of MSH in the medium. This α-MSH-resulted from conversion of parasite ACTH by a neutral endopeptidase (NEP) present on the surface of vertebrate PMN and invertebrate immunocytes (Shipp et al., 1990). MSH α and β were shown to inhibit activation of mammalian PMN and snail immunocytes *in vitro* (Duvaux-Miret et al., 1992). Interestingly the participation of POMC-derived factors in the inhibition of immunocytes obtained from *S. mansoni* infected snails was suggested by the reversal of inhibition in the presence of antibodies directed against ACTH or α MSH (Duvaux-Miret et al., 1992). These findings are highly significant because α MSH has been shown to inhibit adherence and locomotion of PMNs, monocytes and invertebrate immunocytes. ACTH can exert the same immunosuppressive cellular effects through conversion into α MSH by means of NEP (Shipp et al., 1990).

Another possibility of the parasite to acquire resistance is the development of receptor structures for molecules from the host currently used for cellular communication.

It has been recently shown that Tumor Necrosis Factor TNFα could positively influence schistosome worm fecundity (Amiri et al., 1992). While female worm presented a reduced fecundity when infection was performed in SCID mice, a normal level of egg laying was obtained in the same mice after treatment with TNFα. This result strongly favour the hypothesis regarding the use of this cytokine either as a stimulus for fertility or as a embryogenic growth factor. Though the demonstration of the existence of parasite receptors for TNFα has not been already proved, the expression by the parasite of molecular forms showing affinity for host factors able to regulate the reproduction cycle, is more than likely.

Consequently, the co-evolution of the parasite and the immune response of the hosts, and in particular the cytokine production, could be taken under consideration (Sher, 1992; Capron and Dessaint, 1992). Strengthening this hypothesis, the production of TNFα by the intermediate host of *S. mansoni* (*B.glabrata*) and the variation of its production after parasitic infection have recently shown (Ouwe-Missioukem-Boyer et al., 1994).

The discovery that animals and humans develop partial immunity to new schistosome infection, focused attention on these mechanisms which the schistosome deploys to make itself invisible to the host's immune system. As it has been described above, these mechanisms include the schistosome acquiring host-derived macromolecules on its outer membrane. Interestingly, one of the effect of chemotherapy with praziquantel is to blister the surface skin of the schistosome, thereby penetrating the shield that makes it invisible to the host, and rendering it vunerable to imune attack and destruction by the host effector cells.

4.3.3 Effector Mechanisms in Experimental Models to Immunity in Humans

Extensive studies have attempted to a detailed analysis of the humoral components and of the cellular partners involved in vitro killing of target schistosome larvae. Using the rat as a model system in parallel with cytotoxicity assays in humans and in primates, we identified novel ADCC mechanisms involving pro-inflammatory cell populations (macrophages, eosinophils, and platelets) as cellular partners and unusual antibody isotypes such as IgE or subclass of IgG with anaphylactic properties, like rat IgG2a in the particular case of eosinophils (Capron et al., 1984). These observations of IgE dependent cell mediated killing, which could be confirmed in human schistosomiasis and in primates (Joseph et al., 1978; Capron et al., 1984) remained however limited at this stage to *in vitro* experiments thus raising the problem of their *in vivo* relevance. Indeed the essential question brought by these *in vitro* ADCC mechanisms was related to the implication, so far unsuspected, of anaphylactic antibody isotypes and specially IgE in the mechanisms of protection against metazoan parasites. The production of monoclonal anti schistosome antibodies of the rat IgE and IgG2a isotype led to the demonstration of their high protective capacity by passive transfer (Grzych et al., 1982; Verwaerde et al., 1987). The relevance of these experimental finding to human immunity to schistosomes has been recently confirmed by immunoepidemiological studies which have brought convergent evidence for a protective role of IgE antibody in human infection. Studying the rate of reinfection after treatment in community exposed to *S. haematobium* infection in Gambia, Hagan et al. have demonstrated a positive correlation between the specific IgE antibody response to worm antigens and the acquired resistance to reinfection. Multiple regression analyses show in particular that the risk of reinfection is ten times more likely when IgE antibodies are absent or in the lowest quintile (Hagan et al., 1991). Similar findings have been made by Dunne et al. (1992) in a community exposed to *S. mansoni* infection as well as by Rihet et al. in Brazil (1991). Without excluding the possibility of the participation of additional mechanisms, specific IgE antibody response appears therefore as a strong correlate of protective immunity in humans, confirming the views we have expressed for many years regarding the unsuspected functions of this class of antibody (Capron and Dessaint, 1985; Capron et al., 1987).

Among the various mechanisms that regulate the expression of protective immunity, one stems from isotypic regulation itself. Evidence for the selective production of defined antibody classes during the course of experimental schistosome infection in rats raised questions about the functions of other isotypes shown not to be directly implied in killing pathways. The decrease in immunity observed at certain periods of the infection in rats is indeed not related to sharp decrease in antibody production but is concomitant with the appearance of non anaphylactic IgG subclasses. A representative IgG2c monoclonal antibody was shown to inhibit the capacity of an IgG2a monoclonal antibody both to induce eosinophil dependent killing of schistosomula and to confer passive protection *in vivo* (Grzych et al., 1984). The concept of blocking antibody was supported by the observation that this IgG2c monoclonal antibody can inhibit the recognition by the protective IgG2a monoclonal antibody of the carbohydrate moiety of the major surface glycoprotein of schistosomula described as gp 38 (Grzych et al., 1984).

The possibility that a similar phenomenon might be important in humans infected by *S. mansoni* was first indicated by the observation that susceptibility to reinfection after treatment of school children is significantly correlated with the presence of high levels of antibodies that inhibit the binding to the major gp 38 schistosomulum surface antigen of the protective monoclonal IgG2a antibody. In addition, IgM and IgG2 antibodies isolated from the sera of various individuals directly block the eosinophil-dependent killing of schistosomula mediated by IgG antibodies from the same sera. IgM antibodies with specificity for schistosomulum surface antigens are present in higher levels in the young susceptible children than in the order, resistant subjects (Khalife et al., 1986; Butterworth et al., 1987). More recently, analysing the isotypes of antibodies to a recombinant protective protein (28 kDa glutathione S-transferase; 28GST) from *S. mansoni*, and its derived synthetic peptides, we found a significant correlation between susceptibility to reinfection to *S. mansoni* in humans and increased production of IgG4 antibodies to Sm28GST and its defined B cell epitopes (Auriault et al., 1990).

The study of the immunological mechanism, underlying the inhibition of parasite fecundity and of egg viability has revealed the existence of unsuspected mechanism related to the neutralising activity of IgA antibodies. Recent studies performed in human population have revealed a close association between the production of IgA antibodies to Sm28GST, their neutralising activity of the enzymatic function of the molecule and the age dependent decrease in the egg output observed in human population in parallel with acquisition of immunity (Grzych et al., 1993). Monoclonal IgA antibodies have been raised in mice immunised with rSm28GST and their epitopic specificity has been determined. Passive transfer of IgA mAb recognising the central domain (AA115-131) of Sm28GST induced up to 90 % protection. Passive transfer of IgA mAb directed against the enzymatic site of the molecule (AA 10-43 or AA 190-211) induced a reduction in the number of tissue eggs (45 %) without effect on worm burden.

It also appears, both from the study of experimental models and of human populations that at least two distinct immunological mechanisms, for which the cellular components remain to be defined, may account for these two effects. For the first, IgE antibodies appear as a major humoral component of acquired resistance to reinfection whereas for the other, IgA antibodies appear as a major humoral factor affecting parasite fecundity and its pathological consequences. As it could be expected immunity to such complex organisms as schistosomes is obviously multifactorial in nature and there is no *a priori* reason to think that successful immunisation against this pathogen can be achieved though the elicitation of a single effector mechanism. Moreover, in a chronic transmissible disease such schistosomiasis, whereas the antibody profile of potentially protective antibodies might not differ significantly in immune and non-immune populations, susceptibility seems to be controlled by the development of a blocking antibody response and thus it might be crucial to characterise markers of susceptibility rather than putative indicators of protection. Besides, the general concept that a given antigen might selectively be involved in the production of a defined effector isotype and that the same molecule may elicit both effector and blocking isotypes, is obviously of considerable importance in the framework of vaccine strategy.

4.3.4 Target Antigens of Immunity

It therefore appears that in terms of vaccine strategy against schistosomiasis immunisation with specific antigen, might achieve two complementary goals in human populations:

a) a partial but significant reduction of the worm population resulting from infection or reinfection;
b) a significant reduction of pathological consequences by a marked decrease in parasite fecundity and egg viability, this effect affecting directly transmission potentialities of the disease.

The various vaccine candidates could be grouped in various different ways based on the approaches leading to their identification, their biochemical natures or their localisation. We have grouped the main vaccine candidates in two categories; surface membrane antigens and excreted-secreted antigens. This grouping does give some indication of the conceptual approach to their characterisation.

4.3.4.1 Surface Antigens

The approaches used to characterise vaccine antigens toward schistosomes have started with the premise that the antigens presented at the surface of the young larvae would be the best candidates. They were, presumably, targets of the immune mechanisms that were demonstrably only effective against the schistosomula immediately after skin penetration.

Newly-transformed schistosomula present a complex repertoire of molecules (Dissous et al., 1981) among which a 38 kDa glycoprotein (gp38) is immunodominant. A rat IgG2a monoclonal antibody (mAb), specific for gp38, mediated both eosinophil-dependent cytotoxicity toward schistosomula and protection of rats by passive transfer (Grzych et al., 1982). Subsequent studies showed that the epitope involved in the protective response was glycannic (Dissous, et al., 1985) and generated blocking antibody responses (Grzych et al., 1984). This has limited interest in the gp38 molecule as a vaccine candidate.

However, the immunodominant epitope was found to be shared with both the intermediate snail host and a variety of fresh water and marine molluscs, including the keyhole limpet *Megathura crenulata* (Dissous et al., 1986). Moreover, immunisation with Keyhole limpet hemocyanin (KLH) protected rats against a challenge infection (Grzych et al., 1987). Interest in KLH as a possible vaccine component has been rekindled by the finding that it can mediate limited but significant anti-fecundity immunity in cattle (Bushara et al., 1993).

4.3.4.2 Antigens Identified Using Irradiated Cercariae Vaccine Model

Vaccination with radiation-attenuated cercariae elicits a high level of resistance to a challenge infection (Richter et al., 1993). As we have already discussed, a single exposure induces significant levels of protection, and this is dependent on a Thl response involving IFN-γ production (Smythies et al., 1992). Multiple exposure of mice to irradiated cercariae only marginally increases the level of resistance conferred, but different mechanisms are involved since the resistance can be passively transferred using serum from immunised mice (Mangold and Dean, 1986).

A subset of schistosomulum surface antigens is recognised by sera from twice-vaccinated mice (Dalton and Strand, 1987) indicating that they may be targets of antibody-mediated protection in this model. A polyclonal antiserum was prepared that recognised five schistosomula and seven adult worm glycoproteins that were the targets of a unique or enhanced antibody response in the twice-vaccinated mice. cDNA clones encoding two of these molecules have been isolated. One, encodes the protein designated IrV-1 (Hawn et al., 1993) which shows significant sequence identity to both mammalian calnexin and calreticulin, although it is distinct from *S. mansoni* calreticulin (Khalife et al., 1993).

The second cDNA clone encodes a 62 kDa fragment of a 200 kDa protein designated as IrV-5, with extensive sequence identity to mammalian β-myosin heavy chain (Amory-Soisson et al., 1992) and the expressed recombinant protein has been shown to be highly protective in mice. Given the degree of sequence identity (about 50 % overall) with the β-myosin heavy chain of mammals, the reported schistosomular surface location for the protein might be considered surprising. However, the presence of epitopes of the native 200 kDa antigen are unequivocally demonstrated on the surface of live, intact, untreated schistosomula.

4.3.4.3 Integral Membrane Proteins of Schistosomes

Two integral membrane proteins present at both larval and adult stages of schistosomes have been characterised and cloned. At present, their vaccine potential has not been demonstrated. The first of these proteins was initially identified on the surface of lung schistosomula as a 22 kDa protein (Harn et al., 1985). Given the interest in immunity acting at the lung stage in the vaccinated mouse model and the dearth of characterised lung-stage antigens, its potential as vaccine candidate was evident. Following this work, (Dalton et al., 1987) reported the cloning of a cDNA encoding a 18kDa surface protein of schistosomula. (Wright et al., 1990) then characterised the full-length sequence of Sm23, which turned out to be identical to partial amino acid or nucleotide sequences of the 22kDa and 18kDa proteins.

Sm23 and the homologue from *S. japonicum* are members of a superfamily of membrane proteins, the function of which is unknown but which includes a certain number of proteins found on the surface of activated hemopoietic cells including CD9

(Boucheix et al., 1991), CD37 (Classon et al., 1989) and CD53 (Amiot, 1990). It has been suggested that the fact that schistosomes that live in permanent contact with blood cells express on their surface a member of this protein family may not be due to mere coincidence (Reynolds et al., 1992).

The predicted structure of Sm23 suggest the presence of four transmembrane domains and one small and one large hydrophilic extracellular domain (Reynolds et al., 1992). Both the predicted external domains were found to be highly immunogenic and to contain several B cell epitopes. In addition, T cell epitopes were present in the large extracellular domain, one 23 amino acid segment containing both a B and a T cell epitope recognised by all mouse strains tested. Studies on the protective capacity of these Sm23 epitopes against an experimental infection have not yet been reported, however.

Another glycoprotein surface antigen of a similar size (22kDa) has also been characterised independently in several laboratories. Evidence that a protein in this size range may be vaccine candidate came from work contrasting the group of polypeptides recognised by antibodies in serum from twice-infected Fischer rats which conferred protection by passive transfer, to those proteins detected by non-protective sera from twice-infected Wistar-Furth rats (Barker et al., 1985). Differential screening of a cDNA library using the two types of sera led to the characterisation of a partial cDNA clone (El Sherbeini et al., 1990) and a genomic clone covering the entire coding sequence (El Sherbeini et al., 1991).

The predicted amino acid sequence is identical to that of a protein designated as Sm25 (Knight et al., 1989; Omer Ali et al., 1991) and the protein may also be identical to those described by two further groups (Karcz et al., 1988; Rogers et al., 1988). Two promoter sites are utilised on the gene and there are at least three transcription products. All the resulting polypeptides have a hydrophobic trans-membrane segment near the carboxy-terminus and a consensus sequence for thio-esterification by a fatty acid. There is an external domain containing two potential N-glycosylation sites and a leucine-zipper motif, suggesting that the protein may exist as a dimer on the worm surface. As is the case for Sm23, the glycoprotein Sm25 or gp22 as it is variously called is present on all developmental stages and is the main component of adult worm membranes that induce a protective immune response in mice when saponin is used as the adjuvant (Smithers et al., 1989). It is not the only one, however, since another 22 kDa membrane component that is a possible target of this response has been characterised and cloned (Jeffs et al., 1991) and turns out to be identical to a previously described schistosomulum antigen (Stein and David, 1986).

4.3.4.4 Enzymes of the Glycolytic Pathway

Two proteins which would seem unlikely components of surface membrane of schistosome larvae, but which have both been convincingly demonstrated to be present nonetheless, are two enzymes that are components of the Glycolytic pathway, namely glyceraldehyde-3-phosphate dehydrogenase (G3PDH) and triose phosphate isomerase (TPI).

TPI was first identified using a mAb that recognised a 28 kDa antigen found at all schistosome developmental stages (Harn, 1985). This mAb was partially protective against a challenge infection by passive transfer into naive mice. Direct immunisation with immune complexes in which the 28 kDa antigen was bound to the corresponding mAb induced a 38 % reduction in worm burden after challenge (Harn et al., 1987). The M.1 mAb was next used to purify the 28 kDa antigen by immunoaffinity chromatography and tryptic peptides derived from the pure protein were subjected to N-terminal sequencing (Harn et al., 1992). The sequences obtained showed striking homology to mammalian and microbial TPI. A cDNA clone comprising the entire coding sequence has been obtained (Shoemaker et al., 1992) and the expressed recombinant protein processes TPI enzymatic activity.

The glycolytic enzyme, G3PDH, was identified as a 37 kDa antigen associated with resistance to schistosomiasis in humans. An initial study of a Brazilian population infected with *S. mansoni* (Dessein et al., 1988) showed that infection susceptibility of certain individuals who rapidly become reinfected after treatment and develop heavy infections could be related to the failure of IgG antibodies in their sera to recognise a 37 kDa antigen.

This 37 kDa antigen has been identified and cDNA clones obtained (Goudot-Crozel et al., 1989). The predicted amino acid sequence shows more than 70 % identity to chicken an human G3PDH and the residues demonstrated to be involved in substrate and cofactor binding are conserved. Antisera raised against the expressed recombinant protein recognised the surface of living schistosomula.

4.3.4.5 Excreted-Secreted Antigens

The interest in molecules released into the culture medium by newly-transformed schistosomula (which we called schistosomula released products-antigens: SRP-A) was stimulated by the demonstration that these molecules could induce a protective IgE response in the rat and in primates without the need for an adjuvant (Auriault et al., 1984; Damonneville et al., 1986). Among the antigens recognised by anti-SRP-A sera were a 26 kDa glycoprotein that is a specific target of IgE antibodies and was also recognised by a protective rat IgE monoclonal antibody (Verwaerde et al., 1987) and a protein of 29 kDa detected in the *in vitro* translation products of adult worm mRNA and was associated with schistosomulum surface membranes (Pierce et al., 1985). Although some components of SRP-A were surface-derived molecules, this is not a requirement for them to be targets of immune mechanisms. Neither ADCC mechanisms involving macrophages or platelets as the effectors cells, or T cell based immunity in which the effector arm consists of cytokine-induced inflammatory cells require cell/target contact for killing to occur.

Among the antigens identified in the excreted-secreted products are two established vaccine candidates, paramyosin and the 28 kDa glutathione *S*-transferase (Sm28GST).

4.3.4.6 Paramyosin

A target of cell-mediated immunity in mice which is not present on the surface of schistosomula was characterised using a vaccine model in which partial protection is engendered by intradermal immunisation with soluble worm antigens using bacillus Calmette-Guerin (BCG) as adjuvant (James, 1985). Resistance to infection in this model results from the stimulation of specific T cell dependent cell-mediated immunity, and more particularly the production of lymphokines including IFN-γ that activate macrophages for larval schistosome killing (James et al., 1988). An unusually monospecific antibody response is generated by this method of immunisation that is directed against a 97 kDa molecule (Pearce et al., 1986) subsequently cloned and identified as the myofibrillar protein paramyosin (Lanar et al., 1986). Immunisation with recombinant paramyosin produces partial protection (39 %) in the mouse, which suggested that it could be a candidate for inclusion in a multi-component vaccine (Pearce et al., 1988). That paramyosin is secreted by schistosomula was shown by the fact that live schistosomula stimulate primed T cells to produce IFN-γ in *in vitro* cultures. Subsequent work (Matsumoto et al., 1988) showed that paramyosin was present in adult worm tegument in a non-filamentous form, in membrane-bound elongate bodies.

4.3.4.7 Glutathione S-Transferases

The best characterised and most tested vaccine candidate antigen is the 28 kDa glutathione S-transferase of *S. mansoni* (Sm28GST). The protective properties of this molecule have stimulated work both on the GSTs of other schistosome species and on other enzymes involved in detoxification and antioxidant pathways.

Schistosome GSTs were first identified as potential targets of protective immunity by two separate approaches. The identification of a 28 kDa fraction of *S. mansoni* adult worm soluble antigens that elicited a protective antibody response and conferred a high degree of protection to both rats and mice by direct immunisation (Balloul et al., 1985; 1987a) was followed by the optention of a cDNA clone encoding the major antigenic component of the fraction (Balloul et al., 1987b) which was identified as a GST (Taylor et al., 1988). At the same time, a 26 kDa *S. japonicum* GST was identified as a specific target of antibodies in the sera of WEHI 129/J mice that were naturally resistant to infection (Smith et al., 1986). Subsequent work with this antigen, however, has shown that inconsistent levels of protection are obtained in various mouse strains (Mitchell et al. 1988). The equivalent molecule from *S. mansoni* has also been cloned (Trottein et al., 1990; Henkle et al., 1990) and gives significant but modest levels of protection against a challenge infection in rats (Trottein et al., 1992a).

Sm28GST was expressed in both *Escherichia coli* and in yeast in an enzymatically active form that allows its purification by affinity chromatography on glutathione-agarose (Taylor et al., 1988). The recombinant protein elicits protective immunity in rats, hamsters (Balloul et al., 1987b; Grezel et al., 1993), baboons (Balloul et al., 1987c) and in mice (Boulanger et al., 1991; Xu et al., 1993; Pancré et al., 1994). The level of protection conferred against infection varies from a mean of 50-70 % in the rat to 30-45 % in the mouse and 38 % in the baboon. The latter model underlined one of the problems that may occur in the use of single recombinant antigens in outbred populations in that a wide individual variation in response was obtained. The individual level of protection conferred ranged from 0 to 80 %.

Thus far in this review all discussion of protection against schistosome infection has concerned a reduction in the numbers of parasites recovered from immunised animals compared to non-immunised controls. The experiments in baboons using recombinant Sm28GST highlighted a different protective effect of immunisation, which until recently had not been demonstrated with any other antigen. This effect was a reduction in the fecundity of female worms reflected in a reduced excretion of eggs by immunised animals, and a reduction in tissue egg loads (Boulanger et al., 1991). These results have been confirmed in mice and more recently in cattle infected with *S. bovis* (Bushara et al., 1993). In the latter case the homologous native GST fraction (of which more than 95 % corresponds to the 28 kDa form) purified by affinity chromatography was used immunize calves that were subsequently infected with 15000 *S. bovis* cercariae. Mean faecal egg counts (in two separate experiments) between weeks 6-10 post-challenge were reduced by 56 to 82 % compared to controls. In the second experiment tissue egg densities in both the liver and intestine were significantly reduced (57 % and 80 % respectively). Importantly, these reductions in egg levels were not mirrored by a significant reduction in worm numbers and were thus the consequence of a direct effect on fecundity.

The *S. bovis* 28 kDa GST has been cloned (Trottein et al., 1992a), produced in yeast and the recombinant protein is being tested in cattle.

Working from the hypothesis that part of the protective effect of immunisation with Sm28GST might be due to the inhibition of the GST enzymatic activity which is vital to the survival of the parasite in the definitive host (Taylor et al., 1988; Mitchell, 1989) a mAb specific for Sm28GST that inhibited its enzyme activity *in vitro* was used to determine an eventual link between protective immunity and GST activity. This mAb S13 reduced parasite burdens after passive transfer in mice. When S13 was passively transferred at day 15 or 30 post-infection, no effect on worm numbers was produced, but a marked reduction in numbers of tissue eggs was noted, strongly suggesting a link between inhibition of GST activity and anti-fecundity immunity (Xu et al., 1991). A further effect of the S13 mAb was to diminish the viability of the eggs that were produced, both in terms of the numbers of mature eggs present, and the ability of the mature eggs to hatch in an in vitro assay. These effects were also seen after immunisation with Sm28GST.

The S13 mAb was found to interact with peptides derived from both the amino- and carboxy-terminal domains of Sm28GST and more particularly the latter (Xu et al., 1993). This reactivity of the S13 mAb can be explained by the spatial proximity of the amino-and carboxy-terminal domains (Trottein et al., 1992b). The use of the carboxy-terminal peptide (residues 190-211 of Sm28GST) conjugated to ovalbumin as an immunogen led to the production of antibodies that inhibited the enzymatic activity of Sm28GST immunisation of mice with the conjugate protected them against a challenge infection by partially reducing the worm burden compared to controls, but more particularly by markedly reducing tissue egg numbers and viability. The relevance of these observations to human infections is suggested by the capacity of IgA antibodies from infection sera to impair the fecundity of schistosome adult worm pairs maintained in culture (Grzych et al., 1993).

The carboxy-terminal peptide 190-211 corresponds to a major epitope in humans since it stimulates the proliferation of peripheral blood lymphocytes from a high proportion of infected individuals (Wolowczuck, I., personnal communication). Among the other important epitopes of Sm28GST, initially defined in animal models (Auriault et al., 1988; Wolowczuck et al., 1989; 1991) the peptide comprising residues 115-131 is worthy of a particular mention since a multiple antigenic peptide comprising eight copies of the peptide on a lysine core ("octopus") induces protective immunity in rats (Wolowczuck et al., 1991).

The data reviewed so far emphasise the role of the humoral response in anti-fecundity. However, this is not the whole story since T cell responses to Sm28GST can also be important in determining reductions in the numbers of hepatic eggs and hepatic pathology as measured by the collagen content of livers from infected mice (Pancré et al., 1994). In this model, mice immunised with a single dose of recombinant Sm28GST developed CD4+ Th1 and CD8+ antigen-specific responses and the protective effects could be completely abrogated by the administration of anti-IFN-γ antibodies. As well as showing that specific antibodies are not necessary for anti-fecundity immunity, this study underlines the importance of the animal model and the dosing regime in determining the type of response generated.

4.3.5 Conceptual Basis of a Vaccine Strategy

Considered as a whole, these observations and the confrontation of studies in experimental models, particularly the rat, and in infected humans point to three major messages.

The first relates to the selection of potentially protective antigens. A major emphasis has been given in the past to the identification in various parasites of stage-specific antigens, but as far as schistosomiasis is concerned, increasing evidence arising from experimental studies in laboratory animals or epidemiological surveys in human populations supports the original views of Smithers and Terry and their description of concomitant immunity (Smithers and Terry, 1969). The molecular basis of concomitant immunity obviously implies that identical or cross-reactive epitopes are expressed at various stages of the parasite life cycle. Common antigens or common epitopes on both the adult stage of schistosomes, as the major stimulus of the protective immune response, and on invading schistosomula as the privileged targets of its expression appear thus as essential components for the induction of an optimal protection. In this respect, the selection of antigens or crossreactive epitopes common to several stages, rather than stage-specific molecules, has to be favoured (Capron and Dessaint, 1988).

The second message concerns the respective roles played by surface versus excretory-secretory molecules in the induction of protective immune response. Although the efficiencies of effector mechanisms, either cellular or humoral or both, implies the accessibility of surface-associated molecules, considered in terms of their immunogenicity in experimental infection or in natural human infection, schistosomula integral membrane proteins do not necessarily represent the optimal vaccine candidates, contrary to excreted-secreted proteins. It is striking in this respect that by using antibody probes from immunised or infected animals and humans, the majority of over thirty different schistosome genes so far cloned, encode proteins which are either enzymes or non surface structural proteins which are transiently expressed on the worm surface.

Finally, the concept of isotype selection which we have developed since 1976 appears as an essential factor for the expression of optimal resistance to reinfection both in experimental models and in humans. As mentioned above, the dynamic balance in chronic schistosomiasis between effector and blocking antibody isotypes against the same molecule appears to account for the variable expression of immunity. Among effector antibody isotypes, IgE and IgA appear to act on different functional targets of the parasite. The optimal induction through immunisation of the appropriate antibody isotype is therefore of primary importance to the success of any vaccine strategy. Our approach towards this goal is however presently hampered by our limited knowledge of the basic immunological mechanisms driving isotype selection, and by empirical criteria in the choice of appropriate antigen delivery system.

However, in terms of vaccine candidate for a schistosomiasis vaccine strategy, it therefore appears that immunisation with Sm28GST might achieve two complementary goals in human population: 1) a partial but significant reduction of the worm population resulting from infection or reinfection, and 2) a significant reduction of pathological consequences by a marked decrease in parasite fecundity and egg viability, which directly affects the transmission potentialities of the disease.

It also appears, both from the study of experimental models and of human populations, that at least two distinct immunological mechanisms, related to Th2 immune response, may account for these two effects. For the first, IgE antibodies seem to be a major humoral component of acquired resistance to reinfection whereas the other,

IgA antibodies seem to be a major humoral factor affecting parasite fecundity and its pathological consequences. As could be expected, immunity to such complex organisms as schistosomes is obviously multifactorial in nature and there is no a priori reason to think that successful immunisation against this pathogen can be achieved through the elicitation of a single effector mechanism.

To conclude the conceptual basis of vaccine srategy part and in general terms of efficacy of new generation vaccines, the successful modern vaccine development is based on an understanding of the mechanisms of pathogenesis and the immune response in humans and on adequate animal models of response to the infectious organism. We expect a vaccine to have a good efficacy in terms of appropiate immune response with long-lasting protection, a good adaptation to the epidemiological situation in afflicted areas, a good immunogenicity in all individuals, an easy and appropriate delivery systems, no early or delayed side-effects, no disease-potentiating effects, and economic acceptability.

Obviously, there is a general need to bring more basic sciences into vaccine development, for instance to rationalise the entirely empirical immunisation schedules, or to achieve optimal induction of a specialised immune response (i.e. Th2 response). For this reason, the study of vaccine efficacy performed on animal models in the natural context host/related parasite should bring us the preliminary basis for the conception of an adapted vaccine against the urinary or intestinal human schistosomiasis.

4.3.6 New Approaches for Effective Vaccine Against Schistosomiasis

The relevance of our observations to immunisation strategy has been confirmed by vaccination experiments performed in Sudan against cattle schistosomiasis due to *S. bovis*. This model was chosen because of many common features with human infection by *S. mansoni*.

In collaboration with A. Bushara and M. Taylor (Boushara et al., 1993), we could show that immunisation of calves with *S. bovis* 28GST results in a dramatic reduction in egg production and tissue egg count (over 80 %) and acquisition of resistance to a lethal infection. These results confirmed in a natural host that the major effect of immunisation with schistosome GST is to very significantly reduce parasite fecundity. They also open feasible perspectives for a veterinary vaccine against schistosomiasis in a near future and allow to consider the acceptability of Phase I trials in human populations. In more recent experiments, immunisation of goats with the recombinant GST from *S. bovis* (rSb28GST) in presence of complete Freund adjuvant resulted in a significant reduction in the worm burden (48 %). Individual levels of protection could reach up to 80%. Unlike the controls, the immunised animals did not experience weight loss during the acute phase of infection. However, schistosome fecundity was not affected by that protocol (Boulanger et al., 1995). These results while confirming the vaccine potential of the *S. bovis* recombinant GST against veterinary schistosomiasis also clearly illustrate the striking differences in the result of immunisation between distinct animal species.

At the same time, the molecular cloning and the full sequence of the chromosomal gene of Sm28GST (McNair et al., 1993), together with the crystallisation of the Sm28GST (Trottein et al., 1992b) allows the study of its 3D molecular structure and new approaches toward molecular design of an optimal vaccine.

The recent cloning in our laboratory of the GST from *S. haematobium* and *S. bovis* (Trottein et al., 1992a) has allowed a comparative analysis of the GST sequences of the various schistosome species. Although overall predicted aminoacid sequences identities are above 90 % between the species, crucial differences exist compared to defined Sm28GST epitopes. Notably, a single base change determining an aminoacid substitution, leads to the non-recognition of Sh28GST by antibodies directed against the protective B epitope 115-131 of the *S. mansoni* protein.

In contrast, the demonstration of the high degree of conservation of the C terminal domain among the various species of schistosomes (Trottein et al., 1992a) opens the possibility that a cross-specific anti-fecundity vaccine might be achieved.

Recent experiments, in *P. patas* monkeys, appear to support this hypothesis (Boulanger et al., 1995). Indeed, whilst heterologous immunisation of these monkeys with the recombinant Sm28GST did reduce the number of worms after experimental infection with a human strain of *S. haematobium*, urine and faecal egg excretions were significantly reduced during one year of observation respectively by 46 and 53 per cent.

The demonstration of the so far unsuspected function of IgA antibodies in schistosomiasis and their potential role in protective immunity has paved the way to new possibilities of immunisation strategies through mucosal routes. Several approaches have now been designed.

Assessment of muramyl dipeptide (MDP) to enhance specific secretory immune response has been undertaken. Administration of this immunostimulant product by oral route during systemic immunisation with Sm28GST enhanced the production of specific secretory IgA. These antibodies were shown to be effective in inhibiting enzymatic activity of the recombinant protein. In these experiments, anti-Sm28GST secretory IgA production was correlated with the protective activity (60 %) against challenge infection in the mouse model.

New adapted formulations of synthetic vectors for mucosal vaccination including liposomes have been investigated for their capacity to optimise Th2 immune response against the recombinant Sm28GST. Experiments using oral administration indicated that Sm28GST-entrapped liposomes was able to elicit a strong specific mucosal IgA response and a Th2-like type response in serum. This immune response in the mouse was associated with a significant reduction in the worm burden (53 %) and tissue egg count (50 %) (Ivanoff et al., in press).

Other possibilities for the development of the new mucosal vaccines appear to rely in modified or attenuated live vectors which can be used to express genes encoding the candidate vaccine protein. A first model has been developed in collaboration with the laboratory of C. Hormaeche at Newcastle University (UK).

Recombinant Aro-A *Salmonella typhimurium* expressing the Sm28GST as a fusion protein with the C fragment of tetanus toxin (Tet-C) has been used to immunise mice after a single administration by either i.v. or oral route. Both procedures gave an equally strong Th1-type humoral response (Kahn et al., 1994a; Khan et al., 1994b). A marked secretory IgA response to Sm28GST was detected after oral administration and a significant reduction in the worm burden was observed.

The second model developed in our laboratory in collaboration with C. Locht is the construction of a recombinant *Bordetella pertussis* strain in which the cDNA encoding for Sm28GST has been cloned into the *fhaB* region corresponding to the immunodominant C-terminal portion of the mature filamentous hemagglutinin FHA (Renauld et al., submitted). Nasal immunisation with this recombinant *B. pertussis* strain has very recently shown an high protective capacity in terms of reduction in worm burden and worm fecundity (Mielcarek et al., submitted).

Taken altogether, these preliminary results indicate at least the feasibility to a novel and attractive approach towards oral or nasal immunisation against a systemic parasitic disease.

From experimental models to human populations, from the bench to endemic areas, studies performed during the last 15 years have revealed at the level of effector mechanisms of immunity, immunoregulation, and pathogenesis, the interest of which extend far beyond the field of schistosomiasis (Capron and Dessaint 1992; Capron, 1992). These studies seem presently to represent a promising approach towards the possible immunological control of one of the major human parasitic disease through the identification not only of potentially protective antigens and potent vectors, but also of the components of the immune response which vaccination should aim at inducing.

4.3.7 Concluding Remarks

The prospects for an effective vaccine against schistosomiasis are very real. The recombinant 28GST antigen is a candidate molecule for such a vaccine, especially for its capacity to induce an antifecundity effect. Although high levels of protection can be induced by single doses of 28GST, it remains likely that MHC restriction of the immune response in outbred populations will necessitate either a manipulation of the immune response by selective epitopic presentation, or the incorporation of other protective antigens in the definitive vaccine. Several powerful candidates for a vaccine cocktail are on the list, such as paramyosin, TPI and Sm23. However, further work remains to be accomplished, notably on the effects of vaccinated children previously exposed to schistosome antigens *in utero*, and on the possibility of vaccinating individuals after chemotherapy to eliminate an ongoing infection. Indeed, such a combination of chemotherapy with vaccine cocktail strategy could well be the main hope for control of schistosomiasis.

References

Amiot, M. (1990) Identification and analysis of cDNA clone endocing CD53. A pan-leukocyte antigen related to membrane transport proteins. J. Immunol. *145*, 4322-4325.

Amiri, P., Locksley, R. M., Parslow, T. G., Sadick, M., Rector, E., Ritter, D., and Mc Kerrow, J. H. (1992) Tumor necrosis factor α restores granulomas and induces parasite egg-laying in schistosome-infected SCID mice. Nature *356*, 604-607.

Amory-Soisson, L. M., Masterson, C., Tom, T. D., McNally, M. T., Lowell, G. H., and Strand, M. (1992) Induction of protective immunity in mice using a 62-kDa recombinant fragment of a *Schistosoma mansoni* surface antigen. J. Immunol. *149*, 3612-3620.

Auriault, C,. Dammonneville, M., Verwaerde, C., Pierce, R. J., Joseph, M., Capron, M., and Capron, A. (1984) Rat IgE directed against schistosomula released products (SRP-A) is cytotoxic for *Schistosoma mansoni* shistosomula *in vitro*. Eur. J. Immunol. *14*, 132-138.

Auriault, C., Gras-Masse, H., Pierce, R. J., Butterworth, A. E., Wolowczuk, I., Capron, M., Ouma, J. H., Balloul, J. M., Khalife, J., Neyrinck, J.-L., Tartar, A., Koech, D., and Capron, A. (1990) Antibody response of *schistosoma mansoni* infected human subjects to the recombinant P28 glutathione *S*-transferase and to synthetic peptides. J. Clin. Microbiol. *28* , 1918-1924.

Auriault, C., Gras-Masse, H., Wolowczuk, I., Pierce, R. J., Balloul, J. M., Neyrinck, J. L., Drobecq, H., Tartar, A., and Capron, A. (1988) Analysis of the T and B cell epitopes of the *Schistosoma mansoni* P28 antigen in the rat model using synthetic peptides. J. Immunol. *141*, 1687-1694.

Balloul, J. M., Pierce, R. J., Grzych, J. M., and Capron, A. (1985) In vitro synthesis of a 28 kDa antigen present on the surface of the schistsomulum of *Schistosoma mansoni*. Mol. Biochem. Parasitol. *17*, 105-114.

Balloul, J. M., Grzych, J. M., Pierce, R. J., and Capron, A. (1987a) A purified 28 kDa protein from *Schistosoma mansoni* adult worms protects rats and mice against experimental schistosomiasis. J. Immunol. *138*, 3448-3453.

Balloul, J. M., Sondermeyer, P., Dreyer, D., Capron, M., Grzych, J. M., Pierce, R. J., Carvallo, D., Lecocq, J. P., and Capron, A. (1987b) Molecular cloning of protective antigen of schistosomes. Nature *326*, 149-153.

Balloul, J. M., Boulanger, D., Sondermeyer, P., Dreyer, D., Capron, M., Grzych, J. M., Pierce, R. J., Carvallo, D., Lecocq, J. P., and Capron, A. (1987c) Vaccination of baboons with a P28 antigen of *Schistosoma mansoni* expressed in *E. coli*. In: Molecular Paradigms for the Eradication of Helminth Parasites. MacInnis, A. J. (ed.), Alan R Liss, New York, 77-84.

Barker, R. J. Jr., Srivastava, B. S., Suri, P., Goldberg, M., and Knopf, P. M. (1985) Immunopresipitation analysis of radiolabeled protein antigens biosynthetized *in vitro* by *Schistosoma mansoni*: indentification of antigen uniquily recognized by protective antibodies. J. Immunol. *134*, 1192-1201.

Bouchara, H. O., Bashir, M. E. N., Malik, K. H. E., Mukhtar, M. M., Trottein, F., Capron, A., and Taylor, M. G. (1993) Suppression of *S. bovis* egg production in cattle by vaccination with either glutathione *S*-transferase or keyhole limpet hemocyanin. Parasite Immunol. *15*, 383-390.

Boucheix, C., Benoit, P., Frachet, P., Billard, M., Worthington, R. E., Gagnon, J., and Uzan, G. (1991) Molecular cloning of the CD9 antigen. A new family of cell surface proteins. J. Biol. Chem. *226*, 117-122.

Boulanger, D., Warter, A., Trottein, F., Mauny, F., Bremond, P., Audibert, F., Couret, D., Kadri, S., Godin, C., Pierce, R., Lecocq, J. P., Sellin, B., and Capron, A. (1995) Vaccination of patas monkeys experimentally infected with *Schistosoma haematobium* using recombinant glutathione *S*-transferase cloned from *S. mansoni*. Parasite Immunol. *17*, 361-369.

Boulanger, D., Reid, G. D. F., Sturrock, R. F., Wolowczuk, I., Balloul, J. M, Grezel, D., Pierce, R. J., Otieno, M. F., Guerret, S., Grimaud, J. A., Butterworth, A. E., and Capron, A. (1991) Immunization of mice and baboons with the recombinant Sm28GST affects both worm viability and fecundity after experimental infection with *Schistosoma mansoni*. Parasite Immunol. *13*, 473-490.

Butterworth, A. E., Vadas, M. A., Martz, E., and Sher, A. (1979) Cytolytic T lymphocytes recognize alloantigens on schistosomula of *Schistosoma mansoni*, but fail to induce damage. J. Immunol. *122*, 1314-1321.

Butterworth, A. E., Bensted-Smith, R., Capron, A., Capron, M., Dalton, P. R., Dunne, D. W., Grzych, J.-M., Kariuki, H. C., Khalife, J., Koech, D., Mugambi, M., Ouma, J. H., Arap Siongok, T. K., and Sturrock, R. F. (1987) Immunity in human schitosomiasis: prevention by blocking antibodies of the expression of immunity in young children. Parasitology *94*, 281-300.

Capron, A. (1992) Immunity to schistosomes. Current Opinion in Immunilogy *4*, 419-424.

Capron, A. and Dessaint, J.-P. (1985) Effector and regulatory mechanism in immunity to schistosome: a heuristic view. Ann. Rev. Immunol. *3*, 455-476.

Capron, A. Dessaint, J. P., Capron, M., Ouma, J. H., and Butterworth, A. E. (1987) Immunity to schistosome: progress toward vaccine. Science *238*, 1065-1072.

Capron, A. and Dessaint, J. P. (1988) Vaccination against parasitic disease: some alternative concepts for the definition of protective antigens. Ann. Immunol. Inst. Pasteur. *139*, 109.

Capron, A. and Dessaint, J. P. (1992) Immunologic aspects of schistosomiasis. Ann. Rev. Med. *43*, 209-218.

Capron, M., Spiegelberg, H. L., Prin, L., Bennich, H., Butterworth, A. E., Pierce, R. J., Ouaissi, M. A., and Capron, A. (1984) Role of IgE receptors in effector function of human eosinophils. J. Immunol. *232*, 462-468.

Classon, B. J., Williams, A. F., Willis, A. C., Seed, B., and Stamenkovic, I. (1989) The primary structure of the human leukocyte antigens CD37, a species homologue of the rat mrc OX-44 antigen. J. Exp. Med. *169*, 1497-1502.

Dalton, J. P. and Strand, M. (1987) *Schistosoma mansoni* polypeptides immunogenic in mice vaccined with radiation-attenuated cercariae. J. Immunol. *139*, 2474-2481.

Dalton, J. P., Tom, T. D., and Strand, M. (1987) Cloning of a cDNA encoding a surface antigen of *Schistosoma mansoni* schistosomula recognized by sera of vaccinated mice. Proc. Natl. Acad. Sci. USA *84*, 4268-4272.

Damonneville, M., Auriault, C., Verwaerde, C., Delanoye, A., Pierce, R. J., and Capron, A. (1986) Protection against experimental *Schistosoma mansoni* schistosomiasis achieved by immunization with schistosomula released products antigens (SRP-A): role of IgE antibodies. Clin. Exp. Immunol. *65*, 244-252.

Dessein, A. J., Begley, M., Demeure, C., Caillol, D., Fueri, J., Dos Reis, M. G., Andrade, Z. A., Prata, A., and Bina, J. C. (1988) Human resistance of *Schistosoma mansoni* in associated wih IgG reactivity to a 37 kDa larval surface antigen. J. Immunol. *140*, 2727-2736.

Dissous, C. and Capron, A. (1981) Isolation and characterization of surface antigens from *Schistosoma mansoni* schistosomula. Mol. Biochem. Parasitol. *3*, 215-225.

Dissous, C., Grzych, J.-M., and Capron, A. (1985) Biochemical studies on the 30-40 kDa *Schistosoma mansoni* surface antigens. Mol. Biochem. Parasitol. *16*, 277-288.

Dissous, C., Grzych, J.-M., and Capron, A. (1986) *Schistosoma mansoni* shares a protective oligosaccharide epitope with freshwater and marine snails. Nature *323*, 443-445.

Dunne, D., Butterworth, A. E., Fulford, A. J. C., Kariuki, H. C., Langley, J. G., Ouma, J. H., Capron, A., Pierce, R. J., and Sturrock, R. F. (1992) Immunity after treatment of human schistosomiasis *mansoni*: asociation between IgE antibodies to adult worm antigens and resistance to reinfection. Eur. J. Immunol. *22*, 1483-1494.

Duvaux-Miret, O., Dissous, C., Gautron, J. P., Pattou, E., Kordon, C., and Capron, A. (1990) The helminth *Schistosoma mansoni* expresses a peptide highly similar to human β-endorphin and possesses a proopiomelanocortin-related gene. New Biol. *2*, 93-99

Duvaux-Miret, O. and Capron, A. (1991) Proopiomelanocortin gene and gene products: a comparative study. Adv. Neuroimmunol. *1*, 41-57

Duvaux-Miret, O., Stefano, G. B., Smith, E. M., Dissous, C., and Capron, A. (1992) Immunosupression in the definitive and intermediate hosts of human parasite *Schistosoma mansoni* by release of immunoactive neuropeptide. Proc. Natl Acad. Sci. USA. *89*, 778-781.

El-Sherbeini, M., Bostian, K. A., and Knopf, P. M. (1990) *Schistosoma mansoni* cloning of antigen gene sequences in *E. coli*. Exp. Parasitol. *70*, 72-84.

El-Sherbeini, M., Ramadan, N., Bostian, K. A., and Knopf, P. M. (1991) Cloning and sequence analysis of the *Schistosoma mansoni* membrane glycoproteine antigen gene GP22. Mol. Biochem. Parasitol. *49*, 83-98.

Goudot-Crozel, V., Caillol, D., Djabali, M., and Dessein, A. J. (1989) The major parasite surface antigen associated with human resistance to schistosomiasis is a 37 kDa glyceraldehyde-3-phosphate dehydrogenase. J. Exp. Med. *170*, 2065-2080.

Grezel, D., Capron, M., Grzych, J. M., Fontaine, J., Lecocq, J. P., and Capron, A. (1993) Protective immunity induced in rat schistosomiasis by a single dose of the Sm28 GST recombinant antigen: effector mechanisms involving IgE and IgA antibodies. Eur. J. Immunol. *23*, 454-460.

Grzych, J. M., Capron, M., Bazin, H., and Capron, A. (1982) *In vitro* and *in vivo* effector function of rat IgG2a monoclonal anti-S mansoni antibodies. J. Immunol. *129*, 2739-2743.

Grzych, J. M., Capron, M., Dissous, C., and Capron, A. (1984) Blocking activity of rat monoclonal antibodies in experimental schistosomiasis. J. Immunol. *133*, 998-1004.

Grzych, J. M., Dissous, C., Capron, M., Torres, S., Lambert, P. H., and Capron, A. (1987) *Schistosoma mansoni* shares a protective carbohydrate epitope with keyhole limpet hemocyanin. J. Exp. Med. *165*, 865-878.

Grzych, J. M., Grezel, D., Xu, C. B., Neyrinck, J. L., Capron, M., Ouma, J. H., Butterworth, A. E., and Capron, A. (1993) IgA antibodies to a protective antigen in human *Schistosomiasis mansoni*. J. Immunol.. *150*, 527-535.

Hagan, P., Blumenthal, U. J., Dunne, D., Simpson, A. J. G., and Wilkins, H. A. (1991) Human, IgE, IgG4, and resistance to reinfection with *schistosoma haematobium*. Nature *349*, 243-245.

Harn, D. A. (1985) Identification by monoclonal antibody of a major (28 kDa) surface membrane antigen of *Schistosoma mansoni*. Mol. Biochem. Parasilol. *16*, 345-354.

Harn, D. A., Quinn, J. J., Cianci, C. M., and Ko, A. I. (1987) Evidence that a protective membrane epitope is involved in early but not late phase immunity to *Schistosoma mansoni*. J. Immunol. *138*, 1571-1580.

Harn, D. A., Wei, G., Oligino, L. D., Mitsuyama, M., Gebremichael, A., and Richter, D. (1992) A protective monoclonal antibody specifically recognizes and alters the catalytic activity of schistosome triose-phosphate isomerase. J. Immunol. *148*, 562-567.

Hawn, T. R., Tom, T. D., and Strand, M. (1993) Molecular cloning and expresssion of SmIrVl, a *Schistosoma mansoni* antigen with similary to calnexin, calreticulin and OvRa1l. J. Biol. Chem. *268*, 7692-7698.

Henkle, K. J., Davern, K. M., Wright, M. D., Ramos, A. J., and Mitchell, G. F. (1990) Comparison of the cloned genes of the 26 and 28 kDa glutathione *S*-transferase of *Schistosoma japonicum* and *Schistosoma mansoni*. Mol. Biochem. Parasitol. *40*, 23-34.

Ivanoff, N., Phillips, N., Schacht, A.-M., Heydari, C., Capron, A., and Riveau, G. Mucosal vaccination against schistosomiasis using liposome-associated Sm28GST. Vaccine. In press.

James, S. L., Salzmann, C., and Pearce, E. J. (1988) Induction of protective immunity against *Schistosoma mansoni* by a non-living vaccine. VI. Antigen recognition by non-responder mouse strains. Parasite Immunol. *10*, 71-83.

James, S. L. (1985) Induction of protective immunity against *Schistosoma mansoni* by a non-living vaccine is dependent on the method of antigen presentation. J. Immunol. *134*, 1956-1960.

Jeffs, S. A., Hagan, P., Allen, R., Correa-Oliveira, R., Smithers, S. R., and Simpson A. J. G. (1991) Molecular cloning and characterization of the 22-kilodalton adult *Schistosoma mansoni* antigen recognized by antibodies from mice protectively vaccinated with isolated tegumental surface membranes. Mol. Biochem. Parasitol. *46*, 159-167.

Joseph, M., Capron, A., Butterworth, A. E., Sturrock, R. F., and Houba, V. (1978) Cytotoxicity of human and baboon mononuclear phagocytes against schistosomula *in vitro*: induction by immune complex containing IgE and *schistosoma mansoni* antigens. Clin. Exp. Immunol. *33*, 48-56.

Karcz, S. R., Barnard, B. J., and Podesta, R. B. (1988) Biochemical properties of a 24 kDa membrane glycoprotein antigen complex from *Schistosoma mansoni*. Mol. Biochem. Parasitol. *31*, 163-172.

Khalife, J., Capron, M., Capron, A., Grzych, J. M., Butterworth, A. E., Dunne, D. W., and Ouma, J. H. (1986) Immunity in human *Schistosomiasis mansoni*. Regulation of protective immune mechanisms by IgM blocking antibodies. J. Exp. Med. *166*, 1626-1640.

Khalife, J., Trottein, F., Schacht, A. M., Godin, C., Pierce, R. J., and Capron, A. (1993) Cloning of the gene encoding a *Schistosoma mansoni* antigen homologous to human Ro/SS-A autoantigen. Mol. Biochem. Parasitol. *57*, 193-202.

Khan, C. M. A., Villarreal-Ramos, B., Pierce, R., Riveau, G., Demarco de Hormaeche, R., Capron, A., Dougan, G., and Hormaeche C. E. (1994) Construction, expression and immunogenicity of the *Schistosoma mansoni* P28 glutathione *S*-transferase as a genetic fusion to tetanus toxin fragment C in a live Aro attenuated vaccine strain of *Salmonella*. Proc.Nat.Acad. Sc. USA *91*, 11261-11265.

Khan, C. M. A., Villarreal-Ramos, B., Pierce, R., Riveau, G., Demarco de Hormaeche, R., Capron, A., Dougan, G., and Hormaeche C. E. (1994) The construction, expression and immunogenicity of multiple tandem copies of the *S. mansoni* peptide comprising 115-131 of the P28 glutathione *S*-transferase, expressed as C-terminal fusions to tetanus toxin fragment C in live attenuated vaccine strain of *Salmonella*.J. Immunol. *153*, 5634-5642

Knight, M., Kelly, C., Rodrigues, V., Yi, X. Wamachi, A., Smithers, S. R., and Simpson, A. J. G. (1989) A cDNA clone encoding part of major 25000 dalton surface membrane antigen of adult *Schistosoma mansoni*. Parasitol. Res. *75*, 280-286.

Lanar, D., Pearce, E. J., James, S. L., and Sher, A. (1986) Identification of paramyosin as the schistosome antigen recognized by intradermally vaccinated mice. Science *234*, 593-596.

Mangold, B. L. and Dean, D. A. (1986) Passive transfer with serum and IgG antibodies of irradiated cercaria-induced resistance against *Schistosoma mansoni* in mice. J. Immunol. *136*, 2644-2648.

Matsumoto, Y., Perry, G., Levin, R. J. C., Blanton, R., Mahmoud, A. A., and Aikawa, M. (1988) Paramyosin and actin in schistosomal teguments. Nature *333*, 76-78.

McNair, A., Dissous, C., Duvaux-Miret, O., and Capron, A. (1993) Cloning and characterization of the gene encoding the 28 kDa glutathione *S*-transferase of *schistosoma mansoni*. Gene *124*, 245-249.

Mielcarek, N., Cornette, J., Schacht, A. M., Pierce, R. J., Capron, A., Locht, C., Riveau, G. Protection against *Schistosoma mansoni* after nasal immunization with recombinant *Bordetella pertussis*. Submitted

Mitchell, G. F., Garcia, E. G., Davern, K., Tiu, W. U., and Smith, D. B. (1988) Sensitization against the parasite antigen Sj26 is not sufficient for consistent expression of resistance to *Schistosoma japonicum* in mice. Trans. R. Soc. Trop. Med. Hyg. *82*, 885-889.

Mitchell, G. F. (1989) Glutathione *S*-tranferases – potential components of anti-schistosome vaccines. Pararitol. Today *5*, 34-37.

Omer Ali, P., Jeffs, S. A., Meadows, H. M., Hollyer, T., Owen, C. A., Abath, F. G. C., Allen, R., Hackett, F., Smithers, S. R., and Simpson, A. J. G. (1991) Struture of Sm25, an antigenic integral membrane glycoprotein of adult *Schistosoma mansoni*. Mol. Biochem. Parasitol. *45*, 215-222.

Ouwe-Missoukem-Boyer, O., Porchet, E., Capron, A., and Dissos, C. (1994) Characterization of immunoreactive TNFa molecules in the gasteropod *Biophalaria glabrata*. Dev. Comparative Immunol. *18*, 211-218.

Pancré, V., Wolowczuk, I., Guerret, S., Copin, M. C., Delanoye, A., Capron, A., and Auriault, C. (1994) Protective effect of rSm28GST-specific T-cells in schistosomiasis: role of gamma interferon. Infect. Immun. *62*, 3723-3730.

Pearce, E. J., James, S. L., Dalton, J., Barrall, A., Ramos, C., Strand, M., and Sher, A. (1986). Immunological characterization and purification of Sm-97, a *Schistosoma mansoni* antigen monospecifically recognized by antibodies from mice protectivally immunized with a non-living vaccine. J. Immunol. *137*, 3593-3600.

Pearce, E. J., James, S. L., Hieny, S., Lanar, D. E., and Sher, A. (1988) Induction of protective immunity against *Schistosoma mansoni* by vaccination with schistosome paramyosin (Sm-97), a non-surface parasite antigen. Proc. Natl. Acad. Sci. USA *85*, 5678-5682.

Pierce, R. J., Aimar, C., Balloul, J. M., Delarue, M., Grausz, D., Verwaerde, C., and Capron, A. (1985) Translation of *Schistosoma mansoni* antigen in Xenopus oocytes microinjected with mRNA from adult worms. Mol. Biol. Chem. Parasitol. *15*, 171-188.

Renauld, G., Mielcarek, N., Cornette, J., Schacht, A. M., Capron, A., Riveau, G., and Locht, C. Induction of mucosal immune responses against a heterologous antigen fused to filamentous hemagglutinin after intranasal immunization with recombinant Bordetella pertussis. submitted

Reynolds, S. R., Shoemaker, C. B., and Harn, D. (1992) T and B cell epitope mapping of Sm23, an integral membrane protein of *Schistosoma mansoni*. J. Immunol. *149*, 3995-4001.

Richter, D., Incani, R. N., and Harn, D. A. (1993) Isotype responses to candidate vaccine antigens in protective sera obtained from mice vaccinated with cercariae of *Schistosoma mansoni*. Infect. Immun. *61*, 3003-3011.

Rihet, P., Demeure, C. E., Bourgeois, A., Parta, A,. and Dessein, A. J. (1991) Evidence for an association between human resistance to *schistosoma mansoni* and high anti-larval IgE levels. Eur. J. Immunol. *21*, 2679-2686.

Rogers, M. V., Davern, K. M., Smyth, J. A., and Mitchell, J. F. (1988) Immunoblotting analysis of the major integral membrane protein antigens of *Schistosoma mansoni*. Mol. Biochem. Parasitol. *29*, 77-88.

Sher, A., Hall, B. F., and Vadas, M. A. (1978) Acquisition of murine major histocompatibility complex gene products by schistosomula of *Schistosoma mansoni*. J. Exp. Med. *148*, 46-52.

Sher, A., Gazinelli, R. T., Oswald, I. P., Clerici, M., Kullberg, M., Pearce, E. J., Berzofsky, J. A., Mosmann, T. R., James, S. L., Morse III, H. C., and Shearer, G. M. (1992) Role of T-cell derived cytokines in the down-regulation of immune responses in parasitic and retroviral infection. Immunol. Rev. *127*, 183-204.

Shipp, M. A., Stefano, G. B., D'adamio, L., Switzer, S. N., Howard, F. D., Sinisterra, J., Scharrer, B., and Reinherz, E. (1990) Down regulation of enkephalin-mediated inflammatory responses by CD10/neutral endopeptidase 24.11. Nature *347*, 394-396.

Shoemaker, C., Gross, A., Gebremichael, A., and Harn, D. A. (1992) cDNA cloning and functional expression of the *Schistosoma mansoni* protective antigen triose-phosphate isomerase. Proc. Natl. Sci. USA. *89*, 1842-1846.

Smith, J. H. and Christie, J. D. (1986) The pathobiology of *Schistosoma haematobium* infection in humans. Human Pathol. *17*, 333-345.

Smithers, S. R. and Terry, R. J. (1969) Immunity in schistosomiasis. Ann. N.Y. Acad. Sci. *160*, 826-840.

Smithers, S. R., Hackett, F., Omer Ali, P., and Simpson, A. J. G. (1989) Protective immunization of mice against *Schistosoma mansoni* with purified adult worm surface membranes. Parasite Immunol. *11*, 301-318.

Smythies, L. E., Pemberton, R. M., Coulson, P. S., Mountford, A. P., and Wilson, R. A. (1992) T-cell devivated cytokines associated with pulmonary immune mechanism in mice vaccinated with irradiated cercariae of *Schistosoma mansoni*. J. Immunol. *148*, 1512-1518.

Stefano, G. B., Leung, M. K., Zhao, X., and Scharrer, B. (1989) (D-Ala2)-deltrophin I binding and pharmacological evidence for a novel subtype of delta opioid receptor on human and invertebrate immune cells. Proc. Natl. Acad. Sci. USA. *86*, 626-630.

Stein, L. D. and David, J. R. (1986) Cloning of a developmentally regulated tegumental antigen of *Schistosoma mansoni*. Mol. Biochem. Parasitol. *20*, 253-264.

Taylor, J. B., Vidal A., Torpier, G., Meyer, D. J., Roitsch, C., Balloul, J. M., Southan, C., Sondermeyer, P., Pemble, S., Lecocq, J. P., Capron, A., and Ketterer, B. (1988) The gluthathione transferase activity and tissue distribution of a cloned Mr28K protective antigen of Schistosoma mansoni. EMBO J. *7*, 465-472.

Trotein, F., Kieny, M.P., Verwaerde, C., Torpier, G., Pierce, R.J., Balloul, J.M., Schmitt, D., Lecocq, J. P., and Capron, A. (1990) Molecular cloning and tissue distribution of a 26 kDa *Schistosoma mansoni* glutathione *S*-transferase. Mol. Biochem. Parasitol. *41*, 35-44.

Trottein, F. (1992) Contribution à l'étude moléculaire des glutathion S-transferase dans le cadre d'une stratégie vaccinale contre la schistosomiase. Ph.D. Thesis, Université des Sciences et Technologie de Lille.

Trottein, F., Godin, C., Pierce, R. J., Sellin, B., Taylor, M. G., Gorillot, I., Sampaio Silva, M., Lecocq, J. P., and Capron A. (1992a) Inter-species variation of schistosome 28kDa glutathione *S*-transferase. Mol. Biochem. Parasitol. *54*, 63-72.

Trottein, F., Vaney, M. C., Bachet, B., Pierce, R. J., Colloc'h, N., Lecocq, J. P., Capron, A., and Mornon, J. P. (1992b) Crystallization and preliminary X-ray diffraction studies of a protective cloned 28 kDa glutathione *S*-transferase from *Schistosoma mansoni*. J. Mol. Biol. *224*, 515-518.

Verwaerde, C., Joseph, M., Capron, M., Pierce, R. J., Dammonneville, M., Velge, F., Auriault, C., and Capron, A. (1987) Functional properties of a rat monoclonal IgE antibody specific for *Schistosoma mansoni*. J. Immunol. *138*, 4441-4446.

Wolowczuck, I., Auriault, C., Gras-Masse, H., Vendeville, C., Balloul, J. M., Tartar, A., and Capron, A. (1989) Protective immunity in mice vaccinated with the *Schistosoma mansoni* P-28-1 antigen. J. Immunol. *142*, 1342-1350.

Wolowczuck, I., Auriault, C., Bossus, M., Boulanger, D., Gras-Masse, H., Mazingue, C., Pierce, R. J., Grezel, D., Reid, G. D. F., Tartar, A., and Capron, A. (1991) Antigenicity and imunogenicity of a multiple peptidic construction of the *Schistosoma mansoni* Sm-28 GST antigen in rat, mouse and monkey. Jour. Immunol. *146*, 1987-1995.

Wright, M. D., Henkle, K. J., and Mitchell, G. F. (1990) An immunogenic Mr 23,000 integral membrane protein of *Schistosoma mansoni* that closely ressembles a human tumor-associated antigen. J. Immunol. *144*, 3195-3200.

Xu, C.-B., Verwaerde, C., Grzych, J.-M., Fontaine, J., and Capron, A. (1991) A monoclonal antibody blocking the *Schistosoma mansoni* 28-kDa glutathione *S*-transferase activity reduces female worm fecundity and egg viability. Eur. J. Immunol. *21*, 1801-1807.

Xu, C. B., Verwaerde, C., Gras-Masse, H., Fontaine, J., Bossus, M., Trottein, F., Wolowczuk, I., Tartar, A., and Capron, A. (1993) *Schistosoma mansoni* 28-kDa glutathione *S*-transferase and immunity against parasite fecundity and egg viability. Role of the amino- and carboxyl-terminal domains. J. Immunol. *150*, 940-949.

4.4 AIDS Vaccination

Stephen Norley and Reinhard Kurth

4.4.1 Introduction

Although still a major health threat, the dire predictions made ten years ago of a raging AIDS epidemic in Europe have thankfully not been fulfilled. Globally, however, the spread of HIV infection continues at an explosive pace. Over 16 million people are estimated to have been so far infected with this number rising to 30-40 million by the end of the century (Mertens et al., 1994). As 90 % of the infected people live in developing countries it is difficult to imagine how the often rudimentary health-care facilities in these areas will cope with the burden as these infections begin to manifest themselves as AIDS. Even in the USA, where 1 % of the male population is HIV-positive, the problem will be acute.

The 'slow-down' of the epidemic in Europe is almost certainly due to the intense publicity surrounding the risk of HIV-infection and the heightened awareness of those potentially at risk. Such publicity campaigns are difficult to carry out successfully in many of the developing countries and have been relatively ineffectual so far. Similarly, even if an effective anti-HIV therapeutic were to be developed by one of the major Western pharmaceutical companies, the drug would certainly be expensive and would probably need to be taken regularly throughout the life of the infected individual. Neither the citizens, nor the governments of the countries where such drugs are most needed could afford to pay for these treatments.

The biggest hope in the world-wide fight against AIDS is therefore the development of an efficient, one or two shot vaccine which, in much the same way as the smallpox vaccine, would stop the march of the epidemic. Although not helping those millions already HIV-positive such a vaccine would at least prevent further tens of millions from developing AIDS. Unlike smallpox however, the development of an AIDS vaccine has been, and remains, a very difficult and often frustrating task. In this chapter we will review the results of AIDS vaccine efforts in both humans and the various animal models. In particular we will emphasise the novel strategies which are now vigorously being pursued and which offer at least some hope for the future.

4.4.2 The Problem of Variability in the Development of an AIDS Vaccine

A vaccine works by stimulating an immune response to one or more proteins of the infectious agent. More specifically, antibodies, and possibly T-lymphocytes, individually specific for epitopes within the protein are stimulated. One of the extraordinary properties of HIV-1 is it's high rate of variability both within an individual and within a population. Sequence analysis of viruses isolated from an infected patient will normally reveal not one, but many different genotypes – the virus exists as a 'quasispecies' (Cichutek et al., 1992). Moreover, the constituents of the quasispecies change over time, presumably due to adaptation to the host or in response to the immune system. It is this capacity for rapid mutation which permits the almost instant development of drug-resistant variants in patients undergoing treatment with one or more of the available anti-HIV therapeutics and which may, as later described, allow the virus to continually escape immune control.

Globally, HIV does not exist as a homogenous quasispecies. Different areas of the world tend to be predominantly infected with one or more 'clades' of HIV-1 as defined by phylogenetic analysis of the envelope gene (Bachmann et al., 1994; Bobkov et al., 1994; Cernescu et al., 1994; Dumitrescu et al., 1994; Gao et al., 1994; Korber et al., 1994; Louwagie et al., 1995). Viruses belonging to clades A – H, of which clade B is predominant in the USA, Europe and Australia, are collectively referred to Group M viruses. Recently, isolates which are significantly distinct from Group M viruses have been identified and have been classified as Group O ('outliers'). It seems likely that further Group M clades will be identified in the future and possibly even further distinct groups.

To date, the vast majority of preclinical and clinical trials for HIV-1 vaccines have been based on clade B isolates. Even if such vaccines were to prove efficacious against infection with clade B viruses it is unknown how effective the vaccines would be against other clades of Group M or in particular against Group O viruses. In addition, even a vaccine effective against most HIV-1 viruses would probably be ineffective against HIV-2, a virus which although at present appears less pathogenic and widespread than HIV-1, is increasingly a major concern in West Africa.

In addition to the conceptual problems of developing an AIDS vaccine able to prevent infection with the many different forms of HIV, the practical and logistic problems in carrying out phase III vaccine trials in humans are staggering. Although the phase II trials already completed have failed to unequivocally demonstrate efficacy the preparations for phase III trials are already underway in several developed and developing countries (Katongole-Mbidde, 1993). As it is unlikely that the use of a placebo could be ethically justified when dealing with an infection with such a high mortality rate, efficacy of the vaccine would be judged on the basis of the rates of infection within the immunized volunteers compared with that in the general population. However, unless the vaccine was 100 % effective, interpretation of the data will be complicated by the fact that those enrolling in the study will receive counselling on the risks of HIV-infection and advice on prevention, resulting perhaps in a false apparent success of the vaccine on trial.

4.4.3 Possible Protective Immune Responses

As stated above, a vaccine works by stimulating an immune response to viral proteins (tab. 4.4.1). In the design of a vaccine it is important to understand the benefits and weaknesses of each effector mechanism. One of the major unanswered questions for those involved in AIDS vaccine development is which types of immune response against which proteins are needed to achieve a state of protective immunity. This remains a point of some controversy, significantly influencing the choice of vaccine strategy, although the answer may in all likelihood be that the stimulation of multiple and strong immune responses against many proteins will be needed.

Table 4.4.1: Specific Antiviral Immune Mechanisms

Immune Mechanism	Target in HIV	Effect
Neutralizing antibody	Variable and conserved regions of the external and transmembrane envelope glycoproteins	Prevents binding or entry of virus into the host cell
Virolysis	Envelope glycoproteins on the virus particle	Binding of antibody activates the complement cascade resulting in viral disruption
Complement-mediated lysis	Envelope glycoproteins expressed on the surface of infected cells	Binding of antibody activates the complement cascade resulting in cell lysis
Antibody-dependent cellular cytotoxicity (ADCC)	Envelope glycoproteins expressed on the surface of infected cells	Binding of antibody activates Fc-receptor-bearing 'killer' cells which lyse the infected target cell
Cytotoxic T-lymphocytes (CTL)	Short, processed segments of any viral protein presented by cell-surface MHC-I	Recognition of the MHC-I/ peptide complex by the CTL T-cell receptor results in lysis of the infected cell

4.4.3.1 Neutralising Antibody

Antibodies which recognise and neutralise a virus by preventing binding to or entry into the host cell have been shown to be highly effective in providing protective immunity to a number of different viruses, including some retroviruses (Fish et al., 1979; Hunsmann et al., 1975a; Hunsmann et al., 1975b). Indeed, the efficacy of a number of vaccines can be estimated by the measuring the levels of neutralising antibodies stimulated. Once HIV-1 had been identified as the infectious agent causing AIDS, the identification and cloning of the envelope glycoprotein which would stimulate and bind neutralizing antibodies was considered the first crucial step in the development of a vaccine (Hahn et al., 1984; Luciw et al., 1984; Robert-Guroff et al., 1985; Robey et al., 1985; Sarngadharan et al., 1985; Veronese et al., 1985). Since this has been achieved (Putney et al., 1986) the HIV-1 outer and transmembrane envelope glycoproteins (gp120 and gp41) have been analysed in great detail. In particular, the 'third hypervariable region' or 'V3-loop' of HIV-1 gp120 has been the sole subject of scores of publications as early studies showed it to be the 'principal neutralising determinant' (PND) of HIV-1 (Broliden et al., 1991; Javaherian et al., 1990; Javaherian et al., 1989; Matsushita et al., 1988; Nara et al., 1990; Vogel et al., 1994). However, the significance for vaccines of much of this detailed knowledge has been somewhat reduced in the last few years with the realisation that there is a fundamental difference between strains of HIV-1 grown in T-cell lines in laboratories (and usually used in neutralisation assays) and primary isolates of HIV-1 found in infected people (Bou-Habib et al., 1994; Moore et al., 1995; Sawyer et al., 1994). Antibodies raised in human volunteers immunised with recombinant envelope glycoprotein and able to neutralise laboratory strains of HIV have little or no effect on primary isolates (Steimer et al., 1994). This may in fact be a blessing in disguise as one of the major obstacles to vaccine development had been perceived to be the high variability of the V3-loop PND: efforts now tend to concentrate on more conserved regions of the envelope glycoprotein which act as neutralising epitopes in primary isolates (Burton et al., 1994; Chamat et al., 1992; Moore and Ho, 1993; Steimer et al., 1991).

4.4.3.2 Virolysis, Complement Mediated Lysis and Antibody-Dependent Cellular Cytotoxicity (ADCC)

The interaction of many enveloped viruses with complement results in activation of the complement cascade and subsequent destruction of the particle (Bartholomew et al., 1978). HIV, however, appears to be relatively resistant to the effects of antibody-independent virolysis (Banapour et al., 1986; Hoshino et al., 1984), possibly due to the protective effect of the cellular CD55 molecule incorporated into the viral membrane (Marschang et al., 1995). However, binding of specific antibody to a virus particle may also activate the complement cascade and this has been demonstrated to occur with HIV-1 (Spear et al., 1993).

During the replicative process, HIV-infected cells express the envelope glycoproteins on their surface in preparation for the budding progeny virus. These can also be recognised by gp120- and gp41-specific antibodies which should theoretically result in complement-mediated lysis of the cell. As with virolysis however, such lysis has been with a few exceptions (Gregersen et al., 1990; Toth et al., 1989) very difficult to demonstrate (Nara et al., 1987; Posner et al., 1992; Yefenof et al., 1991).

The binding of antibodies to the surface of the infected cell can also result in activation of Fc-receptor bearing 'killer' cells which then lyse the infected target (Norley and Wardley, 1983; Shore et al., 1974). In contrast to complement mediated lysis this mechanism is readily demonstrable in the HIV system (Ljunggren et al., 1987a; Ljunggren et al., 1989b; Ljunggren et al., 1987b). Unlike neutralisation, ADCC is broadly reactive (Ljunggren et al., 1989a) with some sera from HIV-1 infected patients even able to initiate lysis of HIV-2 infected cells (Norley et al., 1990).

4.4.3.3 Cytotoxic T-Lymphocytes (CTL)

CTLs are generally the most crucial immune mechanism in terms of controlling a virus infection (Oldstone, 1994). Recognition via the T-cell receptor of short, processed fragments of viral proteins presented on the surface of the infected cell by the MHC-I molecule by these $CD8^+$ lymphocytes rapidly results in target-cell lysis. HIV-1 specific CTLs are in surprising abundance in healthy seropositive patients (Hoffenbach et al., 1989) and in fact HIV-1 infection is one of the few known cases in humans where CTL activity can be detected without the need for secondary in vitro stimulation (Walker and Plata, 1990). CTLs specific for *env, gag, nef* and *pol* gene products have been demonstrated (Buseyne et al., 1993; Clerici et al., 1991; Culmann et al., 1991; Johnson et al., 1993; Nixon et al., 1988; Walker et al., 1989) and there is evidence that the early control of virus replication is due to the rapid development of virus-specific CTLs (Borrow et al., 1994; Koup et al., 1994).

4.4.4 Primate Models for AIDS Vaccine Development

As mentioned earlier there are a number of phase I and II clinical trials in progress to test the safety and efficacy of potential AIDS vaccines with preparations for large scale phase III trials underway. Before getting to the point of human trials, however, it is crucial that a vaccine strategy be tested in one or more of the animal models which allow classic immunisation and challenge experiments to be performed. There are a number of such models available, each with its advantages and disadvantages and unfortunately the model selected can have a large impact on the result. Which model is the most relevant for AIDS vaccine development is therefore a matter of debate.

4.4.4.1 HIV-1 Infection of Chimpanzees

Perhaps not surprisingly given their close evolutionary relationship to Man, chimpanzees (along with gibbon apes) are the only non-human primates which can be reproducibly infected by HIV-1 (Alter et al., 1984; Kurth et al., 1988; Prince et al., 1988). However, there are of course enormous ethical, financial and practical problems in housing and handling HIV-1 infected chimps and this has limited the use of this model to the bare minimum. Despite it's status as the 'gold-standard' animal model, HIV-1 infection of chimps has fundamental differences to HIV-1 infection of humans. Most importantly, chimps do not appear to develop disease as a result of infection with even chimps infected as early as 1983 remaining healthy today. This is probably due to the fact that the virus does not grow well in these primates, having the in vivo characteristics of an apathogenic attenuated virus (see later). Although it can be argued that this is not a problem if the end-point of the study is whether or not a vaccine can induce 'sterilising immunity' it seems likely that, as with most vaccines, a first generation AIDS vaccine will at best prevent disease development. Attempts to establish a more practical model for HIV-1 infection in lower primates have generally failed with the exception of pig-tailed macaques. These monkeys have been shown to be permissive for HIV-1 infection (Agy et al., 1992; Frumkin et al., 1993; Kent et al., 1995) although it has proved difficult in many primate centres to achieve the level of reproducibility needed for immunization/challenge experiments.

4.4.4.2 SIV Infection of Macaques

The majority of simian immunodeficiency virus (SIV) isolates used in vaccine trials are genetically similar to HIV-2 of humans (Myers et al., 1992; Myers et al., 1991). Indeed, these isolates and HIV-2 are almost certainly derived originally from SIV_{sm} which is apathogenic in its sooty mangaby natural host but causes AIDS in heterologous species (Fultz et al., 1990; Hirsch et al., 1989; Lowenstine et al., 1986; McClure et al., 1991; Villinger et al., 1991). Of the SIV isolates available, SIV_{mac} originally isolated from rhesus macaques is by far the most commonly used. SIV_{mac} infection of macaques is characterised by an initial high virus load which is then brought under control, apparently by the developing immune response (Beer et al., 1995; Reimann et al., 1994), followed by a period of five to 36 months (during which time it is possible to reisolate virus) before a drastic decline in $CD4^+$ cells number occurs and the animal develops the symptoms of human AIDS (Chalifoux et al., 1987; Desrosiers, 1988; Letvin et al., 1985). In these respects, SIV_{mac} infection of macaques mimics more closely HIV-1 infection of humans than does HIV-1 infection of chimpanzees and provides the opportunity to test vaccine strategies which although allowing limited infection to occur can delay or prevent the onset of AIDS.

SIV_{mac} is not the only strain of SIV available for vaccine studies. SIV_{mne} from pig-tailed macaques, SIV_{sm} itself and even the more distantly related SIV_{agm} from African green monkeys have been used in vaccine experiments, each providing additional (and often contrasting) information about vaccine efficacy (Hu et al., 1989; Putkonen et al., 1991; Siegel et al., 1995).

4.4.4.3 HIV-2 Infection of Primates

It has been known for many years that some species of primates can be productively infected with HIV-2 (Castro et al., 1991; Dormont et al., 1989; Putkonen et al., 1989; Stahl-Hennig et al., 1990), but the model had not been used extensively because unlike SIV_{mac} the virus did not appear to cause disease. More recently however, a number of HIV-2 isolates have been obtained by passage from monkey to monkey which do induce the development of AIDS-like symptoms (Morton et al., 1994). The model therefore offers the opportunity to develop and test vaccines which, even if eventually found to have little relevance to HIV-1 in humans might well be effective against the growing epidemic of HIV-2 occurring particularly in Africa.

4.4.5 Vaccine Strategies

Virtually every conceivable form of vaccine, from the 'traditional' to the speculative, has been developed and tested either in animal models for HIV infection or in humans (tab. 4.4.2). Choosing a suitable vaccine strategy is often an exercise in finding the optimal balance between safety and efficacy and in this respect all vaccine preparations have inherent advantages and disadvantages. For the remainder of this review we will examine each of these strategies in turn, discussing the potential benefits and risks in the light of data accumulated so far. As mentioned previously, the efficacy of a vaccine preparation often depends on the system used to test it and this should be kept in mind when evaluating the actual potential for use as an AIDS vaccine.

4.4.5.1 Whole Inactivated Virus

An immunogen comprising of whole inactivated virions is perhaps the simplest form of vaccine available and has been used with success to combat viral epidemics such as influenza and polio. However, the use of a whole inactivated HIV vaccine in humans was initially discounted because of the potential risk of incomplete inactivation: in the age of molecular biology this risk was considered unnecessary. Ironically, the first real successes in reproducibly inducing a state of immunity in a 'vigorous' challenge system such as SIV_{mac}/rhesus were achieved using inactivated virus vaccines (Desrosiers et al., 1989; Murphey-Corb et al., 1989). A large number of groups (including our own) were subsequently able to demonstrate 'sterilizing' immunity against normally lethal challenge (Cranage et al., 1992; Dormont et al., 1995; Hartung et al., 1992; Johnson et al., 1992; Stott et al., 1990). The initial hopes that it was only a matter of time before the protective component(s) of such vaccines and the immune responses stimulated would be identified were dashed by the demonstration that most, if not all, of the protection was mediated by the immune response not to viral proteins but to xenogeneic cellular proteins incorporated into both the immunogen and challenge

Table 4.4.2: Examples of Vaccine Strategies Currently in Pre-Clinical Testing or in Clinical Trial

Particles
Whole inactivated virus
Virus-like particles
Immunostimulatory complexes (ISCOMS)
Pseudovirions

Recombinant proteins
Envelope gp160
Envelope gp120
Core p24

Synthetic Peptides
Monomeric V3 peptides
V3 peptides linked to carriers (PPD, Toxin A)
Multiple antigenic peptides
Multiple epitope peptides
V3-T helper epitope peptides
Lipopetides
Multiple antigenic lipopeptides

Combination Vaccines
Vaccinia-gp160 + rgp160
Vaccinia-gp160 + rgp160 + Env peptides
Canarypox-gp160 + rgp160
rgp160 + V3 peptide

Live Attenuated virus
SIVmac Δ*nef*
SIVmac Δ3 (*nef, NRE, vpr* deleted)
SIVmac Δ*rev*
HIV-1 Δ3

Genetic Immunization
HIV-1 env DNA,
HIV expression vector-coated gold particles

Others
Recombinant CD4
Anti-idiotype responses:
Anti-gp120; anti-CD4

Recombinant Vectors
Vaccinia-gp160
Canarypox-gp160
MVA-*gag/pol/env*
MoMLV-HIV-1 *env, rev*
Vaccinia p55 *gag*/V3
Adenovirus-*env* or -*gag*
Poliovirus-*nef, gag, env*
Mengovirus-*nef*
Salmonella gp120, p24, *nef*
Shigella-V3 peptide
BCG-Env peptides

viruses. Of 140 or so monkeys immunized with inactivated virus (grown in human cells) or fixed infected cells and protected from challenge with live virus grown in human cells, none were protected from subsequent rechallenge with virus grown in monkey cells (Schultz and Stott, 1994). In one study, rhesus macaques immunized with inactivated virus were protected against challenge with homologous virus grown in human cells but not the same virus passaged once through rhesus PBMC (Norley et al., 1996).

That protection could be mediated by cellular proteins alone was formally demonstrated by Stott et al (Stott et al., 1991) who protected 50 % of monkeys by immunisation with uninfected human cells. Although this phenomenon has reduced the significance of the many successful vaccine experiments using inactivated SIV vaccines world-wide there is the possibility that a novel form of vaccination might be developed as a result. One of the protective cellular components would appear to be the major histocompatibility class II (MHC-II) antigen which is incorporated, along with many other cellular proteins, into the virus as it buds from the host cell. Stott et al went on to show that immunisation of monkeys with mouse cells expressing human MHC-II was sufficient to protect 50 % against challenge with the human-grown challenge virus (Stott et al., 1993) and others (Arthur et al., 1995) have shown that immunization with purified MHC-II itself is protective. Furthermore it has been possible to confer immunity by passive transfer of serum from immunized animals, demonstrating that it is the antibody response which is at least partially responsible (E. J. Stott, personal communication, 1995). These experiments were all performed in a xenogeneic system (human/monkey) which by itself would have little relevance to HIV infection of humans. However, Stott and colleagues have recently demonstrated that allogeneic immunisation can also be protective (E. J. Stott, personal communication, 1995). This raises the exciting prospect of developing an AIDS vaccine based not on immunity to viral proteins but to allogeneic MHC molecules. It is worthwhile stressing that the immune response to cellular proteins in the monkey model is far more protective than is the response to viral proteins.

Although inactivated virus vaccines in the SIV/monkey model are ineffective in the absence of an anti-cell response it is unknown whether this is also true for HIV-1 in humans. To circumvent the risk of incomplete inactivation, a number of groups have concentrated on the production of virus-like particles. Human trials are underway using HIV-1 p24 Gag expressed in particulate form with a yeast transposon product (Ty p24-VLP; Weber et al., 1995) which is more immunogenic than the purified p24 itself. Vaccinia viruses expressing both *env* and *gag* gene products have been used to produce particles ('pseudovirions') resembling closely HIV-1 itself but which contain no viral RNA and are therefore not infectious (Haffar et al., 1990; Karacostas et al., 1989). Such strategies allow the production of immunogens which mimic closely the immunogenicity of inactivated virus vaccines but which have none of the associated safety concerns.

4.4.5.2 Purified Recombinant Subunit Vaccines

The vast majority of HIV-1 vaccine candidates developed to date have consisted of HIV-1 gp120 or gp160 produced by recombinant DNA technology in bacteria, yeast or mammalian cells. The rationale for this focus has been the assumption that the stimulation of a neutralising antibody response by such products will be protective. Indeed, a small number of vaccine trials in chimpanzees have demonstrated protective immunity against challenge using such products, more often than not with a correlation between levels of neutralising antibody and protection (Berman et al., 1990). However, the number of failures far outweigh the successes, even though experiments are performed under conditions which optimise as far as possible the likelihood of success.

There are at least seven phase I and II trials in uninfected volunteers completed or currently underway using recombinant gp160 or gp120 (Walker and Fast, 1994) with most showing little or no toxicity and reasonable immunogenicity, including induction of neutralising antibodies (Montefiori et al., 1993; Schwartz et al., 1993). Despite intense pressure from the companies producing the candidate vaccines the decision to continue into phase III trials has been postponed because the sera from immunised vaccinees, although neutralising laboratory strains of HIV-1, appear to have no effect on primary isolates (Steimer et al., 1994). The potential of such vaccines to protect against infection 'in the field' is therefore questionable. The production and testing of recombinant envelope glycoproteins derived from primary isolates of HIV-1 may be needed before their use in phase III trials will be considered.

4.4.5.3 Synthetic Peptide Immunogens

Although often poorly immunogenic compared to large proteins, the use of synthetic peptides offers a number of advantages. First, being totally chemical in origin, there are none of the potential risks of incomplete inactivation or injection of foreign DNA associated with the production and use of some other vaccines. Second, it is possible using peptides to stimulate immune responses to epitopes of choice, leaving out those epitopes which might stimulate enhancing antibodies or even auto-immunity. Third, suitably modified peptides can stimulate cell-mediated immune responses in addition to antibody responses, something which normally require the use of a live replicating vector.

The greatest obstacle to the use of synthetic peptides has been their generally poor immunogenicity and a number of approaches have been taken to overcome this. For example, the synthesis of peptides with neutralisation, CTL and T-helper epitopes in tandem have been shown to be immunogenic (Hart et al., 1990) and even to offer some degree of protection in animal models (Shafferman et al., 1991; Shafferman et al., 1992). Multiple antigenic peptide (MAP) constructs consisting of eight HIV-1 V3 loop peptides bound to a lysine copolymer can stimulate neutralising antibodies to the relevant isolate (Vogel et al., 1994; Wang et al., 1991) and such constructs are currently in phase I and II trials (Walker and Fast, 1994). Similarly, V3 peptides bound to carrier proteins are immunogenic and are being tested in humans (Walker and Fast, 1994). In addition to antibodies, efforts to stimulate CTL responses with peptides have met with some success. Peptides corresponding to known CTL epitopes alone or modified by addition of a tri-palmitic tail has been successfully used to stimulate anti-SIV_{mac} p27 CTL responses in rhesus macaques (Vogel et al., 1996; Yasutomi et al., 1994). In one of these studies the strength of the resulting anamnestic CTL response upon challenge correlated with the reduction in virus load (Vogel et al., 1996).

4.4.5.4 Live Vectors

Most efforts to stimulate CTLs in addition to humoral immune responses have used some form of recombinant, replicating vector to bring the protein of choice into the cellular endogenous processing pathway. The first, and still most commonly used, live vector consists of vaccinia virus carrying one or more HIV or SIV genes. Vaccinia virus is itself relatively safe, having been administered to a significant proportion of the world's population as part of the successful campaign to eradicate smallpox.

However, vaccinia can occasionally cause systemic infection resulting in brain damage or even death, particularly in recipients already immunocompromised, although controlled phase I trials have been performed and are still in progress (Cooney et al., 1991; Graham et al., 1992; Zagury et al., 1988). In order to overcome this danger a number of alternative constructs based on attenuated forms of vaccinia have been made. Canarypox, for example, which cannot complete it's replication in mammalian cells and therefore poses no threat to the vaccinee has been used to produce gp160-bearing recombinants currently being tested in humans (Pialoux et al., 1995). Even more promising is the use of recombinants based on the highly attenuated modified vaccinia Ankara (MVA), a variant of vaccinia derived from 500 passages through chicken cells which like canarypox is unable to replicate in mammalian cells. MVA has the advantage of a late stop in virus assembly which allows the full complement of viral proteins to be expressed in the cells to a high degree (Sutter and Moss, 1995). Comparison of recombinant MVA and other constructs in the SIV/monkey model revealed a higher immunogenicity and more pronounced antiviral effect (Hirsch et al., 1994).

In addition to different pox virus constructs, there are a multitude of live vectors under development and testing in the laboratory. Recombinant MoMuLV vectors carrying HIV-1 *env* and *rev* are being used directly or as a means of expressing proteins in autologous fibroblasts in attempts to boost the immune response in HIV-1 infected volunteers (Ziegner et al., 1995). HIV-1 genes have been engineered into viruses such as adenoviruses and poliovirus in an attempt to specifically stimulate a state of mucosal immunity (Alexander et al., 1994; Chanda et al., 1990; Evans et al., 1989; Lubeck et al., 1994; Natuk et al., 1992; Natuk et al., 1993; Natuk et al., 1995; Vella et al., 1991) and a number of non-viral vectors have also been produced, including salmonella and shigella (Walker and Fast, 1994).

4.4.5.5 Combination Vaccines

Earlier experiments with recombinant vaccinia viruses indicated that although efficient at priming the immune response the actual titres of antibodies was relatively poor. One of the more common approaches to vaccine development has therefore been the 'prime and boost' regime where vaccinees are first infected with the live recombinant virus and then boosted with purified recombinant protein or peptides. Initial experiments in the SIVmne/macaque model showed this to be an efficient means of stimulating both humoral and cellular immune responses and even protection against challenge with the homologous virus clone (Hu et al., 1992a; Hu et al., 1992b; Hu et al., 1992c). This protection unfortunately breaks down partially when animals are challenged with an uncloned 'swarm' of SIVmne (S.-L. Hu, personal communication, 1995). Although it has so far proved difficult to successfully protect animals using the prime and boost approach against the more vigorous SIV_{mac} challenge (Giavedoni et al., 1993; Yilma et al., 1994) it is clear the ensuing viral load in vivo is very much reduced (Israel et al., 1994) and may protect from disease development.

Based partially on the successful protection against challenge achieved in the animal models by the boost and prime approach a number of related vaccine regimes are currently being tested in human volunteers (Cooney et al., 1993; Graham et al., 1993; Pialoux et al., 1995; Picard et al., 1992; Walker and Fast, 1994). All are so far based on the use of vaccinia or canarypox recombinants expressing gp160 for priming followed by boosts with purified recombinant gp160, gp120 or peptides (usually V3). Although immunogenic the efficacy of such vaccines remains to be evaluated.

4.4.5.6 Live Attenuated Virus

Historically, the most successful vaccines have been live attenuated derivatives of the pathogenic virus itself (e.g. vaccinia, Sabin polio vaccine). Understandable fears about the safety of a replicating retrovirus currently rule out the use of a live attenuated HIV-1 vaccine in the general population. Although there is as yet no evidence that integration of the HIV provirus can result in oncogenesis or that HIV can combine with other retroviruses to form new pathogenic variants these possibilities cannot be ruled out. Consequently, no clinical trials with live attenuated virus vaccines are in progress and all of the preliminary work has been performed in the SIV/monkey and, more recently, in the HIV-1 chimpanzee models.

Desrosiers and colleagues initially demonstrated that deletion of the *nef* gene from SIVmac resulted in a highly attenuated phenotype in monkeys, with the virus becoming difficult to reisolate after some months and no signs of disease development apparent after 2 years (Kestler et al., 1991). Moreover, challenge of such 'immunised' monkeys with the pathogenic full-length virus failed to produce any sign of challenge virus replication whatsoever, even when doses as high as 1000 MID_{50} were used or heterologous challenges were performed (Daniel et al., 1992). This degree of protection is in stark contrast to the weak or non-existent immunity afforded by most other vaccines in the SIVmac system. One of the puzzles surrounding these data is the apparent need for a long period (40 weeks or more) of infection with the attenuated virus before a state of immunity is achieved. Most immune responses, particularly CTL responses, develop very quickly after infection (1-4 weeks) and remain relatively constant thereafter (Vogel et al., 1995). In a carefully controlled experiment Wyand et al (Wyand et al., 1994) showed no protection after 8 weeks 'immunisation', partial protection after 20 weeks and full protection at 72 weeks. In contrast, near complete protection has been demonstrated using an attenuated variant of SIVmac (C8) with a small (4 amino acid) deletion in the Nef protein after only 10 and 20 weeks of infection (Norley et al., 1994b). This discrepancy almost certainly reflects the different replication capabilities of the vaccine viruses in vivo: there would appear to be an inverse correlation between the degree of attenuation and the protective capacity of the viruses.

The greatest fear of a live attenuated vaccine concerns the possibility of reversion to the pathogenic wild-type virus. In a virus with a variability as high as HIV-1, this worry is all the more valid. Indeed, a number of monkeys infected with the C8 virus described above have been shown to develop variants which fill in the deletion and evolve over time back to the original full-length sequence (M. Cranage, personal communication, 1995), in one case resulting in AIDS development. Reversion of the large deletions used in Desrosiers' mutants is unlikely to occur, but to reduce the risk further this group has created and tested a range mutants with multiple depletions (Wyand et al., 1994) the best of which, SIVmac Δ3, has deletions in *nef*, the negative regulatory element (NRE) and *vpr*. The chances of reversion in such a construct are minimal, and the HIV-1 equivalent has been tested in chimpanzees.

Reversion, however, is not the only possible danger associated with attenuated vaccines for HIV-1. Recently, Ruprecht's group (Baba et al., 1995) showed that infection of neonatal rhesus macaques with SIVmac Δ3 resulted in high virus loads, a drop in CD4 cell numbers and AIDS without any indication of reversion. This indicates either that the undeveloped immune system of newborn monkeys is unable to cope with even an attenuated strain of SIV or that the deleted genes perform a role in adults which is not needed in neonates. Although it seems unlikely given the low virus load in adult monkeys infected with *nef*-deletion mutants that mother-to-child transmission would occur in the first place the risk of high-level replication in immunosuppressed vaccinees has to be evaluated. In the meantime, live attenuated variants of SIV remain the most potent forms of vaccine available and are being used extensively as tools to identify the immune response(s), if any, responsible for this protection in the hope that such immunity will be achievable using a safer approach.

4.4.5.7 Genetic Immunisation

In the last few years a new means of immunization has been developed which potentially offers the benefits of live recombinant vectors without the associated risks and restrictions. Although it has been known for many years that injection of purified viral DNA can result in the production of live progeny virus it is only recently that direct inoculation of plasmid DNA has been tested extensively as a form of vaccination. Much of the initial work has been carried out in mice, chickens or ferrets using influenza DNA and is reviewed in detail elsewhere in this volume. There are, however, an increasing number of groups attempting to stimulate a state of protective immunity in the various animal models for AIDS.

The intracellular events following immunisation with subgenomic DNA can closely mimic those following infection with the virus itself (fig. 4.4.1). DNA entering the cell will eventually be transcribed and translated to produce the protein or proteins of choice (e.g. HIV-1 gp160) which may be secreted in a near-native form with the correct glycosylation and conformation to stimulate an antibody response. In addition, as occurs with all proteins undergoing *de novo* synthesis, some will be diverted into the endogenous processing pathway and appropriate fragments will associate with

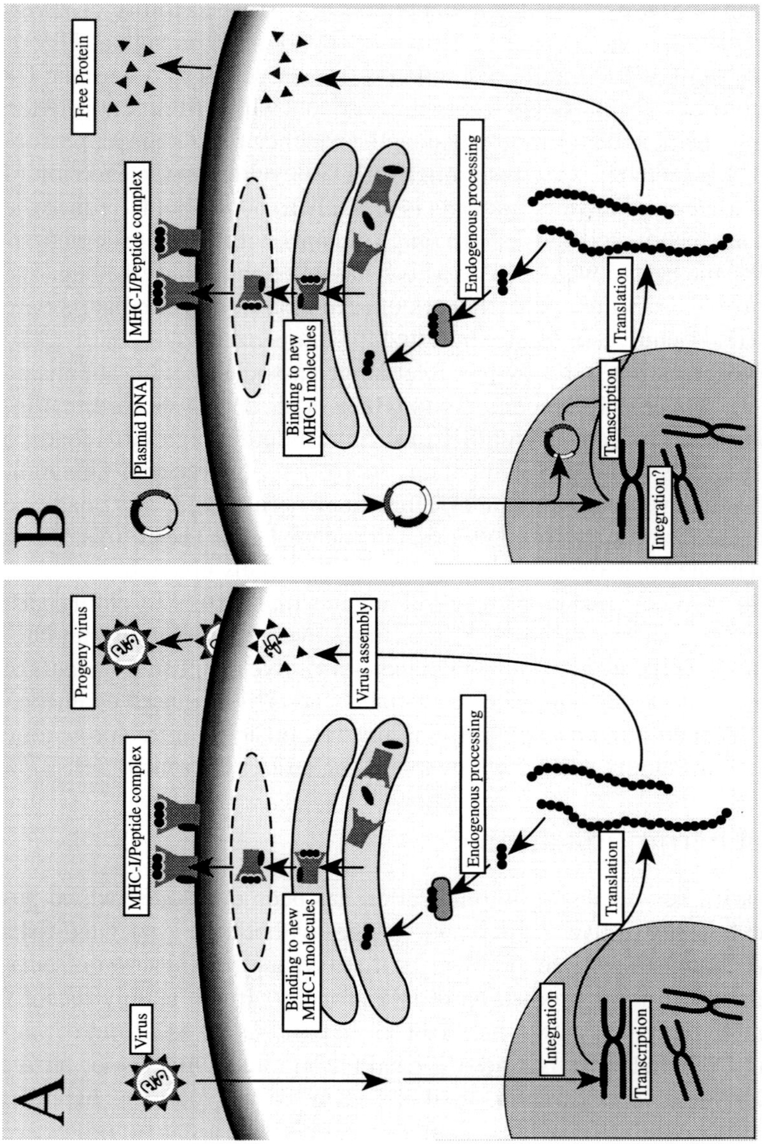

Fig. 4.4.1: Mimicry of natural infection by DNA immunization. A: After infection of a cell by HIV, viral RNA is released, reverse transcribed to DNA and incorporated into the host genome. This provirus is then transcribed and translated to proteins to produce progeny virus which bud from the cell. Proteins are also processed and expressed on the cell surface with MHC-I to stimulate and act as a target for virus specific CTL. B: After inoculation and entry of plasmid DNA, which may or may not become incorporated into the host genome, protein synthesis progresses normally resulting in the release of protein able to stimulate an antibody response and the processing and presentation of protein fragments for stimulation of CTL.

the MHC-I molecule to be expressed on the surface of the infected cell and stimulate a CTL response. In the influenza system, protection has not only been achieved using haemagglutinin DNA (Fynan et al., 1993a; Fynan et al., 1993b; Robinson et al., 1993; Webster et al., 1994), which stimulates neutralisating antibody responses, but also with nucleoprotein DNA against homologous and heterologous challenge in which neutralisation plays no role (Donnelly et al., 1994; Montgomery et al., 1993; Ulmer et al., 1994; Ulmer et al., 1993), presumably through the action of nucleoprotein-specific CTL.

Both neutralising antibody responses and virus-specific CTL have been induced in mice and monkeys following immunisation with HIV-1 *env* DNA (Liu et al., 1994). However, generation of good immune responses in monkeys has proven relatively difficult, and in the small number of vaccination/challenge experiments performed in the SIV/macaque system the degree of protection to challenge has so far been disappointing (Lu et al., 1994). Genetic immunization is at present an extremely new area of vaccine study and hopefully refinement of the vectors and delivery systems over the next few years will allow the generation of the levels of anti-HIV and -SIV immunity in monkeys that are presently being achieved in mice with influenza. One interesting alternative to the use of DNA is to bring messenger RNA itself into the cell, for example via liposomes (Martinon et al., 1993), allowing immediate translation into protein and avoiding some of the potential risks and problems associated with the use of DNA.

4.4.6 Conclusions

Efforts to develop a vaccine against AIDS have been, and remain, thwarted by unexpected obstacles and set-backs. The extreme variability of the virus, it's mode of transmission and the cells it infects make the production of a vaccine conceptually and practically very difficult. After more than a decade of intense research we still do not know which, if any, of the different immune mechanisms have the capability of preventing infection or disease under 'field' conditions. The requirements for a successful vaccine therefore remain a matter of debate.

It is now becoming clear that a vaccine based on the induction of an antiviral neutralising immune response alone will probably be ineffective and recently efforts have in part shifted to the stimulation of the cellular immune response. Indeed, there is circumstantial evidence that an antiviral cellular immune response is protective in humans. There are documented cases of African sex workers who remain HIV-1 seronegative despite hundreds of probable exposures to infection. Of particular interest is the fact that these women have a demonstrable cellular immune response to HIV-1, possibly as a result of prior, abortive infection. Similar immune reactions have been described in seronegative partners of seropositive patients and even in monkeys previously exposed to 'sub-infectious' doses of SIV which are resistant to subsequent infectious challenge. Identification of the protective mechanism which appear to prevent infection under field conditions would provide a goal for development of a vaccine.

In the meantime expectations for a first generation AIDS vaccine have fallen from one able to prevent infection to one able to prevent or delay the onset of disease. Many of the vaccines tested in the various animal models, although allowing infection to occur, seem to suppress the ensuing virus load and to delay, at least, the development of AIDS. If the degree of virus replication in vivo could be reduced to the point that further transmission is unlikely and disease development is considerably delayed the epidemic and it's burden on health care facilities would be greatly diminished.

What then are the prospects for developing an effective AIDS vaccine in the near future? For a variety of reasons described above it is unlikely that most of the vaccine formulations currently in human trials would provide an effective protection in the general population. The only form of vaccine able to induce consistant protection against 'rigorous' challenge in the primate model, i.e. the live attenuated virus vaccine, is at this time considered too dangerous for use as a prophylactic vaccine in humans. All other vaccines are very limited in the strength, duration and breadth of protection induced, even under optimised laboratory conditions. Novel approaches such as live recombinant vectors, combination vaccines and DNA vaccines are promising, but it will be some years before such systems are satisfactorally evaluated in the laboratory and even longer before they have undergone phase I, II and III trials and become, if at all, available to those most in need of an AIDS vaccine.

References

Agy, M. B., Frumkin, L. R., Corey, L., Coombs, R. W., Wolinsky, S. M., Koehler, J., Morton, W. R., and Katze, M. G. (1992) Infection of Macaca nemestrina by human immunodeficiency virus type-1. Science *257,* 103-106.

Alexander, L., Lu, H. H., Gromeier, M., and Wimmer, E. (1994) Dicistronic polioviruses as expression vectors for foreign genes. AIDS Res. Hum. Retroviruses *10* Suppl 2, 57-60.

Alter, H. J., Eichberg, J. W., Masur, H., Saxinger, W. C., Gallo, R., Macher, A. M., Lane, H. C., and Fauci, A. S. (1984) Transmission of HTLV-III infection from human plasma to chimpanzees: an animal model for AIDS. Science *226,* 549-552.

Arthur, L. O., Bess, J. W., Urban, R. G., Strominger, J. L., Morton, W. R., Mann, D. L., Henderson, L. E., and Benveniste, R. E. (1995) Macaques immunized with HLA-DR are protected from challenge with simian immunodeficiency virus. J. Virol. *69,* 3117-3124.

Baba, T. W., Jeong, Y. S., Penninck, D., Bronson, R., Greene, M. F., and Ruprecht, R. M. (1995) Pathogenicity of live attenuated SIV after mucosal infection of neonatal macaques. Science *267,* 1820-1825.

Bachmann, M. H., Delwart, E. L., Shpaer, E. G., Lingenfelter, P., Singal, R., and Mullins, J. I. (1994) Rapid genetic characterization of HIV type 1 strains from four World Health Organization-sponsored vaccine evaluation sites using a heteroduplex mobility assay. WHO Network for HIV Isolation and Characterization. AIDS Res. Hum. Retroviruses *10,* 1345-1353.

Banapour, B., Sernatinger, J., and Levy, J. A. (1986) The AIDS-associated retrovirus is not sensitive to lysis or inactivation by human serum. Virology *152,* 268-271.

Bartholomew, R. M., Esser, A. F., and Mueller-Eberhard, H. J. (1978) Lysis of oncornaviruses by human serum. Isolation of the viral complement (C1) receptor and identification as p15E. J. Exp. Med. *147,* 844-853.

Beer, B., Norley, S., Ennen, J., Plesker, R., Cichutek, K., and Kurth, R. (1995) Control of initial viremia correlates with the development of humoral immune response after SIVmac infection of rhesus macaques. AIDS *(Submitted),*

Berman, P. W., Gregory, T. J., Riddle, L., Nakamura, G. R., Champe, M. A., Porter, J. P., Wurm, F. M., Hershberg, R. D., Cobb, E. K., and Eichberg, J. W. (1990) Protection of chimpanzees from infection by HIV-1 after vaccination with recombinant glycoprotein gp120 but not gp160. Nature *345,* 622-625.

Bobkov, A., Cheingsong, P. R., Garaev, M., Rzhaninova, A., Kaleebu, P., Beddows, S., Bachmann, M. H., Mullins, J. I., Louwagie, J., Janssens, W., van der Groen, G., McCutchan, F., and Weber, J. (1994) Identification of an env G subtype and heterogeneity of HIV-1 strains in the Russian Federation and Belarus. AIDS *8,* 1649-1655.

Borrow, P., Lewicki, H., Hahn, B. H., Shaw, G. M., and Oldstone, M. B. (1994) Virus-specific CD8+ cytotoxic T-lymphocyte activity associated with control of viremia in primary human immunodeficiency virus type 1 infection. J. Virol. *68,* 6103-6110.

Bou-Habib, D. C., Roderiquez, G., Oravecz, T., Berman, P. W., Lusso, P., and Norcross, M. A. (1994) Cryptic nature of envelope V3 region epitopes protects primary monocytotropic human immunodeficiency virus type 1 from antibody neutralization. J. Virol. *68,* 6006-6013.

Broliden, P. A., Maekitalo, B., Akerblom, L., Rosen, J., Broliden, K., Utter, G., Jondal, M., Norrby, E., and Wahren, B. (1991) Identification of amino acids in the V3 region of gp120 critical for virus neutralization by human HIV-1-specific antibodies. Immunol. *73,* 371-376.

Burton, D. R., Pyati, J., Koduri, R., Sharp, S. J., Thornton, G. B., Parren, P., Sawyer, L., Hendry, R. M., Dunlop, N., Nara, P. L., Lamacchia, M., Garratty, E., Stiehm, E. R., Bryson, Y. J., Cao, Y. Z., Moore, J. P., Ho, D. D., and Barbas, C. F. (1994) Efficient neutralization of primary isolates of HIV-1 by a recombinant human monoclonal antibody. Science *266,* 1024-1027.

Buseyne, F., McChesney, M., Porrot, F., Kovarik, S., Guy, B., and Riviere, Y. (1993) Gag-specific cytotoxic T lymphocytes from human immunodeficiency virus type 1-infected individuals: Gag epitopes are clustered in three regions of the p24gag protein. J. Virol. *67,* 694-702.

Castro, B. A., Nepomuceno, M., Lerche, N. W., Eichberg, J. W., and Levy, J. A. (1991) Persistent infection of baboons and rhesus monkeys with different strains of HIV-2. Virology *184,* 219-226.

Cernescu, C. E., Tardei, G., Necula, A., Ruta, S. M., and Pau, C. P. (1994) The serologic significance of F viral genotype for human immunodeficiency virus type 1 epidemic. J. Infect. Dis. *170,* 1043-1044.

Chalifoux, L. V., Ringler, D. J., King, N. W., Sehgal, P. K., Desrosiers, R. C., Daniel, M. D., and Letvin, N. L. (1987) Lymphadenopathy in macaques experimentally infected with the simian immunodeficiency virus (SIV). Am J Pathol *128,* 104-110.

Chamat, S., Nara, P., Berquist, L., Whalley, A., Morrow, W. J., Koehler, H., and Kang, C. Y. (1992) Two major groups of neutralizing anti-gp120 antibodies exist in HIV-infected individuals. Evidence for epitope diversity around the CD4 attachment site. J. Immunol. *149,* 649-654.

Chanda, P. K., Natuk, R. J., Mason, B. B., Bhat, B. M., Greenberg, L., Dheer, S. K., Molnar-Kimber, K. L., Mizutani, S., Lubeck, M. D., Davis, A. R., and Hung, P. P. (1990) High level expression of the envelope glycoproteins of the human immunodeficiency virus type I in presence of rev gene using helper-independent adenovirus type 7 recombinants. Virology *175,* 535-547.

Cichutek, K., Merget, H., Norley, S., Linde, R., Kreuz, W., Gahr, M., and Kurth, R. (1992) Development of a quasispecies of human immunodeficiency virus type 1 in vivo. Proc. Natl. Acad. Sci. USA *89,* 7365-7369.

Clerici, M., Lucey, D. R., Zajac, R. A., Boswell, R. N., Gebel, H. M., Takahashi, H., Berzofsky, J. A., and Shearer, G. M. (1991) Detection of cytotoxic T lymphocytes specific for synthetic peptides of gp160 in HIV-seropositive individuals. J. Immunol. *146,* 2214-2219.

Cooney, E. L., Collier, A. C., Greenberg, P. D., Coombs, R. W., Zarling, J., Arditti, D. E., Hoffman, M. C., Hu, S. L., and Corey, L. (1991) Safety of and immunological response to a recombinant vaccinia virus vaccine expressing HIV envelope glycoprotein. Lancet *337,* 567-572.

Cooney, E. L., McElrath, M. J., Corey, L., Hu, S. L., Collier, A. C., and Arditti, D. (1993) Enhanced immunity to human immunodeficiency virus (HIV) envelope elicited by a combined vaccine regimen consisting of priming with a vaccinia recombinant expressing HIV envelope and boosting with gp160 protein. Proc. Natl. Acad. Sci. USA *90,* 1882-1886.

Cranage, M. P., Ashworth, L. A., Greenaway, P. J., Murphey-Corb, M., and Desrosiers, R. C. (1992) AIDS vaccine developments. Nature *355,* 685-686.

Culmann, B., Gomard, E., Kieny, M. P., Guy, B., Dreyfus, F., Saimot, A. G., Sereni, D., Sicard, D., and Levy, J. P. (1991) Six epitopes reacting with human cytotoxic CD8+ T cells in the central region of the HIV-1 NEF protein. J. Immunol. *146,* 1560-1565.

Daniel, M. D., Kirchhoff, F., Czajak, S. C., Sehgal, P. K., and Desrosiers, R. C. (1992) Protective effects of a live attenuated SIV vaccine with a deletion in the nef gene. Science *258,* 1938-1941.

Desrosiers, R. C. (1988) Simian immunodeficiency viruses. Annu Rev Microbiol *42,* 607-625.

Desrosiers, R. C., Wyand, M. S., Kodama, T., Ringler, D. J., Arthur, L. O., Sehgal, P. K., Letvin, N. L., King, N. W., and Daniel, M. D. (1989) Vaccine protection against simian immunodeficiency virus infection. Proc. Natl. Acad. Sci. USA *86,* 6353-6357.

Donnelly, J. J., Ulmer, J. B., and Liu, M. A. (1994) Immunization with DNA. J. Immunol. Methods. *176,* 145-152.

Dormont, D., Legrand, R., Cranage, M., Greenaway, P., Hunsmann, G., Stahlhennig, C., Rossi, G., Verani, P., Stott, J., Kitchin, P., Osterhaus, A., Devries, P., Kurth, R., Norley, S., Heeney, J., Biberfeld, G., and Putkonen, P. (1995) Protection of macaques against simian immunodeficiency virus infection with inactivated vaccines: Comparison of adjuvants, doses and challenge viruses. Vaccine *13,* 295-300.

Dormont, D., Livartowski, J., Chamaret, S., Guetard, D., Henin, D., Levagueresse, R., van de Moortelle, P. F., Larke, B., Gourmelon, P., Vazeux, R. et al. (1989) HIV-2 in rhesus monkeys: serological, virological and clinical results. Intervirology *30,* 59-65.

Dumitrescu, O., Kalish, M. L., Kliks, S. C., Bandea, C. I., and Levy, J. A. (1994) Characterization of human immunodeficiency virus type 1 isolates from children in Romania: identification of a new envelope subtype. J. Infect. Dis. *169,* 281-288.

Evans, D. J., McKeating, J., Meredith, J. M., Burke, K. L., Katrak, K., John, A., Ferguson, M., Minor, P. D., Weiss, R. A., and Almond, J. W. (1989) An engineered poliovirus chimaera elicits broadly reactive HIV-1 neutralizing antibodies. Nature *339,* 385-388.

Fish, D. C., Gilden, R. V., Bare, R. M., Trimmer, R. M., and Huebner, R. J. (1979) Prevention of spontaneous leukemia in AKR mice by passive immunization and type specificity of this protection. J. Natl. Cancer. Inst. *62,* 943-945.

Frumkin, L. R., Agy, M. B., Coombs, R. W., Panther, L., Morton, W. R., and Koehler, J. (1993) Acute infection of Macaca nemestrina by human immunodeficiency virus type 1. Virology *195,* 422-431.

Fultz, P. N., Gordon, T. P., Anderson, D. C., and McClure, H. M. (1990) Prevalence of natural infection with simian immunodeficiency virus and simian T-cell leukemia virus type I in a breeding colony of sooty mangabey monkeys. AIDS *4,* 619-625.

Fynan, E. F., Robinson, H. L., and Webster, R. G. (1993a) Use of DNA encoding influenza hemagglutinin as an avian influenza vaccine. DNA Cell Biol *12,* 785-789.

Fynan, E. F., Webster, R. G., Fuller, D. H., Haynes, J. R., Santoro, J. C., and Robinson, H. L. (1993b) DNA vaccines: protective immunizations by parenteral, mucosal, and gene-gun inoculations. Proc. Natl. Acad. Sci. USA *90,* 11478-11482.

Gao, F., Yue, L., Craig, S., Thornton, C. L., Robertson, D. L., Mccutchan, F. E., Bradac, J. A., Sharp, P. M., Hahn, B. H., Osmanov, S., Belsey, E. M., Heyward, W., Esparza, J., Galvaocastro, B., Vandeperre, P., Karita, E., Wasi, C., Sempala, S., Tugume, B., Biryahwaho, B., Rubsamenwaigmann, H., Vonbriesen, H., Esser, R., Grez, M., Holmes, H., Newberry, A., Ranjbar, S., Tomlinson, P., Bradac, J., Mccutchan, F., Louwagie, J., Hegerich, P., Lopezgalindez, C., Olivares, I., Dopazo, J., Mullins, J. I., Delwart, E. L., Bachmann, H. M., Goudsmit, J., Dewolf, F., Saragosti, S., Schochetman, G., Kalish, M., Luo, C. C., George, R., Pau, C. P., Weber, J., Cheingsongpopov, R., Kaleebu, P. et al. (1994) Genetic variation of HIV type 1 in four World Health Organization-sponsored vaccine evaluation sites: Generation of functional envelope (glycoprotein 160) clones representative of sequence subtypes A, B, C, and E. AIDS Res. Hum. Retroviruses *10,* 1359-1368.

Giavedoni, L. D., Planelles, V., Haigwood, N. L., Ahmad, S., Kluge, J. D., Marthas, M. L., Gardner, M. B., Luciw, P. A., and Yilma, T. D. (1993) Immune response of rhesus macaques to recombinant simian immunodeficiency virus gp130 does not protect from challenge infection. J. Virol. *67,* 577-583.

Graham, B. S., Belshe, R. B., Clements, M. L., Dolin, R., Corey, L., Wright, P. F., Gorse, G. J., Midthun, K., Keefer, M. C., Roberts, N. J., Agosti, J. M., Fernie, B. F., Stablein, J. S., Montefiori, D. C., Lambert, J. S., Hu, S. L., Esterlitz, J. R., Lawrence, D. N., and Koff, W. C. (1992) Vaccination of vaccinia-naive adults with human immunodeficiency virus type 1 gp160 recombinant vaccinia virus in a blinded, controlled, randomized clinical trial. J. Infect. Dis. *166,* 244-252.

Graham, B. S., Matthews, T. J., Belshe, R. B., Clements, M. L., Dolin, R., Wright, P. F., Gorse, G. J., Schwartz, D. H., Keefer, M. C., Bolognesi, D. P., Corey, L., Stablein, D. M., Esterlitz, J. R., Hu, S. L., Smith, G. E., Fast, P. E., and Koff, W. C. (1993) Augmentation of human immunodeficiency virus type 1 neutralizing antibody by priming with gp160 recombinant vaccinia and boosting with rgp160 in vaccinia-naive adults. The NIAID AIDS Vaccine Clinical Trials Network. J. Infect. Dis. *167,* 533-537.

Gregersen, J. P., Mehdi, S., Baur, A., and Hilfenhaus, J. (1990) Antibody- and complement-mediated lysis of HIV-infected cells and inhibition of viral replication. J Med Virol *30,* 287-293.

Haffar, O., Garrigues, J., Travis, B., Moran, P., Zarling, J., and Hu, S. L. (1990) Human immunodeficiency virus-like, nonreplicating, gag-env particles assemble in a recombinant vaccinia virus expression system. J. Virol. *64,* 2653-2659.

Hahn, B. H., Shaw, G. M., Arya, S. K., Popovic, M., Gallo, R. C., and Wong-Staal, F. (1984) Molecular cloning and characterization of the HTLV-III virus associated with AIDS. Nature *312,* 166-169.

Hart, M. K., Palker, T. J., Matthews, T. J., Langlois, A. J., Lerche, N. W., Martin, M. E., Scearce, R. M., McDanal, C., Bolognesi, D. P., and Haynes, B. F. (1990) Synthetic peptides containing T and B cell epitopes from human immunodeficiency virus envelope gp120 induce anti-HIV proliferative responses and high titers of neutralizing antibodies in rhesus monkeys. J. Immunol. *145,* 2677-2685.

Hartung, S., Norley, S. G., Ennen, J., Cichutek, K., Plesker, R., and Kurth, R. (1992) Vaccine protection against SIVmac infection by high- but not low-dose whole inactivated virus immunogen. J. AIDS *5,* 461-468.

Hirsch, V. M., Olmsted, R. A., Murphey-Corb, M., Purcell, R. H., and Johnson, P. R. (1989) An African primate lentivirus (SIVsm) closely related to HIV-2. Nature *339,* 389-392.

Hirsch, V. M., Sutter, G., Goldstein, S., Lifson, J., Moss, B., Chanock, R., Sissler, J., and Fuerst, T. (1994) Limited virus replication following homologous uncloned SIV challenge of macaques immunized with attenuated MVA vaccinia expressing SIV env and gag-pol. Seventh Annual Meeting of the National Cooperative Vaccine Development Groups for AIDS 117.

Hoffenbach, A., Langlade-Demoyen, P., Dadaglio, G., Vilmer, E., Michel, F., Mayaud, C., Autran, B., and Plata, F. (1989) Unusually high frequencies of HIV-specific cytotoxic T lymphocytes in humans. J. Immunol. *142,* 452-462.

Hoshino, H., Tanaka, H., Miwa, M., and Okada, H. (1984) Human T-cell leukaemia virus is not lysed by human serum. Nature *310,* 324-325.

Hu, S. L., Abrams, K., Barber, G. N., Moran, P., Zarling, J. M., Langlois, A. J., Kuller, L., Morton, W. R., and Benveniste, R. E. (1992a) Protection of macaques against SIV infection by subunit vaccines of SIV envelope glycoprotein gp160. Science *255,* 456-459.

Hu, S. L., Abrams, K., Misher, L., Stallard, V., Moran, P., Zarling, J. M., Langlois, A. J., Kuller, L., Morton, W. R., and Benveniste, R. E. (1992b) Evaluation of protective efficacy of recombinant subunit vaccines against simian immunodeficiency virus infection of macaques. J. Med. Primatol. *21,* 119-125.

Hu, S. L., Travis, B. M., Stallard, V., Abrams, K., Misher, L., Moran, P., Zarling, J. M., Langlois, A. J., Kuller, L., Morton, W. R. et al. (1992c) Immune responses to SIVmne envelope glycoproteins protect macaques from homologous SIV infection. AIDS Res. Hum. Retroviruses *8,* 1489-1494.

Hu, S. L., Zarling, J. M., Chinn, J., Travis, B. M., Moran, P. A., Sias, J., Kuller, L., Morton, W. R., Heidecker, G., and Benveniste, R. E. (1989) Protection of macaques against simian AIDS by immunization with a recombinant vaccinia virus expressing the envelope glycoproteins of simian type D retrovirus. Proc. Natl. Acad. Sci. USA *86,* 7213-7217.

Hunsmann, G., Moennig, V., and Schaefer, W. (1975a) Properties of mouse leukemia viruses. IX. Active and passive immunization of mice against Friend leukemia with isolated viral GP71 glycoprotein and its corresponding antiserum. Virology *66,* 327-329.

Hunsmann, G., Moennig, V., Seifert, E., and Schaefer, W. (1975b) Protection of mice against Friend leukemia by active and passive immunization with isolated viral glycoprotein and its antiserum respectively. Zeitschr. Naturforsch. *30c,* 309-310.

Israel, Z. R., Edmonson, P. F., Maul, D. H., O'Neil, S. P., Mossman, S. P., Thiriart, C., Fabry, L., van Opstal, O., Bruck, C., Bex, F., Burney, A., Fultz, P., Mullins, J. I., and Hoover, E. A. (1994) Incomplete protection, but suppression of virus burden, elicited by subunit simian immunodeficiency virus vaccines. J. Virol. *68,* 1843-1853.

Javaherian, K., Langlois, A. J., LaRosa, G. J., Profy, A. T., Bolognesi, D. P., Herlihy, W. C., Putney, S. D., and Matthews, T. J. (1990) Broadly neutralizing antibodies elicited by the hypervariable neutralizing determinant of HIV-1. Science *250,* 1590-1593.

Javaherian, K., Langlois, A. J., McDanal, C., Ross, K. L., Eckler, L. I., Jellis, C. L., Profy, A. T., Rusche, J. R., Bolognesi, D. P., Putney, S. D., and Matthews, T. J. (1989) Principal neutralizing domain of the human immunodeficiency virus type 1 envelope protein. Proc. Natl. Acad. Sci. USA *86,* 6768-6772.

Johnson, P. R., Montefiori, D. C., Goldstein, S., Hamm, T. E., Zhou, J., Kitov, S., Haigwood, N. L., Misher, L., London, W. T., Gerin, J. L., Allison, A., Purcell, R. H., Chanock, R. M., and Hirsch, V. M. (1992) Inactivated whole-virus vaccine derived from a proviral DNA clone of simian immunodeficiency virus induces high levels of neutralizing antibodies and confers protection against heterologous challenge. Proc. Natl. Acad. Sci. USA *89,* 2175-2179.

Johnson, R. P., Trocha, A., Buchanan, T. M., and Walker, B. D. (1993) Recognition of a highly conserved region of human immunodeficiency virus type 1 gp120 by an HLA-Cw4-restricted cytotoxic T-lymphocyte clone. J. Virol. *67,* 438-445.

Karacostas, V., Nagashima, K., Gonda, M. A., and Moss, B. (1989) Human immunodeficiency virus-like particles produced by a vaccinia virus expression vector. Proc. Natl. Acad. Sci. USA *86,* 8964-8967.

Katongole-Mbidde, E. (1993) Expectations of host countries in HIV vaccine trials. AIDS Res. Hum. Retroviruses *9 (Suppl 1),* S17-S18.

Kent, S. J., Corey, L., Agy, M. B., Morton, W. R., Mcelrath, M. J., and Greenberg, P. D. (1995) Cytotoxic and proliferative T cell responses in HIV-1- infected Macaca nemestrina. J. Clin. Invest. *95,* 248-256.

Kestler, H. W., Ringler, D. J., Mori, K., Panicali, D. L., Sehgal, P. K., Daniel, M. D., and Desrosiers, R. C. (1991) Importance of the nef gene for maintenance of high virus loads and for development of AIDS. Cell *65,* 651-662.

Korber, B. T. M., Osmanov, S., Esparza, J., Myers, G., Belsey, E. M., Heyward, W., Galvaocastro, B., Vandeperre, P., Karita, E., Wasi, C., Sempala, S., Tugume, B., Biryahwaho, B., Rubsamen-Waigmann, H., Vonbriesen, H., Esser, R., Grez, M., Holmes, H., Newberry, A., Ranjbar, S., Tomlinson, P., Bradac, J., McCutchan, F., Louwagie, J., Hegerich, P., Lopezgalindez, C., Olivares, I., Dopazo, J., Mullins, J. I., Delwart, E. L., Bachmann, H. M., Goudsmit, J., Dewolf, F., Hahn, B. H., Gao, F., Yue, L., Saragosti, S., Schochetman, G., Kalish, M., Luo, C. C., George, R., Pau, C. P., Weber, J., Cheingsongpopov, R., Kaleebu, P., Nara, P., Fenyo, E. M., Albert, J., and Korber, B. (1994) The World Health Organization Global Programme on AIDS proposal for standardization of HIV sequence nomenclature. AIDS Res. Hum. Retroviruses *10,* 1355-1358.

Koup, R. A., Safrit, J. T., Cao, Y., Andrews, C. A., McLeod, G., Borkowsky, W., Farthing, C., and Ho, D. D. (1994) Temporal association of cellular immune responses with the initial control of viremia in primary human immunodeficiency virus type 1 syndrome. J. Virol. *68,* 4650-4655.

Kurth, R., Kraus, G., Werner, A., Hartung, S., Centner, P., Baier, M., Norley, S., and Loewer, J. (1988) AIDS: animal retrovirus models and vaccines. J. AIDS *1,* 284-294.

Letvin, N. L., Daniel, M. D., Sehgal, P. K., Desrosiers, R. C., Hunt, R. D., Waldron, L. M., MacKey, J. J., Schmidt, D. K., Chalifoux, L. V., and King, N. W. (1985) Induction of AIDS-like disease in macaque monkeys with T-cell tropic retrovirus STLV-III. Science *230,* 71-73.

Liu, M., Davies, M. E., Perry, H., Yasutami, Y., Letvin, N., Ulmer, J., Donnelly, J., and Shiver, J. (1994) Immune responses to influenza and HIV-1 proteins elicited by DNA vaccination. Seventh Annual Meeting of the National Cooperative Vaccine Development Groups for AIDS 58.

Ljunggren, K., Biberfeld, G., Jondal, M., and Fenyoe, E. M. (1989a) Antibody-dependent cellular cytotoxicity detects type- and strain-specific antigens among human immunodeficiency virus types 1 and 2 and simian immunodeficiency virus SIVmac isolates. J. Virol. *63,* 3376-3381.

Ljunggren, K., Boettiger, B., Biberfeld, G., Karlson, A., Fenyoe, E. M., and Jondal, M. (1987a) Antibody-dependent cellular cytotoxicity-inducing antibodies against human immunodeficiency virus. Presence at different clinical stages. J. Immunol. *139,* 2263-2267.

Ljunggren, K., Chiodi, F., Broliden, P. A., Albert, J., Norkrans, G., Hagberg, L., Jondal, M., and Fenyoe, E. M. (1989b) HIV-1-specific antibodies in cerebrospinal fluid mediate cellular cytotoxicity and neutralization. AIDS Res. Hum. Retroviruses *5,* 629-638.

Ljunggren, K., Fenyoe, E. M., Biberfeld, G., and Jondal, M. (1987b) Detection of antibodies which mediate human immunodeficiency virus-specific cellular cytotoxicity (ADCC) in vitro. J. Immunol. Methods. *104,* 7-14.

Louwagie, J., Janssens, W., Mascola, J., Heyndrickx, L., Hegerich, P., van der Groen, G., Mccutchan, F. E., and Burke, D. S. (1995) Genetic diversity of the envelope glycoprotein from human immunodeficiency virus type 1 isolates of African origin. J. Virol. *69,* 263-271.

Lowenstine, L. J., Pedersen, N. C., Higgins, J., Pallis, K. C., Uyeda, A., Marx, P., Lerche, N. W., Munn, R. J., and Gardner, M. B. (1986) Seroepidemiologic survey of captive Old-World primates for antibodies to human and simian retroviruses, and isolation of a lentivirus from sooty mangabeys (Cercocebus atys). Int J Cancer *38,* 563-574.

Lu, S., Robinson, H., Mustafa, F., Johnson, E., Santoro, J. C., Arthos, J., Winsink, J., Mullins, J. I., Montefiori, D., Yasuhari, Y., Letvin, N., Haynes, J. R., K., M., and Wyand, M. (1994) Simian immunodeficiency virus DNA vaccine trial in macaques. Seventh Annual Meeting of the National Cooperative Vaccine Development Groups for AIDS 60.

Lubeck, M. D., Natuk, R. J., Chengalvala, M., Chanda, P. K., Murthy, K. K., Murthy, S., Mizutani, S., Lee, S. G., Wade, M. S., Bhat, B. M., Bhat, R., Dheer, S. K., Eichberg, J. W., Davis, A. R., and Hung, P. P. (1994) Immunogenicity of recombinant adenovirus-human immunodeficiency virus vaccines in chimpanzees following intranasal administration. AIDS Res. Hum. Retroviruses *10,* 1443-1449.

Luciw, P. A., Potter, S. J., Steimer, K., Dina, D., and Levy, J. A. (1984) Molecular cloning of AIDS-associated retrovirus. Nature *312,* 760-763.

Marschang, P., Sodroski, J., Wurzner, R., and Dierich, M. P. (1995) Decay-accelerating factor (CD55) protects human immunodeficiency virus type 1 from inactivation by human complement. Eur. J. Immunol. *25,* 285-290.

Martinon, F., Krishnan, S., Lenzen, G., Magne, R., Gomard, E. et al. (1993) Induction of virus-specific cytotoxic T-lymphocytes by liposome entrapped messenger RNA. Eur. J. Immunol. *23,* 1719-1722.

Matsushita, S., Robert-Guroff, M., Rusche, J., Koito, A., Hattori, T., Hoshino, H., Javaherian, K., Takatsuki, K., and Putney, S. (1988) Characterization of a human immunodeficiency virus neutralizing monoclonal antibody and mapping of the neutralizing epitope. J. Virol. *62,* 2107-2114.

McClure, H. M., Anderson, D. C., Fultz, P. N., Ansari, A. A., Jehuda-Cohen, T., Villinger, F., Klumpp, S. A., Switzer, W., Lockwood, E., Brodie, A., and Keyserling, H. (1991) Maternal transmission of SIVsmm in rhesus macaques. J. Med. Primatol. *20,* 182-187.

Mertens, T. E., Burton, A., Stoneburner, R. et al (1994) Global estimates and epidemiology of HIV infections and AIDS. AIDS *8 (suppl 1),* S361-S372.

Montefiori, D. C., Graham, B. S., Zhou, J., Zhou, J., Bucco, R. A., Schwartz, D. H., Cavacini, L. A., and Posner, M. R. (1993) V3-specific neutralizing antibodies in sera from HIV-1 gp160-immunized volunteers block virus fusion and act synergistically with human monoclonal antibody to the conformation-dependent CD4 binding site of gp120. NIH-NIAID AIDS Vaccine Clinical Trials Network. J Clin Invest *92,* 840-847.

Montgomery, D. L., Shiver, J. W., Leander, K. R., Perry, H. C., Friedman, A., Martinez, D., Ulmer, J. B., Donnelly, J. J., and Liu, M. A. (1993) Heterologous and homologous protection against influenza A by DNA vaccination: optimization of DNA vectors. DNA Cell Biol *12,* 777-783.

Moore, J. P., Cao, Y. Z., Qing, L., Sattentau, Q. J., Pyati, J., Koduri, R., Robinson, J., Barbas, C. F., Burton, D. R., and Ho, D. D. (1995) Primary isolates of human immunodeficiency virus type 1 are relatively resistant to neutralization by monoclonal antibodies to gp120, and their neutralization is not predicted by studies with monomeric gp120. J. Virol. *69,* 101-109.

Moore, J. P. and Ho, D. D. (1993) Antibodies to discontinuous or conformationally sensitive epitopes on the gp120 glycoprotein of human immunodeficiency virus type 1 are highly prevalent in sera of infected humans. J. Virol. *67,* 863-875.

Morton, W. R., Schmidt, A., McClure, J., Hu, S. L., Thompson, J., Tsai, C. C., and Anderson, D. (1994) HIV-2/EHO287: A pathogenic model in macaca nemestrina, Conference on advances in AIDS vaccine development. Seventh annual meeting of the National Cooperative Vaccine Development Groups for AIDS.

Murphey-Corb, M., Martin, L. N., Davison-Fairburn, B., Montelaro, R. C., Miller, M., West, M., Ohkawa, S., Baskin, G. B., Zhang, J. Y., Putney, S. D., Allison, A. C., and Eppstein, D. A. (1989) A formalin-inactivated whole SIV vaccine confers protection in macaques. Science *246,* 1293-1297.

Myers, G., MacInnes, K., and Korber, B. (1992) The emergence of simian/human immunodeficiency viruses. AIDS Res. Hum. Retroviruses *8,* 373-386.

Myers, G., Rabson, A. B., Berzofsky, J. A., Smith, T. F., and Wong-Staal, F. (1991) Human Retroviruses and AIDS. Los Alamos National Laboratory, Los Alamos.

Nara, P. L., Robey, W. G., Gonda, M. A., Carter, S. G., and Fischinger, P. J. (1987) Absence of cytotoxic antibody to human immunodeficiency virus-infected cells in humans and its induction in animals after infection or immunization with purified envelope glycoprotein gp120. Proc. Natl. Acad. Sci. USA *84,* 3797-3801.

Nara, P. L., Smit, L., Dunlop, N., Hatch, W., Merges, M., Waters, D., Kelliher, J., Gallo, R. C., Fischinger, P. J., and Goudsmit, J. (1990) Emergence of viruses resistant to neutralization by V3-specific antibodies in experimental human immunodeficiency virus type 1 IIIB infection of chimpanzees. J. Virol. *64,* 3779-3791.

Natuk, R. J., Chanda, P. K., Lubeck, M. D., Davis, A. R., Wilhelm, J., Hjorth, R., Wade, M. S., Bhat, B. M., Mizutani, S., Lee, S., Eichberg, J., Gallo, R. C., Hung, P. P., and Robert-Guroff, M. (1992) Adenovirus-human immunodeficiency virus (HIV) envelope recombinant vaccines elicit high-titered HIV-neutralizing antibodies in the dog model. Proc. Natl. Acad. Sci. USA *89,* 7777-7781.

Natuk, R. J., Lubeck, M. D., Chanda, P. K., Chengalvala, M., Wade, M. S., and Murthy, S. C. (1993) Immunogenicity of recombinant human adenovirus-human immunodeficiency virus vaccines in chimpanzees. AIDS Res. Hum. Retroviruses *9,* 395-404.

Natuk, R. J., Wade, M. S., Chengalvala, M., Murthy, K. K., Chanda, P. K., Murthy, S. C. S., Bhat, B. M., Bhat, R., Dheer, S. K., Mizutani, S., Lee, S. G., Davis, A. R., Hung, P. P., Lubeck, M. D., and Eichberg, J. W. (1995) Adenovirus as vector for HIV: Efficacy and safety issues. Non Target Effects of Live Vaccines *84,* 153-156.

Nixon, D. F., Townsend, A. R., Elvin, J. G., Rizza, C. R., Gallwey, J., and McMichael, A. J. (1988) HIV-1 gag-specific cytotoxic T lymphocytes defined with recombinant vaccinia virus and synthetic peptides. Nature *336,* 484-487.

Norley, S., Beer, B., König, H., Jensen, F., and Kurth, R. (1996) SIV_{mac} vaccine studies using whole inactivated virus antigen sequentially depleted of viral proteins. Vaccine *Submitted,*

Norley, S. G., Beer, B., Binninger-Schinzel, D., Cosma, C., and Kurth, R. (1994) Protection from SIVmac challenge following short-term infection with a nef-deficient attenuated virus. Seventh Annual Meeting of the National Cooperative Vaccine Development Groups for AIDS 213.

Norley, S. G., Mikschy, U., Werner, A., Staszewski, S., Helm, E. B., and Kurth, R. (1990) Demonstration of cross-reactive antibodies able to elicit lysis of both HIV-1- and HIV-2-infected cells. J. Immunol. *145,* 1700-1705.

Norley, S. G., and Wardley, R. C. (1983) Effector mechanisms in the pig. Antibody-dependent cellular cytolysis of African swine fever virus infected cells. Res Vet Sci *35,* 75-79.

Oldstone, M. B. A. (1994) Cytotoxic T-lymphocytes in human viral and malaria infections. Current Topics in Microbiology and Immunology *189.*

Pialoux, G., Excler, J. L., Riviere, Y., Gonzalezcanali, G., Feuillie, V., Coulaud, P., Gluckman, J. C., Matthews, T. J., Meignier, B., Kieny, M. P., Gonnet, P., Diaz, I., Meric, C., Paoletti, E., Tartaglia, J., Salomon, H., and Plotkin, S. (1995) A prime-boost approach to HIV preventive vaccine using a recombinant canarypox virus expressing glycoprotein 160 (MN) followed by a recombinant glycoprotein 160 (MN/LAI). AIDS Res. Hum. Retroviruses *11,* 373-381.

Picard, O., Achour, A., Bernard, J., Halbreich, A., Bizzini, B., Boyer, V., Desgranges, C., Bertho, J. M., Lachgar, A., Polliotti, B., Defer, M. C., Lanneval, K., Imbert, J. C., Frottier, J., Salaun, J. J., Burney, A., and Zagury, D. (1992) A 2-year follow-up of an anti-HIV immune reaction in HIV-1 gp160-immunized healthy seronegative humans: evidence for persistent cell-mediated immunity. J. AIDS *5,* 539-546.

Posner, M. R., Elboim, H. S., Cannon, T., Cavacini, L., and Hideshima, T. (1992) Functional activity of an HIV-1 neutralizing IgG human monoclonal antibody: ADCC and complement-mediated lysis. AIDS Res. Hum. Retroviruses *8,* 553-558.

Prince, A. M., Moor-Jankowski, J., Eichberg, J. W., Schellekens, H., Mauler, R. F., Girard, M., and Goodall, J. (1988) Chimpanzees and AIDS research. Nature *333,* 513.

Putkonen, P., Boettiger, B., Warstedt, K., Thorstensson, R., Albert, J., and Biberfeld, G. (1989) Experimental infection of cynomolgus monkeys (Macaca fascicularis) with HIV-2. J. AIDS *2,* 366-373.

Putkonen, P., Thorstensson, R., Ghavamzadeh, L., Albert, J., Hild, K., Biberfeld, G., and Norrby, E. (1991) Prevention of HIV-2 and SIVsm infection by passive immunization in cynomolgus monkeys. Nature *352,* 436-438.

Putney, S. D., Matthews, T. J., Robey, W. G., Lynn, D. L., Robert-Guroff, M., Mueller, W. T., Langlois, A. J., Ghrayeb, J., Petteway, S. R., Weinhold, K. J., Fischinger, P. J., Wong-Staal, F., Gallo, R. C., and Bolognesi, D. P. (1986) HTLV-III/LAV-neutralizing antibodies to an E. coli-produced fragment of the virus envelope. Science *234,* 1392-1395.

Reimann, K. A., Tenner-Racz, K., Racz, P., Montefiori, D. C., Yasutomi, Y., Lin, W., Ransil, B. J., and Letvin, N. L. (1994) Immunopathogenic events in acute infection of rhesus monkeys with simian immunodeficiency virus of macaques. J. Virol. *68,* 2362-2370.

Robert-Guroff, M., Brown, M., and Gallo, R. C. (1985) HTLV-III-neutralizing antibodies in patients with AIDS and AIDS-related complex. Nature *316,* 72-74.

Robey, W. G., Safai, B., Oroszlan, S., Arthur, L. O., Gonda, M. A., Gallo, R. C., and Fischinger, P. J. (1985) Characterization of envelope and core structural gene products of HTLV-III with sera from AIDS patients. Science *228,* 593-595.

Robinson, H. L., Hunt, L. A., and Webster, R. G. (1993) Protection against a lethal influenza virus challenge by immunization with a haemagglutinin-expressing plasmid DNA. Vaccine *11,* 957-960.

Sarngadharan, M. G., diMarzo Veronese, F., Lee, S., and Gallo, R. C. (1985) Immunological properties of HTLV-III antigens recognized by sera of patients with AIDS and AIDS-related complex and of asymptomatic carriers of HTLV-III infection. Cancer Res *45,* 4574-4577.

Sawyer, L. S., Wrin, M. T., Crawford-Miksza, L., Potts, B., Wu, Y., Weber, P. A., Alfonso, R. D., and Hanson, C. V. (1994) Neutralization sensitivity of human immunodeficiency virus type 1 is determined in part by the cell in which the virus is propagated. J. Virol. *68,* 1342-1349.

Schultz, A. M. and Stott, E. J. (1994) Primate models for AIDS vaccines. AIDS *8 (suppl. 1),* 203-212.

Schwartz, D. H., Gorse, G., Clements, M. L., Belshe, R., Izu, A., Duliege, A. M., Berman, P., Twaddell, T., Stablein, D., Sposto, R., Siliciano, R., and Matthews, T. (1993) Induction of HIV-1-neutralising and syncytium-inhibiting antibodies in uninfected recipients of HIV-1IIIB rgp120 subunit vaccine. Lancet *342,* 69-73.

Shafferman, A., Jahrling, P. B., Benveniste, R. E., Lewis, M. G., Phipps, T. J., Eden-McCutchan, F., Sadoff, J., Eddy, G. A., and Burke, D. S. (1991) Protection of macaques with a simian immunodeficiency virus envelope peptide vaccine based on conserved human immunodeficiency virus type 1 sequences. Proc. Natl. Acad. Sci. USA *88,* 7126-7130.

Shafferman, A., Lewis, M. G., McCutchan, F. E., Benveniste, R. E., Jahrling, P. B., Hickman, R. L., Lai, C. Y., Burke, D. S., and Eddy, G. A. (1992) Vaccination of macaques with SIV conserved envelope peptides suppressed infection and prevented disease progression and transmission. AIDS Res. Hum. Retroviruses *8,* 1483-1487.

Shore, S. L., Nahmias, A. J., Starr, S. E., Wood, P. A., and McFarlin, D. E. (1974) Detection of cell-dependent cytotoxic antibody to cells infected with herpes simplex virus. Nature *251,* 350-352.

Siegel, F., Kurth, R., and Norley, S. (1995) Neither whole inactivated virus immunogen nor passive immunoglobulin transfer protect against SIVagm infection in the African green monkey natural host. J. AIDS *8,* 217-226.

Spear, G. T., Takefman, D. M., Sullivan, B. L., Landay, A. L., and Zolla, P. S. (1993) Complement activation by human monoclonal antibodies to human immunodeficiency virus. J. Virol. *67,* 53-59.

Stahl-Hennig, C., Herchenroeder, O., Nick, S., Evers, M., Stille-Siegener, M., Jentsch, K. D., Kirchhoff, F., Tolle, T., Gatesman, T. J., Lueke, W., and Hunsmann, G. (1990) Experimental infection of macaques with HIV-2ben, a novel HIV-2 isolate. AIDS *4,* 611-617.

Steimer, K. S., Scandella, C. J., Skiles, P. V., and Haigwood, N. L. (1991) Neutralization of divergent HIV-1 isolates by conformation-dependent human antibodies to Gp120. Science *254,* 105-108.

Steimer, K. S., Sinangil, F., Kahn, J., Chernoff, D., Duliege, A. M., and Baenziger, J. (1994) Primary isolate neutralizing activity of human antibodies directed to recombinant native HIV-SF2 gp120 (rgp120SF2). AIDS Res. Hum. Retroviruses *10 (suppl 2),* S55.

Stott, E. J., Chan, W. L., Mills, K. H., Page, M., Taffs, F., Cranage, M., Greenaway, P., and Kitchin, P. (1990) Preliminary report: protection of cynomolgus macaques against simian immunodeficiency virus by fixed infected-cell vaccine. Lancet *336,* 1538-1541.

Stott, E. J., Kitchin, P., and Mills, K. (1993) The role of cellular and viral antigens in protection against SIV infection in macaques. IX International Conference on AIDS / IV STD World Congress Abstract WS-A23-2.

Stott, E. J., Kitchin, P. A., Page, M., Flanagan, B., Taffs, L. F., Chan, W. L., Mills, K. H. G., Silvera, P., and Rodgers, A. (1991) Anti-cell antibody in macaques. Nature *353,* 393.

Sutter, G. and Moss, B. (1995) Novel vaccinia vector derived from the host range restricted and highly attenuated MVA strain of vaccinia virus. Non Target Effects of Live Vaccines *84,* 195-200.

Toth, F. D., Kiss, J., Szabo, B., Fust, G., Ujhelyi, E., Hollan, S. R., and Vaczi, L. (1989) Complement-dependent cytotoxicity of antibodies reactive with HIV-induced cell surface antigens in HIV-carrying haemophiliacs. Acta Virol *33,* 521-526.

Ulmer, J. B., Deck, R. R., Dewitt, C. M., Friedman, A., Donnelly, J. J., and Liu, M. A. (1994) Protective immunity by intramuscular injection of low doses of influenza virus DNA vaccines. Vaccine *12,* 1541-1544.

Ulmer, J. B., Donelly, J. J., Parker, S. E., Rhodes, G. H., Feigner, P. L., Dwarki, V. J., Gromkowski, S. H., Deck, R. R., DeWitt, C. M., Friedman, A., Hawe, L. A., Leander, K. R., Martinez, D., Perry, H. C., Shiver, J. W., Montgomery, D. L., and Liu, M. A. (1993) Heterologous protection against influenza by injection of DNA encoding a viral protein. Science *259,* 1745-1749.

Vella, C., Minor, P. D., Weller, I. V., Jenkins, O., Evans, D., and Almond, J. (1991) Recognition of poliovirus/HIV chimaeras by antisera from individuals with HIV infection. AIDS *5,* 425-430.

Veronese, F. D., DeVico, A. L., Copeland, T. D., Oroszlan, S., Gallo, R. C., and Sarngadharan, M. G. (1985) Characterization of gp41 as the transmembrane protein coded by the HTLV-III/LAV envelope gene. Science *229,* 1402-1405.

Villinger, F., Powell, J. D., Jehuda-Cohen, T., Neckelmann, N., Vuchetich, M., De, B., Folks, T. M., McClure, H. M., and Ansari, A. A. (1991) Detection of occult simian immunodeficiency virus SIVsmm infection in asymptomatic seronegative nonhuman primates and evidence for variation in SIV gag sequence between in vivo- and in vitro-propagated virus. J. Virol. *65,* 1855-1862.

Vogel, T., Beer, B., zur Megede, J., Ihlenfeldt, H. G., Jung, G., Kurth, R., and Norley, S. (1996) Induction of anti-SIVmac cellular and humoral immune responses in rhesus macaques by peptide immunogens: Reduction of virus load following challenge. J. Virol *(Submitted),*

Vogel, T., Kurth, R., and Norley, S. (1994) The majority of neutralizing Abs in HIV-1-infected patients recognize linear V3 loop sequences. Studies using HIV-1MN multiple antigenic peptides. J. Immunol. *153,* 1895-1904.

Walker, B. D., Flexner, C., Birch-Limberger, K., Fisher, L., Paradis, T. J., Aldovini, A., Young, R., Moss, B., and Schooley, R. T. (1989) Long-term culture and fine specificity of human cytotoxic T-lymphocyte clones reactive with human immunodeficiency virus type 1. Proc. Natl. Acad. Sci. USA *86,* 9514-9518.

Walker, B. D. and Plata, F. (1990) Cytotoxic T lymphocytes against HIV. AIDS *4,* 177-184.

Walker, M. C. and Fast, P. E. (1994) Clinical trials of candidate AIDS vaccines. AIDS *8 (suppl 1),* S213-S236.

Wang, C. Y., Looney, D. J., Li, M. L., Walfield, A. M., Ye, J., Hosein, B., Tam, J. P., and Wong-Staal, F. (1991) Long-term high-titer neutralizing activity induced by octameric synthetic HIV-1 antigen. Science *254,* 285-288.

Weber, J., Cheinsongpopov, R., Callow, D., Adams, S., Patou, G., Hodgkin, K. and Martin, S. (1995) Immunogenicity of the yeast recombinant p17/p242:Ty virus-like particles (p24-VLP) in healthy volunteers, Vaccine *13*, 831-834.

Webster, R. G., Fynan, E. F., Santoro, J. C., and Robinson, H. (1994) Protection of ferrets against influenza challenge with a DNA vaccine to the haemagglutinin. Vaccine *12,* 1495-1498.

Wyand, M. S., Manson, K., and Desrosiers, R. C. (1994) Vaccine protection against pathogenic SIVmac251 using live-attenuated vaccines SIV239 Δ3 and SIV316 Δnef. Seventh Annual Meeting of the National Cooperative Vaccine Development Groups for AIDS 109.

Yasutomi, Y., Palker, T. J., Gardner, M. B., Haynes, B. F., and Letvin, N. L. (1994) Synthetic peptide in mineral oil adjuvant elicits simian immunodeficiency virus-specific CD8+ cytotoxic T-lymphocytes in rhesus macaques. J. Immunol. *151,* 5096-5105.

Yefenof, E., Asjo, B., and Klein, E. (1991) Alternative complement pathway activation by HIV infected cells: C3 fixation does not lead to complement lysis but enhances NK sensitivity. Int Immunol *3,* 395-401.

Yilma, T., Ahmad, S., McChesney, M., Marthas, M., Kluge, J. D., Bartz, C., Morton, W., Nolte, R., Gardner, M., Luciw, P., and Giavedoni, L. (1994) Protection studies in macaques immunized with vaccinia virus recombinants, subunit antigens, and live attenuated SIVmac239Δnef and SIVmac239Δnef-IFN-gamma. Seventh Annual Meeting of the National Cooperative Vaccine Development Groups for AIDS.

Zagury, D., Bernard, J., Cheynier, R., Desportes, I., Leonard, R., Fouchard, M., Reveil, B., Ittele, D., Lurhuma, Z., Mbayo, K., Wane, J., Salaun, J. J., Goussard, B., Dechazal, L., Burney, A., Nara, P., and Gallo, R. C. (1988) A group specific anamnestic immune reaction against HIV-1 induced by a candidate vaccine against AIDS. Nature *332,* 728-731.

Ziegner, U., Peters, G., Jolly, D. J., Mento, S. J., Galpin, J., Prussak, C. E., Barber, J. R., Hartnett, D. E., Bohart, C., Klump, W., Sajjadi, N., Merchant, B., and Warner, J. F. (1995) Cytotoxic T-lymphocyte induction in asymptomatic HIV-1- infected patients immunized with Retrovector((R)) transduced autologous fibroblasts expressing HIV-1(IIIB) Env/Rev proteins. AIDS *9,* 43-50.

4.5 Vaccination Therapy Against Autoimmune Diseases

Irun R. Cohen and Felix Mor

4.5.1 Introduction

Vaccination is the art of getting the immune system to respond to a particular antigen in a particular way. Note that this definition embodies two distinct concepts: the concept of the antigen epitope as the specific target of vaccination, and the concept of the biologic response, the way we want the immune system to respond to the antigen. Also note that our degree of understanding of each of these two elements differs.

The perception of epitopes by the lymphocytes is now well on the way to being explained; witness our molecular knowledge of the interaction of the T-cell receptor (TCR) with the processed peptide epitope presented in the cleft of a major histocompatibility complex (MHC) molecule (Stern et al., 1994). In contrast, our knowledge of the factors that control the biologic outcome of the immune response is still far from a satisfactory molecular description. Control of the biologic outcome of immunization would appear to be a complex chain of events involving cytokine networks, Th1-Th2 and other lymphocyte interactions, cell adherence, trafficking, and tissue penetration, and many different ligand-receptor interactions between lymphocytes and antigen-presenting cells (Cohen, 1992a). The rational design of new vaccines will likely benefit from a better understanding of the workings of the presently used empirical adjuvants and the more sophisticated future adjuvants. Certainly, control of the biologic response to immunization is at the crux of any possible use of vaccination for therapy of autoimmune diseases. Although vaccination usually implies immunization to prevent future disease, the application of the term to autoimmune disease is meant to include vaccine therapy, the induction of a change in the nature of an autoimmune response that is already underway and causing the disease.

4.5.1.1 Vaccination, Tolerance and Autoimmune Disease

The historic quest of immunologists to use vaccination to turn on an effective immune response would seem, at first glance, to negate vaccination as a therapy for autoimmune disease. Logically, it would seem that therapy for autoimmune diseases would require the induction of tolerance, the turning off of the immune response, the opposite of vaccination. We have been taught that autoimmune diseases result from the activation of lymphocytes whose very existence was forbidden (Burnet, 1959). According to this view, the logical cure for autoimmune diseases should be specific inactivation of the forbidden lymphocytes. The discovery, mentioned above, of the ternary complex of T-cell recognition, the MHC-peptide-TCR complex, and its ancillary CD4 and other ligands has led to the design of peptide blockers or blocking antibodies directed specifically to these molecules (Bach, 1993). Active vaccination as therapy for autoimmune disease is, conceptually, the opposite of such attempts to make the system tolerant by blocking its activity (Cohen, 1995). The basis for vaccination therapy is the appreciation that the expression of autoimmune disease, like the definition of vaccination given above, depends on two components: the first is the focus of an immune response on a particular self-antigen and the second is the biologic nature of the resulting response.

It is clear that perception of a self-antigen by a TCR is not sufficient to cause an autoimmune disease; the disease results from the biology of the immune response that follows recognition of the antigen (Cohen, 1992). Vaccination as the art of inducing a particular biologic response has a place in autoimmune therapy to the extent that a benign or regulatory autoimmune response can be activated to replace or depress the noxious biologic response causing the disease (Cohen, 1995). The substance of this chapter will be the demonstration that several types of vaccination may be preventive and even curative in experimental autoimmune diseases. Clinical applications are underway.

4.5.2 T Cell Vaccination

4.5.2.1 Specific Cell Vaccines

The concept of T cell vaccination as a way to regulate autoimmune responses is deceptively simple: if an autoimmune disease is the work of a forbidden clone of autoimmune T cells, then why not muster the immune system itself to eradicate the offenders by using the autoimmune T cells as a vaccine. Indeed, the first experiments showed that it was possible to prevent experimental autoimmune encephalomyelitis (EAE) in rats by vaccinating them with a single injection of a syngeneic T cell line specific for myelin basic protein (MBP), the target antigen in rat EAE (Ben-Nun et al., 1981). T cell vaccination was then shown to be effective in preventing (Roubaty et al., 1990; Holoshitz et al., 1983; Maron et al., 1983) and even in treating (Lider et al., 1987) other experimentally induced autoimmune diseases. T cell vaccination could also be used to abort the progression of the spontaneous autoimmune diabetes arising in NOD strain mice (Elias et al., 1991; Formby and Shao, 1993). The specificity of the protection induced by T cell vaccination was found to be related to the specificity of the TCR of the T cell line or clone used for vaccination; this suggested that the TCR was probably the target of the anti-autoimmune regulatory T cells (Holoshitz et al., 1983). This idea was supported by the observation that T cell vaccination induced the appearance of clone-specific (anti-idiotypic) T cells of both the CD4 and CD8 types (Lider et al., 1988) and by the observation that the anti-idiotypic T cells were responsible for mediating resistance to the disease (Beraud et al., 1989; Lider et al., 1989). Thus the autoimmune TCR appeared to be the target of the regulatory response induced by T cell vaccination. But, the TCR did not seem to behave as a simple immunogen administered to a naive immune system.

4.5.2.2 Complexities of T Cell Vaccination

At least four observations suggest that autoimmune T cell vaccination might represent a phenomenon beyond the realm of primary immunization. The first observation is that T cell vaccination with whole cells requires a unique "adjuvant": a state of activation of the vaccinating T cells. The T cells can vaccinate when used either alive (Lider et al., 1988; Beraud et al., 1989; Lider et al., 1989), attenuated by irradiation or mitomycin C (Ben-Nun et al., 1981), or killed by chemical cross-linkers (Lider et al., 1987; Lider et al., 1986), but only on condition that the T cells have been activated, for at least 6-8 hours up to several days, before inoculation. This activation can be done using a T-cell mitogen or the specific antigen (Naparstek et al., 1983), but the state of activation appears to be essential; the same T cell line administered in a resting state will not induce resistance to EAE even when inoculated at 10-fold the dose. Therefore, it seems that the mere concentration of TCR is not as critical as is the state of the cells in the vaccine.

Secondly, a single clone can successfully vaccinate against diseases that probably involve many different autoimmune clones and more than one antigen, such as adjuvant arthritis in Lewis rats (Cohen et al., 1985) or spontaneous diabetes in NOD mice (Formby and Shao, 1993). Thus the regulatory response must have some way of spreading to regulate other autoimmune T cells that would seem to be unrelated idiotypically to the vaccine.

Thirdly, the response to T cell vaccination is unusually sensitive; a single injection of very small numbers of a T cell clone, as few as 100, can activate a protective anti-idiotypic response within five days (Beraud et al., 1989). Thus, the immune system acts as if it were already primed to certain autoimmune T cells before the individual has been vaccinated.

This idea is supported by the fourth observation: The induction of an autoimmune disease, such as EAE may itself activate regulatory T cells specific for a peptide epitope of a TCR associated with the disease (Sun et al., 1988; Kumar and Sercarz, 1993). Thus, even without T cell vaccination, the immune system appears outfitted naturally with anti-idiotypic T cells specific for certain autoimmune TCR epitopes. These observations are compatible with the notion that T cell vaccination may work by boosting naturally existing anti-idiotypic T cells specific for common TCR epitopes expressed by particular autoimmune T cells (Cohen, 1992a; Cohen, 1992b). We do not yet know how common TCR idiotopes arise, but such idiotopes may be related to the restricted TCR usage seen to be associated with autoimmunity (Zamvil and Steinman, 1990). We also need to understand how anti-idiotypic T cells become primed naturally to such TCR idiotopes, how the state of activation of a T cell vaccine can act as an "adjuvant", and how the regulatory effects of T cell vaccination might spread to down-regulate idiotypically unrelated autoimmune T cells.

The anti-idiotypic response to T cell vaccination, together with the regularity of expression of autoimmunity, both pathologic and natural, have led to the theory of the immunological homunculus (Cohen, 1992b; Cohen and Young, 1991). The homunculus idea suggests that autoimmunity to a limited and predictable set of self antigens is inherent in the organization of the immune repertoire and that the biologic expression of this potential for autoimmunity is, in health, tightly regulated by elements of the immune system. Autoimmune disease arises through a failure of the regulatory elements to control the biologic expression of the autoimmune lymphocytes normally quiescent in the healthy repertoire. Thus, autoimmune disease can erupt in the wake of any insult, like a virus infection, that might weaken immune regulation. T cell vaccination, by activating the anti-idiotypic T cells, acts to restore the regulatory balance. The phenomenon of T cell vaccination has also stimulated theoretical models of immune regulation (Segel et al., 1995; Borghans and de Boer, 1995). Obviously, the value of these speculations will depend on how much they contribute to the experimental elucidation of the molecular basis of T cell vaccination. It has been suggested that the autoimmune effector T cells responsible for disease are lysed (Sun et al., 1988) by the anti-idiotypic T cells. Recently, we have obtained evidence that T cell vaccination may activate a switch in the cytokines produced by the autoimmune T cells from those causing inflammation and tissue destruction (Th1) to those which can suppress tissue destruction (Th2) (Cohen, 1995). Whatever the molecular mechanism turns out to be, the effectiveness of T cell vaccination in experimental animals supports the idea that autoimmune diseases can be prevented or treated by activating the immune system, by vaccinating.

4.5.2.3 Ergotypic Vaccination

In the course of investigating T cell vaccination with antigen-specific T cells, it was observed that partial resistance to some autoimmune conditions could be induced by vaccinating animals with syngeneic T cells that were not specific for the autoantigen (Lohse et al., 1989). This type of T cell vaccination requires that the T cells be in an activated state. Since the responding T cells seem to be directed to some surface marker of activation on the target T cells and not to the TCR, we termed the response anti-ergotypic, meaning a response against an activation marker (Lohse et al., 1989). Responses of T cells to activated T cells are readily observable in both animals and in humans, but nothing is yet known about the target ergotope, how it is recognized, or how the response inhibits autoimmune disease.

4.5.2.4 TCR Peptide Vaccination

The anti-idiotypic specificity of T cell vaccination using whole T cells led to the discovery that peptides of autoimmune TCR could also be used to induce resistance to experimental autoimmune disease (Broeren et al., 1995; Vandenbark et al., 1989; Howell et al., 1989). In these studies, peptides from the CDR1 (Broeren et al., 1995) and CDR2 (Vandenbark et al., 1989) segment of the TCR β chain (Broeren et al., 1995), a segment outside of the clone-specific idiotype, as well as TCR peptides from the clone-specific CDR3 segment have been used (Howell et al., 1989). It is not yet clear to what extent vaccination with TCR peptides is related to T cell vaccination with intact, activated T cells. TCR peptide vaccination seems to require amounts of peptide orders of magnitude greater than the amounts of TCR present in whole T cell vaccination. Furthermore, TCR peptide vaccination requires strong adjuvants such as complete Freund's adjuvant. It is conceivable that vaccination with certain TCR peptides, in contrast to whole T cell vaccination, may activate a new, primary-type immune response to the TCR. The success of TCR peptide vaccination with CDR1 or CDR2 segments that are not idiotypic indicates that various TCR epitopes might be exploited as targets for regulatory mechanisms, irrespective of whether natural priming has preceded the vaccination. Unfortunately, TCR peptide therapy also has been reported to aggrevate disease (Desquenne et al., 1991). Perhaps a more effective response would be induced by the "right" TCR peptide, the peptide epitope involved in the naturally primed anti-idiotypic network. But this remains to be seen.

4.5.2.5 Clinical T Cell Vaccination

The effectiveness of T cell vaccination and its apparent lack of toxicity in the animal models of autoimmune disease has led to clinical trials in humans. The first trials were undertaken in patients suffering from multiple sclerosis (Hafler et al., 1992) or rheumatoid arthritis (Van Laar et al., 1993). These trials were done using as vaccines T cell clones isolated from inflammatory lesions, but without selection of the T cells for specificity to defined self-antigens. Except for demonstrating the feasibility of raising large numbers of T cells from humans and a lack of toxicity, these studies produced little information about the specificity or immunological effectiveness of T cell vaccination in humans suffering from autoimmune disease. A step towards specificity was made using nickel-specific T cells as vaccines in patients with hypersensitivity to nickel (De Vries et al., 1995). The results were not conclusive.

Recently, Zhang, Raus and their associates have undertaken T cell vaccination in multiple sclerosis patients using defined T cell clones specific for myelin basic protein (Zhang et al., 1993). The results have shown clear immunological specificity: the vaccinated patients developed anti-idiotypic T cells of both CD4 and CD8 types, and the appearance of these T cells was associated with down-regulation of the frequency of the anti-myelin basic protein T cells present in the blood of the patients (reviewed in Zhang and Raus, 1995). More patients and longer follow-up periods are needed to allow conclusions about clinical effectiveness, but the immunological results in hand already indicate that T cell vaccination is achievable in humans.

4.5.3 Vaccination with Autoantigens

4.5.3.1 Mucosal Vaccination

It was discovered many decades ago that cell-mediated immune responses to immunization with a foreign antigen could be inhibited by feeding the antigen to the subject. It was only recently, however, that this procedure of "oral tolerance" was applied to the problem of autoimmune disease (Weiner et al., 1994). Howard Weiner and his associates have elucidated key factors in the basic immunology of the phenomenon and have initiated clinical trials in multiple sclerosis (Weiner et al., 1994; Weiner et al., 1993) and arthritis. The success of this work can be seen to strengthen the concept of vaccination therapy for autoimmune disease because the mechanism of action, to be discussed below, is not "tolerance" in the sense of inactivation, but positive activation of the immune response into a more beneficial channel. The "oral" aspect of "oral tolerance", like the "tolerance" aspect, is also a misnomer because the effect can be induced through other mucosal surfaces, especially those of the respiratory system (Metzler and Wraith, 1993). Thus "oral tolerance" might better be termed "mucosal vaccination".

Infectious agents or toxins that adhere to mucosal surfaces can be repulsed by antibodies that bind to the pathogen and prevent its adherence to mucosal cells. The production of such protective antibodies is aided by T cells that produce cytokines of the Th2 class: IL-4, IL-5 and IL-10 (Romagnani, 1994). For this reason, perhaps, immunization at mucosal surfaces tends to elicit a Th2-type response; antigen presentation at the mucosae may be organized naturally to favor activation of Th2 cells.

But the skewing of mucosal immunization towards Th2-type immunity can also have a medicinal advantage. The damage in many autoimmune diseases, such as multiple sclerosis, rheumatoid arthritis or type I diabetes, is caused by a Th1-type of inflammatory cytokine response in the target tissues: IFNγ, IL-2, TNFα (Liblau et al., 1995). Since Th2 cytokines can suppress the activity of Th1 cells (Romagnani, 1994), activation of Th2 type autoimmunity to a target antigen can suppress the Th1-type cells responsible for causing the damage.

Thus, feeding autoantigens can suppress destructive autoimmune disease, both in animal models and in clinical trials. Immunization at the mucosal surfaces of the gut or respiratory system activates Th2-type T cells that migrate to the organ where their target antigen is normally expressed and there the Th2 cells secrete cytokines which can suppress the activity of the pathogenic Th1 cells (Weiner et al., 1994).

Note that the induction of suppression is specific; the Th2 cells are immunized to a specific antigen. However, the suppressor effect of the Th2 cytokines is not specific and can influence Th1 cells responding to other antigens at the site. This "by-stander suppression" provides a mechanism for the spreading of regulation from T cells that see one antigen to T cells that see other antigens, provided that the interacting cells are attracted to the same anatomical site (Weiner et al., 1994; Miller et al., 1991).

"Oral tolerance" is thus "mucosal vaccination": a damaging Th1 autoimmune reaction is down-regulated by activating a competing Th2-type autoimmune reaction. The "tolerance" is seen at the level of the disease; the autoimmunity persists, albeit modified at the cellular and cytokine levels.

4.5.3.2 Peptide Vaccination

Peptides representing the native sequences of antigens, and even whole antigens themselves, are effective in preventing autoimmune diseases. It has been known for some time that the injection of antigens intravenously or into the thymus can induce specific "unresponsiveness", and intravenous or intrathymic administration of antigen have been used to induce resistance to experimental autoimmune diseases (Tisch et al., 1993; Kaufman et al., 1993). However, the unresponsiveness following such injections has been thought to be due to "tolerance", the deletion or anergy of the autoimmune lymphocytes with receptors for the injected antigen. However the administration of antigen with the intent to delete or anergize lymphocytes is not vaccination, by definition. The concept of vaccination would also not encompass the use of peptides designed to occlude the peptide-presenting clefts of MHC molecules, and thereby block T cell activation (Adorini and Nagy, 1990). But the designs of the experimenter, at least in biology if not in physics (Horgan, 1992), cannot impose themselves on the conduct of nature; the experimenter may have intended to block T cell activation by administering a peptide, but it is still possible that the experimenter has inadvertently activated regulatory T cells. Definitions not withstanding, an intended blockade can be an actual vaccination. Indeed, it seems that peptides are now intentionally being used as vaccines.

Peptide epitopes that have been modified by truncation or amino acid substitution have been found to activate the secretion of cytokines by T cells in the absence of T cell proliferation (Evavold et al., 1993). In other words, the interaction of the TCR with a peptide in an MHC pocket can activate the T cell into variously different pathways of biologic response depending on the fine structure of the peptide. Human T cell clones specific for myelin basic protein were shown to modulate their secretion of cytokines when activated using a peptide ligand modified by a single amino acid substitution (Windhagen et al., 1995). Treatment of rodents with a modified peptide of myelin basic protein was shown to reverse the course of EAE, an effect associated with a reduction in the secretion of Th1-type cytokines (Karin et al., 1994). Thus, T cell activation with modified self-peptides is not merely an in vitro curiosity but can modulate an autoimmune process in the whole animal.

However, effective peptide vaccination can even be achieved using the native, unmodified peptide. We have found that a single subcutaneous injection of 50-100 μg of a peptide designated p277 from the sequence of the human 60KDa heat shock protein (hsp60) can prevent the spontaneous development of type I diabetes in NOD strain mice (Elias et al., 1991). Moreover, a subcutaneous injection of p277 could arrest the progression of advanced β cell destruction that had already become manifest as clinical hyperglycemia (Elias and Cohen, 1994). Some of the NOD mice that were moderately hyperglycemia at the time of p277 peptide vaccination manifested a return to normal blood glucose levels and others survived without insulin treatment although remaining hyperglycemic (Elias and Cohen, 1994). The arrest of the diabetogenic process was accompanied by histologic regression of the intra-islet insulitis and the appearance of normal, uninflamed islets (Elias and Cohen, 1995). Peptide p277 is from the human hsp60 sequence that differs from the mouse peptide by 1 amino acid substitution (Birk et al., 1996). However, both the human and mouse peptides can vaccinate NOD mice with advanced insulitis with equal effectiveness. The peptides, in addition are entirely cross-reactive as determined using a reference T cell clone. Thus, even a unmodified self-peptide can vaccinate.

The mechanism responsible for terminating the destruction of β cells by p277 peptide therapy was found to be a shift in the cytokines produced by the anti-p277 T cells (Elias et al., submitted). The spontaneous progression of β cell destruction was associated with the production of the Th1 cytokine IL-2 and IFNγ, but after vaccination with peptide p277, the anti-p277 T cells secreted the Th2 cytokines IL-4 and IL-10 and not IL-2 or IFNγ. The switch in Th1—>Th2 cytokine production was relatively specific; the spontaneous T cell response to a bacterial peptide remained in the Th1 mode (Elias et al., submitted). However, the spontaneous Th1 autoimmunity to other antigens present in the NOD mice such as glutamic acid decarboxylase (GAD) and insulin was inhibited by the Th2 switch induced to the p277 peptide vaccine. In other words, "bystander suppression", similar to that observed in mucosal vaccination (Weiner et al., 1994), could also be seen following subcutaneous peptide vaccination with p277. This suppression was confirmed in adoptive transfer experiments (Elias and Cohen, 1995).

The "tolerance" induced by mucosal vaccination against EAE was found to wane a month or two after a course of oral administration of myelin basic protein and the rats regained susceptibility to induction of EAE (Weiner et al., 1994). In contrast, subcutaneous vaccination with peptide p277 led to permanent arrest of the spontaneous diabetes in the NOD mice (Elias and Cohen, 1994; Elias and Cohen, 1995). It remains to be resolved whether the apparent difference in duration of suppressive effect is due to the difference between induced and spontaneous diseases, or whether the effects of mucosal vaccination are inherently transitory.

An important feature of subcutaneous p277 peptide vaccination that may explain its long-term effect on NOD diabetes appears to be the further evolution of the induced Th2 response. Several months after treatment, we found that Th2 cytokine secretion was superseded by a return to the low-level reactivity characteristic of the period before the onset of insulitis (Elias et al., submitted). Thus, the Th1—>Th2 switch induced by p277 vaccination was only an interim state of the immune system on the way to a non-autoimmune state of baseline reactivity.

4.5.3.3 The Importance of the Adjuvant

The results using p277 vaccine therapy for spontaneous mouse type I diabetes seem paradoxical. Peptide p277 was identified because it contained a target epitope of a pathogenic T cell clone capable of adoptively transferring diabetes (Elias et al., 1991). How then can the target epitope of a pathogenic T cell function to turn off the pathogenic process? The target epitope is in fact a cause of the disease. The paradox, I think, can be resolved by attending to the distinction made at the opening of this chapter between the antigenic epitope and the biologic nature of the immune response made to that epitope. Vaccination deals with both the target epitope and the signals determine that the biologic response; vaccination involves adjuvant signals. The adjuvant, it turns out, is at least as important for the biology of autoimmunity as is the antigen. Consider the well known and largely ignored fact that experimental autoimmune diseases are difficult, if not impossible to induce using the self-antigen alone. An adjuvant, usually complete Freund's adjuvant (CFA), is essential. The administration of purified myelin basic protein, or any of the other protein self-antigens without CFA will not usually induce disease. In fact, administering the self-antigen in aqueous solution or even in incomplete Freund's adjuvant (mineral oil plus emulsifiers; IFA) will often induce resistance to the disease when CFA is added to the antigen on subsequent immunizations (Swanborg, 1975). Thus there appears to be competition between the biologic responses inducible to a self-antigen; the combination of the antigen with CFA pushes the anti-self response into a damaging effector mode, while the self-antigen in IFA or aqueous solution can push the response into a suppressive mode.

Now p277 therapeutic vaccination can be done in aqueous solution, but is much more efficient in IFA; p277 in IFA is more effective at 50 μg than is p277 in saline at 500 μg (in preparation). Peptide p277 can be used to induce insulitis and hyperglycemia in non-diabetes prone mice such as male C57BL/6 mice, but the disease requires that the p277 peptide be conjugated covalently to a foreign carrier protein such as ovalbumin (Elias et al., 1995). Thus, the adjuvant signals needed to induce a pathogenic mode are complex and include the "helper" signals generated by a carrier molecule as well as the signals present in CFA. Sorting out the various adjuvant signals will make it possible to tailor vaccines to the particular medicinal challenge, be it down-regulation of autoimmunity or allograft immunity, or up-regulation of immunity to an infectious agent.

4.5.4 The Future

Immunology is beginning to appreciate the cognitive potential of the immune system to adjust the nature of its response in the light of the accessory signals that accompany the antigen (Cohen, 1992a; Cohen, 1992b). The accessory signals that influence the biologic response to the antigen include not only the traditional concept of adjuvants (CFA, IFA, etc), but also the anatomic site of administration (mucosal immunization), and the anti-idiotypic circuits to which the system has been primed (T cell vaccination, the immunological homunculus). If experimental autoimmune diseases can be shut down by activating the immune system with an opportune mix of signals, perhaps suitable vaccines will be forthcoming for human autoimmune disease (Cohen, 1995).

Acknowledgements

I. R. Cohen is the incumbent of the Mauerberger Chair in Immunology and the Director of the Robert Koch-Minerva Center for Research in Autoimmune Disease. We thank Mrs. Doris Ohayon for dedicated help in preparing this manuscript.

References

Adorini, L. and Nagy, Z. A. (1990) Peptide competition for antigen presentation. Immunol Today *11*, 21-24.

Bach, J. F. (1993) Immunosuppressive therapy of autoimmune diseases. Immunol. Today *14*, 322-325.

Ben-Nun, A., Wekerle, H., and Cohen, I. R. (1981) Vaccination against autoimmune encephalomyelitis with T-lymphocyte line cells reactive against myelin basic protein. Nature *292*, 60-61.

Beraud, E., Lider, O., Baharav, E., Reshef, T., and Cohen, I. R. (1989) Vaccination against experimental autoimmune encephalomyelitis using a subencephalitogenic dose of autoimmune effector cells (1). Characteristics of vaccination. J Autoimmun *2*, 75-86.

Birk, O. S., Elias, D., Weiss, A. S. et al. (1996) NOD mouse diabetes: the ubiquitous mouse hsp60 is a beta-cell target antigen of autoimmune T cells. J. Autoimmun. in press.

Borghans, J. A. M. and de Boer, R. J. (1995) A minimal model for T-cell vaccination. Proc. Royal Soc. London *259*, 173-179.

Broeren, C. P. M., Luccassen, M. A., and van Eden, W. (1995) In: T cell vaccination and autoimmune disease. Zhang, J. and Raus, J. C. M. (eds.), R.G. Landes Co. 37-60.

Burnet, F. M. (1959) The Clonal Selection Theory of Acquired Immunity, Cambridge University Press.

Cohen, I. R. (1992) The cognitive paradigm and the immunological homunculus. Immunol Today *13*, 490-494.

Cohen, I. R. (1992) The cognitive principal challanges clonal selection. Immunol Today *13*, 441-444.
Cohen, I. R. (1995) In: T Cell Vaccination and Autoimmune Disease. Zhang, J. and Raus, J. C. M. (eds.), R.G. Landes Co, 7-20.
Cohen, I. R. (1995) Treatment of autoimmune disease: to activate or to deactivate? Chem Immunol *60*, 150-160.
Cohen, I. R. and Young, D. B. (1991) Autoimmunity, microbial immunity and the immunological homunculus. Immunol Today *12*, 105-110.
Cohen, I. R., Holoshitz, J., van, E. W., and Frenkel, A. (1985) T lymphocyte clones illuminate pathogenesis and affect therapy of experimental arthritis. Arthritis Rheum *28*, 841-845.
De Vries, R. R. P., Dolhain, R. J. E. M., Van Laar, J. M., and Breedveld, F. C. (1995) In: T Cell Vaccination and Autoimmune Disease. Zhang, J. and Raus, J. C. M. (eds.), R.G. Landes Co. 161-169.
Desquenne, C. L., Esch, T. R., Otvos, L. J., and Heber, K. E. (1991) T-cell receptor peptide immunization leads to enhanced and chronic experimental allergic encephalomyelitis. Proc Natl Acad Sci USA *88*, 7219-7223.
Elias, D. and Cohen, I. R. (1994) Peptide therapy for diabetes in NOD mice. Lancet *343*, 704-706.
Elias, D. and Cohen, I. R. (1995) Treatment of autoimmune diabetes and insulitis in NOD mice with hsp60 peptide p 277. Diabetes, *44*, 1132-1138.
Elias, D., Marcus, H., Reshef, T., Ablumunits, V., and Cohen, I. R. (1995) Induction of diabetes in standard mice by immunization to the p277 peptide of a 60KDa heat shock protein. Eur. J. Immunol. *25*, 2851-2857.
Elias, D., Meilin, A., Birk, O. S. et al. Hsp60 peptide therapy of NOD mouse diabetes downregulates autoimmunity to various antigens associated with a transitory Th1 to Th2 cytokine shift. Submitted for publication
Elias, D., Reshef, T., Birk, O. S. et al. (1991) Vaccination against autoimmune mouse diabetes with a T-cell epitope of the human 65-kDa heat shock protein. Proc Natl Acad Sci USA *88*, 3088-3091.
Evavold, B. D., Sloan, L. J., and Allen, P. M. (1993) Tickling the TCR: selective T-cell functions stimulated by altered peptide ligands. Immunol Today *14*, 602-609.
Formby, B. and Shao, T. (1993) T cell vaccination against autoimmune diabetes in non-obese diabetic mice. Ann. Clin. Lab. Sci. *23*, 137-147.
Hafler, D. A., Cohen, I. R., Benjamin, D. S., and Weiner, H. L. (1992) T cell vaccination in multiple sclerosis: a preliminary report. Clin Immunol Immunopathol *62*, 307-313.
Holoshitz, J., Frenkel, A., Ben-Nun, A., and Cohen, I. R. (1983) Autoimmune (EAE) mediated or prevented by T lymphocyte lines directed against diverse antigenic determinants of myelin basic protein. Vaccination is determinant specific. J Immunol *131*, 2810-2813.
Holoshitz, J., Naparstek, Y., Ben-Nun, A., and Cohen, I. R. (1983) Lines of T lymphocytes mediate or vaccinate against autoimmune arthritis. Science *219*, 56-58.
Horgan, J. (1992) Quantum philosophy. Scientific American *267*, 72-80.
Howell, M. D., Winters, S. T., Olee, T. et al. (1989) Vaccination against experimental allergic encephalomyelitis with T cell receptor peptides. Science *246*, 668-670.
Karin, N., Mitchell, D. J., Brocke, S., Ling, N., and Steinman, L. (1994) Reversal of experimental autoimmune encephalomyelitis by a soluble peptide variant of a myelin basic protein epitope: T cell receptor antagonism and reduction of interferon gamma and tumor necrosis factor alpha production. J Exp Med *180*, 2227-2237.
Kaufman, D. L., Clare, S. M., Tian, J. et al. (1993) Spontaneous loss of T-cell tolerance to glutamic acid decarboxylase in murine insulin-dependent diabetes. Nature *366*, 69-72.

Kumar, V. and Sercarz, E. E. (1993) The involvement of T cell receptor peptide-specific regulatory CD4+ T cells in recovery from antigen-induced autoimmune disease. J. Exp. Med. *178*, 909-916.

Liblau, R. S., Singer, S. M., and McDevitt, H. O. (1995) Th1 and Th2 CD4+ T cells in the pathogenesis of organ-specific autoimmune diseases. Immunol Today *16*, 34-38.

Lider, O., Beraud, E., Reshef, T., Friedman, A., and Cohen, I. R. (1989) Vaccination against experimental autoimmune encephalomyelitis using a subencephalitogenic dose of autoimmune effector T cells. (2). Induction of a protective anti-idiotypic response. J Autoimmun *2*, 87-99.

Lider, O., Karin, N., Shinitzky, M., and Cohen, I. R. (1987) Therapeutic vaccination against adjuvant arthritis using autoimmune T cells treated with hydrostatic pressure. Proc Natl Acad Sci USA *84*, 4577-4580.

Lider, O., Reshef, T., Beraud, E., Ben-Nun, A., and Cohen, I. R. (1988) Anti-idiotypic network induced by T cell vaccination against experimental autoimmune encephalomyelitis. Science *239*, 181-183.

Lider, O., Shinitzky, M., and Cohen, I. R. (1986) Vaccination against experimental autoimmune diseases using T lymphocytes treated with hydrostatic pressure. Ann NY Acad Sci *475*, 267-273.

Lohse, A. W., Mor, F., Karin, N., and Cohen, I. R. (1989) Control of experimental autoimmune encephalomyelitis by T cells responding to activated T cells. Science *244*, 820-822.

Maron, R., Zerubavel, R., Friedman, A., and Cohen, I. R. (1983) T lymphocyte line specific for thyroglobulin produces or vaccinates against autoimmune thyroiditis in mice. J Immunol *131*, 2316-2322.

Metzler, B. and Wraith, D. C. (1993) Inhibition of experimental autoimmune encephalomyelitis by inhalation but not oral administration of the encephalitogenic peptide: influence of MHC binding affinity. Int Immunol *5*, 1159-1165.

Miller, A., Lider, O., and Weiner, H. L. (1991) Antigen-driven bystander suppression after oral administration of antigens. J Exp Med *174*, 791-798.

Naparstek, Y., Ben-Nun, A., Holoshitz, J. et al. (1983) T lymphocyte lines producing or vaccinating against autoimmune encephalomyelitis (EAE). Functional activation induces peanut agglutinin receptors and accumulation in the brain and thymus of line cells. Eur J Immunol *13*, 418-423.

Romagnani, S. (1994) Lymphokine production by human T cells in disease states. Annu. Rev. Immunol. *12*, 227-257.

Roubaty, C., Bedlin, C., and Charreirre, J. (1990) Prevention of experimental autoimmune thyroiditis through the anti-idiotypic network. J. Immunol. *144*, 2157-2172.

Segel, L. A., Jager, E., Elias, D., and Cohen, I. R. (1995) A quantitative model of autoimmune disease and T cell vaccination: Why does more mean less? Immunol. Today *16*, 80-84.

Stern, L. J., Brown, J. H., Jardetzky, T. S. et al. (1994) Crystal structure of the human class II MHC protein HLA-DR1 complexed with an influenza virus peptide. Nature *368*, 215-221.

Sun, D., Qin, Y., Chluba, J., Epplen, J. T., and Wekerle, H. (1988) Suppression of experimentally induced autoimmune encephalomyelitis by cytolytic T-T cell interactions. Nature *332*, 843-845.

Swanborg, R. H. (1975) Antigen-induced inhibition of experimental allergic encephalomyelitis. III. Localization of an inhibitory site distinct from the major encephalitogenic determinant of myelin basic protein. J. Immunol. *114*, 191-195.

Tisch, R., Yang, X. D., Singer, S. M. et al. (1993) Immune response to glutamic acid decarboxylase correlates with insulitis in non-obese diabetic mice. Nature *366*, 72-75.

Van Laar, J., Miltenburg, A. M. M., Verdonk, M. J. A. et al. (1993) Effects of treatment with attenuated, autologous T cells in patients with rheumatoid arthritis. J. Autoimmun. *6*, 159-167.

Vandenbark, A. A., Hashim, G., and Offner, H. (1989) Immunization with a synthetic T-cell receptor V-region peptide protects against experimental autoimmune encephalomyelitis. Nature *341*, 541-544.

Weiner, H. L., Friedman, A., Miller, A. et al. (1994) Oral tolerance: immunologic mechanisms and treatment of animal and human organ-specific autoimmune diseases by oral administration of autoantigens. Annu Rev Immunol. *12*, 809-837.

Weiner, H. L., Mackin, G. A., Matsui, M. et al. (1993) Double-blind pilot trial of oral tolerization with myelin antigens in multiple sclerosis. Science *259*, 1321-1324.

Windhagen, A., Scholz, C., Hollsberg, P. et al. (1995) Modulation of cytokine patterns of human autoreactive T cell clones by a single amino acid substitution of their peptide ligand. Immunity *2*, 373-380.

Zamvil, S. S. and Steinman, L. (1990) The T lymphocyte in experimental allergic encephalomyelitis. Annu Rev Immunol *8*, 579-621.

Zhang, J. and Raus, J. C. M. (1995) In: T Cell Vaccination and Autoimmune Disease. Zhang, J. and Raus, J. C. M. (eds.), R.G. Landes Co. 135-156.

Zhang, J., Medaer, R., Stinissen, P., Hafler, D., and Raus, J. (1993) MHC-restricted depletion of human myelin basic protein-reactive T cells by T cell vaccination. Science *261*, 1451-1454.

Contributors

Dr. *Jeff Alexander*
Cytel Corporation
3525 Johns Hopkins Court
San Diego, CA 92121
USA

Dr. *Roberto Amador*
Instituto de Immunologia
Hospital San Juan de Dios
Carrera 10 CALLE1
Santafé de Bogotà
Columbia

Dr. *C. M. Anjam Khan*
Department of Microbiology
The Medical School
Newcastle University
Framlington Place
Newcastle-upon-Tyne NE2 4HH
England

Dr. *Robert V. Ashley*
Harvard Center for Population and Developmental Studies
Roger and Ellen Revelle Bldg.
9 Bow Street
Cambridge, MA 02138
USA

Dr. *Christian Bogdan*
Universität Erlangen-Nürnberg
Institut für Klin. Mikrobiologie
Wasserturmstr. 3
D-91054 Erlangen
Germany

Prof. Dr. *Peter Bretscher*
Department of Microbiology
University of Saskatchewan
107 Wiggins Road
Saskatoon SK S7N 5E5
Canada

Prof. Dr. *André Capron*
Centre d'Immunologie et de Biologie Parasitaire
1, rue du Prof. A. Calmette
F-59019 Lille Cedex
France

Dr. *Robert W. Chesnut*
Cytel Corporation
3525 Johns Hopkins Court
San Diego, CA 92121
USA

Dr. *Irun R. Cohen*
Department of Cell Biology
The Weizmann Inst. of Science
P.O. Box 26
76100 Rehovot
Israel

Dr. *Geoffrey A. Cole*
Department of Immunology
St. Jude Children's Research Hospital
332 North Lauderdale

Memphis, TN 38105
USA

Dr. *Antonello Covacci*
Biocine S.p.A.
Via Fiorentina 1
I-53100 Siena
Italy

Dr. *John Cox*
The National Veterinary Institute
Department of Virology
Biomedicum, Box 585
Husargatan 3
S-75123 Uppsala
Sweden

Dr. *Cecil Czerkinsky*
Department of Medical Microbiology
and Immunology
University of Göteborg
Guldhedsgatan 10
S-41346 Göteborg
Sweden

Prof. Dr. *Sieghard Dittmann*
Immunization and Vaccine Programme
WHO Regional Office for Europe
Scherfigsvej 8
DK-2100 Copenhagen
Denmark

Prof. Dr. *Peter C. Doherty*
Department of Immunology
St. Jude Children's Research Hospital
332 North Lauderdale
Memphis, TN 38105
USA

Dr. *John J. Donnelly*
Department of Virus and
Cell Biology
Merck Research Laboratories
WP 16-3, Bldg. 16101
West Point, PA 19486
USA

Dr. *Paul W. Ewald*
Department of Biology
Amherst College
Amherst, MA 01002-5000
USA

Dr. *Bruno Gander*
Dutch National Institute of
Public Health and Environmental
Protection
Antonie van Leeuwenhoeklaan 9
P.O. Box 1
NL-3720 Bilthoven
The Netherlands

Dr. *Paolo Ghiara*
Biocine S.p.A.
Via Fiorentina 1
I-53100 Siena
Italy

Dr. *Howard M. Grey*
Cytel Corporation
3525 Johns Hopkins Court
San Diego, CA 92121
USA

Dr. *Carole Heilman*
Chief Respiratory Diseases Branch
NIAID
6003 Executive Boulevard
Solar Building, Room 3B06
Bethesda, MD 20892
USA

Prof. Dr. *Jan Holmgren*
Department of Medical Microbiology
and Immunology
University of Göteborg

Guldhedsgatan 10
S-41346 Göteborg
Sweden

Dr. *Carlos E. Hormaeche*
Department of Microbiology
The Medical School
Newcastle University
Framlington Place
Newcastle-upon-Tyne NE2 4HH
England

Prof. Dr. *Stefan H. E. Kaufmann*
Department of Immunology
University of Ulm
Albert-Einstein-Allee 11
D-89070 Ulm
Germany

Dr. *Gideon Kersten*
Dutch National Institute of
Public Health and Environmental
Protection
Antonie van Leeuwenhoeklaan 9
P.O. Box 1
NL-3720 Bilthoven
The Netherlands

Dr. *David L. Klein*
Respiratory Diseases Branch
NIAID
6003 Executive Boulevard
Solar Building, Room 3B06
Bethesda, MD 20892
USA

Dr. *Ralph T. Kubo*
Cytel Corporation
3525 Johns Hopkins Court
San Diego, CA 92121
USA

Prof. Dr. *Reinhard Kurth*
Präsident des Paul-Ehrlich Instituts
Paul-Ehrlich-Str. 51 – 59
D-63225 Langen
Germany

Dr. *Margaret A. Liu*
Department of Virus and
Cell Biology
Merck Research Laboratories
WP 16-3, Bldg. 16101
West Point, PA 19486
USA

Dr. *Karin Lövgren-Bengtsson*
The National Veterinary Institute
Department of Virology
Biomedicum, Box 585
Husargatan 3
S-75123 Uppsala
Sweden

Dr. *Felix Mor*
Department of Cell Biology
The Weizmann Inst. of Science
P.O. Box 26
76100 Rehovot
Israel

Dr. *Bror Morein*
The National Veterinary Institute
Department of Virology
Biomedicum, Box 585
Husargatan 3
S-75123 Uppsala
Sweden

Prof. Dr. *Christopher J. L. Murray*
Harvard Center for Population and
Developmental Studies
Roger and Ellen Revelle Bldg.
9 Bow Street

Cambridge, MA 02138
USA

Dr. *Stephen Norley*
Paul-Ehrlich Institut
Paul-Ehrlich-Str. 51 – 59
D-63225 Langen
Germany

Dr. *Enzo Paoletti*
Director of Research Virogenetics Corporation
Rensselaer Technology Park
465 Jordan Road
Troy, NY 12180
USA

Dr. *Manuel Patarroyo*
Instituto de Immunologia
Hospital San Juan de Dios
Carrera 10 CALLE1
Santafé de Bogotà
Columbia

Dr. *Marion E. Perkus*
Research Virogenetics Corporation
Rensselaer Technology Park
465 Jordan Road
Troy, NY 12180
USA

Dr. *Mariagrazia Pizza*
Biocine S.p.A.
Via Fiorentina 1
I-53100 Siena
Italy

Dr. *Hans-Georg Rammensee*
Deutsches Krebsforschungszentrum
Abt. Tumorvirus-Immunologie 0620
Im Neuenheimer Feld 242
D-69120 Heidelberg
Germany

Dr. *Rino Rappuoli*
Biocine S.p.A.
Via Fiorentina 1
I-53100 Siena
Italy

Dr. *Gilles Riveau*
Relations Hôte-Parasite et Stratégies Vaccinales
Institut Pasteur – INSERM U 167
1, rue du Prof. A. Calmette
F-59019 Lille Cedex
France

Prof. Dr. *Martin Röllinghoff*
Universität Erlangen-Nürnberg
Institut für Klin. Mikrobiologie
Wasserturmstr. 3
D-91054 Erlangen
Germany

Dr. *Hansjörg Schild*
Deutsches Krebsforschungszentrum
Abt. Tumorvirus-Immunologie 0620
Im Neuenheimer Feld 242
D-69120 Heidelberg
Germany

Dr. *Alessandro Sette*
Cytel Corporation
3525 Johns Hopkins Court
San Diego, CA 92121
USA

Dr. *John W. Shiver*
Department of Virus and Cell Biology
Merck Research Laboratories
WP 16-3, Bldg. 16101
West Point, PA 19486
USA

Dr. *John Sidney*
Cytel Corporation
3525 Johns Hopkins Court
San Diego, CA 92121
USA

Dr. *Jia-Bin Sun*
Department of Medical Microbiology and Immunology
University of Göteborg
Guldhedsgatan 10
S-41346 Göteborg
Sweden

Dr. *Ann-Mari Svennerholm*
Department of Medical Microbiology and Immunology
University of Göteborg
Guldhedsgatan 10
S-41346 Göteborg
Sweden

Dr. *Gudrun Szalay*
Department of Immunology
University of Ulm
Albert-Einstein-Allee 11
D-89070 Ulm
Germany

Dr. *John L. Telford*
Biocine S.p.A.
Via Fiorentina 1
I-53100 Siena
Italy

Dr. *Jeffrey B. Ulmer*
Department of Virus and Cell Biology
Merck Research Laboratories
WP 16-3, Bldg. 16101
West Point, PA 19486
USA

Dr. *Peggy A. Wentworth*
Cytel Corporation
3525 Johns Hopkins Court
San Diego, CA 92121
USA

Dr. *David L. Woodland*
Department of Immunology
St. Jude Children's Research Hospital
332 North Lauderdale
Memphis, TN 38105
USA

Index